INTRODUCTORY
DIFFERENTIAL
EQUATIONS

INTRODUCTORY DIFFERENTIAL EQUATIONS

FOURTH EDITION

MARTHA L. ABELL
Georgia Southern University, Statesboro, Georgia, USA

JAMES P. BRASELTON
Georgia Southern University, Statesboro, Georgia, USA

ELSEVIER

AMSTERDAM • BOSTON • HEIDELBERG • LONDON
NEW YORK • OXFORD • PARIS • SAN DIEGO
SAN FRANCISCO • SINGAPORE • SYDNEY • TOKYO
Academic Press is an imprint of Elsevier

Academic Press is an imprint of Elsevier
225 Wyman Street, Waltham, MA 02451, USA
525 B Street, Suite 1800, San Diego, CA 92101-4495, USA
32 Jamestown Road, London NW1 7BY, UK
The Boulevard, Langford Lane, Kidlington, Oxford OX5 1GB, UK

Fourth edition 2014

ISBN: 978-0-12-417219-7

Library of Congress Cataloging-in-Publication Data
Abell, Martha L., 1962-
  Introductory differential equations/Martha L. Abell, James P. Braselton. – Fourth edition.
    pages cm
  Includes bibliographical references and index.
  ISBN 978-0-12-417219-7 (alk. paper)
1. Differential equations–Textbooks. I. Braselton, James P., 1965- II. Title. III. Title: Differential equations with boundary value problems.
  QA379.A24 2014
  515'.35–dc23
                                                                    2014006809

British Library Cataloguing in Publication Data
A catalogue record for this book is available from the British Library

For information on all Academic Press publications
visit our web site at store.elsevier.com

This book has been manufactured using Print On Demand technology. Each copy is produced to order and is limited to black ink. The online version of this book will show color figures where appropriate.

Working together
to grow libraries in
developing countries

www.elsevier.com • www.bookaid.org

Contents

Please find the companion website at http://booksite.elsevier.com/9780124172197/.
Please find the instructor website at http://textbooks.elsevier.com/web/Manuals.aspx?isbn=9780124172197/.

Preface

Introductory Differential Equations began as the fourth edition of a text originally called Introductory Differential Equations with Boundary Value Problems.

When we were done with the revision, we no longer saw Introductory Differential Equations with Boundary Value Problems but rather a new text that we have titled Introductory Differential Equations.

The first edition of this text, Modern Differential Equations was "modern" because it was one of the first texts that nearly required access to a graphing calculator, computer algebra system, or numerical software package.

Computer algebra systems and sophisticated graphing calculators have changed the ways in which we learn and teach ordinary differential equations. Instead of focusing students' attention only on a sequence of solution methods, we want them to use their minds to understand what solutions mean and how differential equations can be used to answer pertinent questions. Now their use is expected in a standard course, so the term "modern" no longer applies to the text.

Interestingly, this metamorphosis in the teaching of differential equations described occurred relatively quickly and coincided with our professional careers at Georgia Southern University. Our interest in the use of technology in the mathematics classroom began in 1990 when we started to use computer laboratories and demonstrations in our calculus, differential equations, and applied mathematics courses. Over the past years we have learned some ways of how to and how not to use technology in the mathematics curriculum. In the early stages,

we simply wanted to show students how they could solve more difficult problems by using a computer algebra system so that they could be exposed to the technology. However, we soon realized that we were missing the great opportunity of allowing students to discover aspects of the subject matter on their own. We revised our materials to include experimental problems and thought-provoking questions in which students are asked to make conjectures and investigate supporting evidence. We also developed application projects called Differential Equations at Work, not only to emphasize technology, but also to improve the problem-solving and communication skills of our students. To preserve the "wow" aspects of technology, we continue to use it to observe solutions in classroom demonstrations through such things as animating the motion of springs and pendulums. These demonstrations not only grab the attention of students, but also help them to make the connection between a formula and what it represents.

This text is designed to serve as a text for beginning courses in ordinary differential equations. Usually, introductory ordinary differential equations courses are taken by students who have successfully completed a first-year calculus course, a basic linear algebra course, and this text is written at a level readable for them. Previous versions of the texts have included two chapters on boundary value problems and partial differential equations. Often, the first ordinary differential equations course and the first partial differential equations course are completely separated. We believe that there are many more comprehensive texts for a first course in partial differential equations than the previous two chapters on boundary value problems and partial differential equations that were included in this text. Of course, if any instructor or student using this text wants access to those two chapters for their classes we are more than happy to supply them to you.

TECHNOLOGY

The advantages of incorporating technology into mathematics courses are well-known. Some of them include enhancing the ability to solve a variety of problems; helping students work examples; supporting varied, realistic, and illuminating applications; exploiting and improving geometric intuition; encouraging mathematical experiments; and teaching approximation. In addition, technology is implemented throughout this text to promote the following goals in the learning of differential equations:

1. Solving problems: Using different methods to solve problems and generalize solutions;
2. Reasoning: Exploiting computer graphics to develop spatial reasoning through visualization;
3. Analyzing: Finding the most reasonable solution to real problems or observing changes in the solution under changing conditions;
4. Communicating mathematics: Developing written, verbal, and visual skills to communicate mathematical ideas; and
5. Synthesizing: Making inferences and generalizations, evaluating outcomes, classifying objects, and controlling variables.

Students who develop these skills will succeed not only in differential equations, but also in subsequent courses and in the workforce.

The ✔ icon is used throughout the text to indicate those examples in which technology is used in a nontrivial way to develop or visualize

the solution or to indicate the sections of the text, such as those discussing numerical methods, in which the use of appropriate technology is essential or interesting.

APPLICATIONS

Applications in this text are taken from a variety of fields, especially biology, physics, chemistry, engineering, and economics, and they are documented by references. These applications can be found in many of the examples and exercises, in separate sections and chapters of the text, and in the *Differential Equations at Work* subsections at the end of each chapter. Many of these applications are well-suited to exploration with technology because they incorporate real data. In particular, obtaining closed form solutions is not necessarily "easy" (or always possible). These applications, even if not formally discussed in class, show students that differential equations is an exciting and interesting subject with extensive applications in many fields.

STYLE

To keep the text as flexible as possible, addressing the needs of both audiences with different mathematical backgrounds and instructors with varying preferences, *Introductory Differential Equations* is written in an easy-to-read, yet mathematically precise, style. It contains all topics typically included in a standard first course in ordinary differential equations. Definitions, theorems, and proofs are concise but worded precisely for mathematical accuracy Generally, theorems are proved if the proof is instructive or has "teaching value." Of course, discussion of such proofs is optional in the typical classroom for which this text is written. In other cases, proofs of theorems are developed in the exercises or omitted. Theorems and definitions are boxed for easy reference;

key terms are highlighted in boldface. Figures are used frequently to clarify material with a graphical interpretation.

FEATURES

Introductory Differential Equations is an extensive revision of the third edition of *Introductory Differential Equations with Boundary Value Problems*. Particular features include:

- The text's website include background material, proofs of some theorems, solutions to selected exercises, additional exercises, visualizations of certain topics (movies) and podcasts that students can download to their video-capable *iPod* or other compatible mv4 players.
- Because the mathematician's who developed the mathematics discussed in this text were (or *are*) still interesting in their own right, we have tried to include an image and interesting tidbit about their lives whenever possible hoping to help some students become more interested in the course. When credit is not given for a photo, it is because we have reasonable reason to believe that the image is in the public domain. If a copyright applies to an image and appropriate credit has not been given, please alert us so that we can correct the situation promptly
- All graphics have been redone. In each case, the intent of the graphic has been questioned. In some cases, graphics have been eliminated; in other cases, they have been redone to emphasize their purpose.
- We have revised the exercise set considerably. The total number of exercises remains about the same as in previous editions but we have deleted about 250 "outdated" exercises and replaced them with new ones. We continue to believe that a student should be able to solve a basic nontrivial problem by hand.

PEDAGOGICAL FEATURES

Examples

Throughout the text, numerous examples are given, with thorough explanations and a substantial amount of detail. Solutions to more difficult examples are constructed with the help of graphing calculators or a computer algebra system and are indicated by an icon.

"Think About It!"

Many examples are followed by a question indicated by a ⬤ icon. Generally, basic knowledge about the behavior of functions is sufficient to answer the question. Many of these questions encourage students to use technology. Others, focus on the graph of a solution. Thus, "Think about it!" questions help students determine when to use technology and make this text more interactive.

Technology

Many students entering their first differential course have had substantial experience with various sophisticated calculators and computer algebra systems.

Our first ordinary differential equations course attempts to encourage students to *use technology intelligently*. We have italicized the words *use technology intelligently* because they take on different meaning to different instructors because they depend on the instructor's philosophy, institution, and students. Students also interpret the phrase differently depending upon their instructor and exposure to technology.

In any case, many of us have limited resources and would prefer that our students have a good grasp of the fundamentals rather than be "wowed!" by nonsense. We have tried to use technology intelligently here. We believe that it should not be obtrusive so you should not notice when we do. When required in an example or exercise, it should be obvious to an instructor and relatively easy to convince a student that there are two ways to solve a problem: the easy way and the hard way. We choose the hard way when there is instructional value to the approach. The ![icon] icon is intended to alert students that technology is intelligently (and wisely) used to assist in solving the problem. Typically, the technology we have used is a computer algebra system, like *Mathematica* or *Maple*.

Technology is used throughout the text to explore many of the applications and more difficult examples, especially those marked with ![icon] and the problems in the subsections *Differential Equations at Work*.

Answers to most odd exercises are included at the end of the text. More complete answers, solutions, partial solutions, or hints to selected exercises are available separately to students and instructors. *Differential Equations at Work* subsections describe detailed economics, biology, physics, chemistry, and engineering problems documented by references. These problems include real data when available and require students to provide answers based on different conditions. Students must analyze the problem and make decisions about the best way to solve it, including the appropriate use of technology Each *Differential Equations at Work* project can be assigned as a project requiring a written report, for group work, or for discussion in class.

Differential Equations at Work also illustrate how differential equations are used in the real world. Students are often reluctant to believe that the subject matter in calculus, linear algebra, and differential equations classes relates to subsequent courses and to their careers. Each *Differential Equations at Work* subsection illustrates how the material discussed in the

course is used in real life. We keep each *Differential Equations at Work* subsection short because nearly all instructors have enough trouble covering the content expected of them. On the other hand, when a student asks the question "When am I going to use this?" or "How am I going to use this?" these short subsections can give the instructor ideas as to how to handle the question.

The problems in *Differential Equations at Work* are not connected to a specific section of the text; they require students to draw different mathematical skills and concepts together to solve a problem. Because each *Differential Equations at Work* is cumulative in nature, students must combine mathematical concepts, techniques, and experiences from previous chapters and math courses.

Exercises

Numerous exercises, ranging in level from easy to difficult, are included in each section of the text. In particular, the exercise sets for topics that students find most difficult are rich and varied. For the fourth edition of the text, the abundant "routine" exercises have been completely revised and try to encourage students to master basic techniques. Most sections also contain interesting mathematical and applied problems to show that mathematics and its applications are both interesting and relevant. Instructors will find that they can assign a large number of problems, if desired yet still have plenty for review in addition to those found in the review section at the end of each chapter. Answers to most odd-numbered exercises are included at the end of the text; detailed solutions to selected exercises are included in the *Student Resource Manual*.

Chapter Summary and Review Exercises

Each chapter ends with a chapter summary highlighting important concepts, key terms and formulas, and theorems. The Review Exercises following the chapter summary of each chapter offer students extra practice on the topics in that chapter. These exercises are arranged by section so that students having difficulty can turn to the appropriate material for review.

Figures

This text provides an abundance of figures and graphs, especially for solutions to examples. In addition, students are encouraged to develop spatial visualization and reasoning skills, to interpret graphs, and to discover and explore concepts from a graphical point of view. To ensure accuracy, the figures and graphs have been completely computer-generated. Nearly all figures for the fourth edition have been revised. We hope you like the improvement in the graphics.

Historical Material

Nearly every topic is motivated by either an application or an appropriate historical note. We have also included images of paintings, drawings, or photographs of the many famous scientists and descriptions of the mathematics they discovered.

CONTENT

The highlights of each chapter are described briefly below.

Chapter 1 After introducing preliminary definitions, we discuss direction fields not only for first-order differential equations, but also for systems of equations. In this presentation, we establish a basic understanding of solutions and their graphs. We give an overview of some of the applications covered later in the text to point out the usefulness of the topic and some of

the reasons we have for studying differential equations in the exercises.

Chapter 2 In addition to discussing the standard techniques for solving several types of first-order differential equations (separable equations, homogeneous equations, exact equations, and linear equations), we introduce several numerical methods (Euler's Method, Improved Euler's Method, Runge-Kutta Method) and discuss the existence and uniqueness of solutions to first order initial-value problems. Throughout the chapter, we encourage students to build an intuitive approach to the solution process by matching a graph to a solution without actually solving the equation.

Chapter 3 Not only do we cover most standard applications of first order equations in Chapter 3 (orthogonal trajectories, population growth and decay, Newton's law of cooling, free-falling bodies), but we also present many that are not (due to their computational difficulty) in *Differential Equations at Work*.

Chapter 4 This chapter emphasizes the methods for solving homogeneous and nonhomogeneous higher order differential equations. It also stresses the Principle of Superposition and the differences between the properties of solutions to linear and nonlinear equations. After discussing Cauchy-Euler equations, series methods are introduced, which includes a discussion of several special equations and the properties of their solutions/equations important in many areas of applied mathematics and physics.

Chapter 5 Several applications of higher order equations are presented. The distinctive presentation illustrates the motion of spring-mass systems and pendulums graphically to help students understand what solutions represent and to make the applications more meaningful to them.

Chapter 6 The study of systems of differential equations is perhaps the most exciting of all the topics covered in the text. Although we direct most of our attention to solving systems of linear first-order equations with constant coefficients, technology allows us to investigate systems of nonlinear equations and observe phase planes. We also show how to use eigenvalues and eigenvectors to understand the general behavior of systems of linear and nonlinear equations. We have added a section on phase portraits in this edition.

Chapter 7 Several applications discussed earlier in the text are extended to more than one dimension and solved using systems of differential equations, in an effort to reinforce the understanding of these important problems. Numerous applications involving nonlinear systems are discussed as well.

Chapter 8 Laplace transforms are important in many areas of engineering and exhibit intriguing mathematical properties as well. Throughout the chapter, we point out the importance of initial conditions and forcing functions on initial-value problems.

For a one semester course introducing ordinary differential equations, many instructors will choose to cover topics from Chapters 1 to 7 or from Chapters 1 to 6 and Chapter 8. For a two semester course, the instructor will easily be able to cover the remaining chapters of the

text. In our introductory ordinary differential equations course, we typically cover most of Chapters 1, 2, 4, and 6, and instructors choose a variety of applications from Chapters 3, 5, and 7.

- For a one semester course targeted to the computational needs of most engineering majors, cover most topics from Chapters 1, 2, 4, 6, and 8.
- For a casual one semester course directed towards math and math education majors, cover most mathematical topics in Chapters 1, 2, 4, and 6 and selected applications from Chapters 3, 5, and 7.

Finally, we thank those close to us, especially Imogene Abell, Lori Braselton, Ada Braselton, and Mattie Braselton for enduring with us the pressures of a project like this and for graciously accepting our demanding work schedules. We certainly could not have completed this task without their care and understanding.

Martha Abell
(E-mail: martha@georgiasouthern.edu)
James Braselton
(E-mail: jbraselton@georgiasouthern.edu)

Department of Mathematical Sciences
P.O. Box 8093
Georgia Southern University
Statesboro, Georgia
30460

July, 2014

CHAPTER

1

Introduction to Differential Equations

The purpose of *Introductory Differential Equations* is twofold. First, we introduce and discuss the topics covered in an undergraduate course in ordinary differential equations (ODEs). Second, we indicate how certain technologies such as computer algebra systems and graphing calculators are used to enhance the study of differential equations, not only by eliminating some of the computational difficulties that arise in the study of differential equations but also by overcoming some of the visual limitations associated with the solutions of differential equations. The advantages of using technology such as graphing calculators and computer algebra systems in the study of differential equations are numerous, but perhaps the most useful is that of being able to produce the graphics associated with solutions of differential equations. This is particularly beneficial in the discussion of applications because many

physical situations are modeled with differential equations. For example, in Chapter 5, we see that the motion of a pendulum can be modeled by a differential equation. When we solve the problem of the motion of a pendulum, we use technology to watch the pendulum move. The same is true for the motion of a mass attached to the end of a spring, as well as many other problems. In having this ability to use technology, the study of differential equations becomes much more meaningful as well as interesting.

Although this chapter is short in length, the vocabulary introduced here is used throughout the text. To a large extent, this chapter may be read quickly, but it is important to remember that subsequent chapters will take advantage of the terminology and techniques discussed here. Any formal introduction to differential equations should begin with German scientist Gottfried Wilhelm Leibniz (1646-1716) and

British scientist Isaac Newton (1642-1727), the inventors of calculus. In integral calculus, we learn that the area under the graph of a smooth positive function is given by a definite integral, but both Leibniz and Newton were more concerned with solving differential equations than finding areas. Many of the methods of solution we present in this text are from the great Swiss mathematician Leonhard Euler (1707-1783). Subsequently, many problems, such as determining the motion of a plucked string, lead not only to ordinary and partial differential equations but also to other areas of mathematics. Indeed, differential equations are full of rich and exciting applications; interesting applications are included throughout the text to motivate discussions, make the study of differential equations more interesting and pertinent to the real world and indicate how some people use differential equations beyond this course. However, mathematics is also interesting in its own right; mathematical applications are also included throughout the text.

Isaac Newton: We hope this young Isaac Newton was not worried about who discovered calculus first.

Gottfried Wilhelm Leibniz (1646-1716)

Isaac Newton (1642-1727): The controversy as to who (Leibniz or Newton) discovered calculus first became an obsession with Newton during the second half of his life.

Leonhard Euler (1707-1783)

1.1 INTRODUCTION TO DIFFERENTIAL EQUATIONS: VOCABULARY

We begin our study of differential equations by explaining what a differential equation is. From our experience in calculus, we are familiar with some differential equations. For example, suppose that the acceleration $a(t)$ (measured in ft./s^2) of a falling object is $a(t) = -32$. Then, because $a(t) = v'(t)$, where $v(t)$ is the velocity of the object (measured in ft./s), we have $v'(t) = -32$ or

$$\frac{dv}{dt} = -32.$$

An equation like this involving a function of a single variable is called an *ordinary differential equation* (ODE). (If the equation involves partial derivatives, then it is called a *partial differential equation* (PDE).) In this case, the function to be determined is $v = v(t)$, which depends on the variable t, representing time (measured in seconds). The goal in solving an ODE is to find a function that satisfies the equation. We can solve this ODE through integration:

$$v(t) = \int a(t)dt = \int (-32)dt = -32t + C,$$

where C is an arbitrary constant. This result indicates that $v(t) = -32t + C$ is a solution of the ODE for any choice of the constant C. (We call this a *general solution* because it involves an arbitrary constant.) In fact, we have found every solution of the ODE because each is expressed as $-32t + C$. Examples of solutions and the corresponding C values include $v(t) = -32t$ ($C = 0$), $v(t) = -32t + 32$ ($C = 32$), and $v(t) = -32t - 8$ ($C = -8$). This shows that there are an infinite number of solutions to the ODE.

We can *verify* that $v(t) = -32t + C$ is a *general solution* of $dv/dt = -32$ through substitution:

$$\frac{dv}{dt} = \frac{d}{dt}(-32t + C) = -32.$$

When we substitute our solution into the left side of the ODE, we obtain -32, the value on the right side of the ODE. We have verified that $v(t) = -32t + C$ satisfies the ODE for any choice of C.

Many times we are given a particular condition that the solution must satisfy. For example, suppose that the object considered earlier has an initial velocity of -64 ft./s. In other words, the velocity at time $t = 0$ seconds is represented with the *initial condition* $v(0) = -64$. Therefore, we need to find a solution of the ODE that also satisfies the initial condition. We express this *initial value problem* (IVP) as

$$\frac{dv}{dt} = -32, \quad v(0) = -64,$$

where we solve the IVP by first *finding a general solution to the ODE* and then by *applying the initial condition to determine the arbitrary constant*. When we substitute $t = 0$ into the general solution $v(t) = -32t + C$, we obtain $v(0) = -32 \times 0 + C = C$. We conclude that $C = -64$ so that $v(0) = -64$ is satisfied. This means that the solution to the IVP is $v(t) = -32t - 64$. Notice that unlike the ODE, the IVP has only one solution.

Another application of differential equations found in calculus is finding a function when given (a) the slope of the line tangent to the graph of the function at any point (x, y) and (b) a point on the graph.

For example, suppose that the slope of the tangent line at any point on the graph of a function $y = y(x)$ is given by

$$\frac{dy}{dx} = 3x^2 - 4x,$$

and further that the graph passes through the point $(1, 4)$, which means that $y(1) = 4$. In this case, the ODE is given by $dy/dx = 3x^2 - 4x$, where we need to find the function $y = y(x)$ that satisfies the initial condition $y(1) = 4$. Therefore, we solve the IVP

$$\frac{dy}{dx} = 3x^2 - 4x, \quad y(1) = 4.$$

As in the previous example, we find a general solution to $dy/dx = 3x^2 - 4x$ through integration. This yields

$$y = y(x) = x^3 - 2x^2 + C.$$

Now, when we apply the initial condition, we find that

$$y(1) = 1^3 - 2 \times 1^3 + C = -1 + C = 4$$

so that $C = 5$, which means that the solution to the IVP is

$$y = y(x) = x^3 - 2x^2 + 5.$$

Figure 1.1 shows the graph of the solution together with a portion of the tangent line at the point $(1, 4)$. Notice that the slope of the tangent line at the point $(1, 4)$ is the value of dy/dx evaluated if $x = 1$: if $x = 1$, $dy/dx = 3 \times 1^2 - 4 \times 1 = -1$. This observation will be useful in Section 1.2 in helping us better understand the behavior of solutions of differential equations.

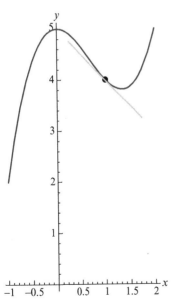

FIGURE 1.1   Solution to $dy/dx = 3x^2 - 4x$, $y(1) = 4$ along with a segment of the tangent line at $(1, 4)$.

The previous examples are similar in that they each involve an ODE in which the highest order derivative is the first derivative. We call equations of this type *first-order ODEs* because the *order* of a differential equation is the order of the highest order derivative appearing in the equation.

EXAMPLE 1.1.1

Determine the order of each of the following differential equations: (a) $dy/dx = x^2/y^2 \cos y$ (b) $u_{xx} + u_{yy} = 0$ (e) $(dy/dx)^4 = y + x$ (d) $d^2x/dt^2 + 2dx/dt + 3x = \sin t$

Solution

(a) This equation is first order because it includes only one first-order derivative, dy/dx. (b) This equation is classified as second order because the highest order derivatives, u_{xx}, representing $\partial^2 u/\partial x^2$, and u_{yy} representing $\partial^2 u/\partial y^2$, are of order two. The equation $u_{xx} + u_{yy} = 0$ arises in many areas of study, which include fluid flows as well as electrostatic and gravitational potential, is often called *Laplace's equation* after Pierre-Simon Laplace (also known as the Marquis de Laplace) (1749-1827) or the *potential equation*. Hence, Laplace's equation is a second-order partial differential equation. (e) This is a first-order equation because the highest order derivative is the first derivative. Raising that derivative to the fourth power does not affect the order of the equation. The expressions

$$\left(\frac{dy}{dx}\right)^4 \quad \text{and} \quad \frac{d^4y}{dx^4}$$

do not represent the same quantities: $(dy/dx)^4$ represents the derivative of y with respect to x, dy/dx, raised to the fourth power; d^4y/dx^4 represents the fourth derivative of y with respect to x. (d) The highest order derivative is d^2x/dt^2, so the equation is second order.

Pierre-Simon Laplace (1749, Normandy, France-
1827, Paris, France): Laplace's name is probably
most remembered in reference to the *Laplace transform* that we will study in Chapter 8.

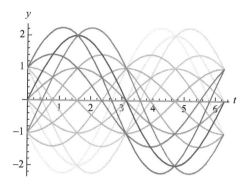

FIGURE 1.2 Graphs of $y = c_1 \cos t + c_2 \sin t$ for various values of c_1 and c_2.

EXAMPLE 1.1.2

(a) Show that $y = c_1 \sin t + c_2 \cos t$ satisfies the second-order ODE (ODE) $y'' + y = 0$ where c_1 and c_2 are arbitrary constants. (b) Find the solution to the IVP $y'' + y = 0$, $y(0) = 0$, $y'(0) = 1$.

Solution

(a) Differentiating, we obtain $y' = dy/dt = c_1 \cos t - c_2 \sin t$ and $y'' = d^2y/dt^2 = -c_2 \cos t - c_1 \sin t$. Therefore,

$$\frac{d^2y}{dt^2} + y = -c_1 \sin t - c_2 \cos t + c_1 \sin t$$
$$+ c_2 \cos t = 0,$$

so the function satisfies the ODE for any choice of c_1 and c_2. We graph the solution for several values of these constants in Figure 1.2. Because the solution depends on at least one constant, we say that the functions shown in Figure 1.3 are members of the *family of solutions* of the ODE.

(b) Evaluating the function at $t = 0$ yields $y(0) = c_1 \sin 0 + c_2 \cos 0 = c_2$. Then, the initial condition, $y(0) = 0$, indicates that $c_2 = 0$.

Similarly, $y'(0) = c_1 \cos 0 - c_2 \sin 0 = c_1$, so $c_1 = 1$ so that the second initial condition, $y'(0) = 1$, is satisfied. Therefore, $y(t) = \sin t$ is the solution to the initial value problem (IVP). We graph this solution in Figure 1.3. Notice that the ODE has an infinite number of solutions while the IVP has only one unique solution. Also observe that the number of initial conditions for the initial value problem (IVP) matches the order of the ODE.

The next level of classification tells us whether an equation is *linear* or *nonlinear* in terms of the *dependent variable*. For an ODE, assuming that the independent variable is x and the dependent variable is y, an ODE (of order n) is called *linear* if it can be written as

$$a_n(x)\frac{d^ny}{dx^n} + a_{n-1}(x)\frac{d^{n-1}y}{dx^{n-1}} + \cdots + a_2(x)\frac{d^2y}{dx^2}$$
$$+ a_1(x)\frac{dy}{dx} + a_0(x)y = f(x), \qquad (1.1)$$

where the functions $a_k(x)$, $k = 0, 1, \ldots, n$, and $f(x)$ are given and $a_n(x)$ is not the zero function. If $f(x)$ is identically the zero function, the linear equation (1.1) is said to be *homogeneous*.

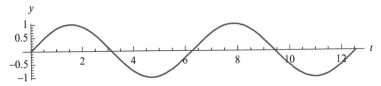

FIGURE 1.3   Graph of $y = y(t) = \sin t$.

You should verify that $y(x) = 0$, which we call the *trivial solution*, is a solution to every linear homogeneous equation. If $f(x)$ is *not* identically the zero function, the linear equation (1.1) is said to be *nonhomogeneous*.

In Chapter 4, we will learn that a general solution of the nth-order linear equation (1.1) is a solution that depends on n arbitrary constants and includes all solutions of the equation.

If the equation under consideration cannot be written in the form given by equation (1.1), the equation is said to be *nonlinear*. Therefore, some of the properties that lead to classifying an equation as linear or nonlinear are *powers of the dependent variable* (or one of its derivatives) and *functions of the dependent variable*.

EXAMPLE 1.1.3

Determine which of the following differential equations are linear: (a) $\dfrac{dy}{dx} = x^3$, (b) $\dfrac{d^2u}{dx^2} + u = e^x$,

(c) $(y - 1)\,dx + x \cos y\,dy = 0$, (d) $\dfrac{d^3y}{dx^3} + y\dfrac{dy}{dx} = x$,

(e) $\dfrac{dy}{dx} + x^2y = x$, and (f) $\dfrac{d^2x}{dt^2} + \sin x = 0$.

Solution

(a) This equation is linear because the nonlinear term x^3 is the function $f(x)$ of the independent variable in the general formula for a linear differential equation. (b) This equation is also linear. Using u as the name of the dependent variable

does not affect the linearity. (c) If y is the dependent variable, solving for dy/dx gives us

$$\frac{dy}{dx} = \frac{1-y}{x \cos y}.$$

Because the right side of this equation includes a nonlinear function of y, the equation is nonlinear (in y). However, if x is the dependent variable, solving for dx/dy yields

$$\frac{dx}{dy} = \frac{\cos y}{1-y}x.$$

This equation is linear in the dependent variable x. (d) The coefficient of the term dy/dx is y instead of an expression involving only the independent variable x. Hence, this equation is nonlinear in the dependent variable y. (e) This equation is linear. The term x^2 is the coefficient function. (f) For this equation, note that x is the dependent variable and t is the independent variable. This equation, known as the *pendulum equation* because it models the motion of a simple pendulum, is nonlinear because it involves a nonlinear function of the dependent variable x, $\sin x$.

If an ODE has the form $dy/dx = f(x)$, we can use integration to determine $y = y(x)$, although the result may be in terms of integrals that cannot be evaluated using standard techniques of integration. The following examples of differential equations of this type illustrate some of the typical methods of integration that may be encountered.

EXAMPLE 1.1.4

Solve the following differential equations: (a) $\dfrac{dy}{dx} = \cos x$, (b) $\dfrac{dy}{dx} = \dfrac{x}{\sqrt{x^2+1}}$, (c) $\dfrac{dy}{dx} = \dfrac{1}{x^2+16}$, (d) $\dfrac{dy}{dx} = xe^x$, (e) $\dfrac{dy}{dx} = \dfrac{1}{4-x^2}$, and (f) $\dfrac{dy}{dx} = \sin x \tan x$.

Solution

In each case, we integrate the indicated function and graph the solution for several values of the constant of integration. Each solution contains a constant of integration so there are infinitely many solutions to each equation. (a) $y = \int \cos x \, dx = \sin x + C$ (see Figure 1.4(a)). (b) To evaluate $y = \int (x/\sqrt{x^2+1}) \, dx$, we let $u = x^2+1$ so that $du = 2x \, dx$ or $(1/2) \, du = x \, dx$. Then,

$$y = \frac{1}{2} \int \frac{2x}{\sqrt{x^2+1}} \, dx$$

$$= \frac{1}{2} \int \frac{1}{\sqrt{u}} \, du = \frac{1}{2} \int u^{-1/2} \, du = u^{1/2} + C$$

$$= \sqrt{x^2+1} + C \quad \text{(see Figure 1.4(b))}.$$

(c) To integrate $y = \int (1/x^2 + 16) \, dx$, we use a trigonometric substitution. Letting $x = 4\tan\theta$, $-\pi/2 < \theta < \pi/2$, so that $dx = 4\sec^2\theta \, d\theta$ gives us

$$y = \int \frac{1}{16 + (4\tan\theta)^2} 4\sec^2\theta \, d\theta$$

$$= \int \frac{1}{16 + 16\tan^2\theta} 4\sec^2\theta \, d\theta$$

$$= \frac{1}{16} \int \frac{1}{\sec^2\theta} 4\sec^2\theta \, d\theta$$

$$= \frac{1}{4} \int d\theta = \frac{1}{4}\theta + C = \frac{1}{4}\tan^{-1}\left(\frac{x}{4}\right) + C.$$

First, we use the identity $1 + \tan^2\theta = \sec^2\theta$ and then resubstitute: $x/4 = \tan\theta$ so $\theta = \tan^{-1}(x/4)$.

(d) To evaluate $y = xe^x$ by hand, we use the Integration by Parts formula with $u = x$ and $dv = e^x \, dx$. Then, $du = dx$ and $v = e^x$. This gives

$$y = \int xe^x \, dx = xe^x - \int e^x \, dx = xe^x - e^x + C.$$

(e) Use partial fractions to evaluate this integral. First, find the partial fraction decomposition of $1/(4 - x^2)$, which is determined by finding constants A and B that satisfy the following equation:

$$\frac{1}{(2-x)(2+x)} = \frac{A}{2-x} + \frac{B}{2+x}.$$

These values are $A = B = 1/4$. (Why?) Thus,

$$y = \frac{1}{4} \int \left(\frac{1}{2-x} + \frac{1}{2+x}\right) dx$$

$$= \frac{1}{4} [-\ln|2-x| + \ln|2+x|] + C$$

$$= \frac{1}{4} \ln\left|\frac{2+x}{2-x}\right| + C \quad \text{(see Figure 1.5)}.$$

The *Integration by Parts formula* states that $\int u \, dv = uv - \int v \, du$.

 Find C if $y(1) = 2$.

Notice that the solutions of $dy/dx = 1/(4 - x^2)$ are undefined if $x = -2$ or $x = 2$. This is because $1/(4-x^2)$ is undefined at these two values of x. Later, we will discuss in greater detail the relationship between the differential equation and its solutions.

(f) In this case, $y = \int \sin x \tan x \, dx = (\sin^2 x / \cos x) \, dx$ because $\tan x = \sin x / \cos x$. Now use the identity $\sin^2 x + \cos^2 x = 1$ or $\sin^2 x = 1 - \cos^2 x$ and divide:

$$y = \int \frac{\sin^2 x}{\cos x} \, dx,$$

$$y = \int \frac{1 - \cos^2 x}{\cos x} \, dx,$$

$$y = \int (\sec x - \cos x) \, dx,$$

$$y = \ln|\sec x + \tan x| - \sin x + C.$$

For $-\pi/2 < x < \pi/2$, $\int \sec x \, dx = \ln(\sec x + \tan x) + C$.

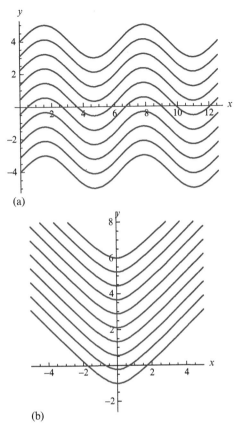

(a)

(b)

FIGURE 1.4 (a) Graph of $y = \sin x + C$ for various values of C. (b) Graph of $y = \sqrt{x^2 + 1} + C$ for various values of C.

In Example 1.1.4, each solution is given as a function $y = y(x)$ of the independent variable. In these cases, the solution is said to be *explicit*. In solving some differential equations, however,

We will see that given an *arbitrary* differential equation, constructing an explicit or implicit solution is nearly always impossible. Consequently, although mathematicians were first concerned with finding analytic (explicit or implicit) solutions to differential equations, they have since (frequently) turned their attention to addressing properties of the solution and finding algorithms to approximate solutions.

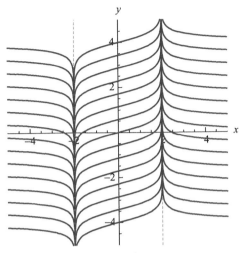

FIGURE 1.5 Graph of $y = \frac{1}{4} \ln |(2 + x)/(2 - x)| + C$ for various values of C.

we can find only an equation involving the independent and dependent variables that the solution satisfies. In this case, we say that we have found an *implicit* solution.

EXAMPLE 1.1.5

Verify that the equation $2x^2 + y^2 - 2xy + 5x = 0$ satisfies the following differential equation:
$$\frac{dy}{dx} = \frac{2y - 4x - 5}{2y - 2x}.$$

Solution

We use implicit differentiation to compute $y' = dy/dx$ if $2x^2 + y^2 - 2xy + 5x = 0$:

$$4x + 2y\frac{dy}{dx} - 2x\frac{dy}{dx} - 2y + 5 = 0,$$

$$\frac{dy}{dx}(2y - 2x) = 2y - 4x - 5,$$

$$\frac{dy}{dx} = \frac{2y - 4x - 5}{2y - 2x}.$$

The equation $2x^2 + y^2 - 2xy + 5x = 0$ satisfies the differential equation $dy/dx = (2y - 4x - 5)/(2y - 2x)$.

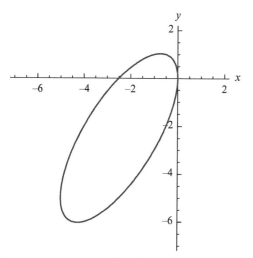

FIGURE 1.6 Graph of $2x^2 + y^2 - 2xy + 5x = 0$.

Although we cannot solve $2x^2 + y^2 - 2xy + 5x = 0$ for y as a *function* of x (see Figure 1.6), we can determine the corresponding y value(s) for a given value of x. For example, if $x = -1$, then

$$2 + y^2 + 2y - 5 = y^2 + 2y - 3 = (y+3)(y-1) = 0.$$

Therefore, the points $(-1, -3)$ and $(-1, 1)$ lie on the graph of $2x^2 + y^2 - 2xy + 5x = 0$ (see Figure 1.7).

Find an equation of the line tangent to the graph of $2x^2 + y^2 - 2xy + 5x = 0$ at the points $(-1, -3)$ and $(-1, 1)$.

In the same manner that we consider systems of equations in algebra, we can also consider systems of differential equations. For example, if x and y represent functions of t, we will learn in Chapter 6 to solve the *system of linear equations*

$$\begin{cases} dx/dt = ax + by, \\ dy/dt = cx + dy, \end{cases}$$

where a, b, c, and d represent constants and differentiation is with respect to t. We will see that systems of differential equations arise naturally in many physical situations that are modeled with more than one equation and involve more than one dependent variable. In addition, we will see that it is often useful to write a differential equation of order greater than one as a system of first-order equations, especially when the original equation is nonlinear.

EXAMPLE 1.1.6

(Duffing's equation). *Duffing's equation* is the second-order nonlinear equation

$$\frac{d^2x}{dt^2} + k\frac{dx}{dt} - x + x^3 = \Gamma \cos \omega t, \qquad (1.2)$$

where k, Γ, and ω are positive constants.

Sources: See texts like Jordan and Smith's *Nonlinear ODEs*, [16].

Write Duffing's equation as a system of first-order equations

Solution

Let $y = x'$. Then, $y' = x''$ and substituting into Duffing's equation gives us

$$x'' + kx' - x + x^3 = \Gamma \cos \omega t,$$
$$y' + ky - x + x^3 = \Gamma \cos \omega t,$$
$$y' = x - x^3 - ky + \Gamma \cos \omega t.$$

Thus, Duffing's equation is equivalent to the nonlinear system

$$\begin{cases} x' = y, \\ y' = x - x^3 - ky + \Gamma \cos \omega t. \end{cases}$$

Note that a system of differential equations can consist of more than two equations. For

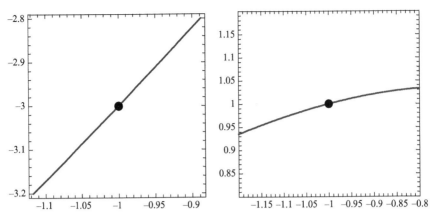

FIGURE 1.7 Notice that near the points $(-1, -3)$ and $(-1, 1)$ the implicit solution *looks* like a function. In fact, when we zoom in near the points $(-1, -3)$ and $(-1, 1)$, we see what appears to be the graph of a function.

example, the basic equations that describe the competition between two organisms, with population densities x_1 and x_2, respectively, in a chemostat are

Sources: See Smith and Waltman's *The Theory of the Chemostat* [27], for a detailed discussion of chemostat models.

$$\begin{cases} S' = 1 - S - \dfrac{m_1 S}{a_1 + S}x_1 - \dfrac{m_2 S}{a_2 + S}x_2 \\ x_1' = x_1\left(\dfrac{m_1 S}{a_1 + S} - 1\right) \\ x_2' = x_2\left(\dfrac{m_2 S}{a_2 + S} - 1\right) \end{cases} \quad , \quad (1.3)$$

where $'$ denotes differentiation with respect to t; $S = S(t)$, $x_1 = x_1(t)$, and $x_2 = x_2(t)$. For Equations (1.3), we remark that S denotes the concentration of the nutrient available to the competitors with population densities x_1 and x_2. We investigate chemostat models in more detail in Chapter 7.

 EXERCISES 1.1

Determine if each of the following equations is an ODE or a partial differential equation. If the equation is an ODE, then determine (a) the order of the ODE and (b) if the equation is linear or nonlinear.

1. $\dfrac{d^2y}{dx^2} + \dfrac{dy}{dx} - 2y = x^3$

2. $y\dfrac{dy}{dx} + y^4 = \sin x$

3. $\dfrac{\partial^2 y}{\partial t^2} = c^2\dfrac{\partial^2 y}{\partial x^2}, c > 0$ constant

4. $y''' - 2y'' + 5y' + y = e^x$,
   $(' = d/dx; y = y(x))$

5. $\left(\dfrac{dy}{dx}\right)^2 + y = 0$

6. $t^2\dfrac{d^2y}{dt^2} + t\dfrac{dy}{dt} + 2y = 0$

7. $\dfrac{1}{c^2}\dfrac{\partial^2 z}{\partial t^2} = \dfrac{\partial^2 z}{\partial x^2} + \dfrac{\partial^2 z}{\partial y^2}, (z = z(t, x, y))$

8. $u\,u_x + u_t = 0, (u = u(t, x))$

9. $x\left(\dfrac{d^2y}{dx^2}\right)^2 + 2y = 2x$

10. $\dfrac{d^2x}{dt^2} + 2\sin x = \sin 2t,\ (x = x(t))$

11. $u_{\widehat{t}} + u\,u_x = \sigma\,u_{xx},\ \sigma$ constant

12. $(2x - 1)\,dx - dy = 0$

13. $(2t - y)\,dt - dy = 0$

14. $\dfrac{\partial u}{\partial x}\dfrac{\partial u}{\partial y} = u,\ (u = u(x, y))$

15. $(2x - y)\,dx - y\,dy = 0$

16. Write each of the following second-order equations as a system of first-order equations:

   (a) $\dfrac{d^2x}{dt^2} - \dfrac{dx}{dt} - 6x = 0$

   (b) $4\dfrac{d^2x}{dt^2} + 4\dfrac{dx}{dt} + 37x = 0$

   (c) $L\dfrac{d^2x}{dt^2} + g\,\sin x = 0,\ L, g$ positive constants, $x = x(t)$

   (d) $\dfrac{d^2x}{dt^2} - \mu(1 - x^2)\dfrac{dx}{dt} + x = 0,\ \mu > 0$ constant

   (e) $t\dfrac{d^2x}{dt^2} + (b - t)\dfrac{dx}{dt} - ax = 0,\ a, b$ constants

In Exercises 17–28, verify that each of the given functions is a solution to the corresponding differential equation. (A, B, and C represent constants.)

17. $dy/dx + 2y = 0,\ y(x) = e^{-2x},\ y(x) = 5e^{-2x}$

18. $dy/dx + xy = 0,\ y(x) = e^{-x^2/2}$

19. $dy/dx + y = \sin x,\ y(x) = e^{-x} - \tfrac{1}{2}\cos x + \tfrac{1}{2}\sin x$

20. $\dfrac{d^2y}{dt^2} - \dfrac{dy}{dt} - 12y = 0,\ y(t) = e^{4t},\ y(t) = e^{-3t}$

21. $y'' + 9y' = 0,\ y = A + Be^{-9t}$ (' denotes d/dt)

22. $x'' + 3x' - 10x = 0,\ x(t) = Ae^{2t} + Be^{-5t}$

23. $x'' + x = t\cos t - \cos t,\ x = A\cos t + B\sin t + \tfrac{1}{4}t^2\sin t - \tfrac{1}{2}t\sin t + \tfrac{1}{4}t\cos t$

24. $y'' - 12y' + 40y = 0,\ y = e^{6x}\cos 2x,\ y = e^{6x}\sin 2x$ (' denotes d/dx)

25. $y''' - 4y' = 0,\ y = A + Be^{2x} + Ce^{-2x}$

26. $y''' - 2y'' = 0,\ y = A + Bt + Ce^{2t}$

27. $x^2y'' - 12xy' + 42y = 0,\ y = Ax^6 + Bx^7$

28. $t^2y'' + 3ty' + 5y = 0,\ y = t^{-1}(A\cos(2\ln t) + B\sin(2\ln t))$

In Exercises 29–33, verify that the given equation satisfies the differential equation. Use the equation to determine y for the given value of x (or t). *Confirm your result by graphing each equation using appropriate technology.*

29. $dy/dx = -x/y,\ x^2 + y^2 = 16,\ x = 0$

30. $3y(t^2 + y)\,dt + t(t^2 + 6y)\,dy = 0,\ t^3y + 3ty^2 = 8,\ t = 2$

31. $dy/dx = -2y/x - 3,\ x^3 + x^2y = 100,\ x = 1$

32. $y\cos t\,dt + (2y + \sin t)\,dy = 0,\ y^2 + y\sin t = 1,\ t = 0$

33. $(y/x + \cos y)\,dx + (\ln x - x\sin y)\,dy = 0,\ y\ln x + x\cos y = 0,\ x = 1$

In Exercises 34–43, use integration to find a solution to the differential equation.

34. $dy/dx = (x^2 - 1)(x^3 - 3x)^3$

35. $dy/d = x\sin x^2$

36. $dy/dx = x/\sqrt{x^2 - 16}$

37. $dy/dx = 1/(x\ln x)$

38. $dy/dx = x\ln x$

39. $dy/dx = xe^{-x}$

40. $\dfrac{dy}{dx} = \dfrac{-2(x + 5)}{(x + 2)(x - 4)}$

41. $\dfrac{dy}{dx} = \dfrac{x - x^2}{(x + 1)(x^2 + 1)}$

42. $dy/dx = \sqrt{x^2 - 16}/x$

43. $dy/dx = (4 - x^2)^{3/2}$

44. $dy/dx = 1/(x^2 - 16)$

45. $dy/dx = \cos x\cot x$

46. $dy/dx = \sin^3 x\tan x$

In Exercises 45-54, use the indicated conditions with the indicated solution to determine the solution to the given problem.

47. $dy/dx + 2y = 0, y(0) = 2, y(x) = Ae^{-2x}$

48. $dy/dt + y = \sin t, y(0) = -1,$
$y(t) = Ae^{-t} - \frac{1}{2}\cos t + \frac{1}{2}\sin t$

49. $y'' - y' - 12y = 0, y(0) = 1, y'(0) = -1,$
$y = Ae^{4x} + Be^{-3x}$

50. $y'' + 9y' = 0, y(0) = 2, y'(0) = -1,$
$y = A + Be^{-9x}$

51. $y''' - 2y'' = 0, y(0) = 0, y'(0) = 1, y''(0) = 3,$
$y = A + Bx + Ce^{2x}$

52. $y''' - 4y' = 0, y(0) = 1, y'(0) = -1, y''(0) = 0,$
$y = A + Be^{2x} + Ce^{-2x}$

53. $t^2 y'' - 12t y' + 42y = 0, y(1) = 0, y'(1) = -1,$
$y(t) = At^6 + Bt^7$

54. $x^2 y'' + 3xy' + 5y = 0, y(1) = 0, y'(1) = 1,$
$y(x) = x^{-1}\left(A\cos(2\ln x) + B\sin(2\ln x)\right)$

In Exercises 55-58, solve the IVP. *Confirm your result by graphing each function on an appropriate interval using appropriate technology.*

55. $dy/dx = 4x^3 - x + 2, y(0) = 1$

56. $dy/dt = \sin 2t - \cos 2t, y(0) = 0$

57. $dy/dx = x^{-2}\cos\left(x^{-1}\right), y(2/\pi) = 1$

58. $dy/dx = (\ln x)/x, y(1) = 0$

59. The velocity of a falling object with mass m that is subjected to air resistance proportional to the instantaneous velocity v of. the object is found by solving the IVP
$$\begin{cases} m\,dv/dt = mg - cv, \\ v(0) = v_0, \end{cases}$$ where $c > 0$ is the proportionality constant.
  (a) Given that a general solution to $m\,dv/dt = mg - cv$ is $v(t) = mg/c + Ke^{-ct/m}$, find the solution of this IVP.
  (b) Determine $\lim_{t\to\infty} v(t)$.

60. The number of cells in a bacteria colony after t hours is determined by solving the
IVP $\begin{cases} dP/dt = kP, \\ P(0) = P_0. \end{cases}$
  (a) Given that a general solution of $dP/dt = kP$ is $P(t) = Ce^{kt}$, use the initial condition to find C.
  (b) Find the value of k so that the population doubles in 8 h.

61. In 1840, the Belgian mathematician-biologist Pierre F. Verhulst (1804-1849) developed the *logistic equation*, $dP/dt = rP - aP^2$, where r and a are positive constants, to predict the population $P(t)$ in certain countries.
  (a) Given that a general solution to this equation is $P(t) = r/(a + Ce^{-rt})$, find the solution that satisfies $P(0) = P_0$, where $P_0 > 0$ is constant.
  (b) Determine $\lim_{t\to\infty} P(t)$.

P. F. VERHULST.

Pierre Francois Verhulst (1804-1849)

62. The differential equation $dS/dt + 3S/(t + 100) = 0$, where $S(t)$ is the number of pounds of salt in a particular tank at time t, is used to approximate the amount of salt in the tank containing a saltwater mixture in which pure water is allowed to flow into the tank while the mixture is allowed to flow out of the tank. If $S(t) = 15,000,000/(t + 100)^3$, show that S satisfies $dS/dt + 3S/(t + 100) = 0$. What is the initial amount of salt in the tank? As $t \to \infty$, what happens to the amount of salt in the tank?

63. The displacement (measured from $x = 0$) of a mass attached to the end of a spring at time t is given by $x(t) = 3\cos 4t + \frac{9}{4}\sin 4t$. Show that x satisfies the ODE $x'' + 16x = 0$. What is the initial position of the mass? What is the initial velocity of the mass?

64. Show that $u(x, y) = \ln\sqrt{x^2 + y^2}$ satisfies *Laplace's equation*, $u_{xx} + u_{yy} = 0$.

65. The temperature in a thin rod of length 2π after t minutes at a position x between 0 and 2π is given by $u(x, t) = 3 - e^{-16kt}\cos 4x$. Show that u satisfies $u_t = k u_{xx}$. What is the initial temperature $(t = 0)$ at $x = \pi$? What happens to the temperature at each point in the wire as $t \to \infty$?

66. The displacement u of a string of length 1 at time t position x, where x is measured from $x = 0$, is given by $u(x, t) = \sin \pi x \cos t$. Show that u satisfies $\pi^2 u_{tt} = u_{xx}$. What is the value of u at the endpoints $x = 0$ and $x = 1$ for all values of t?

67. Find the value(s) of m so that $y = x^m$ is a solution of $x^2 y'' - 2xy' + 2y = 0$.

68. Find the value(s) of k so that $y = e^{kt}$ is a solution of $y'' - 3y' - 18y = 0$.

69. Use the fact that $(d/dx)\left(e^{2x}y\right) = e^{2x}(dy/dx) + 2e^{2x}y$ and integration to solve $e^{2x}\dfrac{dy}{dx} + 2e^{2x}y = e^x$.

70. Use the fact that $(d/dx)\left(e^x y\right) = e^x(dy/dx) + e^x y$ and integration to solve $e^x(dy/dx) + e^x y = xe^x$.

71. The *time-independent Schrödinger equation* is given by

$$-\frac{h^2}{2m}\frac{d^2\psi(x)}{dx^2} + U(x)\psi(x) = E\psi(x).$$

If $U(x) = 0$, find conditions on E so that $\psi(x) = A\sin(n\pi x/L)$ is a solution of the time-independent Schrödinger equation.

The great physicist **Erwin Schrödinger** (1887-1961) received a Nobel prize for his work in 1933.

72. A *singular solution* of a differential equation is a solution that cannot be derived from the general solution of the differential equation. Use implicit differentiation to show that $-1/x + 2/x^2 + 1/y - 1/y^2 = C$ is a general (implicit) solution of the differential equation $dy/dx = (x - 4)y^3/[x^3(y - 2)]$. Is $y = 0$ a solution of this differential equation? Is $y = 0$ a singular solution?

73. Show that $x + x^2/y = C$ is a general (implicit) solution of the differential equation $dy/dx = (y^2 + 2xy)/x^2$. Is $y = 0$ a solution of this differential equation? Is $y = 0$ a singular solution?

74. The current $I(t)$ in an *L-R circuit*, which contains a resistor, an inductor, and a voltage source, satisfies the differential equation $RI + LdI/dt = E(t)$, where R and L

are constants representing the resistance and the inductance and $E(t)$ is the voltage source. Is this equation linear or nonlinear? Determine the order of the equation.

In Exercises 75-77, (a) verify that the indicated function is a solution of the given differential equation and (b) graph the solution on the indicated interval(s).

75. $xy' + y = \cos x$, $y = (\sin x)/x$; $[-2\pi, 0) \cup (0, 2\pi]$

76. $16y'' + 24y' + 153y = 0$, $y = e^{-3t/4} \cos 3t$; $[0, 3\pi/2]$

77. $x^3 y''' + x^2 y'' + xy' - 40y = 0$, $y = x^{-1} \sin(3 \ln x)$; $(0, \pi]$

78. **(a)** Verify that
$$\begin{cases} x = e^{-t} \left(\frac{100\sqrt{3}}{3} \sin \sqrt{3}t + 20 \cos \sqrt{3}t \right) \\ y = e^{-t} \left(-\frac{40\sqrt{3}}{3} \sin \sqrt{3}t + 20 \cos \sqrt{3}t \right) \end{cases}$$
is a solution of the system of differential equations $\begin{cases} dx/dt = 4y, \\ dy/dt = -x - 2y. \end{cases}$

(b) Graph $x(t)$, $y(t)$, and the parametric equations $\begin{cases} x = x(t) \\ y = y(t) \end{cases}$ for $0 \le t \le 2\pi$.

Throughout *Introductory Differential Equations*, we use graphs of solutions of differential equations. In some cases, we are able to predict what the graph of a solution should look like. If the graph of our proposed solution does not appear as predicted, we know that either we made a mistake in constructing our proposed solution or our conjecture about the general shape of the graph of the solution is wrong. In other cases, we will find that it is easier to examine the graph of a solution than it is to examine the solution (if we are able to construct one in the first place).

---

79. **(a)** Show that $(x^2 + y^2)^2 = 5xy$ is an implicit solution of
$$[4x(x^2+y^2)-5y]dx+[4y(x^2+y^2)-5x]dy=0.$$

(b) Graph $(x^2 + y^2)^2 = 5xy$ on the rectangle $[-2, 2] \times [-2, 2]$.

(c) Approximate all points on the graph of $(x^2 + y^2)^2 = 5xy$ with the x-coordinate 1.

(d) Approximate all points on the graph of $(x^2 + y^2)^2 = 5xy$ with the y-coordinate -0.319.

Often the calculus and algebra encountered in solving differential equations can be tedious, if not completely overwhelming or impossible. Today many sophisticated calculators and computer algebra systems are capable of performing the integration and algebraic simplification encountered when solving many differential equations. Having access to these tools can be a great advantage: a large number of problems can be solved quickly, we make conjectures as to the general form of a solution to different forms of differential equations, and these tools allow us to check and verify our work.

---

80. Solve $dy/dx = \sin^4 x$, $y(0) = 0$ and graph the resulting solution on the interval $[0, 4\pi]$.

81. A general solution of
$y^{(4)} + \frac{25}{2}y'' - 5y' + \frac{629}{16}y = 0$ is given by
$y = e^{-x/2}(c_1 \cos 3x + c_2 \sin 3x) + e^{x/2}(c_3 \cos 2x + c_4 \sin 2x)$, where c_1, c_2, c_3, and c_4 are constants. Solve the IVP
$$\begin{cases} y^{(4)} + \frac{25}{2}y'' - 5y' + \frac{629}{16}y = 0 \\ y(0) = 0, \ y'(0) = 1, \ y''(0) = -1, \quad \text{and} \\ y'''(0) = 1 \end{cases}$$
graph the resulting solution.

82. A general solution of the system
$$\begin{cases} dx/dt = 4y \\ dy/dt = -4x \end{cases} \text{is}$$

$$\begin{cases} x = -c_1 \cos 4t + c_2 \sin 4t \\ y = c_2 \cos 4t + c_1 \sin 4t \end{cases}, \text{ where } c_1 \text{ and }$$

c_2 are constants. Solve the IVP

$$\begin{cases} dx/dt = 4y \\ dy/dt = -4x \\ x(0) = 4, \, y(0) = 0 \end{cases} \text{ and then graph } x(t),$$

$y(t)$, and the parametric equations

$$\begin{cases} x = x(t), \\ y = y(t). \end{cases}$$

83. A general solution of the system

$$\begin{cases} dx/dt = -5x + 4y \\ dy/dt = 2x + 2y \end{cases} \text{ is}$$

$$\begin{cases} x = -c_1 e^{3t} - 4c_2 e^{-6t}, \\ y = 2c_1 e^{3t} + c_2 e^{-6t}, \end{cases} \text{ where } c_1 \text{ and } c_2$$

are constants. Solve the IVP

$$\begin{cases} x = -5x + 4y \\ y = 2x + 2y \\ x(0) = 4, \, y(0) = 0 \end{cases} \text{ and then graph } x(t),$$

$y(t)$, and the parametric equations

$$\begin{cases} x = x(t), \\ y = y(t). \end{cases}$$

1.2 A GRAPHICAL APPROACH TO SOLUTIONS: SLOPE FIELDS AND DIRECTION FIELDS

- SYSTEMS OF ODEs AND DIRECTION FIELDS
- RELATIONSHIP BETWEEN SYSTEMS OF FIRST-ORDER AND HIGHER ORDER EQUATIONS

Suppose that we are asked to solve the ODE $dy/dx = e^{-x^2}$. In this case, we do not attempt to solve this equation through integration as we did in the previous section because of the presence of the function $f(x) = e^{-x^2}$ on the right side of the ODE (see Exercise 21 at the end of this section). Instead, we can gain insight into the behavior of solutions of this ODE through a graphical approach by considering the slope of the tangent line to solutions of the ODE. Recall

from Section 1.1 that the differential equation gives the slope of the tangent line to solutions of the ODE at the given point (x, y) in the xy-plane. Therefore, if we wish to determine the slope of the tangent line to solution to the ODE that passes through $(0, 1)$ at this point, we substitute $(0, 1)$ into the right side of $dy/dx = e^{-x^2}$. Because the right side only depends on x, we have slope

$$\frac{dy}{dx} = e^{-(0)^2} = 1.$$

In fact, the slope of the line tangent to solutions at all points of the form $(0, y)$ is 1. At the point $(\sqrt{\ln 2}, 4)$, we find that the slope is

$$\frac{dy}{dx} = e^{-(\sqrt{\ln 2})^2} = e^{-\ln 2} = e^{\ln 2^{-1}} = \frac{1}{2}.$$

Again, we obtain the same slope at all points $(\sqrt{\ln 2}, y) \approx (0.832555, y)$ and $(-\sqrt{\ln 2}, y) \approx (-0.832555, y)$. In Figure 1.8, we draw several short line segments using points of the form $(0, y)$ for $y = 0, \pm 1, \pm 2, \pm 3$. Notice that we use a triangle with base length 1 and height length 1 to assist in sketching the tangent lines with slope 1 at these points. We use a triangle with base length 2 and height length 1 to help us draw the lines of slope $1/2$ that are tangent to solutions at the points $(-\sqrt{\ln 2}, y)$ for $y = 0, \pm 1, \pm 2, \pm 3$. By drawing a set of short line segments representing the tangent lines to solutions of the ODE at numerous points in plane, we construct the *slope field* of the ODE. We show the slope field for $dy/dx = e^{-x^2}$ on the square $[-2, 2] \times [-2, 2]$ in Figure 1.9(a). (Note that because constructing a slope field is time consuming, we usually let a computer algebra system do the work for us.) Observe that at each point along the y-axis, the slope is 1 as we predicted earlier. Notice also that the slope appears to be zero for larger values of $|x|$. This because for large values of x (in absolute value), $dy/dx \approx 0$ because $\lim_{x \to \pm \infty} e^{-x^2} = 0$. Therefore, we expect solutions to "flatten out" as x increases.

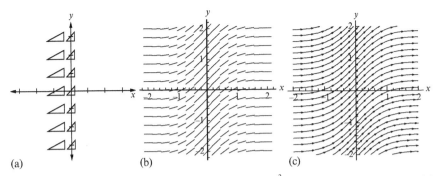

FIGURE 1.8   (a) Several line segments in the slope field for $dy/dx = e^{-x^2}$. (b) One view of the slope field for the equation. (c) A different view of the slope field for the equation.

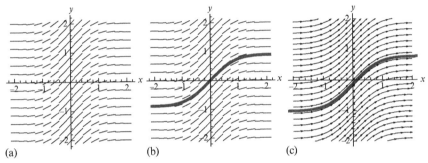

FIGURE 1.9   (a) Slope field for $dy/dx = e^{-x^2}$. (b) Using the slope field to sketch the solution to $dy/dx = e^{-x^2}$, $y(-2) = -1$. (c) A different view of using the slope field to sketch the solution to $dy/dx = e^{-x^2}$, $y(-2) = -1$.

We can use the slope field to investigate the solution to an IVP such as

$$\frac{dy}{dx} = e^{-x^2}, \quad y(-2) = -1$$

by starting at the point $(-2, -1)$ and tracing the solution by following the tangent slopes. This solution is sketched in Figure 1.9(b). Thus, although we were not able to determine explicit formulas for either the general solution of the ODE or the solution to the IVP, we were able to determine some properties of the solutions by using the slope field of the differential equation.

Systems of ODEs and Direction Fields

We can also consider systems of differential equations. In Chapter 6, we will learn how to solve systems of first-order ODEs of the form

$$\begin{cases} dx/dt = ax + by, \\ dy/dt = cx + dy, \end{cases}$$

where a, b, c, and d are given constants. In the case of this system, we solve for $x = x(t)$ and $y = y(t)$. For example, if we consider the system

$$\begin{cases} dx/dt = y, \\ dy/dt = -x, \end{cases}$$

we can verify that the parametric equations

$$\begin{cases} x(t) = \sin t \\ y(t) = \cos t \end{cases}$$

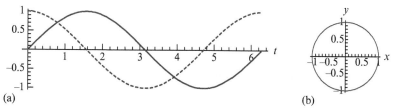

FIGURE 1.10 (a) Graphs of $x(t) = \sin t$ and $y(t) = \cos t$ for $0 \le t \le 2\pi$. (b) Graph of the parametric equations $x(t) = \sin t$ and $y(t) = \cos t$ for $0 \le t \le 2\pi$.

satisfy the system because

$$\frac{dx}{dt} = \frac{d}{dt}(\sin t) = \cos t = y \text{ and}$$

$$\frac{dy}{dt} = \frac{d}{dt}(\cos t) = -\sin t = -x.$$

We can graph each function separately as we do in Figure 1.10(a). (The graph of $x(t)$ is the solid curve; that of $y(t)$ is dashed.) Another option is to graph them as a pair of parametric equations as in Figure 1.10(b). As we recall, the graph of this pair of parametric equations is a circle of radius 1 centered at the origin because $x^2 = \sin^2 t$ and $y^2 = \cos^2 t$, so that $x^2 + y^2 = \sin^2 t + \cos^2 t = 1$. However, we must indicate the *orientation* of the curve (the direction of increasing parameter value t). For this pair of parametric equations, we find that at $t = 0$, $x(0) = \sin 0 = 0$ and $y(0) = \cos 0 = 1$. Therefore, the point $(0,1)$ corresponds to $t = 0$. Similarly, the point $(1,0)$ corresponds to $t = \pi/2$ because $x(\pi/2) = \sin \pi/2 = 1$ and $y(\pi/2) = \cos \pi/2 = 0$. This means that the solution moves from $(0,1)$ to $(1,0)$ as t increases. To determine if the orientation is clockwise or counter-clockwise, we test a t-value between 0 and $\pi/2$. Choosing $t = \pi/4$, we find the $x(\pi/4) = \sin \pi/4 = 1/\sqrt{2}$ and $y(\pi/4) = \cos \pi/4 = 1/\sqrt{2}$ so the orientation is clockwise. The parametric equations $\{x(t) = \sin t, y(t) = \cos t\}$ satisfy the *IVP*

$$\begin{cases} dx/dt = y, & x(0) = 0 \\ dy/dt = -x, & y(0) = 1 \end{cases}$$

because they satisfy the system of differential equations as well as the two initial conditions. Notice that in the case of an IVP involving a system of differential equations, an initial condition is given for each of the variables x and y that depend on t. In the parametric plot, the solution to this IVP passes through the point $(x(0), y(0)) = (0, 1)$.

Another way to view a system of two ODEs is through the use of a *direction field*, which is similar to a slope field. For example, for the first-order system

$$\begin{cases} dx/dt = ax + by, \\ dy/dt = cx + dy, \end{cases}$$

we first write it as a first-order equation with

$$\frac{dy}{dx} = \frac{dy/dt}{dx/dt} = \frac{cx + dy}{ax + by}.$$

Observe that the procedure described here can be used for any system of the form $\begin{cases} dx/dt = f(x,y), \\ dy/dt = g(x,y). \end{cases}$

Then, we can consider the slope field associated with this differential equation. For example, if we refer back to the system

$$\begin{cases} dx/dt = y, \\ dy/dt = -x, \end{cases}$$

we obtain the first-order equation $dy/dx = -x/y$, so we can determine the slope of tangent lines to solutions at points in the xy-plane. For example, at the point $(1/\sqrt{2}, 1/\sqrt{2})$, the solution to this system that passes through this point has slope $(-1/\sqrt{2})/(1/\sqrt{2}) = -1$. In a similar manner, we can find the slope at other points in the plane. However, as we mentioned in our earlier discussion, we must indicate the orientation when we graph parametric equations, so we consider the vector $\langle dx/dt, dy/dt \rangle = dx/dt\,\mathbf{i} + dy/dt\,\mathbf{j}$ with components from the system of differential equations. In the case of this system, we consider $\langle dx/dt, dy/dt \rangle = \langle y, -x \rangle$. At the point $(1/\sqrt{2}, 1/\sqrt{2})$, we obtain the vector $\langle 1/\sqrt{2}, -1/\sqrt{2} \rangle$. This means that the solution through $(1/\sqrt{2}, 1/\sqrt{2})$ has tangent vector $\langle 1/\sqrt{2}, -1/\sqrt{2} \rangle$. The direction field is made up of tangent vectors such as $\langle 1/\sqrt{2}, -1/\sqrt{2} \rangle$ to solutions at points in the plane, so it is similar to the slope field for $dy/dy = -x/y$ shown in Figure 1.11(a), except that vectors are used to indicate the orientation of solutions. In Figure 1.11(b), we show the direction field for this system. The vectors in the direction field indicate that solutions to this system are circles in the xy-plane that are directed clockwise. We graph several solutions along with the direction

---

Generally, we will use standard mathematical notation throughout *Introductory Differential Equations* so $\mathbf{i} = \langle 1, 0 \rangle$ and $\mathbf{j} = \langle 0, 1 \rangle$.

---

field in Figure 1.11(c). A collection of solutions in the xy-plane is called the *phase portrait* of the system. Notice that at points in the first quadrant, where $x > 0$ and $y > 0$, $dx/dt = y > 0$ and $dy/dt = -x < 0$. This means that $x(t)$ increases and $y(t)$ decreases along solutions in the first quadrant. In the second quadrant, where $x < 0$ and $y > 0$, $dx/dt = y > 0$ and $dy/dt = -x > 0$. Therefore, $x(t)$ and $y(t)$ increase along solutions in the second quadrant. We can perform a similar analysis for points in the other two quadrants.

Relationship Between Systems of First-Order and Higher Order Equations

Again, consider the system $\begin{cases} dx/dt = y, \\ dy/dt = -x. \end{cases}$ If we differentiate the equation $dx/dt = y$ with respect to t, we obtain $d^2x/dt^2 = dy/dt$. Therefore, if we equate $dy/dt = -x$ and $dy/dt = d^2x/dt^2$, we have $d^2x/dt^2 = -x$ or $d^2x/dt^2 + x = 0$. We say that the system of two first-order ODEs is *equivalent* to the second-order ODE $d^2x/dt^2 + x = 0$. Often, however, we begin with a second-order ODE of the form $d^2x/dt^2 + b\,dx/dt + cx = f(t)$ and would like to write the equation as a system of first-order ODEs. We do this by letting $dx/dt = y$ and by differentiating this equation with respect to t to obtain $d^2x/dt^2 = dy/dt$. Solving the second-order ODE for d^2x/dt^2, we find $d^2x/dt^2 = -b\,dx/dt - cx + f(t)$ so that $dy/dt = -b\,dx/dt - cx + f(t)$. Replacing dx/dt

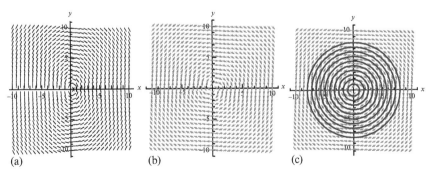

FIGURE 1.11 (a) Slope field for $dy/dx = -x/y$. (b) Direction field for $dx/dt = y$, $dy/dt = -x$. (c) Direction field for $dx/dt = y$, $dy/dt = -x$ and several solution curves.

in this equation with y, we obtain the following equivalent system of first-order ODEs:

$$\begin{cases} dx/dt = y, \\ dy/dt = -by - cx + f(t). \end{cases}$$

By writing the second-order ODE as a system of first-order ODEs, we investigate the behavior of the solution of the second-order ODE by observing the behavior in the direction field and phase portrait of the corresponding system. A similar procedure is used to transform an ODE of degree greater than two into a system of first-order ODEs.

EXERCISES 1.2

In Exercises 1-4, use the slope field to determine if the indicated path is that of a solution to the differential equation.

1. $\dfrac{dy}{dx} = -y/x$

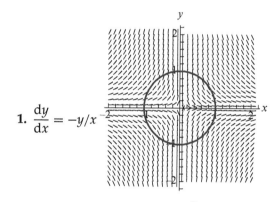

2. $\dfrac{dy}{dx} = x^2 + y^2$

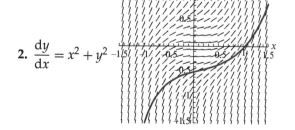

3. $\dfrac{dy}{dx} = x/y$

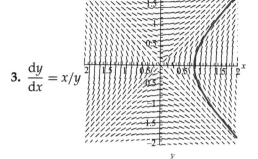

4. $\dfrac{dy}{dx} = x^2 - y$

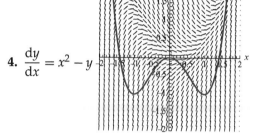

In Exercises 5-8, use the slope field to sketch the solutions of the differential equation that pass through the given points.

5. $\dfrac{dy}{dx} = x/y$

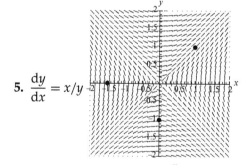

6. $\dfrac{dy}{dx} = x^2 - y$

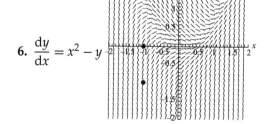

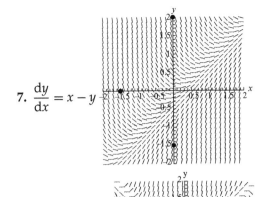

7. $\dfrac{dy}{dx} = x - y$

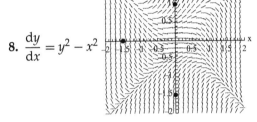

8. $\dfrac{dy}{dx} = y^2 - x^2$

9. Graph the slope field for $dy/dt = \sin y$. Determine $\lim_{t\to\infty} y(t)$ if $y(0) = y_0$ where (a) $y_0 = -3\pi/2$; (b) $y_0 = -\pi/2$; (c) $y_0 = \pi/2$; (d) $y(0) = 3\pi/2$.

10. Graph the slope field for $dy/dx = \sin x$. Does $\lim_{x\to\infty} y(x)$ exist for any initial condition $y(0) = y_0$? Solve the ODE and find $\lim_{x\to\infty} y(x)$. Does this match the graphical result?

11. Graph the slope field for $dy/dx = e^{-x}$. Does $\lim_{x\to\infty} y(x)$ exist for any initial condition $y(0) = y_0$? Solve the ODE and find $\lim_{x\to\infty} y(x)$. Does this match the graphical result?

12. Graph the slope field for $dy/dx = 1/(x^2 + 1)$. Does $\lim_{x\to\infty} y(x)$ exist for any initial condition $y(0) = y_0$? Solve the ODE and find $\lim_{x\to\infty} y(x)$. Does this match the graphical result?

In Exercises 13-14, use the direction field of the given system to sketch the graph of the solution that satisfies the indicated initial conditions. Determine $\lim_{t\to\infty} x(t)$ and $\lim_{t\to\infty} y(t)$ in each case (if they exist).

13. System: $\begin{cases} dx/dt = y \\ dy/dt = x \end{cases}$; (a) $x(0) = 0$, $y(0) = 2$; (b) $x(0) = -2$, $y(0) = 0$; (c) $x(0) = -2$, $y(0) = 2$;

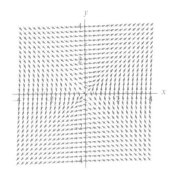

14. System: $\begin{cases} dx/dt = 3x - 2y \\ dy/dt = 4x - y \end{cases}$; $x(0) = 0$, $y(0) = 5$;

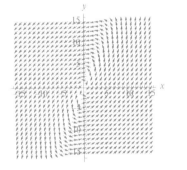

In Exercises 15-20, write the second-order equation as a system of first-order equations.

15. $d^2x/dt^2 + 4x = 0$
16. $d^2x/dt^2 - 5\,dx/dt = 0$
17. $d^2x/dt^2 + 4\,dx/dt + 13x = 0$
18. $x'' - 6x' + 7x = 0$ $('= d/dt; x = x(t))$
19. $x'' + 16x = \sin t$
20. $x'' + 4x' + 13x = e^{-t}$
21. (a) Use a computer algebra system to solve $dy/dx = e^{-x^2}$; (b) Graph the solution to the IVP $dy/dx = e^{-x^2}$, $y(0) = a$ for $a = -2, -1, 0, 1, 2$. (c) Graph the slope field of the differential equation together with the solutions in (b). (d) Do the solutions

appear to match the results described at the beginning of the section?

22. (a) Use a computer algebra system to graph the slope field for $dy/dx = \sin(2x - y)$. (b) Use the slope fie to graph the solution $y(x)$ that satisfies the initial condition $y(0) = 5$. Does $\lim_{x \to \infty} y(x)$ appear to exist?

23. Consider the systems

$$\begin{cases} dx/dt = -y \\ dy/dt = x \end{cases} \quad \text{and} \quad \begin{cases} dx/dt = -y, \\ dy/dt = -x. \end{cases}$$

(a) For any given initial conditions, in a brief paragraph explain why you *think* that the solutions of the systems are similar or different.

(b) Figure 1.12 shows the direction field associated with each system. Use the direction field to help you graph the solutions that satisfy these initial conditions

   (i) $x(0) = 0.5, y(0) = 0$;
   (ii) $x(0) = -0.25, y(0) = 0$;
   (iii) $x(0) = 0, y(0) = 0.75$; and
   (iv) $x(0) = 0, y(0) = -0.5$.

(c) How do your graphs affect your conjecture in (a)?

24. Consider the systems

$$\begin{cases} dx/dt = x/2 \\ dy/dt = y \end{cases} \quad \text{and} \quad \begin{cases} dx/dt = -x/2, \\ dy/dt = -y. \end{cases}$$

(a) For any given initial conditions, in a brief paragraph explain why you *think* that the solutions of the systems are similar or different.

(b) Figure 1.13 shows the direction field associated with each system. Use the direction field to help you graph the solutions that satisfy these initial condition

   (i) $x(0) = 0.5, y(0) = 0.25$;
   (ii) $x(0) = -0.25, y(0) = -0.5$;
   (iii) $x(0) = -0.5, y(0) = 0.75$; and
   (iv) $x(0) = 0.75, y(0) = -0.5$.

(c) How do your graphs affect your conjecture in (a)?

25. (*Competing Species*) The system of equations,
$$\begin{cases} dx/dt = x(a - b_1x - b_2y) \\ dy/dt = y(c - d_1x - d_2y) \end{cases}$$
where a, b_1, b_2, c, d_1, and d_2 represent positive constants, can be used to model the size of the population of two species, represented by $x(t)$ and $y(t)$, competing for a common food supply.

(a) Figure 1.14(a) shows the direction field for the system if $a = 1, b_1 = 2, b_2 = 1, c = 1, d_1 = 0.75$, and $d_2 = 2$. (i) Use the direction field to graph various solutions if both $x(0)$ and $y(0)$ are positive. (ii) Use

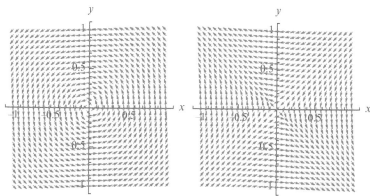

FIGURE 1.12   Figure for Exercise 23.

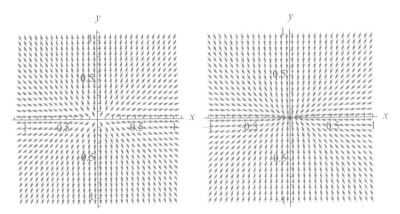

FIGURE 1.13   Figure for Exercise 24.

the direction field and your graphs to approximate $\lim_{t\to\infty} x(t)$ and $\lim_{t\to\infty} y(t)$.

(b) Figure 1.14(b) shows the direction field for the system if $a = 1, b_1 = 1, b_2 = 1, c = 0.67, d_1 = 0.75$, and $d_2 = 1$. (i) Use the direction field to graph various solutions if both $x(0)$ and $y(0)$ are positive. (ii) Use the direction field and your graphs to determine the fate of the species with population $y(t)$. What happens to the species with population $x(t)$?

CHAPTER 1 SUMMARY: ESSENTIAL CONCEPTS AND FORMULAS

Differential Equation (DE) An equation that contains the derivative or differentials of one or more dependent variables with respect to one or more dependent variables.

Ordinary Differential Equation (ODE) If a differential equation contains only ordinary derivatives (of one or more dependent variables) with respect to a single independent variable, the equation is called an *ODE*.

Partial Differential Equation (PDE) A differential equation that contains the partial derivatives or differentials of one or more independent variables with respect to more than one independent variable is called a *partial differential equation.*

Linear ODE A *linear ODE* is an equation that can be written in the form
$a_n(x)y^{(n)} + a_{n-1}(x)y^{(n-1)} + \cdots + a_2(x)y'' + a_1(x)y + a_0y = f(x)$, where $y = y(x)$.

Order of an Equation The order of the highest order derivative in a differential equation is called the *order* of the equation.

Solution A *solution* of a differential equation on a given interval is a function that is continuous on the interval and has all the necessary derivatives that are present in the differential equation such that when substituted into the equation yields an identity for all values on the interval.

Explicit Solution A solution given as a function of the independent variable.

Implicit Solution A solution given as a relation such as $f(x, y) = 0, f(t, x) = 0$, or $f(t, y) = 0$.

Trivial Solution $y = 0$ is *always* a solution of the nth-order linear homogeneous equation.

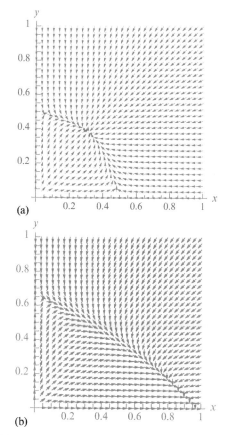

(a)

(b)

FIGURE 1.14 (a) Figure for Exercise 25 (a). (b) Figure for Exercise 25 (b).

Slope Field A collection of line segments that indicate the slope of the tangent line to the solution(s) of a differential equation.

Phase Portrait A collection of graphs of solutions to the system $\begin{cases} dx/dt = f(x,y) \\ dy/dt = g(x,y) \end{cases}$ in the xy-plane.

Direction Field A collection of vectors that indicate the slope and direction of the tangent line to the solutions of a system of differential equations.

CHAPTER 1 REVIEW EXERCISES

In Exercises 1-5, determine (a) if the equation is an ordinary differential or partial differential equation; (b) the order of the differential equation; and (c) if the equation is linear or nonlinear.

1. $dy/dt = y$
2. $a\,u_x + u_t = 0$, $u = u(x,t)$, a constant
3. $d^2y/dx^2 + 2\,dy/dx + y = 0$
4. $m\,x'' + kx = \sin t$, m and k positive constants; $x = x(t)$
5. $\dfrac{\partial \phi}{\partial x}\dfrac{\partial^2 \phi}{\partial x^2} = \dfrac{\partial^2 \phi}{\partial y^2}$

In Exercises 6-13, verify that the given function is a solution of the corresponding differential equation. (A and B denote constants.)

6. $dy/dx + y\cos x = 0$, $y = e^{-\sin x}$
7. $dy/dx - y = \sin x$, $y = (e^x - \cos x - \sin x)/2$
8. $y'' + 4y' - 5y = 0$, $y = e^{-5x}$, $y = e^x$
9. $y'' - 6y' + 45y = 0$, $y = e^{3x}(\cos 6x - \sin 6x)$
10. $xy'' - xy' - 16y = 0$, $y = Ax^5 + Bx^{-3}$
11. $x^2 y'' + 3xy' + 2y = 0$, $y = x^{-1}(\cos(\ln x) - \sin(\ln x))$
12. $d^2y/dx^2 + 2\,dy/dx + 2y = x$, $y = (x-1)/2$
13. $y'' - 7y' + 12y = 2$, $y = Ae^{3x} + Be^{4x} + 1/6$

In Exercises 14 and 15, verify that the given implicit function satisfies the differential equation.

14. $(2x - 3y)dx + (2y - 3x)dy = 0$, $x^2 - 3xy + y^2 = 1$
15. $(y\cos(xy) + \sin x)dx + x\cos(xy)dy = 0$, $\sin(xy) - \cos x = 0$

In Exercises 16-19, find a solution of the differential equation.

16. $dy/dx = xe^{-x^2}$
17. $dy/dx = x^2 \sin x$
18. $\dfrac{dy}{dx} = \dfrac{2x^2 - x + 1}{(x-1)(x^2+1)}$
19. $dy/dx = x^2/\sqrt{x^2 - 1}$

Exercises 20-22, use the indicated initial or boundary conditions with the given general solution to determine the solution(s) to the given problem.

20. $dy/dx + 2y = x^2$, $y(0) = 1$,
    $y = \frac{1}{4} - \frac{1}{2}x + \frac{1}{2}x^2 + Ae^{-2x}$
21. $y'' + 4y = t$, $y(0) = 1$, $y(\pi/4) = \pi/16$,
    $y = t/4 + A\cos 2t + B\sin 2t$
22. $x^2y'' + 5xy' + 4y = 0$, $y(1) = 1$, $y'(1) = 0$,
    $y = Ax^{-2} + Bx^{-2}\ln x$

Exercises 23 and 24, solve the initial-value problem. Graph the solution on an appropriate interval.

23. $dy/dx = \cos^2 x \sin x$, $y(0) = 0$
24. $dy/dx = \frac{1}{3}(4x - 9)(x - 3)^{-2/3}$, $y(0) = 0$
25. The temperature on the surface of a steel ball at time t is given by $u(t) = 70e^{-kt} + 30$ (in °F) where k is a positive constant. Show that u satisfies the first-order equation $du/dt = -k(u - 30)$. What is the initial temperature ($t = 0$) on the surface of the ball? What happens to the temperature as $t \to \infty$?
26. The displacement (measured from $x = 0$) of a mass attached to the end of a spring at time t is given by $x(t) = \frac{1}{4}e^{-t}\left(\cos\sqrt{35}t + \frac{9}{\sqrt{35}}\sin\sqrt{35}t \right)$. Show that x satisfies the ODE $x'' + 2x' + 36x = 0$. What is the initial displacement of the mass? What is the initial velocity of the mass?
27. For a particular wire of length 1 ft., the temperature at time t hours at a position of x feet from the end ($x = 0$) of the wire is estimated by $u(x, t) = e^{-\pi^2 kt}\sin\pi x - e^{-4\pi^2 kt}\sin 2\pi x$. Show that u satisfies the equation $u_t = ku_{xx}$. What is the initial temperature ($t = 0$) at $x = 1$? What happens to the temperature at each point in the wire as $t \to \infty$?
28. The height u of a long string at time t and position x where x is measured from the

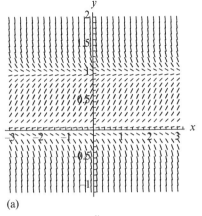

(a)

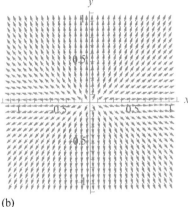

(b)

FIGURE 1.15 (a) Figure for Exercise 30. (b) Figure for Exercise 31.

middle of the string ($x = 0$) is given by $u(x, t) = \sin x \cos 2t$. Show that u satisfies the *wave equation* $u_{tt} = 4u_{xx}$. What is the initial height ($t = 0$) at $x = 0$?
29. Show that $u(x, y) = \tan^{-1}(y/x)$ satisfies Laplace's equation $u_{xx} + u_{yy} = 0$.
30. The slope field of $y' = 2(y - y^2)$ is shown in Figure 1.15(a). Sketch the graph of the solution that satisfies the indicated initial condition. Also, determine if $\lim_{x\to\infty} y(x)$ exists. (a) $y(0) = 0.5$, (b) $y(0) = 1.5$, (c) $y(0) = -0.5$.

31. The direction field for
$\{dx/dt = x,\ dy/dt = 2y\}$ is shown in
Figure 1.15(b). Is it possible to select initial
conditions so that $\lim_{t\to\infty} x(t) = 0$ and
$\lim_{t\to\infty} y(t) = 0$? (Briefly explain.)

32. The function
$J_1(x) = \sum_{k=0}^{\infty}((-1)^k/k!(k+1)!2^{2k+1})x^{2k+1}$ is
called the *Bessel function of order 1*. Verify
that $J_1(x)$ is a solution of *Bessel's equation of
order 1*, $x^2y'' + xy + (x^2 - 1)y = 0.$

CHAPTER

2

First-Order Equations

OUTLINE

We will devote a considerable amount of time in this course to developing explicit, implicit, numerical, and graphical solutions of differential equations. In this chapter, we discuss first-order ordinary differential equations (ODEs) and some methods used to construct explicit, implicit, numerical, and graphical solutions of them. Several of the equations and methods of solution discussed here will be used in later chapters.

2.1 INTRODUCTION TO FIRST-ORDER EQUATIONS

- SOME DIFFERENCES BETWEEN LINEAR AND NONLINEAR EQUATIONS

To better understand the solution or solutions to an initial-value problem (IVP), consider the following IVP:

$$dy/dt = t/y, \quad y(0) = 0.$$

Introductory Differential Equations
http://dx.doi.org/10.1016/B978-0-12-417219-7.00002-8

In Section 2.2, we will learn to solve the equation $dy/dt = t/y$ and see that its general solution is $y^2 - t^2 = C$. If $C \neq 0$, this solution is represented graphically by a family of hyperbolas. For $C < 0$, we have hyperbolas that intersect the t-axis (Figure 2.1(a)), and for $C < 0$, the hyperbolas intersect the y-axis (Figure 2.1(b)). In the case of this IVP, we are interested in finding the solution or solutions, if any, that satisfy $y(0) = 0$. In other words, we require the solution curve(s) pass through $(0,0)$ if the solution exists. When we substitute $(0,0)$ into $y^2 - t^2 = C$, we find that $C = 0$, so $y^2 - t^2 = 0$ satisfies the IVP. Factoring indicates that $(y - t)(y + t) = 0$ so the IVP has *two* solutions, $y = t$ and $y = -t$ (Figure 2.1(c)). A question we may ask at this point is whether the IVP has more than one solution if we use another initial condition. For example, if we consider

$$dy/dt = t/y, \quad y(1) = 0,$$

we have the same general solution, $y^2 - t^2 = C$, as before. However, when we apply the initial condition $y(1) = 0$, we find that $(0)^2 - (1)^2 = C$ so $C = -1$. Therefore, $y^2 - t^2 = -1$ so $y^2 = t^2 - 1$, which means that $y = \sqrt{t^2 - 1}$, $t \geq 1$ or $y = -\sqrt{t^2 - 1}$, $t \geq 1$. Both of these functions satisfy the IVP so this IVP also has two solutions. On the other hand, the IVP

$$dy/dt = t/y, \quad y(\sqrt{2}) = 1,$$

also leads to $y = \sqrt{t^2 - 1}$, $t \geq 1$ or $y = -\sqrt{t^2 - 1}$, $t \geq 1$ but because $y = \sqrt{t^2 - 1}$ is the only one of these two functions that satisfies $y(\sqrt{2}) = 1$, this IVP has the unique solution $y = \sqrt{t^2 - 1}$. As we see from this example, an IVP may have one or more solutions depending on how the problem is stated.

Charles Emile Picard (July 24, 1856, Paris, France-December 11, 1941, Paris, France): According to O'Connor [19], "Picard and his wife had three children, a daughter and two sons, who were all killed in World War I. His grandsons were wounded and captured in World War II."

In regards to Picard's teaching, the famous French mathematician Jacques Salomon Hadamard (1865-1963) wrote in Picard's obituary "A striking feature of Picard's scientific personality was the perfection of his teaching, one of the most marvelous, if not the most marvelous that I have known." [19]

Ernst Leonard Lindelöf (March 7, 1870, Helsingors, Russian Empire-June 4, 1946, Helsinki, Finland): Both Ernst Lindelöf and his father, Leonard Lorenz Lindelöf, were professors of mathematics at Helsinki. According to O'Connor [19], "Later in his life Lindelöf gave up research to devote himself to teaching and writing his excellent textbooks. ...he wrote the textbook *Differential and integral calculus and their applications* which was published in four volumes between 1920 and 1946. Another fine textbook *Introduction to function theory* was published in 1936."

In fact, an IVP may have no solutions, a unique solution, several solutions, or infinitely many solutions. The following theorem, which is

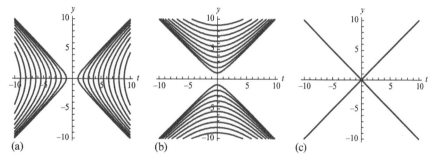

FIGURE 2.1   (a) Graphs of $y^2 - t^2 = C$ for several values of $C < 0$. (b) Graphs of $y^2 - t^2 = C$ for several values of $C > 0$. (c) Graphs of $y = t$ and $y = -t$.

the Picard-Lindelöf Theorem, Picard's Existence Theorem, or the Existence and Uniqueness Theorem, helps us understand the types of IVPs that have unique solutions.

Theorem 1 (Picard-Lindelöf   Theorem). *Consider the IVP*

$$\mathrm{d}y/\mathrm{d}t = f(t, y), \quad y(t_0) = y_0.$$

If f and $\partial f / \partial y$ are continuous functions on the rectangular region R: $a < t < b, c < y < d$ containing the point (t_0, y_0) then there exists an interval $|t - t_0| < h$ centered at t_0 on which there exists one and only one solution to the differential equation that satisfies the initial condition.

EXAMPLE 2.1.1

Does the fact that the IVP $\mathrm{d}y/\mathrm{d}t = t/y, y(0) = 0$ has two solutions contradict the Existence and Uniqueness Theorem?

Solution

In this case, $f(t, y) = t/y$ and $(t_0, y_0) = (0, 0)$. The hypothesis of the Existence and Uniqueness Theorem is not satisfied because f is not continuous at $(0, 0)$. Therefore, the fact that the IVP has two solutions does not contradict the Existence and Uniqueness Theorem.

The theorems proved by Picard in 1890 and Lindelöf in 1894 are stated more precisely than stated here in their original works and in more advanced texts like Refs. [6, 7, 2]. The technicalities are usually discussed in a more advanced differential equations course.

EXAMPLE 2.1.2

Verify that the IVP $\mathrm{d}y/\mathrm{d}t = y, y(0) = 1$ has a unique solution.

Solution

In this case, $f(t, y) = y$, $t_0 = 0$, and $y_0 = 1$. Hence, both $f(t, y) = y$ and $\partial f / \partial y = 1$ are continuous on all rectangular regions containing the point $(t_0, {}_0) = (0, 1)$. By the Picard-Lindelöf Theorem, there exists a unique solution to the differential equation that satisfies the initial condition $y(0) = 1$. In Sections 2.2 and 2.3, we will learn to solve first-order separable and linear equations so that we are able to find a general solution of $\mathrm{d}y/\mathrm{d}t = y$ is $y = Ce^t$ and the solution that satisfies the initial condition $y(0) = 1$ is $y = e^t$.

The Picard-Lindelöf Theorem gives sufficient, but not necessary, conditions for the existence of a unique solution to a first-order IVP. If an IVP does not satisfy the hypotheses of the theorem, we cannot conclude that a unique solution does not exist. In fact, the problem may have a unique solution, no solution, or many solutions, as illustrated in the examples and exercises.

EXAMPLE 2.1.3

Discuss the existence and uniqueness of solutions to the IVPs (a) $\begin{cases} dy/dt = ty^{-2/3} \\ y(0) = 0 \end{cases}$ and

(b) $\begin{cases} 2(ty^3 - t^5y)dt + (t^6 - t^2y^2)dy = 0, \\ y(0) = 0, \end{cases}$

if possible.

Solution

The Existence and Uniqueness Theorem does not guarantee the existence of a solution to either problem because both $ty^{-2/3}$ and $-2(ty^3 - t^5y)/(t^6 - t^2y^2)$ are discontinuous at the point $(0,0)$ specified by the initial condition.

(a) The equation $dy/dt = ty^{-2/3}$ is separable (see Section 2.2) and has general solution $y = \left(\frac{5}{6}t^2 + C\right)^{3/5}$. Application of the initial condition leads to the *unique* solution $y = (5/6)^{3/5}t^{6/5}$, which is graphed in Figure 2.2(a).

(b) Dividing the equation $2(ty^3 - t^5y)dt + (t^6 - t^2y^2)dy = 0$ by $(t^4 + y^2)^2$ leads to an exact equation (see Section 2.4) that has general solution $t^2y = C(t^4 + y^2)$. Applying the initial condition $y(0) = 0$ results in the identity $0 = 0$, which means that $t^2y = C(t^4 + y^2)$ is a solution to the IVP for *any* value of C. There are infinitely many solutions to the IVP, so the solution is *nonunique*. Several solutions are graphed in Figure 2.2(b).

Are the hypotheses of the Existence and Uniqueness Theorem satisfied for the IVP $dy/dt = \sqrt{y^2 - 1}$, $y(0) = 1$? Given that $\int (1/\sqrt{y^2 - 1})dy = \ln\left|y + \sqrt{y^2 - 1}\right| + C$, does a unique solution to this problem exist? Does more than one solution exist?

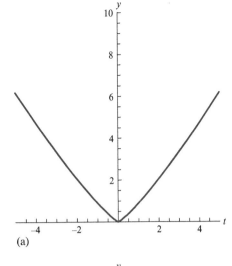

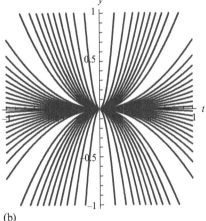

FIGURE 2.2 (a) Plot of $y = (5/6)^{3/5}t^{6/5}$. (b) Graph of $t^2y = C(t^4 + y^2)$ for various values of C.

Some Differences Between Linear Equations and Nonlinear Equations

While solving both linear and nonlinear first-order equations throughout this chapter, we will notice differences between the two types of equations.

One difference we will notice is that IVPs involving first-order linear equations will have unique solutions under much simpler situations than those stated in the Existence and Uniqueness Theorem.

Theorem 2 (Existence and Uniqueness: First-Order Linear Equations). *Consider the first-order IVP*

$$y' + p(t)y = q(t), \quad y(t_0) = y_0.$$

If $p(t)$ and $q(t)$ are continuous on an interval I that contains $t = t_0$, then there exists one and only one function that satisfies the differential equation and initial condition for each value of t in I.

Note that if the initial condition was $y(-2) = 4$, then we would have looked for a solution on the interval $-\infty < t < 0$.

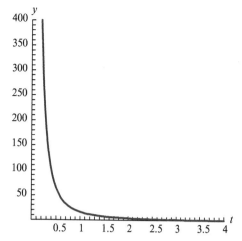

FIGURE 2.3   Solution of $ty' + 2y = t\cos t$, $y(2) = 4$.

EXAMPLE 2.1.4

Find the largest interval on which the IVP $ty' + 2y = t\cos t$, $y(2) = 4$ has a unique solution.

Solution

First, we write the ODE in the form $y' + (2/t)y = \cos t$ to identify $p(t) = 2/t$ and $q(t) = \cos t$ The function $p(t)$ is continuous for $t < 0$ or for $t > 0$; $q(t)$ is continuous for all values of t. Therefore, we must determine the interval containing $t_0 = 2$ for which both functions are continuous. This interval is $0 < t < \infty$.

When we solve this IVP (see Section 2.3 or use a computer algebra system (CAS)), we find that

$$y = \frac{1}{t^2}\left(2t\cos t - 2(-8 + 2\cos 2 + \sin 2)\right.$$

$$\left. + \left(-2 + t^2\right)\sin t\right).$$

We graph this function in Figure 2.3 to observe that the solution is valid only on $0 < t < \infty$. The solution becomes unbounded as $t \to 0^+$; it approaches zero as $t \to \infty$.

As we see, the interval on which the solution to a first-order IVP is unique depends on the functions $p(t)$ and $q(t)$ in the general form of the ODE. This is not necessarily true with an IVP involving a nonlinear ODE. In this case, the interval of definition of the solution may depend on the initial condition, as we illustrate in the following example.

EXAMPLE 2.1.5

Determine the interval of definition of the solution to $dy/dt = y^3$, $y(0) = a$, $a > 0$ constant.

Solution

In Section 2.2, we will learn that this equation is *separable*. Using the techniques discussed in Section 2.2 or a CAS, we find that a general solution of $dy/dt = y^3$ is

$$y = \frac{1}{\sqrt{C - 2t}} \quad \text{or} \quad y = 0, \, C \text{ an arbitrary constant.}$$

Applying the initial condition $y(0) = a$ indicates that $C = a^{-2}$, so $y = 1/\sqrt{a^{-2} - 2t}$. This solution is defined only if $a^{-2} - 2t > 0$ or $t < 1/(2a^2)$. In Figure 2.4, we graph the solution for

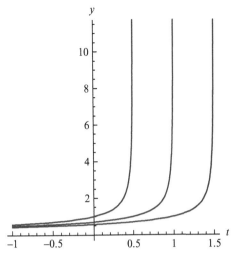

FIGURE 2.4   Graph of $y = 1/\sqrt{C - 2t}$ for $C = 1, 2, 3$.

$C = a^{-2} = 1, 2, 3$. In this graph, notice that if $a^{-2} = 1$, the solution is defined only for $t < 1/2$. The other two solutions are defined on $t < 1$ and $t < 2/3$, respectively. Notice also that each becomes unbounded as t approaches $1/(2a^2)$ from the left.

Another difference between linear and nonlinear equations that you will notice while working problems and studying examples in this chapter is that a linear equation has an explicit solution because of the way we solve linear ODEs (see Section 2.3). On the other hand, a nonlinear equation may have an implicit solution that cannot be solved analytically for the dependent variable to find an explicit form of the solution.

 EXERCISES 2.1

1. Does the Existence and Uniqueness Theorem guarantee a unique solution to the following IVPs on some interval? Explain.
   (a) $dy/dt + t^2 = y^2$, $y(0) = 0$.
   (b) $dy/dt + t^2 = y^{-2}$, $y(0) = 0$.
   (c) $dy/dt = y + 1/(1 - t)$, $y(1) = 0$.
2. According to the Existence and Uniqueness Theorem, the IVP $dy/dt = |y|$, $y(1) = 0$ has a unique solution. (a) Must the solution be unique? (b) Given that a general solution of $dy/dt = y$ is $y = Ce^t$ and a general solution of $dy/dt = -y$ is $y = Ce^{-t}$, solve the IVP by hand. Is the solution unique? *Hint:* Separate the IVP into two parts. First solve the problem for $y \geq 0$ and then solve it for $y < 0$.
3. The Existence and Uniqueness Theorem implies that at least one solution to the IVP $dy/dt = y^{1/5}$, $y(0) = 0$ exists. (a) Must the solution be unique according to the theorem? (b) Given that a general solution of $dy/dt = y^{1/5}$ is $y = \left(\frac{4}{5}t + C\right)^{5/4}$ or $y = 0$, solve the problem by hand. Is the solution unique?
4. Show that $y = 0$ and $y = \frac{1}{16}t^4$ both satisfy the IVP $1/t \, dy/dt = \sqrt{y}$, $y(0) = 0$. Does this contradict the Existence and Uniqueness Theorem?
5. Show that $y = 0$ and $y = t \, |t|$ both satisfy the IVP $dy/dt = 2\sqrt{|y|}$, $y(0) = 0$. Does this contradict the Existence and Uniqueness Theorem?
6. Does the IVP $dy/dt = 4t^2 - ty^2$, $y(2) = 1$ have a unique solution on an interval containing $t = 2$?
7. Does the IVP $dy/dt = y\sqrt{t}$, $y(1) = 1$ have a unique solution on an interval containing $t = 1$?
8. Does the IVP $dy/dt = 6y^{2/3}$, $y(1) = 0$ have a unique solution on an interval containing $t = 1$? In the next section, we will learn that by separating variables, we obtain that y and t satisfy $3y^{1/3} = 6t + C$. Is $y = 0$ obtained from this solution? Is $y = 0$ a solution to the IVP?
9. Does the IVP $y' = \sin y - \cos t$, $y(\pi) = 0$ have a unique solution on an interval containing $t = \pi$?

10. Does the IVP $ty' = y, y(0) = 1$ have a unique solution on an interval containing $t = 0$?

11. Show that $y = \sec t$ satisfies the IVP $y' = y \tan t, y(0) = 1$. What is the largest open interval containing $t = 0$ over which $y = \sec t$ is a solution? Explain.

12. Show that $y = \tan^{-1} t$ satisfies the IVP $y' = 1/(1 + t^2), y(0) = 0$. What is the largest open interval containing $t = 0$ over which $y = \tan t$ is a solution? Explain.

13. Using the Existence and Uniqueness Theorem, determine if $dy/dt = \sqrt{y^2 - 1}$ is guaranteed to have a unique solution passing through the point (a) $(0, 2)$, (b) $(4, -1)$, (c) $(0, 1/2)$, and (d) $(2, 1)$.

14. Using the Existence and Uniqueness Theorem, determine if $dy/dt = \sqrt{25 - y^2}$ is guaranteed to have a unique solution passing through the point (a) $(-4, 3)$, (b) $(0, 5)$, (c) $(3, -6)$, and (d) $(4, -5)$.

In Exercises 15-22, use the Existence and Uniqueness Theorem for Linear IVPs to determine the largest interval on which the solution is guaranteed to exist.

15. $ty' + y = t^3, y(1) = 0$.

16. $t^3 y' + t^4 y = 2t^3, y(0) = 0$.

17. $2y' + ty = \ln t, y(e) = 0$.

18. $y' + y \sec t = t, y(0) = 0$.

19. $y' + \dfrac{1}{t-3} y = \dfrac{1}{t-1}, y(-1) = 0$.

20. $(t - 2)y' + (t^2 - 4)y = \dfrac{1}{t+2}, y(0) = 3$.

21. $y' + y/\sqrt{4 - t^2} = t, y(0) = 0$.

22. $y' + y/\sqrt{4 - t^2} = t, y(3) = -1$.

23. (a) Over what interval is the solution to the IVP $ty' + y = t \sin t, y(\pi) = 1$ certain to exist? (b) Use a CAS to solve this IVP and graph the solution. (c) Is the solution valid on a larger interval than what is guaranteed by the Existence and Uniqueness Theorem for Linear IVPs?

24. (a) Over what interval is the solution to the IVP $y' + y \tan t = \sin t, y(0) = 0$ certain to

exist? (b) Use a CAS to solve this IVP and graph the solution. (c) Is the solution valid on a larger interval than what is guaranteed by the Existence and Uniqueness Theorem for Linear IVPs?

In Exercises 25-28, determine the interval of definition of the solution to each IVP. Use appropriate software to graph the solution for several values of $a > 0$ to verify your result. (Use values of a such as $a = 1/2, 1, 3/2$, and 2.)

25. $dy/dt = y^2, y(0) = a$.

26. $dy/dt = ty^2, y(0) = a$.

27. $dy/dt = -t/y, y(0) = a$.

28. $dy/dt = -y^3, y(0) = a$.

29. The proof of the Picard-Lindelöf Theorem relies on the proof of the following theorem.

Theorem 3. *Let $f(t, y)$ be continuous. A function $\phi(t)$, defined on an interval I, is a solution of $y' = f(t, y), y(t_0) = y_0$ if and only if it is a continuous solution of $y = y_0 + \int_{t_0}^{t} f(s, y) ds$.*

Prove this theorem. *Hints:* First assume that $\phi(t)$ is a solution of $y' = dy/dt = f(t, y), y(t_0) = y_0$. This means $\phi' = f(t, \phi)$. Now integrate both sides of the equation $\int_{t_0}^{t} \phi'(s) ds = \int_{t_0}^{t} f(s, \phi(s)) ds$ and simplify using the Fundamental Theorem of Calculus. For the converse, differentiate $\phi(t) = y_0 + \int_{t_0}^{t} f(s, \phi(s)) ds$ and simplify.

2.2 SEPARABLE EQUATIONS

Many interesting problems involving populations are solved through the use of first-order ordinary differential equations. For example, let $y(t)$ be the fish population size (in tons) at time t (years) and suppose that the population has birth rate $by(t)$ and mortality rate $my(t)$, where we are assuming that these rates are proportional to $y(t)$ and that b and m are constant. If there are no other factors affecting the rate of change in the population, dy/dt, then dy/dt equals the rate at which the population increases (the birth rate) minus the rate at which it decreases (the mortality rate), or

$$\frac{dy}{dt} = \text{(birth rate)} - \text{(mortality rate)}.$$

Mathematically, we represent this relationship with the differential equation

$$\frac{dy}{dt} = by - my = (b - m)y \quad \text{or} \quad \frac{dy}{dt} = ky,$$

where $k = b - m > 0$. If $y(0) = y_0$ is the initial population size, then we find $y(t)$ by solving the IVP

$$dy/dt = ky, \quad y(0) = y_0.$$

This is known as the *Malthus model* and is due to the work of the English clergyman and economist Thomas Robert Malthus (1766-1834). Notice that the ODE can be written in *differential form* as

$$\frac{1}{y}dy = k\,dt.$$

In 1798, **Thomas Robert Malthus** (1766-1834) published *An Essay on the Principle of Population*, which can be downloaded at *Electronic Scholarly Publishing* [23], where he noticed that the human population would double every 25 years unless it were kept under control by external factors like food supply and disease. For more information regarding early population models, point your browser to O'Neil's Web site [24]. You can learn more about "Bob" Malthus by pointing your browser to the *Malthus Family Homepage* [22].

We say that this ODE is *separable* because we are able to collect all of the terms involving y on one side of the equation and all the terms involving t on the other side of the equation.

Definition 1 (Separable Differential Equation). *A first-order differential equation that can be written in the form $g(y)y' = f(t)$ or $g(y)dy = f(t)dt$ is called a **separable differential equation**.*

After we place the ODE in this form, we solve through integrating each side with respect to the indicated variable. For the Malthus equation, this gives

$$\int \frac{1}{y}dy = \int k\,dt.$$

Integration yields

$$\ln|y| = kt + C_1,$$

where C_1 is a constant of integration. When possible to obtain an explicit solution, we like to solve for $y(t)$. In this case, we use the exponential function and its properties to simplify:

$$e^{\ln|y|} = e^{kt+C_1} = e^{kt}e^{C_1},$$

$$|y| = Ce^{kt},$$

$$y = Ce^{kt},$$

where we replace the constant e^{C_1}, with the arbitrary constant $C > 0$. Finally, because $y(t)$ represents population size, $y(t) \geq 0$ for all t, and we eliminate the absolute value to obtain a general solution of the ODE, $y(t) = Ce^{kt}$.

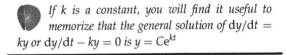
If k is a constant, you will find it useful to memorize that the general solution of $dy/dt = ky$ or $dy/dt - ky = 0$ is $y = Ce^{kt}$

Note: If we had not assumed that $y(t) \geq 0$ for all t, then we simplify $|y| = Ce^{kt}$ with $y = \pm Ce^{kt}$. Therefore, by letting $C_2 = \pm C$ or 0, we obtain $y = C_2e^{kt}$, which is equivalent to the obtained result.

To solve the IVP, we must choose C so that $y(0) = y_0$. Applying the initial condition, we find that $y(0) = Ce^{k \cdot 0} = C$, which indicates that $C = y_0$. Therefore, the solution to the IVP is $y(t) = y_0 e^{kt}$.

 Is $t^2 + y^2 = 0$ an implicit solution of $dy/dt = -t/y$? Briefly explain.

EXAMPLE 2.2.1

Show that the equation $dy/dt = -t/y$ is separable and solve this equation.

Solution

In this case, we see that the equation is separable by expressing it as $t \, dt = -y \, dy$. Integration yields

$$\frac{1}{2}t^2 = -\frac{1}{2}y^2 + C_1.$$

Multiplying by 2 and simplifying give us $t^2 + y^2 = 2C_1$. If we let $k = 2C_1$, then the solutions satisfy the equation $t^2 + y^2 = k$, so that an implicit solution is the family of circles centered at the origin with radius $\sqrt{k}$, as shown in Figure 2.5.

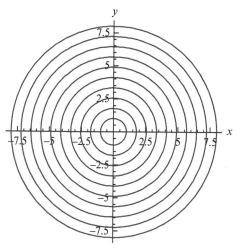

FIGURE 2.5 Graph of $t^2 + y^2 = k$ for various values of $k > 0$.

In the previous example, the relation that satisfies the differential equation is not an explicit function of the dependent variable. Graphs of equations that are solutions to a differential equation but are not (usually) explicit functions of the independent variable are often called *integral curves*.

As we have seen, separable differential equations can be stated along with an initial condition.

Theorem 4 (Existence and Uniqueness: First-Order Separable Equations). *Consider the IVP*

$$g(y)dy = f(t)dt, \quad y(t_0) = y_0.$$

If $f(t)$ is continuous on (t_1, t_2) with t_0 in (t_1, t_2) and $g(y)$ is continuous on (y_1, y_2) with y_0 in (y_1, y_2), then there exists one and only one function that satisfies the differential equation and initial condition for each value of t on an interval containing t_0.

In this situation, we find a general solution to the differential equation using the separation of variables technique and we then apply the initial condition to determine the unknown constant in the general solution.

EXAMPLE 2.2.2

Solve the IVP $dy/dt = e^{2t+y}$, $y(0) = 0$.

Solution

If we write the equation as $dy/dt = e^{2t}e^y$, we see that the equation is separable and that the variables are separated by writing the equation as $e^{-y} dy = e^{2t} dt$. In this case, $f(t) = e^{2t}$ is continuous for $-\infty < t < \infty$ and $g(y) = e^{-y}$ is continuous on $-\infty < y < \infty$. So, the Existence and Uniqueness Theorem guarantees a unique solution valid on an interval containing $t = 0$.

Integration gives us $-e^{-y} = \frac{1}{2}e^{2t} + C_1$ and simplifying results in $e^{-y} = -\frac{1}{2}e^{2t} - c_1$. Application of the natural logarithm to both sides of this equation yields

$$-y = \ln\left(-\frac{1}{2}e^{2t} - C_1\right) \quad \text{or}$$

$$y = -\ln\left(-\frac{1}{2}e^{2t} - C_1\right).$$

To find the value of C_1, so that the solution satisfies the condition $y(0) = 0$, we use the equation $-e^{-y} = \frac{1}{2}e^{2t} + C_1$. Applying $y(0) = 0$ yields $1 = -1/2 - C_1$, so $C_1 = -3/2$ and thus the solution to the IVP is

$$y = -\ln\left(-\frac{1}{2}e^{2t} + \frac{3}{2}\right).$$

The domain of the natural logarithm is $(0, \infty)$, so this solution is valid only for the values of t such that $3 - e^{2t} > 0$, which is equivalent to $e^{2t} < 3$ and has solution $t < \ln(3)/2 \approx 0.5493$. Notice that the graph of the solution shown in Figure 2.6 passes through the point $(0, 0)$. If it had not, we would check for mistakes in our solution procedure. Observe also that the domain of the solution is contained in the domain of $f(t)$.

In some cases, a separable differential equation may be difficult to solve because of the integration that is required.

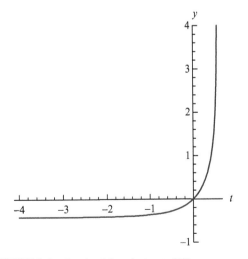

FIGURE 2.6   Graph of the solution to IVP.

EXAMPLE 2.2.3

 Solve $dy/dt = (t^4 + 1)(y^4 + 1)$, $y(0) = 0$.

Solution

When we separate variables, we obtain

$$\frac{1}{y^4 + 1}dy = (t^4 + 1)dt.$$

The right side of the equation is easily integrated. However, the left side presents a challenge. Using a CAS, we find that

$$\int \frac{1}{y^4 + 1}dy = \frac{1}{4\sqrt{2}}\left(-2\tan^{-1}(1 - \sqrt{2}y)\right.$$

$$\left. +2\tan^{-1}(\sqrt{2}y + 1) + \ln\left|\frac{y^2 + \sqrt{2}y + 1}{-y^2 + \sqrt{2}y - 1}\right|\right) + C.$$

Therefore, we find that

$$\frac{1}{4\sqrt{2}}\left(-2\tan^{-1}(1 - \sqrt{2}y) + 2\tan^{-1}(\sqrt{2}y + 1)\right.$$

$$\left. + \ln\left|\frac{y^2 + \sqrt{2}y + 1}{-y^2 + \sqrt{2}y - 1}\right|\right) + C = \frac{1}{5}t^5 + t$$

is a general solution of the equation. Substitution of $t = 0$ and $y = 0$ (from the initial condition) into this equation indicates that $C = 0$ because $\tan^{-1}(-1/\sqrt{2}) = -\pi/4$, $\tan^{-1}(1/\sqrt{2}) = \pi/4$, and $\ln|-1| = 0$. This means that the solution to the IVP is

$$\frac{1}{4\sqrt{2}}\left(-2\tan^{-1}(1 - \sqrt{2}y) + 2\tan^{-1}(\sqrt{2}y + 1)\right.$$

$$\left. + \ln\left|\frac{y^2 + \sqrt{2}y + 1}{-y^2 + \sqrt{2}y - 1}\right|\right) = \frac{1}{5}t^5 + t.$$

We can gain insight into the behavior of the solution by observing the slope field of the ODE. For example, the slope of the tangent line to the solution passing through $(0, 0)$ at $(0, 0)$ is $dy/dt = (0^4 + 1)(0^4 + 1) = 1$. The slope at $(0, 1)$ is $dy/dt = (0^4 + 1)(1^4 + 1) = 2$ and that at $(0, 2)$ is $dy/dt = (0^4 + 1)(2^4 + 1) = 17$. As we see, slopes become steeper as the y-coordinate increases (or decreases because we obtain the same values for dy/dt at $(0,-1)$ and $(0,-2)$). Along the t-axis, we obtain the same slopes at $(1,0)$ and $(2,0)$ as we did at $(0,1)$ and $(0,2)$ because dy/dt is symmetric in t and y. We generate the slope field with a CAS in Figure 2.7(a). To understand the solution of the IVP, we begin at the point $(0,0)$ and follow the tangent line slopes shown in the slope field for $t > 0$ and for $t < 0$. Doing so, we obtain the curve shown in Figure 2.7(b). We notice that y approaches $+\infty$ as t approaches a value near 0.9, and y approaches $-\infty$ as t approaches a value near -0.9. We also note that given any initial condition, the solution to the corresponding IVP becomes unbounded (i.e., y approaches $\pm\infty$) as the t-coordinate moves away from 0.

Notice that both $dy/dt = ky$, k constant, and $dy/dt = f(y)$ are *always* separable. See Exercises 13 and 20.

Equilibrium Solutions of $dy/dt = f(y)$

Malthus's equation, $dy/dt = ky$ is a member of a special category of separable equations because it has the form $y' = dy/dt = f(y)$. Such equations are called *autonomous equations* because the independent variable t does not appear in the equation on the right. We find that, although we are able to solve many equations of this form, we can learn much about the behavior of solutions without actually solving the differential equation.

An *equilibrium solution* of the autonomous first-order equation $dy/dt = f(y)$ is a constant solution of the ODE: a function $y(t) = C$, C constant for which $f(y) = 0$. Consider the equation

$$\frac{dy}{dt} = 2y - y^2 \quad \text{or} \quad \frac{1}{2y - y^2}dy = dt.$$

(Note that this is a nonlinear ODE because it involves the term y^2. We learn more about the differences in linear and nonlinear equations throughout this chapter.) Instead of solving by separating variables, let us use a graphical approach to determining properties of solutions. First, we locate the equilibrium solutions of the equation by solving $2y - y^2 = 0$ or $y(2 - y) = 0$.

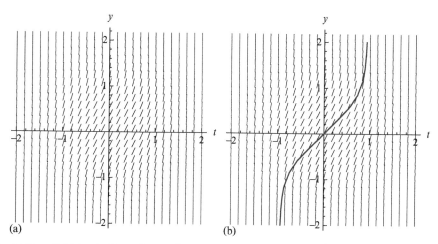

FIGURE 2.7 (a) Slope field for $dy/dt = (t^4 + 1)(y^4 + 1)$. (b) Solution to $dy/dt = (t^4 + 1)(y^4 + 1)$, $y(0) = 0$.

FIGURE 2.8   Phase line for $dy/dt = 2y - y^2$.

The roots of this equation are $y = 0$ and $y = 2$, so the equilibrium solutions are $y(t) = 0$ and $y(t) = 2$. Notice that each of these functions satisfies $dy/dt = 2y - y^2$ because $dy/dt = 0$ and $y(2 - y) = 0$.

Next, we investigate the behavior of solutions on the intervals $y < 0$, $0 < y < 2$, and $y > 2$ by sketching the *phase line*. After marking the two equilibrium solutions on the vertical line in Figure 2.8, we determine the sign of dy/dt on each of the three intervals listed above. This can be done by substituting a value of y on each interval into $f(y)v2y - y^2$. For example, $f(-1) = -3$, so $dy/dt < 0$ if $y < 0$. We use an arrow directed downward to indicate that solutions decrease on this interval. In a similar manner, we find that $f(1) = 1$ and $f(3) = -3$, so $dy/dt > 0$ if $0 < y < 2$, and $dy/dt < 0$ if $y > 2$. We include arrows on the phase line directed upward and downward, respectively, to indicate this behavior. Based on the orientation, solutions that satisfy $y(0) = a$, where either $0 < a < 2$ or $a > 2$ approaches $y = 2$ as $t \rightarrow \infty$, so we classify $y = 2$ as *asymptotically stable*. Solutions that satisfy $y(0) = a$ where $a < 0$ or $0 < a < 2$ move away from $y = 0$ as $t \rightarrow \infty$. We say that $y = 0$ is *unstable* because there are solutions that begin near $y = 0$ and move away from $y = 0$ as $t \rightarrow \infty$. In Figure 2.9, we graph several solutions to this equation (in the ty-plane) along with the phase portrait to illustrate this behavior. Notice that solutions with initial value near $y = 0$ move away from this line while solutions that begin on $0 < y < 2$ or $y > 2$ move toward $y = 2$ as $t \rightarrow \infty$.

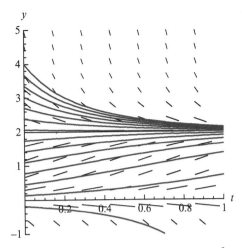

FIGURE 2.9   Several solutions of $dy/dt = 2y - y^2$.

Note: Sometimes, we refer to equilibrium solutions as *steady-state solutions*, because over time, the solution approaches a constant value that no longer depends on time t. For example, if we interpret the solution $y(t)$ of the differential equation to represent the size of a population, then if $y(0) = a$ where either $0 < a < 2$ or $a > 2$, we expect the population to approach 2 as $t \rightarrow \infty$.

Notice that the equation $dy/dt = f(y)$ is always separable because it can be written in the form $(1/f(y))dy = dt$.

EXAMPLE 2.2.4

Solve $dy/dt = 2y - y^2$.

Solution

After separating variables, we use partial fractions

$$\frac{1}{2y - y^2}dy = dt,$$

$$\frac{1}{2}\left(\frac{1}{y} - \frac{1}{y - 2}\right)dy = dt.$$

Now we integrate and apply properties of the logarithm to simplify

$$\frac{1}{2}\left(\ln|y| - \ln|y - 2|\right) = t + C_1,$$

$$\ln\left|\frac{y}{y - 2}\right| = 2t + C_2,$$

$$\frac{y}{y - 2} = C_3 e^{2t}.$$

Solving for y gives us

$$y = \frac{2C_3 e^{2t}}{C_3 e^{2t} - 1} = \frac{2}{1 + Ce^{-2t}},$$

where C is arbitrary.

Note that from the general solution, we can obtain the equilibrium solution $y = 2$ by setting $C = 0$. However, there is no value of C for which this general solution gives us the solution $y = 0$. Thus, $y = 0$ is a singular solution for this equation. Combining these two solutions together means that all solutions of the equation $dy/dt = 2y - y^2$ can be written as

$$y = \frac{2}{1 + Ce^{-2t}} \quad \text{or} \quad y = 0,$$

where C is constant.

 EXERCISES 2.2

In Exercises 1-30, solve each equation.

1. $dy/dx = x/y^2$.

2. $\frac{1}{2}t^{-1/2} dt + y^2 dy = 0$.

3. $dy/dx = \sqrt{y}/x^2$.

4. $dy/dt = (1 + y^2)/y$.

5. $(6 + 4t^3)dt + (5 + 9y^{-8})dy = 0$.

6. $(6t^{-9} - 6t^{-3} + t^7)dt + (9 + s^{-2} - 4s^8)ds = 0$. (Solve for $s = s(t)$.)

7. $4\sinh 4y\, dy = 6\cosh 3x\, dx$.

8. $dy/dt = (y + 1)/(t + 1)$.

9. $dy/dt = (y + 2)/(2t + 1)$.

10. $3t^{-2}\, dt = (y^{-1/2} + y^{1/2})\, dy$.

11. $3\sin x\, dx - 4\cos y\, dy = 0$.

12. $\cos y\, dy = 8\sin 8t\, dt$.

13. $y' + ky = 0$ (k constant).

14. $(5x^5 - 4\cos x)dx + (2\cos 9t + 2\sin 7t)dt = 0$.

15. $(\cosh 6t + 5\sinh 4t)dt + 20\sinh y\, dy = 0$.

16. $dy/dt = e^{2y+10t}$.

17. $dy/dt = e^{3y+2t}$.

18. $\sin^2\theta\, d\theta = \cos^2\phi\, d\phi$.

19. $(3\sin\theta - \sin 3\theta)d\theta = (\cos 4y - 4\cos y)dy$.

20. $dx/dt = \sec^2 t/(\sec x \tan x)$.

21. $(2 - 5/y^2)dy + 4\cos^2 x\, dx = 0$.

22. $\dfrac{dy}{dt} = \dfrac{t^3}{y\sqrt{(1 - y^2)(t^4 + 9)}}$.

23. $\tan y\, \sec^2 y\, dy + \cos^3 2x\, \sin 2x\, dx = 0$.

24. $\dfrac{dy}{dt} = \dfrac{1 + 2e^y}{e^y\, t\ln t}$.

25. $x\sin\left(x^2\right) dx = \dfrac{\cos\sqrt{y}}{\sqrt{y}}dy$.

26. $\dfrac{x - 2}{x^2 - 4x + 3}dx = \left(1 - \dfrac{1}{y}\right)^2 \dfrac{1}{y^2}dy$.

27. $\dfrac{\cos y}{(1 - \sin y)^2}dy = \sin^3 x\cos x\, dx$.

28. $\dfrac{d\phi}{d\theta} = \dfrac{(5 - 2\cos\theta)^3\,\sin\theta\,\cos^4\phi}{\sin\phi}$.

29. $\dfrac{\sqrt{\ln x}}{x}dx = \dfrac{e^{3/y}}{y^2}dy$.

30. $dy/dt = 5^{-t}/y^2$.

31. $dy/dt = t^2 y^2 + y^2 - t^2 - 1$.

32. $dy = (y^2 - 3y + 2)\, dx$.

33. $4(x - 1)^2 dy - 3(y + 3)^2 dx = 0$.

34. $dy/dt = \sin(t - y) + \sin(t + y)$.

35. $dy/dt = y^3 + 1$.

36. $dy/dt = y^3 - 1$.

37. $dy/dt = y^3 + y$.

38. $dy/dt = y^3 - y^2$.

39. $dy/dt = y^3 - y$.

40. $dy/dt = y^3 + y$.

In Exercises 36-40, draw the phase line for the equation and classify the equilibrium solutions.

In Exercises 41-52, solve the IVP. Graph the solution on an appropriate interval.

41. $dy/dx = x^3$, $y(0) = 0$.

42. $dy/dt = \cos t$, $y(\pi/2) = -1$.

43. $dx = \cos y\, dy$, $x(0) = 2$.

44. $\sin^2 y\, dy = dx$, $x(0) = 0$.

45. $dy/dt = \sqrt{t}/y$, $y(0) = 2$.

46. $d\phi/d\theta = \sqrt{\phi/\theta}$, $\phi(1) = 2$.

47. $dy/dt = e^t/(y+1)$, $y(0) = -2$.

48. $dy/dy = e^{t-y}$, $y(0) = 0$.

49. $dy/dx = y/\ln y$, $y(0) = e$.

50. $dy/dt = t\sin(t^2)$, $y(\sqrt{\pi}) = 0$.

51. $dy/dx = 1/(1+x^2)$, $y(0) = 1$.

52. $\dfrac{dy}{d\theta} = \dfrac{\sin\theta}{\cos y + 1}$, $y(0) = 0$.

53. $dy/dxn = (y+3)/(3x+1)$, $y(0) = 1$.

54. $dy/dx = e^{x-y}$, $y(0) = 1$.

55. $dy/dx = e^{2x-y}$, $y(0) = 1$.

56. $dy/dx = (3y+1)/(x+3)$, $y(0) = 1$.

57. Solve each of the following IVPs and graph the results.

(a) $\begin{cases} dy/dt = y\cos t, \\ y(0) = 1. \end{cases}$

(b) $\begin{cases} dy/dt = y^2\cos t, \\ y(0) = 1. \end{cases}$

(c) $\begin{cases} dy/dt = \sqrt{y}\cos t, \\ y(0) = 1. \end{cases}$

58. Carefully show that the solution to the IVP $y' + f(t)y = 0$, $y(0) = 0$ is $y = y_0 e^{-\int f(t)\, dt}$.

59. Find an equation of the curve that passes through the point $(0,0)$ and has slope $dy/dx = -(y-2)/(x-2)$ on each point on the curve (x,y).

60. In some cases, substitutions can convert an equation to a form that we recognize and can solve.

For example, consider the first-order equation

$$\frac{dy}{dx} = \frac{ax + ay + c}{bx + by + d}, \qquad (2.1)$$

Note that if $ad - bc = 0$, $ad = bc$ and Equation (2.1) reduces to $dy/dx = a/b$.

---

where $a, b \neq 0, c$, and d are constants. Let $Y = x + y$. Then, $dY/dx = 1 + dy/dx$ and substituting into Equation (2.1) gives us

$$\frac{dY}{dx} - 1 = \frac{aY + c}{bY + d},$$

$$\frac{dY}{dx} = 1 + \frac{aY + c}{bY + d},$$

$$\frac{dY}{dx} = \frac{(a+b)Y + (c+d)}{bY + d},$$

which is separable. Separating, solving for Y, and replacing Y with $x + y$ gives an implicit solution of the equation. Use this substitution to solve

(a) $\dfrac{dy}{dx} = \dfrac{x + y + 3}{3x + 3y + 1}$,

(b) $\dfrac{dy}{dx} = \dfrac{x - y + 2}{2x - 2y - 1}$.

61. A differential equation of the form $dy/dx = f(ax + by + k)$ is separable if $b = 0$. However, if $b \neq 0$, the substitution $u(x) = ax + by + k$ yields a separable equation. Use this substitution to solve

(a) $dy/dx = (x + y - 4)^2$,

(b) $dy/dx = (3y + 1)^4$.

62. Let $\omega > 0$ be constant. (a) Show that the system

$$\begin{cases} dx/dt = x(1-r) - \omega y, \\ dy/dt = y(1-r) + \omega x, \end{cases} \quad r = \sqrt{x^2 + y^2} \text{ can}$$

be rewritten as the system

$$\begin{cases} dr/dt = r(1-r) \\ d\theta/dt = \omega \end{cases} \quad \text{by changing to polar}$$

coordinates $\begin{cases} x = r\cos\theta, \\ y = r\sin\theta. \end{cases}$ (b) Show that

the solution to $\begin{cases} dr/dt = r(1-r) \\ d\theta/dt = \omega \\ r(0) = r_0, \theta(0) = \theta_0 \end{cases}$ is

$$\begin{cases} r(t) = r_0 e^t / ((1 - r_0) + r_0 e^t), \\ \theta(t) = \omega t + \theta_0, \end{cases} \quad \text{and the}$$

solution to

$$\begin{cases} dx/dt = x(1-r) - \omega y, \\ dy/dt = y(1-r) + \omega x, \end{cases} \quad r = \sqrt{x^2 + y^2} \text{ is}$$

$$\begin{cases} x = r(t)\cos\theta(t), \\ y = r(t)\sin\theta(t). \end{cases} \quad \text{(c) How does the}$$

solution change for various initial conditions? (*Hint:* First determine how ω and θ_0 affect the solution. Then, determine how r_0 affects the solution. Try graphing the solution if $\omega = 2$, $\theta_0 = 0$, and $r_0 = 1/2, 1$, and $3/2$. What happens if you increase ω? What happens if you increase θ_0?)

In Exercises 63-68, draw the phase line for each equation then find and classify (as asymptotically stable or unstable) the equilibrium solutions of the first-order equation.

63. $dy/dt = 3y$.

64. $dy/dt = -y$.

65. $dy/dt = y^2 - y$.

66. $dy/dt = 16y - 8y^2$.

67. $dy/dt = 12 + 4y - y^2$.

68. $dy/dt = y^2 - 5y + 4$.

69. (*Lasers*) The variables that characterize a two-level laser are listed in the following table:

ϕ	Total number of photons in optical resonator
l	Length of resonator
N_1	Number of atoms per unit volume in Level 1
N_2	Number of atoms per unit volume in Level 2
V	Volume of optical resonator
n	Total inversion given by $(N_2 - N_1)V$
n_i	Initial inversion
n_t	Total inversion at threshold
t_c	Decay time constant for photons in passive resonator
c	Phase velocity of light wave
v	Light wave frequency
h	Magnetic field

Suppose that a light wave of frequency v and intensity I_v propagates through an atomic medium with N_2 atoms per unit volume in Level 2 and N_1 in Level 1. If $N_2 > N_1$, then the medium is amplifying as is the case with lasers. On the other hand, if $N_1 > N_2$, then the medium is absorbing. These two situations are pictured in Figure 2.10. Optical resonators are used to build up large field intensities with moderate power inputs. A measure of this property is the quality factor Q, where

$$Q = \omega \times \frac{\text{field energy stored by resonator}}{\text{power dissipated by resonator}}.$$

A technique called "Q-switching" is used to generate short and intense bursts of oscillation from lasers. This is done by lowering the quality factor Q during the pumping so that the inversion $N_2 - N_1$ builds up to a high value without oscillation. When the inversion reaches its peak, the factor Q is suddenly restored to its ordinary value. At this point, the laser medium is well above its threshold, which causes a large buildup of the oscillation and a simultaneous exhaustion of the inversion

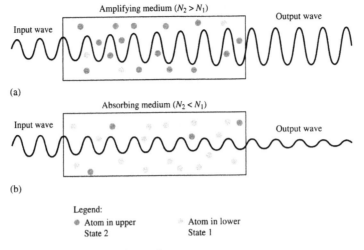

(a)

(b)

Legend:
● Atom in upper   ⟶ Atom in lower
    State 2   State 1

FIGURE 2.10   Light wave propagating through a medium.

by simulated transitions from Level 2 to Level 1. This process converts most of the energy that was stored by atoms pumped into Level 2 into photons, which are now in the optical resonator. These photons bounce back and forth between the reflectors with a fraction $1 - R$ escaping from the resonator with each bounce. This causes a decay of the pulse with a photon lifetime

$$t_c \approx \frac{nl}{c(1 - R)}.$$

The quantities of ϕ and n are related by the differential equation

$$\frac{d\phi}{dn} = \frac{n_t}{2n} - \frac{1}{2}.$$

(a) Solve this equation to determine the total number of photons in the optical resonator ϕ as a function of n (where ϕ depends on an arbitrary constant).

(b) Assuming that the initial values of ϕ and n are ϕ_0 and n_0 at $t = 0$, show that ϕ can be simplified to obtain

$$\phi = \frac{1}{2}n_t \ln \frac{n}{n_0} + \frac{1}{2}(n_0 + n) + \phi_0.$$

(c) If the instantaneous power output of the laser is given by $P = \phi h v / t_c$, show that the maximum power output occurs when $n = n_t$. (Hint: Solve $\partial P / \partial n = 0$.)

70. (Kinetic Reactions) Suppose that substances A and B, called the reactants, combine to form substance C. The reaction depends on the concentration of the reactants because the rate of the reaction is proportional to the product of their concentrations. When substances A and B are placed in the test tube and begin to combine to form C, the amounts of A and B diminish and the reaction stops when either A or B is no longer present. Let $y = y(t)$ represent the amount of substance C in the test tube at time t. If the initial amounts of substances A and B are A_0 and B_0 and if the volume in the test tube is V, then the concentration of A at time t is $(A_0 - y)/V$ and that of B is $(B_0 - y)/V$. Therefore, we determine $y(t)$ by solving

$$\frac{dy}{dt} = K\left[\frac{1}{V}(A_0 - y)\right]\left[\frac{1}{V}(B_0 - y)\right]$$

$$= \frac{K}{V^2}(A_0 - y)(B_0 - y).$$

(a) Identify the equilibrium solutions of the equation. Assuming that $B_0 > A_0$, classify each as stable or unstable. Generate the direction field for the equation with $A_0 = 1$, $B_0 = 2$, $K = 1$, and $V = 1$ to verify your results. What happens when either $y(t) = A_0$ or $y(t) = B_0$?

(b) Assume that $y(0) = y_0$ and solve the IVP to determine $y(t)$. Calculate $\lim_{t\to\infty} y(t)$. How does this limiting value relate to the physical situation described by the problem?

(c) If B_0 is much larger than A_0, then we may approximate the solution to this problem by solving the simplified IVP

$$\frac{dy}{dt} = K\left[\frac{1}{V}(A_0 - y)\right]\frac{B_0}{V}$$

$$= \frac{KB_0}{V^2}(A_0 - y), \quad y(0) = y_0.$$

(In other words, the reaction stops when substance A is exhausted.) Solve this IVP. Calculate $\lim_{t\to\infty} y(t)$.

(d) Suppose that $K = 0.2$, $V = 1$, $A_0 = 1$, $B_0 = 10$, and $y_0 = 0.1$. Compare the solution found in (b) to that found in (c). Do the same for the parameter values $K = 0.2$, $V = 1$, $A_0 = 1$, $B_0 = 50$, and $y_0 = 0.1$. How does the value of B_0 affect the accuracy of the approximation in (c)?

71. How does the graph of the solution to the initial value problem

$$dy/dt = cy/t^2, \quad y(1) = 1$$

changes as c varies from -2 to 2?

72. (*Tumor Growth*) Ludwig von Bertalanffy (1901-1972) made valuable contributions to the study of organism growth based on the relationship between body size and the metabolic rate of the organism. He theorized that weight is directly proportional to volume, V, and that the metabolic rate is proportional to surface area, S. This would mean that surface area is

Ludwig von Bertalanffy (September 19, 1901, near Vienna-June 12, 1972, Buffalo, New York): Many mathematical biologists consider Von Bertalanffy to be one of the greatest geniuses of mathematical biology ever.

According to Mark Davidson in his 1983 text *Uncommon Sense: The Life and Thought of Ludwig Von Bertalanffy*, "Ludwig von Bertalanffy may well be the least known intellectual titan of the twentieth century."

proportional to $V^{2/3}$ (because V is measured in cubic units and S in square units). Therefore, Bertalanffy studied the IVP

$$\frac{dV}{dt} = aV^{2/3} - bV, \quad V(0) = V_0,$$

where $V(t)$ represents volume; a, b, and V_0 are positive constants.

(a) Solve the IVP to show that volume is given by

$$V(t) = \left[\frac{a}{b} - \left(\frac{a}{b} - V_0^{1/3}\right)e^{-bt/3}\right]^3.$$

(b) Find $\lim_{t\to\infty} V(t)$ and explain what the limit represents.

73. If r_B (the *von Bertalanffy growth rate*) and L_∞ (the *ultimate length of the individual*) represent positive constants, solve the *von Bertalanffy growth equation*

$$L'(t) = r_B(L_\infty - L(t)), \quad L(0) = 0.$$

74. **(a)** Assume that $y(t) > 0$ and $\int_1^t f(u)du$ exists for $t \geq 1$. Show that the solution to the IVP

$$\begin{cases} y' = f(t)y \\ y(1) = 1 \end{cases} \quad \text{is } y(t) = 2\exp(\int_1^t f(u)du).$$

(b) Find three functions $f(t)$ so that (i) $y(t)$ is periodic, (ii) $\lim_{t\to\infty} y(t) = 0$, and (iii) $\lim_{t\to\infty} y(t) = \infty$. Confirm your result by graphing each solution. (c) If $c > 0$ is given, is it *always* possible to choose $f(t)$ so that $\lim_{t\to\infty} y(t) = c$? Explain.

2.3 FIRST-ORDER LINEAR EQUATIONS

In addition to the assumptions made in the fish population model at the beginning of Section 2.2, now suppose that we harvest the fish population at the constant rate h. In other words, we remove fish from the population by fishing at the rate h, where h is constant. By modifying the Malthus model developed in Section 2.2, we find that the rate at which the population size changes satisfies $dy/dt = ky - h$. Therefore, we determine the population size by solving the IVP

$$\frac{dy}{dt} = ky - h, \quad y(0) = y_0,$$

where y_0 represents the initial population size. This differential equation is classified as a *first-order linear differential equation*.

Definition 2 (First-Order Linear Differential Equation). *A differential equation that can be written in the form*

$$\frac{dy}{dt} + p(t)y = q(t) \qquad (2.2)$$

*is a **first-order linear differential equation**.*

The *corresponding homogeneous equation* for the first-order linear equation (2.2) is

$$\frac{dy}{dt} + p(t)y = 0. \qquad (2.3)$$

For simplicity, we assume that h is constant. Of course, h might depend on many external factors that could lead to interesting problems well beyond the scope of this text.

Equation (2.3) is separable and has general solution

$$y_h = Ce^{-\int p(t)dt}. \qquad (2.4)$$

(See Exercise 58 in Section 2.2 Exercises.)

A *particular solution* of a differential equation is a specific function that does not contain any arbitrary constants that is a solution of the differential equation.

EXAMPLE 2.3.1

Show that $y_p = 1/(1 + t^2)$ is a particular solution of

$$\frac{dy}{dt} + \frac{3t}{1+t^2}y = \frac{t}{1+t^2}.$$

Solution

We substitute y_p into the differential equation,

$$\frac{dy_p}{dt} + \frac{3t}{1+t^2}y_p = \frac{-2t}{(1+t^2)^2} + \frac{3t}{1+t^2}\frac{1}{1+t^2} = \frac{t}{1+t^2}$$

and see that the result is an identity. Thus, $y_p = 1/(1 + t^2)$ is a particular solution of the ODE.

Now assume that $y = y(t)$ is the general solution of the first-order linear equation (2.2) and y_p is a particular solution of the first-order linear equation. Then,

$$(y - y_p)' + p(t)(y - y_p) = (y' + p(t)y)$$
$$- (y_p' + p(t)y_p) = q(t) - q(t) = 0,$$

which means that $y - y_p$ is a solution of the corresponding homogeneous equation (2.3). Thus, $y - y_p = y_h$ so $y = y_h + y_p$. That is, a general solution of the first-order linear equation is the sum of the general solution to the corresponding homogeneous equation and a particular solution to the nonhomogeneous equation.

One approach to finding a particular solution of a nonhomogeneous linear equation is to "vary" the parameter C in the general solution to the corresponding homogeneous equation (2.4). In other words, we assume that a particular solution of the nonhomogeneous linear equation has the form $y_p(t) = u(t)y_h(t)$, where $u(t)$ is a function to be determined. For convenience, we now omit the arguments and assume that $y_p = uy_h$, where $u = u(t)$ is a function of t to be determined. Differentiating, substituting into the nonhomogeneous equation, and regrouping give us

$$y'_p + p(t)y_p = (u'y_h + uy'_h) + p(t)uy_h$$
$$= u'y_h + u(y'_h + p(t)y_h)$$
$$= u'y_h = q(t)$$

because $y'_h + p(t)y_h = 0$ since y_h is a solution of the corresponding homogeneous equation. Solving this equation for u gives us

$$u' = q(t)(y_h)^{-1},$$
$$u = \int q(t)(y_h)^{-1}dt,$$
$$u = \int q(t)e^{\int p(t)dt}dt$$

so

$$y_p = e^{-\int p(t)dt}\int q(t)e^{\int p(t)dt}dt. \qquad (2.5)$$

Now, when we include our arbitrary constant when evaluating $\int q(t)e^{\int p(t)dt}$ and multiply the result by $e^{\int p(t)dt}$ we see that y_h is included in Equation (2.5) so a general solution of the first-order linear equation is given by Equation (2.5): a general solution of the first-order linear equation is

$$y = e^{-\int p(t)dt}\int q(t)e^{\int p(t)dt}dt. \qquad (2.6)$$

The term $\mu(t) = e^{\int p(t)dt}$ in Equation (2.6) is of particular interest and called an *integrating factor* for the first-order linear equation because if we multiply the first-order linear equation (2.2) by the integrating factor, we obtain

$$e^{\int p(t)dt}\frac{dy}{dt} + p(t)e^{\int p(t)dt}y = q(t)e^{\int p(t)dt}. \qquad (2.7)$$

However, by the product rule and Fundamental Theorem of Calculus,

$$\frac{d}{dt}\left(e^{\int p(t)dt}y\right) = e^{\int p(t)dt}\frac{dy}{dt} + p(t)e^{\int p(t)dt}y. \qquad (2.8)$$

This expression is the left side of Equation (2.8), so we can rewrite Equation (2.7) as

$$\frac{d}{dt}\left(e^{\int p(t)dt}y\right) = q(t)e^{\int p(t)dt}. \qquad (2.9)$$

We solve this equation by integrating each side and solving for y, which results in the general solution given by Equation (2.6):

$$\frac{d}{dt}\left(e^{\int p(t)dt}y\right) = q(t)e^{\int p(t)dt}$$
$$e^{\int p(t)dt}y = \int q(t)e^{\int p(t)dt}\,dt$$
$$y = e^{-\int p(t)dt}\int q(t)e^{\int p(t)dt}\,dt. \qquad (2.10)$$

Recall that under reasonable conditions (see Theorem 2), IVPs involving first-order linear equations have unique solutions.

We illustrate the solution method in the following example.

EXAMPLE 2.3.2

Consider the IVP $dy/dt = y - 1/2$, $y(0) = a$. In this case, the fish population grows at the rate of 1 (ton per year) and fish are harvested (or fished) at the rate of 1/2 (ton per year). (a) Determine the behavior of solutions by drawing the phase line. (b) Find the solution to the IVP. (c) Investigate the behavior of solutions of the IVP using $y(0) = 1/4$, $y(0) = 1$, and $y(0) = 1/2$.

Solution

(a) By solving $y - 1/2 = 0$, we find that the equilibrium solution is $y = 1/2$. We draw the phase line in Figure 2.11. We include only nonnegative

FIGURE 2.11 Phase line for $dy/dt - y = -1/2$.

values on the phase line because we are interested in $y(t) \geq 0$ because $y(t)$ represents population size. Note that if $y > 1/2$, then $dy/dt > 0$ while if $y < 1/2$, then $dy/dt < 0$. Therefore, on the interval $y > 1/2$, the arrow is directed upward and on $0 \leq y < 1/2$, the arrow is directed downward. The phase line indicates that solutions to the IVP $dy/dt = y - 1/2$, $y(0) = a$ with $a > 1/2$ increase as $t \to \infty$. Conversely, solutions to this IVP with $0 < a < 1/2$ decrease as $t \to \infty$.

(b) We solve $dy/dt = y - 1/2$ as a first-order linear equation by rewriting the equation as $dy/dt - y = -1/2$. The integrating factor is

$$\mu(t) = e^{\int p(t)\,dt} = e^{\int (-1)\,dt} = e^{-t}.$$

Multiplying the equation by the integrating factor gives us

$$e^{-t}\frac{dy}{dt} - e^t y = -\frac{1}{2}e^{-t} \quad \text{or} \quad \frac{d}{dt}\left(e^{-t}y\right) = -\frac{1}{2}e^{-t}.$$

Integration yields $e^{-t}y = \frac{1}{2}e^{-t} + C$, so a general solution of the ODE is $y(t) = \frac{1}{2} + Ce^t$.

To solve the IVP, we solve $y(0) = \frac{1}{2} + C = a$ for C to find that $C = a - \frac{1}{2}$ so the solution to the IVP is $y = \frac{1}{2} + \left(a - \frac{1}{2}\right)e^t$.

(c) If $y(0) = 1/4$, then $y(t) = \frac{1}{2} - \frac{1}{4}e^t$. We graph $y(t)$ in Figure 2.12(a). Notice that the population becomes extinct when $\frac{1}{2} - \frac{1}{4}e^t = 0$, or $t = \ln 2 \approx 0.693$ year. (We ignore the portion of the solution curve where $y(t) < 0$.)) We should note that any solution to this IVP with $y(0) < 1/2$ means that the species will become extinct at some time. If $y(0) = 1$, $y(t) = \frac{1}{2} + \frac{1}{2}e^t$, so $y(t) > 0$ for all t, as we see in Figure 2.12(b). In fact, the population grows without bound. If $y(0) = 1/2$, then $y(t) = 1/2$ as expected because $y(t) = 1/2$ is an equilibrium solution. If the population is initially $1/2$ ton, then it remains at that level (see Figure 2.12(c)).

EXAMPLE 2.3.3

Solve the IVP $(dy/dt) + 5t^4 y = t^4$, $y(0) = -7$.

Solution

We begin by multiplying the linear equation $\dfrac{dy}{dt} + 5t^4 y = t^4$ by the integrating factor $\mu(t) = e^{\int 5t^4\,dt} = e^{t^5}$. Then, the equation can be written as

$$\frac{d}{dt}\left(e^{t^5}y\right) = t^4 e^{t^5}$$

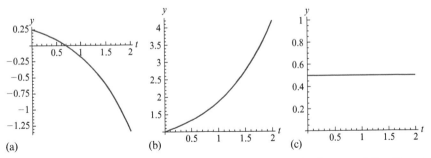

FIGURE 2.12   (a) Solution to $dy/dt = y - 1/2$, $y(0) = 1/4$. (b) Solution to $dy/dt = y - 1/2$, $y(0) = 1$. (c) Solution to $dy/dt = y - 1/2$, $y(0) = 1/2$.

and integration of both sides of the equation yields

$$e^{t^5}y = \frac{1}{5}e^{t^5} + C.$$

A general solution is

$$y = \frac{1}{5} + Ce^{-t^5}.$$

We find the unknown constant C by substituting the initial condition $y(0) = -7$ into the general solution and solving for C. This gives $-7 = 1/5 + C$, so $C = -36/5$. Therefore, the solution to the IVP is

$$y = \frac{1}{5} - \frac{36}{5}e^{-t^5}.$$

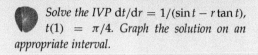 *Figure 2.13 shows the graph of $y = \frac{1}{5} + Ce^{-t^5}$ for various values of C. Identify the graph of the solution that satisfies the condition $y(0) = -7$.*

In some cases, we must rewrite an equation to place it in the form of a linear first-order differential equation.

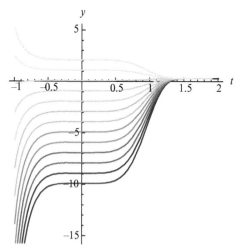

FIGURE 2.13 Graph of $y = \frac{1}{5} + Ce^{-t^5}$ for various values of C.

EXAMPLE 2.3.4

Solve $dt/dr = 1/(\sin t - r\tan t), 0 < t < \pi/2$.

Solution

Notice that if t is the dependent variable, the equation is nonlinear (in t). (Why?) However, solving the equation for dr/dt yields

$$\frac{dr}{dt} = \sin t - r\tan t \quad \text{or} \quad \frac{dr}{dt} + (\tan t)r = \sin t,$$

which is a *linear* equation in the dependent variable r. With $p(t) = \tan t$, the integrating factor is

$$\mu(t) = e^{\int \tan t\,dt} = e^{-\ln|\cos t|} = \frac{1}{\cos t}, \quad 0 < t < \pi/2.$$

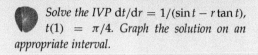 *Solve the IVP $dt/dr = 1/(\sin t - r\tan t)$, $t(1) = \pi/4$. Graph the solution on an appropriate interval.*

Then,

$$\frac{d}{dt}(\mu(t)r) = \mu(t)\sin t,$$

$$\frac{d}{dt}\left(\frac{1}{\cos t}r\right) = \frac{\sin t}{\cos t},$$

$$\frac{1}{\cos t}r = -\ln|\cos t| + C,$$

$$r = -(\cos t)\ln(\cos t) + C\cos t.$$

Note that $|\cos t| = \cos t$ for $0 < t < \pi/2$.

EXAMPLE 2.3.5

Show that the IVP $t(dy/dt) - y = t^2\cos t$, $y(0) = 0$ has infinitely many solutions.

Solution

Writing $ty' - y = t^2\cos t$ in the form $y' = f(t,y)$ results in

$$\frac{dy}{dt} = \frac{t^2\cos t + y}{t}.$$

and because $f(t,y) = \left(t^2 \cos t + y\right)/t$ is *not* continuous on an interval containing $t = 0$, The Existence and Uniqueness Theorem for Linear IVPs *does not* guarantee the existence or uniqueness of a solution to the IVP.

Rewriting the equation in standard form shows us that the equation is linear and gives us

$$\frac{dy}{dt} - \frac{1}{t}y = t \cos t$$

so an integrating factor is $\mu(t) = \exp\left(\int -1/t\, dt\right) = \exp\left(-\ln t\right) = 1/t$. Multiplying the equation by $1/t$, rewriting and integrating give us

$$\frac{1}{t}\frac{dy}{dt} - \frac{1}{t^2}y = \cos t,$$

$$\frac{d}{dt}\left(\frac{1}{t}y\right) = \cos t,$$

$$\frac{1}{t}y = \sin t + C,$$

$$y = t \sin t + tC.$$

Thus, a general solution of $ty' - y = t^2 \cos t$ and every solution of the equation satisfy $y(0) = 0$ so the IVP has infinitely many solutions. A few solutions are graphed in Figure 2.14.

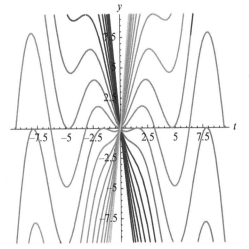

FIGURE 2.14   Every solution of the differential equation satisfies $y(0) = 0$.

EXAMPLE 2.3.6

If a drug is introduced into the bloodstream in dosages of $D(t)$ and is removed at a rate proportional to the concentration, the concentration $C(t)$ at time t is given by

$$\frac{dC}{dt} = D(t) - kC, \quad C(0) = 0,$$

where $k > 0$ is the constant of proportionality.[1]

(a) Solve this IVP.

(b) Suppose that over a 24-h period, a drug is introduced into the bloodstream at a rate of $24/t_0$ for exactly t_0 hours and then stopped so that

$$D_{t_0}(t) = \begin{cases} 24/t_0, & \text{if } 0 \le t_0 \\ 0, & \text{if } t > t_0 \end{cases}.$$

What is the total dosage and average dosage over a 24-h period?

(c) Calculate and then graph $C(t)$ on the interval $[0, 30]$ if $k = 0.05, 0.10, 0.15,$ and 0.20 for $t_0 = 4, 8, 12, 16,$ and 20. How does increasing t_0 affect the concentration of the drug in the bloodstream? How does increasing k affect it?

Solution

(a) After rewriting the equation as $dC/dt + kC = D(t)$, we find the integrating factor to be $\mu(t) = e^{\int k\, dt} = e^{kt}$. Therefore, after multiplying the equation by the integrating factor, we obtain that $(d/dt)(e^{kt}C(t)) = e^{kt}D(t)$ and integrating yields

$$e^{kt}C(t) = \int_0^t e^{ks}D(s)\,ds \quad \text{or}$$

$$C(t) = e^{-kt}\int_0^t e^{ks}D(s)\,ds.$$

(b) In each case, the total dosage over a 24-h period is

$$\int_0^{24} D_{t_0}(t)\,dt = \int_0^{t_0} \frac{24}{t_0}\,dt = 24;$$

the average dosage is

$$\frac{1}{24 - 0}\int_0^{24} D_{t_0}(t)\,dt = \frac{1}{24}\int_0^{t_0}\frac{24}{t_0}\,dt = 1.$$

The average value of $y = f(x)$ on $[a,b]$ is
$\frac{1}{b-a}\int_a^b f(x)dx$.

(c) To compute $C(t) = e^{-kt}\int_0^t e^{ks}D(s)ds$, we must keep in mind that $D_{t_0}(t)$ is a piecewise-defined function:

$$C(t) = e^{-kt}\int_0^t e^{ks}D(s)\,ds$$

$$= \begin{cases} e^{-kt}\int_0^t e^{ks}\dfrac{24}{t_0}\,ds, & 0 \le t \le t_0 \\ e^{-kt}\int_0^{t_0} e^{ks}\dfrac{24}{t_0}\,ds, & t > t_0 \end{cases}$$

$$= \begin{cases} \dfrac{24}{k_0}\left(1-e^{-kt}\right), & 0 \le t \le t_0 \\ \dfrac{24}{kt_0}\left(e^{-k(t-t_0)}-e^{-kt}\right), & t > t_0. \end{cases}$$

We graph $C(t)$ on the interval $[0,30]$ if $k = 0.05$, 0.10, 0.15, and 0.20 for $t_0 = 4, 8, 12, 16,$ and 20 in Figure 2.15. From the graphs, we see that as t_0 is increased, the maximum concentration level decreases and occurs at later times, while increasing k increases the rate at which the drug is removed from the bloodstream.

[1]J.D. Murray, Mathematical Biology, Springer-Verlag, pp. 645-649, 1990.

EXERCISES 2.3

In Exercises 1-25, solve each equation.

1. $dy/dt - y = 10$.
2. $dy/dt - y = 2e^{-t}$.
3. $dy/dt - y = 2\cos t$.
4. $dy/dt - y = t^2 - 2t$.
5. $dy/dt - y = 4te^{-t}$.
6. $ty' + y = t^2$.
7. $ty' + y = t$.
8. $x\,dy/dx + y = xe^x$.
9. $x\,dy/dx + y = e^{-x}$.

10. $dy/dt - 2t/(1+t^2)y = 2$.
11. $dy/dt - 4t/(1+4t^2)y = 4t$.
12. $dy = \left(2x + \dfrac{xy}{x^2-1}\right)dx$.
13. $dy/dt + y\cot t = \cos t$.
14. $dy/dt - 3t/(t^2-4)y = t$.
15. $dy/dt - 4t/(4t^2-9)y = t$.
16. $dy/dx - 9x/(9x^2+49)y = x$.
17. $dy/dx + 2y\cot x = \cos x$.
18. $dt/ds + st = s^3$.
19. $d\theta/dr - r\theta = r$.
20. $dy/dx = 1/(y^2+x)$.
21. $dx/dy - x = y$.
22. $y\,dx - (x+3y^2)dy = 0$.
23. $\dfrac{dx}{dt} = \dfrac{3xt^2}{1-t^3}$.
24. $dp/dt = t^3 + p/t$.
25. $dv/ds + v = e^{-s}$.

In Exercises 26-35, solve the IVP. *Graph your solution on an appropriate interval to help you verify that your answer is correct.*

26. $dy/dt - y = 4e^t$, $y(0) = 4$.
27. $dy/dt + y = e^{-t}$, $y(0) = -1$.
28. $dy/dt + 3t^2y = e^{-t^3}$, $y(0) = 2$.
29. $dy/dt + 2ty = 2t$, $y(0) = -1$.
30. $t\,dy/dt + y = \cos t$, $y(\pi/2) = 4/\pi$, $t > 0$.
31. $t\,dy/dt + y = 2te^t$, $y(1) = -1$, $t > 0$.
32. $(e^t+1)dy/dt + e^ty = t$, $y(0) = -1$.
33. $(t^2+4)dy/dt + 2ty = 2t$, $y(0) = -4$.
34. $dx/dt = x + t + 1$, $x(0) = 2$.
35. $d\theta/dt = e^{2t} + 2\theta$, $\theta(0) = 0$.
36. If $y = y_1(t)$ satisfies the homogeneous equation $dy/dt + p(t)y = 0$ and $y = y_2(t)$ satisfies the nonhomogeneous equation $dy/dt + p(t)y = r(t)$, show that $y(t) = y_1(t) + y_2(t)$ satisfies the nonhomogeneous equation $dy/dt + p(t)y = r(t)$.
37. (a) Show that if $y = y_1(t)$ is a solution of $dy/dt + p(t)y = r(t)$ and $y = y_2(t)$ is a solution of $dy/dt + p(t)y = q(t)$, then $y(t) = y_1(t) + y_2(t)$ is a solution of $dy/dt + p(t)y = r(t) + q(t)$.

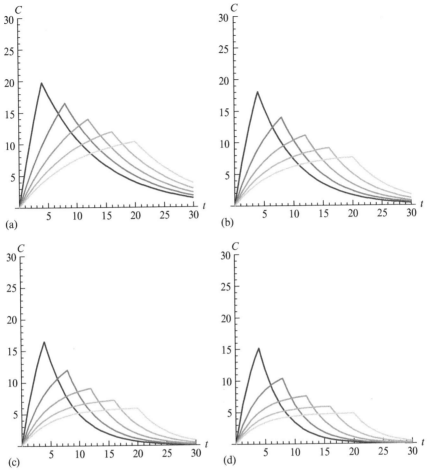

FIGURE 2.15   (a) $k = 0.05$, (b) $k = 0.10$, (c) $k = 0.15$, and (d) $k = 0.20$.

(b) Use the result obtained in (a) to solve
$$dy/dt + 2y = e^{-t} + \cos t.$$
38. (a) Show that if $y = y_1(t)$
is a solution of $dy/dt + p(t)y = q(t)$ then
$y(t) = y_1(t)$ is a solution of
$d^2y/dt^2 + p'(t)y + p(t)y' = q'(t).$
(b) Solve $dy/dt - (1/t)y = \ln t, t > 0.$
(c) Use the results obtained in (a) and (b)
to solve
$$\frac{d^2y}{dt^2} - \frac{1}{t}\frac{dy}{dt} + \frac{1}{t^2}y = \frac{1}{t}, \quad t > 0.$$

It is *very important* to understand that a general solution of the inhomogeneous linear equation $y' + p(t)y = q(t)$ is the sum of a general solution, y_h, to the corresponding homogeneous equation $y' + p(t)y = 0$ and a particular solution, y_p, of the nonhomogeneous equation $y' + p(t)y = q(t)$. In an earlier exercise, you showed that if k is a constant, a general solution of $y' + ky = 0$ is $y_h = Ce^{-kt}$. In the following exercises, find y_h and then guess the form of y_p and form a general solution of the differential equation, $y = y_h + y_p$.

39. $y' - y = -1;$
$y' + y = t + 1.$

40. $y' + y = \cos t + \sin t;$
$y' - y = -2e^{-t}.$

41. Observe Figure 2.15. Describe how the value of k affects the concentration of the drug in the system. Do larger or smaller values cause the drug to remain in the system longer? How would the chemist working for the pharmaceutical company design the drug?

In Exercises 42-45, solve each of the IVPs that involve piecewise-defined functions.

42. $dy/dt + y = q(t)$, where
$$q(t) = \begin{cases} 4, & 0 \le t < 2 \\ 0, & t \ge 2 \end{cases}, y(0) = 0.$$

43. $dy/dt + y = q(t)$, where
$$q(t) = \begin{cases} t, & 0 \le t < 1 \\ 0, & t \ge 1 \end{cases}, y(0) = 1.$$

44. $dy/dt + p(t)y = 0$, where
$$p(t) = \begin{cases} 1, & 0 \le t < 2 \\ -1, & t \ge 2 \end{cases}, y(0) = 2.$$

45. $dy/dt + p(t)y = 0$, where
$$p(t) = \begin{cases} 2, & 0 \le t < 1 \\ 4, & t \ge 1 \end{cases}, y(0) = 1.$$

46. (*Method of Undetermined Coefficients*) Consider the first-order nonhomogeneous equation $y' + y = \cos t$. We can solve the problem in two parts. First, we can solve the corresponding homogeneous equation $y' + y = 0$ for y_h. (a) Show that $y_h(t) = Ce^{-t}$ is a general solution of this equation. A general solution of the inhomogenous equation is $y = y_h + y_p$, where y_h is a general solution of the corresponding homogeneous equation and y_p is a particular solution of the inhomogeneous equation. (b) Next, we attempt to find a

particular solution of the inhomogenous equation by guessing the form of the solution. Suppose that this particular solution is $y_p(t) = A \cos t + B \sin t$. Substitute y_p into $y' + y = \cos t$ to show that the undetermined coefficients are $A = 1/2$ and $B = 1/2$. Then, a general solution of the inhomogenous equation is
$y(t) = y_h(t) + y_p(t) = Ce^{-t} + \frac{1}{2}\cos t + \frac{1}{2}\sin t.$
(What integral would need to be evaluated if we used an integrating factor to solve the ODE?)

47. Solve $y' - y = \sin 2t$.

48. Solve $y' + y = 5e^{2t}$ by assuming that $y_p(t) = Ae^{2t}$.

49. Solve $y' + y = e^{-t}$ by assuming that $y_p(t) = Ate^{-t}$. Why don't we choose $y_p(t) = Ae^{-t}$?

50. Solve $y' + y = 2 - e^{2t}$ by assuming that $y_p(t) = A + Be^{2t}$

51. Solve $y' - 5y = t$ by assuming that $y_p(t) = At + B$.

52. Solve $y' + 3y = 27t^2 + 9$ by assuming that $y_p(t) = At^2 + Bt + C$.

53. Solve $y' - \frac{1}{2}y = 5\cos t + 2e^t$ by assuming that $y_p(t) = A\cos t + B\sin t + Ce^t$.

In Exercises 54-59, solve each of the equations using the Method of Undetermined Coefficients.

54. $y' + 4y = 8\cos 4t$.

55. $y' + 10y = 2e^t$.

56. $y' - 3y = 27t^2$.

57. $y' - y = 2e^t$.

58. $y' + y = 4 + 3e^t$.

59. $y' + y = 2\cos t + t$.

60. (*First-Order Linear with Periodic Forcing Function*) Consider the differential equation $dy/dt + cy = f(t)$, where $f(t)$ is a periodic function and c is a constant. The goal of this exercise is to determine if equations of this form have a periodic solution. (a) Solve the IVP $y' + y/2 = \sin t, y(0) = a$. For what values of a does the IVP have a periodic solution? Graph the slope field for this ODE. Describe the behavior of the other solutions. Do they approach the periodic solution found in (a) as $t \to \infty$ or $t \to -\infty$? (b) Solve the IVP $y' - y/2 = \sin t$, $y(0) = a$. For what values of a does the IVP have a periodic solution? Graph the slope field for this ODE. Describe the behavior of the other solutions. Do they approach the periodic solution found in (b) as $t \to \infty$ or $t \to -\infty$? (c) Based on your findings in (a) and (b), does the ODE $y' + cy = f(t)$, where $f(t)$ is a periodic function and c is a constant, have a periodic solution? How does the value of c affect the other solutions?

61. Suppose that a drug is added to the body at a rate $r(t)$, and let $y(t)$ represent the concentration of the drug in the bloodstream at time t hours. In addition, suppose that the drug is used by the body at the rate ky, where k is a positive constant. Then, the net rate of change in $y(t)$ is given by the equation $dy/dt = r(t) - ky$. If at $t = 0$, there is no drug in the body, we determine $y(t)$ by solving the IVP

$$dy/dt = r(t) - ky, \quad y(0) = 0.$$

(a) Suppose that $r(t) = r$, where r is a positive constant. In this case, the drug is added at a constant rate. Sketch the phase line for $dy/dt = r - ky$. Solve the IVP. Determine $\lim_{t \to \infty} y(t)$. How does this limit correspond to the phase line?

(b) Suppose that $r(t) = 1 + \sin t$ and $k = 1$. In this case, the drug is added at a periodic rate. Solve the IVP. Determine $\lim_{t \to \infty} y(t)$, if it exists. Describe what happens to the drug concentration over time.

(c) Suppose that $r(t) = e^{-t}$ and $k = 1$. In this case, the rate at which the drug is added decreases over time. Solve the IVP. Determine $\lim_{t \to \infty} y(t)$, if it exists. Describe what happens to the drug concentration over time.

62. Find a general solution of the equation $t \, dy/dt + y = t \cos t$. Graph various solutions on the rectangle $[0, 2\pi] \times [-10, 10]$.

One advantage of using technology is that often when a large number of problems are solved, their solutions are compared and conjectures about general patterns can be discovered and tested. Many CASs are capable of solving a variety of linear equations, particularly those that are frequently encountered in an elementary differential equations course.

---

63. Compare the solutions of $dy/dt + y = f(t)$ subject to $y(0) = 0$, where $f(t) = t$, $\sin t$, $\cos t$, and e^t.

64. Compare the solutions of $dy/dt + ky = t$, $y(0) = 1$ for $k = -2, -1, 0, 1$, and 2.

65. Compare the solutions of $dy/dt + y = t$, $y(0) = k$ for $k = -2, -1, 0, 1$, and 2.

66. (a) Graph the slope field for the differential equation $dy/dt = y - t$. (b) Solve and graph the solution to the following IVPs (i)

$$\begin{cases} y' = y - t \\ y(0) = 1 \end{cases} \; ; \text{(ii)} \; \begin{cases} y' = y - t \\ y(0) = 1.1 \end{cases} \; ; \text{and (iii)}$$

$$\begin{cases} y' = y - t \\ y(0) = 0.9 \end{cases} \; . \text{(c) Comment on this}$$

statement: if we slightly change the initial conditions in a linear IVP, the solution also slightly changes.

67. (*Destruction of Microorganisms*) Microorganisms can be removed from fluids by mechanical methods such as filtration, centrifugation, or flotation. However, they can also be destroyed by heat, chemical agents, or electromagnetic waves. The fermentation industry is interested in improving this process of sterilizing media. An important component in the design of a sterilizer is the kinetics of the death of microorganisms.

The destruction of microorganisms by heat indicates loss of viability as opposed to physical destruction. This destruction follows the rate of reaction $dN/dt = -kN$, where k is the reaction constant (min^{-1}) and is a function of temperature, N is the number of viable organisms, and t is time.[1]

Microbiologists use the term *decimal reduction time D* to indicate the time of exposure to heat during which the original number of viable microbes is reduced by one-tenth. If $N(0) = N_0$, find $N(t)$. Use the fact that $N(D) = \frac{1}{10}N_0$ to find D in terms of k.

2.4 EXACT DIFFERENTIAL EQUATIONS

We now turn our attention as to why $\mu(t) = e^{\int p(t)dt}$, in Equation (2.6), is of particular interest and called an *integrating factor* for the first-order linear equation (2.2).

We begin with an example: the first-order differential equation

$$(\sin y + y \cos t)dt + (\sin t + t \cos y)dy = 0$$

is neither separable or linear nor can it be reduced to a separable equation or linear equation by an appropriate substitution. Nevertheless, a general solution of this equation can be calculated, as we will see in this section.

Definition 3 (Exact Differential Equation). *A first-order differential equation that can be written in the form*

$$M(t,y)dt + N(t,y)dy = 0,$$

where

$$M(t,y)dt + N(t,y)dy = \frac{\partial f}{\partial t}(t,y)dt + \frac{\partial f}{\partial y}(t,y)dy$$

*for some function $f(t,y)$ is called an **exact differential equation**.*

In calculus, we learn that the *total differential* of the function $z = f(t,y)$ is

$$df = \frac{\partial f}{\partial t}(t,y)dt + \frac{\partial f}{\partial y}(t,y)dy.$$

Therefore, the equation $M(t,y)dt + N(t,y)dy = 0$ is exact if there exists a function $z = f(t,y)$ such that $M(t,y)dt + N(t,y)dy$ is the total differential of $z = f(t,y)$. In physics classes, the function $z = f(t,y)$ is often called the *potential function*.

In multivariable calculus, we learn that if f, $\partial f/\partial t$, $\partial f/\partial y$, $\partial^2 f/(\partial t\, \partial y)$, and $\partial^2 f/(\partial y \partial t)$ are continuous on an open region R, then $\partial^2 f/(\partial t \partial y) = \partial^2 f/(\partial y \partial t)$ on R. Hence, if $M(t,y)dt + N(t,y)dy = 0$ is exact and $M(t,y)dt + N(t,y)dy$ is the total differential of $z = f(t,y)$,

$$\frac{\partial N}{\partial t} = \frac{\partial}{\partial t}\left(\frac{\partial f}{\partial y}\right) = \frac{\partial}{\partial y}\left(\frac{\partial f}{\partial t}\right) = \frac{\partial M}{\partial y}.$$

In fact, we can prove the following (see Exercise 56 in this section).

Theorem 5 (Test for Exactness). *The first-order differential equation $M(t,y)dt + N(t,y)dy = 0$ is exact if and only if $\partial M/\partial y = \partial N/\partial t$.*

EXAMPLE 2.4.1

Show that the equation $2ty^3\, dt + (1 + 3t^2y^2)dy = 0$ is an exact equation and that the equation $t^2y\, dt + 5ty^2\, dy = 0$ is not exact.

Solution

The equation $2ty^3\, dt + (1 + 3t^2y^2)dy = 0$ is an exact equation because

$$\frac{\partial}{\partial y}(2ty^3) = 6ty^2 = \frac{\partial}{\partial t}(1 + 3t^2y^2).$$

Conversely, the equation $t^2y\, dt + 5ty^2\, dy = 0$ is not exact because

$$\frac{\partial}{\partial y}(t^2y) = t^2 \neq 5y^2 = \frac{\partial}{\partial t}(5ty^2).$$

---

[1] S. Aiba, A.E. Humphrey, N.E Millis, Biochemical Engineering, second ed., Academic Press, pp. 240-242, 1973.

If the equation $M(t,y)dt + N(t,y)dy = 0$ is exact, we can find a function $z = f(t,y)$ such that $M(t,y) = (\partial f/\partial t)(t,y)$ and $N(t,y) = (\partial f/\partial y)(t,y)$. Then the differential equation becomes

$$M(t,y)dt + N(t,y)dy = 0$$
$$df = 0.$$

A general solution of the equation is $f(t,y) = C$, where C is an arbitrary constant.

EXAMPLE 2.4.2

Find a general solution of $(\sin y + y \cos t)dt + (\sin t + t \cos y)dy = 0$.

Solution

The equation is exact because

$$\frac{\partial}{\partial y}(\sin y + y \cos t) = \cos y + \cos t$$

$$= \frac{\partial}{\partial t}(\sin t + t \cos y).$$

Let $f(t,y)$ be a function with $\partial f/\partial t = \sin y + y \cos t$ and $\partial f/\partial y = \sin t + t \cos y$. Integrating $\partial f/\partial t = \sin y + y \cos t$ with respect to t results in

$$\int (\sin y + y \cos t)dt = t \sin y + y \sin t + g(y),$$

where $g(y)$ denotes an arbitrary function of y. We must include this arbitrary function $g(y)$ because the derivative of a function of y with respect to t is zero. That is, the general form of the function $f(t,y)$ whose partial derivative with respect to t is $\partial f/\partial t = \sin y + y \cos t$ is given by

$$f(t,y) = t \sin y + y \sin t + g(y).$$

Because we are looking for a function $f(t,y)$ that satisfies $\partial f/\partial y = \sin t + t \cos y$ and differentiating $f(t,y) = t \sin y + y \sin t + g(y)$ with respect to y results in $\partial f/\partial y = \sin t + t \cos y + g'(y)$. Now, we must have that

$$\sin t + t \cos y + g'(y) = \sin t + t \cos y,$$

which indicates that $g'(y) = 0$. This means that $g(y) = k$ for some constant k. Thus,

$$f(t,y) = t \sin y + y \sin t + k.$$

Therefore, the implicit function $t \sin y + y \sin t + k = C_1$ or $t \sin y + y \sin t = C$, where $C = C_1 - k$ represents an arbitrary constant, is a general solution of $(\sin y + y \cos t)dt + (\sin t + t \cos y)dy = 0$. (Note that there is no need to include an arbitrary constant in $f(t,y)$ because it is included in $f(t,y) = C$.) Several members of the family of solutions are graphed in Figure 2.16 by graphing several level curves of the function $f(t,y) = t \sin y + y \sin t$.

As with the other IVPs involving first-order equations that we have discussed, under reasonable conditions, IVPs involving the first-order exact differential equation $M(t,y)dt + N(t,y)dy = 0$ will have a unique solution.

Theorem 6 (Existence and Uniqueness: First-Order Exact Equations). *Consider the IVP*

$$M(t,y)dt + N(t,y)dy = 0, \quad y(t_0) = y_0,$$

where $M(t,y)dt + N(t,y)dy = 0$ is exact. If M, N, $\partial M/\partial t$, and $\partial N/\partial y$ are continuous on an open region R containing (t_0, y_0) and $N(t_0, y_0) \neq 0$, then

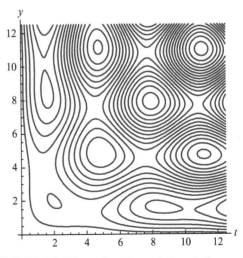

FIGURE 2.16 We graph various solutions to the equation by graphing several level curves of the function $f(t,y) = t \sin y + y \sin t$.

there exists one and only one function that satisfies the differential equation and initial condition for each value of t on an interval containing t_0.

EXAMPLE 2.4.3

Solve $2t \sin y\, dt + (t^2 \cos y - 1)dy = 0$ subject to $y(0) = 1/2$.

Solution

The equation is exact because

$$\frac{\partial}{\partial y}(2t \sin y) = 2t \cos y = \frac{\partial}{\partial t}(t^2 \cos y - 1).$$

Let $f(t,y)$ be a function with $\partial f/\partial t = 2t \sin y$ and $\partial f/\partial y = t^2 \cos y - 1$. Integrating $\partial f/\partial t$ with respect to t yields

$$f(t,y) = \int 2t \sin y\, dt = t^2 \sin y + g(y).$$

Notice that the arbitrary function $g(y)$ serves as a "constant" of integration with respect to t. From the differential equation, we have $\partial f/\partial y = t^2 \cos y - 1$, and differentiating $f(t,y) = t^2 \sin y + g(y)$ with respect to y gives us $\partial f/\partial y = t^2 \cos y + g'(y)$. Thus,

$$t^2 \cos y - 1 = t^2 \cos y + g'(y),$$
$$g'(y) = -1$$

and $g(y) = -y + C_1$ so that substitution into $f(t,y) = t^2 \sin y + g(y)$ yields

$$f(t,y) = t^2 \sin y - y + C_1.$$

A general solution of the exact equation is then $t^2 \sin y - y + C_1 = C$. Simplifying, we have $t^2 \sin y - y = k$, where $k = C - C_1$ is an arbitrary constant. Our solution must satisfy $y(0) = 1/2$, so we must find the solution that passes through the point $(0, 1/2)$. Substituting $t = 0$ and $y = 1/2$ into the general solution, we obtain $0^2 \times \sin(1/2) - 1/2 = k$ so that $k = -1/2$ and the solution to the IVP is $t^2 \sin y - y = -1/2$. The solution is graphed in Figure 2.17. We see that the graph passes through the point $(0, 1/2)$ as required by the initial condition.

Solving the Exact Differential Equation
$M(t,y)dt + N(t,y)dy = 0$

A similar algorithm can be stated so that in step 2, $N(t,y)$ is integrated with respect to y as we show in the next example.

1. Assume that $M(t,y) = \partial f/\partial t(t,y)$ and $N(t,y) = \partial f/\partial y(t,y)$.

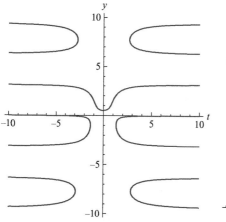
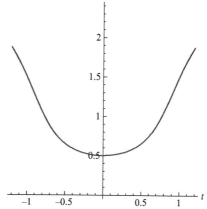

FIGURE 2.17 On the left, a graph of the equation $t^2 \sin y - y = -1/2$. On the right, we see that the solution is unique only on an interval containing approximately $-1.2 < t < 1.2$.

2. Integrate $M(t, y)$ with respect to t. (For the arbitrary constant, add an arbitrary function of y, $g(y)$.)

3. Differentiate the result in step 2 with respect to y and set the result equal to $N(t, y)$. Solve for $g'(y)$.

4. Integrate $g'(y)$ with respect to y to obtain an expression for $g(y)$. (There is no need to include an arbitrary constant at this time because it will be taken into account when the general solution is formed.)

5. Substitute $g(y)$ into the result obtained in step 2 for $f(t, y)$.

6. A general solution is $f(t, y) = C$, where C is an arbitrary constant.

7. Apply the initial condition if given.

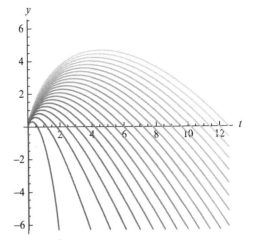

FIGURE 2.18 Graph of $te^{y/t} + \tan^{-1} t = C$ for various values of C.

EXAMPLE 2.4.4

Solve $(e^{y/t} - y/t\, e^{y/t} + 1/(1+t^2))dt + e^{y/t}dy = 0$.

Solution

This equation is exact because

$$\frac{\partial}{\partial y}\left(e^{y/t} - \frac{y}{t}e^{y/t} + \frac{1}{1+t^2}\right) = -\frac{y}{t^2}e^{y/t} = \frac{\partial}{\partial t}(e^{y/t}).$$

Let $f(t, y)$ be a function such that $\partial f/\partial t = e^{y/t} - y/t e^y y/t + 1/(1 + t^2)$ and $\partial f/\partial y = e^{y/t}$. Integrating $\partial f/\partial y$ with respect to y because it is a less complicated expression than $\partial f/\partial t$ gives us

$$f(t, y) = \int e^y y/t\, dy = \frac{1}{1/t}e^{y/t} + g(t) = te^{y/t} + g(t),$$

where $g(t)$ is an arbitrary function of t. Differentiating $f(t, y)$ with respect to t leads to

$$\frac{\partial f}{\partial t} = e^{y/t} + t\left(-\frac{y}{t^2}e^y y/t\right) + g'(t) = e^{y/t} - \frac{y}{t}e^{y/t} + \frac{1}{1+t^2},$$

so $g'(t) = 1/(1+t^2)$. This implies that $g(t) = \tan^{-1} t$ so $f(t, y) = te^{y/t} + \tan^{-1} t$. Therefore, a general solution of the exact equation is $te^{y/t} + \tan^{-1} t = C$. Several solutions are graphed in Figure 2.18.

Most IVPs that we have considered have had a unique solution. However, it is important to remember that this need not be true.

EXAMPLE 2.4.5

Find a value of y_0 so that there is *not* a unique solution to the IVP

$$\begin{cases} (\cos t \cos y - \sin t)dt + (\cos y - \sin t \sin y)dy = 0 \\ y(0) = y_0. \end{cases}$$

Solution

The equation $(\cos t \cos y - \sin t)dt + (\cos y - \sin t \sin y)dy = 0$ is exact because

$$\frac{\partial}{\partial y}(\cos t \cos y - \sin t) = -\cos t \sin y$$

$$= \frac{\partial}{\partial t}(\cos y - \sin t \sin y).$$

A general solution is found to be $\cos t + \sin t \cos y + \sin y = C$, which we graph for various values of C in Figure 2.19.

From the graphs, we see that if $C = 0$, it is possible to find y_0 such that there is *not* a unique solution to the IVP. Indeed an exact value of y_0 is $3\pi/2$ (Why?) (see Figure 2.20).

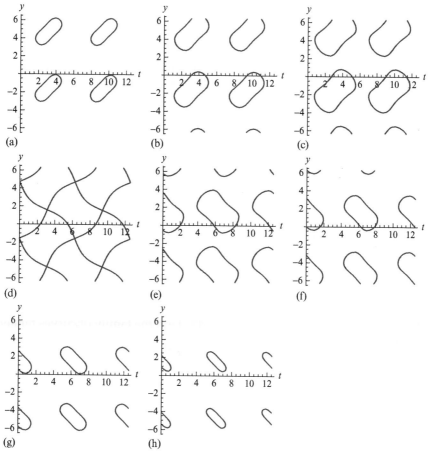

FIGURE 2.19   (a) $C = -3/2$, (b) $C = -1$, (c) $C = -1/2$, (d) $C = 0$, (e) $C = 1/2$, (f) $C = 1$, (g) $C = 3/2$, and (h) $C = 9/5$.

EXERCISES 2.4

In Exercises 1-10, determine if the equation is exact.

1. $(y^2 - y/(2\sqrt{t}))dt + (2ty - \sqrt{t} + 1)dy = 0$.
2. $t/\sqrt{t^2 + y^2}\, dt + y/\sqrt{t^2 + y^2}\, dy = 0$.
3. $y\cos(ty)dt + t\cos(ty)dy = 0$.
4. $(y\sec^2 t + 2t)dt + \tan t\, dy = 0$.
5. $3ty^2\, dt + y^3\, dy = 0$.
6. $(t - y\sin t)dt + (y^6 + \cos t)\, dy = 0$.
7. $y\sin 2t\, dt - (\sqrt{y} + \cos 2t)dy = 0$.
8. $(e^{2t} - y)\, dt - (e^y - t)dy = 0$.
9. $\ln(ty)\, dt + t/y\, dy = 0$.
10. $e^{ty}\, dt + te^{ty}/y\, dy = 0$.

In Exercises 11-30, solve each equation.

11. $3t^2\, dt - dy = 0$.
12. $-dt + 3y^2\, dy = 0$.
13. $y^2\, dt + 2ty\, dy = 0$.
14. $3t^2/y\, dt - t^3/y^2\, dy = 0$.
15. $(2t + y^3)dt + (3ty^2 + 4)dy = 0$.
16. $-1/y\, dt + (t/y^2 + 3y^2)\, dy = 0$.
17. $2ty\, dt + (t^2 + y^2)dy = 0$.
18. $2ty^3\, dt + (1 + 3t^2y^2)dy = 0$.

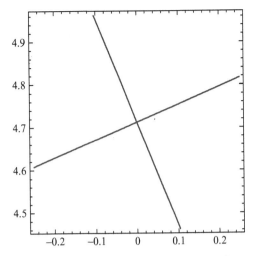

FIGURE 2.20 The solution is *not unique* near this point because when we zoom in, we see the graph of more than one function passing through the point.

19. $\sin^2 y\, dt + t \sin 2y\, dy = 0.$
20. $(3t^2 + 3y^2)dt + 6ty\, dy = 0.$
21. $e^t \sin y\, dt + (1 + e^t \cos y)dy = 0.$
22. $(3t^2 y + 3y^2 - 1)dt + (t^3 + 6ty)dy = 0.$
23. $-2ty^2 \sin(t^2)dt + 2y \cos(t^2)\, dy = 0.$
24. $(2t - y^2 \sin ty)dt + (\cos ty - ty \sin ty)dy = 0.$
25. $(1 + y^2 \cos ty)dt + (ty \cos ty + \sin ty)dy = 0.$
26. $(2t \sin y - 2ty \sin(t^2))dt + (t^2 \cos y + \cos(t^2))dy = 0.$
27. $((3 + t)\cos(t + y) + \sin(t + y))dt + (3 + t)\cos(t + y)dy = 0.$
28. $\dfrac{1}{t^2}(2t^2 y \cos t^2 - y \sin t^2)dt + \dfrac{1}{t}(2ty + \sin t^2)\, dy = 0.$
29. $(-t^{-2}y^2 e^{y/t} + 1)dt + e^{y/t}(1 + y/t)dy = 0.$
30. $(2t \sin(y/t) - y \cos(y/t))dt + t \cos(y/t)dy = 0.$

In Exercises 31-40, solve the IVP. Graph the solution on an appropriate interval.

31. $2ty^2\, dt + 2t^2 y\, dy = 0,\ y(1) = 1.$
32. $(1 + y/t^2)\, dt - 1/t\, dy = 0,\ y(2) = 1.$
33. $(2ty + 3t^2)dt + (t^2 - 1)dy = 0,\ y(0) = 1.$
34. $(1 + 5t - y)dt - (t + 2y)dy = 0,\ y(0) = 0.$
35. $(e^y - 2ty)dt + (te^y - t^2)\, dy = 0,\ y(0) = 0.$

36. $\left(2tye^{t^2} + 2te^{-y}\right)dt + \left(e^{t^2} - t^2 e^{-y} + 1\right)dy = 0,\ y(0) = 0.$
37. $(y^2 - 2 \sin 2t)\, dt + (1 + 2ty)\, dy = 0,\ y(0) = 1.$
38. $(\cos^2 t - \sin^2 t + y)dt + (\sec y \tan y + t)dy = 0,\ y(0) = 0.$
39. $(1/(1 + t^2) - y^2)dt - 2ty\, dy = 0,\ y(0) = 0.$
40. $(2t/(1 + t^2) + y)dt + (e^y + t)dy = 0,\ y(0) = 0.$
41. (a) Solve
    $(-2x - y \cos(xy))dx + (2y - x \cos(xy))dy = 0,$
    $y(0) = 0.$ (b) Explain why the results do or do not contradict the Picard-Lindelöf Theorem or the Existence and Uniqueness Theorem for First-Order Exact Equations.
42. (a) Solve $\left(-4x^3 + 6y \sin(6xy)\right)dx + \left(4y^3 + 6x \sin(6xy)\right)dy,\ y(0) = 0.$ (b) Explain why the results do or do not contradict the Picard-Lindelöf Theorem or the Existence and Uniqueness Theorem for First-Order Exact Equations.
43. Find the family of curves tangent to the force field
$$F(x,y) = \underbrace{\frac{2xy}{\sqrt{x^2 + y^2}}}_{dx/dt}\,\mathbf{i} - \underbrace{\frac{y^2 - x}{\sqrt{x^2 + y^2}}}_{dy/dt}\,\mathbf{j}.$$

44. Find the family of curves tangent to the force field
$$F(x,y) = \underbrace{\frac{x^2 - y^2}{\sqrt{x^2 + y^2}}}_{dx/dt}\,\mathbf{i} - \underbrace{\frac{2xy}{\sqrt{x^2 + y^2}}}_{dy/dt}\,\mathbf{j}.$$

45. Show that an equation of the form $g(y)\, dy - h(t)dt = 0$ is exact.

If $u = u(t)$ and $v = v(t)$, then the product rule tells us that $(d/dt)(uv) = v\,(du/dt) + u(dv/dt).$

---

46. (*Integrating Factors*) If the differential equation $M(t,y)dt + N(t,y)\, dy = 0$ is not exact, multiplying it by an *appropriate* function $\mu(t,y)$ sometimes yields an exact

equation. To find $\mu(t, y)$, we use the fact that if the equation $\mu(t, y) M(t, y)dt + \mu(t, y) N(t, y) dy = 0$ is exact $(\mu M)_y = (\mu N)_t$.

(a) Use the product rule to show that μ must satisfy the differential equation
$$\mu_y M - \mu_t N + (M_y - N_t) \mu = 0.$$

(b) Use this equation to show that if $\mu = \mu(t)$, μ satisfies the differential equation
$$\frac{d\mu}{dt} = \frac{M_y - N_t}{N} \mu, \quad \text{where} \quad \frac{M_y - N_t}{N}$$
is a function of y only.

(c) Show that if $\mu = \mu(t)$,
$$\mu(t) = \exp\left(\frac{1}{N(t, y)}\left[\frac{\partial M(t, y)}{\partial y} - \frac{\partial N(t, y)}{\partial t}\right] dt\right).$$

(d) If $\mu = \mu(y)$, find a differential equation that μ must satisfy and a restriction on $(N_t - M_y)/M$. Show that
$$\mu(y) = \exp\left(\frac{1}{M(t, y)}\left[\frac{\partial N(t, y)}{\partial t} - \frac{\partial M(t, y)}{\partial y}\right] dy\right).$$

47. (a) Find restrictions on p and q so that the first-order equation $dy/dt + p(t) y = q(t)$ is exact.

(b) Consider the first-order linear equation $dy/dt + p(t) y = q(t)$. Write it in the form $M(t, y)dy + N(t, y)dt = 0$, multiplying by the integrating factor $\mu(t)e^{\int p(t)\, dt}$ results in an exact equation.

In Exercises 48-55, use an integrating factor to solve each differential equation (see Exercise 46).

48. $t^2 y \, dt + t^3 \, dy = 0$.

49. $y(2e^t + 4t)dt + 3(e^t + t^2)dy = 0$.

50. $y \, dt + (2t - ye^y) \, dy = 0$.

51. $(2ty + y^2)dt - t^2 \, dy = 0$.

52. $(y + 2t^2)dt + (t^2 y - t)dy = 0$.

53. $(5ty + 4y^2 + 1)dt + (t^2 + 2ty)dy = 0$.

54. $(2ty^2 + y) \, dt + (2t^3 - t)dy = 0$.

55. $(2t + \tan y)dt + (t - t^2 \tan y)dy = 0$.

56. Suppose that $M(t, y)dt + N(t, y) \, dy = 0$ is an equation for which $\partial M/\partial y = \partial N/\partial t$.

(a) Let
$$g(y) = \int (N(t, y) - (\partial/\partial y) \int M(t, y)dt)dy.$$
Show that g is a function of y. *Hint:*
Show that $N(t, y) - \dfrac{\partial}{\partial y} \int M(t, y)dt$ is a function of y by showing that
$$(\partial/\partial t)(N(t, y) - (\partial/\partial y) \int M(t, y)dt) = 0.$$

(b) Let $f(x, y) = g(y) + \int M(t, y)dt$. Show that $M(t, y)dt + N(t, y)dy = (\partial f/\partial x)dx + (\partial f/\partial y)dy$.

As with other types of equations, technology is useful in solving exact equations or in performing the steps necessary to solve an exact equation.

---

57. Find a general solution of the equation $(2t - y^2 \sin ty) \, dt + (\cos ty - ty \sin ty)dy = 0$. Graph various solutions on the rectangle $[0, 3\pi] \times [0, 3\pi]$.

58. Find a general solution of $(-1 + e^{ty}y + y \cos ty) \, dt + (1 + e^{ty}t + t \cos(ty)) \, dy = 0$. Graph several solutions.

59. (a) Find a general solution for each of the following differential equations:
    (i) $(2t + 2y)dt + (2t + 2y)dy = 0$
    (ii) $(1.8t + 2y) \, dt + (2t + 2y)dy = 0$
    (iii) $(2t + 1.9y)dt + (1.9t + 2y) \, dy = 0$

(b) Graph the direction field for each equation along with several solutions.

(c) Comment on this statement: "If we slightly change a differential equation, the solutions also slightly change."

60. How does the graph of the solution to the IVP $(\sin cy - yc \sin ct)dt + (ct \cos cy + \cos ct) \, dy = 0$, $y(0) = 1$ change as c takes on values from -2 to 2.

2.5 SUBSTITUTION METHODS AND SPECIAL EQUATIONS

Substitutions are used often in integral calculus to transform integrals into forms that can be computed easily. Similarly, with some differential equations, we can perform substitutions that transform a given differential equation into an equation that is easier to solve. For example, we have considered population problems modeled with the first-order linear IVPs:

$$dy/dt = ky, \qquad y(0) = y_0 \quad \text{and}$$
$$dy/dt = ky - h, \quad y(0) = y_0.$$

However, we can consider a more complicated model that involves a nonlinear equation with

$$dy/dt = ky - ay^2, \quad y(0) = y_0.$$

The equation in this model, known as the *logistic equation*, (see Exercise 61 in Section 1.1), is almost linear because it can be written as $dy/dt - ky = -ay^2$. (The y^2 term on the right side causes the equation to be nonlinear.) It can also be expressed as $dy/dt = (k - ay)y$ to show that it differs from $dy/dt = ky$ in that the rate of growth (the coefficient of y), $(k - ay)$, is not constant. Instead, this rate depends on y. Although this is a separable equation, we can solve this equation through the use of the substitution

$$w = y^{1-2} = y^{-1}.$$

We will transform the logistic equation from an ODE that depends on y and t into one that depends on w and t. To do so, we start by finding dw/dt with the chain rule:

$$\frac{dw}{dt} = \frac{dw}{dy}\frac{dy}{dt} = -y^{-2}\frac{dy}{dt}.$$

Therefore,

$$\frac{dy}{dt} = -y^2\frac{dw}{dt}.$$

Substitution of this expression into $dy/dt - ky = -ay^2$ gives us

$$-y^2\frac{dw}{dt} - ky = -ay^2 \quad \text{or} \quad \frac{dw}{dt} + ky^{-1} = a.$$

Finally, with the substitution $w = y^{-1}$, we obtain the first-order linear ODE

$$dw/dt + kw = a,$$

which we solve by multiplying by the integrating factor $\mu(t) = e^{\int k\,dt} = e^{kt}$. Therefore, $\frac{d}{dt}\left(e^{kt}w\right) = ae^{kt}$, so

$$e^{kt}w = \frac{a}{k}e^{kt} + C_1 \quad \text{or} \quad w = \frac{a}{k} + C_1e^{kt}.$$

Returning to the original dependent variable, we have

$$y^{-1} = \frac{a}{k} + C_1e^{kt}.$$

Applying the initial condition, $y(0) = y_0$, we find that $1/y_0 = a/k + C_1$, so that $C_1 = 1/y_0 - a/k$, or $C_1 = (k - ay_0)/(ky_0)$. Therefore,

$$y^{-1} = \frac{a}{k} + \frac{k - ay_0}{ky_0}e^{-kt}.$$

Solving for y, we find that

$$y = \frac{ky_0}{ay_0 + (k - ay_0)e^{-kt}}.$$

In Figure 2.21, we graph this function using the parameter values $y_0 = 1/4$, $a = 1$, and $k = 3$. Notice that unlike the solutions to the Malthus model, solutions to the logistic equation are bounded.

Equations such as the logistic equation are classified as *Bernoulli equations*, and named after the theologian, mathematician, and business man, Jacob (Jacques) Bernoulli.

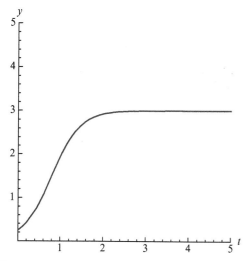

FIGURE 2.21 Solution to the logistic equation ($y_0 = 1/4$, $a = 1$, and $k = 3$).

Jakob (Jacques) Bernoulli (December 27, 1654–August, 16, 1705) In 1696, Bernoulli solved the equation, $y' = p(t) y + q(t) y^n$. Inscribed on his tomb at Münster, Basel is *Eadem mutata resurgo*, which translates to "Though changed I shall rise the same."

Definition 4 (Bernoulli Equation). *A **Bernoulli** equation is a first-order equation of the form*

$$\frac{dy}{dt} + p(t)\, y = q(t)\, y^n.$$

Of course, if $n = 0$ or $n = 1$, this equation is linear, so we would not need to make a change of variable to solve the equation. However, for other values of n, such as $n = 2$ as we considered in the logistic equation example, we solve Bernoulli equations with the substitution $w = y^{1-n}$. In doing so, we transform the nonlinear equation into the linear equation

$$\frac{dw}{dt} + (1 - n)p(t)\, w = (1 - n)\, q(t).$$

After solving this equation for w by using an integrating factor, we return to the original variable with the substitution $w = y^{1-n}$ and by solving for y. Of course, Bernoulli equations may involve dependent and independent variables other than y or t, but we follow the same solution method.

EXAMPLE 2.5.1

Solve $dy/dx + y/x = 1/(xy^2)$, $x > 0$.

Solution

In this case, the independent variable is x (instead of t as it was in the previous problem), and the right side of the equation can be written as y^{-2}/x, so $p(x) = 1/x$, $q(x) = 1/x$, and $n = -2$. Therefore, we make the substitution $w = y^{1-(-2)} = y^3$. Then, $dw/dx = 3y^2\, dy/dx$ so that $dy/dx = \frac{1}{3}y^{-2}\, dw/dx$. With this substitution, we obtain

$$\frac{1}{3}y^{-2}\frac{dw}{dx} + \frac{1}{x}y = \frac{1}{x}y^{-2}.$$

Multiplication by $3y^2$ gives us the first-order ODE

$$\frac{dw}{dx} + \frac{3}{x}y^3 + \frac{1}{x}y^{-2} \quad \text{or} \quad \frac{dw}{dx} + \frac{3}{x}w = \frac{3}{x}$$

because $w = y^3$. We solve this equation with the integrating factor

$$\mu(x) = e^{\int (3/x)\, dx} = e^{3\ln x} = x^3, \quad x > 0.$$

Multiplying the ODE by $\mu(x)$ and simplifying yields $(d/dx)(x^3 w) = 3x^2$, so that $x^3 w + x^3 + C$. Solving for w, we find that $w = 1 + Cx^{-3}$. Substituting $w = y^3$ and solving for y give us

$$y = (1 + Cx^{-3})^{1/3}.$$

Now, consider the differential equation $dy/dt = -(t + y)/(y - t)$, which can be written as

$$(y - t)dy + (t + y)\, dt = 0.$$

Because this equation is not of the form $f_1(t)g_1(y)dt+f_2(t)g_2(y)dy = 0$, it is not separable. However, if we let $y = ut$, then we use the product rule to obtain $dy = u\,dt + t\,du$. Substituting these expressions into $(y - t)\,dy + (t + y)dt = 0$ and simplifying results in

$$(ut - t)(u\,dt + t\,du) + (t + ut)\,dt = 0 \quad \text{or}$$

$$t(u^2 + 1)\,dt + t^2(u - 1)\,dyu = 0.$$

This equation is separable and can be written as

$$\frac{1}{t}\,dt = \frac{1 - u}{u^2 + 1}\,du.$$

Because the right side of this equation is equivalent to $((1/(u^2 + 1)) - u/(u^2 + 1))du$, we find that

$$\ln|t| = \tan^{-1} u - \frac{1}{2}\ln(u^2 + 1) + C.$$

Notice that the absolute value is not needed in the term containing $\ln(u^2 + 1)$ because $u^2 + 1 > 0$ for all u. Now, $u = y/t$ because $y = ut$, so resubstitution gives us

$$\ln|t| = \tan^{-1}(y/t) - \frac{1}{2}\ln((y/t)^2 + 1) + C$$

as a general solution of $(y - t)\,dy + (t + y)\,dt = 0$.
    The equation $(y - t)\,dy + (t + y)\,dt = 0$ is called a *homogeneous equation*. We can always reduce a homogeneous equation to a separable equation by a suitable substitution.
    Definition 5 (Homogeneous Differential Equation). *A first-order differential equation that can be written in the form $M(t, y)\,dt + N(t, y)\,dy = 0$, where $M(xt, xy) = x^n M(t, y)$ and $N(xt, xy) = x^n N(t, y)$ is called a **homogeneous differential equation** (of degree n).*

EXAMPLE 2.5.2

Show that the equation $(t^2 + ty)\,dt - y^2\,dy = 0$ is homogeneous.

Solution

Let $M(t, y) = t^2 + ty$ and $N(t, y) = -y^2$. The equation $(t^2 + ty)\,dt - y^2\,dy = 0$ is homogeneous of degree 2 because

$$M(xt, xy) = (xt)^2 + (xt)(xy) = x^2(t^2 + ty)$$
$$= x^2 M(t, y)$$

and

$$N(xt, xy) = -x^2 y^2 = x^2 N(t, y).$$

Homogeneous equations are reduced to separable equations by either of the substitutions

$$y = ut \quad \text{or} \quad t = vy.$$

Use the substitution $y = ut$ if $N(t, y)$ is less complicated than $M(t, y)$, and use $t = vy$ if $M(t, y)$ is less complicated than $N(t, y)$. If a difficult integration problem is encountered after a substitution is made, try the other substitution to see if it yields an easier problem. As with the separation of variables technique, this technique was also discovered by Leibniz.

EXAMPLE 2.5.3

Solve the IVP $dy/dt = y/(t + \sqrt{t^2 + y^2})$, $y(0) = 1, y \geq 0$.

Solution

When we write the differential equation in differential form, $y\,dt - (t + \sqrt{t^2 + y^2})\,dy = 0$, we recognize that the equation is homogeneous of order 1. Then, because $M(t, y) = y$ is less complicated than $N(t, y) = -(t + \sqrt{t^2 + y^2})$, we let $t = vy$ so that $dt = v\,dy + y\,dv$. Substitution into the homogeneous equation yields

$$y(v\,dy + y\,dv) - (vy + \sqrt{v^2 y^2 + y^2})dy = 0,$$

$$vy\,dy + y^2\,dv - vy\,dy - y\sqrt{v^2 + 1}\,dy = 0.$$

Notice that by writing $\sqrt{y^2 v^2 + y^2} = |y|\sqrt{v^2 + 1} = y\sqrt{v^2 + 1}$ we are taking advantage of our assumption that $y \geq 0$. The initial condition includes a positive value of y, so we can solve the IVP with this assumption. We simplify and separate variables in this equation to obtain

$$y^2 \, dv = y\sqrt{v^2+1} \, dy \quad \text{or} \quad \frac{1}{\sqrt{v^2+1}} \, dv = \frac{1}{y} \, dy.$$

Using a trigonometric substitution, table of integrals, or a CAS to evaluate $\int 1/\sqrt{v^2+1}\,dv$, we obtain

$$\ln\left|v+\sqrt{v^2+1}\right| = \ln y + C_1.$$

Returning to the original variables with $v = t/y$, we find that

$$\ln\left|\frac{t}{y}+\sqrt{\left(\frac{t}{y}\right)^2+1}\right| = \ln y + C_1$$

$$\frac{t}{y}+\sqrt{\left(\frac{t}{y}\right)^2+1} = C_2 y$$

$$\frac{t+\sqrt{t^2+y^2}}{y} = C_2 y$$

$$t+\sqrt{t^2+y^2} = C_2 y^2$$

$$y = \pm\frac{1}{C_2}\sqrt{1+2C_2 t}.$$

Applying the initial conditions gives us $(1/C_2)\sqrt{1+2C_2 \times 0} = 1$ so $C_2 = 1$. Therefore, the solution to the IVP is $y = \sqrt{1+2t}$. We graph this solution in Figure 2.22. Notice that the curve passes through the point $(0,1)$ as required by the initial condition.

Solve the equation $2xy\,dx + (x^2 - y^2)\,dy = 0$ by viewing it as a homogeneous equation of degree 2 rather than as an exact equation as done in Exercise 44 in Section 2.4.

In addition to showing that an equation is homogeneous of degree n by verifying that $M(xt, xy) = x^n M(t, y)$ and $N(xt, xy) = x^n N(t, y)$, there are

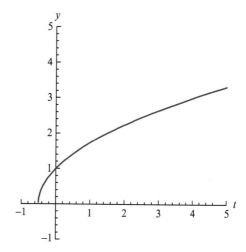

FIGURE 2.22   Graph of $y = \sqrt{1+2t}$.

other ways to determine if an equation is homogeneous or not. For example, solving the differential equation $4ty^2\,dt + (t^3+y^3)\,dy = 0$ for dy/dt, we obtain

$$\frac{dy}{dt} = \frac{-4ty^2}{t^3+y^3} = \frac{\frac{1}{t^3}(-4ty^2)}{\frac{1}{t^3}(t^3+y^3)} = \frac{-4\left(\frac{y}{t}\right)^2}{1+\left(\frac{y}{t}\right)^3} = F\left(\frac{y}{t}\right),$$

where $F(x) = -4x^2/(1+x^3)$. Similarly, we find that

$$\frac{dy}{dt} = \frac{-4ty^2}{t^3+y^3} = \frac{\frac{1}{y^3}(-4ty^2)}{\frac{1}{y^3}(t^3+y^3)} = \frac{-4\frac{t}{y}}{\left(\frac{t}{y}\right)^3+1} = G\left(\frac{t}{y}\right),$$

where $G(x) = -4x/(x^3+1)$. This indicates (although we have not shown it in general) that an equation is homogeneous if we can write it in either of the forms $dy/dt = F(y/t)$ or $dy/dt = G(t/y)$ (see Exercise 46).

EXAMPLE 2.5.4

Find values of a_0, a_1, a_2, and a_3, none of which are zero so that the nontrivial solutions to $dy/dt = (a_0 t + a_1 y)/(a_2 t + a_3 y)$ are periodic.

Solution

If none of a_0, a_1, a_2, and a_3 are zero, the equation is homogenous because

$$\frac{dy}{dt} = \frac{a_0 t + a_1 y}{a_2 t + a_3 y} = \frac{dy}{dt} = \frac{a_0 + a_1 \frac{y}{t}}{a_2 + a_3 \frac{y}{t}} = F\left(\frac{y}{t}\right),$$

where $F(x) = (a_0 + a_1 x)/(a_2 + a_3 x)$. Letting $y = ut$ leads to

$$-t\left(a_3 u^2 + (a_2 - a_1)u - a_0\right) dt - t^2 (a_3 u + a_2) du = 0,$$

$$\frac{a_3 u + a_2}{a_3 u^2 + (a_2 - a_1)u - a_0} du = -\frac{1}{t} dt.$$

Integrating and substituting $u = y/t$ yields a general solution of the equation

$$\ln \sqrt{a_0 + (a_1 - a_2)u - a_3 u^2} - \frac{a_1 + a_2}{\sqrt{(a_2 - a_1)^2 + 4a_0 a_3}}$$

$$\times \tanh^{-1}\left(\frac{a_2 - a_1 + 2a_3 u}{\sqrt{(a_2 - a_1)^2 + 4a_0 a_3}}\right) = -\ln t + C$$

or

$$\ln \sqrt{a_0 + (a_1 - a_2)\frac{y}{t} - a_3 \frac{y^2}{t^2}} - \frac{a_1 + a_2}{\sqrt{(a_2 - a_1)^2 + 4a_0 a_3}}$$

$$\times \tanh^{-1}\left(\frac{a_2 - a_1 + 2a_3 y}{t\sqrt{(a_2 - a_1)^2 + 4a_0 a_3}}\right) = -\ln t + C.$$

This solution is complicated and nearly impossible to analyze analytically, so we proceed graphically. We begin by generating *random* values of a_0, a_1, a_2, and a_3, none of which are zero, and then graphing the direction field associated with the equation $dy/dt = (a_0 t + a_1 y)/(a_2 t + a_3 y)$. Several of the results we obtain are shown in Figure 2.23. (Use each direction field to graph several solutions to each equation.)

From these graphs, we see that the choices $a_0 = 3$, $a_1 = -1$, $a_2 = 1$, and $a_3 = 2$ *may* lead to period solutions. A general solution of $dy/dt = (3t - y)/(t - 2y)$ is $\frac{3}{2}t^2 - ty + y^2 = C$ (why?), which is a family of ellipses as shown in Figure 2.24. We see that the choices $a_0 = 3$, $a_1 = -1$, $a_2 = 1$, and $a_3 = 2$ yield periodic solutions to the equation

 Show that the nontrivial solutions to $dy/dt = (a_0 t + a_1 y)/(a_2 t + a_3 y)$ are periodic if $a_2 = -a_1$ and $a_1^2 + a_0 a_3 < 0$.

 EXERCISES 2.5

In Exercises 1-10, solve the Bernoulli equation

1. $dy/dt - \frac{1}{2}y = ty^{-1}$.
2. $y' + y = ty^2$.
3. $2ty' - y = 2ty^3 \cos t$.
4. $ty' - y = ty^3 \sin t$.
5. $y' - 2y = y^{-1/2} \cos t$.
6. $y' + 3y = \sqrt{y} \sin t$.
7. $\dfrac{dy}{dt} - \dfrac{1}{t}y = ty^2$.
8. $\dfrac{dy}{dt} - \dfrac{1}{t}y = y^2/t^2$.
9. $\dfrac{dy}{dt} - \dfrac{1}{t}y = y^2/t$.
10. $\dfrac{dy}{dt} - \dfrac{1}{t}y = t^2 y^{3/2}$.

In Exercises 11-16, determine if the first-order differential equation is homogeneous of degree n. If so, determine its degree.

11. $\cos(t/(t + y))dt + e^{2y/t} dy = 0$.
12. $y \ln(t/y) dt + t^2/(t + y) dy = 0$.
13. $2 \ln t \, dt - \ln(4y^2)dy = 0$.
14. $(2/t + 1/y) dt + t/y^2 dy = 0$.
15. $\dfrac{\sin 2t}{\cos 2y} dt + \dfrac{\ln y}{\ln t} dy = 0$.
16. $\sqrt{t^2 + 1} \, dt + y \, dy = 0$.

In Exercises 17-32, solve each equation.

17. $2t \, dt + (y - 3t)dy = 0$.
18. $(2y - 3t)dt + t \, dy = 0$.
19. $(ty - y^2)dt + t(t - 3y)dy = 0$.
20. $(t^2 + ty + y^2)dt - ty \, dy = 0$.
21. $(t^3 + y^3)dt - ty^2 \, dy = 0$.
22. $dy/dt = (t + 4y)/(4t + y)$.
23. $(t - y) \, dt + t \, dy = 0$.

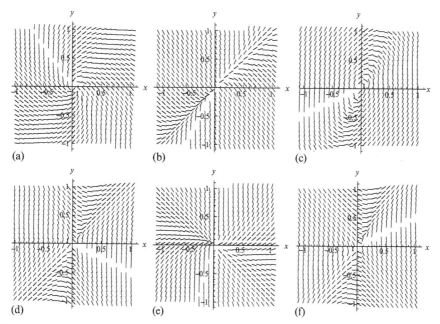

(a)   (b)   (c)

(d)   (e)   (f)

FIGURE 2.23 First Row: (a)-(c): (a) $a_0 = -4$, $a_1 = 3$, $a_2 = 7$, $a_3 = 5$; (b) $a_0 = 4$, $a_1 = -5$, $a_2 = -6$, $a_3 = 3$; and (c) $a_0 = 3$, $a_1 = -1$, $a_2 = 1$, $a_3 = -2$. Second Row: (d)-(f) (d) $a_0 = 3$, $a_1 = -1$, $a_2 = 1$, $a_3 = 2$; (e) $a_0 = -2$, $a_1 = -7$, $a_2 = -7$, $a_3 = 2$; and (f) $a_0 = -5$, $a_1 = 1$, $a_2 = 2$, $a_3 = -4$.

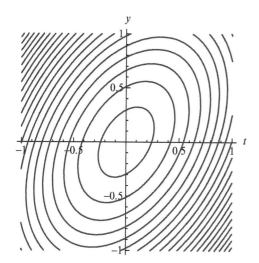

FIGURE 2.24 Graph of $\frac{3}{2}t^2 - ty + y^2 = C$ for various values of C.

24. $y\,dt + (t + y)dy = 0$.

25. $(2t^2 - 7ty + 5y^2)dy + ty\,dy = 0$.

26. $\left(y + 2\sqrt{t^2 + y^2}\right)dt - t\,dy = 0$.

27. $y^2\,dt = (ty - 4t^2)dy$.

28. $y\,dt - (3\sqrt{ty} + t)dy = 0$.

29. $(t^2 - y^2)\,dy + (y^2 + ty)dt = 0$.

30. $ty\,dy - (t^2 e^{-y/t} + y^2)dt = 0$.

31. $dt/dy = 2ye^{-t/y}/t + t/y$.

32. $t(\ln t - \ln y)dy = y\,dt$.

In Exercises 33-41, solve the IVP. Graph the solution on an appropriate interval.

33. $dy/dt + 2y = t^2\sqrt{y}$, $y(0) = 1$.

34. $dy/dt - 2y = t^2\sqrt{y}$, $y(0) = 1$.

35. $dy/dt = (4y^2 - t^2)/(2ty)$, $y(1) = 1$.

36. $(t + y)\,dt - t\,dy = 0$, $y(1) = 1$.

37. $t\,dy - (y + \sqrt{t^2 + y^2})dt = 0$, $y(1) = 0$.

38. $(t^3 + y^2\sqrt{t^2 + y^2})\,dt - ty\sqrt{t^2 + y^2}dy = 0$, $y(1) = 1$.

39. $(y^3 - t^3)dt - ty^2\, dy = 0$, $y(1) = 3$.

40. $ty^3\, dt - (t^4 + y^4)\, dy = 0$, $y(1) = 1$.

41. $y^4\, dt + (t^4 - ty^3)dy = 0$, $y(1) = 2$.

42. First-order equations of the form

$$(a_1 t + b_1 y + c_1)dt + (a_2 t + b_2 y + c_2)\, dy = 0$$

can sometimes be transformed into a homogeneous equation with a transformation. If $a_2/a_1 \neq b_2/b_1$, the transformation $t = T + h$, $y = Y + k$, where (h, k) satisfies the linear system

$$\begin{cases} a_1 h + b_1 k + c_1 = 0 \\ a_2 h + b_2 k + c_2 = 0 \end{cases} \quad \text{reduces this equation}$$

to a homogeneous equation in the variables T and Y. If $a_2/a_1 = b_2/b_1 = C$, for some constant C, the transformation $z = a_1 t + b_1 y$ also reduces this equation to a homogeneous equation in the variables t and z.

  Use the appropriate transformation described here to solve the following equations:

 (a) $(t - 2y + 1)dt + (4t - 3y - 6)dy = 0$.
 (b) $(5t + 2y + 1)\, dt + (2t + y + 1)dy = 0$.
 (c) $(3t - y + 1)dt - (6t - 2y - 3)\, dy = 0$.
 (d) $(2t + 3y + 1)dt + (4t + 6y + 1)dy = 0$.

43. Find an equation of the curve that passes through the point $(\sqrt{e}, \sqrt{e})$ and has slope $y/t + t/y$ at each point (t, y) on the curve.

44. Find the family of curves tangent to the force field

$$\mathbf{F}(x, y) = \underbrace{-\frac{xy + y^2}{\sqrt{x^2 + y^2}}}_{dx/dt}\, \mathbf{i} + \underbrace{\frac{y^2}{\sqrt{x^2 + y^2}}}_{dy/dt}\, \mathbf{j}.$$

45. Find the family of curves tangent to the force field

$$\mathbf{F}(x, y) = \underbrace{\frac{x^2 - xy}{\sqrt{x^2 + y^2}}}_{dx/dt}\, \mathbf{i} - \underbrace{\frac{y^2}{\sqrt{x^2 + y^2}}}_{dy/dt}\, \mathbf{j}.$$

46. The Bernoulli equation
$(dy/dx) - (2/x)y = -x^2 y$ is also known as *Ricatti's equation*. Show that the solutions of Ricatti's equation are $y = x^2/(\frac{1}{5}x^5 + C)$ and $y = 0$.

47. Solve $y' + y \cot x = y^4 \sin x$, $y(0) = 0$. Explain why the result does or does not contradict the Existence and Uniqueness Theorem.

48. Show that if the differential equation $dy/dt = f(t, y)$ is homogeneous, the equation can be written as $dy/dt = F(y/t)$. (*Hint:* Let $x = 1/t$.)

49. If $M(t, y)dt + N(t, y)dy = 0$ is a homogeneous equation, show that the change in variables $x = r\cos\theta$ and $y = r\sin\theta$ transforms the homogeneous equation into a separable equation.

Alexis Clairaut (1713-1765): As with many other mathematicians, there is a crater on the moon named after Clairaut, which is in the center of the following photograph.

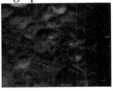

50. Equations of the form $f(ty' - y) = g(y')$ are called *Clairaut equations* after the French mathematician Alexis Clairaut (1713-1765), who studied equations of this type in 1734.

Solutions to this equation are determined by differentiating each side of the equation with respect to t.

(a) Use the chain rule to show that the derivative of $f(ty' - y)$ is

$$f'(ty' - y)(ty'' + y' - y') = f'(ty' - y)(ty''),$$

where $'$ denotes differentiation with respect to the argument of the function, t.

(b) Show that the equation

$$f'(ty' - y)(ty'') = g'(y')y'',$$

which is equivalent to

$$[f'(ty' - y)t - g'(y')]y'' = 0,$$

is obtained by differentiating both sides of the Clairaut equation with respect to t.

(c) This result indicates that $y'' = 0$ or $f'(ty' - y)t - g'(y') = 0$. If $y'' = 0$, $y' = c$ where c is constant. Substitute $y' = c$ into the differential equation $f(ty' - y) = g(y')$ to find that a general solution is $f(tc - y) = g(c)$. If $f'(ty' - y)t - g'(y') = 0$, this equation can be used along with $f(ty' - y) = g(y')$ to determine another solution by eliminating y'. This is called the *singular solution* of the Clairaut equation.

51. Use the following steps to solve the Clairaut equation $ty' - (y')^3 = y$.

(a) Place the equation in the appropriate form to find that $f(t) = t$ and $g(t) = t^3$.

(b) Use the form of a general solution to find that $tc - y = c^3$ and solve this equation for y.

(c) Find the singular solution by differentiating $ty' - (y')^3 = y$ with respect to t to obtain $ty'' + y' - 3(y')^2y'' = y'$, which can be

simplified to $[t - 3(y')^2]y'' = 0$. Since $y'' = 0$ was used to find the general solution, solve $t - 3(y')^2 = 0$ for y' to obtain $y' = (t/3)^{1/2}$. Substitute this expression for y' into $ty' - y = (y')^3$ to obtain a relationship between t and y.

In Exercises 52-55, solve the Clairaut equation.

52. $ty' - y - 2(ty' - y)^2 = y' + 1.$
53. $ty' - y - 1 = (y')^2 - y'.$
54. $1 + y - ty' = \ln(y').$
55. $1 - 2(ty' - y) = (y')^{-2}.$
56. *(Pollution)* Under normal atmospheric conditions, the density of soot particles $N(t)$ satisfies the differential equation

$$\frac{dN}{dt} = -k_cN^2 + k_dN,$$

where k_c, called the *coagulation constant*, is a constant that relates how well particles stick together; and k_d, called the *dissociation constant*, is a constant that relates to how well particles fall apart. Both of these constants depend on temperature, pressure, particle size, and other external forces.[2]

(a) Find a general solution for this Bernoulli equation.

(b) Find the solution that satisfies the initial condition $N(0) = N_0$. ($N_0 > 0$)
The following table lists typical values of k_c and k_d.

k_c	k_d
163	5
125	26
95	57
49	85
300	26

(c) For each pair of values in the previous table, sketch the graph of $N(t)$ if

[2] Chr. Feldermann, H. Jander, H.Gg. Wagner, Soot Particle Coagulation of Premixed Ethylene/Air Flames at 10 bar, International Journal of Research in Physical Chemistry and Chemical Physics, **186** (Part 11), 1994, 127-140.

$N(0) = N_0$ for $N_0 = 0.01, 0.05, 0.1, 0.5,$ 0.75, 1, 1.5, and 2. Regardless of the initial condition $N(0) = N_0$, what do you notice in each case? Do pollution levels seem to be more sensitive to $k_c y$ or k_d? Does your result make sense? Why?

(d) Show that if $k_d y > 0$, $\lim_{t \to \infty} N(t) = k_d/k_c$. Why is the assumption that $k_d > 0$ reasonable?

(e) For each pair in the table, calculate $\lim_{t \to \infty} N(t) = k_d/k_c$. Which situation results in the highest pollution levels? How could the situation be changed?

Joseph-Louis Lagrange (January 25, 1736, Turin, Italy–April 10, 1813, Paris, France): As with Clairaut, a crater on the moon is named after Lagrange. Some people classify Lagrange as a French mathematician but others consider him an Italian mathematician as he was born in Turin, Italy. Lagrange is famous for his contributions to analysis and number theory in mathematics as well as his contributions to analytical and celestial mechanics.

"If I had been rich, I probably would not have devoted myself to mathematics."

57. Equations of the form $y = tf(y') + g(y')$ are called *Lagrange equations*. These equations are solved by making the substitution $p = y'(t)$.

(a) Differentiate $y = tf(y') + g(y')$ with respect to t to obtain

$$y' = tf'(y')y'' + f(y') + g'(y')y''.$$

(b) Substitute p into the equation to obtain

$$p = tf'(p)\frac{dp}{dt} + f(p) + g'(p)\frac{dp}{dt}$$

$$= f(p) + \left(tf'(p) + g'(p)\right)\frac{dp}{dt}.$$

(c) Solve this equation for dt/dp to obtain the linear equation

$$\frac{dt}{dp} = \frac{tf'(p) + g'(p)}{p - f(p)},$$

which is equivalent to

$$\frac{dt}{dp} + \frac{f'(p)}{f(p) - p}t = \frac{g'(p)}{p - f(p)}.$$

This first-order linear equation can be solved for t in terms of p. Then, $t(p)$ can be used with $y = tf(p) + g(p)$ to obtain an equation for y.

58. Solve the Lagrange equation $y = -ty' + \frac{1}{5}(y')^5$ using the following steps. In this case, $f(y') = -y'$ and $g(y') = \frac{1}{5}(y')^5$.

(a) Differentiate the equation with respect to t to obtain $y' = -y' - ty'' + (y')^4 y''$.

(b) Substitute $y' = p$ to obtain $p = -p - t\,dp/dt + p^4\,dp/dt$. Simplify this equation to obtain $dp/dt = 2p/(p^4 - t)$.

(c) Rewrite this equation as $dt/dp = (p^4 - t)/(2p)$ and solve this first-order linear equation for t to obtain $t = \frac{1}{9}p^4 + Cp^{-1/2}$.

(d) Substitute $y' = p$ and $t = \frac{1}{9}p^4 + Cp^{-1/2}$ into the differential equation $y = -ty' + \frac{1}{5}(y')^5$ to obtain a formula for y.

(e) Graph the integral curves for various values of C.

In Exercises 59-61, solve the Lagrange equation. (In each case, $' = d/dt$.)

59. $y = t(y')^2 + 3(y')^2 - 2(y')^3$.
60. $y = t(y' + 1) + (2y' + 1)$.
61. $y = t(2 - y') + (2(y')^2 + 1)$.

Leonhard Euler (April 15, 1707-September 18, 1783, St. Petersburg, Russia): According to O'Connor and Robertson [19], "Euler served as a medical lieutenant in the Russian navy from 1727 to 1730. In St. Petersburg, he lived with Daniel Bernoulli, who, already unhappy in Russia, had requested that Euler bring him tea, coffee, brandy, and other delicacies from Switzerland. Euler became professor of physics at the Academy in 1730 and, since this allowed him to become a full member of the Academy, he was able to give up his Russian navy post."

Euler was married twice and had 13 children. All but five of the children died when they were very young, but according to legend, Euler did most of his science with young children at his feet.

62. Euler was the first mathematician to take advantage of integrating factors to solve linear differential equations.[3] Euler used the following steps to solve the differential equation

$$\frac{dz}{dv} - 2z + \frac{z}{v} = \frac{1}{v}.$$

(a) Multiply the equation by the integrating factor $e^{-2v}v$.
(b) Show that $(d/dv)(e^{-2v}vz) = e^{-2v}v(dz/dv) - 2e^{-2z}vz + e^{-2z}z$.
(c) Express the equation as $(d/dv)(e^{-2v}vz) = e^{-2v}$ and solve this equation for z.

63. Consider the solution to the logistic equation $y = ky_0/(ay_0 + (k - ay_0)e^{-kt})$. Find $\lim_{t\to\infty} y(t)$. Sketch the phase line for $dy/dt = (k - ay)y$ and compare it to the value of $\lim_{t\to\infty} y(t)$.

64. Solve the equation
$(t^{1/3}y^{2/3} + t)dt + (t^{2/3}y^{1/3} + y)\, dy = 0$.
Graph several solutions.

65. Solve the IVP $y' = (y^2 - t^2)/(ty)$, $y(4) = 0$ and graph the solution for $0 < t \leq 4$.

66. (a) The *sine integral function*, Si(t), is defined by

$$\mathrm{Si}(t) = \int_0^t \frac{\sin x}{x}\, dx.$$

(i) Evaluate Si$'(t)$ and $\lim_{t\to\infty}$ Si(t). (ii) Graph Si(t) on the interval $[0, 6\pi]$. (iii) Approximate the maximum value of Si(t).

(b) Graph the direction field associated with the equation $y \sin(t/y)\, dt - (t + t\sin(t/y))dy = 0$.

(c) Solve the IVP $y \sin(t/y)\, dt - (t + t\sin(t/y))dy = 0$, $y(1) = 2$ and graph the solution.

In the same manner that CASs and graphing utilities can be used to help find solutions of other first-order equations, then can also be used to help solve and graph solutions of many other special first-order equations.

2.6 NUMERICAL METHODS FOR FIRST-ORDER EQUATIONS

- EULER'S METHOD
- IMPROVED EULER'S METHOD
- ERRORS

[3] V.J. Katz, A History of Mathematics: An Introduction, HarperCollins, 1993, p. 503.

- RUNGE-KUTTA METHOD
- COMPUTER-ASSISTED SOLUTIONS USING
  COMMERCIAL SOFTWARE

In many cases, we cannot obtain a formula for the solution to an IVP of the form

$$dy/dt = f(t, y), \quad y(t_0) = y_0.$$

For example, we cannot find an exact solution of $dy/dx = \sin(2x - y)$, $y(0) = 0.5$ using any of the methods we have discussed to this point. Of course, we can determine certain properties of the solution by looking at the slope field. However, we sometimes would like to find numerical values of y at particular values of t. For that reason, we now discuss numerical methods for approximating solutions to IVPs. We begin with a method based on tangent line approximations.

Euler's Method

Suppose that we wish to approximate the solution to

$$dy/dx = f(x, y), \quad y(x_0) = y_0$$

over the interval $x_0 \le x \le X$, where X is a specified value of x. For example, we may wish to approximate the solution to an IVP on an interval like $0 \le x \le 1$. This indicates that the initial condition is specified at $x_0 = 0$ and that $X = 1$ is the end of the interval. Now, we partition $x_0 \le x \le X$ into N subintervals of equal length h, where $h = (X - x_0)/N$. Therefore, the nth value of x in the partition is $x_n = x_0 + nh$, $n = 1, 2, \ldots, N$. This means that $x_1 = x_0 + h$,

---

In this text, we frequently use t as our independent variable. In this section, we exclusively use x to represent the independent variable and y to represent the dependent variable to make programming these numerical methods into most graphing calculators easier for those students who wish to do so.

$x_2 = x_0 + 2h, \ldots, x_N = x_0 + Nh = x_0 + N \times (X - x_0)/N = X$. If we are solving the IVP $dy/dx = f(x, y)$, $y(x_0) = y_0$, then the point (x_0, y_0) is on the solution curve. Therefore, as we discussed in relation to slope fields, we can follow the line tangent to the solution curve at (x_0, y_0) to approximate the value of the solution at the next value of x, $x = x_1$. Because the slope of this tangent line is $f(x_0, y_0)$, we find that an equation for the tangent line is $y - y_0 = f(x_0, y_0)(x - x_0)$ or $y = f(x_0, y_0)(x - x_0) + y_0$. Therefore, the approximate value of the solution at $x = x_1$ is

$$y_1 = f(x_0, y_0)(x_1 - x_0) + y_0 = hf(x_0, y_0) + y_0.$$

Notice that we refer to the approximate value of y at $x = x_i$ as y_i while we refer to the exact value of the solution at $x = x_i$ as $y(x_i)$. This means that y_i approximates $y(x_i)$ (see Figure 2.25(a)). We remember from calculus that this gives a good approximation if h is a small number. Next, we assume that the point (x_1, y_1) is on the solution curve. If we determine the equation of the line with slope $f(x_1, y_1)$ that passes through (x_1, y_1), we obtain

$$y - y_1 = f(x_1, y_1)(x - x_1),$$

so we find the approximate value of y at $x = x_2$ is

$$y_2 = f(x_0, y_0)(x_2 - x_1) + y_1 = hf(x_0, y_0) + y_1.$$

We continue this process until we reach y_N, the approximate value of $y(x_N) = y(X)$. In doing so, we obtain a sequence of points of the form (x_n, y_n), $n = 1, 2, \ldots, N$. We show several points of this type along with actual values of y in Figure 2.25(b).

Theorem 7 (Euler's Method). *The solution of the IVP*

$$dy/dx = f(x, y), \quad y(x_0) = y_0$$

is approximated at the sequence of points (x_n, y_n), $n = 1, 2, \ldots$, where y_n is the approximate value of $y(x_n)$ by computing

$$y_n = hf(x_{n-1}, y_{n-1}) + y_{n-1}, \quad n = 1, 2, \ldots,$$

where $x_n = x_0 + nh$ and h is the selected stepsize.

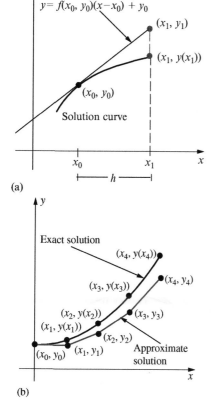

FIGURE 2.25 (a) Approximating (x_1, y_1). (b) Successive approximations using Euler's method.

EXAMPLE 2.6.1

Use Euler's method with $h = 0.1$ and with $h = 0.05$ to approximate the solution of $y' = xy$, $y(0) = 1$ on $0 \leq x \leq 1$. Determine the exact solution and compare the results.

Solution

First, we note that $f(x,y) = xy$, $x_0 = 0$, and $y_0 = 1$. With $h = 0.1$, we have the formula

$$y_n = hf(x_{n-1}, y_{n-1}) + y_{n-1} = 0.1x_{n-1}y_{n-1} + y_{n-1}.$$

Then, for $x_1 = x_0 + h = 0.1$, we have

$$y_1 = 0.1x_0y_0 + y_0 = 0.1 \times 0 \times 1 + 1 = 1.$$

Similarly, for $x_2 = x_0 + 2h = 0.2$,

$$y_2 = 0.1x_1y_1 + y_1 = 0.1 \times 0.1 \times 1 + 1 = 1.01.$$

In Table 2.1, we show the results of this sequence of approximations. From this, we see that $y(1)$ is approximately 1.54711.

First, $h = 0.05$, we use

$$y_n = hf(x_{n-1}, y_{n-1}) + y_{n-1} = 0.05x_{n-1}y_{n-1} + y_{n-1}$$

to obtain the values given in Table 2.2. With this stepsize, the approximate value of $y(1)$ is 1.59594.

The exact solution to the IVP, which is found with separation of variables, is $y = e^{x^2/2}$ so the exact value of $y(1)$ is $e^{1/2} \approx 1.64872$. The smaller value of h, therefore, yields a better approximation. We graph the approximations obtained with $h = 0.1$ and $h = 0.05$ as well as the graph of $y = e^{x^2/2}$ in Figure 2.26. Notice from these graphs that the approximation is more accurate when h is decreased.

TABLE 2.1   Euler's Method with $h = 0.1$

x_n	y_n	x_n	y_n
0.0	1.0	0.6	1.15873
0.1	1.0	0.7	1.22825
0.2	1.01	0.8	1.31423
0.3	1.0302	0.9	1.41937
0.4	1.06111	1.0	1.54711
0.5	1.10355		

TABLE 2.2   Euler's Method with $h = 0.05$

x_n	y_n	x_n	y_n	x_n	y_n
0.0	1.0	0.35	1.05361	0.70	1.2523
0.05	1.0	0.40	1.07204	0.75	1.29613
0.10	1.0025	0.45	1.09348	0.80	1.34474
0.15	1.00751	0.50	1.11809	0.85	1.39853
0.20	1.01507	0.55	1.14604	0.90	1.45796
0.25	1.02522	0.60	1.17756	0.95	1.52357
0.30	1.03803	0.65	1.21288	1.00	1.59594

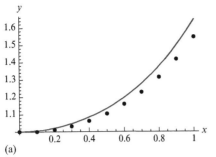

(a)

(b)

FIGURE 2.26   (a) $h = 0.1$ and (b) $h = 0.05$.

Improved Euler's Method

Euler's method is improved by using an average slope over each interval. Using the tangent line approximation of the curve through (x_0, y_0), $y = f(x_0, y_0)(x - x_0) + y_0$, we find the approximate value of y at $x = x_1$, which we now call y_1^*. Then, $y_1^* = hf(x_0, y_0) + y_0$, and with the differential equation $y' = f(x, y)$, we find that the approximate slope of the tangent line at $x = x_1$ is $f(x_1, y_1^*)$. The average of the two slopes, $f(x_0, y_0)$ and $f(x_1, y_1^*)$ is $\frac{1}{2}\left(f(x_0, y_0) + f(x_1, y_1^*)\right)$, and an equation of the line through (x_0, y_0) with slope $\frac{1}{2}\left(f(x_0, y_0) + f(x_1, y_1^*)\right)$ is

$$y = \frac{1}{2}\left(f(x_0, y_0) + f(x_1, y_1^*)\right)(x - x_0) + y_0.$$

We illustrate the determination of this average slope and the position of these lines in Figure 2.27.

At $x = x_1$, the approximate value of y is given by

$$y = \frac{1}{2}\left(f(x_0, y_0) + f(x_1, y_1^*)\right)(x - x_0) + y_0.$$

Continuing in this manner, the approximation in each step in the improved Euler's method depends on the following two calculations:

$$y_n^* = hf\left(x_{n-1}, y_{n-1}\right) + y_{n-1},$$

$$y_n = \frac{1}{2}h\left(f(x_{n-1}, y_{n-1}) + f(x_n, y_n^*)\right) + y_{n-1},$$

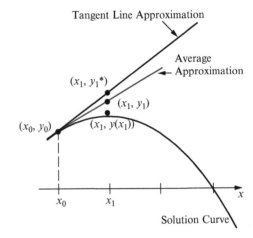

FIGURE 2.27   Improving Euler's method.

where $x_n = x_0 + nh$.

Theorem 8 (Improved Euler's Method). *The solution of the IVP*

$$y' = f(x, y), \quad y(x_0) = y_0$$

is approximated at the sequence of points (x_n, y_n), $n = 1, 2, \ldots$, where y_n is the approximate value of $y(x_n)$ by computing at each step the following two calculations:

$$y_n^* = hf\left(x_{n-1}, y_{n-1}\right) + y_{n-1},$$

$$y_n = \frac{1}{2}h\left(f(x_{n-1}, y_{n-1}) + f(x_n, y_n^*)\right) + y_{n-1},$$

where $x_n = x_0 + nh$ and h is the selected stepsize.

EXAMPLE 2.6.2

Use the improved Euler's method to approximate the solution of $y' = xy$, $y(0) = 1$ on $0 \leq x \leq 1$ for $h = 0.1$. Compare the results to the exact solution.

Solution

In this case, $f(x, y) = xy$, $x_0 = 0$, and $y_0 = 1$. Therefore, we use the equations

$$y_n^* = h x_{n-1} y_{n-1} + y_{n-1}$$

$$y_n = \frac{1}{2} h \left(x_{n-1} y_{n-1} + x_n y_n^* \right) + y_{n-1},$$

for $n = 1, 2, \ldots, 10$. For example, if $n = 1$, we have

$$y_1^* = h x_0 y_0 + y_0 = 0.1 \times 0 \times 1 + 1 = 1,$$

$$y_1 = \frac{1}{2} h \left(x_0 y_0 + x_1 y_1^* \right) + y_0$$

$$= \frac{1}{2}(0 \times 1 + 0.1 \times 1) \times 0.1 + 1 = 1.005.$$

Then,

$$y_2^* = h x_1 y_1 + y_1 = 0.1 \times 0.1 \times 1.005$$

$$+1.005 = 1.01505,$$

$$y_2 = \frac{1}{2} h \left(x_1 y_1 + x_2 y_2^* \right) + y_1 = \frac{1}{2}(0.1 \times 1.005$$

$$+0.2 \times 1.01505) \times 0.1 + 1.005 = 1.0201755.$$

In Table 2.3, we list the approximations obtained with this improved method and compare them to those obtained with Euler's method in Example 2.6.1 (see Figure 2.28). Which method yields a better approximation?

Errors

When approximating the solution of an IVP, there are several sources of error. One of these

TABLE 2.3 Improved Euler's Method with $h = 0.1$

x_n	y_n (IEM)	y_n (EM)	Actual Value
0.0	1.0	1.0	1.0
0.1	1.005	1.0	1.00501
0.2	1.0201755	1.01	1.0202
0.3	1.0459859	1.0302	1.04603
0.4	1.083223	1.06111	1.08329
0.5	1.1330513	1.10355	1.13315
0.6	1.1970687	1.15873	1.19722
0.7	1.277392	1.22825	1.27762
0.8	1.3767731	1.31423	1.37713
0.9	1.4987552	1.41937	1.49930
0.10	1.6478813	1.54711	1.64872

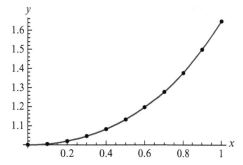

FIGURE 2.28 Using the Improved Euler's Method to approximate the solution to an IVP.

sources is *round-off error* because computers and calculators use only a finite number of digits in all calculations. Of course, the error is compounded as rounded values that are used in subsequent calculations. Therefore, one way to minimize round-off error is to reduce the number of calculations.

Another source of error is *truncation error*, which results from using an approximate formula. We begin our discussion of error by considering the error associated with Euler's method, which uses only the first two terms of the Taylor series expansion.

Recall *Taylor's Formula with Remainder* and *Taylor's Theorem* from introductory calculus. Provided that the necessary derivatives exist and are continuous on an interval containing $x = x_0$ for the function $y = y(x)$ and other reasonable conditions are met, Taylor's Theorem tells us that

$$y(x) = y(x_0) + y'(x_0)(x - x_0) + \frac{y''(x_0)}{2!}(x - x_0)^2$$

$$+ \cdots + \frac{y^{(n)}(x_0)}{n!}(x - x_0)^n + \frac{y^{(n+1)}(c)}{(n + 1)!}(x - x_0)^{n+1},$$

where c is a number between x and x_0. The remainder term is

$$R_n(x) = \frac{y^{(n+1)}(c)}{(n + 1)!}(x - x_0)^{n+1}$$

so the accuracy of the approximation obtained by using the first n terms of the Taylor series depends on the size of $R_n(x)$. For Euler's method, the remainder is

$$R_1(x) = \frac{y''(c)}{2!}(x - x_0)^2$$

so at $x = x_1 = x_0 + h$, the remainder is

$$R_1(x_1) = R(x_0 + h) = \frac{y''(c)}{2!}h^2.$$

Therefore, a *bound* on the error is given by

$$|R_1(x_1)| \leq \frac{1}{2}h^2 \max_{x_0 \leq x \leq x_1} |y''(x)|.$$

EXAMPLE 2.6.3

Find a bound for the local truncation error when Euler's method is used to approximate the solution of the IVP $y' = xy$, $y(0) = 1$ at $x = 0.1$.

Solution

We found in Example 2.6.1 that the exact solution of this problem is $y = e^{x^2/2}$. Therefore, a bound on the error is

$$|R_1(x_1)| \leq \frac{1}{2}(0.1)^2 \max_{x_0 \leq x \leq x_1} |y''(x)|,$$

where $y' = xe^{x^2/2}$ and $y'' = e^{x^2/2} + x^2 e^{x^2/2}$. Then,

$$|R_1(x_1)| \leq \frac{1}{2}(0.1)^2 \max_{x_0 \leq x \leq x_1} |y''(x)|$$

$$= \frac{1}{2}(0.1)^2 \max_{x_0 \leq x \leq x_1} |e^{x^2/2} + x^2 e^{x^2/2}|$$

$$\leq (0.005)\left(\max_{x_0 \leq x \leq x_1} |e^{x^2/2}| + \max_{x_0 \leq x \leq x_1} |t^2 e^{x^2/2}|\right)$$

$$\leq (0.005)\left(e^{(0.1)^2/2} + (0.1)^2 e^{(0.1)^2/2}\right)$$

$$\approx 0.005075.$$

Notice that the value of $y = e^{x^2/2}$ at $x = 0.1$ is 1.005012521 and the approximate value obtained with Euler's method is $y_1 = 1$. Therefore, the error is $1.005012521 - 1.0 = 0.005012521$, which is less than the bound $|R_1(x_1)| \leq 0.005075$.

Runge-Kutta Method

To improve on the approximate solutions obtained with Euler's method and to avoid the analytic differentiation of the function $f(t, y)$ to obtain y'', y''', $\ldots$, we introduce the *Runge-Kutta method*. We begin with the *Runge-Kutta method* of order 2.

Suppose that we know the value of y at x_n. We now use the point (x_n, y_n) to approximate the value of y at a nearby value $x = x_n + h$ by assuming that

$$y_{n+1} = y_n + Ak_1 + Bk_2,$$

where

$$k_1 = hf(x_n, y_n) \quad \text{and} \quad k_2 = hf(x_n + ah, y_n + bk_1).$$

We also use the Taylor series expansion of y to obtain another representation of $y_{n+1} = y(x_n + h)$ as follows:

$$y(x_n + h) = y(x_n) + hy'(x_n) + \frac{h^2}{2!}y''(x_n) + \cdots$$

$$= y_n + hy'(x_n) + \frac{h^2}{2!}y''(x_n) + \cdots .$$

Now, because

$$y_{n+1} = y_n + Ak_1 + Bk_2 = y_n + Ahf(x_n, y_n)$$
$$+ Bhf(x_n + ah, y_n + bhf(x_n, y_m)),$$

we wish to determine values of $A, B, a,$ and b such that these two representations of y_{n+1} agree. Notice that if we let $A = 1$ and $B = 0$, then the relationships match up to order h. However, we can choose these parameters more wisely so that agreement occurs up through terms of order h^2. This is accomplished by considering the Taylor series expansion of a function $z = F(x, y)$ of two variables about (x_0, y_0) which is given by

$$F(x_0, y_0) + \frac{\partial F}{\partial x}(x_0, y_0)(x - x_0)$$
$$+ \frac{\partial F}{\partial y}(x_0, y_0)(y - y_0) + \cdots .$$

In our case, we have

$$f(x_n + ah, y_n + bhf(x_n, y_m)) = f(x_n, y_n)$$
$$+ ah\frac{\partial f}{\partial x}(x_n, y_n) + bhf(x_n, y_n)$$
$$\frac{\partial f}{\partial y}(x_n, y_n)(y - y_0) + O(h^2).$$

The power series is then substituted into the following expression and simplified to yield:

$$y_{n+1} = y_n + Ahf(x_n, y_n)$$
$$+ Bhf(x_n + ah, y_n + bhf(x_n, y_m))$$
$$= y_n + (A + B)hf(x_n, y_n) + aBh^2\frac{\partial f}{\partial x}(x_n, y_n)$$
$$+ bBh^2 f(x_n, y_n)\frac{\partial f}{\partial x}(x_n, y_n) + O(h^3).$$

Comparing this expression to the following power series obtained directly from the Taylor series of y,

$$y(x_n + h) = y(x_n) + hf(x_n, y_n)$$
$$+ \frac{1}{2}h^2\frac{\partial f}{\partial x}(x_n, y_n)$$
$$+ \frac{1}{2}h^2\frac{\partial f}{\partial y}(x_n, y_n) + O(h^3)$$

Wilhelm Kutta's (1867-1944) thesis published in 1900 contains the Runge-Kutta method for solving differential equations. Although Kutta's theorem is famous now, he may have been more interested in subject areas outside of mathematics as he took many courses in music, art, and languages.

Carle Runge's (1856-1927) is more famous for his contributions to physics than for his contributions to mathematics. Although active and fit, Runge died of a heart attack at the age of 70.

or

$$y_{n+1} = y_n + hf(x_n, y_n) + \frac{1}{2}h^2 \frac{\partial f}{\partial x}(x_n, y_n)$$
$$+ \frac{1}{2}h^2 \frac{\partial f}{\partial y}(x_n, y_n) + O(h^3),$$

we see that A, B, a, and b must satisfy the following system of nonlinear equations:

$$A + B = 1, \quad aA = \frac{1}{2}, \quad \text{and} \quad bB = \frac{1}{2}.$$

Therefore, choosing $a = b = 1$, the Runge-Kutta method of order 2 uses the following equation:

$$y_{n+1} = y(x_n + h) = y_n + \frac{1}{2}hf(x_n, y_n)$$
$$+ \frac{1}{2}hf(x_n + h, y_n + hf(x_n, y_n))$$
$$= y_n + \frac{1}{2}(k_1 + k_2),$$

where $k_1 = hf(x_n, y_n)$ and $k_2 = hf(x_n + h, y_n + k_1)$. Notice that this method is equivalent to the improved Euler method.

Theorem 9 (Runge-Kutta Method of Order 2). *The solution of the IVP*

$$y' = f(x, y), \ y(x_0) = y_0$$

is approximated at the sequence of points (x_n, y_n) $(n = 1, 2, \ldots)$, where y_n is the approximate value of $y(t_n)$ by computing at each step

$$y_{n+1} = y(x_n + h) = y_n + \frac{1}{2}hf(x_n, y_n)$$
$$+ \frac{1}{2}hf(x_n + h, y_n + hf(x_n, y_n))$$
$$= y_n + \frac{1}{2}(k_1 + k_2) \ (n = 0, 1, 2, \ldots),$$

where $k_1 = hf(x_n, y_n)$, $k_2 = hf(x_n + h, y_n + k_1)$, $x_n = x_0 + nh$, and h is the selected stepsize.

EXAMPLE 2.6.4

Use the Runge-Kutta method of order 2 with $h = 0.1$ to approximate the solution of the IVP $y' = xy$, $y(0) = 1$ on $0 \le x \le 1$.

Solution

In this case, $f(x, y) = xy$, $x_0 = 0$, and $y_0 = 1$. Therefore, on each step we use the following three equations:

$$k_1 = hf(x_n, y_n) = 0.1x_n y_n,$$
$$k_2 = hf(x_n + h, y_n + k_1)$$
$$= 0.1(x_n + 0.1)(y_n + k_1),$$

and

$$y_{n+1} = y_n + \frac{1}{2}(k_1 + k_2).$$

For example, if $n = 0$, then

$$k_1 = 0.1x_0 y_0 = 0.1 \times 0 \times 1 = 0,$$
$$k_2 = 0.1(x_0 + 0.1)(y_0 + k_1)$$
$$= 0.1 \times 0.1 \times 1 = 0.01,$$

and

$$y_1 = y_0 + \frac{1}{2}(k_1 + k_2) = 1 + \frac{1}{2} \times 0.01 = 1.005.$$

Therefore, the Runge-Kutta method of order 2 approximates that the value of y at $x = 0.1$ is 1.005. Similarly, if $n = 1$, then

$$k_1 = 0.1x_1 y_1 = 0.1 \times 0.1 \times 1.005 = 0.01005,$$
$$k_2 = 0.1(x_1 + 0.1)(y_1 + k_1) = 0.1$$
$$\times 0.2 \times 1.01505 = 0.020301,$$

and

$$y_2 = y_1 + \frac{1}{2}(k_1 + k_2) = 1.005 + \frac{1}{2}$$
$$\times (0.01005 + 0.020301) = 1.0201755.$$

In Table 2.4, we display the results obtained for the other values on $0 \le x \le 1$ using the Runge-Kutta method of order 2.

The terms of the power series expansions used in the derivation of the Runge-Kutta method of order 2 can be made to match up to order 4. These computations are rather complicated, so they will not be discussed here. However, after much work, the *fourth-order Runge-Kutta method* approximation at each step is found to be made with

TABLE 2.4 Runge-Kutta Method of Order 2 with $h = 0.1$

x_n	y_n (RK)	Actual Value
0.0	1.0	1.0
0.1	1.005	1.00501
0.2	1.0201755	1.0202
0.3	1.0459859	1.04603
0.4	1.083223	1.08329
0.5	1.1330513	1.13315
0.6	1.1970687	1.19722
0.7	1.277392	1.27762
0.8	1.3767731	1.37713
0.9	1.4987552	1.4993
1.0	1.6478813	1.64874

$$y_{n+1} = y_n + \frac{1}{6}h\left(k_1 + 2k_2 + 2k_3 + k_4\right),$$
$$n = 0, 1, 2, \ldots,$$

where

$$k_1 = f\left(x_n, y_n\right),$$

$$k_2 = f\left(x_n + \frac{1}{2}h, y_n + \frac{1}{2}hk_1\right),$$

$$k_3 = f\left(x_n + \frac{1}{2}h, y_n + \frac{1}{2}hk_2\right),$$

and

$$k_4 = f\left(x_{n+1}, y_n + hk_3\right).$$

Theorem 10 (Runge-Kutta Method of Order 4). *The solution of the IVP*

$$y' = f(x, y), \quad y(x_0) = y_0$$

is approximated at the sequence of points (x_n, y_n) ($n = 1, 2, \ldots$), where y_n is the approximate value of $y(x_n)$, by computing at each step

$$y_{n+1} = y_n + \frac{1}{6}h\left(k_1 + 2k_2 + 2k_3 + k_4\right)$$
$$(n = 0, 1, 2, \ldots),$$

where $k_1 = f\left(x_n, y_n\right)$, $k_2 = f\left(x_n + \frac{1}{2}h, y_n + \frac{1}{2}hk_1\right)$, $k_3 = f\left(x_n + \frac{1}{2}h, y_n + \frac{1}{2}hk_2\right)$, $k_4 = f\left(x_{n+1}, y_n + hk_3\right)$, $x_n = x_0 + nh$, and h is the selected stepsize.

EXAMPLE 2.6.5

Use the fourth-order Runge-Kutta method with $h = 0.1$ to approximate the solution of the problem $y' = xy$, $y(0) = 1$ on $0 \le x \le 1$.

Solution

With $f(x, y) = xy$, $x_0 = 0$, and $y_0 = 1$, the formulas are

$$y_{n+1} = y_n + \frac{0.1}{6}\left(k_1 + 2k_2 + 2k_3 + k_4\right),$$
$$n = 0, 1, 2, \ldots,$$

where

$$k_1 = f\left(x_n, y_n\right) = x_n y_n,$$

$$k_2 = f\left(x_n + \frac{1}{2}h, y_n + \frac{1}{2}hk_1\right)$$
$$= \left(x_n + \frac{1}{2} \times 0.1\right)\left(y_n + \frac{1}{2} \times 0.1k_1\right),$$

$$k_3 = f\left(x_n + \frac{1}{2}h, y_n + \frac{1}{2}hk_2\right)$$
$$= \left(x_n + \frac{1}{2} \times 0.1\right)\left(y_n + \frac{1}{2} \times 0.1k_2\right),$$

and

$$k_4 = f\left(x_{n+1}, y_n + hk_3\right) = x_{n+1}\left(y_n + 0.1k_3\right).$$

For $n = 0$, we have

$$k_1 = x_0 y_0 = 0 \times 1 = 0$$

$$k_2 = \left(x_0 + \frac{1}{2} \times 0.1\right)\left(y_0 + \frac{1}{2} \times 0.1k_1\right)$$
$$= 0.05 \times 1 = 0.05$$

$$k_3 = \left(x_0 + \frac{1}{2} \times 0.1\right)\left(y_0 + \frac{1}{2} \times 0.1k_2\right)$$
$$= 0.05 \times (1 + 0.0025) = 0.050125$$

and

$$k_4 = x_1\left(y_0 + 0.1k_3\right) = 0.1$$
$$\times (1 + 0.0050125) = 0.10050125.$$

Therefore,

$$y_1 = y_0 + \frac{0.1}{6}(k_1 + 2k_2 + 2k_3 + k_4) = 1.005012521.$$

In Table 2.5, we list the results for the Runge-Kutta method of order 4-5 decimal places. Notice that this method yields the most accurate approximation of the methods used to this point.

 Computer-Assisted Solutions Using Commercial Software

CAS like *Mathematica* and *Maple*, numerical software like *MatLab*, as well as many other commercially available software programs and your graphing calculator can help you generate numerical solutions to differential equations.

Most of you are familiar with some of these programs from your calculus courses, so refer to appropriate resources to see how your software can help you solve problems that require numerical methods.

TABLE 2.5 Fourth-Order Runge-Kutta Method with $h = 0.1$

x_n	y_n (RK Order 4)	y_n (IEM)	Actual Value
0.0	1.0	1.0	1.0
0.1	1.00501	1.005	1.00501
0.2	1.0202	1.0201755	1.0202
0.3	1.04603	1.0459859	1.04603
0.4	1.08329	1.083223	1.08329
0.5	1.13315	1.1330513	1.13315
0.6	1.19722	1.1970687	1.19722
0.7	1.27762	1.277392	1.27762
0.8	1.37713	1.3767731	1.37713
0.9	1.4993	1.4987552	1.4993
1.0	1.64872	1.6478813	1.64874

Detailed discussions of error analysis as well as other numerical methods can be found in most numerical analysis texts.[4]

EXAMPLE 2.6.6

Graph the solution to the IVP $dy/dx = \sin(2x - y)$, $y(0) = 0.5$. What is the value of $y(1)$?

Solution

Using a CAS, we generated the figure of the solution to the IVP that is shown in Figure 2.29(a). Note that the figure has been enhanced by a graphics specialist so might look better in this text than if you generate your own. We also generate several solutions together with the slope field for the equation in Figure 2.29(b).

Using the same software, we determined that $y(1) \approx 0.875895$. To confirm that the graph and y-value are reasonably correct, we verified the result with a different CAS.

 EXERCISES 2.6

In Exercises 1-8, use Euler's method with $h = 0.1$ and $h = 0.05$ to approximate the solution of the IVP at the given value of x.

1. $y' = 4y + 3x + 2$, $y(0) = 1$, $x = 1$.
2. $y' = 4x - y + 1$, $y(0) = 0$, $x = 1$.
3. $y' - x = y^2 - 1$, $y(0) = 1$, $x = 1$.
4. $y' + x = 5\sqrt{y}$, $y(0) = 1$, $x = 1$.
5. $y' = \sqrt{xy} + 5y$, $y(1) = 1$, $x = 2$.
6. $y' = xy^{1/3} - y$, $y(1) = 1$, $x = 2$.
7. $y' = \sin y$, $y(0) = 1$, $x = 1$.
8. $y' = \sin(y - x)$, $y(0) = 0$, $x = 1$.

---

[4] See, for example, R.L. Burden, J.D. Faires, Numerical Analysis, third ed., PWS Publishers, 1985.

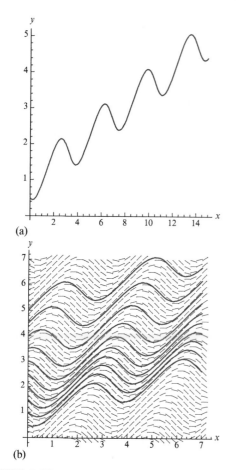

(a)

(b)

FIGURE 2.29 (a) Graph of the solution to the IVP. (b) Several slopes together with the slope field for the ODE.

In Exercises 9-16, use the improved Euler's method with $h = 0.1$ and $h = 0.05$ to approximate the solution of the corresponding exercise above at the given value of x. Compare these results with those obtained in Exercises 1-8.

9. Exercise 1.
10. Exercise 2.
11. Exercise 3.
12. Exercise 4.
13. Exercise 5.
14. Exercise 6.
15. Exercise 7.
16. Exercise 8.

In Exercises 17-24, use the Runge-Kutta method with $h = 0.1$ and $h = 0.05$ to approximate the solution of the corresponding exercise above. Compare these results with those obtained in Exercises 1-16.

17. Exercise 1.
18. Exercise 2.
19. Exercise 3.
20. Exercise 4.
21. Exercise 5.
22. Exercise 6.
23. Exercise 7.
24. Exercise 8.
25. Consider the IVP $dy/dx = -y, y(0) = 1$. (a) Find the exact solution to the problem. What is $y(1)$? (b) Approximate $y(1)$ with Euler's method using $h = 0.1, h = 0.05$, and $h = 0.025$. (c) Do the values in (b) approach the exact value of $y(1)$ as h approaches zero?
26. Repeat Exercise 25 using the improved Euler's method. Do the approximate values of $y(1)$ obtained with the improved Euler's method converge more quickly than those obtained with Euler's method?
27. Repeat Exercise 25 using the Runge-Kutta method of order 4. Do the approximate values of $y(1)$ obtained with this method converge more quickly than those obtained with the improved Euler's method and Euler's method?

As indicated in our examples, computers and CASs are of great use in implementing numerical techniques. In addition to being able to implement the algorithm illustrated earlier, many CASs contain built-in commands that you can use to implement various numerical methods.

28. Graph the solution to the IVP $dy/dx = \cos(y - 2x), y(0) = 0.5$ on the interval $[0, 15]$.
29. Graph the solution of $y' = \sin(xy)$ subject to the initial condition $y(0) = i$ on the interval

[0,7] for $i = 0.5, 1.0, 1.5, 2.0$, and 2.5. In each case, approximate the value of the solution if $x = 0.5$.

30. **(a)** Graph the direction field associated with $dy/dx = x^2 + y^2$ for $-1 \leq x \leq 1$ and $-1 \leq y \leq 1$.
    (b) Graph the solution to the IVP $dy/dx = x^2 + y^2, y(0) = 0$ on the interval $[-1,1]$.
    (c) Approximate $y(1)$.

31. Solve $dy/dt = y^2 + 1, y(0) = 1$ for $0 \leq t \leq \pi/2$. Can you resolve any problems with your solution that occur near $t = \pi/4$?

CHAPTER 2 SUMMARY: ESSENTIAL CONCEPTS AND FORMULAS

Existence and Uniqueness Theorems: In general, if f and $\partial f/\partial y$ are continuous functions on the rectangular region R: $a < t < b, c < y < d$ containing the point (t_0, y_0), then there exists an interval $|t - t_0| < h$ centered at t_0 on which there exists one and only one solution to the IVP $y' = f(t,y), y(t_0) = y_0$. Special cases of the theorem are stated in Section 2.2 for separable equations, Section 2.3 for linear equations, and Section 2.4 for exact equations.

Separable Differential Equation: A differential equation that can be written in the form $g(y)y' = f(t)$ or $g(y)\,dy = f(t)\,dt$ is called a *separable differential equation*.

First-Order Linear Differential Equation: A differential equation that can be written in the form $dy/dt + p(t)y = q(t)$ is called a *first-order linear differential equation*.

Integrating Factor: An *integrating factor* for the first-order linear differential equation is $\mu(t) = e^{\int p(t)\,dt}$.

Exact Differential Equation: A differential equation that can be written in the form

$M(t,y)\,dt + N(t,y)\,dy = 0$ where $M(t,y)\,dt + N(t,y)\,dy = (\partial f/\partial t)(t,y)dt + (\partial f/\partial y)(t,y)dy$ for some function $z = f(t,y)$ is called an *exact differential equation*.

Bernoulli Equation: A differential equation of the form $y' + p(t)y = q(t)y^n$.

Homogeneous Equation: A differential equation that can be written in the form $M(t,y)\,dt + N(t,y)\,dy = 0$ is homogeneous of degree n if both $M(t,y)$ and $N(t,y)$ are homogeneous of degree n.

Numerical Methods Numerical methods include computer-assisted solution, Euler's method, the improved Euler's method, and the Runge-Kutta method. Other methods are typically discussed in more advanced numerical analysis courses.

CHAPTER 2 REVIEW EXERCISES

In Exercises 1-23, solve each equation.

1. $dy/dt = 2t^5/(5y^2)$.
2. $\cos 4\theta \, d\theta - 8 \sin r \, dr = 0$.
3. $y' - y/t = y^2/t, t > 0$.
4. $dy/dt = e^{8y}/t$.
5. $dy/dt = e^{5t}/y^4$.
6. $(-\theta^{-5} + \theta^{-3})\,d\theta = (2\phi^4 - 6\phi^9)\,d\phi$.
7. $dy/dt = y/(e^{2t} \ln y)$.
8. $\dfrac{dy}{dx} = \dfrac{(4 - 7x)(2y - 3)}{(x - 1)(2x - 5)}$.
9. $dy/dt + 3y = -10 \sin t$.
10. $3t\,dt + (t - 4y)\,dy = 0$.
11. $(y - t)\,dt + (t + y)\,dy = 0$.
12. $(y - x)dx + dy = 0$.
13. $y^2\,dt + (ty + t^2)\,dy = 0$.
14. $dr/d\theta = (r^2 + \theta^2)/(r\theta)$.
15. $dx/dt = 5tx/(x^2 + t^2)$.
16. $(t^2 - y)dt + (y - t)\,dy = 0$.

17. $(t^2 y + \sin t) dt + (\frac{1}{3} t^3 - \cos y) \, dy = 0.$

18. $(\tan y - t) \, dt + (t \sec^2 y + 1) \, dy = 0.$

19. $t \ln y \, dt + (t^2/(2y) + 1) \, dy = 0.$

20. $y' + y = 5.$

21. $y' + ty = t.$

22. $dx/dy + x/y = y^2.$

23. $\theta r' + r = \theta \cos \theta.$

24. $dy/dt - y = ty^3.$

25. $dy/dt + y = e^t y^{-2}.$

Solve the following Clairaut equations (see Exercise 50, Section 2.5).

26. $y = ty' + 3(y')^4.$

27. $y - ty' = 2y^2 \ln(t).$

Solve the following Lagrange equations (see Exercise 57, Section 2.5).

28. $y - 2ty' = -2(y')^3.$

29. $y - 2ty' = -4(y')^2.$

In Exercises 30-35, solve each IVP. (*Check your solution by graphing the result on an appropriate interval.*)

30. $(2x - y - 2) \, dx + (2y - x) \, dy = 0, y(0) = 1.$

31. $\cos(t - y) \, dt + (1 - \cos(t - y)) \, dy = 0,$
$y(\pi) = \pi.$

32. $(ye^{ty} - 2t) dt + te^{ty} \, dy = 0, y(0) = 0.$

33. $(\sin y - y \cos t) \, dt + (t \cos y - \sin t) \, dy = 0,$
$y(\pi) = 0.$

34. $y^2 \, dt + (2ty - 2 \cos y \sin y) \, dy = 0, y(0) = \pi.$

35. $(y/t + \ln y) \, dt + (t/y + \ln t) dy = 0, y(1) = 1.$

For each of the following IVPs, use (a) Euler's method, (b) the improved Euler's method, and (c) the Runge-Kutta method of order 4 with $h = 0.05$ to approximate the solution to the IVP on the indicated interval.

36. $y' = y^2 - x, y(0) = 0, [0, 1].$

37. $y' = \sqrt{x - y}, y(1) = 1, [1, 2].$

38. $y' = x + y^{1/3}, y(1) = 1, [1, 2].$

39. $y' = \sin(x^2 y), y(0) = 1, [0, 1].$

40. Does the Existence and Uniqueness Theorem guarantee a unique solution to the following IVPs?
(a) $dy/dt = ty^3, y(0) = 0.$
(b) $dy/dt = ty^{-3}, y(0) = 0.$
(c) $dy/dt = -y/(t - 2), y(2) = 0.$

41. (*Higher-Order Methods with Taylor Series Expansion*) Consider the IVP $y' = f(x, y)$, $y(x_0) = y_0$. Recall that the Taylor series expansion of y about $x = x_0$ is given by

$$\sum_{k=0}^{\infty} \frac{y^{(k)}(x_0)}{k!} (x - x_0)^k.$$

We know the value of y at the initial value of $x = x_0$, so we use this value to approximate y at $x_1 = x_0 + h$, which is near x_0, in the following manner. Assuming that the Taylor series converges to $y(x)$ on an appropriate interval containing x_0, we first evaluate the Taylor series at $x_1 = x_0 + h$ to yield

$$y(x_0 + h) = y(x_0) + y'(x_0)h + \frac{y''(x_0)}{2!} h^2$$

$$+ \frac{y'''(x_0)}{3!} h^3 + \cdots .$$

Substituting $y' = f(x, y)$, $y'' = f'(x, y)$, and $y''' = f''(x, y)$ into this expansion, using $y(x_0) = y_0$ and calling this new value y_1, we have

$$y_1 = y(x_0 + h) = y(x_0) + f(x_0, y_0)h$$

$$+ \frac{1}{2!} f'(x_0, y_0)h^2 + \frac{1}{3!} f''(x_0, y_0)h^3 + \cdots$$

$$= y_0 + f(x_0, y_0)h + \frac{1}{2!} f'(x_0, y_0)h^2$$

$$+ \frac{1}{3!} f''(x_0, y_0)h^3 + \cdots .$$

Hence, the initial point (x_0, y_0) is used to determine y_1. A first-order approximation is obtained from this series by disregarding the terms of order h^2 and higher. In other

words, we determine y_1 from
$y_1 = y_0 + f(x_0, y_0)h$. We next use the point
(x_1, y_1) to approximate the value of y at
$x_2 = x_1 + h$. Calling this value y_2, we have

$$y(x) = y(x_1) + y'(x_1)(x - x_1)$$
$$+ \frac{y''(x_1)}{2!}(x - x_1)^2 + \cdots$$

so that an approximate value of $y(x_2)$ is

$$y_2 = y(x_1) + f(x_1, y_1)h = y_1 + f(x_1, y_1)h.$$

Continuing this procedure, we have that the
approximate value of y at $x = x_n = x_{n-1} + h$
is given by

$$y_n = y(x_{n-1}) + f(x_{n-1}, y_{n-1})h$$
$$= y_{n-1} + f(x_{n-1}, y_{n-1})h.$$

If we use the first three terms of the Taylor
series expansion instead, then we obtain the
approximation

$$y_n = y(x_{n-1}) + f(x_{n-1}, y_{n-1})h$$
$$+ \frac{1}{2}h^2 f'(x_{n-1}, y_{n-1})$$
$$= yy_{n-1} + f(x_{n-1}, y_{n-1})h$$
$$+ \frac{1}{2}h^2 f'(x_{n-1}, y_{n-1}).$$

We call the approximation that uses the first
three terms of the expansion the *three-term
Taylor method*.

(a) Use the three-term Taylor method with
$h = 0.1$ to approximate the solution of
the IVP $y' = xy, y(0) = 1$ on $0 \leq x \leq 1$. A
four-term approximation is derived in a
similar manner to be

$$y_n = y(x_{n-1}) + f(x_{n-1}, y_{n-1})h$$
$$+ \frac{1}{2}h^2 f'(x_{n-1}, y_{n-1})$$
$$+ \frac{1}{6}h^3 f''(x_{n-1}, y_{n-1})$$

$$= y_{n-1} + f(x_{n-1}, y_{n-1})h$$
$$+ \frac{1}{2}h^2 f'(x_{n-1}, y_{n-1})$$
$$+ \frac{1}{6}h^3 f''(x_{n-1}, y_{n-1}).$$

(b) Use the four-term method with $h = 0.1$
to approximate the solution of
$y' = 1 + y + x^2, y(0) = 0$ on $0 \leq x \leq 1$.

42. *(Running Shoes)* The rate of change of
energy absorption in a running shoe is
given by $dE/dt = F\,du/dt$, where
$F(t) = \frac{1}{2}F_0(1 - \cos \omega t)$. represents the force
magnitude and pulse duration exerted on
the shoe and $du/dt = \frac{1}{2}u_0\omega \sin(\omega t - \delta)$
represents the rate of change of the vertical
displacement of the midsole.[5] The constant
F_0 represents the maximum magnitude of
the input force in Newtons (N), ω represents
the frequency of the input profile in radians
per second (rad/s), u_0 represents the
maximum rate of change of the vertical
displacement of the midsole in meters (m), δ
represents the phase angle between $F(t)$ and
$u(t)$ in radians (rad), and t represents time
in seconds (s). Thus, the rate of change of
energy absorbed by the shoe is given by

$$\frac{dE}{dt} = \frac{1}{4}F_0 u_0 \omega (1 - \cos \omega t) \sin(\omega t - \delta).$$

(a) Find the energy absorbed by the shoe
from $t = t_1$ to $t = t_2$.

(b) The *maximum energy absorbed by the shoe*,
ME, is found by calculating the energy
absorbed if $t_1 = \delta/\omega$ and $t_2 = (\pi + \delta)/\omega$.
Show that

$$\text{ME} = \frac{1}{2}F_0 u_0 \left(1 + \frac{\pi}{4} \sin \delta\right).$$

43. *(Fermentation)* In the fermentation industry,
one of the goals of the molecular biologist is
to control the environment and regulate the
fermentation. To achieve meaningful

---

[5] J.F. Swigart, A.G. Erdman, P.J. Cain, An Energy-Based Method for Testing Cushioning Durability of Running Shoes, *Journal of Applied Biomechanics*, **9**, 1993, 47-65.

environmental control, fermentation research must be carried out on fully monitored environmental systems, the environmental observations must be correlated with existing knowledge of cellular control mechanisms, and environmental control conditions must be reproduced through continuous computer monitoring, analysis, and feedback control of the fermentation environment.

One component of environmental control is the measurement of dissolved oxygen with a steam sterilizable dissolved oxygen sensor made up of a polymer membrane-covered electrode. Suppose that the electrode is immersed in a liquid medium and that the oxygen is reduced according to the chemical reaction

$$O_2 + 2H_2O \rightarrow 4OH^-.$$

The rate of change $d\overline{C}/dt$ in dissolved oxygen concentration at a particular point in a fermentor vessel is given by

$$\frac{d\overline{C}}{dt} = k_L a\,(C^* - \overline{C}) - Q_{O_2}X,$$

where C^* is the concentration of dissolved oxygen that is in equilibrium with partial pressure $\overline{p}$ in bulk gas phase, $\overline{C}$ is the concentration of dissolved oxygen in bulk liquid, $k_L a$ is the volumetric oxygen transfer coefficient, Q_{O_2} is the specific rate of oxygen uptake (microbial respiration), and X is the cell mass concentration. If there are no cells present, the equation becomes

$$\frac{d\overline{C}}{dt} = k_L a\,(C^* - \overline{C}).^6$$

(a) If $\overline{C}(0) = 0$, determine $\overline{C}(t)$ as a function of C^* and $k_L a$.
(b) Suppose that $\overline{C}_p$ is the concentration of dissolved oxygen that corresponds to the sensor reading and k_p is the sensor

constant that depends on the conductance of the membrane and the liquid film outside the sensor. If $\overline{C}_p$ satisfies the equation
$d\overline{C}/dt = k_p\,(\overline{C} - \overline{C}_p)$ and $\overline{C}_p(0) = 0$,
determine $\overline{C}_p(t)$ as a function of k_p, $k_L a$, and C^*. By knowing k_p in advance and by observing $\overline{C}_p$ experimentally, the value of $k_L a$ can be estimated with the solution of this IVP. Thus, the amount of dissolved oxygen can be determined.

DIFFERENTIAL EQUATIONS AT WORK

A. Modeling the Spread of a Disease

Suppose that a disease is spreading among a population of size $N = N(t)$ at time t. In some diseases, like chicken pox, once an individual has had the disease, the individual becomes immune to the disease. In other diseases, like most venereal diseases, once an individual has had the disease and recovers from it, the individual does not become immune to the disease. Subsequent encounters can lead to recurrence of the infection.

Let $S(t)$ denote the percentage of the population susceptible to a disease at time t, $I(t)$ the percentage of the population infected with the disease, and $R(t)$ the percentage of the population unable to contract the disease. For example, $R(t)$ could represent the percentage of persons who have had a particular disease, recovered, and have subsequently become immune to the disease.

To model the spread of various diseases, we begin by making several assumptions and introducing some notation.

1. Susceptible and infected individuals die at a rate proportion to the number of susceptible

[6] S. Aiba, A.E Humphrey, N.F. Millis, Biochemical Engineering, second ed., Academic Press, 1973, pp. 317-336.

and infected individuals with proportionality constant μ called the *daily death removal rate*; the number $1/\mu$ is the *average lifetime* or *life expectancy*.

2. The constant λ represents the *daily contact rate*. On average, an infected person will spread the disease to λ people per day.

3. Individuals recover from the disease at a rate proportional to the number infected with the disease with proportionality constant γ. The constant γ is called the *daily recovery removal rate*; the *average period of infectivity* is $1/\gamma$.

4. The *contact number* $\sigma = \lambda/(\gamma + \mu)$ represents the average number of contacts an infected person has with both susceptible and infected persons.

If a person becomes susceptible to a disease after recovering from it (such as gonorrhea, meningitis, or streptococcal sore throat), then the percentage of people susceptible to becoming infected with the disease, $S(t)$, and the percentage of people in the population infected with the disease, $I(t)$, can be modeled by the system

$$\begin{cases} dS/dt = -\lambda IS + \gamma I + \mu - \mu S \\ dI/dt = \lambda IS - \gamma I - \mu I \\ S(0) = S_0, \ I(0) = I_0, \ S(t) + I(t) = 1. \end{cases}$$

This model is called an *SIS* (susceptible-infected-susceptible) model because once an individual has recovered from the disease, the individual again becomes susceptible to the disease.[7]

Since $S(t) = 1 - I(t)$, we can write $I'(t) = \lambda IS - \gamma I + \mu I$ as $dI/dt = \lambda I(1 - I) - (\gamma + \mu)I$ and thus we need to solve the IVP

$$dI/dt = [\lambda - (\gamma + \mu)]I - \lambda I^2, \quad I(0) = I_0.$$

1. Convert the Bernoulli equation $I'(t) = \lambda I(1 - I) - (\gamma + \mu)I$ to a linear equation and solve the result with the substitution $y = I^{-1}$.

2. (a) Show that the solution to the IVP
$I'(t) = \lambda I(1 - I) - (\gamma + \mu)I, \ I(0) = I_0$ is

$$I(t) = \begin{cases} \dfrac{e^{(\lambda+\mu)(\sigma-1)t}}{\sigma\left(e^{(\lambda+\mu)(\sigma-1)t} - 1\right)/(\sigma-1) + 1/I_0}, & \text{if } \sigma \neq 1 \\[4mm] \dfrac{1}{\lambda t + 1/I_0}, & \text{if } \sigma = 1 \end{cases}$$

and graph various solutions if (b) $\lambda = 3.6$, $\gamma = 2$, and $\mu = 1$; (c) $\lambda = 3.6$, $\gamma = 2$, and $\mu = 2$. In each case, find the contact number. How does the contact number affect $I(t)$ for large values of t?

3. Evaluate $\lim_{t \to \infty} I(t)$.

The incidence of some diseases, such as measles, rubella, and gonorrhea, oscillates seasonally. To model these diseases, we may wish to replace the constant contact rate λ by a periodic function $\lambda(t)$.

4. Graph various solutions if (a) $\lambda(t) = 5 - 2\sin 6t$, $\gamma = 1$, and $\mu = 4$; and (b) $\lambda(t) = 5 - 2\sin 6t$, $\gamma = 1$, and $\mu = 2$. In each case, calculate the average contact number. How does the average contact number affect $I(t)$ for large values of t?

5. Explain why diseases such as gonorrhea, meningitis, and streptococcal sore throat continue to persist in the population. Do you think there is any way to eliminate these diseases completely from the population? Why or why not?

B. Linear Population Model with Harvesting

1. Consider: $y' = ay - h$, $y(0) = y_0$, where a and h are positive constants. The value of a represents the growth rate, while that of h represents the fishing (or harvesting) rate. The solution $y(t)$ represents the size of a population at time t.

---

[7] H.W., Hethcote, Three Basic Epidemiological Models, Applied Mathematical Ecology, S.A. Levin, T.G. Hallan, L.J. Gross (Eds.), Springer-Verlag, 1989, pp. 119-143.

(a) Determine the equilibrium solution.

(b) Sketch the phase line.

(c) Solve the IVP.

(d) If $y_0 > h/a$, determine $\lim_{t\to\infty} y(t)$.

(e) If $y_0 = h/a$, determine $\lim_{t\to\infty} y(t)$.

(f) If $y_0 < h/a$, determine $\lim_{t\to\infty} y(t)$.

2. $y' = \frac{1}{2}y - 1$, $y(0) = y_0$

(a) Determine the equilibrium solution.

(b) Sketch the phase line.

(c) Solve the IVP.

(d) If $y_0 = 3$, determine $\lim_{t\to\infty} y(t)$.

(e) If $y_0 = 2$, determine $\lim_{t\to\infty} y(t)$.

(f) If $y_0 = 1.5$, determine the value of t for which $y(t) = 0$. What does this mean in terms of the population?

3. Consider the model $y' = y - 1$, $y(0) = 0.5$, and the model $y' = \frac{1}{2}y - 1$. In which model do you predict that the population will become extinct more quickly? Verify your response.

4. Consider the model $y' = \frac{1}{2}y - \frac{1}{2}$, $y(0) = 0.5$ and the model $y' = \frac{1}{2}y - 1$, $y(0) = 0.5$. In which model do you predict that the population will become extinct more quickly? Verify your response.

5. Suppose that you own a "fish farm" and that during the first year, the growth rate of the population is $a = 1/2$ and the harvesting rate is $h = 1/2$. However, during the second year, instead of harvesting, you decide to add fish at a rate of $r = 1/2$. If $y(0) = 0.5$, find $y(2)$. How does this value compare to the original size of the population? What (approximate) value of r should be used so that $y(2) = y(0)$?

6. Suppose that in problem 5 above, you do not fish or add to the population after the first year. If $y(0) = 0.5$, find $y(2)$. How does this value compare to the original size of the population? How much time T is required so that $y(T) = y(0)$?

7. Suppose that fishing occurs on a seasonal basis so that the situation is modeled with

$y' = ay - h(1 - \sin(\pi t/6))$, $y(0) = y_0$. What is the maximum value, minimum value, and period of the harvesting function $h(1 - \sin(\pi t/6))$? Using the indicated parameter values, describe what happens to the size of the population over 1 year.

(a) $a = 1/10$, $h = 1$, $y(0) = 20$.

(b) $a = 1/10$, $h = 2$, $y(0) = 20$.

(c) $a = 1/10$, $h = 4$, $y(0) = 20$.

(d) $a = 1/10$, $h = 10$, $y(0) = 20$.

8. Suppose that a situation is modeled with $y' = ay - h\cos(\pi t/6)$, $y(0) = y_0$. That is, at some times, fishing occurs while at others fish are added. During what time intervals does fishing occur? During what time intervals are fish added? What is the maximum value, minimum value, and period of the function, $h\cos(\pi t/6)$? Using the indicated parameter values, describe what happens to the size of the population over 1 year.

(a) $a = 1/10$, $h = 1$, $y(0) = 50$.

(b) $a = 1/10$, $h = 4$, $y(0) = 50$.

(c) $a = 1/10$, $h = 10$, $y(0) = 50$.

(d) $a = 1/10$, $h = 20$, $y(0) = 50$.

C. Logistic Model with Harvesting

1. Consider the logistic model with harvesting, $y' = ay - cy^2 - h$, $y(0) = y_0$, where a represents the growth rate, c the inhibitive factor, and h the harvesting rate. Each of these constants is positive.

(a) Find the equilibrium solutions by solving $ay - cy^2 - h = 0$. (What condition must hold so that there are two real solutions to this equation? What condition must hold so that there is only one equilibrium solution? What condition must hold so that there is no equilibrium solution?)

(b) Sketch the phase line assuming that there are two equilibrium solutions.

(c) Sketch the phase line assuming that there is only one equilibrium solution.

(d) What happens to the population if there is no equilibrium solution?

2. Consider the problem $y' = \frac{7}{10}y - \frac{1}{10}y^2 - 1$, $y(0) = y_0$.

   (a) What are the two equilibrium solutions?

   (b) Sketch the phase line.

   (c) If $y(0) = 3$, then determine $\lim_{t \to \infty} y(t)$.

   (d) If $y(0) = 6$, then determine $\lim_{t \to \infty} y(t)$.

   (e) If $y(0) = 1$, what do you expect to happen to the population?

3. Consider the problem $y' = \frac{7}{10}y - \frac{1}{10}y^2 - h$, $y(0) = y_0$.

   (a) What value of h, $h = h_0$, causes this model to have only one equilibrium solution? (This value is called a *bifurcation point* because the dynamics of the model change for values slightly larger and smaller than this value.)

   (b) Describe solutions to the model with $h = h_0$.

   (c) Describe solutions to the model with $h < h_0$.

   (d) Describe solutions to the model with $h > h_0$.

4. Suppose that over the first year, the rate of change in the size of a fish population is given by $y' = \frac{7}{10}y - \frac{1}{10}y^2 - \frac{1}{2}$, where $y(0) = 0.5$. After the first year, however, the growth rate remains the same while no fishing is allowed. What is the size of the population after 5 years? Describe what happens to the population size on the interval $0 \le t \le 5$. Compare this result to the population that follows $y' = \frac{7}{10}y - \frac{1}{10}y^2 - \frac{1}{2}$, where $y(0) = 0.5$ over the entire interval $0 \le t \le 5$. (That is, fishing continues after the first year.)

5. Suppose that over the first year, the rate of change in the size of a fish population is given by $y' = \frac{13}{10}y - \frac{1}{10}y^2 - \frac{1}{2}$, where $y(0) = 0.5$. After the first year, however, the growth rate remains the same while no fishing is allowed. What is the size of the population after 5 years? Describe what happens to the population size on the interval $0 \le t \le 5$. Compare this result to population that follows $y' = \frac{13}{10}y - \frac{1}{10}y^2 - \frac{1}{2}$, where $y(0) = 0.5$ over the entire interval $0 \le t \le 5$. (That is, fishing continues after the first year.)

6. Consider the model
   $y' = \frac{13}{10}y - \frac{1}{10}y^2 - \frac{1}{2}\sin\frac{\pi t}{6}$, $y(0) = a$, in which fishing occurs at certain times while at other times, fish are added. Investigate the solution to this IVP over $0 \le t \le 12$ for $a = 2, 4, 6, 8, 10, 12, 14$. What eventually happens to the size of the population in each case?

7. Consider the model
   $y' = \frac{13}{10}y - \frac{1}{10}y^2 - \frac{1}{2}\left(1 - \sin\frac{\pi t}{6}\right)$, $y(0) = a$, in which fishing occurs at different rates. Investigate the solutio to this IVP over $0 \le t \le 12$ for $a = 2, 4, 6, 8, 10, 12, 14$. What eventually happens to the size of the population in each case?

8. Consider the model
   $y' = \frac{7}{10}y - \frac{1}{10}y^2 - \frac{1}{2}\sin\frac{\pi t}{6}$, $y(0) = a$, in which fishing occurs at certain times while at other times, fish are added. Investigate the solution to this IVP over $0 \le t \le 12$ for $a = 2, 4, 6, 8, 10, 12, 14$. What eventually happens to the size of the population in each case?

9. Consider the model
   $y' = \frac{7}{10}y - \frac{1}{10}y^2 - \frac{1}{2}\left(1 - \sin\frac{\pi t}{6}\right)$, $y(0) = a$, in which fishing occurs at different rates. Investigate the solution to this IVP over $0 \le t \le 10$ for $a = 2, 4, 6, 8, 10, 12, 14$. What eventually happens to the size of the population in each case?

D. Logistic Model with Predation

One of the more predominant pests in Canadian forests is the spruce budworm. These insects, which are moths in the adult stage, lay eggs in the needles of the trees. After hatching, the larvae eventually tunnel into the old foliage of the tree. Later, they spin a webbing in the needles and devour new growth until they are interrupted. Although they do not destroy the tree, budworms weaken the trees and make them susceptible to disease and forest fires. We can model the rate of change in the size of the budworm population by building on the logistic model. Let $W(t)$ represent the size of the budworm population at time t. One factor that helps control the budworm population is predation by birds (that is, birds eat budworms). Consider the model,

$$\frac{dW}{dt} = rW\left(1 - \frac{1}{k}W\right) - P(W),$$

where $P(W)$ is a function of W describing the rate of predation. Notice that for low population levels, the model demonstrates exponential growth rate r. Based on the logistic model, k represents the carrying capacity of the forest. In the absence of predation, this gives a maximum size of the budworm population.

Properties of the Predation Function: To develop a formula for $P(W)$, we make two assumptions. First, if $W = 0$, then $P(W) = 0$ (that is, if there are no budworms, the birds have none to eat). Second, $P(W)$ has a limiting value (that is, the birds have a limited appetite. Even if the budworm population is large, the birds eat only what they need). A function that satisfies both of these properties is $P(W) = aW^2/(b^2 + W^2)$. Therefore, our model becomes:

$$\frac{dW}{dt} = rW\left(1 - \frac{1}{k}W\right) - \frac{aW^2}{b^2 + W^2}, \quad W(0) = W_0.$$

1. What is the limiting value of $P(W) = aW^2/(b^2 + W^2)$ as $W \to \infty$?

2. Using the indicated parameter values, approximate the equilibrium solutions. Sketch a phase line in each case.
   (a) $r = 1, k = 15, a = 5, b = 2$.
   (b) $r = 1, k = 20, a = 5, b = 2$.

3. Using the values $r = 1, a = 5, b = 2$, approximate the value of k that leads to three equilibrium solutions. (Do this experimentally by selecting values for k and then plotting. This value is a bifurcation point.) Sketch the phase line.

4. Using various initial conditions, generate numerical solutions to the two situations in problem 2 above and to that in problem 3. Compare these results to the corresponding phase line.

5. Suppose that your job is to determine the carrying capacity of the forest. How important is it that you do a good job? (Use the three cases above for guidance.)

6. Discuss the differences in the following three models:
   (a) $y' = 0.48y\left(1 - \frac{1}{15}y\right) - \frac{2y^2}{4 + y^2}$.

   (b) $y' = 0.48y\left(1 - \frac{1}{17}y\right) - \frac{2y^2}{4 + y^2}$.

   (c) $y' = 0.48y\left(1 - \frac{1}{15.5}y\right) - \frac{2y^2}{4 + y^2}$.

References

[1] Laurentian Forestry Centre Success Stories, July 30, 1997.
[2] E.K. Yeargers, R.W. Shonkwiler, J.V. Herod, An Introduction to the Mathematics of Biology with Computer Algebra Models, Birkhauser, Boston, 1996.
[3] T. Howard, Eating Worms, Columbus State University, 1994.

3

Applications of First-Order
Differential Equations

When a space shuttle was launched from the Kennedy Space Center, the minimum initial velocity needed for the shuttle to escape the Earth's atmosphere is determined by solving a first-order differential equation. The same can be said for finding the flow of electromagnetic forces, the temperature of a cup of coffee, the population of a species, and numerous other applications. In this chapter, we show how these problems can be expressed as first-order equations. We focus our attention on setting up problems and explaining the meaning of the subsequent solutions because techniques for solving the first-order equations were discussed in Chapter 2.

3.1 POPULATION GROWTH AND DECAY

- Logistic Equation
- Population Model with a Threshold

As we discussed in Section 2.2, the *Malthus model*,

$$dy/dt = ky, \quad y(0) = y_0,$$

can be used to investigate population growth if $k > 0$. It can also be applied to problems involving the decay of radioactive material if $k < 0$.

Henri Becquerel (1852-1908) observed but did not understand radioactive decay.

Marie Sklodowska Curie (1867-1934): Marie Curie is the only woman who has won two Nobel prizes. She was also the first person to be awarded a second Nobel prize. Of the four individuals who have received the Nobel prize twice, the others are Linus Pauling (in 1954 and 1962), John Bardeen (in 1956 and 1972), and Frederick Sange (in 1958 and 1980).

Forms of a given element with different numbers of neutrons are called *nuclides*. Some nuclides are not stable. For example, potassium-40 (^{40}K) naturally decays to reach argon-40 (^{40}Ar). This decay that occurs in some nuclides was first observed, but not understood, by Henri Becquerel (1852-1908) in 1896. Marie Curie, however, began studying this decay in 1898, named it *radioactivity*, and discovered the radioactive substances polonium and radium. Marie Curie

(1867-1934), along with her husband Pierre Curie (1859-1906) and Henri Becquerel, received the Nobel Prize in Physics in 1903 for their work on radioactivity. Marie Curie subsequently received the Nobel Prize in Chemistry in 1910 for discovering polonium and radium.

Pierre Curie (1859-1906): Marie and Pierre's daughter, Irene, won the Nobel prize in Chemistry in 1935 by showing that radioactivity can be artificially produced.

Given a sample of ^{40}K of sufficient size, after 1.2×10^9 years, approximately half of the sample will have decayed to ^{40}Ar. The *half-life* of a nuclide is the time for half the nuclei in a given sample to decay (see Table 3.1). We see that the rate of decay of a nuclide is proportional to the amount present because the half-life of a given nuclide is constant and independent of the sample size.

EXAMPLE 3.1.1

If the half-life of polonium, ^{209}Po, is 100 years, determine the percentage of the original amount of ^{209}Po that remains after 50 years.

Solution

Let y_0 represent the original amount of ^{209}Po that is present. The amount present after t years

TABLE 3.1   Half-Lives of Various Nuclides

Element	Nuclide	Half-Life	Element	Nuclide	Half-Life
Aluminum	^{26}Al	7.4×10^5 years	Polonium	^{209}Po	100 years
Beryllium	^{10}Be	1.51×10^6 years	Polonium	^{210}Po	138 days
Carbon	^{14}C	5730 years	Radon	^{222}Rn	3.82 days
Chlorine	^{36}Cl	3.01×10^5 years	Radium	^{226}Ra	1700 years
Iodine	^{131}I	8.05 days	Thorium	^{230}Th	75,000 years
Potassium	^{40}K	1.2×10^9 years	Uranium	^{238}U	4.51×10^9 years

is $y(t) = y_0 e^{kt}$. Using $y(100) = \frac{1}{2}y_0$ and $y(100) = y_0 e^{100k}$, we solve $y_0 e^{100k} = \frac{1}{2}y_0$ for e^k:

$$e^{100k} = \frac{1}{2},$$

$$(e^k)^{100} = \frac{1}{2},$$

$$e^k = \left(\frac{1}{2}\right)^{1/100}.$$

Hence,

$$y(t) = y_0 e^{kt} = y_0 (e^k)^t = y_0 \left(\frac{1}{2}\right)^{t/100}.$$

To determine the percentage of y_0 that remains, we evaluate

$$y(50) = y_0 \left(\frac{1}{2}\right)^{50/100} \approx 0.7071 y_0.$$

Therefore, 70.71% of the original amount of ^{209}Po remains after 50 years.

In Example 3.1.1, we determined the percentage of the original amount of ^{209}Po that remains even though we do not know the value of y_0, the initial amount of ^{209}Po. Instead of letting $y(t)$ represent the amount of the substance present after time t, we can let it represent the fraction (or percentage) of y_0 that remains after time t. In doing so, we use the initial condition $y(0) = 1 = 1.00$ to indicate that 100% of y_0 is present at $t = 0$.

Roman sarcophagi at Worms, Germany. (Copyright released by Mike Chapman.)

EXAMPLE 3.1.2

The wood in a Roman sarcophagus (burial case) is found to contain 79% of the carbon-14, ^{14}C, found in a present-day sample. What is the age of the sarcophagus?

Solution

From Table 3.1, we see that the half-life of ^{14}C is 5730 years. Let $y(t)$ be the percent of ^{14}C in the sample after t years. Then $y(0) = 1$. Now, $y(t) = y_0 e^{kt}$, so $y(5730) = e^{5730k} = 0.5$. Solving for k yields:

$$\ln(e^{5730k}) = \ln(0.5),$$

$$5730k = \ln(0.5),$$

$$k = \frac{\ln(0.5)}{5730} = -\frac{\ln 2}{5730}.$$

Thus, $y(t) = e^{kt} = e^{-(\ln 2/5730)t} = 2^{-t/5730}$. (An alternate approach to obtain the solution is to solve $e^{5730k} = 0.5$ for e^k instead of for k, as we did in Example 3.1.1. This yields $e^k = (0.5)^{1/5730} = \left(\frac{1}{2}\right)^{1/5730} = 2^{-1/5730}$. Substitution of this expression into $y(t) = y_0 e^{kt} = y_0 (e^k)^t$ gives the same solution as found previously.)

We use the fact that $\ln(a/b) = -\ln(b/a)$.

In this problem, we must find the value of t for which $y(t) = 0.79$. Solving this equation results in:

$$2^{-t/5730} = 0.79 = \frac{79}{100},$$

$$\ln(2^{-t/5730}) = \ln\frac{79}{100},$$

$$-\frac{t}{5730}\ln 2 = \ln\frac{79}{100},$$

$$t = -\frac{5730}{\ln 2}\ln\frac{79}{100}$$

$$= \frac{5730}{\ln 2}\ln\frac{100}{79} \approx 1948.63.$$

We conclude that the sarcophagus is approximately 1949-year old.

We can use the Malthus model to predict the size of a population at a given time if the rate of growth of the population is proportional to the present population.

EXAMPLE 3.1.3

Suppose that the number of cells in a bacteria culture doubles after 3 days. Determine the number of days required for the initial population to triple.

Solution

In this case, $y(0) = y_0$, so the population is given by $y(t) = y_0 e^{kt}$ and $y(3) = 2y_0$ because the population doubles after 3 days. Substituting this value $y(t) = y_0 e^{kt}$, we have

$$y(3) = y_0 e^{3k} = 2y_0.$$

Solving for e^k, we find that $e^{3k} = 2$ or $(e^k)^3 = 2$. Therefore, $e^k = 2^{1/3}$. Substitution into $y(t) = y_0 e^{kt}$ then yields

$$y(t) = y_0 (e^k)^t = y_0 2^{t/3}.$$

We find when the population triples by solving $y(t) = y_0 2^{t/3} = 3y_0$ for t. This yields $2^{t/3} = 3$ or $\frac{1}{3}t\ln 2 = \ln 3$. Therefore, the population triples in $t = (3\ln 3)/(\ln 2) \approx 4.755$ days.

To observe some of the limitations of the Malthus model, we consider a population problem in which the rate of growth of the population does not exclusively depend on the present population.

 Determine the number of days required for the culture of bacteria considered in Example 3.1.3 to reach nine times its initial size.

EXAMPLE 3.1.4

The population of the United States was recorded as 5.3 million in 1800. Use the Malthus model to approximate the population for years after 1800 if $k = 0.03$. Compare these results to the actual population. Is this a good approximation for years after 1800?

Solution

In this example, $k = 0.03$, $y_0 = 5.3$, and our model for the population of the United States at time t (where t is the number of years from 1800) is $y(t) = 5.3e^{0.03t}$. To compare this model with the actual population of the United States, census figures for the population of the United States for various years are listed in Table 3.2 along with the corresponding value of $y(t)$. A graph of $y(t)$ with the corresponding points is shown in Figure 3.1.

TABLE 3.2   U.S. Population and Values of $y(t)$

Year (t)	Actual Population (in millions)	Value of $y(t) =$ $5.3\,e^{0.03t}$	Year (t)	Actual Population (in millions)	Value of $y(t) =$ $5.3\,e^{0.03t}$
1800 (0)	5.30	5.30	1870 (70)	38.56	43.28
1810 (10)	7.24	7.15	1880 (80)	50.19	58.42
1820 (20)	9.64	9.66	1890 (90)	62.98	78.86
1830 (30)	12.68	13.04	1900 (100)	76.21	106.45
1840 (40)	17.06	17.60	1910 (110)	92.23	143.70
1850 (50)	23.19	23.75	1920 (120)	106.02	193.97
1860 (60)	31.44	32.06	1930 (130)	123.20	261.83

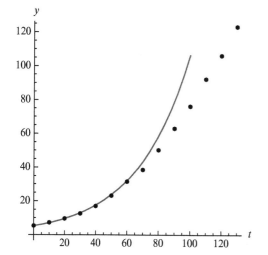

FIGURE 3.1   A population with limited resources cannot grow exponentially forever.

Pierre Verhulst (1804-1849) derived the Verhulst equation in 1846 and predicted that the upper limit of the Belgian population would be 9,400,000. Interestingly, as of July, 2006, the population of Belgium was estimated to be 10,379,067 [5].

Although the model appears to approximate the data for several years after 1800, the accuracy of the approximation diminishes over time because the population of the United States does not exclusively increase at a rate proportional to the population. Another model that better approximates the population would take other factors into account.

The Logistic Equation

Because the approximation obtained with the Malthus model is less than desirable in the previous example, we see that another model is needed. The *logistic equation* (or *Verhulst equation*), which was mentioned in Sections 1.1 (see Exercise 61) and 2.5, is the equation

$$y'(t) = (r - ay(t))y(t), \qquad (3.1)$$

where r and a are constants, subject to the condition $y(0) = y_0$. This equation was first introduced by the Belgian mathematician Pierre Verhulst to study population growth. The logistic equation differs from the Malthus model in that the term $r - ay(t)$ is not constant. This equation can be written as $dy/dt = (r - ay)y = ry - ay^2$ where the term $-y^2$ represents an inhibitive factor. Under these assumptions, the population is neither allowed to grow out of control nor grow or decay constantly as it was with the Malthus model.

Here, we solve the logistic equation as a separable equation. For convenience, we write the equation as

$$y' = (r - ay)y \quad \text{or} \quad \frac{dy}{dt} = (r - ay)y.$$

Sometimes the logistic equation is written in the form $\dfrac{dy}{dt} = \alpha y\left(1 - (1/K)y\right)$. In this form, α is called the *growth rate* and K is called the *carrying capacity*.

Separating variables and using partial fractions to integrate with respect to t and y, we have

$$\frac{1}{(r - ay)y}dy = dt,$$

$$\left(\frac{a/r}{r - ay} + \frac{1/r}{y}\right)dy = dt,$$

$$\left(\frac{a}{r - ay} + \frac{1}{y}\right)dy = r\,dt,$$

$$-\ln|r - ay| + \ln|y| = rt + C.$$

We solve this expression for y using properties of logarithms and exponentials:

$$\ln\left|\frac{y}{r - ay}\right| = rt + C,$$

$$\frac{y}{r - ay} = \pm e^{rt+C} = Ke^{rt} \quad (K = \pm e^C),$$

$$y = r\left(\frac{1}{K}e^{-rt} + a\right)^{-1}.$$

Applying the initial condition $y(0) = y_0$ and solving for K, we find that

$$K = \frac{y_0}{r - ay_0}.$$

After substituting this value into the general solution and simplifying, the solution can be written as

$$y = \frac{ry_0}{ay_0 + (r - ay_0)e^{-rt}}.$$

Use a computer algebra system to solve the logistic equation. If the result you obtain is not in the same form as that given here, show (by hand) that the two are the same.

Notice that $\lim_{t\to\infty} y(t) = r/a$ because $\lim_{t\to\infty} e^{-rt} = 0$ if $r > 0$. This makes the solution to the logistic equation different from that of the Malthus model in that the solution to the logistic equation approaches a finite nonzero limit as $t \to \infty$ while that of the Malthus model approaches either infinity or zero as $t \to \infty$.

EXAMPLE 3.1.5

Use the logistic equation to approximate the population of the United States using $r = 0.03$, $a = 0.0001$, and $y_0 = 5.3$. Compare this result with the actual census values given in Table 3.2. Use the model obtained to predict the population of the United States in the year 2010. How does your approximation compare with the population of the United States given by the Census Bureau to be 308.7 million at that time?

Solution

We substitute the indicated values of r, a, and y_0 into $y = ry_0/(ay_0 + (r - ay_0)e^{-rt})$ to obtain the approximation of the population of the United States at time t, where t represents the number of years since 1800,

$$y(t) = \frac{0.03 \times 5.3}{0.0001 \times 5.3 + (0.03 - 0.0001 \times 5.3)e^{-0.03t}}$$

$$= \frac{0.159}{0.00053 + 0.02947e^{-0.03t}}.$$

In Table 3.3, we compare the approximation of the population of the United States given by the approximation $y(t)$ with the actual population obtained from census figures. Note that this model appears to approximate the population more closely over a longer period of time than the Malthus model did (Example 3.1.4), as we can see in Figure 3.2.

To predict the population in the year 2010 with this model, we evaluate $y(210)$ to obtain

$$y(210) = \frac{0.159}{0.00053 + 0.02947e^{-0.03 \times 210}} \approx 300.$$

Thus, we predict that the population will be approximately 300 million in the year 2010. Note that the estimation of the population of the United States in the year 2010 made by the Bureau of the Census was 308.7 million, so the approximation obtained with the logistic equation appears to be sensible.

Interestingly, the *Factbook's* [5] estimate of the size of the population of the United States in July, 2006 was 298,444,215. Remember that the results we obtain with the logistic equation do not take into account many variables, including but not limited to immigration patterns, as well as changes and developments in the distribution and availability of food and health resources.

Before moving on, we remark that the logistic equation (3.1) has many different but equivalent formulations. For example, If α and K represent positive constants, the logistic equation can be written as

$$\frac{dy}{dt} = \alpha y \left(1 - \frac{1}{K}y\right), \quad y(0) = y_0. \quad (3.2)$$

TABLE 3.3   U.S. Population and Values of $y(t)$

Year (t)	Actual Population (in millions)	Value of $y(t)$	Year (t)	Actual Population (in millions)	Value of $y(t)$
1800 (0)	5.30	5.30	1900 (100)	76.21	79.61
1810 (10)	7.24	7.11	1910 (110)	92.23	98.33
1820 (20)	9.64	9.52	1920 (120)	106.02	119.08
1830 (30)	12.68	12.71	1930 (130)	123.20	141.14
1840 (40)	17.06	16.90	1940 (140)	132.16	163.59
1850 (50)	23.19	22.38	1950 (150)	151.33	185.45
1860 (60)	31.44	29.44	1960 (160)	179.32	205.82
1870 (70)	38.56	38.42	1970 (170)	203.30	224.05
1880 (80)	50.19	49.63	1980 (180)	226.54	239.78
1890 (90)	62.98	63.33	1990 (190)	248.71	252.94

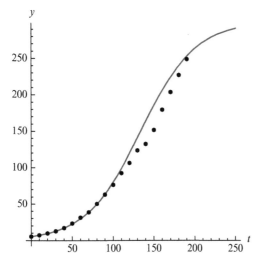

FIGURE 3.2 With these parameter values, the upper limit on the population of the United States is approximately 300 million.

In this formulation, α is called the *growth rate* and K is called the *carrying capacity* because $\lim_{t \to \infty} y(t) = K$, unless $y(0) = 0$. We assume that $y(0)$ is a positive number because our interpretation is that $y(t)$ represents a population (a nonnegative number) at time t. Of course we can solve Equation (3.2) by viewing at as separable equation as we did with its equivalent form (3.1). To illustrate a different approach, we solve it as a Bernoulli equation. First multiply through by αy and subtract:

$$\frac{dy}{dt} = \alpha y - \frac{\alpha}{K} y^2, \qquad (3.3)$$

$$\frac{dy}{dt} - \alpha y = -\frac{\alpha}{K} y^2.$$

After rewriting in this form, we see a Bernoulli equation with $n = 2$ so we let $w = y^{1-2} = y^{-1}$. Differentiating with respect to t gives us $dw/dt = -y^{-2} dy/dt$ or $-y^2\, dw/dt = dy/dt$. With this substitution, Equation (3.3) becomes

$$\frac{dy}{dt} - \alpha y = -\frac{\alpha}{K} y^2,$$

$$-y^2 \frac{dw}{dt} - \alpha y = -\frac{\alpha}{K} y^2, \qquad (3.4)$$

$$\frac{dw}{dt} + \alpha y^{-1} = \frac{\alpha}{K},$$

$$\frac{dw}{dt} + \alpha w = \frac{\alpha}{K}.$$

Equation (3.4) is best solved using the method of undetermined coefficients. The corresponding homogeneous equation is $(dw/dt) + \alpha w = 0$, which has general solution $w_h = Ce^{-\alpha t}$. We search for a particular solution of the nonhomogeneous equation for $w_p = A$, where A is a constant to be determined. Substituting into the nonhomogeneous equation gives us

$$\frac{dw_p}{dt} + \alpha w_p = \frac{\alpha}{K},$$

$$0 + \alpha \cdot A = \frac{\alpha}{K},$$

$$A = \frac{1}{K}.$$

Thus, $w_p = 1/K$ and $w = w_h + w_p = Ce^{-\alpha t} + 1/K = (1 + Ce^{-\alpha t})/K$. Because $w = y^{-1}$, it follows that $y = K/(1 + Ce^{-\alpha t})$. Solving $y(0) = y_0$ for C gives us $C = (K - y_0)/y_0$ so the solution to the initial-value problem (IVP) is

$$y = \frac{K}{1 + \dfrac{K - y_0}{y_0} e^{-\alpha t}}.$$

Population Model with a Threshold

Suppose that when the population of a species falls below a certain level (or threshold), the species cannot sustain itself, but otherwise, the population follows logistic growth. To describe this situation mathematically, we introduce the differential equation

$$\frac{dy}{dt} = -r\left(1 - \frac{1}{A}y\right)\left(1 - \frac{1}{B}y\right)y,$$

where $0 < A < B$. In this case, the equilibrium solutions are $y = 0$, $y = A$, and $y = B$.

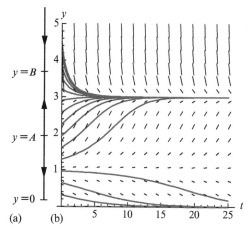

FIGURE 3.3 (a) Phase line for $dy/dt = -r(1 - (1/A)y)$ $(1 - (1/B)y)y$. (b) Several solutions together with the slope field for $dy/dt = -0.25(1 - y)(1 - y/3)y$. The phase portrait shows that all nontrivial solutions other than if $y_0 = 1$, $y(t) = 1$. If $y_0 < 1$, then $\lim_{t\to\infty} y(t) = 0$. If $y_0 > 1$, then $\lim_{t\to\infty} y(t) = 3$.

To understand the dynamics of the equation, we generate the corresponding phase line. If $0 < y < A$, then $y' < 0$; if $A < y < B$, then $y' > 0$; and if $y > B$, then $y' < 0$. From the phase line in Figure 3.3(a), we see that $y = 0$ is asymptotically stable, $y = A$ is unstable, and $y = B$ is asymptotically stable. Notice that we expect the population to die out when $y < A$, so we call $y = A$ the *threshold of the population*. As with the logistic equation, $y = B$ is the *carrying capacity*. In Figure 3.3(b), we graph several solutions to $y' = -0.25(1 - y)(1 - y/3)y$ using different values for the initial population y_0 together with the slope field. Notice that if $0 < y_0 < 1$, then $\lim_{t\to\infty} y(t) = 0$, and if $1 < y_0 < 3$ or $y_0 > 3$, then $\lim_{t\to\infty} y(t) = 3$. In this case, the threshold is $y = 1$ and the carrying capacity is $y = 3$.

 EXERCISES 3.1

Solve the following problems. Unless otherwise stated, use the Malthus model.

1. Suppose that a culture of bacteria has an initial population of $n = 100$. If the population doubles every 3 days, determine the number of bacteria present after 30 days. How much time is required for the population to reach 4250 in number?

2. Suppose that the population in a yeast culture triples every 7 days. What is the population after 35 days? How much time is required for the population to be 10 times the initial population?

3. Suppose that two-thirds of the cells in a culture remain after 1 day. Use this information to determine the number of days until only one-third of the initial population remains.

4. Consider a radioactive substance with half-life 10 days. If there are initially 5000 g of the substance, how much remains after 365 days?

5. Suppose that the half-life of an element is 1000 h. If there are initially 100 g, how much remains after 1 h? How much remains after 500 h?

6. Suppose that the population of a small town is initially 5000. Due to the construction of an interstate highway, the population doubles over the next year. If the rate of growth is proportional to the current population, when will the population reach 25,000? What is the population after 5 years?

7. Suppose that mold grows at a rate proportional to the amount present. If there are initially 500 g of mold and 6 h later there are 600 g, determine the amount of mold present after 1 day. When is the amount of mold 1000 g?

8. Suppose that the rabbit population on a small island grows at a rate proportional to the number of rabbits present. If this population doubles after 100 days, when does the population triple?

9. In a chemical reaction, chemical A is converted to chemical B at a rate proportional to the amount of chemical A

present. If half of chemical A remains after 5 h, when does 1/6 of the initial amount of chemical A remain? How much of the initial amount remains after 15 h?

10. If 90% of the initial amount of a radioactive element remains after 1 day, what is the half-life of the element?

11. If $y(t)$ represents the percent of a radioactive element that is present at time t and the values of $y(t_1)$ and $y(t_2)$ are known, show that the half-life H is given by

$$H = \frac{(t_2 - t_1)\ln 2}{\ln(y(t_1)) - \ln(y(t_2))}.$$

12. The half-life of ^{14}C is 5730 years. If the original amount of ^{14}C in a particular living organism is 20 g and that found in a fossil of that organism is 0.01 g, determine the approximate age of the fossil.

13. After 10 days, 800 g of a radioactive element remain, and after 15 days, 560 g remain. What is the half-life of this element?

14. After 1 week, 10% of the initial amount of a radioactive element decays. How much decays after 2 weeks? When does half of the original amount decay?

15. Determine the percentage of the original amount of ^{226}Ra that remains after 100 years.

16. If an artifact contains 40% of the amount of ^{230}Th as a present-day sample, what is the age of the artifact?

17. On an archeological dig, scientists find an ancient tool near a fossilized human bone. If the tool and fossil contain 65% and 60% of the amount of ^{14}C as that in present-day samples, respectively, determine if the tool could have been used by the human.

18. A certain group of people with initial population 10,000 grows at a rate proportional to the number present. The population doubles in 5 years. In how many years will the population triple?

19. Solve the logistic equation, $dy/dt = \alpha y(1 - (1/K)y)$, by viewing it as a Bernoulli equation and then solve the resulting linear equation by using an integrating factor rather than the method of undetermined coefficients that is illustrated in the examples.

20. What is the limiting population, $\lim_{t\to\infty} y(t)$, of the United States population using the result obtained in Example 3.1.5?

21. Solve the logistic equation if $r = 1/100$ and $a = 10^{-8}$ given that $y(0) = 100,000$. Find $y(25)$. What is the limiting population?

22. Five college students with the flu virus return to an isolated campus of 2500 students. If the rate at which this virus spreads is proportional to the number of infected students y and to the number not infected $2500 - y$, solve the IVP $dy/dt = ky(2500 - y)$, $y(0) = 5$ to find the number of infected students after t days if 25 students have the virus after 1 day. How many students have the flu after 5 days?

23. One student in a college organization of 200 members proceeds to spread a rumor. If the rate at which this rumor spreads is proportional to the number of students y that know about the rumor as well as the number that do not know, then solve the IVP to find the number of students informed of the rumor after t days if 50 students are informed after 1 day. How many students know the rumor after 2 days? Will all of the students eventually be informed of the rumor? (see Exercise 22).

24. Suppose that glucose enters the bloodstream at the constant rate of r grams per minute while it is removed at a rate proportional to the amount y present at any time. Solve the IVP $dy/dt = r - ky$, $y(0) = y_0$ to find $y(t)$. What is the eventual concentration of glucose in the bloodstream according to this model?

25. What is the concentration of glucose in the bloodstream after 10 min if $r = 5\,\text{g/min}$, $k = 5$, and the initial concentration is $y(0) = 500$? After 20 min? Does the concentration appear to reach its limiting value quickly or slowly? (see Exercise 24).

26. Suppose that we deposit a sum of money in a money market fund that pays interest at an annual rate k, and let $S(t)$ represent the value of the investment at time t. If the compounding takes place continuously, then the rate at which the value of the investment changes is the interest rate times the value of the investment, $dS/dt = kS$. Use this equation to find $S(t)$ if $S(0) = S_0$.

27. Banks use different methods to compound interest. If the interest rate is k, and if interest is compounded m times per year, then $S(t) = S_0 (1 + k/m)^{mt}$. When interest is compounded continuously, then $m \to \infty$. Compare

$$\lim_{m \to \infty} S_0 (1 + k/m)^{mt}$$

to the formula obtained in Exercise 26.

28. (*Dating Works of Art*) We can determine if a work of art is more than 100 years old by determining if the lead-bearing materials contained in the work were manufactured within the last 100 years. The half-life of lead-210 (^{210}Pb) is 22 years, while the half-life of radium-226 (^{226}Ra) is 1700 years. Let SF denote the ratio of ^{210}Pb to ^{226}Ra per unit mass of lead. The approximate value of SF for works of art created in the last 80 years is 100. Then, the quantity of lead $(1 - \text{Ra})/(\text{Po})$ at time t is given by

$$\frac{1 - \text{Ra}}{\text{Po}} = \frac{(\text{SF} - 1)e^{-\lambda t}}{1 + (\text{SF} - 1)e^{-\lambda t}},$$

where λ is the disintegration constant for ^{210}Pb.[1] On the other hand, for very old paintings $(1 - \text{Ra})/(\text{Po}) \approx 0$.

(a) Determine the disintegration constant λ for ^{210}Pb where the amount of ^{210}Pb at time t is $y = y_0 e^{-\lambda t}$.

(b) Graph $(1 - \text{Ra})/(\text{Po})$ for $0 \le t \le 250$ using SF = 100.

The following table shows the ratio of $(1 - \text{Ra})/(\text{Po})$ for various famous paintings.

The last two paintings, *Lace Maker* and *Laughing Girl*, were painted by the Dutch painter Jan Vermeer, who lived from 1632 to 1675.

Painting	^{210}Po Concentration (dpm/g of Pb)	^{226}Ra Concentration (dpm/g of Pb)	$\frac{1 - \text{Ra}}{\text{Po}}$
Washing of Feet	12.6	0.26	0.98
Woman Reading Music	10.3		0.97
		0.30	
Woman Playing Mandolin	8.2	0.17	0.98
Woman Drinking	8.3	0.1	0.99
Disciples of Emmaus		0.8	0.91
Boy Smoking	4.8	0.31	0.94
Lace Maker	1.5	1.4	0.07
Laughing Girl	5.2	6.0	−0.15

[1] B. Keisch, Dating works of art through their natural radioactivity: improvements and applications, Science, **160**, 1968, 413-415.

(c) Determine if it is likely that any of the first six paintings were also painted by Vermeer (which would make them very valuable!). If not, approximate when they were painted.

The Gompertz equation is named after the English mathematician and statistician **Benjamin Gompertz** (1779-1865). He is most famous for *Gompertz's Law of Mortality* that was published in 1825.

29. Consider the differential equation $dy/dt = -r(1 - y/A)y$, where r and A are positive constants. (a) Find the equilibrium solutions, sketch the phase line, and classify the equilibrium solutions. (b) Describe how this equation differs from the logistic equation. (c) If $A = 2$ and $y(0) = 1$, what is $\lim_{t \to \infty} y(t)$? (d) If $A = 2$ and $y(0) = 3$, what is $\lim_{t \to \infty} y(t)$?

30. (*Gompertz's Law of Mortality*) (a) Find and classify the equilibrium solutions of the *Gompertz equation*, $dy/dt = y(r - a \ln y)$, where r and a are positive constants. (b) *Gompertz's Law of Mortality* states that $N'(t) = rN(t) \ln(K/N(t))$, where $N(t)$ is the size of the population at time t, r the growth rate, and K the equilibrium population size. Show that $dy/dt = y(r - a \ln y)$ is equivalent to $N'(t) = rN(t) \ln(K/N(t))$.

31. The equilibrium solution $y = c$ is classified as *semistable* if solutions on one side of $y = c$ approach $y = c$ as $t \to \infty$ but on the other side of $y = c$, they move away from $y = c$ as $t \to \infty$. Use this definition to classify the equilibrium solutions (as asymptotically stable, semistable, or unstable) of the following differential equations.
(a) $y' = y(y - 2)^2$.
(b) $y' = y^2(y - 1)$.
(c) $y' = y - \sqrt{y}$.
(d) $y' = y^2(9 - y^2)$.
(e) $y' = 1 + y^3$.
(f) $y' = y - y^3$.
(g) $y' = y^2 - y^3$.

32. Consider the Malthus population model with $k = 0.01, 0.05, 0.1, 0.5$, and 1.0 using $y_0 = 1$. Solve the model, plot the solution with these values, and compare the results. How does the value of k affect the solution?

33. Consider the logistic equation with $r = 0.01, 0.05, 0.1, 0.5$, and 1.0 using $y_0 = 1$ and $a = 1$. Solve the model, plot the solution with these values, and compare the results. How does the value of r affect the solution?

34. (*Tumor and Organism Growth*) Ludwig von Bertalanffy (1901-1972) made valuable contributions to the study of organism growth, including the growth of tumors, based on the relationship between body size of the organism and the metabolic rate. He theorized that weight is directly proportional to volume and that the metabolic rate is proportional to surface area. This indicates that surface area is proportional to $V^{2/3}$ because V is measured in cubic units and S in square units. Therefore, Bertalanffy studied the IVP $dV/dt = aV^{2/3} - bV, V(0) = V_0$.

(a) Solve this IVP to show that volume is given by

$$V(t) = [(a/b) - ((a/b) - V_0^{1/3})e^{-bt/3}]^3.$$

(b) Find $\lim_{t\to\infty} V(t)$ and explain what the limit represents.

(c) This IVP is similar to that involving the logistic equation $dV/dt = aV^2 - bV$, $V(0) = V_0$. Compare the limiting volume between the logistic equation and the model proposed by Bertalanffy in the cases when $a/b > 1$ and when $a/b < 1$.

35. (*Harvesting*) If we wish to model a population of size $P(t)$ at time t and consider a constant harvest rate h (like hunting, fishing, or disease), then we might modify the logistic equation and use the equation $P' = rP - aP^2 - h$ to model the population under consideration. Assume that $h \geq r^2/(4a)$.

(a) Show that if $h \geq r^2/(4a)$, a general solution of $P' = rP - aP^2 - h$ is

$$P(t) = \frac{1}{2a}\left[r + \sqrt{4ah - r^2}\,\tan\right.$$
$$\left. \times \left(\frac{1}{2a}(C - at)\sqrt{4ah - r^2}\right)\right].$$

(b) Suppose that for a certain species it is found that $r = 0.03$, $a = 0.0001$, $h = 2.26$, and $C = -1501.85$. At what time will the species become extinct?

(c) If $r = 0.03$, $a = 0.0001$, and $P(0) = 5.3$, graph $P(t)$ if $h = 0, 0.5, 1.0, 1.5, 2.0, 2.25$ and 2.5.

(d) What is the maximum allowable harvest rate to assure that the species survives?

(e) Generalize your result from (d). For arbitrary a and r, what is the maximum allowable harvest rate that ensures survival of the species?

36. (*Radiation Poisoning*) Shortly after the Chernobyl accident in the Soviet Union in 1986, several nations reported that the level of ^{131}I in milk wasfive times that considered

safe for human consumption. Make a table of the level of ^{131}I in milk as a multiple of that considered safe for human consumption for the first 3 weeks following the accident. After how long did the milk become safe for human consumption?

37. Consider a solution to the logistic equation with initial population y_0 where $0 < y_0 < r/a$. Show that this solution is concave up for $0 < y < r/(2a)$ and concave down for $r/(2a) < y < r/a$. Hint: Differentiate the right side of the logistic equation and set the result equal to 0. Describe the behavior of dy/dt based on this result.

38. From the early 1800s to the mid-1800s, the passenger pigeon population was thriving. However, due to hunting, the population size was reduced dramatically by the late 1800s. Unfortunately, the passenger pigeon requires a large number of cohorts to achieve successful reproduction. Having fallen below this level, the population size continued to decrease and the bird is now extinct. Which population model should be used to describe this situation?

39. (a) Find the solution to the IVP $dy/dt = -r(1 - y/A)y$, $y(0) = y_0$, where r and A are positive constants. (b) If $y_0 < A$, then determine $\lim_{t\to\infty} y(t)$. (c) If $y_0 > A$, show that $\lim_{t\to\infty} y(t) = \infty$. (d) Show that if $y_0 > A$, then $y(t)$ has a vertical asymptote at $t = (1/r)\ln(y_0/(y_0 - A))$.

40. Solve the IVP $dy/dt = r(1 - (1/A)y)(1 - (1/B)y)$, $y(0) = y_0$.

41. (*The Logistic Difference Equation*) Given x_0, the *logistic difference equation* is

$$x_{n+1} = rx_n(1 - x_n). \qquad (3.5)$$

Assume that $x_0 = 0.5$.

(a) If $r = 3.83$, calculate (n, x_n) for $n = 1, 2, \ldots, 50$, plot the resulting set of points and connect consecutive points with line segments. Why do you think this is called a "three-cycle?"

(b) Compute $(r, x_n(r))$ for 250 equally spaced values of r between 2.8 and 4.0 and $n = 101, \dots, 300$. Plot the resulting set of points. This famous image is called the "Pitchfork diagram."

(c) Repeat (b) for 250 equally spaced values of r between 3.7 and 4.0 and $n = 101, \dots, 300$.

(d) The logistic difference equation exhibits "chaos." Approximate the r-values between 2.8 and 4.0 for which the logistic difference equation exhibits chaos. Explain your reasoning. In your explanation, approximate those r-values that lead to a two-cycle, four-cycle, and so on.

See Smith and Waltman's, *The Theory of the Chemostat: Dynamics of Microbial Competition* [27], for a detailed discussion of various chemostat models.

---

42. *(Growth in the Chemostat)* The *scaled* equations for the growth of a population in a chemostat are

$$\frac{dS}{dt} = 1 - S - \frac{mS}{a + S}x,$$

$$\frac{dx}{dt} = x\left(\frac{mS}{a + S} - 1\right), \qquad (3.6)$$

$$S(0) \geq 0, \quad x(0) > 0,$$

where $S(t)$ denotes the concentration of the nutrient at time t for the organism with concentration $x(t)$ at time t.

(a) If $\Sigma = 1 - S - x$, show that $\Sigma' = -\Sigma$. Then, system (3.6) can be written as

$$\frac{d\Sigma}{dt} = -\Sigma,$$

$$\frac{dx}{dt} = x\left(\frac{m(1 - \Sigma - x)}{a + (1 - \Sigma - x)} - 1\right), \qquad (3.7)$$

$$\Sigma(0) > 0, \quad x(0) > 0.$$

Because $\Sigma(t) = \Sigma(0)e^{-t}, \lim_{t \to \infty} \Sigma(t) = 0$ so system (3.7) can be rewritten as the single first-order equation

$$\frac{dx}{dt} = x\left(\frac{m(1 - x)}{a + (1 - x)} - 1\right) \quad \text{or}$$

$$\frac{dx}{dt} = x\left[\frac{m(1 - x)}{1 + a - x} - 1\right], \quad 0 \leq x \leq 1,$$

$$(3.8)$$

where $x(0) > 0$.

(b) Find the equilibrium solutions of Equation (3.8).

(c) Write the nontrivial solutions found in (a) in the form $1 - \lambda$. (λ is called the *break-even* concentration.)

(d) Find conditions on λ so that a nontrivial equilibrium solution exists. Confirm your result by graphing various solutions for various values of m and a. Illustrate that if m is sufficiently small, the organism becomes extinct, regardless of the initial conditions. On the other hand, if a nontrivial equilibrium solution exists, all nontrivial solutions tend to it.

3.2 NEWTON'S LAW OF COOLING AND RELATED PROBLEMS

- Newton's Law of Cooling
- Mixture Problems

Newton's Law of Cooling

First-order linear differential equations can be used to solve a variety of problems that involve temperature. For example, a medical examiner can find the time of death in a homicide case, a chemist can determine the time required for a plastic mixture to cool to a hardening temperature, and an engineer can design the cooling and heating system of a manufacturing facility. Although distinct, each of these problems depends on a basic principle, *Newton's Law of Cooling* that is used to develop the differential equation associated with each problem.

> *Newton's Law of Cooling*: The rate at which the temperature $T(t)$ changes in a cooling body at time t is proportional to the difference between the temperature of the body, $T(t)$, and the constant temperature T_s of the surrounding medium.

This situation is represented as the first-order IVP

$$\frac{dT}{dt} = k(T - T_s), \quad T(0) = T_0,$$

where T_0 is the initial temperature of the body and k is the constant of proportionality. The equation $dT/dt = k(T - T_s)$ is separable, and separating variables gives us

Newton's Law of Cooling is named after the English scientist and mathematician **Sir Isaac Newton** (1643-1727). "Truth is ever to be found in the simplicity, and not in the multiplicity and confusion of things." Newton probably discovered the Law of Cooling when performing experiments on thermometers that he had invented in the late 1690s. He first reported his discovery in May, 1701.

$$\frac{1}{T - T_s} dT = k \, dt,$$

followed by integrating yields $\ln|T - T_s| = kt + C$. Using the properties of natural logarithms and simplifying yields $T = C_1 e^{kt} + T_s$, where $C_1 = e^C$. Applying the initial condition implies that $T_0 = C_1 + T_s$, so $C_1 = T_0 - T_s$. Therefore, the solution of the IVP is

$$T = (T_0 - T_s)e^{kt} + T_s. \tag{3.9}$$

If $k < 0$, $\lim_{t \to \infty} e^{kt} = 0$. Therefore, in this case, $\lim_{t \to \infty} T(t) = T_s$, so the temperature of the body approaches that of its surroundings.

Instead of solving the ordinary differential equation (ODE) as a separable equation, we could have solved this equation by viewing it as a first-order linear equation by writing the equation in standard form as $T' - kT = -kT_s$.

Some teachers prefer using the method of undetermined coefficients to solve this equation. To do so, we first keep in mind that the solution of $y' + ky = 0$, where k is a constant, is $y = Ce^{-kt}$. To use this remark, we rewrite $dT/dt = k(T - T_s)$, $T(0) = T_0$ in the form $(dT/dt) - kT = -kT_s$, $T(0) = T_0$. The corresponding homogeneous equation is $(dT/dt) - kT = 0$, which has solution $T_h = Ce^{kt}$. A particular solution of the nonhomogeneous equation $(dT/dt) - kT = -kT_s$ will take the form $T_P = A$, where A is a constant to be determined. With $T_P = A$, differentiating gives us $T_P' = 0$, because the derivative of a constant is 0. With this substitution into Newton's equation for his Law of Cooling, we have

$$T_P' - kT_P = -kT_s,$$
$$0 - k \cdot A = -kT_s,$$
$$A = T_s.$$

 In Equation (3.10), k is assumed to be positive. Explain why the proportionality constant is negative?

This means that a particular solution of Newton's Law of Cooling is $T_P = T_s$. Thus, a general solution of Newton's Law of Cooling is $T = T_h + T_P = Ce^{kt} + T_s$. Solving this equation for $T(0) = T_0$ gives us

$$T = (T_0 - T_s)\, e^{kt} + T_s, \qquad (3.10)$$

which, not surprising, is the same as we obtained in Equation (3.9). In Equation (3.11), k is negative. Thus, we generally write the solution of Newton's Law of Cooling as

$$T = (T_0 - T_s)\, e^{-kt} + T_s, \qquad (3.11)$$

where k is a positive constant.

To better understand the model, suppose that $k < 0$. If $T > T_s$, then $dT/dt < 0$, so $T(t)$ is a decreasing function. Similarly, if $T < T_s$, then $dT/dt > 0$, so $T(t)$ is an increasing function. Therefore, the object cools off if the surrounding temperature is less than the object's temperature, whereas it warms up if the temperature of the surroundings is cooler than the object's temperature.

EXAMPLE 3.2.1

A pie is removed from a 350 °F oven and placed to cool in a room with temperature 75 °F. In 15 min, the pie has a temperature of 150 °F Determine the time required to cool the pie to a temperature of 80 °F.

Solution

In this example, $T_0 = 350$ and $T_s = 75$. Substituting these values into $T = (T_0 - T_s)e^{kt} + T_s$, we obtain $T(t) = (350 - 75)e^{kt} + 75 = 275\, e^{kt} + 75$. To solve the problem we must find e^k (or k). We know that $T(15) = 150$, so $T(15) = 275e^{15k} + 75 = 150$. Solving this equation for e^k gives us

$$275\, e^{15k} = 75,$$

$$e^{15k} = 3/11 \quad \text{or} \quad (e^k)^{15} = 3/11 \quad \text{so}$$

$$e^k = (3/11)^{1/15}.$$

Thus, $T(t) = 275\left(e^k\right)^t + 75 = 275(3/11)^{t/15} + 75$. To find the value of t for which $T(t) = 80$, we solve the equation $275(3/11)^{t/15} + 75 = 80$ for t:

$$275(3/11)^{t/15} = 5,$$

$$(3/11)^{t/15} = 1/55,$$

$$\ln(3/11)^{t/15} = \ln 1/55 = -\ln 55,$$

$$\frac{t}{15}\ln(3/11) = -\ln 55,$$

$$t = \frac{-15 \times \ln 55}{\ln(3/11)} \approx 46.264.$$

Thus, the pie will reach a temperature of 80 °F after about 46 min.

An interesting problem associated with this example is to determine if the pie ever reaches room temperature. From the formula $T(t) = 275(3/11)^{t/15} + 75$, we note that $275(3/11)^{t/15} > 0$, so $T(t) = 275(3/11)^{t/15} + 75 > 75$. Therefore, the pie never actually reaches room temperature according to our model. However, we see its temperature approaches 75 °F as t increases because

$$\lim_{t \to \infty} (275(3/11)^{t/15} + 75) = 75.$$

In the investigation of a homicide, the time of death can be important. In some situations, using the fact that the normal body temperature of most healthy people is 98.6 °F together with Newton's Law of Cooling can be used to approximate the time of death.

EXAMPLE 3.2.2

Suppose that when Miss Scarlett discovered Mr. Boddy's body in the conservatory at noon, its temperature was 82 °F. Two hours later, the temperature of the corpse was 72 °F. If the temperature of the conservatory was 65 °F, what was the approximate time of Mr. Boddy's death?

Solution

This problem is solved like the previous example. Let $T(t)$ denote the temperature of the body at time t, where $T(0)$ represents the temperature of the body when it is discovered and $T(2)$ represents the temperature of the body 2 h after it is discovered. In this case we have $T(0) = 82$ and $T_s = 65$, and substituting these values into $T(t) = (T_0 - T_s)e^{kt} + T_s$, yields $T(t) = (82 - 65)e^{kt} + 65 = 17e^{kt} + 65$. Using $T(2) = 72$, we solve the equation $T(2) = 17e^{2k} + 65 = 72$ for e^k to find $e^k = (7/17)^{1/2}$, So $T(t) = 17(e^k)^t + 65 = 17(7/17)^{t/2} + 65$. To find the value of t for which $T(t) = 98.6$, we solve the equation $17(7/17)^{t/2} + 65 = 98.6$ for t and obtain

$$t = \frac{2\ln(1.97647)}{\ln 7 - \ln 17} = -1.53569.$$

This result means that the time of death occurred approximately 1.53 h before the body was discovered, as we observe in Figure 3.4. Therefore, the time of death was approximately 10:30 a.m. because the body was discovered at noon.

In each of the previous cases, the temperature of the surroundings was assumed to be constant. However, this does not have to be the case. For example, determining the temperature inside a building over the span of a 24-h day is

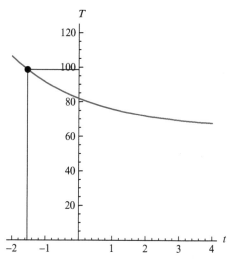

FIGURE 3.4   Graph of $T(t) = 17(7/17)^{t/2} + 65$.

more complicated because the outside temperature varies. If we assume that a building has no heating or air conditioning system, the differential equation that needs to be solved to find the temperature $u(t)$ at time t inside the building is

$$\frac{du}{dt} = k(C(t) - u(t)),$$

where $C(t)$ is a function that describes the outside temperature and $k > 0$ is a constant that depends on the insulation of the building. According to this equation, if $C(t) > u(t)$, then $du/dt > 0$, which implies that u increases. However, if $C(t) < u(t)$, then $du/dt < 0$, which means that u decreases.

EXAMPLE 3.2.3

Suppose that during the month of April in Atlanta, Georgia, the outside temperature in °F is given by $C(t) = 70 - 10\cos(\pi t/12)$, $0 \le t \le 24$. (This implies that the average value of $C(t)$ is 70 °F.) Determine the temperature in a building that has an initial temperature of 60 °F if $k = 1/4$.

Solution

The IVP that we must solve is

$$\frac{du}{dt} = \frac{1}{4}\left(70 - 10\cos\left(\frac{\pi}{12}t\right) - u\right), \quad u(0) = 60.$$

The differential equation can be solved if we write it as $(du/dt) + \frac{1}{4}u = \frac{1}{4}(70 - 10\cos((\pi/12)t))$ and use an integrating factor. This gives us

$$u(t) = \frac{10}{9 + \pi^2}\left(63 + 7\pi^2 - 9\cos\left(\frac{\pi}{12}t\right)\right.$$
$$\left. - 3\pi\sin\left(\frac{\pi}{12}t\right)\right) + Ce^{-t/4}.$$

We then apply the initial condition $u(0) = 60$ to determine the arbitrary constant C and obtain the solution

$$u(t) = \frac{10}{9 + \pi^2}\left(63 + 7\pi^2 - 9\cos\left(\frac{\pi}{12}t\right)\right.$$
$$\left. - 3\pi\sin\left(\frac{\pi}{12}t\right)\right) - \frac{10\pi^2}{9 + \pi^2}e^{-t/4},$$

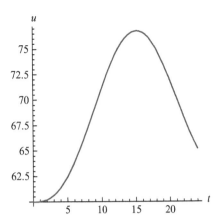

FIGURE 3.5   Modeling the temperature in a building over the course of a 24-h day.

which is graphed in Figure 3.5. From the graph, we see that the temperature reaches its maximum (approximately 72°F) near $t = 15.5$ h, which corresponds to 3:30 p.m.

At what time during the day is the temperature in the building increasing (decreasing) at the fastest rate?

In many situations, a heating or cooling system is installed to control the temperature in a building. Another factor in determining the temperature, which we ignored in prior calculations, is the generation of heat from the occupants of the building, including people and machinery. We investigate the inclusion of these factors in the exercises at the end of this section.

Mixture Problems

An application that involves a differential equation similar to the one encountered with Newton's Law of Cooling is that associated with *mixture problems*. In this case, suppose that a tank contains V_0 gallons of a brine solution, a mixture of salt and water. A brine solution of concentration S_1 pounds per gallon is allowed to flow into the tank at a rate R_1 gallons per minute while a well-stirred mixture flows out of the tank at the rate of R_2 gallons per minute. If $y(t)$ represents the amount (in pounds) of salt in the tank at time t minutes, then the equation that describes the net rate of change in the amount of salt in the tank is

$$\frac{dy}{dt} = \text{(Rate salt enters the tank)}$$
$$- \text{(Rate salt leaves the tank)}$$
$$= \left(S_1 \frac{\text{lb}}{\text{gal}}\right)\left(R_1 \frac{\text{gal}}{\text{min}}\right)$$
$$- \left(\frac{y(t)}{V(t)} \frac{\text{lb}}{\text{gal}}\right)\left(R_2 \frac{\text{gal}}{\text{min}}\right),$$

where $V(t)$ is the volume of solution (liquid) in the tank at any time t and $y(t)/V(t)$ represents the salt concentration of the well-stirred mixture. To determine $V(t)$, we solve the equation that gives the net rate of change in the amount of liquid in the tank,

$$\frac{dV}{dt} = \text{(Rate liquid enters the tank)}$$
$$- \text{(Rate liquid leaves the tank)}$$
$$= \left(R_1 \frac{\text{gal}}{\text{min}}\right) - \left(R_2 \frac{\text{gal}}{\text{min}}\right)$$

using the initial condition $V(0) = V_0$. We then use the solution $V(t)$ to solve the previous ODE for $y(t)$ where $y(0) = y_0$ is the initial amount of salt in the tank.

EXAMPLE 3.2.4

Suppose that $S_1 = 1$, $R_1 = 4$, and $R_2 = 3$. In addition, suppose that the tank initially holds 500 gal of liquid and 250 lb of salt. Find the amount of salt contained in the tank at any time. Determine how much time is required for 1000 lb of salt to accumulate. If the tank holds a maximum of

800 gal, can 1000 lb of salt be present in the tank at any time before the tank reaches its maximum capacity?

Solution

We notice that the volume (of liquid) does not remain constant in the tank because $R_1 \neq R_2$. Therefore, we determine $V(t)$ by solving the IVP $dV/dt = 4 - 3 = 1$, $V(0) = 500$ with the general solution $V(t) = t + C_1$. Applying the initial condition yields $V(t) = t + 500$, so that the IVP used to find $y(t)$ is

$$\frac{dy}{dt} = 4 - \frac{3y}{t + 500}, \quad y(0) = 250.$$

Rewriting the ODE as $dy/dt + 3y/(t + 500) = 4$, we use the integrating factor

$$\mu(t) = e^{\int 3/(t+500)dt} = e^{3\ln(t+500)}$$

$$= (t + 500)^3, \quad t \geq 0$$

to obtain $(d/dt)[(t + 500)^3 y] = 4(t + 500)^3$ so that $(t + 500)^3 y = (t + 500)^4 + C_2$. Therefore, $y(t) = t + 500 + C_2(t + 500)^{-3}$, so that $y(0) = 250$ indicates that $C_2 = -250 \times 500^3 = -31,250,000,000$. After graphing $y(t) = t + 500 - 31,250,000,000(t+500)^{-3}$, we can use a numerical solver to find that the approximate solution to $t + 500 - 31,250,000,000(t + 500)^{-3} = 1000$ is $t = 528.7$ min. However, if $t = 528.7$, then $V(528.7) = 528.7 + 500 = 1028.7$ gal, so the tank would overflow before 1000 lb of salt could be present in the tank.

 EXERCISES 3.2

1. A hot cup of tea is initially 100 °C when poured. How long does it take for the tea to reach a temperature of 50 °C if it is at 80 °C

after 15 min and the room temperature is 30 °C?

2. Suppose that the tea in Exercise 1 is allowed to cool at room temperature for 20 min. It is then placed in a cooler with temperature 15 °C. What is the temperature of the tea after 60 min if it is 60 °C after 10 min in the cooler?

3. A can of orange soda is removed from a refrigerator having temperature 40 °F. If the can is at 50 °F after 5 min, how long does it take for the can to reach a temperature of 60 °F if the surrounding temperature is 75 °F?

4. Suppose that a container of tea is placed in a refrigerator at 35 °F to cool. If the tea is initially at 75 °F and it has a temperature of 70 °F after 1 h, then when does the temperature of the tea reach 55 °F?

5. Determine the time of death if the temperature of a corpse is 79 °F when discovered at 3:00 p.m. and 68 °F 3 h later. Assume that the temperature of the surroundings is 60 °F (normal body temperature is 98.6 °F).

6. At the request of his children, a father makes homemade popsicles. At 2:00 p.m., one of the children asks if the popsicles are frozen (0 °C), at which time the father tests the temperature of a popsicle and finds it to be 5 °C. If the father placed the popsicles with a temperature of 15 °C in the freezer at 12:00 p.m. and the temperature of the freezer is −2 °C, when will the popsicles be frozen?

7. A thermometer that reads 90 °F is placed in a room with temperature 70 °F. After 3 min, the thermometer reads 80 °F. What does the thermometer read after 5 min?

8. A thermometer is placed outdoors with temperature 80 °F. After 2 min, the thermometer reads 68 °F, and after 5 min, it reads 72 °F. What was the initial temperature reading of the thermometer?

9. A casserole is placed in a microwave oven to defrost. It is then placed in a conventional oven at 300 °F and bakes for 30 min, at which time its temperature is 150 °F. If after baking an additional 30 min its temperature is 200 °F, what was the temperature of the casserole when it was removed from the microwave?

10. A bottle of wine at room temperature (70 °F) is placed in ice to chill at 32 °F. After 20 min, the temperature of the wine is 58 °F. When will its temperature be 50 °F?

11. When a cup of coffee is poured its temperature is 200 °F. Two minutes later, its temperature is 170 °F. If the temperature of the room is 68 °F, when is the temperature of the coffee 140 °F?

12. After dinner, a couple orders two cups of coffee. Upon being served, the gentleman immediately pours one container of cream into his cup of coffee. His companion waits 4 min before adding the same amount of cream to her cup. Which person has the hotter cup of coffee when they both take a sip of coffee after she adds the cream to her cup? (Assume that the cream's temperature is less than that of the coffee.) Explain.

13. Suppose that during the month of February in Washington, D.C., the outside temperature in °F is given by
$C(t) = 40 - 5\cos(\pi t/12), 0 \le t \le 24$.
Determine the temperature in a building that has an initial temperature of 50 °F if $k = 1/4$. (Assume that the building has no heating or air conditioning system.)

14. Suppose that during the month of August in Savannah, Georgia, the outside temperature in °F is given by $C(t) = 85 - 10\cos(\pi t/12)$, $0 \le t \le 24$. Determine the temperature in a building that has an initial temperature of 60 °F if $k = 1/4$. (Assume that the building has no heating or air conditioning system.)

15. Suppose that during the month of October in Los Angeles, California, the outside temperature in °F is given by

$C(t) = 70 - 5\cos(\pi t/12), 0 \le t \le 24$.
Determine the temperature in a building that has an initial temperature of 65 °F if $k = 1/4$. (Assume that the building has no heating or air conditioning system.)

16. Suppose that during the month of January in Dayton, Ohio, the outside temperature in °F is given by $C(t) = 20 - 5\cos(\pi t/12)$, $0 \le t \le 24$. Determine the temperature in a building that has an initial temperature of 40 °F if $k = 1/4$. (Assume that the building has no heating or air conditioning system.)

17. Find the amount of salt $y(t)$ in a tank with initial volume V_0 gallons of liquid and y_0 pounds of salt using the given conditions.
   (a) $R_1 = 4\,\text{gal/min}$, $R_2 = 4\,\text{gal/min}$, $S_1 = 2\,\text{lb/gal}$, $V_0 = 400$, $y_0 = 0$.
   (b) $R_1 = 4\,\text{gal/min}$, $R_2 = 2\,\text{gal/min}$, $S_1 = 1\,\text{lb/gal}$, $V_0 = 500$, $y_0 = 20$.
   (c) $R_1 = 4\,\text{gal/min}$, $R_2 = 6\,\text{gal/min}$, $S_1 = 1\,\text{lb/gal}$, $V_0 = 600$, $y_0 = 100$.
   Describe how the values of R_1 and R_2 affect the volume of liquid in the tank.

18. A tank contains 100 gal of a brine solution in which 20 lb of salt is initially dissolved. (a) Water (containing no salt) is then allowed to flow into the tank at a rate of 4 gal/min and the well-stirred mixture flows out of the tank at an equal rate of 4 gal/min. Determine the amount of salt $y(t)$ at any time t. What is the eventual concentration of the brine solution in the tank? (b) If instead of water a brine solution with concentration 2 lb/gal flows into the tank at a rate of 4 gal/min, what is the eventual concentration of the brine solution in the tank?

19. A tank contains 200 gal of a brine solution in which 10 lb of salt is initially dissolved. A brine solution with concentration 2 lb/gal is then allowed to flow into the tank at a rate of 4 gal/min and the well-stirred mixture flows out of the tank at a rate of 3 gal/min. Determine the amount of salt $y(t)$ at any time t. If the tank can hold a maximum of

400 gal, what is the concentration of the brine solution in the tank when the volume reaches this maximum?

20. A tank contains 300 gal of a brine solution in which 300 lb of salt is initially dissolved. A brine solution with concentration 4 lb/gal is then allowed to flow into the tank at a rate of 3 gal/min, and the well-stirred mixture flows out of the tank at a rate of 4 gal/min. Determine the amount of salt $y(t)$ at any time t. What is the concentration of the brine solution after 10 min? What is the eventual concentration of the brine solution in the tank? For what values of t is the solution defined? Why?

The temperature $u(t)$ inside a building can be based on three factors: (1) the heat produced by people or machinery inside the building, (2) the heating (or cooling) produced by the furnace (or air conditioning system), and (3) the temperature outside the building based on Newton's Law of Cooling. If the rate at which these factors affect (increase or decrease) the temperature is given by $A(t)$, $B(t)$, and $C(t)$, respectively, the differential equation that models this situation is

$$\frac{du}{dt} = k(C(t) - u(t)) + A(t) + B(t),$$

where the constant $k > 0$ depends on the insulation of the building.

21. Find the temperature (and the maximum temperature) in the building with $k = 1/4$ if the initial temperature is 70 °F, and (a) $A(t) = 0.25$, $C(t) = 75$, and $B(t) = 0$; (b) $A(t) = 0.25$, $C(t) = 70 - 10\cos(\pi t/12)$, and $B(t) = 0$; and (c) $A(t) = 1$, $C(t) = 70 - 10\cos(\pi t/12)$, and $B(t) = 0$.

If a heating or cooling system is considered, we must model the system with an appropriate function. Of course, the system could run constantly, but we know that most are controlled by a thermostat. Suppose that the desired temperature is u_d. Then $B(t) = k_d(u_d - u(t))$, where k_d is a constant approximately equal to two for most systems.

22. Determine the temperature (and the maximum temperature) in a building with $k = 1/4$ and initial temperature 70 °F if (a) $A(t) = 0.25$, $B(t) = 1.75(68 - u(t))$, and $C(t) = 70 - 10\cos(\pi t/12)$; (b) $A(t) = 1$, $B(t) = 1.75(68 - u(t))$, and $C(t) = 70 - 10\cos(\pi t/12)$; and (c) $A(t) = 0.25$, $B(t) = 1.75(68 - u(t))$, and $C(t) = 80 - 10\cos(\pi t/12)$.

23. If $k = 1/4$, $A(t) = 0.25$, $B(t) = 1.75(u_d - u(t))$, and $C(t) = 70 - 10\cos(\pi t/12)$, determine the value of u_d needed so that the average temperature in the building over a 24-h period is 70 °F.

3.3 FREE-FALLING BODIES

The motion of some objects can be determined through the solution of a first-order equation. We begin by explaining some of the theory that is needed to set up the differential equation that models the situation.

Newton's Second Law of Motion: The rate at which the momentum of a body changes with respect to time is equal to the resultant force acting on the body.

A body's *momentum* is defined as the product of its mass and velocity, so this statement is modeled as

$$\frac{d}{dt}(mv) = F,$$

where m and v represent the body's mass and velocity, respectively, and F is the sum of the forces acting on the body. The mass m of the body is constant, so differentiation leads to the differential equation

$$m\frac{dv}{dt} = F.$$

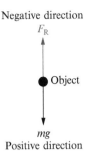

FIGURE 3.6   Force diagram.

Although Newton had formed early versions of his three laws of motion by 1666, they appear in his crowning achievement, the *Principia* that he published in 1687. "Be courageous and steady to the Laws and you cannot fail."

If the body is subjected to the force due to gravity, then its velocity is determined by solving the differential equation

$$m\frac{\mathrm{d}v}{\mathrm{d}t} = mg \quad \text{or} \quad \frac{\mathrm{d}v}{\mathrm{d}t} = g,$$

where $g \approx 32$ ft./s^2 (English system) or $g \approx 9.8$ m/s^2 (international system). See the summary of units in Table 3.4.

This differential equation is applicable only when the resistive force due to the medium (such as air resistance) is ignored. If this offsetting resistance is considered, we must discuss all of the forces acting on the object. Mathematically, we write the equation as

$$m\frac{\mathrm{d}v}{\mathrm{d}t} = \sum(\text{forces acting on the object}),$$

TABLE 3.4   Units Useful in Solving Problems Associated with Newton's Second Law of Motion

	English	International
Mass	Slug (1b s^2/ft.)	Kilogram (kg)
Force	Pound (1b)	Newton (m kg/s^2)
Distance	Foot (ft.)	Meter (m)
Time	Second (s)	Second (s)

where the direction of motion is taken to be the positive direction.

We use the force diagram in Figure 3.6 to set up the differential equation that models the situation. Air resistance acts against the object as it falls and g acts in the same direction of the motion. We state the differential equation in the form:

$$m\frac{\mathrm{d}v}{\mathrm{d}t} = mg + (-F_R) \quad \text{or} \quad m\frac{\mathrm{d}v}{\mathrm{d}t} = mg - F_R,$$

where F_R represents this resistive force. Note that *down* is assumed to be the positive direction. The resistive force is typically proportional to the body's velocity (v) or a power of its velocity. Hence, the differential equation is linear or nonlinear based on the resistance of the medium taken into account.

EXAMPLE 3.3.1

(a) Determine the velocity and the distance traveled by an object with mass $m = 1$ slug that is thrown downward with an initial velocity of 2 ft./s from a height of 1000 ft. Assume that the object is subjected to air resistance that is equivalent to the instantaneous velocity of the object. (b) Determine the time at which the object strikes the ground and its velocity when it strikes the ground.

Solution

(a) First, we set up the IVP to determine the velocity of the object. The air resistance is equivalent to the instantaneous velocity, so $F_R = v$. The formula $m\,dv/dt = mg - F_R$ then gives us $dv/dt = 32 - v$ and imposing the initial velocity $v(0) = 2$ yields the IVP $dv/dt = 32 - v$, $v(0) = 2$, which can be solved through several methods. We choose to solve it as a linear first-order equation and use an integrating factor. (It also can be solved by separating variables.) With the integrating factor e^t, we have $(d/dt)(e^{tv}) = 32e^t$. Integrating both sides gives us $e^{tv} = 32e^t + C_1$, so $v = 32 + C_1e^{-t}$, and applying the initial velocity gives us $v(0) = 32 + C_1e^0 = 32 + C_1 = 2$, so $C_1 = -30$. Therefore, the velocity of the object is $v = 32 - 30\,e^{-t}$.

Notice that the velocity of the object cannot exceed 32 ft./s, called the *limiting velocity*, which is found by evaluating $\lim_{t\to\infty} v(t)$.

To determine the distance traveled at time t, $s(t)$, we solve the first-order equation $ds/dt = v = 32 - 30e^{-t}$ with initial condition $s(0) = 0$. This differential equation is solved by integrating both sides of the equation to obtain $s(t) = 32t + 30e^{-t} + C_2$. Application of $s(0) = 0$ then gives us $s(0) = 32 \times 0 + 30e^0 + C_2 = 30 + C_2 = 0$, so $C_2 = -30$, and the distance traveled by the object is given by $s(t) = 32t + 30e^{-t} - 30$.

(b) The object strikes the ground when $s(t) = 1000$. Therefore, we must solve the equation $s(t) = 32t + 30e^{-t} - 30 = 1000$ for t. The solutions of this equation can be approximated with numerical methods like Newton's method. From the graph of this function, shown in Figure 3.7, we see that $s(t) = 1000$ near $t = 35$. Numerical methods show that $s(t) = 32t + 30e^{-t} - 30 = 1000$ when $t \approx 32.1875$ s.

The velocity at the point of impact is found to be 32.0 ft./s by evaluating the derivative at the time at which the object strikes the ground, given by $s'(32.1875)$.

EXAMPLE 3.3.2

Suppose that the object in Example 3.3.1 of mass 1 slug is thrown downward with an initial

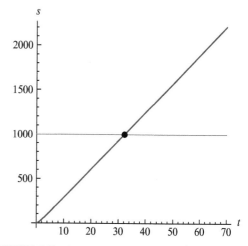

FIGURE 3.7   Graph of $s(t) = 32t + 30e^{-t} - 30$.

velocity of 2 ft./s and that the object is attached to a parachute, increasing this resistance so that it is given by v^2. Find the velocity at any time t and determine the limiting velocity of the object.

Solution

This situation is modeled by the IVP

$$\frac{dv}{dt} = 32 - v^2, \quad v(0) = 2.$$

We solve the differential equation by separating the variables and using partial fractions:

$$\frac{1}{32 - v^2}\,dv = dt,$$

$$\frac{1}{\left(4\sqrt{2} + v\right)\left(4\sqrt{2} - v\right)}\,dv = dt,$$

$$\frac{1}{8\sqrt{2}}\left(\frac{1}{v + 4\sqrt{2}} - \frac{1}{v - 4\sqrt{2}}\right)dv = dt,$$

$$\ln\left|v + 4\sqrt{2}\right| - \ln\left|v - 4\sqrt{2}\right| = 8\sqrt{2}t + C_1,$$

$$\ln\left|\frac{v + 4\sqrt{2}}{v - 4\sqrt{2}}\right| = 8\sqrt{2}t + C_1,$$

$$\left|\frac{v + 4\sqrt{2}}{v - 4\sqrt{2}}\right| = C_2e^{8\sqrt{2}t}$$

$$\left(C_2 = e^{C_1}\right),$$

$$\frac{v + 4\sqrt{2}}{v - 4\sqrt{2}} = C_3 e^{8\sqrt{2}t}$$

$$(C_3 = \pm C_2 \text{ or } 0).$$

Solving for v, we find that

$$v + 4\sqrt{2} = C_3 e^{8\sqrt{2}t}(v - 4\sqrt{2}) \quad \text{or}$$

$$(1 - C_3 e^{8\sqrt{2}t})v = -4\sqrt{2}(C_3 e^{8\sqrt{2}t} + 1),$$

so $v = -4\sqrt{2}(C_3 e^{8\sqrt{2}t} + 1)/(1 - C_3 e^{8\sqrt{2}t})$. Application of the initial condition yields $C_3 = (1 + 2\sqrt{2})/(1 - 2\sqrt{2})$. The limiting velocity of the object is found with L'Hopital's rule to be $\lim_{t\to\infty} v(t) = -4\sqrt{2}C_3/(-C_3) = 4\sqrt{2}$ ft./s.

In Example 3.3.1, the limiting velocity is 32 ft./s, so the parachute causes the velocity of the object to be reduced (see Figure 3.8). This shows that the object does not have to endure as great an impact as it would without the help of the parachute.

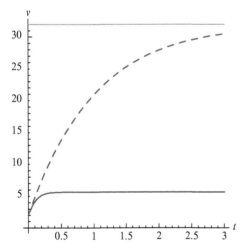

FIGURE 3.8 The velocity functions from Example 3.3.1 (dashed) and Example 3.3.2. Notice how the different forces due to air resistance affect the velocity of the object.

EXAMPLE 3.3.3

Determine a solution (for the velocity and the height) of the differential equation that models the motion of an object of mass m when directed upward with an initial velocity of v_0 from an initial position s_0, assuming that the air resistance equals cv where c is a positive constant.

Solution

The motion of the object is upward, so g and F_R act against the upward motion of the object as shown in Figure 3.9.

Therefore, the differential equation that must be solved in this case is the linear equation $dv/dt = -g - (c/m)v$. We solve the IVP

$$\frac{dv}{dt} = -g - \frac{c}{m}v, \quad v(0) = v_0 \qquad (3.12)$$

by first rewriting the equation $dv/dt = -g - (c/m)v$ as $dv/dt + (c/m)v = -g$ and then calculating the integrating factor $e^{\int (c/m)dt} = e^{ct/m}$. Multiplying each side of the equation by $e^{ct/m}$ gives us $e^{ct/m}dv/dt + (c/m)e^{ct/m}v = -g e^{ct/m}$ so that $(d/dt)(e^{ct/m}v) = -g e^{ct/m}$. Integrating we obtain $e^{ct/m}v = -gm/c\, e^{ct/m} + C$ and, consequently, the general solution is

$$v(t) = -\frac{gm}{c} + Ce^{-ct/m}.$$

Applying the initial condition $v(0) = v_0$ and solving for C yields $C = (cv_0 + gm)/c$ so that the solution to the IVP is

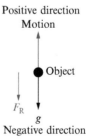

Positive direction
Motion

Object

F_R

g
Negative direction

FIGURE 3.9 By drawing a force diagram, we see that g and F_R are in the negative direction.

$$v(t) = -\frac{gm}{c} + \frac{cv_0 + gm}{c}e^{-ct/m}. \qquad (3.13)$$

For example, the velocity function for the case with $m = 1/128$ slugs, $c = 1/160$, $g = 32$ ft./s^2, and $v_0 = 48$ ft./s is $v(t) = 88e^{-4t/5} - 40$. This function is graphed in Figure 3.10(a). Notice where $v(t) = 0$. This value of t represents the time at which the object reaches its maximum height and begins to fall toward the ground.

Similarly, this function (3.13) can be used to investigate numerous situations without solving the differential equation (3.12) each time the parameter values are changed.

The height function $s(t)$, which represents the distance above the ground at time t, is determined by integrating the velocity function:

$$s(t) = \int v(t)dt = \int \left(-\frac{gm}{c} + \frac{cv_0 + gm}{c}e^{-ct/m} \right) dt$$

$$= -\frac{gm}{c}t + \frac{cmv_0 + gm^2}{c^2}e^{-ct/m} + C.$$

If the initial height is given by $s(0) = s_0$, solving for C results in $C = (gm^2 + c^2s_0 + cmv_0)/c^2$, so that

$$s(t) = -\frac{gm}{c}t + \frac{cmv_0 + gm^2}{c^2}e^{-ct/m} \qquad (3.14)$$

$$+ \frac{gm^2 + c^2s_0 + cmv_0}{c^2}.$$

The height and velocity functions are shown in Figure 3.10(b) using the parameters $m = 1/128$ slugs, $c = 1/160$, $g = 32$ ft./s^2, and $v_0 = 48$ ft./s as well as $s_0 = 0$.

The time at which the object reaches its maximum height occurs when the derivative of the height function is equal to zero. From Figure 3.10(b), we see that $s'(t) = v(t) = 0$ when $t \approx 1$. Solving $s'(t) = 0$ for t yields the better approximation $t \approx 0.985572$.

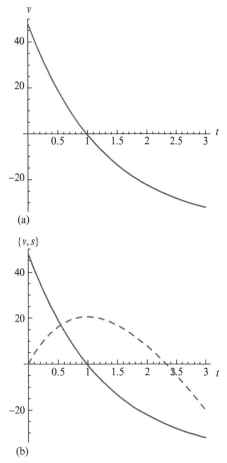

(a)

(b)

FIGURE 3.10   (a) Graph of $v(t) = 88e^{-4t/5} - 40$. (b) Graph of $v(t)$ together with $s(t) = 110 - 40t - 110e^{-4t/5}$ (dashed).

Weight and Mass Notice that in the English system, *pounds* describe *force*. Therefore, when the weight W of an object is given, we must calculate its *mass* with the relationship $W = mg$ or $m = W/g$. Conversely, in the international system, the mass of the object (in kilograms) is typically given.

We now combine several of the topics discussed in this section to solve the following problem.

EXAMPLE 3.3.4

A 32-lb object is dropped from a height of 50 ft. above the surface of a small pond. While the object is in the air, the force due to air resistance is v. However, when the object is in the pond it is subjected to a buoyancy force equivalent to $6v$. Determine how much time is required for the object to reach a depth of 25 ft. in the pond.

Solution

The mass of this object is found using the relationship $W = mg$, where W is the weight of the object. With this, we find that $32 \text{ lb} = m(32 \text{ ft./s}^2)$, so $m = 1 (\text{lb s}^2)/\text{ft}$ (slug).

This problem must be broken into two parts: an IVP for the object above the pond and an IVP for the object below the surface of the pond. Using techniques discussed in previous examples, the IVP above the pond's surface is found to be

$$dv/dt = 32 - v, \quad v(0) = 0.$$

However, to determine the IVP that yields the velocity of the object beneath the pond's surface, the velocity of the object when it reaches the surface must be known. Hence, the velocity of the object above the surface must be determined first.

The equation $dv/dt = 32 - v$ is separable and rewriting it yields $1/(32 - v)dv = dt$. Integrating and applying the initial condition results in $v(t) = 32 - 32e^{-t}$. To find the velocity when the object hits the pond's surface, we must know the time at which the object has fallen 50 ft. Thus, we find the distance traveled by the object by solving $ds/dt = v(t), s(0) = s_0$, obtaining $s(t) = 32e^{-t} + 32t - 32$. From the graph of $s(t)$ shown in Figure 3.11(a), we see that the value of t at which the object has traveled 50 ft. appears to be about 2.5 s.

A more accurate value of the time at which the object hits the surface is $t \approx 2.47864$. The velocity at this time is then determined by substitution into the velocity function, resulting in $v(2.47864) \approx$

29.3166. Note that this value is the initial velocity of the object when it hits the surface of the pond.

Thus, the IVP that determines the velocity of the object beneath the surface of the pond is given by $dv/dt = 32 - 6v, v(0) = 29.3166$. The solution of this IVP is $v(t) = 16/3 + 23.9833e^{-6t}$, and solving $ds/dt = 16/3 + 23.9833e^{-6t}, s(0) = 0$ we obtain

$$s(t) = 3.99722 - 3.99722\,e^{-6t} + \frac{16}{3}t,$$

which gives the depth of the object at time t. From the graph of this function, shown in Figure 3.11(b), we see that the object is 25 ft. beneath the surface of the pond after approximately 4 s.

A more accurate approximation of the time at which the object is 25 ft. beneath the pond's surface is $t \approx 3.93802$.

Finally, the time required for the object to reach the pond's surface is added to the time needed for it to travel 25 ft. beneath the surface to see that approximately 6.41667 s are required for the object to travel from a height of 50 ft. above the pond to a depth of 25 ft. below the surface.[1]

[1] L.N. Long, H. Weiss, The velocity dependence of aerodynamic drag: a primer for mathematicians, The American Mathematics Monthly, **106**(2), 127-135.

Note The model presented in this section in which the force due to air resistance is proportional to the velocity $mdv/dt = mg - cv$ is a simplified model used to illustrate how differential equations are used to model physical situations. In most cases, the equation in which the force due to air resistance is assumed to be proportional to the square of the velocity, $mdv/dt = mg - cv^2$, does a much better job in predicting the velocity of a falling object. In fact, the first model typically works only when an object is dropped in a highly viscous medium or when the object has negligible mass.

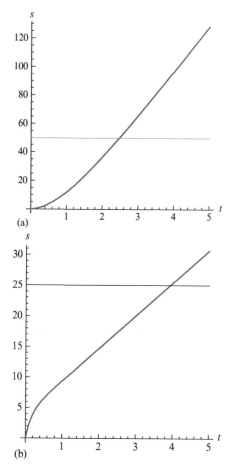

(a)

(b)

FIGURE 3.11 (a) Graph of $s(t) = 32e^{-t} + 32t - 32$. (b) Graph of $s(t) = 3.99722 - 3.9972e^{-6t} + \frac{16}{3}t$.

⊕ **EXERCISES 3.3**

1. A rock that weighs 32 lb is dropped from rest from the edge of a cliff. (a) Find the velocity of the rock at time t if the air resistance is equivalent to the instantaneous velocity v. (b) What is the velocity of the rock at $t = 2$ s?

2. An object that weighs 4 lb is dropped from the top of a tall building. (a) Find the velocity of the object at time t if the air

resistance is equivalent to the instantaneous velocity v. (b) What is the velocity of the object at $t = 2$ s? How does this compare to the result in Exercise 1?

3. An object weighing 1 lb is thrown downward with an initial velocity of 8 ft./s. (a) Find the velocity of the object at time t if the air resistance is equivalent to twice the instantaneous velocity. (b) What is the velocity of the object at $t = 1$ s?

4. An object weighing 16 lb is dropped from a tall building. (a) Find the velocity of the object at time t if the air resistance is equivalent to twice the instantaneous velocity. (b) What is the velocity of the object at $t = 1$ s? How does this compare to the result in Exercise 3?

5. A ball of weight 4 oz is tossed into the air with an initial velocity of 64 ft./s. (a) Find the velocity of the object at time t if the air resistance is equivalent to 1/16 the instantaneous velocity. (b) When does the ball reach its maximum height?

6. A tennis ball weighing 8 oz is hit vertically into the air with an initial velocity of 128 ft./s. (a) Find the velocity of the object at time t if the air resistance is equivalent to half of the instantaneous velocity. (b) When does the ball reach its maximum height?

7. A rock of weight 0.5 lb is dropped (with zero initial velocity) from a height of 300 ft. If the air resistance is equivalent to 1/64 times the instantaneous velocity, find the velocity and distance traveled by the object at time t. Does the rock hit the ground before 4 s elapse?

8. An object of weight 0.5 lb is thrown downward with an initial velocity of 16 ft./s from a height of 300 ft. If the air resistance is equivalent to 1/64 times the instantaneous velocity, find the velocity of and distance traveled by the object at time t. Compare these results to those in Exercise 7.

9. An object of mass 10 kg is dropped from a great height. (a) If the object is subjected to

air resistance equivalent to 10 times the instantaneous velocity, find the object's velocity (b) What is the limiting velocity of the object?

10. Suppose that an object of mass 1 kg is thrown with a downward initial velocity of 5 m/s and is subjected to an air resistance equivalent to the instantaneous velocity. (a) Find the velocity of the object and the distance fallen at time t. (b) How far does the object drop after 5 s?

11. A projectile of mass 100 kg is launched vertically from ground level with an initial velocity of 100 m/s. (a) If the projectile is subjected to air resistance equivalent to 1/10 times the instantaneous velocity, determine the velocity of and the height of the projectile at any time t. (b) What is the maximum height attained by the projectile?

12. In a carnival game, a participant uses a mallet to project an object up a 20-m pole. If the mass of the object is 1 kg and the object is subjected to resistance equivalent to 1/10 times the instantaneous velocity, determine if an initial velocity of 20 m/s causes the object to reach the top of the pole.

13. Assuming that air resistance is ignored, find the velocity and height functions if an object with mass m is thrown vertically up into the air with an initial velocity v_0 from an initial height s_0.

14. Use the results of Exercise 13 to find the velocity and height functions if $m = 1/128$ slugs, $g = 32$ ft./s^2, $v_0 = 48$ ft./s, and $s_0 = 0$ ft. What is the shape of the height function? What is the maximum height attained by the object? When does the object reach its maximum height? When does the object hit the ground? How do these results compare to those of Example 3?

15. Consider the situation described in Exercise 13. What is the velocity of the object when it hits the ground, assuming that the object is thrown from ground level?

16. Consider the situation described in Exercise 13. If the object reaches its maximum height after T seconds, when does the object hit the ground, assuming that the object is thrown from ground level?

17. Suppose that an object of mass 10 kg is thrown vertically into the air with an initial velocity of v_0 m/s. If the limiting velocity is -19.6 m/s, what can be said about $c > 0$ (positive constant), in the force due to air resistance $F_R = cv$ acting on the object?

18. If the limiting velocity of an object of mass m that is thrown vertically into the air with an initial velocity v_0 m/s is -9.8 m/s, what can be said about the relationship between m and c in the force due to air resistance $F_R = cv$ acting on the object?

19. A parachutist weighing 192 lb falls from a plane (that is, $v_0 = 0$). When the parachutist's speed is 60 ft./s, the parachute is opened and the parachutist is then subjected to air resistance equivalent to $F_R = 3v^2$. Find the velocity $v(t)$ of the parachutist. What is the limiting velocity of the parachutist?

20. A parachutist weighing 60 kg falls from a plane (that is, $v_0 = 0$) and is subjected to air resistance equivalent to $F_R = 10v$. After one minute, the parachute is opened so that the parachutist is subjected to air resistance equivalent to $F_R = 100v$. (a) What is the parachutist's velocity when the parachute is opened? (b) What is the parachutist's velocity $v(t)$ after the parachute is opened? (c) What is the parachutist's limiting velocity? How does this compare to the limiting velocity if the parachute does not open?

21. (*Escape Velocity*) Suppose that a rocket is launched from the Earth's surface. At a great (radial) distance r from the center of the Earth, the rocket's acceleration is not the constant g. Instead, according to Newton's law of gravitation, $a = k/r^2$, where k is the constant of proportionality ($k > 0$ if the

rocket is falling toward the Earth; $k < 0$ if the rocket is moving away from the Earth). (a) If $a = -g$ at the Earth's surface (when $r = R$), find k and show that the rocket's velocity is found by solving the IVP $dv/dt = -gR^2/r^2$, $v(0) = v_0$. (b) Show that $dv/dt = v\,dv/dr$ so that the solution to the IVP $v\,dv/dr = -gR^2/r^2$, $v(R) = v_0$ is $v^2 = 2gR^2/r + v_0^2 - 2gR$. (c) If $v > 0$ (so that the rocket does not fall to the ground), show that the minimum value of v_0 for which this is true (even for very large values of r) is $v_0 = \sqrt{2gR}$. This value is called the *escape velocity* and signifies the minimum velocity required so that the rocket does not return to the Earth.

22. If $R \approx 3960\,\text{mile}$ and $g \approx 32\,\text{ft./s}^2$ $\approx 0.006\,\text{mile/s}^2$, use the results of Exercise 21 to compute the escape velocity of the Earth.

23. (*See Exercise 21*) Determine the minimum initial velocity needed to launch the lunar module (used on early space missions) from the surface of the Earth's moon given that the moon's radius is $R \approx 1080\,\text{mile}$ and the acceleration of gravity of the moon is 16.5% of that of the Earth.

24. (*See Exercise 21*) Compare the Earth's escape velocity to those of Venus and Mars if for Venus $R \approx 3800\,\text{mile}$ and acceleration of gravity is 85% of that of the Earth; and for Mars $R \approx 2100\,\text{mile}$ and acceleration of gravity is 38% of that of the Earth. Which planet has the largest escape velocity? Which has the smallest?

25. In an electric circuit with one loop that contains a resistor R, a capacitor C, and a voltage source $E(t)$, the charge Q on the capacitor is found by solving the IVP $R\,dQ/dt + Q/C = E(t)$, $Q(0) = Q_0$. Solve this problem to find $Q(t)$ if $E(t) = E_0$ where E_0 is constant. Find $I(t) = Q'(t)$ where $I(t)$ is the current at any time t.

26. If in the $R - C$ circuit described in Exercise 25 $C = 10^{-6}\,\text{F}$, $R = 4000\,\Omega$, and $E(t) = 200\,\text{V}$,

find $Q(t)$ and $I(t)$ if $Q(0) = 0$. What eventually happens to the charge and the current as $t \to \infty$?

27. An object that weighs 48 lb is released from rest at the top of a plane metal slide that is inclined $30°$ to the horizontal. Air resistance (lb) is numerically equal to one-half the velocity (ft./s), and the coefficient of friction is $\mu = 1/4$. Using Newton's Second Law of Motion by summing the forces along the surface of the slide, we find the following forces:

(a) the component of the weight parallel to the slide: $F_1 = 48 \sin 30° = 24$;

(b) the component of the weight perpendicular to the slide: $N = 48 \cos 30° = 24\sqrt{3}$;

(c) the frictional force (against the motion of the object): $F_2 = -\mu N = -\frac{1}{4} \times 24\sqrt{3} = -6\sqrt{3}$; and

(d) the force due to air resistance (against the motion of the object): $F_3 = -\frac{1}{2}v$. Because the mass of the object is $m = 48/32 = 3/2$, we find that the velocity of the object satisfies the IVP $m\,dv/dt = F_1 + F_2 + F_3$ or

$$\frac{3}{2}\frac{dv}{dt} = 24 - 6\sqrt{3} - \frac{1}{2}v, \quad v(0) = 0.$$

Solve this problem for $v(t)$. Determine the distance traveled by the object at time t, $x(t)$, if $x(0) = 0$.

28. A boat weighing 150 lb with a single rider weighing 170 lb is being towed in a particular direction at a rate of 20 mph. At $t = 0$, the tow rope is cut and the rider begins to row in the same direction, exerting a constant force of 12 lb in the direction that the boat is moving. The resistance is equivalent to twice the instantaneous velocity (ft./s). The forces acting on the boat are $F_1 = 12$ in the direction of motion and the force due to resistance in the opposite direction, $F_2 = -2v$. Because the total weight (boat

and rider) is 320 lb, $m = 320/32 = 10$. Therefore, the velocity satisfies the differential equation $m\,dv/dt = F_1 + F_2$ or $10\,dv/dt = 12 - 2v$ with initial velocity 20 mph, which is equivalent to

$$v(0) = 20\frac{\text{mile}}{\text{h}} \times \frac{5280\,\text{ft.}}{1\,\text{mile}} \times \frac{1\,\text{h}}{3600\,\text{s}} = \frac{88}{3}\,\text{ft./s.}$$

Find $v(t)$ and the distance traveled by the boat, $x(t)$, if $x(0) = 0$.

29. What is the equilibrium solution of $dv/dt = 32 - v$? How does this relate to the solution $v = 32 + Ce^{-t}$?

30. What is the equilibrium solution of $dv/dt = 32 - v^2$? How does this relate to the solution
$$v = (-4\sqrt{2}(Ce^{8t\sqrt{2}} + 1))/(1 - Ce^{8t\sqrt{2}})?$$

31. Find the equilibrium solution to $dv/dt = -g - (c/m)v$. What is the limiting velocity?

32. Find the equilibrium solution to $dv/dt = -g - (c/m)v^2$. What is the limiting velocity?

33. Suppose that a falling body is subjected to air resistance assumed to be $F_R = cv$. Use the values of $c = 0.5, 1$, and 2; and plot the velocity function with $m = 1, g = 32$, and $v_0 = 0$. How does the value of c affect the velocity?

34. Compare the effects that air resistance has on the velocity of a falling object of mass $m = 0.5$ that is released with an initial velocity of $v_0 = 16$. Consider $F_R = 16v^2$ and $F_R = 16v$.

35. Compare the effects that air resistance has on the velocity of a falling object of mass $m = 0.5$ that is released with an initial velocity of $v_0 = 0$. Consider $F_R = 16v^3$ and $F_R = 16\sqrt{v}$.

36. Consider the velocity and height functions found in Example 3.3.3 with $m = 1/128$, $c = 1/160$, and $g = 32$ ft./s^2.
    (a) Suppose that on the first toss, the object is thrown with $v_0 = 48$ ft./s from an initial height of $s_0 = 0$ ft. and on the

second toss with $v_0 = 36$ ft./s and $s_0 = 6$. On which toss does the object reach the greater maximum height?
    (b) If $s_0 = 0$ ft., compare the effect that the initial velocities $v_0 = 48$ ft./s, $v_0 = 64$ ft./s, and $v_0 = 80$ ft./s have on the height function.
    (c) If $v_0 = 48$ ft./s, compare the effect that the initial heights $s_0 = 0$ ft., $s_0 = 10$ ft., and $s_0 = 20$ ft. have on the height function.

37. A woman weighing 125 lb falls from an airplane at an altitude of 4000 ft. and opens her parachute after 5 s. If the force due to air resistance is $F_R = v$ before she opens her parachute and $F_R = 10v$ afterward, how long does it take for the woman to reach the ground?

38. Consider the problem discussed in Example 3.3.4. Instead of a buoyancy force equivalent to $6v$, suppose that when the object is in the pond, it is subjected to a buoyancy force equivalent to $6v^2$. Determine how much time is required for the object to reach a depth of 25 ft. in the pond.

CHAPTER 3 SUMMARY: ESSENTIAL CONCEPTS AND FORMULAS

Malthus model The IVP $dy/dt = ky$, $y(0) = y_0$ has the solution $y = y_0 e^{kt}$.

Logistic equation (or Verhulst equation): The IVP $dy/dt = (r - ay)y$, $y(0) = y_0$ has the solution $y = ry_0/(ay_0 + (r - ay_0)e^{-rt})$.

Newton's Law of Cooling: The IVP $dT/dt = k(T - T_s)$, $T(0) = T_0$ has the solution $T = (T_0 - T_s)e^{kt} + T_s$.

Newton's Second Law of Motion: The rate at which the momentum of a body changes with respect to time is equal to the resultant force acting on the body: $(d/dt)(mv) = F$.

The velocity of the falling body is found by solving the differential equation determined with $m(dv/dt) = \sum(\text{forces acting on the object})$.

CHAPTER 3 REVIEW EXERCISES

In Exercises 1-6, classify the equilibrium solutions.

1. $dy/dt = y(1 - 2y)$.
2. $dy/dt = -y$.
3. $dy/dt = -\frac{1}{4}y(y - 4)$.
4. $dy/dt = \frac{1}{8}y(y - 2)$.
5. $dy/dt = -y(1 - \frac{1}{2}y)(1 - \frac{1}{4}y)$.
6. $dy/dt = y(1 - \frac{1}{3}y)(1 - \frac{1}{6}y)$.
7. The initial population in a bacteria culture is y_0. Suppose that after 4 days the population is $3y_0$. Assuming exponential growth so that the Malthus model applies, when is the population $5y_0$?
8. Suppose that a bacteria culture contains 200 cells. After 1 day the culture contains 600 cells. Assuming exponential growth so that the Malthus model applies, how many cells does the culture contain after 2 days?
9. What percentage of the original amount of the element ^{226}Ra remains after 50 years? (Refer to Table 3.1.)
10. If an artifact contains 10% of the amount of ^{14}C as that of a present-day sample, how old is the artifact? (Refer to Table 3.1.)
11. Suppose that in an isolated population of 1000 people, 250 initially have a virus. If after 1 day, 500 have the virus, how many days are required for three-fourths of the population to acquire the virus? (Use the Verhulst equation.)
12. The *Gompertz equation* given by $dy/dt = y(r - a \ln y)$ is used by actuaries to predict trends of certain populations. If $y(0) = y_0$, then find $y(t)$. Find $\lim_{t\to\infty} y(t)$ if $a > 0$.

13. A bottle that contains water with a temperature of 40 °F is placed on a tennis court with temperature 90 °F. After 20 min, the water is 65 °F. What is the water's temperature after 30 min?
14. A can of diet cola at room temperature of 70 °F is placed in a cooler with temperature 40 °F. After 30 min, the can's temperature is at 60 °F. When is the temperature of the can at 45 °F?
15. A frozen turkey breast is placed in a microwave oven to defrost. It is then placed in a conventional oven at 325 °F and bakes for 1 h, at which time its temperature is 100 °F. If after baking an additional 45 min its temperature is 150 °F, what was the temperature of the turkey when it was removed from the microwave?
16. Suppose that during the month of July in Statesboro, Georgia, the outside temperature in °F is given by $C(t) = 85 - 10\cos(\pi t/12)$, $0 \le t \le 24$. Find the temperature in a parked car that has an initial temperature of 70 °F if $k = 1/4$. (Assume that the car has no heat or air conditioning system.)
17. A rock weighing 4 lb is dropped from rest from a large height and is subjected to air resistance equivalent to $F_R = v$. Find the velocity $v(t)$ of the rock at any time t. What is the velocity of the rock after 3 s? How far has the rock fallen after 3 s?
18. A container of waste weighing 6 lb is accidentally released from an airplane at an altitude of 1000 ft. with an initial velocity of 6 ft./s. If the container is subjected to air resistance equivalent to $F_R = 2v/3$, find the velocity $v(t)$ of the container at any time t. What is the velocity of the container after 5 s? How far has the container fallen after 5 s? Approximately when does the container hit the ground?
19. An object of mass 5 kg is thrown vertically in the air from ground level with an initial velocity of 40 m/s. If the object is subjected

to air resistance equivalent to $F_R = 5v$, find the velocity of the object at any time t. When does the object reach its maximum height? What is its maximum height?

20. A ball weighing 0.75 lb is thrown vertically in the air with an initial velocity of 20 ft./s. If the ball is subjected to air resistance equivalent to $F_R = v/64$, find the velocity of the object at any time t. When does the object reach its maximum height? What is its maximum height if it is thrown from an initial height of 5 ft.?

21. A parachutist weighing 128 lb falls from a plane ($v_0 = 0$). When the parachutist's speed is 30 ft./s, the parachute is opened and the parachutist is then subjected to air resistance equivalent to $F_R = 2v^2$. Find the velocity, $v(t)$, of the parachutist. What is the limiting velocity of the parachutist?

22. A relief package weighing 256 lb is dropped from a plane ($v_0 = 0$) over a war-ravaged area and is subjected to air resistance equivalent to $F_R = 16v$. After 2 s the parachute opens and the package is then subjected to air resistance equivalent to $F_R = 4v^2$. Find the velocity, $v(t)$, of the package. What is the limiting velocity of the package? Compare this to the limiting velocity if the parachute does not open.

23. Atomic waste is placed in sealed canisters and dumped in the ocean. It has been determined that the seal will not break and leak the waste when the canister hits the bottom of the ocean as long as the velocity of the canister is less than 12 m/s when it hits the bottom of the ocean. Using Newton's second law, the velocity satisfies the equation $m \, dv/dt = W - B - kv$, where $v(0) = 0$, W is the weight of the canister, B is the buoyancy force, and the drag is given by $-kv$. Solve this first-order linear equation for $v(t)$ and then integrate to find the position $y(t)$. If $W = 2254$ Newtons, $B = 2090$ Newtons, and $k = 0.637$ kg/s,

determine the time at which the velocity is 12 m/s. Determine the depth H of the ocean so that the seal will not break when the canister hits the bottom.

24. According to the *law of mass action*, if the temperature is constant, then the velocity of a chemical reaction is proportional to the product of the concentrations of the substances that are reacting. The chemical reaction $A + B \rightarrow M$ combines a moles per liter of substance A and b moles per liter of substance B. If $y(t)$ is the number of moles per liter that have reacted after time t, the reaction rate is $dy/dt = k(a - y)(b - y)$. Find $y(t)$ if $y(0) = 0$. Find $\lim_{t \to \infty} y(t)$ first if $a > b$ and then if $b > a$.

25. Solve the following equations for $r(\theta)$. Graph the polar equation that results.
    (a) $r \, dr/d\theta + 4 \sin 2\theta = 0$, $r(0) = 2$,
    (b) $dr/d\theta = r$, $r(1) = 0$,
    (c) $dr/d\theta = \theta$, $r(0) = 0$,
    (d) $dr/d\theta = 2 \sec \theta \tan \theta$, $r(0) = 4$,
    (e) $dr/d\theta = 6 \sin \theta/2 \cos \theta/2$, $r(0) = 0$.

The Italian scientist **Evangelista Torricelli** (1608-1647) made contributions in many areas of science. Not only did he make contributions to calculus and differential equations, also he discovered (or invented) the barometer.

26. A cylindrical tank 1.50 m high stands on its circular base of radius $r = 0.50$ m and is

initially filled with water. At the bottom of the tank, there is a hole of radius $r = 0.50$ cm that is opened at some instant so that draining starts due to gravity. According to *Torricelli's law*, $v = 0.600\sqrt{2gh}$, where $g = 980$ cm/s and h is the height of the water. By determining the rate at which the volume changes, we find that $dh/dt = -0.600A\sqrt{2gh}/B$, where A is the cross-sectional area of the outlet and B is the cross-sectional area of the tank. In this case, $A = 0.500^2\pi$ cm^2 and $B = 50.0^2\pi$ cm^2, so $dh/dt = -0.00266 - \sqrt{h}$. Find $h(t)$ if $h(0) = 150$.

27. (*Fishing*)[2] Consider a population of fish with population size at time t given by $x(t)$. Suppose the fish are harvested at a rate of $h(t)$. If the fish are sold at price p and δ is the interest rate, the present value P of the harvest is given by the improper integral,

$$P = \int_0^\infty e^{-\delta t}\left(p - \frac{c}{qx(t)}\right)h(t)dt, \quad (3.15)$$

where c and q are constants related to the cost of the effort of catching fish (q is called the *catchability*).

(a) Evaluate P if $\delta = 0.05$, $x(t) = 1$, and $h(t) = 1/2$. If we assume that the harvesting rate $h(t)$ is proportional to the population of the fish, then $h(t) = qEx(t)$, where E represents the effort in catching the fish, and, under certain assumptions, the size of the population of the fish $x(t)$ satisfies the differential equation

$$dx/dt = (r - ax)x - h \quad \text{or}$$
$$dx/dt = (r - ax)x - qEx.$$

This equation can be rewritten in the form

$$dx/dt = ((r - qE) - ax)x.$$

(b) Solve the equation $dx/dt = ((r - qE) - ax)x$ and find the solution that satisfies the initial condition $x(0) = x_0$. What is $h(t)$? Suppose that $x_0 = 1$, $r = 1$, and $a = 1/2$.

(c) Graph $x(t)$ if there is no harvesting for $0 \le t \le 10$. *Hint:* If there is no harvesting, $qE = 0$.

(d) Graph $x(t)$ and $h(t)$ for $0 \le t \le 20$ using $qE = 0, 0.1, 0.2, \ldots, 2$. What is the maximum sustainable harvest rate? In other words, what is the highest rate at which the fish can be harvested without becoming extinct? At what rate should the fish be harvested to produce the largest overall harvest? How does this result compare to (a)?

In 1965, the values of r, a, p, c, and q for the Antarctic whaling industry, were determined to be $r = 0.05$, $a = 1.25 \times 10^{-7}$, $P = 7000$, $c = 5000$, and $q = 1.3 \times 10^{-5}$. Assume that $t = 0$ corresponds to the year 1965 and that $x(0) = 78,000$. Assume that a typical firm expects a return of 10% on their investment, so that $\delta = 0.10$.

(e) Approximate the improper integral given by Equation (3.15) if (i) $E = 5000$ and (ii) $E = 7000$. What happens to the whale population in each case? What advice would you give to the whaling industry?

[2] J.N. Kapur, Some problems in biomathematics, International Journal of Mathematical Education in Science and Technology, 9(3), 1978, 287-306; and C.W. Clark, Bioeconomic Modeling and Resource Management, in: Applied Mathematical Ecology, S.A. Levin, T. Hallam, L.J. Goss (Eds.), Springer-Verlag, New York, 1980, pp. 11-57. For more information, see R.M. May, J.R. Beddington, C.W. Clark, S.J. Holt, R.M. Laws, Management of Multispecies Fisheries, Science, Vol. 205 (4403), July 20, 1979, pp. 267-277.

(f) Approximate the improper integral (3.15) using the values in the following table.

E	Approximation of $P = \int_0^\infty e^{\delta t}\left[p - \frac{c}{qx}\right]h(t)\,dt$
1000	
1500	
2000	
2500	
3000	
3500	
4000	
4500	

What value of E produces the maximum profit? What happens to the whale population in this case?

(g) Some reports have indicated that the optimal stock level of whales should be about 227,500. How does this number compare to the maximum number of whales that the environment can sustain? *Hint:* Evaluate $\lim_{t\to\infty} x(t)$ if there is no harvesting. How can the whaling industry make a profit and maintain this number of whales in the ocean?

28. *(Family of Orthogonal Trajectories)* Two curves C_1 and C_2 are orthogonal at a point of intersection if their respective tangent lines to the curves at that point are orthogonal. For example, consider the curves $y = x$ and $y = \sqrt{1 - x^2}$. (a) Show that the derivatives of these functions are $y' = 1$ and $y' = -x/\sqrt{1 - x^2}$, respectively. (b) Show that the curves are orthogonal at the point $(1/\sqrt{2}, 1/\sqrt{2})$ because the derivatives evaluated at this point are 1 and -1, respectively. Therefore, the tangent lines are perpendicular at $(1/\sqrt{2}, 1/\sqrt{2})$.

29. Determine the orthogonal trajectories of the given family of curves. (*Graph the orthogonal trajectories and curves simultaneously.*)

(a) $y + 2x = c$.
(b) $y = e^{cx}$.
(c) $y^2 = x^2 + c$.
(d) $y^2 = x^2 + cx$.

30. Given a family of curves F, we would like to find a set of curves that are orthogonal to each curve in F. We refer to this set of orthogonal curves as the *family of orthogonal trajectories*. Suppose that a family of curves is defined as $F(x, y) = c$ and that the slope of the tangent line at any point on these curves is $dy/dx = f(x, y)$ when it is obtained by differentiating $F(x, y) = c$ with respect to x and solving for dy/dx. Then the slope of the tangent line of the orthogonal trajectory is $dy/dx = -1/f(x, y)$. Therefore, the family of orthogonal trajectories is found by solving $dy/dx = -1/f(x, y)$. Find the family of orthogonal trajectories of the set of curves $y = cx^2$ by carrying out the following steps. (a) Show that the slope of the tangent line at any point on the parabola $y = cx^2$ is $dy/dx = 2cx$. (b) Solve $y = cx^2$ for c and substitute this result into $dy/dx = 2cx$ to obtain $dy/dx = 2y/x$. (c) Solve the equation $dy/dx = -1/f(x, y) = -x/(2y)$ to find that the family of orthogonal trajectories is $x^2/2 + y^2 = k$, $k > 0$, a family of ellipses.

31. Let $T(x, y)$ represent the temperature at the point (x, y). The curves given by $T(x, y) = c$, where c is a constant, are called *isotherms*. The orthogonal trajectories are curves along which heat flows. Show that the isotherms are $x^2/2 + xy - y^2/2 = c$ if the curves of heat flow are given by $y^2 + 2xy - x^2 = k$. *Hint:* On the heat flow curves,
$$dy/dx = (x - y)/(x + y), \text{ so we must solve}$$
$$dy/dx = (x + y)/(x - y), \text{ which is an exact}$$
differential equation.

32. A family of curves is *self-orthogonal* if the family of orthogonal trajectories is the same as the original family of curves. Is $y^2 - 2cx = c^2$ a self-orthogonal family of curves (parabolas)?

33. Find a value of c so that the two families of curves $y = k_1 x^2 + c$ and $x^2 + 2y^2 = k_2 + y$, where k_1 and k_2 are constants, are orthogonal.

34. Suppose that an electrical current is flowing in a wire along the z-axis. Then, the equipotential lines in the xy-plane are concentric circles centered at the origin. If the electric lines of force are the orthogonal trajectories of these circles, find the electric lines of force.

35. The path along which a fluid particle flows is called a *streamline*, and the orthogonal trajectories are called *equipotential lines*. If the streamlines are $y = k/x$, find the equipotential lines.

36. (*Oblique Trajectories*) Let $\mathcal{L}_1$ and $\mathcal{L}_2$ denote two lines, not perpendicular to each other, with slopes m_1 and m_2, respectively; and let θ denote the angle between them as shown in Figure 3.12(a). Then,
$\tan\theta = (m_2 - m_1)/(1 + m_2 m_1)$.

   (a) Show that
   $m_2 = (m_1 + \tan\theta)/(1 - m_1 \tan\theta)$.

   (b) Suppose we are given a family of curves that satisfies the differential equation $dy/dx = f(x,y)$. Use (a) to show that if we want to find a family of curves that intersects this family at a constant angle θ, we must solve the differential equation $dy/dx = (f(x,y) \pm \tan\theta)/(1 \mp f(x,y)\tan\theta)$. *Hint:* Studying Figure 3.12(b) might help.

   (c) Find a family of curves that intersects the family of curves $x^2 - y^2 = c^2$ at an angle of $\pi/4$. Graph several members of both families to confirm your result.

   (d) Find a family of curves that intersects the family of curves $x^2 + y^2 = c^2$ at an angle of $\pi/6$. Graph several members of both families to confirm your result.

37. (a) Determine the orthogonal trajectories of the family of curves $y^2 = 2cx + 2c^2$.

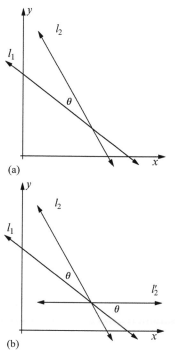

FIGURE 3.12 (a) Calculating the angle of intersection of two non-parallel lines. (b) Drawing the parallel can help compute the angle.

   (b) Graph several members of both families of curves on the same set of axes. (c) What is your reaction to the graphs?

38. (*Barometric Formula*) Consider a point in the atmosphere. The change in pressure P associated with a small change in height h at that point can be determined in terms of the weight of the air. This yields $\Delta P = -\rho g A \Delta h / A = -\rho g \Delta h$ where A represents the area of the face of cross-section of the atmosphere with thickness Δh and ρ the density. ($\rho = m/V = n N_A m / nRT / P$ where m is the mass of one molecule, V is the volume. T is the temperature, R is the gas constant, and N_A is Avogadro's number.) If $k = R/N_A$, then $\rho = \frac{mP}{kT}$ where k is Boltzmann's

constant. Therefore, $(\Delta P/\Delta h) = -mgP/kT$ so that $(dP/dh) = -mg/kTP$. (Notice that $(dP/dt) < 0$ because P decreases as h increases.) Assuming that $P(0) = P_0$, find $P(h)$. What is the limiting value of P as $h \to \infty$?

39. (*Stress in a Bar with a Non-constant Cross section*) Suppose that a bar has circular cross-sections, is supported at the top and has a load L. The bar has length p and density ρ, and it is assumed that the stress at every point in the bar is the constant σ.
    (a) Find the formula for the volume of the bar at a height x from the base of the bar. Also calculate the weight of this section of the bar.
    (b) Find a formula for the stress in the bar at height x.
    (c) Show that the ODE that describes this situation is $\rho\pi y^2 = \sigma\pi 2y\, dy/dx$ or $(2/y)(dy/dx) = \rho/\sigma$. Solve this ODE if $y(0) = 1$.

40. (*Selling Smartphone Apps*)
    A company is trying to sell a smartphone app to a potential audience of P customers through a marketing campaign. If approximately $1/4$ of the potential audience knows about the app after 6 months, how many will know about the app after 1 year?

41. (*Chemical Reaction*) Suppose that in a chemical reaction Substance X is converted to substance Y at a rate proportional to the square of the amount of substance X present in the reaction. Determine the ODE that models this situation and then find X(t), the amount of Substance X at time t, if initially there are $100\,g$ of Substance X and only $60\,g$ remain after $1\,h$. How much remains after $3\,h$?

DIFFERENTIAL EQUATIONS AT WORK

A. Mathematics of Finance

Suppose that P dollars are invested in an account at an annual rate of r% compounded continuously. To find the balance of the account $x(t)$ at time t, we must solve the IVP

$$dx/dt = rx, \quad x(0) = P, \quad \text{for } x = x(t).$$

1. Show that $x(t) = Pe^{rt}$.
2. If \$1000 is deposited into an account with an annual interest rate of 8% compounded continuously, what is the balance of the account at the end of 5, 10, 15, and 20 years?

    If we allow additions or subtractions of sums of money from the account, the problem becomes more complicated. Suppose that an account like a savings account, home mortgage loan, student loan or car loan, has an initial balance of P dollars and r denotes the interest rate per compounding period. Let $p(t)$ denote the money flow per unit time and $\delta = \ln(1 + r)$. Then the balance of the account at time t, $x(t)$, satisfies the IVP[3]

$$\frac{dx}{dt} - \delta x = p(t), \quad x(0) = P.$$

3. Show that the balance of the account at time t_0 is given by

$$x(t_0) = Pe^{\delta t_0} + e^{\delta t_0} \int_0^{t_0} p(t)e^{-\delta t}\, dt$$

4. Suppose that the initial balance of a student loan is \$12,000 and that monthly payments are made in the amount of \$130. If the annual interest rate, compounded

[3] T.C.I. Kotiah, Difference and differential equations in the mathematics of finance, International Journal of Mathematics in Education, Science, and Technology, **22**(5), 1991, 783-789.
E.W. Herold, Inflation Mathematics for the Professional, in: Mathematical Modeling: Classroom Notes in Applied Mathematics, M.S. Klamkin (Ed.), SIAM, 1987, pp. 206-209.

monthly, is 9%, then $P = 12,000$,
$p(t) = -130, r = 0.09/12 = 0.0075$, and
$\delta = \ln(I + r) = \ln(1.0075)$.

(a) Show that the balance of the loan at time t, in months, is given by

$$x(t) = 17398.3 - 5398.25 \times 1.0075^t.$$

(b) Graph $x(t)$ on the interval $[0, 180]$, corresponding to the loan balance for the first 15 years.

(c) How long will it take to pay off the loan?

5. Suppose that the initial balance of a home mortgage loan is $80,000 and that monthly payments are made in the amount of $599. If the annual interest rate, compounded monthly, is 8%, how long will it take to pay off the mortgage loan?

6. Suppose that the initial balance of a home mortgage loan is $80,000 and that monthly payments are made in the amount of $599. If the annual interest rate, compounded monthly, is 8%, how long will it take to pay off the mortgage loan if the monthly payment is increased at an annual rate of 3%, which corresponds to a monthly increase of 1/4%?

7. If an investor invests $250 per month in an account paying an annual interest rate of 10%, compounded monthly, how much will the investor have accumulated at the end of 10, 20, and 30 years?

8. Suppose an investor begins by investing $250 per month in an account paying an annual interest rate of 10%, compounded monthly, and in addition increases the amount invested by 6% each year for an increase of 1/2% each month. How much will the investor have accumulated at the end of 10, 20, and 30 years?

9. Suppose that a 25-year-old investor begins investing $250 per month in an account paying an annual interest rate of 10%, compounded monthly. If at the age of 35 the investor stops making monthly payments, what will be the account balance when the

investor reaches 45, 55, and 65 years of age? Suppose that a 35-year-old friend begins investing $250 per month in an account paying an annual interest rate of 10%, compounded monthly, at the same time the first investor stops. Who has a larger investment at the age of 65?

10. If you are given a choice between saving $150 a month beginning when you first start working and continuing until you retire, or saving $300 per month beginning 10 years after you first start working and continuing until you retire, which should you do to help ensure a financially secure retirement?

From Exercises 8 and 9, we see that consistent savings beginning at an early age can help assure that a large sum of money will accumulate by the age of retirement. How much money does a person need to have accumulated to help ensure a financially secure retirement? For example, corporate pension plans and social security generally provide a relatively small portion of living expenses, especially for those with above average incomes. In addition, inflation erodes the buying power of the dollar; large sums of money today will not necessarily be large sums of money tomorrow.

As an illustration, we see that if inflation were to average a modest 3% per year for the next 20 years and a person has living expenses of $20,000 annually, then after 20 years the person would have living expenses of $20,000 \times 1.03^{20} \approx \$36,122$.

Let t denote years and suppose that a person's after-tax income as a function of years is given by $I(t)$ and $E(t)$ represents living expenses. Here $t = 0$ might represent the year a person enters the work force. Generally, during working years, $I(t) > E(t)$; during retirement years, when $I(t)$ represents income from sources such as corporate pension plans and social security, $I(t) < E(t)$.

Suppose that an account has an initial balance of S_0 and the after-tax return on the account is $r\%$. We assume that the amount deposited into the account each year is $I(t) - E(t)$. What is the balance of the account at year t, $S(t)$? S must satisfy the IVP

$$\frac{dS}{dt} = rS(t) + I(t) - E(t), \quad S(0) = S_0.$$

11. Show that the balance of the account at time $t = t_0$ is

$$S(t_0) = e^{rt_0} \left(S_0 + \int_0^{t_0} (I(t) - E(t)) e^{-rt} \, dt \right).$$

Assume that inflation averages an annual rate of i. Then, in terms of $E(0)$, $E(t) = E(0)e^{it}$. Similarly, during working years we assume that annual raises are received at an annual rate of j. Then, in terms of $I(0)$, $I(t) = I(0)e^{jt}$. However, during retirement years, we assume that $I(t)$ is given by a fixed sum F, such as a corporate pension or annuity, and a portion indexed to the inflation rate V, such as social security. Thus, during retirement years

$$I(t) = F + Ve^{i(t-T)},$$

where T denotes the number of working years. Therefore,

$$I(t) = \begin{cases} I(0)e^{jt}, & 0 \le t \le T \\ F + Ve^{i(t-T)}, & t > T \end{cases}.$$

12. Suppose that a person has an initial income of $I(0) = 20,000$ and receives annual average raises of 5%, so that $j = 0.05$ and initial living expenses are $E(0) = 18,000$. Further, we assume that inflation averages 3%, so that $i = 0.03$, while the after tax return on the investment is 6%, so that $r = 0.06$. Upon retirement after T years of work, we assume that the person receives a fixed pension equal to 20% of his living expenses at that time, so that

$$F = 0.2 \times 18,000 \times e^{0.03T},$$

while social security provides 30% of his living expenses at that time, so that

$$V = 0.3 \times 18,000 \times e^{0.03T}.$$

(a) Find the smallest value of T so that the balance in the account is zero after 30 years of retirement. Sketch a graph of S for this value of T.

(b) Find the smallest value of T so that the balance in the account is never zero. Sketch a graph of S for this value of T.

13. What is the relationship between the results you obtained in Exercises 9 and 10 and that obtained in Exercise 12?

14. (a) How would you advise a person of 22 years of age first entering the work force to prepare for a financially secure retirement?

(b) How would you advise a person of 50 years of age with no savings who hopes to retire at 65 years of age?

(c) When should you start saving for retirement?

B. Algae Growth

When wading in a river or stream, you may notice that microorganisms like algae are frequently found on rocks. Similarly, if you have a swimming pool, you may notice that without maintaining appropriate levels of chlorine and algaecides, small patches of algae take over the pool surface, sometimes overnight. Underwater surfaces are attractive environments for microorganisms because water movement removes wastes and provides a continuous supply of nutrients. The organisms, however, must spread over the surface without being washed away. If conditions become unfavorable, they must be able to free themselves from the surface and recolonize on a new surface.

The rate at which cells accumulate on a surface is proportional to the rate of growth of the cells and the rate at which the cells attach to the surface. An equation describing this situation is given by

$$\frac{dN}{dt} = \mu(N + A),$$

where N represents the cell density, μ the growth rate, A the attachment rate, and t the time.[4]

1. If the attachment rate A is constant, solve the IVP

$$\frac{dN}{dt} = \mu(N + A), \quad N(0) = 0$$

for N and then solve the result for μ.

2. In a colony of cells, it was observed that $A = 3$. The number of cells N at the end of t hours is shown in the following table. Estimate the growth rate at the end of each hour.

t	N	μ
1	3	
2	9	
3	21	
4	45	

3. Using the growth rate obtained in Problem 2, estimate the number of cells at the end of 24 and 36 h.

C. Dialysis

The primary purpose of the kidney is to remove waste products like urea, creatinine, and excess fluid from blood. When the kidneys are not working properly, wastes accumulate in the blood; when toxic levels are reached, death is certain. The leading causes of chronic kidney failure in the United States are hypertension (high blood pressure) and diabetes mellitus. In fact, one-fourth of all patients requiring *kidney dialysis* have diabetes. Fortunately, kidney dialysis removes waste products from the blood of patients with improperly working kidneys. During the hemodialysis process (see Figure 3.13), the patient's blood is pumped through a *dialyzer*, usually at a rate of 1-3 dL/min. The patient's blood is separated from the "cleaning fluid" by a semipermeable membrane, which permits wastes (but not blood cells) to diffuse to the cleaning fluid. The cleaning fluid contains some substances beneficial to the body which diffuse to the blood. The cleaning fluid, called the *dialyzate*, is flowing in the opposite direction as the blood, usually at a rate of 2-6 dL/min. Waste products from the blood diffuse to the dialyzate through the membrane at a rate proportional to the difference in concentration of the waste products in the blood and dialyzate. If we let $u(x)$ represent the concentration of wastes in blood and $v(x)$ represent the concentration of wastes in the dialyzate, where x is the distance along the dialyzer, Q_{Dy} is the flow rate of the dialyzate through the machine, and Q_B is the flow rate of the blood through the machine, then

$$Q_B u' = -k(u - v),$$
$$-Q_D v' = k(u - v),$$

where k is the proportionality constant.[5]

We let L denote the length of the dialyzer, and the initial concentration of wastes in the blood is $u(0) = u_0$, while the initial concentration of wastes in the dialyzate is $v(L) = 0$. Then, we must solve the IVP

---

[4] D.E. Caldwell, Microbial Colonization of Solid-Liquid Interfaces, in: Biochemical Engineering V, Annals of the New York Academy of Sciences, Vol. 56, New York Academy of Sciences, 1987, pp. 274-280.

[5] D.N. Burghess, M.S. Borrie, Modeling with Differential Equations, Ellis Horwood Limited, pp. 41-45. J.M. Black, E. Matassarin-Jacobs, Luckman and Sorensen's Medical-Surgical Nursing: A Psychophysiologic Approach, fourth ed., W.B. Saunders Company, 1993, pp. 1509-1519, 1775-1808.

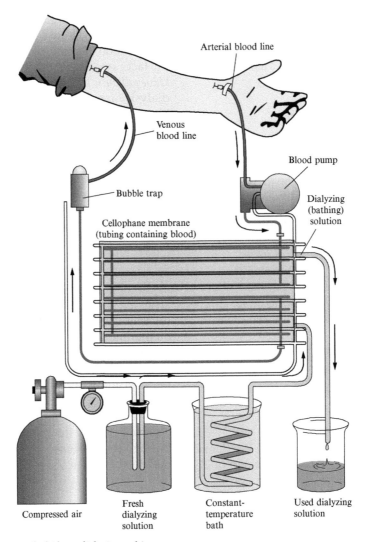

FIGURE 3.13   Diagram of a kidney dialysis machine.

$$Q_B u' = -k(u - v),$$
$$-Q_D v' = k(u - v),$$
$$u(0) = u_0, \quad v(L) = 0.$$

is

$$u(0) = u_0, \quad v(L) = 0.$$

1. Show that the solution to

$$Q_B u' = -k(u - v),$$
$$-Q_D v' = k(u - v),$$

$$u(x) = u_0 \frac{Q_B e^{\alpha x} - Q_D e^{\alpha L}}{e^{\alpha x} \left(Q_B - Q_D e^{\alpha L}\right)}$$

$$v(x) = u_0 \frac{e^{\alpha L} - e^{\alpha x}}{e^{\alpha x} \frac{Q_D}{Q_B} \left(e^{\alpha L} - 1\right)}$$

where $\alpha = k/Q_B - k/Q_D$. (*Hint:* First add the equations $Q_B u' = -k(u - v)$ and $-Q_D v' = k(u - v)$ to obtain the linear equation

$$\frac{d}{dx}(u - v) = -\frac{k}{Q_B}(u - v) + \frac{k}{Q_D}(u - v),$$

then let $z = u - v$. Next solve for z and subsequently solve for u and v.)

In healthy adults, typical urea nitrogen levels are 11-23 mg/dL (1 dL = 100 mL), while serum creatinine levels range from 0.6-1.2 mg/dL, and the total volume of blood is 4-5 L (1 L = 1000 mL).

2. Suppose that hemodialysis is performed on a patient with urea nitrogen level of 34 mg/dL and serum creatinine level of 1.8 using a dialyzer with $k = 2.25$ and $L = 1$. If the flow rate of blood, Q_B, is 2 dL/min while the flow rate of the dialyzate, Q_D, is 4 dL/min, will the level of wastes in the patient's blood reach normal levels after dialysis is performed? For what waste levels would dialysis have to be performed twice?

3. The *amount of waste removed* is given by $\int_0^L k\,(u(x) - v(x))\,dx$. Show that

$$\int_0^L k\,(u(x) - v(x))\,dx = Q_B\,(u_0 - u(L)).$$

4. The *clearance of a dialyzer*, CL, is given by $CL = (Q_B/u_0)(u_0 - u(L))$. Use the solution obtained in Problem 1 to show that

$$CL = Q_B \frac{1 - e^{-\alpha L}}{1 - \dfrac{Q_B}{Q_D}e^{-\alpha L}}.$$

Typically, hemodialysis is performed 3-4 h at a time three or four times per week. In some cases, a kidney transplant can free patients from the restrictions of dialysis. Of course, transplants have other risks not necessarily faced by those on dialysis; the number of available kidneys also affects the number of transplants performed. For example, in 1991 over 130,000 patients were on dialysis while only 7000 kidney transplants had been performed.

D. Antibiotic Production

When you are injured or sick, your doctor may prescribe antibiotics to prevent or cure infections. In the journal article "Changes in the Protein Profile of *Streptomyces griseus* during a Cycloheximide Fermentation," we see that production of the antibiotic cycloheximide by *Streptomyces* is typical of antibiotic production. During the production of cycloheximide, the mass of Streptomyces grows relatively quickly and produces little cycloheximide. After approximately 24 h, the mass of *Streptomyces* remains relatively constant and cycloheximide accumulates. However, once the level of cyclo-heximide reaches a certain level, extracellular cycloheximide is degraded (feedback inhibited). One approach to alleviating this problem and to maximize cycloheximide production is to remove extracellular cycloheximide continu-ously. The rate of growth of *Streptomyces* can be described by the following equation:

$$\frac{dX}{dt} = \mu_{max}\left(1 - \frac{1}{X_{max}}X\right)X,$$

where X represents the mass concentration in g/L, μ_{max} is the maximum specific growth rate, and X_{max} represents the maximum mass concen-tration.[6]

1. Find the solution to the IVP

$$\frac{dX}{dt} = \mu_{max}\left(1 - \frac{1}{X_{max}}X\right)X, \quad X(0) = 1$$

[6] K.H. Dykstra, H.Y. Wang, Changes in the Protein Profile of *Streptomyces griseus* during a Cycloheximide Fermentation, in: Biochemical Engineering V, Annals of the New York Academy of Sciences, Vol. 56, New York Academy of Sciences, 1987, pp. 511-522.

by first converting the equation
$(dX/dt) = \mu_{max}(1 - (1/X_{max})X)X$ to a linear
equation with the substitution $y = X^{-1}$.

2. Experimental results have shown that
$\mu_{max} = 0.3h^{-1}$ and $X_{max} = 10\,g/L$. Substitute
these values into the result obtained in
Problem 1. (a) Graph $X(t)$ on the interval
$[0, 24]$. (b) Find the mass concentration at the
end of 4, 8, 12, 16, 20, and 24 h.

The rate of accumulation of cycloheximide
is the difference between the rate of synthesis
and the rate of degradation:

$$\frac{dP}{dt} = R_s - R_d.$$

It is known that $R_d = K_d P$, where
$K_d \approx 5 \times 10^{-3}h^{-1}$, so $dP/dt = R_s - R_d$ is
equivalent to $dP/dt = R_s - K_d P$.
Furthermore,

$$R_s = Q_{po}EX\left(1 + \frac{P}{K_I}\right)^{-1},$$

where Q_{po} represents the specific enzyme
activity with value $Q_{po} \approx 0.6$ and K_I
represents the inhibition constant. E
represents the intracellular concentration of
an enzyme, which we will assume is
constant. For large values of K_I and t,
$X(t) \approx 10$ and $(1 + P/K_I)^{-1} \approx 1$. Thus,
$R_s \approx 10Q_{po}E$, so

$$\frac{dP}{dt} = 10Q_{po}E - K_d P.$$

3. Solve the IVP

$$\frac{dP}{dt} = 10Q_{po}E - K_d P, \quad p(24) = 0.$$

4. Graph $(1/E)P(t)$ on the interval $[24, 1000]$.

5. What happens to the net accumulation of the
antibiotic as time increases?

6. If instead the antibiotic is removed from the
solution so that no degradation occurs, what
happens to the net accumulation of the
antibiotic as time increases?

In Chapters 2 and 3, we saw that first-order differential equations can be used to model a variety of physical situations. However, many physical situations need to be modeled by higher order differential equations. For example, in 1735, Daniel Bernoulli's (1700-1782) study of the vibration of an elastic beam led to the fourth-order differential equation

$$k^4 \frac{\mathrm{d}^4 y}{\mathrm{d}x^4} = y,$$

Introductory Differential Equations
http://dx.doi.org/10.1016/B978-0-12-417219-7.00004-1

which describes the displacement of the *simple modes*. This equation can be rewritten in the form

$$y - k^4 \frac{d^4 y}{dx^4} = 0.$$

Both Bernoulli and Euler realized that a solution to the equation is $y = e^{x/k}$ but that other solutions to the equation must also exist (see Exercise 71 in "Chapter 4 Review Exercises" section).

Daniel Bernoulli's (1700-1782) study of the vibrations of an elastic beam probably occurred while working with Euler at St. Petersburg, Russia, around 1728. According to O'Connor and Robertson [19], "While in St. Petersburg he made one of his most famous discoveries when he defined the simple modes and the frequencies of oscillation of a system. He showed that the movements of strings of musical instruments are composed of an infinite number of harmonic vibrations all superimposed on the string."

4.1 SECOND-ORDER EQUATIONS: AN INTRODUCTION

- THE GENERAL CASE
- REDUCTION OF ORDER

In this chapter, we turn our attention on solving ordinary differential equations of order 2 or higher. We focus on linear differential equations because there are many instances in which we encounter equations of this type. For example, we see them in the study of vibrations and sound, and as we mentioned in Chapter 1, there is a relationship between a system of two first-order differential equations and a second-order differential equation. We'll discuss nonlinear equations more in some of the exercises in this chapter and at the end of Chapter 8, where we discuss how rewriting a nonlinear equation as a system of first-order nonlinear equations can help us understand the underlying behavior and structure of the system.

Let's begin exploring higher order ordinary differential equations by considering the second-order equation $y'' + y = 0$ or $y'' = -y$. To find a solution, we must determine a function with the property that the second derivative of the function is the negative of the function itself. When we consider functions familiar to us, such as polynomials, trigonometric functions, exponential functions, and natural logarithms, we can conclude that $y_1(t) = \sin t$ is a solution because

$$y_1' = \cos t \quad \text{and} \quad y_1'' = -\sin t; \quad y_1'' = -y_1.$$

Similarly, $y_2(t) = \cos t$ is a solution because

$$y_2' = -\sin t \quad \text{and} \quad y_2'' = -\cos t; \quad y_2'' = -y_2.$$

Of course, we are not always able to determine solutions of a differential equation by inspection as we have in this case. Therefore, we now investigate a more general approach to solving second-order equations. We begin this discussion by considering a problem studied in physics.

The Second-Order Linear Homogeneous Equation with Constant Coefficients

Suppose that we attach an object to the end of a spring that is mounted to a horizontal rod (see Figure 4.1). The object comes to rest. We call this the equilibrium position. Let $x(t)$ represent the displacement of the object from equilibrium.

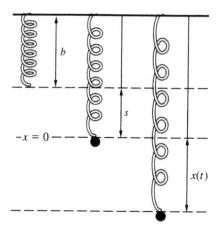

FIGURE 4.1   A spring-mass system.

If $x(t) > 0$, then the object is below equilibrium; if $x(t) < 0$, it is above equilibrium. If we assume that there is a damping force acting on the object that impedes motion, then we can find $x(t)$ by solving a second-order linear differential equation such as

$$\frac{d^2x}{dt^2} + 3\frac{dx}{dt} + 2x = 0.$$

This is a *second-order linear ordinary differential equation* because it has the form

$$a_2(t)\frac{d^2x}{dt^2} + a_1(t)\frac{dx}{dt} + a_0(t)x = f(t),$$

where $a_2(t)$ is not identically the zero function and the highest order derivative in the equation is of order 2. Of course in this equation, $a_2(t) = 1$, $a_1(t) = 3$, and $a_0(t) = 2$, so the coefficient functions are constant. In addition, $f(t) = 0$ for all t, so this equation is *homogeneous*. (We give more details on this spring-mass problem in Chapter 5.) For now, we attempt to find a solution of this differential equation by assuming that solutions have the form $x(t) = e^{rt}$, where r is a constant to be determined, because e^{rt} and its derivatives are constant multiples of one another. We find the value(s) of r that lead to a solution by substituting $x(t) = e^{rt}$ into the differential equation. Notice that $dx/dt = re^{rt}$ and $d^2x/dt^2 = r^2e^{rt}$.

Substitution into $d^2x/dt^2 + 3(dx/dt) + 2x = 0$ then gives us $r^2e^{rt} + 3re^{3r} + 2e^{rt} = 0$ or

$$e^{rt}(r^2 + 3r + 2) = 0.$$

This algebraic equation tells us that either $e^{rt} = 0$ or $r^2 + 3r + 2 = 0$. We know that there are no values of t so that $e^{rt} = 0$. Therefore, we solve the *characteristic equation*

$$r^2 + 3r + 2 = 0$$

to find the values of r that satisfy the equation. In the case of a second-order equation, the characteristic equation is quadratic, so we can either factor or use the quadratic formula. Factoring yields $(r + 1)(r + 2) = 0$, so $r = -1$ or $r = -2$. This means that $x_1(t) = e^{-t}$ and $x_2(t) = e^{-2t}$ each satisfy the differential equation. We verify this by substituting these two functions into the differential equation. If $x_1(t) = e^{-t}$, then $dx_1/dt = -e^{-t}$ and $d^2x_1/dt^2 = e^{-t}$. Consequently,

$$\frac{d^2x_1}{dt^2} + 3\frac{dx_1}{dt} + 2x_1 =$$
$$e^{-t} + 3\left(-e^{-t}\right) + 2e^{-t} = 0.$$

Both $x_1'' + 3x_1' + 2x_1 = 0$ and $x_2'' + 3x_2' + 2x_2 = 0$ because x_1 and x_2 are solutions of the linear homogeneous equation.

Because simplification of the left-hand side of the differential equation yields zero (the right-hand side of the equation), $x_1(t) = e^{-t}$ satisfies the differential equation. In a similar manner, we verify that $x_2(t) = e^{-2t}$ also satisfies the differential equation. Notice that these two functions are not constant multiples of one another, so we say that $x_1(t) = e^{-t}$ and $x_2(t) = e^{-2t}$ are *linearly independent*. Notice also that for arbitrary constants c_1 and c_2, the function $x(t) = c_1x_1(t) + c_2x_2(t) = c_1e^{-t} + c_2e^{-2t}$, called a *linear combination* of $x_1(t) = e^{-t}$ and $x_2(t) = e^{-2t}$, is also a solution of the differential equation. We verify this through substitution as well. If $x(t) = c_1e^{-t} + c_2e^{-2t}$, then $dx/dt = -c_1e^{-t} - 2c_2e^{-2t}$ and $d^2x/dt^2 = c_1e^{-t} + 4c_2e^{-2t}$. Therefore,

$$\frac{d^2x}{dt^2} + 3\frac{dx}{dt} + 2x = \left(c_1e^{-t} + 4c_2e^{-2t}\right)$$
$$+ 3\left(-c_1e^{-t} - 2c_2e^{-2t}\right) + 2\left(c_1e^{-t} + c_2e^{-2t}\right)$$
$$= c_1 \underbrace{\left(e^{-t} - 3e^{-t} + 2e^{-t}\right)}_{x_1'' + 3x_1' + 2x_1 = 0}$$
$$+ c_2 \underbrace{\left(4e^{-2t} - 6e^{-2t} + 2e^{-2t}\right)}_{x_2'' + 3x_2' + 2x_2 = 0}$$
$$= 0.$$

This calculation illustrates an important property of linear homogeneous equations that is called the *Principle of Superposition*: if two functions are solutions of a linear homogeneous differential equation, then any linear combination of the two functions is also a solution of the differential equation.

Theorem 11 (Principle of Superposition). *If $\phi_1(t)$ and $\phi_2(t)$ are solutions of $a_2(t)y'' + a_1(t)y' + a_0(t)y = 0$, then for all values of the constants c_1 and c_2, $\phi(t) = c_1\phi_1(t) + c_2\phi_2(t)$ is also a solution of $a_2(t)y'' + a_1(t)y' + a_0(t)y = 0$.*

You should learn the *Principle of Superposition*: "Every linear combination of solutions to a linear homogeneous equation is also a solution."

Proof. When we say that "$\phi_i(t)$, $i = 1, 2$, is a solution of $a_2(t)y'' + a_1(t)y' + a_0(t)y = 0$" we mean that $a_2(t)\phi_i'' + a_1(t)\phi_i' + a_0(t)\phi_i$ simplifies to identically the zero function: $a_2(t)\phi_i'' + a_1(t)\phi_i' + a_0(t)\phi_i = 0$.

Assume that c_1 and c_2 are constants and that $\phi(t) = c_1\phi_1(t) + c_2\phi_2(t)$. Then,

$$a_2(t)\phi'' + a_1(t)\phi' + a_0(t)\phi = a_2(t)(c_1\phi_1(t)$$
$$+ c_2\phi_2(t))'' + a_1(t)(c_1\phi_1(t) + c_2\phi_2(t))$$
$$+ a_0(t)(c_1\phi_1(t) + c_2\phi_2(t))$$

$$= c_1 \underbrace{\left(a_2(t)\phi_1'' + a_1(t)\phi_1' + a_0(t)\phi_1\right)}_{\text{equals 0 because } \phi_1 \text{ is a solution of the homogeneous equation}}$$
$$+ c_2 \underbrace{\left(a_2(t)\phi_2'' + a_1(t)\phi_2' + a_0(t)\phi_2\right)}_{\text{equals 0 because } \phi_2 \text{ is a solution of the homogeneous equation}}$$
$$= 0.$$

In the specific case of the initial example, $x_1(t) = e^{-t}$ and $x_2(t) = e^{-2t}$ are two linearly independent solutions of the second-order linear homogeneous equation and so by the Principle of Superposition $x(t) = c_1e^{-t} + c_2e^{-2t}$ is also a solution and we will learn that it is a *general solution* of this second-order differential equation because all solutions of initial-value problems (IVPs) for this differential equation are obtained from it. In Figure 4.2(a), we graph this solution for several different choices of c_1 and c_2.

If we are given additional information about the spring-mass system such as the initial displacement and the initial velocity of the object, then we have a second-order IVP. For example, suppose that the object is released from equilibrium ($x = 0$) $x(0) = 0$ with initial velocity $dx/dt(0) = -1$. Therefore, we need to solve

$$x'' + 3x' + 2x = 0, \quad x(0) = 0, \quad x'(0) = -1.$$

After finding a general solution of the differential equation, we apply the two initial conditions. In this case, we have $x(0) = c_1 + c_2$ and $dx/dt(0) = -c_1 - 2c_2$, so we solve the system $c_1 + c_2 = 0$, $-c_1 - 2c_2 = -1$ for c_1 and c_2.

Adding these equations, we obtain $-c_2 = -1$, so $c_2 = 1$. Substituting this value into $c_1 + c_2 = 0$ then indicates that $c_1 = -1$. Therefore, the unique solution to this IVP is $x(t) = e^{-2t} - e^{-t}$, which we graph in Figure 4.2(b). Notice that the curve passes through $(0, 0)$ because of the initial condition $x(0) = 0$. The curve then moves in the downward direction because $x'(0) = -1$. Notice also that $\lim_{t \to \infty} x(t) = 0$ (as with any solution, which is indicated in

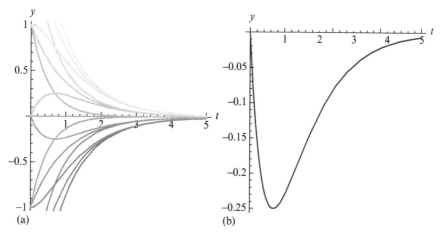

FIGURE 4.2 (a) Graph of general solution for various values of c_1 and c_2. (b) Graph of the solution $x(t) = e^{-2t} - e^{-t}$ to the IVP.

Figure 4.2(a)). This means that the object eventually comes to rest.

In the previous problem, a general solution is $x(t) = c_1 e^{-t} + c_2 e^{-2t}$. Notice that if we select $c_1 = c_2 = 0$, we obtain the *trivial solution* $x(t) = 0$. This shows us that the trivial solution always satisfies a linear homogeneous differential equation. Therefore, when solving equations, we generally search for *nontrivial solutions*.

The General Case

Now that we have discussed a representative problem considered in this chapter, we can state more precisely several theorems and definitions that were mentioned earlier. Let's begin with the definition of linear dependence and linear independence.

Definition 6 (Linearly Dependent). *Let* $S = \{f_1(t), f_2(t)\}$ *be a set of two functions. The set S is **linearly dependent** on an interval I if there are constants c_1 and c_2 not both zero, so that*

$$c_1 f_1(t) + c_2 f_2(t) = 0$$

*for every value of t in the interval I. The set S is **linearly independent** if S is not linearly dependent.*

We can rewrite $c_1 f_1(t) + c_2 f_2(t) = 0$ as $f_2(t) = -(c_1/c_2) f_1(t)$, so we say that the set of *two* functions $S = \{f_1(t), f_2(t)\}$ is linearly dependent if the two functions are constant multiples of each other. They are linearly independent otherwise.

EXAMPLE 4.1.1

Determine if the following sets of functions are linearly dependent or linearly independent: (a) $S = \{2t, 4t\}$, (b) $S = \{e^t, e^{-2t}\}$, (c) $S = \{\sin 2t, 5 \sin t \cos t\}$, and (d) $S = \{t, 0\}$.

Solution

(a) Notice that $4t = 2 \times 2t$, so the functions are linearly dependent. (b) These functions are not constant multiples of one another, so they are linearly independent. (c) We use the identity $\sin 2t = 2 \sin t \cos t$ to show that $\frac{5}{2} \sin 2t = \frac{5}{2}(2 \sin t \cos t) = 5 \sin t \cos t$, so the functions are linearly dependent. (d) This set contains the zero function. Because we can obtain the zero function by multiplying the second function by 0, $0 \times t = 0$, the functions are linearly dependent. Therefore, if 0 is a member of a set of two functions, the set is linearly dependent.

Consider the second-order linear differential equation in its general form

$$a_2(t)y'' + a_1(t)y' + a_0(t)y = g(t),$$

where the coefficient functions $a_2(t)$, $a_1(t)$, $a_0(t)$, and $g(t)$ are continuous on the open interval I. Also, assume that $a_2(t) \neq 0$ for each value of t in I so that we may divide the equation by $a_2(t)$ to write the equation in *standard form* (or *normal form*) given by

$$y'' + p(t)y' + q(t)y = f(t),$$

where $p(t) = a_1(t)/a_2(t)$, $q(t) = a_0(t)/a_2(t)$, and $f(t) = g(t)/a_2(t)$. If $f(t) = 0$, that is, if $f(t)$ is identically the zero function, then the equation is *homogeneous* and is represented by

$$y'' + p(t)y' + q(t)y = 0.$$

If $f(t)$ is not identically the zero function, the equation, $y'' + p(t)y' + q(t)y = f(t)$, is *nonhomogeneous*.

Recall the Principle of Superposition for linear homogeneous equations.

Theorem 12 (Principle of Superposition). *If $\phi_1(t)$ and $\phi_2(t)$ are solutions of $a_2(t)y'' + a_1(t)y' + a_0(t)y = 0$, then for all values of the constants c_1 and c_2, $\phi(t) = c_1\phi_1(t) + c_2\phi_2(t)$ is also a solution of $a_2(t)y'' + a_1(t)y' + a_0(t)y = 0$.*

For nonhomogeneous equations, we have a corresponding theorem, although we need to be careful when simplifying.

Theorem 13 (Principle of Superposition for Nonhomogeneous Equations). *If $\phi_1(t)$ and $\phi_2(t)$ are solutions of the linear nonhomogeneous differential equations $y'' + p(t)y' + q(t)y = f_1(t)$ and $y'' + p(t)y' + q(t)y = f_2(t)$, respectively, on the interval I and c_1 and c_2 are constants, then $y = c_1\phi_1(t) + c_2\phi_2(t)$ is also a solution of $y'' + p(t)y' + q(t)y = c_1f_1(t) + c_2f_2(t)$ on the interval I.*

Proof. Suppose that $\phi_1(t)$ and $\phi_2(t)$ are solutions of the linear nonhomogeneous differential equations $y'' + p(t)y' + q(t)y = f_1(t)$ and $y'' + p(t)y' + q(t)y = f_2(t)$, respectively, on the interval I and assume that c_1 and c_2 are constants. Then, $\phi_1'' + p(t)\phi_1' + q(t)\phi_1 = f_1(t)$ and $\phi_2'' + p(t)\phi_2' +$

$q(t)\phi_2 = f_2(t)$. Let $\phi(t) = c_1\phi_1(t) + c_2\phi_2(t)$. Differentiating, $\phi' = c_1\phi_1' + c_2\phi_2'$ and $\phi'' = c_1\phi_1'' + c_2\phi_2''$. Substitution into the left-hand side of the differential equation yields

$$
\begin{aligned}
y'' + p(t)y' + q(t)y &= c_1\phi_1'' + c_2\phi_2'' + p(t)(c_1\phi_1' \\
&\quad + c_2\phi_2') + q(t)(c_1\phi_1 + c_2\phi_2) \\
&= c_1\left(\phi_1'' + p(t)\phi_1' + q(t)\phi_1\right) \\
&\quad + c_2\left(\phi_2'' + p(t)\phi_2' + q(t)\phi_2\right) \\
&= c_1 \times f_1(t) + c_2 \times f_2(t) = c_1f_1(t) + c_2f_2(t).
\end{aligned}
$$

Therefore, $\phi(t) = c_1\phi_1(t) + c_2\phi_2(t)$ is a solution of $y'' + p(t)y' + q(t)y = c_1f_1(t) + c_2f_2(t)$.

As stated earlier, one way to express Theorem 11 is to say "every linear combination of two or more solutions of $y'' + p(t)y' + q(t)y = 0$ is also a solution of the differential equation."

For Theorem 13, we encourage you to be more careful when expressing it in words: "every linear combination of solutions $\phi_1(t)$ of $y'' + p(t)y' + q(t)y = f_1(t)$ and $\phi_2(t)$ of $y'' + p(t)y' + q(t)y = f_2(t)$, $\phi(t) = c_1\phi_1(t) + c_2\phi_2(t)$, is also a solution of $y'' + p(t)y' + q(t)y = c_1f_1(t) + c_2f_2(t)$."

EXAMPLE 4.1.2

 Duffing's equation with a forcing term is the equation

$$y'' + ky' + cy + \epsilon y^3 = F\cos\omega t.$$

Assume that $k = \epsilon = F = 0$ and $c = 4$. For these parameter values, Duffing's equation is the second-order linear homogeneous differential equation $y'' + 4y = 0$.

Show that $y_1(t) = \cos 2t$ and $y_2(t) = \sin 2t$ are solutions of this differential equation. Use the Principle of Superposition to find another solution of this differential equation.

Solution

For $y_1(t) = \cos 2t$, we have $y_1'(t) = -2\sin 2t$ and $y_1''(t) = -4\cos 2t$. Then, $y'' + 4y = -4\cos 2t + 4\cos 2t = 0$ so y_1 satisfies the differential equation. In a similar manner, we can show that y_2 is a solution of the differential equation. (We leave this task to you as an exercise.) Therefore, by the Principle of Superposition, any linear combination of these two functions, $y(t) = c_1 \cos 2t + c_2 \sin 2t$, is also a solution of the differential equation. Notice that the two functions $y_1(t) = \cos 2t$ and $y_2(t) = \sin 2t$ are linearly independent because they are not constant multiples of each other. Then, because the differential equation is second order and we know two linearly independent solutions, a *general solution* of the differential equation is a linear combination of these two functions. That is, every solution of the IVP

$$y'' + 4y = 0, \quad y(t_0) = a, \quad y'(t_0) = b$$

can be obtained from $y(t) = c_1 \cos 2t + c_2 \sin 2t$. In other words, there are c_1 and c_2 so that the initial conditions are satisfied.

As with first-order IVPs, we state an existence and uniqueness theorem for an IVP involving a second-order linear ordinary differential equation.

Theorem 14 (Existence and Uniqueness for Second-Order Linear Equations). *Suppose that $p(t)$, $q(t)$, and $f(t)$ are continuous functions on an open interval I that contains $t = t_0$. Then, the IVP*

$$y'' + p(t)y' + q(t)y = f(t), \quad y(t_0) = y_0, \quad y'(t_0) = y_0'$$

has a unique solution on I.

EXAMPLE 4.1.3

Find the solution to the IVP $y'' + 4y = 0$, $y(0) = 4, y'(0) = -4$.

Solution

We found a general solution of the differential equation $y'' + 4y = 0$ to be $y = c_1 \cos 2t + c_2 \sin 2t$ in Example 4.1.2. Application of the initial condition $y(0) = 4$ gives us $y(0) = c_1 \times \cos 0 + c_2 \times \sin 0 = c_1 = 4$, so $c_1 = 4$. To apply the condition $y'(0) = -4$, we compute $y'(t) = -2c_1 \sin 2t +$

$2c_2 \cos 2t$ so that $y'(0) = -2c_1 \times \sin 0 + 2c_2 \times \cos 0 = 2c_2$. Therefore, $2c_2 = -4$, so $c_2 = -2$. The unique solution to the IVP is $y(t) = -4 \cos 2t - 2 \sin 2t$. This solution is unique over the interval $(-\infty, \infty)$ because the coefficient functions, $p(t) = 0$ and $q(t) = 4$, and the forcing function $f(t) = 0$ are continuous on $(-\infty, \infty)$, and the existence and uniqueness theorem guarantees that the solution is as well.

Now we discuss why we can write every solution of the differential equation, $y'' + p(t)y' + q(t)y = 0$, with linearly independent solutions $y_1(t)$ and $y_2(t)$, as the linear combination $y(t) = c_1 y_1(t) + c_2 y_2(t)$. So, we call the form $y(t) = c_1 y_1(t) + c_2 y_2(t)$ *a general solution* of this differential equation. We begin by stating the following definition.

Definition 7 (Wronskian of a Set of Two Functions). *Let $S = \{f_1(t), f_2(t)\}$ be a set of two differentiable functions. The **Wronskian** of S, denoted by $W(S) = W(\{f_1(t), f_2(t)\})$, is the determinant*

$$W(S) = \begin{vmatrix} f_1(t) & f_2(t) \\ f_1'(t) & f_2'(t) \end{vmatrix} = f_1(t)f_2'(t) - f_1'(t)f_2(t).$$

The Wronskian is named after **Josef Ho'ené de Wronski** (1778, Poland-1853, France). Wronski published numerous incorrect results and is believed to have been quite conceited.

Thomas Muir (1844, Scotland-1934, South Africa) named the determinant defined here after Wronski.

Theorem 15 (Wronskian of Solutions). *Consider the second-order linear homogeneous equation $y'' + p(t)y' + q(t)y = 0$, where $p(t)$ and $q(t)$ are continuous on the open interval I. Suppose that $y_1(t)$ and $y_2(t)$ are solutions of this differential equation on I. If $y_1(t)$ and $y_2(t)$ are linearly dependent, then $W(\{y_1(t), y_2(t)\}) = 0$ for all values of t in I. If $y_1(t)$ and $y_2(t)$ are linearly independent, then $W(\{y_1(t), y_2(t)\}) \neq 0$ for all values of t in I.*

We omit the proof of this theorem here. We leave it as part of the section exercises (see Abel's formula).

EXAMPLE 4.1.4

Compute the Wronskian of each of the following sets of functions: (a) $S = \{\cos t, \sin t\}$, (b) $S = \{e^t, te^t\}$, and (c) $S = \{2t, 4t\}$.

Solution

We compute a 2×2 determinant with $\begin{vmatrix} a & b \\ c & d \end{vmatrix} = ad - bc$.

(a)

$$W(S) = \begin{vmatrix} \cos t & \sin t \\ \frac{d}{dt}(\cos t) & \frac{d}{dt}(\sin t) \end{vmatrix} = \begin{vmatrix} \cos t & \sin t \\ -\sin t & \cos t \end{vmatrix}$$

$$= \cos^2 t - \left(-\sin^2 t\right) = \cos^2 t + \sin^2 t = 1.$$

(b)

$$W(S) = \begin{vmatrix} e^t & te^t \\ \frac{d}{dt}(e^t) & \frac{d}{dt}(te^t) \end{vmatrix} = \begin{vmatrix} e^t & te^t \\ e^t & (t+1)e^t \end{vmatrix}$$

$$= (t+1)e^{2t} - te^{2t} = e^{2t}.$$

(c) $W(S) = \begin{vmatrix} 2t & 4t \\ 2 & 4 \end{vmatrix} = 8t - 8t = 0.$

EXAMPLE 4.1.5

Show that $y_1(t) = t^{2/3}$ and $y_2(t) = t$ are solutions of $3t^2y'' - 2ty' + 2y = 0$. Compute the Wronskian of $S = \{t^{2/3}, t\}$. Does your result contradict Theorem 15?

Solution

First, we find that $y_1' = \frac{2}{3}t^{-1/3}$ and $y_1'' = -\frac{2}{9}t^{-4/3}$. Then, $3t^2y_1'' - 2ty_1' + 2y_1 = 3t^2 \times -\frac{2}{9}t^{-4/3} - 2t \times \frac{2}{3}t^{-1/3} + 2t^{2/3} = \left(-\frac{2}{3} - \frac{4}{3} + 2\right)t^{2/3} = 0$, so $y_1(t) = t^{2/3}$ satisfies the differential equation. In a similar manner, we can show that $y_2(t) = t$ is also a solution. Now,

$$W(S) = \begin{vmatrix} t^{2/3} & t \\ \frac{2}{3}t^{-1/3} & 1 \end{vmatrix} = t^{2/3} - \frac{2}{3}t^{2/3} = \frac{1}{3}t^{2/3}.$$

Notice that $W(S) \neq 0$ for all values of t except $t = 0$. However, this does not contradict Theorem 15. After writing the differential equation in normal form, $y'' - \frac{2}{3t}y' + \frac{2}{3t^2}y = 0$, we see that the coefficient functions $p(t) = -2/(3t)$ and $q(t) = 2/(3t^2)$ are not continuous at $t = 0$. The theorem holds only on an open interval I where these functions are continuous.

 Show that $y_2(t) = t$ is a solution of $3t^2y'' - 2ty' + 2y = 0$.

In part (c) of Example 4.1.4, we found that $W(S) = 0$. Recall also that we concluded in Example 4.1.1 that the functions in $S = \{2t, 4t\}$ are linearly dependent.

Theorem 16 (General Solution). *Consider the second-order linear homogeneous equation $y'' + p(t)y' + q(t)y = 0$, where $p(t)$ and $q(t)$ are continuous on the open interval I. Suppose that $y_1(t)$ and $y_2(t)$ are linearly independent solutions of this differential equation on I. Then, if Y is any solution of this differential equation, there are constants c_1 and c_2 so that $Y(t) = c_1 y_1(t) + c_2 y_2(t)$ for all t in I.*

This theorem states that if we have two linearly independent solutions of a second-order linear homogeneous equation, then we have all solutions, called a *general solution*, given by the linear combination of the two solutions. We call the set of two linearly independent solutions $S = \{y_1(t), y_2(t)\}$ a *fundamental set of solutions*, and we note that a fundamental set of solutions for the second-order equation $y'' + p(t)y' + q(t)y = 0$ contains two linearly independent solutions. Later in this chapter, we will generalize this result to see that the number of functions in the fundamental set for a linear homogeneous equation equals the order of the differential equation and these functions must be linearly independent.

Proof of Theorem 16 Let $t = t_0$ be a value on I where $W(\{y_1(t_0), y_2(t_0)\}) \neq 0$. Then, $Y(t_0) = c_1 y_1(t_0) + c_2 y_2(t_0)$ and $Y'(t_0) = c_1 y_1'(t_0) + c_2 y_2'(t_0)$. We can write this system of two equations in matrix form as

$$\begin{pmatrix} y_1(t_0) & y_2(t_0) \\ y_1'(t_0) & y_2'(t_0) \end{pmatrix} \begin{pmatrix} c_1 \\ c_2 \end{pmatrix} = \begin{pmatrix} Y(t_0) \\ Y'(t_0) \end{pmatrix}, \quad (4.1)$$

where the determinant of the coefficient matrix at $t = t_0$ is $W(\{y_1(t_0), y_2(t_0)\}) = y_1(t_0)y_2'(t_0) - y_1'(t_0)y_2(t_0)$ is not zero because $y_1(t)$ and $y_2(t)$ are linearly independent. This means that system (4.1) has a nontrivial solution. Using Cramer's rule or elimination, the unique solution of system (4.1) is given by

$$c_1 = \frac{Y(t_0)y_2'(t_0) - Y'(t_0)y_2(t_0)}{y_1(t_0)y_2'(t_0) - y_1'(t_0)y_2(t_0)} \quad \text{and}$$

$$c_2 = \frac{-Y(t_0)y_1'(t_0) + Y'(t_0)y_1(t_0)}{y_1(t_0)y_2'(t_0) - y_1'(t_0)y_2(t_0)}. \quad (4.2)$$

(Notice that these values make sense only if $y_1(t_0)y_2'(t_0) - y_1'(t_0)y_2(t_0) \neq 0$, as we mentioned earlier in connection to the linear independence of $y_1(t)$ and $y_2(t)$.) Using the values of c_1 and c_2 given by Equations (4.2), let $Z(t) = c_1 y_1(t) + c_2 y_2(t)$ be a solution of $y'' + p(t)y' + q(t)y = 0$. Then, in addition to satisfying the differential equation, $Z(t_0) = c_1 y_1(t_0) + c_2 y_2(t_0) = Y(t_0)$ and $Z'(t_0) = c_1 y_1'(t_0) + c_2 y_2'(t_0) = Y'(t_0)$. By the existence and uniqueness theorem, the IVP

$$y'' + p(t)y' + q(t)y = 0, \ y(t_0) = Y(t_0), \ y'(t_0) = Y'(t_0) \quad (4.3)$$

has a unique solution. Because $Z(t)$ and $Y(t)$ each satisfy the differential equation and have the same initial conditions, IVP (4.3), they must be equal (i.e., $Z(t) = Y(t)$ on I). There is only one solution to the IVP.

EXAMPLE 4.1.6

Show that $y = c_1 e^{-5t} + c_2 e^{2t}$ is a general solution of $y'' + 3y' - 10y = 0$.

Solution

First, we must verify that $y_1(t) = e^{-5t}$ and $y_2(t) = e^{2t}$ satisfy the differential equation. With $y_1'(t) = -5e^{-5t}$ and $y_1''(t) = 25e^{-5t}$, we have

$$y_1'' + 3y_1' - 10y_1 = 25e^{-5t} + 3 \times -5e^{-5t} - 10e^{-5t} = 0,$$

so $y_1(t) = e^{-5t}$ is a solution. In a similar manner, we can show that $y_2(t) = e^{2t}$ is also a solution. Next, we verify that these functions are linearly independent. We can establish this by noting that the two are not constant multiples, or we can compute

$$W(\{y_1, y_2\}) = \begin{vmatrix} e^{-5t} & e^{2t} \\ -5e^{-5t} & 2e^{2t} \end{vmatrix}$$

$$= 2e^{-3t} - \left(-5e^{-3t}\right) = 7e^{-3t}.$$

The Wronskian is not identically zero, so the functions are linearly independent. Therefore, because we have two linearly independent solutions to the second-order differential equation, a general solution is the linear combination of the solutions, $y = c_1 e^{-5t} + c_2 e^{2t}$. In addition, we say that $\{e^{-5t}, e^{2t}\}$ is a fundamental set of solutions for $y'' + 3y' - 10y = 0$.

Reduction of Order

In the next section, we learn how to find solutions of homogeneous equations with constant coefficients. In doing so, we will find it necessary to determine a second linearly independent solution from a known solution. We illustrate this procedure, called *reduction of order*, by considering the second-order equation in normal (or standard) form $y'' + p(t)y' + q(t)y = 0$ and assuming that we are given one solution, $y_1(t) = f(t)$. We know from our previous discussion that to find a general solution of this second-order equation, we must have two linearly independent solutions. We must determine a second linearly independent solution. We accomplish this by attempting to find a solution of the form

$$y_2(t) = v(t)f(t)$$

One would never look for a second linearly independent solution of the form $y_2 = cf(t)$ for a constant c because $f(t)$ and $cf(t)$ are linearly dependent.

---

and solving for $v(t)$. Differentiating with the product rule, we obtain

$$y_2' = v'f + vf' \quad \text{and} \quad y_2'' = v''f + 2v'f' + vf''.$$

For convenience, we have omitted the argument of these functions. We now substitute y_2, y_2', and y_2'' into the equation $y'' + p(t)y' + q(t)y = 0$. This gives us

$$y_2'' + p(t)y_2' + q(t)y_2 = v''f + 2v'f'$$
$$+ vf'' + p(t)(v'f + vf') + q(t)vf$$

$$= \underbrace{(f'' + p(t)f' + q(t)f)}_{=0} v + v''f + 2v'f' + p(t)v'f$$

$$= fv'' + (2f' + p(t)f)v'.$$

Therefore, we have the equation $fv'' + (2f' + p(t)f)v' = 0$, which can be written as a first-order equation by letting $w = v'$. Making this substitution gives us the linear first-order equation

$$fw' + (2f' + p(t)f)w = 0 \quad \text{or}$$

$$f\frac{dw}{dt} + (2f' + p(t)f)w = 0$$

which is separable, so we obtain the separated equation

$$\frac{1}{w}dw = \left(-\frac{2f'}{f} - p(t)\right)dt.$$

We solve this equation by integrating both sides of the equation to yield

$$\ln|w| = \ln\left(\frac{1}{f^2}\right) - \int p(t)dt.$$

If $y_1 = f(t)$ is a known solution of the differential equation $y'' + p(t)y' + q(t)y = 0$, we can obtain a second linearly independent solution of the form $y_2 = v(t)f(t)$, where $v(t) = \int \frac{1}{(f(t))^2} e^{-\int p(t)dt} dt$.

This means that $w = \frac{1}{f^2}e^{-\int p(t)dt}$, so we have the formula $\frac{dv}{dt} = \frac{1}{f^2}e^{-\int p(t)dt}$ or

$$v(t) = \int \frac{1}{(f(t))^2} e^{-\int p(t)dt} dt. \qquad (4.4)$$

We leave the proof that

$$y_1(t) = f(t) \quad \text{and}$$

$$y_2(t) = v(t)f(t) = f(t)\int \frac{1}{(f(t))^2} e^{-\int p(t)dt} dt$$

are linearly independent as an exercise.

EXAMPLE 4.1.7

Determine a second linearly independent solution to the differential equation $y'' + 6y' + 9y = 0$ given that $y_1 = e^{-3t}$ is a solution.

Solution

First we identify the functions $p(t) = 6$ and $f(t) = e^{-3t}$. Then we determine the function $v(t)$ so that $y_2(t) = v(t)f(t)$ is a second linearly independent solution of the equation with the formula

$$v(t) = \int \frac{1}{(f(t))^2} e^{-\int p(t)dt} dt$$

$$= \int \frac{1}{(e^{-3t})^2} e^{-\int 6 dt} dt$$

$$= \int \frac{1}{e^{-6t}} e^{-6t} = \int dt = t.$$

A second linearly independent solution is $y_2 = v(t)f(t) = te^{-3t}$; a general solution of the differential equation is $y = (c_1 + c_2t)e^{-3t}$; and a fundamental set of solutions for the equation is $\{e^{-3t}, te^{-3t}\}$.

EXAMPLE 4.1.8

Determine a second linearly independent solution to the differential equation $4t^2y'' + 8ty' + y = 0$, $t > 0$, given that $y_1 = t^{-1/2}$ is a solution.

Solution

In this case, we must first write the equation in normal (or standard) form to use formula (4.4) so we divide by $4t^2$ to obtain $y'' + 2t^{-1}y' + \frac{1}{4}t^{-2}y = 0$. Therefore, $p(t) = 2t^{-1}$ and $f(t) = t^{-1/2}$. Using the formula for v, we obtain

$$v(t) = \int \frac{1}{(f(t))^2} e^{-\int p(t)dt} dt$$

$$= \int \frac{1}{(t^{-1/2})^2} e^{-\int 2/t \, dt} dt$$

$$= \int te^{-2\ln t} dt = \int t^{-1} dt = \ln t, \quad t > 0.$$

A second linearly independent solution is $y_2 = v(t)f(t) = t^{-1/2} \ln t$; a general solution of the differential equation is $y = t^{-1/2}(c_1 + c_2 \ln t)$; and a fundamental set of solutions for the equation is $\{t^{-1/2}, t^{-1/2} \ln t\}$.

 EXERCISES 4.1

In Exercises 1-8, calculate the Wronskian of the indicated set of functions. Classify each set of functions as linearly independent or linearly dependent.

1. $S = \{t, 4t - 1\}$
2. $S = \{t, e^t\}$
3. $S = \{e^{-6t}, e^{-4t}\}$
4. $S = \{\cos 2t, \sin 2t\}$
5. $S = \{t^{-1}, t^{-2}\}$
6. $S = \{t^{-2}, t^{-2} \ln t\}$
7. $S = \{t, |t|\}$
8. $S = \{t, t - |t|\}$

In Exercises 9-12, show that $y = c_1y_1(t) + c_2y_2(t)$ is a general solution of the given differential equation by first showing that $y = c_1y_1(t) + c_2y_2(t)$ satisfies the differential equation and then that $\{y_1(t), y_2(t)\}$ is a linearly independent set of functions.

9. $y = c_1e^t + c_2e^{-t}, y'' - y = 0$
10. $y = c_1e^{-t} + c_2te^{-t}, y'' + 2y' + y = 0$
11. $y = c_1t^{-1/2} + c_2t^3, t > 0, 2t^2y'' - 3ty' - 3y = 0$
12. $y = c_1 \cos 3t + c_2 \sin 3t, y'' + 9y = 0$

In Exercises 13-17, use the given solution on the indicated interval to find the solution to the IVP.

13. $y'' - y' - 2y = 0, y(0) = -1, y'(0) = -5$,
    $y = c_1e^{-t} + c_2e^{2t}, -\infty < t < \infty$
14. $y'' + 9y = 0, y(0) = 1, y'(0) = -3$,
    $y = c_1 \cos 3t + c_2 \sin 3t, -\infty < t < \infty$

15. $3t^2 y'' - 5ty' - 3y = 0, y(1) = 1, y'(1) = 17/3$,
$y = c_1 t^{-1/3} + c_2 t^3, 0 < t < \infty$

16. $t^2 y'' + 7ty' - 7y = 0, y(1) = 2, y'(1) = -22$,
$y = c_1 t^{-7} + c_2 t, 0 < t < \infty$

17. $y'' + y = 2 \cos t, y(0) = 1, y'(0) = 1$,
$y = c_1 \cos t + c_2 \sin t + t \sin t, -\infty < t < \infty$

18. Use the Wronskian to show that if y_1 and y_2 are solutions of the first-order equation $y' + p(t)y = 0$, then y_1 and y_2 are linearly dependent.

19. Consider the hyperbolic trigonometric functions $\cosh t = (e^t + e^{-t})/2$ and $\sinh t = (e^t - e^{-t})/2$. Show that
   (a) $d/dt \, (\cosh t) = \sinh t$
   (b) $d/dt \, (\sinh t) = \cosh t$
   (c) $\cosh^2 t - \sinh^2 t = 1$
   (d) $\cosh t$ and $\sinh t$ are linearly independent functions.

In Exercises 20-23, show that S is a fundamental set of solutions for the given equation by showing that each function in S is a solution of the equation and then that the functions in S are linearly independent.

20. $S = \{e^{-6t}, e^{-4t}\}, y'' + 10y' + 24y = 0$

21. $S = \{\cos 4t, \sin 4t\}, y'' + 16y = 0$

22. $S = \{e^{-3t} \cos 3t, e^{-3t} \sin 3t\}$,
$y'' + 6y' + 18y = 0$

23. $S = \{t^{-1}, t\}, t^2 y'' + ty' - y = 0$

24. Let $ay'' + by' + cy = 0$ be a second-order linear homogeneous equation with constant real coefficients and let m_1 and m_2 be the solutions of the equation $am^2 + bm + c = 0$.
   (a) If $m_1 \neq m_2$ and both m_1 and m_2 are real, show that $\{e^{m_1 t}, e^{m_2 t}\}$ is a fundamental set of solutions of $ay'' + by' + cy = 0$.
   (b) If $m_1 = m_2 = m$, show that $\{e^{mt}, te^{mt}\}$ is a fundamental set of solutions of $ay'' + by' + cy = 0$.
   (c) If $m_{1,2} = \alpha \pm \beta i, \beta \neq 0$, show that $\{e^{\alpha t} \cos \beta t, e^{\alpha t} \sin \beta t\}$ is a fundamental set of solutions of $ay'' + by' + cy = 0$.

In Exercises 25-33, (a) use reduction of order with the solution $y_1(t)$ to find a second linearly independent solution of the given differential equation. (b) What is a general solution of the differential equation? (c) What is a fundamental set of solutions of the differential equation? (d) If initial conditions are specified, solve the IVP.

25. $y_1(t) = e^{3t}, y'' - 5y' + 6y = 0$

26. $y_1(t) = e^{-2t}, y'' + 6y' + 8y = 0$

27. $y_1(t) = e^{2t}, y'' - 4y' + 4y = 0, y(0) = 0,$
$y'(0) = 1$

28. $y_1(t) = e^{-5t}, y'' + 10y' + 25y = 0$

29. $y_1(t) = \cos 3t, y'' + 9y = 0, y(0) = 1,$
$y'(0) = -4$

30. $y_1(t) = \sin 7t, y'' + 49y = 0$

31. $y_1(t) = t^{-4}, t^2 y'' + 4ty' - 4y = 0$

32. $y_1(t) = t^{-2}, t^2 y'' + 6ty' + 6y = 0$

33. $y_1(t) = \dfrac{\cos t}{\sqrt{t}}, t^2 y'' + ty' + \left(t^2 - \tfrac{1}{4}\right)y = 0$

34. $y_1(t) = t^{-1}, t^2 y'' + 3ty' + y = 0$

35. Find integer constants $a, b,$ and c for which the greatest common factor of $a, b,$ and c is one so that $S = \{e^{-t/2}, e^{t/3}\}$ is a fundamental set of solutions for $ay'' + by' + cy = 0$.

36. Find constants a and b so that $S = \{t^{-1/2}, t^{1/2}\}$ is a fundamental set of solutions for $t^2 y'' + aty' + by = 0$.

37. Find constants a and b so that $S = \{e^{-t} \cos 2t, e^{-t} \sin 2t\}$ is a fundamental set of solutions for $y'' + ay' + by = 0$.

38. If $\phi(t)$ is a solution of $y' + p(t)y = q(t)$, what is a solution of $y'' + p'(t)y + p(t)y' = q'(t)$?

39. Suppose the $f(t)$ is a solution to the equation $y'' + p(t)y' + q(t)y = 0$. Show that $f(t)$ and the solution $f(t) \int \frac{1}{(f(t))^2} e^{-\int p(t)dt} dt$ obtained by reduction of order are linearly independent. *Hint:* Use the Wronskian.

40. Show that a general solution of $y'' - k^2 y = 0$ is $y = c_1 \cosh kt + c_2 \sinh kt$ (see Exercise 19 for a reminder of the definitions of $\cosh t$ and $\sinh t$).

41. *(Abel's Formula)* Suppose that $y_1(t)$ and $y_2(t)$ are two linearly independent solutions of $y'' + p(t)y' + q(t)y = 0$ on an open interval I, where $p(t)$ and $q(t)$ are continuous on I.

Then, the Wronskian of y_1 and y_2 is $W(\{y_1, y_2\}) = Ce^{-\int p(t)dt}$. Prove Abel's formula using the following steps. Begin with the system

$$y_1'' + p(t)y_1' + q(t)y_1 = 0,$$
$$y_2'' + p(t)y_2' + q(t)y_2 = 0.$$

(a) Multiply the first equation by $-y_2$ and the second equation by y_1. Add the resulting equations to obtain $(y_1 y_2'' - y_1'' y_2) + p(t)(y_1 y_2' - y_1' y_2) = 0.$
(b) Show that $d/dt \left(W \left(\{y_1, y_2\} \right) \right) = y_1 y_2'' - y_1'' y_2.$ (c) Use the results in (a) and (b) to obtain the first-order equation $dW/dt + p(t)W = 0.$ Solve this equation for W.

Niels Henrik Abel (1802, Norway-1829, Norway): Although Abel's parents were probably deadbeats, he overcame the obstacle to become a famous mathematician at a young age. He is more famous for his contributions to algebra than his contributions to differential equations. In 1824, he proved that the *general* equation $p(x) = 0$, where $p(x) = a_n x^n + a_{n-1}x^{n-1} + \cdots + a_1 x + a_0$ is a polynomial of degree $n \geq 5$, is not solvable by algebraic methods.

42. Can the Wronskian be zero at only one value of t on I? *Hint:* Use Abel's formula.
43. (a) Find conditions on the constants $c_1, c_2,$ $c_3,$ and c_4 so that $y(t)=c_1+c_2 \tan (c_3 + c_4 \ln t)$

is a solution of the nonlinear second-order equation $ty'' - 2yy' = 0$. (b) Is the Principle of Superposition valid for this equation? Explain.
44. Find a general solution of $4t^2 y'' + 4ty' + (36t^2 - 1)y = 0$ given that $y = t^{-1/2} \cos 3t$ is one solution.
45. Given that $y = t^{-1} \sin 4t$ is a solution of $ty'' + 2y' + 16ty = 0$, find and graph the solution of the equation that satisfies $y(\pi/8) = 0$ and $y'(\pi/8) = 32$.
46. Find a general solution of $y'' + b(t)y' + c(t)y = 0$ given that $y = t^{-2} \sin t$ is one solution.
47. Given that $y = t^{-3} \cos t$ is a solution of $y'' + b(t)y' + c(t)y = 0$, find and graph the solution(s), if any, that satisfy the boundary conditions $y(\pi) = y(2\pi) = 0$.
48. Show that if y_1 and y_2 are linearly independent solutions of $y'' + p(t)y' + q(t)y = 0$ then $y_1 + y_2$ and $y_1 - y_2$ are also linearly independent solutions of $y'' + p(t)y' + q(t)y = 0$.
49. If $S = \{f_1(t), f_2(t)\}$ is a set of differentiable functions on an interval I and $W(S) \neq 0$ on I, determine $p(t)$ and $q(t)$ so that S is a fundamental set of solutions for $y'' + p(t)y' + q(t)y = 0$.
50. Find $p(t)$ and $q(t)$ so that $y = c_1 \tan t + c_2 \sec t, -\pi/2 < t < \pi/2$, is a general solution of $y'' + p(t)y' + q(t)y = 0$.

4.2 SOLUTIONS OF SECOND-ORDER LINEAR HOMOGENEOUS EQUATIONS WITH CONSTANT COEFFICIENTS

- TWO DISTINCT REAL ROOTS
- REPEATED ROOTS
- COMPLEX CONJUGATE ROOTS

In Section 4.1, we learned the theory related to solving second-order linear homogeneous equations. Here, we restrict our attention to second-order linear homogeneous equations with constant coefficients,

$$ay'' + by' + cy = 0,$$

where $a \neq 0$, b, and c are constant, and we try to solve the equation by assuming that $y = e^{rt}$ is a solution for some value(s) of r. We try to find r by substituting the function $y = e^{rt}$ into the differential equation and assume that it is a solution. With $y' = re^{rt}$ and $y'' = r^2 e^{rt}$, we obtain $ar^2 e^{rt} + bre^{rt} + ce^{rt} = 0$, or $e^{rt}(ar^2 + br + c) = 0$. Therefore, we solve the *characteristic equation*

$$ar^2 + br + c = 0$$

to determine r (because $e^{rt} > 0$ for all values of t). The roots of the characteristic equation are

$$r = \frac{1}{2a}\left(-b \pm \sqrt{b^2 - 4ac}\right), \qquad (4.5)$$

so the roots of the characteristic equation depend on the values of a, b, and c. There are three possibilities: two distinct real roots when $b^2 - 4ac > 0$, one repeated real root when $b^2 - 4ac = 0$, and two complex conjugate roots when $b^2 - 4ac < 0$. We now consider each case.

Two Distinct Real Roots

Suppose that the characteristic equation of a linear homogeneous differential equation with constant coefficients has two real roots r_1 and r_2, where $r_1 \neq r_2$, so two solutions to the differential equation are $y_1(t) = e^{r_1 t}$ and $y_2(t) = e^{r_2 t}$. We show that these solutions are linearly independent by computing the Wronskian:

$$W(\{y_1, y_2\}) = \begin{vmatrix} e^{r_1 t} & e^{r_2 t} \\ r_1 e^{r_1 t} & r_2 e^{r_2 t} \end{vmatrix}$$

$$= r_2 e^{(r_1+r_2)t} - r_1 e^{(r_1+r_2)t}$$

$$= (r_2 - r_1)e^{(r_1+r_2)t}.$$

These functions are linearly independent because $W(\{y_1, y_2\}) \neq 0$ since $r_2 - r_1 \neq 0$ and $e^{(r_1-r_2)t} > 0$. Therefore, a fundamental set of solutions for the differential equation is $S = \{e^{r_1 t}, e^{r_2 t}\}$ and a general solution of the

second-order linear differential equation with constant coefficients is $y(t) = c_1 y_1(t) + c_2 y_2(t) = c_1 e^{r_1 t} + c_2 e^{r_2 t}$.

Theorem 17 (Distinct Real Roots). *If $r_1 \neq r_2$ are real solutions of $ar^2 + br + c = 0$, then a general solution of $ay'' + by' + cy = 0$ is $y(t) = c_1 e^{r_1 t} + c_2 e^{r_2 t}$.*

EXAMPLE 4.2.1

Solve $y'' + 3y' - 4y = 0$.

Solution

The characteristic equation for this differential equation is $r^2 + 3r - 4 = (r+4)(r-1) = 0$, with roots $r_1 = -4$ and $r_2 = 1$. Therefore, a fundamental set of solutions for the equation is $S = \{e^{-4t}, e^t\}$ and a general solution of the differential equation is $y(t) = c_1 e^{-4t} + c_2 e^t$.

Next, suppose that the characteristic equation of the differential equation (4.5) has a repeated root $r = r_{1,2} = r_1 = r_2$. In other words, $b^2 - 4ac = 0$ in the quadratic formula so that the root is $r = -b/(2a)$ or $(r + b/(2a))^2$ is a factor in the characteristic equation. Because $r = r_{1,2} = -b/(2a)$ is repeated, we only obtain one solution of the differential equation from the characteristic equation, $y_1(t) = e^{rt} = e^{-bt/(2a)}$. We use this solution to obtain a second linearly independent solution through reduction of order with the formula

$$y_2(t) = y_1(t) \int \frac{1}{(f(t))^2} e^{-\int p(t)dt} dt,$$

where $p(t)$ appears in the normal (or standard) form of the equation $y'' + p(t)y' + q(t)y = 0$. In this case, the differential equation is $ay'' + by' + cy = 0$. Dividing by a yields $y'' + (b/a)y' + (c/a)y = 0$, so $p(t) = b/a$. Returning to the reduction of order formula, we find that

$$y_2(t) = e^{-bt/(2a)} \int \frac{1}{(f(t))^2} e^{-\int p(t)dt}$$

$$dt = e^{-bt/(2a)} \int \frac{1}{e^{-bt/(2a)}} e^{-bt/(2a)}$$

$$dt = e^{-bt/(2a)} \int dt = te^{-bt/(2a)}.$$

Therefore, a fundamental set of solutions for the differential equation is $S = \{e^{rt}, te^{rt}\}$ and a general solution is $y(t) = c_1e^{rt} + c_2te^{rt} = c_1e^{-bt/(2a)} + c_2te^{-bt/(2a)}$.

Theorem 18 (Repeated Root). *If $r = r_1 = r_2$ are real repeated solutions of $ar^2 + br + c = 0$, then a general solution of $ay'' + by' + cy = 0$ is $y(t) = c_1e^{rt} + c_2te^{rt}$.*

$$\cos t = \sum_{k=0}^{\infty} \frac{(-1)^k}{(2k)!} t^{2k}$$

$$= 1 - \frac{1}{2!}t^2 + \frac{1}{4!}t^4 - \frac{1}{6!}t^6 + \cdots,$$

and

$$\sin t = \sum_{k=0}^{\infty} \frac{(-1)^k}{(2k+1)!} t^{2k+1}$$

$$= t - \frac{1}{3!}t^3 + \frac{1}{5!}t^5 - \frac{1}{7!}t^7 + \cdots.$$

EXAMPLE 4.2.2

Solve $y'' + 2y' + y = 0$.

Solution

The characteristic equation for this differential equation is $r^2 + 2r + 1 = (r+1)^2 = 0$ with roots $r_1 = r_2 = -1$. Therefore, a fundamental set of solutions is $S = \{e^{-t}, te^{-t}\}$ and a general solution is $y(t) = c_1e^{-t} + c_2te^{-t}$

Note: Often, we denote the roots as $r_{1,2} = \alpha \pm \beta i$.

Complex Conjugate Roots

Suppose that the characteristic equation of a linear homogeneous differential equation with constant coefficients has the complex conjugate roots $r_1 = \alpha + \beta i$ and $r_2 = \alpha - \beta i$ where α and β ($\beta > 0$) are real numbers and $i = \sqrt{-1}$.

To construct a real-valued general solution, we use Euler's formula,

$$e^{i\theta} = \cos\theta + i\sin\theta,$$

which can be obtained through the use of the Maclaurin series:

$$e^t = \sum_{k=0}^{\infty} \frac{1}{k!}t^k = 1 + t + \frac{1}{2!}t^2 + \frac{1}{3!}t^3$$

$$+ \frac{1}{4!}t^4 + \frac{1}{5!}t^5 + \cdots,$$

Leonhard Euler (1707-1783): Euler and Daniel Bernoulli worked together at St. Petersburg. Not only was Euler the most prolific mathematician of all time (his unpublished works continued to be published for 50 years after his death), he was prolific in other ways as well: he and his wife Katharina Gsell had 13 children.

We derive Euler's formula using these Maclaurin series and substitution:

$$e^{i\theta} = \sum_{k=0}^{\infty} \frac{1}{k!}(i\theta)^k = 1 + (i\theta) + \frac{1}{2!}(i\theta)^2$$

$$+ \frac{1}{3!}(i\theta)^3 + \frac{1}{4!}(i\theta)^4 + \frac{1}{5!}(i\theta)^5 + \cdots$$

$$= 1 + i\theta - \frac{1}{2!}\theta^2 - \frac{1}{3!}i\theta^3 + \frac{1}{4!}\theta^4$$

$$+ \frac{1}{5!}i\theta^5 - \frac{1}{6!}\theta^6 - \frac{1}{7!}i\theta^7 + \cdots$$

$$= \left(1 - \frac{1}{2!}\theta^2 + \frac{1}{4!}\theta^4 - \frac{1}{6!}\theta^6 + \cdots\right)$$

$$+ i\left(\theta - \frac{1}{3!}\theta^3 + \frac{1}{5!}\theta^5 - \frac{1}{7!}\theta^7 + \cdots\right)$$

$$= \cos\theta + i\sin\theta.$$

Using the properties of sine (sine is an odd function so $\sin(-x) = -\sin x$) and cosine (cosine is an even function so $\cos(-x) = \cos x$), Euler's formula also implies that $e^{-i\theta} = \cos\theta - i\sin\theta$. The roots of the characteristic equation are $r_{1,2} = \alpha \pm \beta i$, so

$$y_1 = e^{(\alpha+\beta)t} = e^{\alpha t}(\cos\beta t + i\sin\beta t) \quad \text{and}$$

$$y_2 = e^{(\alpha-\beta)t} = e^{\alpha t}(\cos\beta t - i\sin\beta t)$$

are both solutions to the differential equation. By the Principle of Superposition, any linear combination of y_1 and y_2 is also a solution. For example, $z_1 = \frac{1}{2}(y_1+y_2) = e^{\alpha t}\cos\beta t$ and $z_2 = -\frac{i}{2}(y_1-y_2) = e^{\alpha t}\sin\beta t$ are both solutions. You should verify that $S = \{e^{\alpha t}\cos\beta t, e^{\alpha t}\sin\beta t\}$ is linearly independent. This means that $y_1 = e^{\alpha t}\cos\beta t$ and $y_2 = e^{\alpha t}\sin\beta t$ are linearly independent solutions of $ay'' + by' + cy = 0$ so a fundamental set of solutions for $ay'' + by' + cy = 0$ is $S = \{e^{\alpha t}\cos\beta t, e^{\alpha t}\sin\beta t\}$ and a general solution of $ay'' + by' + cy = 0$ is $y = e^{\alpha t}(c_1\cos\beta t + c_2\sin\beta t)$.

Theorem 19 (Complex Conjugate Roots). *If $r_{1,2} = \alpha \pm \beta i$, where α and β ($\beta > 0$) are real numbers, are solutions of $ar^2 + br + c = 0$, then a general solution of $ay'' + by' + cy = 0$ is $y = e^{\alpha t}(c_1\cos\beta t + c_2\sin\beta t)$.*

In Exercise 48 of Section 4.1, we show that if y_1 and y_2 are linearly independent solutions of $y'' + p(t)y' + q(t)y = 0$, then $y_1 + y_2$ and $y_1 - y_2$ are also linearly independent solutions of $y'' + p(t)y' + q(t)y = 0$.

EXAMPLE 4.2.3

Solve $\quad y'' + 4y' + 20y = 0, \quad y(0) = 3, \quad$ and $y'(0) = -1$.

Solution

The characteristic equation is $r^2 + 4r + 20 = 0$ with roots

$$r_{1,2} = \frac{1}{2}\left(-4 \pm \sqrt{4^2 - 4 \times 20}\right)$$

$$= \frac{1}{2}\left(-4 \pm \sqrt{-64}\right)$$

$$= \frac{1}{2}(-4 \pm 8i) = -2 \pm 4i.$$

Therefore, in our earlier notation, $\alpha = -2$ and $\beta = 4$, so that a fundamental set of solutions for the differential equation is $S = \{e^{-2t}\cos 4t, e^{-2t}\sin 4t\}$ and a general solution of the differential equation is $y = e^{-2t}(c_1\cos 4t + c_2\sin 4t)$ because the solutions of the characteristic equation are complex conjugates. To find the solution for which $y(0) = 3$ and $y'(0) = -1$, we first calculate

$$y' = 2e^{-2t}[(-c_1 + 2c_2)\cos 4t + (-2c_1 - c_2)\sin 4t]$$

with the product rule and then evaluate both $y(0) = c_1$ and $y'(0) = 2(2c_2 - c_1)$ obtaining the system of equations

$$\begin{cases} c_1 & = 3, \\ 2(2c_2 - c_1) & = -1. \end{cases}$$

Substituting $c_1 = 3$ into the second equation results in $c_2 = 5/4$. Thus, the solution of $y'' + 4y' + 20y = 0$ for which $y(0) = 3$ and $y'(0) = -1$ is $y = e^{-2t}\left(3\cos 4t + \frac{5}{4}\sin 4t\right)$, which is graphed in Figure 4.3.

In Example 4.2.3, evaluate $\lim_{t\to\infty} y(t)$. Does this limit depend on the initial conditions? Does this result agree with the graph in Figure 4.4?

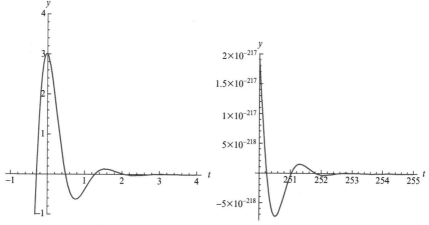

FIGURE 4.3   Graph of $y = e^{-2t}\left(3\cos 4t + \frac{5}{4}\sin 4t\right)$. Note that y is not identically the zero function but converges to the zero function quickly so it appears that y and the zero function are identical in the plot as t increases, even though the function oscillates forever.

The theorems stated here can be summarized as follows:

Solving Second-Order Equations with Constant Coefficients

Let $ay'' + by' + cy = 0$ be a linear homogeneous second-order equation with constant real coefficients and let r_1 and r_2 be the solutions of the characteristic equation $ar^2 + br + c = 0$.

1. If $r_1 \neq r_2$ and both r_1 and r_2 are real, a general solution of $ay'' + by' + cy = 0$ is $y = c_1 e^{r_1 t} + c_2 e^{r_2 t}$; a fundamental set of solutions for the equation is $S = \{e^{r_1 t}, e^{r_2 t}\}$.
2. If $r_1 = r_2$, where r_1 is real, a general solution of $ay'' + by' + cy = 0$ is $y = c_1 e^{r_1 t} + c_2 t e^{r_1 t}$; a fundamental set of solutions for the equation is $S = \{e^{r_1 t}, t e^{r_1 t}\}$.
3. If $r_1 = \alpha + i\beta$, $\beta > 0$, and $m_2 = \overline{m_1} = \alpha - i\beta$, a general solution of $ay'' + by' + cy = 0$ is $y = e^{\alpha t}(c_1 \cos \beta t + c_2 \sin \beta t)$; a fundamental set of solutions for the equation is $S = \{e^{\alpha t} \cos \beta t, e^{\alpha t} \sin \beta t\}$.

 EXERCISES 4.2

In Exercises 1-15, find a general solution of each equation.

1. $y'' = 0$
2. $y'' - 4y' - 12y = 0$
3. $y'' + y' = 0$
4. $y'' + 3y' - 4y = 0$
5. $y'' + 8y' + 12y = 0$
6. $6y'' + 5y' + y = 0$
7. $8y'' + 6y' + y = 0$
8. $4y'' + 9y = 0$
9. $y'' + 16y = 0$
10. $y'' + 8y = 0$
11. $y'' + 7y = 0$
12. $4y'' + 21y' + 5y = 0$
13. $7y'' + 4y' - 3y = 0$
14. $4y'' + 4y' + y = 0$
15. $y'' - 6y' + 9y = 0$

In Exercises 16-31, solve the IVP. Graph the solution on an appropriate interval.

16. $y'' - y' = 0, y(0) = 3, y'(0) = 2$
17. $3y'' - y' = 0, y(0) = 0, y'(0) = 7$
18. $y'' + y' - 12y = 0, y(0) = 0, y'(0) = 7$
19. $y'' - 7y' + 12y = 0, y(0) = 3, y'(0) = -2$
20. $2y'' - 7y' - 4y = 0, y(0) = 0, y'(0) = 1$
21. $y'' - 7y' + 10y = 0, y(0) = 1, y'(0) = 5$
22. $y'' + 36y = 0, y(0) = 2, y'(0) = -6$
23. $y'' + 100y = 0, y(0) = 1, y'(0) = 10$
24. $y'' - 2y' + y = 0, y(0) = 4, y'(0) = 0$
25. $y'' + 4y' + 4y = 0, y(0) = 1, y'(0) = 3$
26. $y'' + 2y' + 5y = 0, y(0) = 1, y'(0) = 0$
27. $y'' + 4y' + 20y = 0, y(0) = 2, y'(0) = 0$
28. $y'' + y' - y = 0, y(0) = 1, y'(0) = 0$
29. $y'' + y' + y = 0, y(0) = 1, y'(0) = 0$
30. $y'' - y' + y = 0, y(0) = 0, y'(0) = 1$
31. $y'' - y' - y = 0, y(0) = 0, y'(0) = 1$
32. *(Operator Notation)* Assume that y is a function of t, $y = y(t)$ and that "prime" denotes differentiation with respect to t, $' = d/dt$. We define the *differential operator D* by $D = d/dt$. With this notation,
$D^2 = \frac{d}{dt} \times \frac{d}{dt} = \frac{d^2}{dt^2}, D^3 = \frac{d}{dt} \times \frac{d}{dt} \times \frac{d}{dt} = \frac{d^3}{dt^3}$,
and so on. Then, in *operator notation* the second-order linear homogeneous equation with constant coefficients,
$ay'' + by' + cy = 0, a \neq 0$, becomes

$$a\frac{d^2y}{dt^2} + b\frac{dy}{dt} + cy = 0$$

$$a\frac{d^2}{dt^2}(y) + b\frac{d}{dt}(y) + c(y) = 0$$

$$aD^2(y) + bD(y) + c(y) = 0$$

$$(aD^2 + bD + c)(y) = 0.$$

Observe that the characteristic equation for $ay'' + by' + cy = 0$ is $ar^2 + br + c = 0$. Write two or three sentences that describe the relationship between the second-order linear homogeneous equation using operator notation, $(aD^2 + bD + c)(y) = 0$, and the characteristic equation for the second-order linear homogeneous equation, $ar^2 + br + c = 0$.

In Exercises 33-35, the differential equation is given to you in factored operator notation. Use Exercise 32 to help you solve the differential equation. Plot the solution that satisfies the initial conditions $y(0) = 1$ and $y'(0) = 1$.

33. $6y'' + 5y' + 1 = 0, (3D + 1)(2D + 1)(y) = 0$
34. $9y'' + 6y' + y = 0, (3D + 1)^2(y) = 0$
35. $y'' + 4y' + 20y = 0$,
    $(D - (-2 + 4i))(D - (-2 - 4i))(y) = 0$
36. Find conditions on a and b so that the real part of the solutions of $r^2 + ar + b = 0$ is negative. If these conditions are satisfied what can you say about *all* solutions $\phi(t)$ of $y'' + ay' + by = 0$ as $\lim_{t \to \infty} \phi(t)$?
37. Use the substitution $t = e^x$ to solve
    (a) $3t^2y'' - 2ty' + 2y = 0, t > 0$ and
    (b) $t^2y'' - ty' + y = 0, t > 0$. *Hint:* Show that
    if $t = e^x$, $\frac{dy}{dt} = \frac{1}{t}\frac{dy}{dx}$ and $\frac{d^2y}{dt^2} = \frac{1}{t^2}\left(\frac{d^2y}{dx^2} - \frac{dy}{dx}\right)$.
38. *(Diabetes)* Under certain circumstances,[1] the equation

$$y'' + 2ay' + by = 0,$$

where $y(t) = Y(t) - Y_0$ is the difference between the concentration of glucose ($Y(t)$) in the bloodstream at time t hours after a glucose injection and Y_0 is the ideal concentration of glucose in the bloodstream, can help determine whether an individual has diabetes or not. Braun notices that if $2\pi/\sqrt{b} < 4$, the patient is unlikely to have diabetes while $2\pi/\sqrt{b} > 4$ indicates at least mild diabetes in some patients. Assume that $y(0) = 12$ and $y'(0) = 0$. Solve this equation if for a patient it is determined that $b = 4$ and then solve this equation for a patient if it is determined that $b = 1$. Plot the solution in each case. Is either patient likely to have diabetes? Explain.
39. Show that a general solution of the differential equation $ay'' + 2by' + cy = 0$, where $b^2 - ac > 0$, can be written as

---

[1]M. Braun, Differential Equations and Their Applications, second ed., Springer-Verlag, New York, 1978.

$$y = e^{-bt/a} \left[c_1 \cosh\left(\frac{\sqrt{b^2 - ac}}{a} t \right) \right.$$

$$\left. + c_2 \sinh\left(\frac{\sqrt{b^2 - ac}}{a} t \right) \right]$$

40. Express the solution to each differential equation in terms of the hyperbolic trigonometric function (see Exercise 39).
(a) $y'' + 6y' + 2y = 0$
(b) $y'' - 5y' + 6y = 0$
(c) $y'' - 6y' - 16y = 0$
(d) $y'' - 16y = 0$

41. Show that the boundary-value problem $y'' + 2y' + 5y = 0, y(0) = 0, y(\pi/2) = 0$ has infinitely many solutions, the boundary value $y'' + 2y' + 5y = 0, y(0) = 0, y(\pi/4) = 0$ has no nontrivial solutions, and that the boundary-value problem $y'' + 2y' + 5y = 0$, $y(0) = 1, y(\pi/4) = 0$ has one nontrivial solution.

42. Use factoring to solve each of the following nonlinear equations. For which equations, if any, is the Principle of Superposition valid? Explain
(a) $(y'')^2 - 5y''y + 4y^2 = 0$
(b) $(y'')^2 - 2y''y + y^2 = 0$

43. Find conditions on a and b, if possible, so that the solution to the IVP $y'' + 4y' + 3y = 0, y(0) = a, y'(0) = b$ has (a) neither local maxima nor local minima, (b) exactly one local maximum, and (c) exactly one local minimum on the interval $[0, \infty)$.

44. (a) Show that the roots of the characteristic equation of $x'' + a_1x' + a_0x = 0$ are

$$r_{1,2} = \frac{1}{2}\left(-a_1 \pm \sqrt{a_1^2 - 4a_0} \right).$$

(b) Graph $a_0 = \frac{1}{4}a_1^2$ and $a_1 = 0$, using the horizontal axis to represent a_0 and the vertical axis to represent a_1. *Randomly generate three points on the* $a_0 = \frac{1}{4}a_1^2$, *three points in the region* $a_0 < \frac{1}{4}a_1^2$, and *three points in the region* $a_0 > \frac{1}{4}a_1^2$.

(c) Write the second-order homogeneous equation $x'' + a_1x' + a_0x = 0$ as a system by letting $y = x' = dx/dt$.

(d) For each pair of points obtained in (b), graph the phase plane associated with the system for $-1 \le x \le 1$ and $-1 \le y \le 1$. If possible, compare your results with your classmates and note similarities and differences between your results.

(e) How does the phase plane associated with the system change as the roots of the characteristic equation change?

45. (a) Is it possible for $y = c_1t\cos t + c_2t\sin t$ to be a general solution of $ay'' + by' + cy = 0$, where a, b, and c are constants? (b) Find $p(t)$ and $q(t)$ so that $y = c_1t\cos t + c_2t\sin t$ is a general solution of $y'' + p(t)y' + q(t)y = 0$.

46. If y_p is a particular solution of $y'' + a_1(t)y' + a_0(t)y = f(t)$, what is a particular solution of $y''' + a_1'(t)y' + a_1(t)y'' + a_0'(t)y + a_0(t)y' = f'(t)$?

4.3 SOLVING SECOND-ORDER LINEAR EQUATIONS: UNDETERMINED COEFFICIENTS

- BASIC THEORY
- METHOD OF UNDETERMINED COEFFICIENTS

Basic Theory

We now turn our attention to the second-order linear nonhomogeneous equation

$$\frac{d^2y}{dt^2} + p(t)\frac{dy}{dt} + q(t)y = f(t) \qquad (4.6)$$

that has corresponding homogeneous equation

$$\frac{d^2y}{dt^2} + p(t)\frac{dy}{dt} + q(t)y = 0. \qquad (4.7)$$

From the previous section, we know that a general solution of Equation (4.7) is $y_h = c_1y_1 + c_2y_2$, where $S = \{y_1, y_2\}$ is a *fundamental set* of solutions of Equation (4.7): the elements of S are both solutions of the homogeneous equation and both elements of S are linearly independent.

A *particular solution* of a differential equation is a *specific* function that does not contain arbitrary constants that is a solution of the differential equation.

EXAMPLE 4.3.1

Verify that $y_p(t) = \frac{3}{5}\sin t$ is a particular solution of $y'' - 4y' = -3\sin t$.

Solution

First we compute $y_p'(t) = \frac{3}{5}\cos t$ and $y_p''(t) = -\frac{3}{5}\sin t$. Substituting into $y'' - 4y$ results in

$$y_p'' - 4y_p = -\frac{3}{5}\sin t - 4 \times \frac{3}{5}\sin t = -3\sin t.$$

We conclude that $y_p(t) = \frac{3}{5}\sin t$ is a particular solution of $y'' - 4y' = -3\sin t$ because y_p satisfies the equation $y'' - 4y' = -3\sin t$ and contains no arbitrary constants.

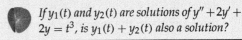 *Show that $y_p(t) = \frac{1}{5}e^{-2t}(15-10e^{4t}+3e^{2t}\sin t)$ is also a particular solution of $y'' - 4y' = -3\sin t$.*

Suppose that y is the general solution of Equation (4.6), y_p is a particular solution of Equation (4.6), and y_h is the general solution of Equation (4.7). Then,

$$(y - y_p)'' + p(t)(y - y_p)' + q(t)(y - y_p)$$

$$= y'' + p(t)y' + q(t)y - \left(y_p'' + p(t)y_p' + q(t)y_p\right)$$

$$= f(t) - f(t) = 0,$$

which shows us that $y - y_p$ is a solution of the corresponding homogeneous equation (4.7). Thus,

$$y - y_p = y_h \quad \text{or} \quad y = y_h + y_p.$$

That is, a general solution of the nonhomogeneous second-order linear equation $y'' + p(t)y' + q(t)y = f(t)$ is $y = y_h + y_p$, where y_p is a particular solution of $y'' + p(t)y' + q(t)y = f(t)$ and y_h is a general solution of the corresponding homogeneous equation, $y'' + p(t)y' + q(t)y = 0$.

EXAMPLE 4.3.2

Show that $y_p = \frac{1}{2}t^3 - \frac{3}{2}t^2 + \frac{3}{2}t$ is a particular solution of $y'' + 2y' + 2y = t^3$ and that $y_h = e^{-t}(c_1\cos t + c_2\sin t)$ is a general solution of the corresponding homogeneous equation $y'' + 2y' + 2y = 0$. What is a general solution of $y'' + 2y' + 2y = t^3$?

Solution

We first show that $y_p = \frac{1}{2}t^3 - \frac{3}{2}t^2 + \frac{3}{2}t$ is a particular solution of $y'' + 2y' + 2y = t^3$. Calculating $y_p' = \frac{3}{2}t^2 - 3t + \frac{3}{2}$, $y_p'' = 3t - 3$, and $y_p'' + 2y_p' + 2y_p$ gives us

$$y_p'' + 2y_p' + 2y_p = (3t - 3)$$

$$+ 2\left(\frac{3}{2}t^2 - 3t + \frac{3}{2}\right)$$

$$+ 2\left(\frac{1}{2}t^3 - \frac{3}{2}t^2 + \frac{3}{2}t\right) = 3t - 3$$

$$+ 3t^2 - 6t + 3 + t^3 - 3t^2 + 3t = t^3.$$

The corresponding homogeneous equation of $y'' + 2y' + 2y = t^3$ is $y'' + 2y' + 2y = 0$ with characteristic equation $r^2 + 2r + 2 = 0$ with roots $r_{1,2} = -1 \pm i$. Thus, a general solution of the corresponding homogeneous equation $y'' + 2y' + 2y = 0$ is $y_h = e^{-t}(c_1\cos t + c_2\sin t)$ so that a general solution of the nonhomogeneous equation $y'' + 2y' + 2y = t^3$ is

$$y = y_h + y_p = e^{-t}(c_1\cos t + c_2\sin t) + \frac{1}{2}t^3 - \frac{3}{2}t^2 + \frac{3}{2}t.$$

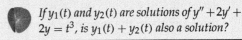 *If $y_1(t)$ and $y_2(t)$ are solutions of $y'' + 2y' + 2y = t^3$, is $y_1(t) + y_2(t)$ also a solution?*

Generally, to solve the nonhomogeneous equation, we begin by solving the corresponding homogeneous equation to find y_h and then find a particular solution, y_p, of the nonhomogeneous equation. Then, a general solution of the nonhomogeneous equation is the sum of the general solution to the corresponding homogeneous equation and a particular solution of the nonhomogeneous equation: $y = y_h + y_p$.

Method of Undetermined Coefficients

We now turn our attention to determining how to find a particular solution, y_p, of the nonhomogeneous equation

$$ay'' + by' + cy = f(t),$$

where a, b, and c are constants and the *forcing function*, $f(t)$, is a linear combination of functions of the form

$$1, t, t^2, \ldots,$$
$$e^{\alpha t}, te^{\alpha t}, t^2 e^{\alpha t}, \ldots,$$
$$\cos \beta t, \sin \beta t, t \cos \beta t, t \sin \beta t,$$
$$t^2 \cos \beta t, t^2 \sin \beta t, \ldots$$

or

$$e^{\alpha t} \cos \beta t \, e^{\alpha t} \sin \beta t, te^{\alpha t} \cos \beta t, te^{\alpha t} \sin \beta t,$$
$$t^2 e^{\alpha t} \cos \beta t, t^2 e^{\alpha t} \sin \beta t, \ldots. \qquad (4.8)$$

After determining the general solution, y_h, of the corresponding homogeneous equation, the *Method of Undetermined Coefficients* can be used to determine a particular solution, y_p, of the nonhomogeneous equation $ay'' + by' + cy = f(t)$. A general solution of the nonhomogeneous equation is then given by $y = y_h + y_p$.

 Outline of the Method of Undetermined Coefficients to Solve

$$ay'' + by' + cy = f(t),$$

where $f(t)$ is a linear combination of the functions in Equation (4.8).

1. Solve the corresponding homogeneous equation for $y_h(t)$.
2. Determine the form of a particular solution $y_p(t)$ (see *Determining the Form of* $y_p(t)$ next).
3. Determine the unknown coefficients in $y_p(t)$ by substituting $y_p(t)$ into the nonhomogeneous equation and equating the coefficients of like terms.
4. Form a general solution with $y(t) = y_h(t) + y_p(t)$.

Determining the Form of **y_p(t)** *(Step 2)*

Suppose that $f(t) = b_1 f_1(t) + b_2 f_2(t) + \cdots + b_j f_j(t)$, where $b_1, b_2, \ldots, b_j$ are constants and each $f_i(t)$, $i = 1, 2, \ldots, j$, is a function of the form t^m, $t^m e^{\alpha t}$, $t^m e^{\alpha t} \cos \beta t$, or $t^m e^{\alpha t} \sin \beta t$.

1. If $f_i(t) = t^m$, the associated set of functions is

$$F = \left\{ 1, t, t^2, \ldots, t^m \right\}.$$

2. If $f_i(t) = t^m e^{\alpha t}$, the associated set of functions is

$$F = \left\{ e^{\alpha t}, te^{\alpha t}, t^2 e^{\alpha t}, \ldots, t^m e^{\alpha t} \right\}.$$

3. If $f_i(t) = t^m e^{\alpha t} \cos \beta t$ or $f_i(t) = t^m e^{\alpha t} \sin \beta t$, the associated set of functions is

$$F = \{ e^{\alpha t} \cos \beta t, \, te^{\alpha t} \cos \beta t, \, t^2 e^{\alpha t} \cos \beta t, \ldots,$$
$$t^m e^{\alpha t} \cos \beta t,$$
$$e^{\alpha t} \sin \beta t, \, te^{\alpha t} \sin \beta t, \, t^2 e^{\alpha t} \sin \beta t, \ldots,$$
$$t^m e^{\alpha t} \sin \beta t \}.$$

For each function $f_i(t)$ in $f(t)$, determine the associated set of functions F_i. If any of the functions in F_i appears in the general solution to the corresponding homogeneous equation, $y_h(t)$, multiply each function in F_i by t^r to obtain a new set $t^r F_i$, where r is the smallest positive integer so that each function in $t^r F_i$ is not a solution of the corresponding homogeneous equation. A particular solution is obtained by taking the linear combination of all functions in the associated sets where repeated functions should appear only once in the particular solution.

The Method of Undetermined Coefficients works because linear combinations of the functions and their derivatives stated in *Step 2 are* the forcing functions, $f(t)$, in Equation (4.8).

EXAMPLE 4.3.3

For each of the following functions, determine the associated set(s) of functions: (a) $f(t) = t^4$, (b) $f(t) = t^3 e^{-2t}$, (c) $f(t) = t^2 e^{-t} \cos 4t$, and (d) $f(t) = \sin 2t + t^2 e^{2t}$.

Solution

(a) Using (1), we have $F = \{t^4, t^3, t^2, t, 1\}$. (b) In this case we use (2): $F = \{t^3 e^{-2t}, t^2 e^{-2t}, te^{-2t}, e^{-2t}\}$. (c) According to (3),

$$F = \left\{ t^2 e^{-t} \cos 4t, \ t^2 e^{-t} \sin 4t, \ te^{-t} \cos 4t, \right.$$

$$\left. te^{-t} \sin 4t, e^{-t} \cos 4t, \ e^{-t} \sin 4t \right\}.$$

(d) Using (2) and (3), we have two associated sets:

$$F_1 = \{\cos 2t, \sin 2t\} \quad \text{and} \quad F_2 = \{t^2 e^{2t}, te^{2t}, e^{2t}\}.$$

In general, *do not* allow the associated sets to overlap. For example, if $f(t) = 7te^{-t} + 4e^{-t} - 3e^t$, the associated set for $7te^{-t}$ is $F_1 = \{te^{-t}, e^{-t}\}$. The associated set for $4e^{-t}$ is $F_{1,1} = \{e^{-t}\}$. Because the associated set for $4e^{-t}$ is contained in the associated set for $7te^{-t}$, the associated set for the two terms is the union (or "larger") of the two associated sets: $F_1 = \{te^{-t}, e^{-t}\}$. The associated set for $-3e^t$ is $F_2 = \{e^t\}$. Thus, for $f(t) = 7te^{-t} + 4e^{-t} - 3e^t$ we have two associated sets: $F_1 = \{te^{-t}, e^{-t}\}$ and $F_2 = \{e^t\}$.

EXAMPLE 4.3.4

Solve the nonhomogeneous equations (a) $y'' + 5y' + 6y = 2e^t$, (b) $y'' + 5y' + 6y = 2t^2 + 3t$, (c) $y'' + 5y' + 6y = 3e^{-2t}$, and (d) $y'' + 5y' + 6y = 4 \cos t$.

In other words, no element of F is an element of S.

Solution

For all four equations, the corresponding homogeneous equation is $y'' + 5y' + 6y = 0$, which has general solution $y_h = c_1 e^{-2t} + c_2 e^{-3t}$. (Why?) A fundamental set of solutions for the corresponding homogeneous equation is $S = \{e^{-2t}, e^{-3t}\}$. (a) Now, we determine the form of y_p. Because $f(t) = 2e^t$, the associated set is $F = \{e^t\}$. Notice that e^t is not a solution to the homogeneous equation, so we take y_p to be the linear combination of the functions in F. Therefore, we assume that a particular solution of the nonhomogeneous equation has the form $y_p = Ae^t$, where A is a constant to be determined. Substituting y_p into the left-hand side of $y'' + 5y' + 6y = 2e^t$ and setting the result equal to $2e^t$ gives us

$$Ae^t + 5Ae^t + 6Ae^t = 12Ae^t = 2e^t.$$

Equating the coefficients of e^t then gives us $12A = 2$ so $A = 1/6$. A particular solution of the nonhomogeneous equation is $y_p = \frac{1}{6}e^t$, so a general solution of $y'' + 5y' + 6y = 2e^t$ is

$$y = y_h + y_p = c_1 e^{-2t} + c_2 e^{-3t} + \frac{1}{6}e^t.$$

(b) Here, we see that $f(t)$ is a linear combination of the functions given in Equation (4.8): $f(t) = a_1 f_1(t) + a_2 f_2(t) = 2t^2 + 3t$. The set of functions associated with $f_1(t) = 2t^2$ is $F_1 = \{t^2, t, 1\}$ and that associated with $f_2(t) = 3t$ is $F_2 = \{t, 1\}$. First, notice that none of these functions are solutions of the corresponding homogeneous equation. Further, notice that F_1 and F_2 have (two) elements in common (t and 1). If we take the linear combination of the functions $F = F_1 \cup F_2 = F_1$, the set consisting of t^2, t, and 1, then we look for a particular solution of the form

$$y_p = At^2 + Bt + C,$$

where A, B, and C are constants to be determined.

Substitution of y_p and the derivatives $y_p' = 2At + B$ and $y_p'' = 2A$ into the nonhomogeneous equation $y'' + 5y' + 6y = 2t^2 + 3t$ yields

$$y_p'' + 5y_p' + 6y_p = 6At^2 + (10A + 6B)t$$
$$+ (2A + 5B + 6C)$$
$$= 2t^2 + 3t.$$

Equating coefficients of the linearly independent terms (t^2, t, and 1) then yields, $6A = 2$, $10A + 6B = 3$, and $2A + 5B + 6C = 0$. This system of equations has the solution $A = 1/3$, $B = -1/18$, and $C = -7/108$, so $y_p = \frac{1}{3}t^2 - \frac{1}{18}t - \frac{7}{108}$. A general solution of the nonhomogeneous equation is

$$y = y_h + y_p = c_1 e^{-2t} + c_2 e^{-3t} + \frac{1}{3}t^2 - \frac{1}{18}t - \frac{7}{108}.$$

(c) In this case, we see that $f(t) = 3e^{-2t}$ with associated set $F = \{e^{-2t}\}$. However, because e^{-2t} is a solution to the corresponding homogeneous equation, we must multiply each element of F by t^r where r is the smallest positive integer so that *no function* in $t^r F$ is a solution to the corresponding homogeneous equation. We multiply by $t^1 = t$ to obtain $tF = \{te^{-2t}\}$ because te^{-2t} is not a solution of $y'' + 5y' + 6y = 0$. Differentiating $y_p = Ate^{-2t}$ twice and substituting into the nonhomogeneous equation yields:

$$y_p'' + 5y_p' + 6y_p = -4Ae^{-2t} + 4Ate^{-2t}$$
$$+ 5\left(Ae^{-2t} - 2Ate^{-2t}\right)$$
$$+ 6Ate^{-2t}$$
$$= Ae^{-2t} = 3e^{-2t}.$$

Thus, $A = 3$ so $y_p = 3te^{-2t}$, and a general solution of the nonhomogeneous equation is

$$y = y_h + y_p = c_1 e^{-2t} + c_2 e^{-3t} + 3te^{-2t}.$$

(d) In this case, we see that $f(t) = 4\cos t$, so the associated set is $F = \{\cos t, \sin t\}$. Because neither of these functions is a solution of the corresponding homogeneous equation, we take the

particular solution to be the linear combination $y_p = A\cos t + B\sin t$. We determine A and B by differentiating y_p twice and substituting into the nonhomogeneous equation

$$y_p'' + 5y_p' + 6y_p = -A\cos t - B\sin t$$
$$+ 5(B\cos t - A\sin t) + 6(A\cos t + B\sin t)$$
$$= (5A + 5B)\cos t + (-5A + 5B)\sin t$$
$$= 4\cos t.$$

Hence, $5A + 5B = 4$ and $-5A + 5B = 0$, so $A = B = 2/5$. Therefore, $y_p = \frac{2}{5}\cos t + \frac{2}{5}\sin t$ and a general solution of the nonhomogeneous equation is

$$y = y_h + y_p = c_1 e^{-2t} + c_2 e^{-3t} + \frac{2}{5}\cos t + \frac{2}{5}\sin t.$$

EXAMPLE 4.3.5

Determine the *form* of a particular solution to each of the following nonhomogeneous equations: (a) $4y'' + 4y' + y = t$, (b) $4y'' + 4y' + y = t^2 e^{-t/2}$, and (c) $4y'' + 4y' + y = t^3 e^{-t/2} + t^2 + 1$.

Solution

For all three equations, the corresponding homogeneous equation is $4y'' + 4y' + y = 0$, which has characteristic equation $4r^2 + 4r + 1 = (2r+1)^2 = 0$ so $r_{1,2} = -1/2$ has multiplicity two. Therefore, two linearly independent solutions to the corresponding homogeneous equation are $y_1 = e^{-t/2}$ and $y_2 = te^{-t/2}$; a fundamental set of solutions is $S = \{e^{-t/2}, te^{-t/2}\}$; and a general solution of the corresponding homogeneous equation is $y_h = c_1 e^{-t/2} + c_2 te^{-t/2}$.

(a) For $f(t) = t$, the associated set is $F = \{t, 1\}$. Because no function in F is a solution to the corresponding homogeneous equation, we assume that a particular solution has the form $y_p = At + B$, where A and B are constants to be determined.

(b) For $f(t) = t^2 e^{-t/2}$, the associated set is $F = \{t^2 e^{-t/2}, te^{-t/2}, e^{-t/2}\}$. Because F contains solutions to the corresponding homogeneous equation, we multiply each function in F by t^r where r is the smallest positive integer so that *no function* in $t^r F$ is a solution to the corresponding homogeneous equation. In this case, we multiply by t^2 to obtain $t^2 F = \{t^4 e^{-t/2}, t^3 e^{-t/2}, t^2 e^{-t/2}\}$. Therefore, we assume that a particular solution has the form $y_p = A t^4 e^{-t/2} + B t^3 e^{-t/2} + C t^2 e^{-t/2}$.

(c) At first glance, it appears as though $f(t) = f_1(t) + f_2(t) + f_3(t) = t^3 e^{-t/2} + t^2 + 1$ has three associated sets. The set of functions associated with $f_1(t) = t^3 e^{-t/2}$ is $F_1 = \{t^3 e^{-t/2}, t^2 e^{-t/2}, t e^{-t/2}, e^{-t/2}\}$, the set of functions associated with $f_2(t) = t^2$ is $F_2 = \{t^2, t, 1\}$, and the set of functions associated with $f_3(t) = 1$ is $F_3 = \{1\}$. But, because the set of functions associated with the function $f_3(t) = 1$ is contained in the set of functions associated with the function $f_2(t) = t^2$ ("the larger set"), we need only consider the two associated sets

$$F_1 = \{t^3 e^{-t/2}, t^2 e^{-t/2}, t e^{-t/2}, e^{-t/2}\} \quad \text{and}$$

$$F_2 = \{t^2, t, 1\}.$$

We see that F_1 contains solutions of the corresponding homogeneous equation so we multiply F_1 by t^r where r is the smallest positive integer so that *no function* in $t^r F_1$ is a solution to the corresponding homogeneous equation. In this case, we multiply by t^2 to obtain $t^2 F_1 = \{t^5 e^{-t/2}, t^4 e^{-t/2}, t^3 e^{-t/2}, t^2 e^{-t/2}\}$. We see that F_2 does not contain any solutions to the corresponding homogeneous equation. Therefore, we assume that a particular solution has the form

$$y_p = \underbrace{A t^5 e^{-t/2} + B t^4 e^{-t/2} + C t^3 e^{-t/2} + D t^2 e^{-t/2}}_{\text{terms contributed by } t^2 F_1}$$

$$+ \underbrace{E t^2 + F t + G}_{\text{terms contributed by } F_2} .$$

To solve an IVP, first determine a general solution and then use the initial conditions to solve for the unknown constants in the general solution.

EXAMPLE 4.3.6

Solve $9y'' + 6y' + 37y = 36 t e^{-t/3} + 18 e^{-t/3} \cos 2t$, $y(0) = 0$, $y'(0) = 1$.

Solution

The corresponding homogeneous equation $9y'' + 6y' + 37y = 0$ has characteristic equation $9r^2 + 6r + 37 = 0$ with solutions $r_{1,2} = -\frac{1}{3} \pm 2i$. Therefore, a general solution of the corresponding homogeneous equation is $y_h = c_1 e^{-t/3} \cos 2t + c_2 e^{-t/3} \sin 2t$.

Looking at the two terms in the forcing function, the set of functions associated with $f_1(t) = 36 t e^{-t/3}$ is $F_1 = \{t e^{-t/3}, e^{-t/3}\}$. No function in F_1 is a solution to the corresponding homogeneous equation. The set of functions associated with $f_2(t) = 18 e^{-t/3} \cos 2t$ is $F_2 = \{e^{-t/3} \cos 2t, e^{-t/3} \sin 2t\}$. F_2 contains functions that are solutions to the corresponding homogeneous equation so we multiply F_2 by t to obtain $t F_2 = \{t e^{-t/3} \cos 2t, t e^{-t/3} \sin 2t\}$. Therefore, we assume that a particular solution has the form

$$y_p = \underbrace{A t e^{-t/3} + B e^{-t/3}}_{\text{terms contributed by } F_1}$$

$$+ \underbrace{C t e^{-t/3} \cos 2t + D t e^{-t/3} \sin 2t}_{\text{terms contributed by } t F_2} .$$

Differentiating y_p twice gives us

$$y_p' = -\frac{1}{3} A t e^{-t/3} + \left(A - \frac{1}{3} B \right) e^{-t/3}$$

$$+ \left(-\frac{1}{3} C + 2D \right) t e^{-t/3} \cos 2t$$

$$+ \left(2C - \frac{1}{3} D \right) t e^{-t/3} \sin 2t + C e^{-t/3} \cos 2t$$

$$+ D e^{-t/3} \sin 2t$$

and

$$y_P'' = \frac{1}{9}Ate^{-t/3} + \left(-\frac{2}{3}A + \frac{1}{9}B\right)e^{-t/3}$$

$$+ \left(-\frac{35}{9}C - \frac{4}{3}D\right)te^{-t/3}\cos 2t$$

$$+ \left(\frac{4}{3}C - \frac{35}{9}\right)te^{-t/3}\sin 2t$$

$$+ \left(-\frac{2}{3}C + 4D\right)e^{-t/3}\cos 2t$$

$$+ \left(-4C - \frac{2}{3}D\right)e^{-t/3}\sin 2t.$$

Substituting y_P and its derivatives into the non-homogeneous equation and simplifying the result leads to

$$9y_P'' + 6y_P' + 37y_P = 36Ate^{-t/3} + 36Be^{-t/3}$$
$$+ 36De^{-t/3}\cos 2t - 36Ce^{-t/3}\sin 2t = 36te^{-t/3}$$
$$+ 18e^{-t/3}\cos 2t.$$

Equating coefficients, $36A = 36$, $36B = 0$, $36D = 18$, and $-36C = 0$ so $A = 1$, $B = 0$, $D = 1/2$, and $C = 0$ resulting in $y_P = te^{-t/3} + \frac{1}{2}e^{-t/3}\sin 2t$. A general solution of the nonhomogeneous equation is then

$$y = y_h + y_p = c_1 e^{-t/3}\cos 2t + c_2 e^{-t/3}\sin 2t$$

$$+ te^{-t/3} + \frac{1}{2}te^{-t/3}\sin 2t.$$

To solve the IVP, we compute y',

$$y' = -\frac{1}{3}te^{-t/3} + e^{-t/3} + te^{-t/3}\cos 2t$$

$$- \frac{1}{6}te^{-t/3}\sin 2t$$

$$+ \left(-\frac{1}{3}c_1 + 2c_2\right)e^{-t/3}\cos 2t$$

$$+ \left(-2c_1 - \frac{1}{3}c_2\right)e^{-t/3}\sin 2t$$

and then apply the initial conditions

$$y(0) = c_1 = 0 \quad \text{and} \quad y'(0) = 1 - \frac{1}{3}c_1 + 2c_2 = 1$$

to see that $c_1 = c_2 = 0$. Therefore, the solution to the IVP is $y = te^{-t/3} + \frac{1}{2}e^{-t/3}\sin 2t$, which is

plotted in Figure 4.4(c). Keep in mind that usually two initial values are required for a second-order initial value to have a unique solution, as illustrated in Figure 4.4(a) and (b).

EXAMPLE 4.3.7

If $\omega \geq 0$ is constant, solve $y'' + y = \cos\omega t$, $y(0) = 0$, $y'(0) = 1$.

Solution

A general solution of the corresponding homogeneous equation $y'' + y = 0$ is $y_h = c_1\cos t + c_2\sin t$. We see that if $\omega \neq 1$, we use Undetermined Coefficients to find a particular solution of the nonhomogeneous equation of the form $y_p = A\cos\omega t + B\sin\omega t$, while if $\omega = 1$ we use Undetermined Coefficients to find a particular solution of the nonhomogeneous equation of the form $y_p = At\cos\omega t + Bt\sin\omega t$. Solving for a particular solution, forming a general solution, and applying the initial conditions yields the solution

$$y = \begin{cases} \dfrac{1}{\omega^2 - 1}(\cos t - \cos\omega t) + \sin t, & \text{if } \omega \neq 1 \\ \dfrac{1}{2}(t + 2)\sin t, & \text{if } \omega = 1 \end{cases}.$$

Notice that the behavior of the solution changes dramatically when $\omega = 1$: if $0 \leq \omega < 1$ or $\omega > 1$, the solution is periodic and bounded (Why?), but if $\omega = 1$, the solution is unbounded (see Figure 4.5).

 EXERCISES 4.3

For each function in Exercises 1-12, determine the associated set(s) of functions.

1. $f(t) = 2t + 5$
2. $f(t) = 3 - 8t^2$
3. $f(t) = 7e^{2t} + 2$

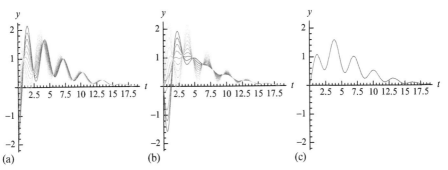

FIGURE 4.4 (a) Plots of some solutions satisfying $y'(0) = 1$. (b) Plots of some solutions satisfying $y(0) = 0$. (c) Plot of *the* solution satisfying $y(0) = 0$ and $y'(0) = 1$.

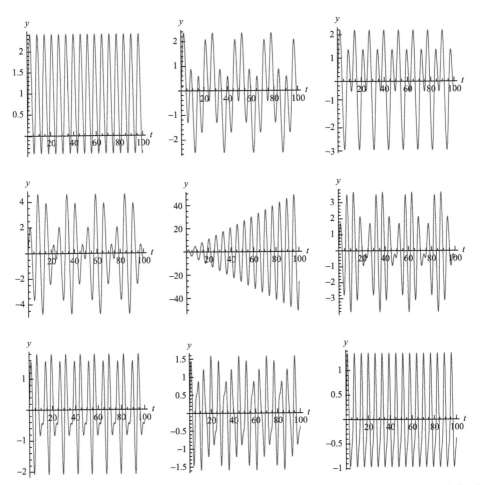

FIGURE 4.5 From left to right, $\omega = 0, 1/4, 1/2, 3/4, 1, 5/4, 3/2, 7/4$, and 2. Notice that the solution is bounded unless $\omega = 1$.

4. $f(t) = -2te^{-t}$
5. $f(t) = 4e^{-t} - 2t^4 + t$
6. $f(t) = 4t - 8t^2 e^{-3t}$
7. $f(t) = 6\sin 2t + 3e^{-4t}$
8. $f(t) = 2t\sin t$
9. $f(t) = 2\sin 3t + 4\sin 2t - 8\cos 2t$
10. $f(t) = 2te^t \sin t - 7t^3 + 9$
11. $f(t) = e^{-t}\cos 2t + 1$
12. $f(t) = te^{2t}\sin t + \cos 4t + 3t^2 - 8e^{-4t}$

In Exercises 13-24, determine the *form* of a particular solution for the given nonhomogeneous equation. (Do not solve for the unknown coefficients.)

13. $y'' + y = 8e^{2t}$
14. $y'' - 4y' + 3y = -e^{-9t}$
15. $y'' - 4y' + 3y = 2e^{3t}$
16. $y'' - y = 2t - 4$
17. $y'' - 2y' + y = t^2$
18. $y'' + 2y' = 3 - 4t$
19. $y'' + y = \cos 2t$
20. $y'' + 4y = 4\cos t - \sin t$
21. $y'' + 4y = \cos 2t + t$
22. $y'' + 4y = 3te^{-t}$
23. $y'' = 3t^4 - 2t$
24. $y'' - 4y' + 13y = 2te^{-2t}\sin 3t$

In Exercises 25-45, find a general solution of each equation.

25. $y'' + y' - 2y = -1$
26. $5y'' + y' - 4y = -3$
27. $y'' - 2y' - 8y = 32t$
28. $16y'' - 8y' - 15y = 75t$
29. $y'' + 2y' + 26y = -338t$
30. $y'' + 3y' - 4y = -32t^2$
31. $8y'' + 6y' + y = 5t^2$
32. $y'' - 6y' + 8y = -256t^3$
33. $y'' - 2y' = 52\sin 3t$
34. $y'' - 6y' + 13y = 25\sin 2t$
35. $y'' - 9y = 54t\sin 2t$
36. $y'' - 5y' + 6y = -78\cos 3t$
37. $y'' + 4y' + 4y = -32t^2\cos 2t$
38. $y'' - y' - 20y = -2e^t$
39. $y'' - 4y' - 5y = -648t^2 e^{5t}$
40. $y'' - 7y' + 12y = -2t^3 e^{4t}$
41. $y'' + 4y' = 8e^{4t} - 4e^{-4t}$
42. $y'' - 3y' = t^2 - e^{3t}$
43. $y'' + 4y' = -24t - 6 - 4te^{-4t} + e^{-4t}$
44. $y'' - 3y' = t^2 - e^{3t}$
45. $y'' = t^2 + e^t + \sin t$

In Exercises 46-58, solve the IVP

46. $y'' + 3y' = 18, y(0) = 0, y'(0) = 3$
47. $y'' - y = 4, y(0) = 0, y'(0) = 0$
48. $y'' - 4y = 32t, y(0) = 0, y'(0) = 6$
49. $y'' + 2y' - 3y = -2, y(0) = 2/3, y'(0) = 8$
50. $y'' + y' - 6y = 3t, y(0) = 23/12, y'(0) = -3/2$
51. $y'' + 8y' + 16y = 4, y(0) = 5/4, y'(0) = 0$
52. $y'' + 7y' + 10y = te^{-t}, y(0) = -15/48,$
    $y'(0) = 9/16$
53. $y'' + 6y' + 25y = -1, y(0) = -1/25, y'(0) = 7$
54. $y'' - 3y' = -e^{3t} - 2t, y(0) = 0, y'(0) = 8/9$
55. $y'' - y' = -3t - 4t^2 e^{2t}, y(0) = -7/2, y'(0) = 0$
56. $y'' - 2y' = 2t^2, y(0) = 3, y'(0) = 3/2$
57. $y'' + 4y' = -24t - 6 - 4te^{-4t} + e^{-4t}, y(0) = 0, y'(0) = 0$
58. $y'' - 3y' = e^{-3t} - e^{3t}, y(0) = 1, y'(0) = 1$

In Exercises 59-61, the forcing function is piecewise defined. Solve each IVP and graph the solution on an appropriate interval.

59. $y'' + 9y = f(t), y(0) = y'(0) = 0,$
    where $f(t) = \begin{cases} 2t, 0 \le t < \pi \\ 0, t \ge \pi \end{cases}$
60. $y'' + 9\pi^2 y = f(t), y(0) = 0, y'(0) = 0,$
    where $f(t) = \begin{cases} 2t, 0 \le t < \pi \\ 2(t-\pi), \pi \le t < 2\pi \\ 0, t \ge 2\pi \end{cases}$
61. $y'' + 4y = f(t), y(0) = 0, y'(0) = 2,$
    where $f(t) = \begin{cases} 0, 0 \le t < \pi \\ 10, \pi \le t < 2\pi \\ 0, t \ge 3\pi \end{cases}$
62. The method of undetermined coefficients can be used to solve first-order linear equations with constant coefficients if the forcing function is of the form given by

Equation (4.8). Use this method to solve the following problems.

(a) $y' - 4y = t^2$
(b) $y' + y = \cos 2t$, $y(0) = 0$
(c) $y' - y = e^{4t}$
(d) $y' + 4y = e^{-4t}$, $y(0) = 1$
(e) $y' + 4y = te^{-4t}$

63. (a) Suppose that y_1 and y_2 are solutions of $ay'' + by' + cy = f(t)$, where $a \neq 0$, b, and c are positive constants. Show that
$$\lim_{t\to\infty} (y_1(t) - y_2(t)) = 0.$$

(b) Is the result of (a) true if $b = 0$?
(c) Suppose that $f(t) = k$ where k is a constant. Show that $\lim_{t\to\infty} y(t) = k/c$ for every solution $y(t)$ of $ay'' + by' + cy = k$.
(d) Determine the solution of $ay'' + by' + cy = k$ and use the solution to calculate $\lim_{t\to\infty} y(t)$.
(e) Determine the solution of $y'' = k$ and calculate $\lim_{t\to\infty} y(t)$.

64. Let $y_p = \frac{3}{5}\sin t$. Show that if $y(t)$ is any solution of $y'' - 4y = -3\sin t$ then $y(t) - y_p(t)$ is a solution of $y'' - 4y = 0$. Explain why there are constants c_1 and c_2 so that $y(t) - y_p(t) = c_1 e^{-2t} + c_2 e^{2t}$ and, thus, $y(t) = c_1 e^{-2t} + c_2 e^{2t} + y_p(t)$.

65. Show that if y_1 is a solution of $y'' + p(t)y' + q(t)y = f_1(t)$ and y_2 is a solution of $y'' + p(t)y' + q(t)y = f_2(t)$, then $y_1 + y_2$ is a solution of $y'' + p(t)y' + q(t)y = f_1(t) + f_2(t)$. Generalize this result.

66. (a) Find conditions on ω so that $y_p(t) = A\cos\omega t + B\sin\omega t$, where A and B are constants to be determined, is a particular solution of $y'' + y = \cos\omega t$.
(b) For the value(s) of ω obtained in (a), show that every solution of the nonhomogeneous equation $y'' + y = \cos\omega t$ is bounded (and periodic).

67. (a) Find conditions on ω so that $y_p(t) = t(A\cos\omega t + B\sin\omega t)$, where A and B are constants to be determined, is a

particular solution of $y'' + 4y = \sin\omega t$. **(b)** For the value(s) of ω obtained in (a), show that every solution of the nonhomogeneous equation $y'' + 4y = \sin\omega t$ is unbounded.

68. Show that the solution to the IVP
$$y'' + y' - 2y = f(t), \quad y(0) = 0, \quad y'(0) = a$$
is
$$y(t) = \frac{1}{3}ae^t - \frac{1}{3}ae^{-2t} + \frac{1}{3}e^t \int_0^t f(x)e^{-x}dx$$
$$- \frac{1}{3}e^{-2t} \int_0^t f(x)e^{2x}dx$$
(as long as the integrals can be evaluated).
(a) Use the Fundamental Theorem of Calculus to verify this result.
(b) If possible, find a function $f(t)$ and a value of a so that (i) the solution is periodic, (ii) the limit of the solution as t approaches infinity is 0, and (iii) the limit of the solution as t approaches infinity does not exist. Confirm your results graphically.

69. How does the solution to the IVP $x'' + 9x = \sin ct$, $x(0) = 0$, $x'(0) = 0$ change as c assumes values from 0 to 6?

70. (a) Show that the boundary-value problem $4y'' + 4y' + 37y = \cos 3t$, $y(0) = y(\pi)$ has infinitely many solutions. Confirm your results graphically.
(b) If possible, find conditions on $y(0)$ and $y(\pi)$ so that the solution to the boundary-value problem is unique.

4.4 SOLVING SECOND-ORDER LINEAR EQUATIONS: VARIATION OF PARAMETERS

In Section 4.3 we learned that to solve
$$y'' + p(t)y' + q(t)y = f(t) \tag{4.9}$$
we need a general solution of the corresponding homogeneous equation
$$y'' + p(t)y' + q(t)y = 0, \tag{4.10}$$

$y_h = c_1 y_1 + c_2 y_2$, where $S = \{y_1, y_2\}$ is a fundamental set of solutions for the corresponding homogeneous equation (4.10), and a particular solution, y_p, of the nonhomogeneous equation (4.9).

In the special case when the coefficients are constants and the forcing function $f(t)$ has the form given by Equation (4.8), the method of undetermined coefficients can be used to find a particular solution of the nonhomogeneous equation (4.9).

When undetermined coefficients is not practical to use to find a particular solution of Equation (4.9), such as when the coefficient functions are *not* constant or when $f(t)$ *does not* have the form given by Equation (4.8), we can "vary" the parameters c_1 and c_2 in much the same way as we did in Section 2.3 to find a particular solution of Equation (4.9). However, instead of "varying" one parameter as we did in Section 2.3 for the first-order linear nonhomogeneous equation, for the second-order linear nonhomogeneous equation we now vary two parameters.

Starting with the second-order linear nonhomogeneous equation, $a_2(t)y'' + a_1(t)y' + a_0(t)y = g(t)$, we begin by dividing by $a_2(t)$ to write it in the standard (or normal) form given by Equation (4.9). Let $y_h = c_1 y_1 + c_2 y_2$ be a general solution of the corresponding homogeneous equation, where $S = \{y_1, y_2\}$ is a fundamental set of solutions for the corresponding homogeneous equation (4.10). We now assume that a particular solution has a form similar to the general solution by "varying" the parameters c_1 and c_2 and let

$$y_p(t) = u_1(t)y_1(t) + u_2(t)y_2(t). \qquad (4.11)$$

We determine the functions $u_1 = u_1(t)$ and $u_2 = u_2(t)$ by requiring that their derivatives, u_1' and u_2', satisfy a linear system of equations with a unique nontrivial solution. Integrating will then yield nonconstant u_1 and u_2. A particular solution of Equation (4.9) will then be given by $y_p = u_1 y_1 + u_2 y_2$.

For convenience, in the following calculations we omit the arguments of y_p, u_1, u_2, y_1, and y_2. Differentiating y_p, we obtain

$$y_p'(t) = u_1' y_1 + u_1 y_1' + u_2' y_2 + u_2 y_2',$$

which is simplified to

$$y_p' = u_1 y_1' + u_2 y_2',$$

with the assumption that

$$y_1 u_1' + y_2 u_2' = 0, \qquad (4.12)$$

which is our *first equation* for determining u_1 and u_2. The assumption in Equation (4.12) is made to eliminate the derivatives of u_1 and u_2, u_1' and u_2', from y_p', which will simplify the process of finding u_1' and u_2' for us. Using our simplified expression for y_p', the second derivative of y_p is

$$y_p'' = u_1 y_1'' + u_1' y_1' + u_2 y_2'' + u_2' y_2'.$$

At this point, assuming that $y_1' u_1' + y_2' u_2' = 0$ would lead to concluding that $u_1' = 0$ and $u_2' = 0$ so that u_1 and u_2 would be constant functions, which would not give us a particular solution of the nonhomogeneous equation (4.9). (Why?) Substitution of y_p, y_p', and y_p'' into the nonhomogeneous equation (4.9), then yields

---

Remember that for the second-order linear homogeneous equation $y'' + a_1(t)y' + a_0(t)y = 0$, a fundamental set of solutions, $S = \{y_1, y_2\}$, is a set of two linearly independent solutions. Then, a general solution of $y'' + a_1(t)y' + a_0(t)y = 0$ is given by $y_h = c_1 y_1 + c_2 y_2$.

Hint: Every linear combination of solutions to a linear homogeneous equation is a solution.

$$y_p'' + p(t)y_p' + q(t)y_p = u_1\left(y_1'' + p(t)y_1' + q(t)y_1\right)$$

$$+ u_2\left(y_2'' + p(t)y_2' + q(t)y_2\right) + y_1'u_1' + y_2'u_2'$$

$$= y_1'u_1' + y_2'u_2' = f(t),$$

where we have used the fact that both $y_1'' + p(t)y_1' + q(t)y_1 = 0$ and $y_2'' + p(t)y_2' + q(t)y_2 = 0$ because y_1 and y_2 are solutions of the corresponding homogeneous equation (4.10). Therefore, our *second equation* for determining u_1' and u_2' is

$$y_1'u_1' + y_2'u_2' = f(t). \qquad (4.13)$$

Combining Equations (4.12) and (4.13) gives us the linear system in u_1' and u_2',

$$y_1 u_1' + y_2 u_2' = 0,$$

$$y_1' u_1' + y_2' u_2' = f(t),$$

which is written in matrix form as $\begin{pmatrix} y_1 & y_2 \\ y_1' & y_2' \end{pmatrix}\begin{pmatrix} u_1' \\ u_2' \end{pmatrix}$

$= \begin{pmatrix} 0 \\ f(t) \end{pmatrix}$. In linear algebra, we learn that this system has a unique solution if and only if $\begin{vmatrix} y_1 & y_2 \\ y_1' & y_2' \end{vmatrix} \neq 0$. Notice that the determinant, $W(S) = \begin{vmatrix} y_1 & y_2 \\ y_1' & y_2' \end{vmatrix}$, is the Wronskian of the set $S = \{y_1, y_2\}$, $W(S)$. We stated in Section 4.1 that $W(S) \neq 0$ if the functions y_1 and y_2 in the set S are linearly independent. Because $S = \{y_1, y_2\}$ represents a fundamental set of solutions of the corresponding homogeneous equation, $W(S) \neq 0$. Hence, this system has a unique solution that can be found by Cramer's rule to be

$$u_1'(t) = \frac{1}{W(S)}\begin{vmatrix} 0 & y_2(t) \\ f(t) & y_2'(t) \end{vmatrix} = -\frac{1}{W(S)}y_2(t)f(t)$$

and

$$u_2'(t) = \frac{1}{W(S)}\begin{vmatrix} y_1(t) & 0 \\ y_1'(t) & f(t) \end{vmatrix} = \frac{1}{W(S)}y_1(t)f(t).$$

Note: If you are not familiar with Cramer's rule, these formulas can be found with elimination.

The functions $u_1(t)$ and $u_2(t)$ are then found by integrating $u_1'(t)$ ($u_1(t) = \int u_1'(t)dt$) and $u_2'(t)$ ($u_2(t) = \int u_2'(t)dt$) and a particular solution of the nonhomogeneous equation (4.9) is given by $y_p = u_1y_1 + u_2y_2$.

EXAMPLE 4.4.1

Solve $y'' - 2y' + y = e^t \ln t, t > 0$.

Solution

The corresponding homogeneous equation, $y'' - 2y' + y = 0$, has characteristic equation $r^2 - 2r + 1 = (r-1)^2 = 0$ so $y_h = c_1 \underbrace{e^t}_{y_1} + c_2 \underbrace{te^t}_{y_2}$ and a fundamental set of solutions for the corresponding homogeneous equation is $S = \{e^t, te^t\}$. Now, we compute

$$W(S) = \begin{vmatrix} e^t & te^t \\ e^t & te^t + e^t \end{vmatrix} = te^{2t} + e^{2t} - te^{2t} = e^{2t}.$$

Through the use of integration by parts,

$$u_1(t) = \int \frac{1}{e^{2t}} \times -te^t \times e^t \ln t\, dt = -\int t \ln t\, dt$$

$$= \frac{1}{4}t^2 - \frac{1}{2}t^2 \ln t$$

and

$$u_2(t) = \int \frac{1}{e^{2t}} \times e^t \times e^t \ln t\, dt$$

$$= \int \ln t\, dt = t \ln t - t.$$

Then, a particular solution of the nonhomogeneous equation is

$$y_p = u_1(t)y_1(t) + u_2(t)y_2(t)$$

$$= \left(\frac{1}{4}t^2 - \frac{1}{2}t^2 \ln t\right)e^t + (t \ln t - t)\,te^t$$

$$= \frac{1}{2}t^2 e^t \ln t - \frac{3}{4}t^2 e^t.$$

Combining the general solution of the corresponding homogeneous equation with the particular solution of the nonhomogeneous equation gives us a general solution of the nonhomogeneous equation,

$$y = y_\text{h} + y_\text{p} = c_1 e^t + c_2 t e^t + \frac{1}{2} t^2 e^t \ln t - \frac{3}{4} t^2 e^t, \quad t > 0.$$

To solve an IVP, first determine a general solution and then use the initial conditions to solve for the unknown constants in the general solution.

EXAMPLE 4.4.2

Solve $y'' + 4y = \sec 2t$, $y(0) = 1$, $y'(0) = 1$, $-\pi/4 < t < \pi/4$.

Solution

The characteristic equation of the corresponding homogeneous equation, $y'' + 4y = 0$, is $r^2 + 4 = 0$ with the complex conjugate roots $r_{1,2} = \pm 2i$ so a general solution of the corresponding homogeneous equation is $y_\text{h} = c_1 \underbrace{\cos 2t}_{y_1} + c_2 \underbrace{\sin 2t}_{y_2}$ and the fundamental set of solutions $S = \{\cos 2t, \sin 2t\}$ so $W(S) = \begin{vmatrix} \cos 2t & \sin 2t \\ -2\sin 2t & 2\cos 2t \end{vmatrix} = 2$. Then,

$$u_1' = \frac{1}{2}\begin{vmatrix} 0 & \sin 2t \\ \sec 2t & 2\cos 2t \end{vmatrix} = -\frac{1}{2}\tan 2t \quad \text{and}$$

$$u_2' = \frac{1}{2}\begin{vmatrix} \cos 2t & 0 \\ -2\sin 2t & \sec 2t \end{vmatrix} = \frac{1}{2}$$

so $u_1 = \frac{1}{4}\ln(\cos 2t)$ and $u_2 = \frac{1}{2}t$. Then, a particular solution of the nonhomogeneous equation is given by $y_\text{p} = u_1 y_1 + u_2 y_2$, which simplifies to $y_\text{p} = \frac{1}{4}\cos 2t \ln \cos 2t + \frac{1}{2}t \sin 2t$, and a general solution of the nonhomogeneous equation is given by $y = y_\text{h} + y_\text{p} = c_1 \cos 2t + c_2 \sin 2t + \frac{1}{4}\cos 2t \ln \cos 2t + \frac{1}{2}t \sin 2t$.

To solve the IVP, we first differentiate to obtain $y' = -2c_1 \sin 2t + 2c_2 \cos 2t - \sin 2t \ln \cos 2t + t \cos 2t$. Application of the initial conditions yields

$\{y(0) = c_1 = 1, y'(0) = 2c_2 = 1\}$ so the solution to the IVP is $y = \cos 2t + \frac{1}{2}\sin 2t + \frac{1}{4}\cos 2t \ln \cos 2t + \frac{1}{2}t \sin 2t$. Figure 4.6 shows the graph of the solution of $y'' + 4y = \sec 2t$, $y(0) = 1$, $y'(0) = c$ for various values of c.

Variation of parameters can be used to find a particular solution of the linear nonhomogeneous equations that were solved in the preceding section by using the method of undetermined coefficients to find a particular solution of the linear nonhomogeneous equation. From a practical point of view, in cases when both methods can be used to find a particular solution of a linear nonhomogeneous equation, *usually* undetermined coefficients will be easier to implement.

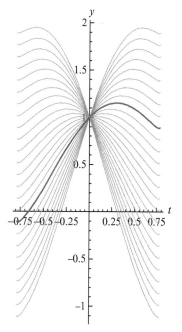

FIGURE 4.6 Plots of $y'' + 4y = \sec 2t$, $y(0) = 1$, $y'(0) = c$ for various values of c. Which is the plot of the solution satisfying $y'(0) = 1$?

EXAMPLE 4.4.3

Solve the IVP $y'' + 4y = \sin 2t$, $y(0) = 0$, $y'(0) = 1$ using variation of parameters.

Solution

The corresponding homogeneous equation is $y'' + 4y = 0$, with general solution $y_h = c_1 \cos 2t + c_2 \sin 2t$, fundamental set $S = \{\cos 2t, \sin 2t\}$, and Wronskian $W(S) = 2$. Through the use of the trigonometric identity $\sin^2 \theta = \frac{1}{2}(1 - \cos 2\theta)$ with $\theta = 2t$ and standard techniques of integration, we obtain

$$u_1(t) = -\frac{1}{2} \int \sin^2 2t \, dt = -\frac{1}{4}t + \frac{1}{16} \sin 4t$$

and

$$u_2(t) = \frac{1}{2} \int \cos 2t \sin 2t \, dt = -\frac{1}{8} \cos^2 2t.$$

Simplifying and using the identity $\sin 2\theta = 2 \sin \theta \cos \theta$ with $\theta = 2t$ leads to the particular solution

$$\begin{aligned}
y_p(t) &= u_1(t)y_1(t) + u_2(t)y_2(t) \\
&= \left(-\frac{1}{4}t + \frac{1}{16} \sin 4t\right) \cos 2t \\
&\quad - \frac{1}{8} \cos^2 2t \times \sin 2t \\
&= -\frac{1}{4}t \cos 2t + \frac{2}{16} \sin 2t \cos 2t \times \cos 2t \\
&\quad - \frac{1}{8} \cos^2 2t \times \sin 2t \\
&= -\frac{1}{4}t \cos 2t
\end{aligned}$$

and, thus, a general solution of the nonhomogeneous equation is

$$y = y_h + y_p = c_1 \cos 2t + c_2 \sin 2t - \frac{1}{4}t \cos 2t.$$

To solve the IVP, we first compute

$$y' = 2c_2 \cos 2t - 2c_1 \sin 2t + \frac{1}{2}t \sin 2t - \frac{1}{4} \cos 2t.$$

We then evaluate $y(0) = c_1 = 0$ and $y'(0) = 2c_2 - 1/4 = 1$. Hence, $c_1 = 0$ and $c_2 = 5/8$ so the solution to the IVP is $y = \frac{5}{8} \sin 2t - \frac{1}{4}t \cos 2t$.

 Figure 4.7 *shows the graph of* $y = c_1 \cos 2t + c_2 \sin 2t - \frac{1}{4}t \cos 2t$ *for various values of* c_1 *and* c_2. *Identify the graphs of the solutions that satisfy the initial conditions* (a) $y(0) = 0$ *and* $y'(0) = 1$ *and* (b) $y(0) = 0$ *and* $y'(0) = -1$.

Note: If we had used the method of undetermined coefficients, we should not expect a particular solution of the same form as that obtained through variation of parameters until we simplify $u_1(t)y_1(t) + u_2(t)y_2(t)$. (Why?)

An advantage that the method of variation of parameters has over the method of undetermined coefficients is particular solutions of nonhomogeneous equations with nonconstant coefficients can be considered.

EXAMPLE 4.4.4

A general solution of $t^2 y'' + ty' = 0$, $t > 0$, is $y_h = c_1 + c_2 \ln t$. Solve the nonhomogeneous equation $t^2 y'' + ty' = 2t^2$, $t > 0$.

Solution

In this case, a fundamental set of solutions of the corresponding homogeneous equation is $S = \{y_1, y_2\} = \{1, \ln t\}$, so $W(S) = \begin{vmatrix} 1 & \ln t \\ 0 & t^{-1} \end{vmatrix} = t^{-1}$. Recall that we divided the second-order equation $a_2(t)y'' + a_1(t)y' + a_0(t)y = g(t)$ by the coefficient function of y'', $a_2(t)$, to place it in the standard form $y'' + p(t)y' + q(t)y = f(t)$ before deriving the formulas for u_1 and u_2. Doing so for this equation, dividing by $a_2(t) = t^2$ gives us $y'' + t^{-1}y' = 2$ so $p(t) = t^{-1}$ and $f(t) = 2$. Therefore, using integration by parts we obtain

$$\begin{aligned}
u_1(t) &= -2 \int 1/t^{-1} \ln t \, dt = -2 \int t \ln t \, dt \\
&= -t^2 \ln t + \frac{1}{2}t^2
\end{aligned}$$

and

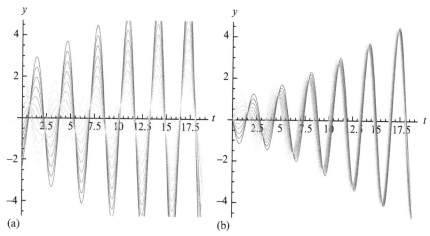

(a)   (b)

FIGURE 4.7   (a) Various solutions satisfying $y'(0) = 1$. (b) Various solutions satisfying $y(0) = 0$.

$$u_2(t) = 2 \int 1/t^{-1} dt = 2 \int t\, dt = t^2,$$

so $y_p(t) = u_1(t)y_1(t) + u_2(t)y_2(t) = \left(-t^2 \ln t + \frac{1}{2}t^2\right) \times$
$1 + t^2 \times \ln t = \frac{1}{2}t^2$, and a general solution to the nonhomogeneous equation is

$$y(t) = y_h(t) + y_p(t) = c_1 + c_2 \ln t + \frac{1}{2}t^2, \quad t > 0.$$

Integration by Parts: $\int u\, dv = uv - \int v\, du$.

Summary of Variation of Parameters for Second-Order Equations

Given the second-order equation $a_2(t)y'' + a_2(t)y' + a_0(t)y = g(t)$.

1. Divide by $a_2(t)$ to rewrite the equation in standard form, $y'' + p(t)y' + q(t)y = f(t)$.
2. Find a general solution, $y_h = c_1 y_1 + c_2 y_2$, and fundamental set of solutions, $S = \{y_1, y_2\}$, of the corresponding homogeneous equation $y'' + p(t)y' + q(t)y = 0$.
3. Let $W(S) = \begin{vmatrix} y_1 & y_2 \\ y_1' & y_2' \end{vmatrix}$.

4. Let $u_1' = -\dfrac{y_2 f(t)}{W(S)}$ and $u_2' = \dfrac{y_1 f(t)}{W(S)}$.
5. Integrate to obtain u_1 and u_2.
6. A particular solution of $a_2(t)y'' + a_2(t)y' + a_0(t)y = g(t)$ is given by $y_p = u_1 y_1 + u_2 y_2$.
7. A general solution of $a_2(t)y'' + a_2(t)y' + a_0(t)y = g(t)$ is given by $y = y_h + y_p$.

EXERCISES 4.4

In Exercises 1-38, solve each differential equation using variation of parameters. *If an equation can be solved using undetermined coefficients, do so. If both methods can be used to find y_p, which method is easier when both can be used?*

1. $y'' + 4y = 1$
2. $y'' + 16y' = t$
3. $y'' - 7y' + 10y = e^{3t}$
4. $y'' + 16y = 2\cos 4t$
5. $y'' + 4y' + 20y = 2te^{-2t}$
6. $y'' + \frac{1}{4}y = \sec(t/2) + \csc(t/2), 0 < t < \pi$
7. $y'' + 16y = \csc 4t, 0 < t < \pi/4$
8. $y'' + 16y = \cot 4t, 0 < t < \pi/4$

9. $y'' + 2y' + 50y = e^{-t} \csc 7t$

10. $y'' + 6y' + 25y = e^{-3t}(\sec 4t + \csc 4t)$

11. $y'' - 2y' + 26y = e^t(\sec 5t + \csc 5t)$

12. $y'' + 12y' + 37y = e^{-6t} \csc t, t > 0$

13. $y'' - 6y' + 34y = e^{3t} \tan 5t$

14. $y'' - 10y' + 34y = e^{5t} \cot 3t, t > 0$

15. $y'' - 12y' + 37y = e^{6t} \sec t$

16. $y'' - 8y' + 17y = e^{4t} \sec t$

17. $y'' - 9y = (1 + e^{3t})^{-1}$

18. $y'' - 25y = (1 - e^{5t})^{-1}$

19. $y'' - y = 2 \sinh t$

20. $y'' - 2y' + y = t^{-1}e^t, t > 0$

21. $y'' - 4y' + 4y = t^{-2}e^{2t}, t > 0$

22. $y'' + 8y' + 16y = t^{-4}e^{-4t}, t > 0$

23. $y'' + 6y' + 9y = t^{-1}e^{-3t}, t > 0$

24. $y'' + 6y' + 9y = e^{-3t} \ln t, t > 0$

25. $y'' + 3y' + 2y = \cos(e^t)$

26. $y'' + 4y' + 4y = e^{-2t}\sqrt{1 - t^2}, -1 \le t \le 1$

27. $y'' - 2y' + y = e^t\sqrt{1 - t^2}, -1 \le t \le 1$

28. $y'' - 10y' + 25y = e^{5t} \ln 2t, t > 0$

29. $y'' - 4y' + 4y = e^{2t} \tan^{-1} t$

30. $y'' + 8y' + 16y = e^{-4t}(1 + t^2)^{-1}$

31. $4y'' + y = \sec(t/2) + \csc(t/2), 0 < t < \pi$

32. $y'' + 9y = \tan^2 3t, -\pi/6 < t < \pi/6$

33. $y'' + 9y = \sec 3t, -\pi/6 < t < \pi/6$

34. $y'' + 9y = \tan 3t, -\pi/6 < t < \pi/6$

35. $y'' + 4y = \tan 2t, -\pi/4 < t < \pi/4$

36. $y'' + 16y = \tan 2t, -\pi/4 < t < \pi/4$

37. $y'' + 4y = \tan t, -\pi/2 < t < \pi/2$

38. $y'' + 9y = \sec 3t \tan 3t, -\pi/6 < t < \pi/6$

39. $y'' + 4y = \sec 2t \tan 2t, -\pi/4 < t < \pi/4$

In Exercises 40-45, solve the IVP.

40. $y'' + 9y = \frac{1}{2} \csc 3t, y(\pi/4) = \sqrt{2}, y'(\pi/4) = 0,$
$0 < t < \pi/3$

41. $y'' + 4y = \sec^2 2t, y(0) = 0, y'(0) = 1$

42. $y'' - 16y = 16te^{-4t}, y(0) = 0, y'(0) = 0$

43. $y'' + y = \tan^2 t, y(0) = 0, y'(0) = 1,$
$-\pi/2 < t < \pi/2$

44. $y'' + 4y = \sec 2t + \tan 2t, y(0) = 1, y'(0) = 0,$
$-\pi/4 < t < \pi/4$

45. $y'' + 9y = \csc 3t, y(\pi/12) = 0, y'(\pi/12) = 1,$
$0 < t < \pi/6$

46. Find a particular solution of
$y'' + 4y' + 3y = 65 \cos 2t$ by (a) the method

of undetermined coefficients; and (b) the method of variation of parameters. Which method is more easily applied to solve the problem?

47. Show that the solution of the IVP
$y'' + a_1(t)y' + a_0(t)y = f(t), y(t_0) = y_0,$
$y'(t_0) = y'_0$ can be written as
$y(t) = u(t) + v(t)$ where $u = u(t)$ is the
solution of $y'' + a_1(t)y' + a_0(t)y = 0,$
$y(t_0) = y_0, y'(t_0) = y'_0$ and $v = v(t)$ is the
solution of $y'' + a_1(t)y' + a_0(t)y = f(t),$
$y(t_0) = 0, y'(t_0) = 0.$

48. Let y_1 and y_2 be two linearly independent
solutions of the second-order linear
differential equation $a_2(t)y'' + a_1(t)y' +$
$a_0(t)y = 0.$ What is a general solution of
$a_2(t)y'' + a_1(t)y' + a_0(t)y = g(t)$? (*Hint:* In
order to apply the variation of parameters
formulas, the lead coefficient must be 1.
Divide by $a_2(t)$ and then apply the variation
of parameters formula, being careful with
notation.)

Green's Functions

Suppose that we are trying to solve the
IVP $y'' + p(t)y' + q(t)y = f(t), y(t_0) = 0,$
$y'(t_0) = 0.$ Based on the method of variation
of parameters, we write a particular solution
to this equation as $y = u_1(t)y_1(t) + u_2(t)y_2(t),$
where $u_1(t) = \int_{t_0}^t -\frac{1}{W(S)}y_2(x)f(x)dx$ and $u_2(t) =$
$\int_{t_0}^t \frac{1}{W(S)}y_1(x)f(x)dx;$ y_1 and y_1 are linearly
independent solutions of $y'' + p(t)y' + q(t)y = 0.$
Here, $W(S)$ represents the Wronskian of y_1 and
y_1 as a function of $x.$ Therefore,

$$y(t) = u_1(t)y_1(t) + u_2(t)y_2(t)$$

$$= -y_1(t) \int_{t_0}^t \frac{1}{W(S)}y_2(x)f(x)dx$$

$$+ y_2(t) \int_{t_0}^t \frac{1}{W(S)}y_1(x)f(x)dx$$

$$= \int_{t_0}^t \frac{1}{W(S)}(y_1(x)y_2(t) - y_1(t)y_2(x))f(x)dx.$$

Thus, we can solve the nonhomogeneous problem using an integral involving the function $f(t)$. The function

$$G(t,x) = \frac{1}{W(S)}\left(y_1(x)y_2(t) - y_1(t)y_2(x)\right)$$

is called the *Green's function*.

49. Show that $u_1(t) = \int_{t_0}^{t} -\frac{1}{W(S)}y_2(x)f(x)dx$ and $u_2(t) = \int_{t_0}^{t}\frac{1}{W(S)}y_1(x)f(x)dx$ indicate that $u_1(t_0) = 0$ and $u_2(t_0) = 0$. Recall that the method of variation of parameters is based on the two equations $y(t) = u_1(t)y_1(t) + u_2(t)y_2(t)$ and $y'(t) = u_1'(t)y_1(t) + u_2'(t)y_2(t)$ (because we assume that $u_1y_1' + u_2y_2' = 0$). Show that $y(t_0) = y'(t_0) = 0$, so that the solution obtained with the Green's function satisfies the IVP.

Suppose that we wish to solve $y'' + 4y = f(t)$, $y(0) = 0$, $y'(0) = 0$. In this case, $y_h(t) = c_1\cos 2t + c_2\sin 2t$, so $y(t) = \frac{1}{2}\int_0^t(\sin 2t \cos 2x - \cos 2t \sin 2x)$ $f(x)dx = \frac{1}{2}\int_0^t \sin 2(t-x)\sin x\,dx$. Therefore, $G(t,x) = \frac{1}{2}\sin 2(t-x)$. For the IVP, $y'' + 4y = \sin t$, $y(0) = 0$, $y'(0) = 0$, the solution is $y(t) = \frac{1}{2}\int_0^t \sin 2(t-x)$ $\sin x\,dx = \frac{1}{3}\sin t - \frac{1}{3}\sin 2t$.

50. Use the Green's function obtained above to solve $y'' + 4y = \sin \omega t$, $y(0) = 0$, $y'(0) = 0$ for (a) $\omega = 3$, (b) $\omega = 2.5$, (c) $\omega = 2.1$, and (d) $\omega = 2$. What happens to the solution as $\omega \to 2$?

51. Use a Green's function to solve $y'' + k^2y = f(t)$, $k > 0$ a positive constant, $y(0) = 0$, $y'(0) = 0$.

52. If we use the solution in Exercise 51 to solve $y'' + k^2y = \sin \omega t$, $y(0) = 0$, $y'(0) = 0$, what value of ω leads to the behavior seen in Exercise 50 (d)?

53. Use a Green's function to solve $y'' - k^2y = f(t)$, $k > 0$ a positive constant, $y(0) = 0$, $y'(0) = 0$.

54. Use the solution in Exercise 53 to solve the IVP $y'' - 4y = f(t)$, $y(0) = 0$, $y'(0) = 0$ for (a) $f(t) = 1$, (b) $f(t) = e^t$, (c) $f(t) = e^{2t}$, (d) $f(t) = \cos t$, (e) $f(t) = e^{-2t}\cos t$, and (f) $f(t) = e^{2t}\cos t$. What type of function $f(t)$ (if any) yields a periodic solution to this IVP?

55. Given that $y = c_1t^{-1} + c_2t^{-1}\ln t$ is a general solution to $t^2y'' + 3ty' + y = 0$, $t > 0$, use variation of parameters to solve $t^2y'' + 3ty' + y = \ln t$, $t > 0$.

56. Given that $y = c_1\sin(2\ln t) + c_2\cos(2\ln t)$ is a general solution to $t^2y'' + ty' + 4y = 0$, $t > 0$, use variation of parameters to solve $t^2y'' + ty' + 4y = t$, $t > 0$.

57. Given that $y = c_1t^6 + c_2t^{-1}$ is a general solution to $t^2y'' - 4ty' - 6y = 0$, $t > 0$, use variation of parameters to solve $t^2y'' - 4ty' - 6y = 2\ln t$, $t > 0$.

58. Consider the IVP $4y'' + 4y' + y = e^{-t/2}$, $y(0) = a$, $y'(0) = b$.
 (a) Show that regardless of the choices of a and b, the limit as $t \to \infty$ of every solution of the IVP is 0.
 (b) Find conditions on a and b, if possible, so that $y(t)$ has (i) no local minima or maxima; (ii) exactly one local minimum or local maximum; and (iii) two distinct local minima and maxima.
 (c) If possible, find a and b so that $y(t)$ has local extrema if $t = 1$ and $t = 3$. Graph $y(t)$ on the interval $[0, 5]$.
 (d) If possible, find a and b so that $y(t)$ has exactly one local extremum at $t = 3$. Graph $y(t)$ on the interval $[0, 6]$.
 (e) Is it possible to determine a and b so that $y(t)$ has local extrema if $t = t_0$ and $t = t_1$ ($t_0 \neq t_1$, both arbitrary)? Support your conclusion with several randomly generated values of t_0 and t_1.

59. Solve the equation

$$e^{-2t}\left[y\frac{d^2y}{dt^2} - \left(\frac{dy}{dt}\right)^2\right] - 2t(1+t)y^2 = 0$$

by making the substitution $y = e^{u(t)}$. Graph the solution for various values of the constants.

60. (a) Show that the solution to the IVP
$y'' + 4y = f(t), y(0) = 0, y'(0) = 2$ is

$$y(t) = \sin 2t + \frac{1}{2} \sin 2t \int_0^t f(x)$$

$$\cos 2x \, dx - \frac{1}{2} \cos 2t$$

$$\int_0^t f(x) \sin 2x \, dx$$

(as long as the integrals can be evaluated). (b) Show that if $f(t)$ is constant, the resulting solution is periodic and find its period. Confirm your result with a graph. (c) If $f(t)$ is periodic, is the resulting solution periodic? As in (b), confirm your result with a graph.

61. One solution of $t^2 y'' - 4ty' + (t^2 + 6)y = 0$ is $y = t^2 \cos t$. (a) Solve $t^2 y'' - 4ty' + (t^2 + 6)y = t^3 + 2t$. (b) Find the solutions, if any, that satisfy the initial conditions $y(0) = 0$ and $y'(0) = 1$. (c) Does the result found in (b) contradict the existence and uniqueness theorem? Explain.

62. One solution of $ty'' + 2y' + ty = 0, t > 0$, is $y = t^{-1} \cos t$. (a) Solve $ty'' + 2y' + ty = -t$. (b) Find the solutions, if any, that satisfy the initial conditions $y(\pi) = -1$ and $y'(\pi) = -1/\pi$. (c) Does the result found in (b) contradict the existence and uniqueness theorem? Explain.

63. One solution of $4t^2 y'' + 4ty' + (16t^2 - 1)y = 0, t > 0$, is $y = t^{-1/2} \sin 2t$. (a) Solve $4t^2 y'' + 4ty' + (16t^2 - 1)y = 16t^{3/2}$. (b) Find the solutions, if any, that satisfy the boundary conditions $y(\pi) = y(3\pi/2) = 0$. (c) Under what conditions, if any, do infinitely many solutions satisfy the boundary conditions $y(\pi) = y'(2\pi) = 0$?

64. A fundamental set of solutions for $t^2 (\ln t - 1)y'' - ty' + y = 0$ is $S = \{t, \ln t\}$. Use this information to solve

$t^2 (\ln t - 1)y'' - ty' + y = -\frac{3}{4}(1 + \ln t)t^{-1/2}$, $y(1) = 0, y'(1) = 0$.

65. A fundamental set of solutions for $(\sin t - t \cos t)y'' - t \sin t \, y' + \sin ty = 0$ is $S = \{t, \sin t\}$. (a) Use this information to solve $(\sin t - t \cos t)y'' - t \sin t \, y' + \sin ty = t$ for $0 < t < x_0$. (b) Find x_0. (x_0 is the largest value so that the solution exists on the interval $0 < x < x_0$.) (c) Find conditions on c and d so that $y(0) = c$ and $y'(0) = d$ has a solution. (d) Find *the* solution of the differential equation that satisfies the initial conditions $y(\pi/4) = 0$ and $y'(\pi/4) = 0$.

4.5 SOLVING HIGHER ORDER LINEAR HOMOGENEOUS EQUATIONS

- BASIC THEORY
- CONSTANT COEFFICIENTS

Basic Theory

The ideas presented for second-order linear homogeneous equations in Sections 4.1 and 4.2 can be extended to those of order 3 or higher. We begin with the general form of an nth-order linear equation.

Definition 8 (*n*th-order Ordinary Linear Differential Equation). *An ordinary differential equation of the form*

$$a_n(t)y^{(n)}(t) + a_{n-1}(t)y^{(n-1)}(t) +$$

$$\cdots + a_1(t)y'(t) + a_0(t)y(t) = g(t), \quad (4.14)$$

*where $a_n(t) \neq 0$, is called an **nth-order ordinary linear differential equation**. If $g(t)$ is identically the zero function, the equation is said to be **homogeneous**; if $g(t)$ is not the zero function, the equation is said to be **nonhomogeneous**; and if the functions $a_i(t), i = 0, 1, 2, \ldots, n$ are constants, the equation is said to have **constant coefficients**. An nth-order equation accompanied by the conditions*

$$y(t_0) = y_0, \; y'(t_0) = y_0', \; \ldots, \; y^{(n-1)}(t_0) = y_0^{(n-1)}$$

where $y_0, y_0', \ldots, y_0^{(n-1)}$ are constants is called an **nth-order initial-value problem.**

On the basis of this definition, the equation $y''' - 8y'' - 3y = \cos t$ is a third-order linear nonhomogeneous equation with constant coefficients, while $y^{(4)} - t^2 y = 0$ is a fourth-order linear homogeneous equation. We assume that the coefficient functions $a_n(t), \ldots, a_0(t)$ and $g(t)$ in Equation (4.14) are continuous on an open interval I. Therefore, if we also assume that $a_n(t) \neq 0$ for all t in I, we can divide by $a_n(t)$ to place the differential equation (4.14) in standard (or normal) form,

$$y^{(n)}(t) + p_{n-1}(t)y^{(n-1)}(t) + \cdots + p_1(t)y'(t)$$
$$+ \, p_0(t)y(t) = f(t) \qquad (4.15)$$

where $p_i(t) = a_i(t)/a_n(t)$, $i = 0, 1, \ldots, n - 1$ and $f(t) = g(t)/a_n(t)$.

In Sections 4.1 and 4.2, we discussed that we needed two linearly independent solutions to solve a second-order linear homogeneous differential equation. Then, we gave a definition for the linear dependence or linear independence of a set of two functions. We now define this concept for a set of n functions because we need to understand this property to solve higher order linear equations.

Definition 9 (Linearly Dependent and Linearly Independent Sets of Functions). *Let $S = \{f_1(t), f_2(t), f_3(t), \ldots, f_{n-1}(t), f_n(t)\}$ be a set of n functions. S is **linearly dependent** on an interval I if there are constants $c_1, c_2, c_3, \ldots, c_{n-1}, c_n$, not all zero, so that*

$$c_1 f_1(t) + c_2 f_2(t) + c_3 f_3(t) + \cdots$$
$$+ \, c_{n-1} f_{n-1}(t) + c_n f_n(t) = 0$$

for every value of t in the interval I containing t_0.

*S is **linearly independent** if S is not linearly dependent.*

Thus, $S = \{f_1(t), f_2(t), f_3(t), \ldots, f_{n-1}(t), f_n(t)\}$ is linearly independent if

$$c_1 f_1(t) + c_2 f_2(t) + c_3 f_3(t) + \cdots$$
$$+ \, c_{n-1} f_{n-1}(t) + c_n f_n(t) = 0$$

implies that $c_1 = c_2 = \cdots = c_n = 0$.

EXAMPLE 4.5.1

Classify the sets of functions as linearly independent or linearly dependent. (a) $S = \{t, 3t - 6, 1\}$, (b) $S = \{\cos 2t, \sin 2t, \sin t \cos t\}$, and (c) $S = \{1, t, t^2\}$

Solution

Here, we must find constants c_1, c_2, and c_3 such that

$$c_1 t + c_2(3t - 6) + c_3 \times 1 = (c_1 + 3c_2)t + (c_3 - 6c_2) = 0,$$

Equating each of the coefficients to zero leads to the system of equations $\{c_1 + 3c_2 = 0, \; c_3 - 6c_2 = 0\}$. This system has infinitely many solutions of the form $\{c_1 = -3c_2, \; c_3 = 6c_2, \; c_2 \text{ arbitrary}\}$. With the choice $c_1 = -3$, $c_2 = 1$, and $c_3 = 6$, the equation is satisfied. Therefore, the set $S = \{t, 3t - 6, 1\}$ is linearly dependent. (b) For this set of functions, we apply the identity $\sin 2\theta = 2 \sin \theta \cos \theta$. Doing so, we consider the equation

$$c_1 \cos 2t + c_2 \sin 2t + c_3 \sin t \cos t = c_1 \cos 2t$$
$$+ \, 2c_2 \sin t \cos t + c_3 \sin t \cos t = 0.$$

The choices of $c_1 = 0$, $c_2 = 1$, and $c_3 = -2$ lead to a solution of this equation. At least one of these constants is not zero, so the set of functions $S = \{\cos 2t, \sin 2t, \sin t \cos t\}$ is linearly dependent.

(c) For the set $S = \{1, t, t^2\}$, we consider the equation $c_1 \times 1 + c_2 \times t + c_3 \times t^2 = 0$. Differentiating twice gives us the system of equations

$$c_1 + c_2 t + c_3 t^2 = 0$$
$$c_2 + 2c_3 t = 0$$
$$2c_3 = 0.$$

Then, $c_3 = 0$, so $c_2 = 0$, and, consequently, $c_1 = 0$. The only constants that satisfy the equation are $c_1 = c_2 = c_3 = 0$, so the set is linearly independent.

We also extend the Principle of Superposition to include more than two functions.

Theorem 20 (Principle of Superposition). *If $f_1(t)$, $f_2(t)$, ..., $f_m(t)$ are solutions of the linear homogeneous differential equation $y^{(n)}(t) + p_{n-1}(t)y^{(n-1)}(t) + \cdots + p_1(t)y'(t) + p_0(t)y(t) = 0$ on the interval I, and if $c_1, c_2, \ldots, c_m$ are arbitrary constants, then $y = c_1 f_1(t) + c_2 f_2(t) + \cdots c_m f_m(t)$ is also a solution of this differential equation on I.*

Another way to express Theorem 20 is to say that every linear combination of solutions to $y^{(n)}(t) + p_{n-1}(t)y^{(n-1)}(t) + \cdots + p_1(t)y'(t) + p_0(t)y(t) = 0$ is also a solution to this differential equation.

EXAMPLE 4.5.2

Show that $y_1(t) = e^{-t}$, $y_2(t) = \cos t$, and $y_3(t) = \sin t$ are solutions of the third-order linear homogeneous differential equation $y''' + y'' + y' + y = 0$. Use the Principle of Superposition to find another solution to this differential equation.

Solution

For $y_1(t) = e^{-t}$, we have $y_1'(t) = -e^{-t}$, $y_1''(t) = e^{-t}$, and $y_1'''(t) = -e^{-t}$.

Then, $y_1''' + y_1'' + y_1' + y_1 = -e^{-t} + e^{-t} - e^{-t} + e^{-t} = 0$, so y_1 satisfies the differential equation. In a similar manner, we can show that y_2 and y_3 are solutions of the differential equation. (We leave this to the reader.) Therefore by the Principle of Superposition, any linear combination of these three functions, $y(t) = c_1 e^{-t} + c_2 \cos t + c_3 \sin t$, is also a solution of the differential equation.

As with first-order and second-order IVPs, we can state an existence and uniqueness theorem for an IVP involving a higher order linear ordinary differential equation.

Theorem 21 (Existence and Uniqueness). *Suppose that $p_{n-1}(t), \ldots, p_0(t)$, and $f(t)$ are continuous functions on an open interval I that contains $t = t_0$. Then, the IVP*

$$y^{(n)}(t) + p_{n-1}(t)y^{(n-1)}(t) + \cdots + p_1(t)y'(t)$$
$$+ p_0(t)y(t) = f(t)$$

$$y(t_0) = y_0, \, y'(t_0) = y_0', \ldots, y^{(n-1)}(t_0) = y_0^{(n-1)}$$

has a unique solution on I.

The proof of this theorem is well beyond the scope of this text but can be found in advanced differential equations texts.[2]

At this point, we need to discuss why we can write every solution of the differential equation $y^{(n)}(t) + p_{n-1}(t)y^{(n-1)}(t) + \cdots + p_1(t)y'(t) + p_0(t)y(t) = 0$ as a linear combination of solutions if we have n linearly independent solutions of the nth-order equation. (We call this a *general solution* of the differential equation.) Before moving on, we must define the Wronskian of a set of n functions. This definition is particularly useful because determining whether a set of functions is linearly independent or dependent becomes more difficult as the number of functions in the set under consideration increases.

Definition 10 (Wronskian). *Let $S = \{f_1(t), f_2(t), f_3(t), \ldots, f_{n-1}(t), f_n(t)\}$ be a set of n functions for which each is differentiable at least $n - 1$ times. The* **Wronskian** *of S, denoted by*

$$W(S) = W(\{f_1(t), f_2(t), f_3(t), \ldots, f_{n-1}(t), f_n(t)\}),$$

is the determinant

$$W(S) = \begin{vmatrix} f_1(t) & f_2(t) & \cdots & f_n(t) \\ f_1'(t) & f_2'(t) & \cdots & f_n'(t) \\ \vdots & \vdots & \vdots & \vdots \\ f_1^{(n-1)}(t) & f_2^{(n-1)}(t) & \cdots & f_n^{(n-1)}(t) \end{vmatrix}.$$

---

[2]For example, see Chapter 3 of C. Corduneanu, Principles of Differential and Integral Equations, Chelsea Publishing Company, 1971.

Although most computer algebra systems can quickly find and simplify the Wronskian of a set of functions, the following example illustrates how to compute the Wronskian for a set of three functions by hand.

EXAMPLE 4.5.3

Compute the Wronskian for each of the following sets of functions. (a) $S = \{t, 3t - 6, 1\}$; (b) $S = \{1, t, t^2\}$.

Show that three functions contained in $S = \{t, 3t-6, 1\}$ are solutions of $y'' = 0$ and that the three function in $S = \{1, t, t^2\}$ are solutions of $y''' = 0$.

Solution

In linear algebra, we learn that we can compute a determinant by expanding along any row or column. So, for example, we compute a 3×3 determinant with

$$\begin{vmatrix} a_1 & a_2 & a_3 \\ b_1 & b_2 & b_3 \\ c_1 & c_2 & c_3 \end{vmatrix} = a_1 \begin{vmatrix} b_2 & b_3 \\ c_2 & c_3 \end{vmatrix} - a_2 \begin{vmatrix} b_1 & b_3 \\ c_1 & c_3 \end{vmatrix} + a_3 \begin{vmatrix} b_1 & b_2 \\ c_1 & c_2 \end{vmatrix}$$

(a) $W(S) = \begin{vmatrix} t & 3t-6 & 1 \\ 1 & 3 & 0 \\ 0 & 0 & 0 \end{vmatrix}$. We can carry out the

step given above for computing a 3×3 determinant or by expanding along the third row to see that $W(S) = 0$.

(b) Because the third row of the Wronskian matrix contains the most 0's, we compute the Wronskian by expanding along the third row,

$$W(S) = \begin{vmatrix} 1 & t & t^2 \\ 0 & 1 & 2t \\ 0 & 0 & 2 \end{vmatrix} = 2 \begin{vmatrix} 1 & t \\ 0 & 1 \end{vmatrix} = 2 \times (1 - 0) = 2.$$

In addition, we could have observed that all nonzero entries lie on the diagonal or above, so the determinant is the product of the entries on the diagonal.

Note: This approach also works if the nonzero entries lie on the diagonal or below. Matrices having these properties are called *upper triangular* and *lower triangular*, respectively.

In part (a) of Example 4.5.3, we found that $W(S) = 0$. Recall that we concluded that the functions in $S = \{t, 3t - 6, 1\}$ are linearly dependent because $3t - 6 = 3 \times t - 6 \times 1$, which means that one of the functions in the set can be written as a linear combination of other members of the set. This is no coincidence. The Wronskian of a set of linearly dependent functions is identically zero when the functions are solutions to the same linear homogeneous differential equation (for these three functions, $y'' = 0$), whereas the Wronskian of a set of linearly independent functions is not equal to zero for at least one value when they are. In part (b), we found that $W(S) = 2 \neq 0$ *and* that the functions in S are solutions to a linear homogeneous differential equation (for these three functions, $y''' = 0$). Therefore, the functions are linearly independent as we discussed earlier.

Theorem 22 (Wronskian of Solutions). *Consider the nth-order linear homogeneous equation* $y^{(n)}(t) + p_{n-1}(t)y^{(n-1)}(t) + \cdots + p_1(t)y'(t) + p_0(t)y(t) = 0$, *where* $p_{n-1}(t), \ldots, p_0(t)$ *are continuous functions on an open interval I. Suppose that* $y_1(t), y_2(t), \ldots, y_n(t)$ *are solutions of this differential equation on I. If* $y_1(t), y_2(t), \ldots, y_n(t)$ *are linearly dependent, then* $W(\{y_1(t), y_2(t), \ldots, y_n(t)\}) = 0$ *on I. If* $y_1(t), y_2(t), \ldots, y_n(t)$ *are linearly independent, then* $W(\{y_1(t), y_2(t), \ldots, y_n(t)\}) \neq 0$ *for all values of t on I.*

We omit the proof of this theorem here. We leave it as part of the section exercises regarding Abel's formula.

EXAMPLE 4.5.4

The functions $y_1 = t^{-4}$, $y_2 = t^{-1}$, $y_3 = t$, and $y_4 = t \ln t$ are solutions of $t^4 y^{(4)} + 9t^3 y''' + 11t^2 y'' - 4ty' + 4y = 0$. Show that these solutions are linearly independent on the interval $t > 0$.

Solution

Notice that we have assumed that $t > 0$, so we can divide each term by t^4 to place the differential equation in normal (or standard) form,

$$y^{(4)} + 9t^{-1}y''' + 11t^{-2}y'' - 4t^{-3}y' + 4t^{-4}y = 0,$$

for use with Theorem 22. (Note that we assumed $t > 0$ so that the coefficient functions in the differential equation are continuous.) Then, if $S = \{t^{-4}, t^{-1}, t, t\ln t\}$, we expand along the third column and use a computer algebra system to check the algebra to find that

$$W(S) = \begin{vmatrix} t^{-4} & t^{-1} & t & t\ln t \\ -4t^{-5} & -t^{-2} & 1 & 1+\ln t \\ 20t^{-6} & 2t^{-3} & 0 & t^{-1} \\ -120t^{-7} & -6t^{-4} & 0 & -t^{-2} \end{vmatrix} = 300t^{-9},$$

so $W(S) \neq 0$ for $t > 0$. Therefore, the functions are linearly independent on the interval $t > 0$.

Theorem 23 (General Solution). *Consider the nth-order linear homogeneous equation $y^{(n)}(t) + p_{n-1}(t)y^{(n-1)}(t) + \cdots + p_1(t)y'(t) + p_0(t)y(t) = 0$, where $p_{n-1}(t), \ldots, p_0(t)$ are continuous functions on an open interval I. Suppose that $y_1(t), y_2(t), \ldots, y_n(t)$ are solutions of this differential equation on I. If $y_1(t), y_2(t), \ldots, y_n(t)$ are linearly dependent solutions of this differential equation, there are constants $c_1, c_2, \ldots, c_n$ so that $y(t) = c_1y_1(t) + c_2y_2(t) + \cdots + c_ny_n(t)$ for all t in I.*

Theorem 23 tells us that if we have n linearly independent solutions of an nth-order linear homogeneous equation, then we have all solutions, called a *general solution*, given by the linear combination of the solutions. In addition, if the n differentiable functions, $S = \{y_1, y_2, \ldots, y_n\}$, are solutions of the differential equation, we call the set a *fundamental set of solutions*, and we note that a fundamental set of solutions for $y^{(n)}(t) + p_{n-1}(t)y^{(n-1)}(t) + \cdots + p_1(t)y'(t) + p_0(t)y(t) = 0$ must contain n linearly independent solutions.

(In other words, the number of functions in the fundamental set of solutions equals the order of the linear homogeneous differential equation, and these functions must be linearly independent.)

The proof of this theorem follows the same steps as those used in the proof of the case for second-order equations discussed in Section 4.2 in the proof of Theorem 16. To illustrate the similarities between the second-order case and the general situation, we briefly outline the proof here, leaving the details for you to fill in.

One such set is the set of functions $S = \{y_1, y_2, \ldots, y_n\}$, where each y_i satisfies the differential equation and y_1 satisfies $y_1(t_0) = 1$, $y_1'(t_0) = 0, \ldots, y^{(n-1)}(t_0) = 0$, y_2 satisfies $y_2(t_0) = 0$, $y_1'(t_0) = 1, \ldots, y^{(n-1)}(t_0) = 0$, and so on. Briefly explain why.

Begin with a set of n functions, $S = \{y_1, y_2, \ldots, y_n\}$, that are solutions of the differential equation, $y^{(n)}(t) + p_{n-1}(t)y^{(n-1)}(t) + \cdots + p_1(t)y'(t) + p_0(t)y(t) = 0$, for which $W(S) \neq 0$ on an interval I containing $t = t_0$. Now let Y represent any solution of the differential equation. Then, on I

$$Y^{(n)}(t) + p_{n-1}(t)Y^{(n-1)}(t) + \cdots + p_1(t)Y'(t) + p_0(t)Y(t) = 0$$

and evaluated at $t = t_0$, we want to find $c_1, c_2, \ldots, c_n$ so that

$$Y(t_0) = c_1y_1(t_0) + c_2y_2(t_0) + \cdots + c_ny_n(t_0)$$
$$Y'(t_0) = c_1y_1'(t_0) + c_2y_2'(t_0) + \cdots + c_ny_n'(t_0)$$
$$\vdots$$
$$Y^{(n-1)}(t_0) = c_1y_1^{(n-1)}(t_0) + c_2y_2^{(n-1)}(t_0) + \cdots + c_ny_n^{(n-1)}(t_0).$$

In matrix form, $\mathbf{Ax} = \mathbf{b}$, this linear system is written as

$$\underbrace{\begin{pmatrix} y_1(t_0) & y_2(t_0) & \cdots & y_n(t_0) \\ y_1'(t_0) & y_2'(t_0) & \cdots & y_n'(t_0) \\ \vdots & \vdots & \cdots & \vdots \\ y_1^{(n-1)}(t_0) & y_2^{(n-1)}(t_0) & \cdots & y_n^{(n-1)}(t_0) \end{pmatrix}}_{\mathbf{A}} \underbrace{\begin{pmatrix} c_1 \\ c_2 \\ \vdots \\ c_n \end{pmatrix}}_{\mathbf{x}}$$

$$= \underbrace{\begin{pmatrix} Y(t_0) \\ Y'(t_0) \\ \vdots \\ Y^{(n-1)}(t_0) \end{pmatrix}}_{\mathbf{b}}. \qquad (4.16)$$

Notice that the determinant of the coefficient matrix is $W(S(t_0)) \neq 0$ so Equation (4.16) has a unique solution and using Cramer's Rule, c_i is given by

$$c_i = \frac{1}{W(S(t_0))} W_i(S(t_0)),$$

where $W_i(S(t_0))$ is the determinant of the matrix obtained by replacing the ith column of the Wronskian matrix, $W(S(t_0))$, with the vector

$$\mathbf{b} = \begin{pmatrix} Y(t_0) \\ Y'(t_0) \\ \vdots \\ Y^{(n-1)}(t_0) \end{pmatrix}.$$

EXAMPLE 4.5.5

Given that the four functions in Example 4.5.4 are solutions of $t^4 y^{(4)} + 9t^3 y''' + 11t^2 y'' - 4ty' + 4y = 0$, determine a general solution of $t^4 y^{(4)} + 9t^3 y''' + 11t^2 y'' - 4ty' + 4y = 0$.

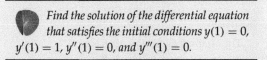

Find the solution of the differential equation that satisfies the initial conditions $y(1) = 0$, $y'(1) = 1$, $y''(1) = 0$, and $y'''(1) = 0$.

Solution

We showed in Example 4.5.4 that the four functions $y_1 = t^{-4}$, $y_2 = t^{-1}$, $y_3 = t$, and $y_4 = t \ln t$ are linearly independent. Therefore, because we have four linearly independent solutions of this fourth-order linear homogeneous differential equation, we can write a general solution as the linear combination

$$y = c_1 y_1 + c_2 y_2 + c_3 y_3 + c_4 y_4$$
$$= c_1 t^{-4} + c_2 t^{-1} + c_3 t + c_4 t \ln t.$$

The set $S = \{t^{-4}, t^{-1}, t, t \ln t\}$ is a fundamental set of solutions for this differential equation.

Constant Coefficients

Theorem 23 tells us that we need n linearly independent solutions (a fundamental set of solutions) of the nth-order linear homogeneous equation to form a general solution. When the coefficient functions are constant, we may be able to find a fundamental set and form a general solution of the equation.

Suppose that we wish to solve

$$a_n y^{(n)}(t) + a_{n-1} y^{(n-1)}(t) + \cdots + a_1 y'(t) + a_0 y(t) = 0, \qquad (4.17)$$

where $a_n, a_{n-1}, \ldots, a_1, a_0$ are real constants with $a_n \neq 0$. Following the same procedure we used to solve $ay'' + by' + cy = 0$, we assume that $y = e^{rt}$ is a solution, and we substitute this function into the differential equation to determine the value(s) of r that lead to solutions. Differentiating, we find that $y' = re^{rt}$, $y'' = r^2 e^{rt}$, $y''' = r^3 e^{rt}$ and continuing we see that the kth derivative is $y^{(k)} = r^k e^{rt}$. Substitution into the linear homogeneous equation with constant coefficients (Equation 4.17) gives us

$$a_n r^n e^{rt} + a_{n-1} r^{n-1} e^{rt} + \cdots + a_1 r e^{rt} + a_0 e^{rt} = 0$$

or

$$e^{rt}\left(a_n r^n + a_{n-1}r^{n-1} + \cdots + a_1 r + a_0\right) = 0,$$

so the *characteristic equation* is

$$a_n r^n + a_{n-1}r^{n-1} + \cdots + a_1 r + a_0 = 0 \qquad (4.18)$$

because $e^{rt} \neq 0$ for all values of t.

Distinct Real Roots

Suppose that the characteristic equation (4.18) has the roots $r_1, r_2, \ldots, r_m$ and they are distinct so that no two listed here are the same. Then, m linearly independent solutions of Equation (4.17) are $y_1 = e^{r_1 t}, y_2 = e^{r_2 t}, \ldots, y_m = e^{r_m t}$.

Theorem 24. *If the numbers $r_1, r_2, \ldots, r_m$ are real distinct solutions of the characteristic equation* (4.18), *m linearly independent solutions of Equation* (4.17) *are $y_1 = e^{r_1 t}, y_2 = e^{r_2 t}, \ldots, y_m = e^{r_m t}$.*

EXAMPLE 4.5.6

Solve $y''' - y' = 0, y(0) = 0, y'(0) = -2, y''(0) = 2$.

Solution

In this case, the characteristic equation is $r^3 - r = r(r+1)(r-1) = 0$. The three distinct roots are $r_1 = 0, r_2 = -1$, and $r_3 = 1$ with corresponding linearly independent solutions $y_1 = e^{0 \times t}$, $y_2 = e^{-1 \times t}$, and $y_3 = e^{1 \times t}$, respectively. Thus, $S = \{1, e^t, e^{-t}\}$ is a fundamental set of solutions for this third-order linear homogeneous equation, $y''' - y' = 0$. Therefore a general solution is the linear combination of these function, $y = c_1 + c_2 e^t + c_3 e^{-t}$.

To solve the IVP, we apply the initial conditions: $y' = c_2 e^t - c_3 e^{-t}$ and $y'' = c_2 e^t + c_3 e^{-t}$. Then, $y(0) = c_1 + c_2 + c_3$, $y'(0) = c_2 - c_3$, and $y''(0) = c_2 + c_3$, so we obtain the linear system of three equations and three unknowns:

$$c_1 + c_2 + c_3 = 0$$

$$c_2 - c_3 = -2$$

$$c_2 + c_3 = 2.$$

Adding the last two equations indicates that $c_2 = 0$. Substitution of this value into either of the last two equations gives $c_3 = 2$. Then, substitution of $c_2 = 0$ and $c_3 = 2$ into the first equation gives us $c_1 = -2$. Therefore, the unique solution to this IVP is $y = -2 + 0 \times e^t + 2 \times e^{-t} = -2 + 2e^{-t}$ (see Figure 4.8).

Repeated Real Roots

When solving second-order equations in Section 4.2, we encountered equations in which the root of the characteristic equation was repeated. For example, the characteristic equation for $y'' - 2y' + y = 0$ is $r^2 - 2r + 1 = 0$ or $(r - 1)^2 = 0$. In this case, we say that the root $r = 1$ has *multiplicity* two, and we found a general solution to be $y(t) = c_1 e^t + c_2 t e^t$. As we increase the order of the differential equation, we can have roots of multiplicity greater than two, so we present the following theorem.

Theorem 25. *Suppose that the root $r = r_1$ of the characteristic equation* (4.18), *has multiplicity k. (That is, $(r - r_1)^k$ is a factor of the characteristic equation but $(r - r_1)^{k+1}$ is not.) Then, k linearly independent solutions of Equation* (4.17) *associated with $r = r_1$ are $y_1 = e^{r_1 t}, y_2 = t e^{r_1 t}, y_3 = t^2 e^{r_1 t}, \ldots, y_k = t^{k-1} e^{r_1 t}$.*

For example, applying this theorem to a second-order linear homogeneous equation with constant coefficients with $k = 2$ and $r = -1$, we obtain $y = c_1 e^{-t} + c_2 t e^{-t}$ as we found in Example 4.2.2 in Section 4.2.

EXAMPLE 4.5.7

Solve $2y^{(6)} - 7y^{(5)} - 4y^{(4)} = 0$.

Solution

The characteristic equation is

$$2r^6 - 7r^5 - 4r^4 = r^4(2r^2 - 7r - 4) = r^4(2r+1)(r-4) = 0$$

Therefore, the roots are $r = 0$ (of multiplicity $k = 4$), $r = -1/2$ (of multiplicity $k = 1$), and

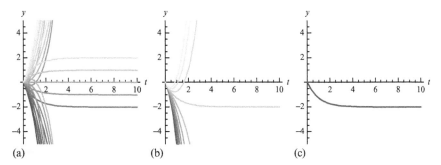

FIGURE 4.8 (a) Various solutions of the differential equation satisfying $y(0) = 0$. (b) Various solutions of the differential equation satisfying $y(0) = 0$ and $y'(0) = -2$. (c) The solution of the differential equation satisfying $y(0) = 0$, $y'(0) = -2$, and $y''(0) = 2$.

TABLE 4.1 Linearly Independent Solutions of $2y^{(6)} - 7y^{(5)} - 4y^{(4)} = 0$

Root	Multiplicity	Corresponding Solution(s)
$r = 0$	$k = 4$	$y_1 = 1, y_2 = t, y_3 = t^2, y_4 = t^3$
$r = -1/2$	$k = 1$	$y_5 = e^{-t/2}$
$r = 4$	$k = 1$	$y_6 = e^{4t}$

$r = 4$ (of multiplicity $k = 1$). Then, the *four* linearly independent solutions corresponding to $r = 0$ are $y_1 = e^{0 \times t} = 1$, $y_2 = te^{0 \times t} = t$, $y_3 = t^2 e^{0 \times t} = t^2$, and $y_4 = t^3 e^{0 \times t} = t^3$. Notice that the number of solutions associated with $r = 0$ equals the multiplicity of $r = 0$.

The solution associated with $r = -1/2$ is $y_5 = e^{-t/2}$ and that associated with $r = 4$ is $y_6 = e^{4t}$ (see Table 4.1). We form a general solution of $2y^{(6)} - 7y^{(5)} - 4y^{(4)} = 0$ by taking the linear combination of the six (linearly independent) solutions. The general solution is

$$y = c_1 + c_2 t + c_3 t^2 + c_4 t^3 + c_5 e^{-t/2} + c_6 e^{4t}.$$

Note that in Example 4.5.7, a fundamental set of solutions for the differential equation is $\{1, t, t^2, t^3, e^{-t/2}, e^{4t}\}$. The number of solutions in a fundamental set equals the order of the differential equation. *Therefore, to form a general*

solution of an nth-order linear homogeneous differential equation, we take the linear combination of the n functions in a fundamental set.

Complex Conjugate Roots

Theorem 26 (Complex Conjugate Roots). *Suppose that $r_1 = \alpha + \beta i$ and $r_2 = \alpha - \beta i$ (α and $\beta > 0$ are real) are roots of the characteristic equation (4.18) with multiplicity k. Then, $2k$ linearly independent solutions of Equation (4.17) associated with $r_{1,2} = \alpha \pm \beta i$ are $y_1 = e^{\alpha t} \cos \beta t$, $y_2 = e^{\alpha t} \sin \beta t$, $y_3 = te^{\alpha t} \cos \beta t$, $y_4 = te^{\alpha t} \sin \beta t, \ldots, y_{2k-1} = t^{k-1} e^{\alpha t} \cos \beta t$, $y_{2k} = t^{k-1} e^{\alpha t} \sin \beta t$.*

EXAMPLE 4.5.8

Solve $y^{(4)} - y = 0$.

Solution

In this case, the characteristic equation is

$$r^4 - 1 = (r^2 - 1)(r^2 + 1) = (r - 1)(r + 1)(r^2 + 1) = 0,$$

so the roots are $r_1 = 1$, $r_2 = -1$, $r_3 = i$, and $r_4 = -i$. The solutions corresponding to the roots $r_1 = 1$ and $r_2 = -1$ are $y_1 = e^t$ and $y_2 = e^{-t}$, respectively, while the solutions corresponding to the complex conjugate pair $r_{3,4} = \pm i$ (where $\alpha = 0$ and $\beta = 1$) are $y_3 = e^{0 \times t} \cos t = \cos t$ and $y_4 = e^{0 \times t} \sin t = \sin t$ (see Table 4.2). Therefore, a general solution of the equation is the linear combination of these four functions given by

$$y = c_1 e^t + c_2 e^{-t} + c_3 \cos t + c_4 \sin t.$$

TABLE 4.2   Linearly Independent Solutions of $y^{(4)} - y = 0$

Root	Multiplicity	Corresponding Solution(s)
$r_1 = 1$	$k = 1$	$y_1 = e^t$
$r_2 = -1$	$k = 1$	$y_2 = e^{-t}$
$r_{3,4} = \pm i$	$k = 1, k = 1$	$y_3 = \cos t, y_4 = \sin t$

We summarize the following rules for finding a general solution of an nth-order linear homogeneous equation with constant coefficients because of the many situations that are encountered.

Figure 4.9 shows the graph of $y = c_1 e^t + c_2 e^{-t} + c_3 \cos t + c_4 \sin t$ for various values of $c_1, c_2, c_3,$ and c_4. Identify those graphs for which (a) $c_1 = c_3 = c_4 = 0$; (b) $c_1 = c_2 = 0$; and (c) $c_3 = c_4 = 0$

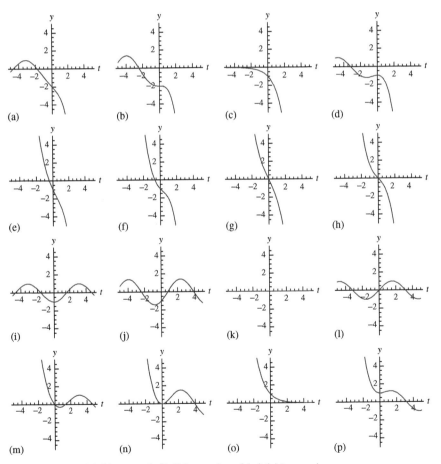

FIGURE 4.9   (a)-(d) in row 1, (e)-(h) in row 2, (i)-(l) in row 3, and (m)-(p) in row 4.

Determining a General Solution of a Higher Order Equation

1. Let r be a real root of the characteristic equation of an nth-order homogeneous linear differential equation with real constant coefficients. Then, e^{rt} is the solution associated with the root r.

   If r is a root of multiplicity k where $k \geq 2$ of the characteristic equation, then the k solutions associated with r are

   $$e^{rt}, \ te^{rt}, \ t^2 e^{rt}, \ \ldots, t^{k-1} e^{rt}$$

2. Suppose that r and $\bar{r}$ represent the complex conjugate pair $r_{1,2} = \alpha \pm \beta i$. Then the two solutions associated with these two roots are

   $$e^{\alpha t} \cos \beta t \quad \text{and} \quad e^{\alpha t} \cos \beta t.$$

   If the values $r_{1,2} = \alpha \pm \beta i$ are each a root of multiplicity k of the characteristic equation, then the other solutions associated with this pair are

   $$te^{\alpha t} \cos \beta t, \ te^{\alpha t} \sin \beta t, \ t^2 e^{\alpha t} \cos \beta t, \ t^2 e^{\alpha t} \sin \beta t,$$

   $$\ldots, t^{k-1} e^{\alpha t} \cos \beta t, \ t^{k-1} e^{\alpha t} \sin \beta t$$

   A general solution to the nth-order linear homogeneous differential equation with constant coefficients is the linear combination of the solutions obtained for all values of r.

   Note that if $r_1, r_2, \ldots, r_j$ are the roots of the characteristic equation of multiplicity $k_1, k_2, \ldots, k_j$, respectively, then $k_1 + k_2 + \cdots + k_j = n$, where n is the order of the differential equation.

Solution

The solutions that correspond to the complex conjugate pair $r_{1,2} = 2 \pm 3i$ are $y_1 = e^{2t} \cos 3t$ and $y_2 = e^{2t} \sin 3t$. These roots are repeated so the two other solutions corresponding to the pair are $y_3 = te^{2t} \cos 3t$ and $y_4 = te^{2t} \sin 3t$. For the single root $r_5 = -5$, the corresponding solution is $y_5 = e^{-5t}$. Finally, the root of multiplicity three $r_{6,7,8} = 2$ yields the three solutions $y_6 = e^{2t}$, $y_7 = te^{2t}$, and $y_8 = t^2 e^{2t}$. A fundamental set of solutions for the equation is

$$S = \left\{ e^{2t} \cos 3t, \ e^{2t} \sin 3t, \ te^{2t} \cos 3t, \ te^{2t} \sin 3t, \right.$$
$$\left. e^{-5t}, e^{2t}, te^{2t}, t^2 e^{2t} \right\}$$

Therefore, a general solution is

$$y = (c_1 + c_2 t)e^{2t} \cos 3t + (c_3 + c_4 t)e^{2t} \sin 3t$$
$$+ c_5 e^{-5t} + (c_6 + c_7 t + c_8 t^2)e^{2t}. \quad (4.19)$$

Note that the characteristic equation for the corresponding eighth-order homogeneous linear differential equation with constant coefficients is

$$[r - (2+3i)]^2[r - (2-3i)]^2(r+5)(r-2)^3 = r^8$$
$$- 9r^7 + 32r^6 + 50r^5 - 939r^4 + 4207r^3$$
$$- 10130r^2 + 12948r - 6760 = 0.$$

Hence,

$$y^{(8)} - 9y^{(7)} + 32y^{(6)} + 50y^{(5)} - 939y^{(4)} + 4207y'''$$
$$+ 10130y'' + 12948y' - 6760y = 0$$

is a linear homogeneous differential equation with general solution given by Equation (4.19).

EXAMPLE 4.5.9

Determine a general solution of the eighth-order homogeneous linear differential equation with constant coefficients if the roots and corresponding multiplicities of the characteristic equation are $r_{1,2} = 2 - 3i, \ k = 2; \ r_{3,4} = 2 + 3i, \ k = 2;$ $r_5 = -5, k = 1;$ and $r_{6,7,8} = 2, k = 3$.

 EXERCISES 4.5

In Exercises 1-6, calculate the Wronskian of the indicated set of function. Classify each set of functions as linearly independent or linearly dependent.

1. $S = \{3t^2, t, 2t - 2t^2\}$
2. $S = \{\cos 2t, \sin 2t, 1\}$
3. $S = \{e^{-t}, e^{3t}, te^{3t}\}$
4. $S = \{e^t, e^{-2t}, e^{-4t}\}$
5. $S = \{1, t, t^2, t^3, t^4\}$
6. $S = \{e^{3t}, e^{-2t}, e^{5t}, te^{5t}\}$

In Exercises 7-12, given the list of roots and multiplicities of the characteristic equation, form a general solution. What is the order of the corresponding differential equation?

7. $r = -2, k = 3; r = 2, k = 1$
8. $r = 1, k = 2; r = -10, k = 2$
9. $r = 0, k = 2; r = 3i, k = 1, r = -3i, k = 1$
10. $r = 4, k = 3; r = 1 + i, k = 1; r = 1 - i, k = 1$
11. $r = -3 + 4i, k = 2; r = -3 - 4i, k = 2;$
    $r = -5, k = 1; r = -1/3, k = 1$
12. $r = i/2, k = 3; r = -i/2, k = 3; r = -1, k = 2$
13. Can $\{1, t, t^2, 2t - 8\}$ be a fundamental set of solutions for a fourth-order linear homogeneous differential equation with real constant coefficients?
14. Can $\{e^{-t}, e^t, \cos t, \sin t, \cos 2t\}$ be a fundamental set of solutions for a fifth-order linear homogeneous differential equation with real constant coefficients?
15. Can $\{e^{2t}, e^{t/4}, e^{-t}\cos t, e^{-t}\sin t\}$ be a fundamental set of solutions for a fifth-order linear homogeneous differential equation with real constant coefficients?
16. Can $\{e^{2t}, e^{t/4}, e^{-t}, te^{-t}, t^2e^{-t}\}$ be a fundamental set of solutions for a fourth-order linear homogeneous differential equation with real constant coefficients?

In Exercises 17-39, find a general solution of each equation. (A computer algebra system may be useful in solving some of the equations.)

17. $y''' = 0$
18. $y^{(n)} = 0$
19. $y''' - 10y'' + 25y' = 0$

20. $8y''' + y'' = 0$
21. $y^{(4)} + 16y'' = 0$
22. $y''' - 2y'' - y' + 2y = 0$
23. $3y''' - 4y'' - 5y' + 2y = 0$
24. $6y''' - 5y'' - 2y' + y = 0$
25. $y''' - 5y' + 2y = 0$
26. $4y''' - 15y' + 11y = 0$
27. $y^{(4)} + y^{(3)} = 0$
28. $y^{(4)} - 9y'' = 0$
29. $y^{(4)} - 16y = 0$
30. $y^{(4)} - 6y''' - y'' + 54y' - 72y = 0$
31. $y^{(4)} + 7y''' + 6y'' - 32y' - 32y = 0$
32. $y^{(4)} + 2y''' - 2y'' + 8y = 0$
33. $y^{(5)} + 4y^{(4)} = 0$
34. $y^{(5)} + 4y^{(3)} = 0$
35. $y^{(5)} + 3y^{(4)} + 3y''' + y'' = 0$
36. $y^{(4)} + 2y'' + y = 0$
37. $y^{(4)} + 8y'' + 16y = 0$
38. $y^{(6)} + 3y^{(4)} + 3y'' + y = 0$
39. $y^{(6)} + 12y^{(4)} + 48y'' + 64y = 0$

In Exercises 40-50, solve the IVP. Graph the solution on an appropriate interval.

40. $y''' - 2y'' = 0, y(0) = 1, y'(0) = 2, y''(0) = 0$
41. $y''' - y = 0, y(0) = 0, y'(0) = 0, y''(0) = 3$
42. $y^{(4)} + 16y''' = 0, y(0) = 0,$
    $y'(0) = 1, y''(0) = 0, y'''(0) = 1$
43. $y^{(4)} - 8y'' + 16y = 0, y(0) = 0, y'(0) = 0,$
    $y''(0) = 8, y'''(0) = 0$
44. $24y''' - 26y'' + 9y' - y = 0, y(0) = 0,$
    $y'(0) = 1, y''(0) = 0$
45. $y^{(4)} - 5y'' + 4y = 0, y(0) = -1,$
    $y'(0) = 3, y''(0) = -7, y'''(0) = 15$
46. $y^{(4)} - 16y = 0, y(0) = 1, y'(0) = 2, y''(0) = 4,$
    $y'''(0) = -24$
47. $8y^{(5)} + 4y^{(4)} + 66y''' - 41y'' - 37y' = 0,$
    $y(0) = 4, y'(0) = -14, y''(0) = -14,$
    $y'''(0) = 139, y^{(4)}(0) = -29/4$
48. $2y^{(5)} + 7y^{(4)} + 17y''' + 17y'' + 5y' = 0,$
    $y(0) = -3, y'(0) = 15/2, y''(0) = 17/4,$
    $y'''(0) = -385/8, y^{(4)}(0) = 1217/16$
49. $y^{(5)} + 8y^{(4)} = 0, y(0) = 8, y'(0) = 4,$
    $y''(0) = 0, y'''(0) = 48, y^{(4)}(0) = 0$

50. $y^{(6)} - 3y^{(4)} + 3y'' - y = 0$, $y(0) = 16$,
$y'(0) = 0$, $y''(0) = 0$, $y'''(0) = 0$, $y^{(4)}(0) = 0$,
$y^{(5)}(0) = 0$

51. Suppose that the roots r_1, r_2, and r_3 of the
characteristic equation of a third-order
linear homogeneous differential equation
with constant coefficients are real and
distinct, and consider a general solution
$y = c_1 e^{r_1 t} + c_2 e^{r_2 t} + c_3 e^{r_3 t}$. Show that the
functions $e^{r_1 t}$, $e^{r_2 t}$, and $e^{r_3 t}$ are linearly
independent using the following steps: (a)
Assume that the functions are linearly
dependent. Then, there are constants c_1, c_2,
and c_3 (not all zero) such that
$c_1 e^{r_1 t} + c_2 e^{r_2 t} + c_3 e^{r_3 t} = 0$. (b) Multiply this
equation by $e^{-r_1 t}$ to obtain
$c_1 + c_2 e^{(r_2 - r_1)t} + c_3 e^{(r_3 - r_1)t} = 0$.
(c) Differentiate this equation with respect
to t to obtain $c_2(r_2 - r_1)e^{(r_2 - r_1)t} +$
$c_3(r_3 - r_1)e^{(r_3 - r_1)t} = 0$. (d) Multiply this
equation by $e^{-(r_2 - r_1)t}$ and differentiate the
resulting equation to obtain
$c_3(r_3 - r_2)(r_2 - r_1)e^{(r_3 - r_2)t} = 0$. This indicates
that $c_3 = 0$. (e) Follow similar steps to show
that $c_1 = c_2 = 0$, which contradicts the
assumption. (f) How could you apply a
similar argument to show that the functions
$e^{r_1 t}$, $e^{r_2 t}$, ..., $e^{r_n t}$ are linearly independent if
the n roots r_1, r_2, ..., r_n are real and
distinct?

52. *(Operator Notation)* The nth-order derivative
of a function $y = y(t)$ is given in operator
notation by $D^n y = d^n y/dt^n$. Then, the left
side of the nth-order linear homogeneous
differential equation with constant
coefficients (Equation 4.17) can be
expressed as

$$a_n D^n y + a_{n-1} D^{n-1} y + \cdots + a_1 D y + a_0 y$$

or

$$\left(a_n D^n + a_{n-1} D^{n-1} + \cdots + a_1 D + a_0 \right) y,$$

where $p(D) = a_n D^n + a_{n-1} D^{n-1} + \cdots +$
$a_1 D + a_0$ is called an *nth-order linear
differential operator* with constant
coefficients. Therefore, the differential
equation can be written as $p(D)y = 0$.
For example, we can write $y'' + 2y' -$
$8y = 0$ as $(D^2 - 2D - 8)y = 0$ or in either
of the forms $(D + 2)(D - 4)y = 0$ or
$(D - 4)(D + 2)y = 0$. Write each of the
following equations in differential operator
notation.
(a) $y'' + y' + 7y = 0$
(b) $y'' + 16y = 0$
(c) $y''' + 8y' - y = 0$
(d) $16y^{(4)} + y = 0$

53. Consider the linear differential operator
$p(D) = (2D - 1)$. When we apply $p(D)$ to the
function $f(t) = t^2$, we obtain $(2D - 1)(t^2)$
$= 2D(t^2) - t^2 = 2 \times 2t - t^2 = 4t - t^2$. Apply
the given differential operator to the given
function.
(a) $p(D) = 3D, f(t) = t$
(b) $p(D) = (1 - D), f(t) = t$
(c) $p(D) = D^2, f(t) = t^2$
(d) $p(D) = (D^2 + 1), f(t) = \cos t$

54. The linear differential operator $p(D)$ is
said to *annihilate* a function $f(t)$ if $p(D)$
$(f(t)) = 0$ for all t. In fact, if $P_1 = p_1(D)$
annihilates $f(t)$ and if $P_2 = p_2(D)$ is another
linear differential operator, then
$P_1 P_2 = P_2 P_1$ annihilates $f(t)$. For example,
$(D - 1)$ annihilates e^t because
$(D - 1)(e^t) = e^t - e^t = 0$. Therefore, if we
apply another linear differential
operator such as D, we have
$D(D - 1)(e^t) = D(e^t - e^t) = D(0) = 0$. Show
that $(D - 1)D(e^t) = 0$. In terms of
differential equations, this means that a
solution to an equation like
$(D - 1)y = y' - y = 0$ is also a solution to the
equation $D(D - 1)y = y'' - y' = 0$.

55. *(Repeated Roots)* Suppose that the
characteristic equation $r^n + a_{n-1}r^{n-1}$

$+\cdots+a_1 r+a_0 = 0$ has two distinct roots, r_1, of multiplicity one and r_2 of multiplicity $k = n-1$. Then, the characteristic equation can be written as $(r-r_1)(r-r_2)^k = 0$. Similarly, the differential equation can be written in differential operator notation as $(D-r_1)(D-r_2)^k y = 0$. Therefore, every solution of $(D-r_2)^k y = 0$ is also a solution of $(D-r_1)(D-r_2)^k y = 0$ (see Exercise 54). To find solutions of $(D-r_2)^k y = 0$, assume that $y = v(t)e^{r_2 t}$. (a) Use the product rule to show that $(D-r_2)\left(v(t)e^{r_2 t}\right) = (Dv)e^{r_2 t}$ so that, by induction, $(D-r_2)^k \left(v(t)e^{r_2 t}\right) = (D^k v)e^{r_2 t}$ for any function $v(t)$. (b) Therefore, a solution of $(D-r_2)^k y = 0$ must satisfy $(D^k v)e^{r_2 t} = 0$ or $D^k v = 0$. Solve $D^k v = 0$ (or $v^{(k)} = 0$) to find $v(t)$ and to conclude that the portion of the solution corresponding to $r = r_2$ is $y = v(t)e^{r_2 t} = (c_1 + c_2 t + \cdots + c_k t^{k-1})e^{r_2 t}$.

56. Let y_1 and y_2 be linearly independent solutions of $y''' + p_2(t)y'' + p_1(t)y' + p_0(t)y = 0$. Derive a "reduction of order formula" for a third linearly independent solution of $y''' + p_2(t)y'' + p_1(t)y' + p_0(t)y = 0$. If you are given $n-1$ linearly independent solutions of an nth-order linear homogeneous equation, how do you generalize your result to determine an nth linearly independent solution?

57. (*Abel's Formula for Higher Order Equations*) Consider the nth-order linear homogeneous equation $y^{(n)} + p_{n-1}(t)y^{(n-1)} + \cdots + p_0(t)y = 0$ with n linearly independent solutions $y_1, y_2, \ldots, y_n$ on an interval I. We can show that the Wronskian of these n solutions satisfies the same identity as that presented in Exercises 4.1. We do this for the third-order differential equation, $y''' + p_2(t)y'' + p_1(t)y' + p_0(t)y = 0$, with linearly independent solutions y_1, y_2, and y_3 on an interval I.

(a) Show that
$$\frac{d}{dt}\left(W(\{y_1, y_2, y_3\})\right) = \begin{vmatrix} y_1 & y_2 & y_3 \\ y_1' & y_2' & y_3' \\ y_1''' & y_2''' & y_3''' \end{vmatrix}.$$
(b) Use the differential equation to solve for y_1''', y_2''', and y_3'''. Substitute these values to obtain $dW/dt + p_2(t)W = 0$.
(c) Solve this differential equation to find that $W(\{y_1, y_2, y_3\}) = Ce^{-\int p_2(t)dt}$.
(d) Indicate how to generalize this result to the nth-order linear homogeneous equation.

In Exercises 58 and 59 show that S is a fundamental set of solutions for the given equation.

58. $S = \{e^t, e^{-5t}\sin t, e^{-5t}\cos t\}$; $y''' + 9y'' + 16y' - 26y = 0$
59. $S = \{e^{-3t}, e^{-t}, e^{-4t}\cos 3t, e^{-4t}\sin 3t\}$; $y^{(4)} + 12y''' + 60y'' + 124y' + 75y = 0$
60. Is it possible to choose values of $c_1, c_2, \ldots, c_7$, not all zero, so that $f(t) = c_1 t^{-7/2} + c_2 t^{-5/2} + c_3 t^{-3/2} + c_4 t^{-1/2} + c_5 t^{1/2} + c_6 t^{3/2} + c_7 t^{5/2}$ is the zero function?
61. Use the symbolic manipulation capabilities of a computer algebra system to help you prove the following theorem. Suppose that $f(t)$ is a differentiable function on an interval I and that $f(t) \neq 0$ for all t in I. (a) Prove that $f(t)$ and $tf(t)$ are linearly independent on I. (b) Prove that $f(t)$, $tf(t)$, and $t^2 f(t)$ are linearly independent on I. (c) Generalize your result.
62. Find a general solution of
(a) $y''' + 2y'' + 5y' - 26y = 0$
(b) $0.9y''' + 18.78y' - 0.2987y = 0$
(c) $8.9y^{(4)} - 2.5y'' + 32.0y' + 0.773y = 0$
Graph the solution for various initial conditions.
63. Solve each of the following IVPs. Verify that your result satisfies the initial conditions by graphing it on an appropriate interval.
(a) $y''' + 3y'' + 2y' + 6y = 0$, $y(0) = 0$, $y'(0) = 1$, $y''(0) = -1$

(b) $y^{(4)} - 8y''' + 30y'' - 56y' + 49y = 0$,
$y(0) = 1, y'(0) = 2, y''(0) = -1$,
$y'''(0) = -1$

(c) $0.31y''' + 11.2y'' - 9.8y' + 5.3y = 0$,
$y(0) = -1, y'(0) = -1, y''(0) = 0$

64. Use a computer algebra system to find a general solution of $y - k^4 \frac{d^4 y}{dx^4} = 0$. Show that the result you obtain is equivalent to

$$y = Ae^{-x/k} + Be^{x/k} + C\sin\frac{x}{k} + D\cos\frac{x}{k}.$$

Find conditions on A, B, C, and D (if possible) so that y has (a) neither local maxima nor local minima; (b) exactly one local maximum; and (c) exactly one local minimum on the interval $[0, \infty)$.

65. Show that if $y_0 \neq 0$ the initial boundary-value problem
$4.02063y''' - 0.224975y'' + 4.486y' - 2.48493y = 0, y(0) = 0, y'(0) = y_0, y(4) = 0$ has a unique solution.

66. Let $S = \{y_1, y_2, y_3\}$ be a set of differentiable functions for which $W(S) \neq 0$ on an interval I. Determine formulas for p_2, p_1, and p_0 so that S is a fundamental set for
$y''' + p_2(t)y' + p_1(t)y' + p_0(t)y = 0$. Indicate how to generalize your result for an nth-order linear homogeneous equation.

67. Solve $2yy'' + y^2 = (y')^2$. Is the Principle of Superposition valid for this nonlinear equation? Explain.

4.6 SOLVING HIGHER ORDER LINEAR EQUATIONS: UNDETERMINED COEFFICIENTS AND VARIATION OF PARAMETERS

- GENERAL SOLUTION OF A NONHOMOGENEOUS EQUATION
- UNDETERMINED COEFFICIENTS
- VARIATION OF PARAMETERS

General Solution of a Nonhomogeneous Equation

The methods discussed in Sections 4.3 and 4.4 for solving second-order linear nonhomogeneous equations can be extended to solving higher order equations.

Definition 11 (Particular Solution). *A particular solution, $y_p(t)$, of the nonhomogeneous differential equation $a_n(t)y^{(n)} + a_{n-1}(t)y^{(n-1)} + \cdots + a_1(t)y' + a_0(t)y = g(t)$ is a specific function that contains no arbitrary constants and satisfies the differential equation.*

Suppose that y_p is a particular solution of the nonhomogeneous equation

$$a_n(t)y^{(n)} + a_{n-1}(t)y^{(n-1)} + \cdots + a_1(t)y' + a_0(t)y = g(t) \tag{4.20}$$

and $S = \{f_1, f_2, \ldots, f_n\}$ is a fundamental set of solutions of the corresponding homogeneous equation

$$a_n(t)y^{(n)} + a_{n-1}(t)y^{(n-1)} + \cdots + a_1(t)y' + a_0(t)y = 0 \tag{4.21}$$

so that

$$y_h = c_1 f_1(t) + c_2 f_2(t) + \cdots + c_n f_n(t) \tag{4.22}$$

is a general solution of the corresponding homogeneous equation (4.21). Now let $y = y(t)$ be *any* solution of the nth-order linear nonhomogeneous equation (4.20). Then,

$$a_n(t)(y - y_p)^{(n)} + a_{n-1}(t)(y - y_p)^{(n-1)} + \cdots + $$
$$a_1(t)(y - y_p)' + a_0(t)(y - y_p)$$
$$= \left(a_n(t)y^{(n)} + a_{n-1}(t)y^{(n-1)} + \cdots + a_1(t)y' + a_0(t)y \right)$$
$$- \left(a_n(t)y_p^{(n)} + a_{n-1}(t)y_p^{(n-1)} + \cdots + \right.$$
$$\left. a_1(t)y_p' + a_0(t)y_p \right)$$
$$= g(t) - g(t) = 0.$$

Thus, $y - y_p$ is a solution of the corresponding homogeneous equation (4.21). So there are c_1, c_2, ..., c_n so that

$$y - y_p = c_1 f_1(t) + c_2 f_2(t) + \cdots + c_n f_n(t)$$

$$y = \underbrace{c_1 f_1(t) + c_2 f_2(t) + \cdots + c_n f_n(t)}_{\text{This is } y_h.} + y_p \quad \text{or}$$

$$y = y_h + y_p.$$

This means that every solution to the nonhomogeneous equation (4.20) can be written as the sum of a general solution, y_h, to the corresponding homogeneous equation (4.21) and a particular solution, y_p, of the nonhomogeneous equation (4.20). These observations lead us to the following definition.

Definition 12 (General Solution of a Nonhomogeneous Equation). *A **general solution** of the nth-order linear nonhomogeneous differential equation* (4.20) *is* $y = y_h(t) + y_p(t)$, *where* $y_h(t)$ *is a general solution of the corresponding homogeneous equation* (4.21) *and* $y_p(t)$ *is a particular solution of the nonhomogeneous equation* (4.20).

EXAMPLE 4.6.1

A fundamental set of solutions for $t^3 y''' + t y' - y = 0$, $t > 0$, is $S = \{t, t \ln t, t(\ln t)^2\}$ and a particular solution of $t^3 y''' + t y' - y = 3 - \ln t$, $t > 0$ is $y_p = \ln t$.

Use this information to solve the IVP $t^3 y''' + t y' - y = 3 - \ln t$, $y(1) = 0$, $y'(1) = 0$, $y''(1) = 1$.

Solution

From the given information, a general solution of the corresponding homogeneous equation of $t^3 y''' + t y' - y = 3 - \ln t$ is $y_h = c_1 t + c_2 t \ln t + c_3 t (\ln t)^2$. Therefore, a general solution of the nonhomogeneous equation is $y = y_h + y_p = c_1 t + c_2 t \ln t + c_3 t (\ln t)^2 + \ln t$ (see Figure 4.10(a)).

To solve the IVP we first compute

$$y' = c_1 + c_2 + t^{-1} + c_2 \ln t + 2c_3 \ln t + c_3 (\ln t)^2$$

and

$$y'' = -t^{-2} + c_2 t^{-1} + 2c_3 t^{-1} + 2c_3 t^{-1} \ln t.$$

Then,

$$\begin{aligned} y(1) &= c_1 & &= 0 \\ y'(1) &= c_1 + c_2 & &= -1 \\ y''(1) &= c_2 + 2c_3 & &= 2 \end{aligned}$$

so $c_1 = 0$, $c_2 = -1$, and $c_3 = 3/2$. Therefore, the solution to the IVP is $y = -t \ln t + \frac{3}{2} t (\ln t)^2 + \ln t$ (see Figure 4.10(b)).

Undetermined Coefficients

In the special case that the corresponding homogeneous equation has constant coefficients and the forcing function is a linear combination of functions of the form

$$1, t, , t^2, \ldots,$$

$$e^{\alpha t}, t e^{\alpha t}, t^2 e^{\alpha t}, \ldots,$$

$$\cos \beta t, \sin \beta t, t \cos \beta t, t \sin \beta t, \tag{4.23}$$

$$t^2 \cos \beta t, t^2 \sin \beta t, \ldots$$

or

$$e^{\alpha t} \cos \beta t, e^{\alpha t} \sin \beta t, t e^{\alpha t} \cos \beta t, t e^{\alpha t} \sin \beta t,$$

$$t^2 e^{\alpha t} \cos \beta t, t^2 e^{\alpha t} \sin \beta t, \ldots,$$

then the method of undetermined coefficients can be used to determine the form of a particular solution of the nonhomogeneous equation in the same way as that discussed in Section 4.2.

EXAMPLE 4.6.2

Solve $y^{(5)} + 4y''' = 48t - 6 - 10e^{-t}$.

Solution

The corresponding homogeneous equation is $y^{(5)} + 4y''' = 0$, which has characteristic equation $r^5 + 4r^3 = r^3(r^2 + 4) = 0$ so $r_{1,2,3} = 0$ has multiplicity three and $r_{4,5} = \pm 2i$ each have multiplicity one. Thus, a fundamental set for the corresponding homogeneous equation is $S = \{1, t, t^2, \cos 2t, \sin 2t\}$ and a general solution

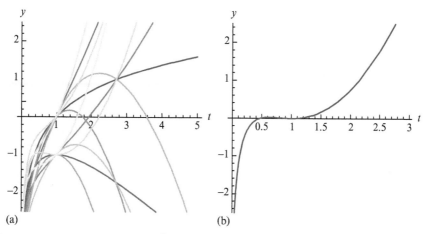

FIGURE 4.10 (a) Plots of various solutions of $t^3 y''' + ty' - y = 3 - \ln t$. (b) Plot of *the* solution of $t^3 y''' + ty' - y = 3 - \ln t$ that satisfies the initial conditions $y(1) = 0$, $y'(1) = 0$, and $y''(1) = 1$.

of the corresponding homogeneous equation is
$$y_h = c_1 + c_2 t + c_3 t^2 + c_4 \cos 2t + c_5 \sin 2t.$$

For the forcing function, $f(t) = 48t - 6 - 10e^{-t}$ we have two associated sets:

$$F_1 = \{e^{-t}\} \quad \text{and} \quad F_2 = \{t, 1\},$$

corresponding to the terms $-10e^{-t}$ and $48t - 6$, respectively, in the forcing function. Note that no element of F_1 is a solution of the corresponding homogeneous equation. On the other hand, functions in F_2 *are* solutions to the corresponding homogeneous equation. So, following the outline in Section 4.2, we multiply F_2 by t^n where n is the smallest positive integer so that no function in $t^n F_2$ is a solution of the corresponding homogeneous equation. In this case, we multiply F_2 by t^3 to obtain $t^3 F_2 = \{t^4, t^3\}$.

Thus, we assume that a particular solution to the nonhomogeneous equation has the form $y_p = At^4 + Bt^3 + Ce^{-t}$. Differentiating we have,

$$y_p' = 4At^3 + 3Bt^2 - Ce^{-t},$$

$$y_p'' = 12At^2 + 6Bt + Ce^{-t},$$

$$y_p''' = 24A + Ce^{-t} \quad \text{and} \quad y_p^{(4)} = -Ce^{-t}.$$

Substituting into the nonhomogeneous equation, simplifying the result, and equating coefficients gives us

$$24B + 96At - 5Ce^{-t} = 48t - 6 - 10e^{-t},$$

so $24B = -6$, $96A = 48$, and $-5C = -10$. Thus, $A = 1/2$, $B = -1/4$, and $C = 2$ so a particular solution of the nonhomogeneous equation is $y_p = \frac{1}{2}t^4 - \frac{1}{4}t^3 + 2e^{-t}$.

A general solution of the nonhomogeneous equation is then given by

$$y = y_h + y_p = c_1 + c_2 t + c_3 t^2 + c_4 \cos 2t$$
$$+ c_5 \sin 2t + \frac{1}{2}t^4 - \frac{1}{4}t^3 + 2e^{-t}.$$

 What is the form of a particular solution of $y^{(5)} + 4y''' = t \cos 2t$?

To solve an IVP, first determine a general solution and then use the initial conditions to solve for the unknown constants in the general solution.

EXAMPLE 4.6.3

Solve the IVP

$$4y''' + 4y'' + 65y' = e^{-t/3}\left(-\frac{286}{3}\cos 2t + \frac{577}{27}\sin 2t\right),$$

$y(0) = 4/3$, $y'(0) = -5/2$, $y''(0) = -173/12$.

Solution

The corresponding homogeneous equation is $4y''' + 4y'' + 65y' = 0$ with characteristic equation $4r^3 + 4r^2 + 65r = r(4r^2 + 4r + 65) = 0$ so $r_1 = 0$ and using the quadratic formula to solve $4r^2 + 4r + 65 = 0$ gives us $r_{2,3} = -\frac{1}{2} \pm 4i$. Thus, a fundamental set of solutions for the corresponding homogeneous equation is $S = \{1, e^{-t/2}\cos 4t, e^{-t/2}\sin 4t\}$ and a general solution of the corresponding homogeneous equation is $y_h = c_1 + e^{-t/2}(c_2\cos 4t + c_3\sin 4t)$.

The associated set of functions for the forcing function

$$g(t) = e^{-t/3}\left(-\frac{286}{3}\cos 2t + \frac{577}{27}\sin 2t\right)$$

is $F = \{e^{-t/3}\cos 2t, e^{-t/3}\sin 2t\}$. No function in F is a solution of the corresponding homogeneous equation so we assume that a particular solution of the nonhomogeneous equation has the form $y_p = Ae^{-t/3}\cos 2t + Be^{-t/3}\sin 2t$, where A and B are constants to be determined. Differentiating y_p three times gives us

$$y_p' = \left(-\frac{1}{3}A + 2B\right)e^{-t/3}\cos 2t$$
$$+ \left(2A - \frac{1}{3}B\right)e^{-t/3}\sin 2t,$$

$$y_p'' = \left(-\frac{35}{9}A - \frac{4}{3}B\right)e^{-t/3}\cos 2t$$
$$+ \left(\frac{4}{3}A - \frac{35}{9}\right)e^{-t/3}\sin 2t,$$

and

$$y_p''' = \left(\frac{107}{27}A - \frac{22}{3}B\right)e^{-t/3}\cos 2t$$
$$+ \left(\frac{22}{3}A + \frac{107}{27}B\right)e^{-t/3}\sin 2t.$$

Substituting y_p and its derivatives into the nonhomogeneous equation and simplifying the result gives us

---

When algebra becomes unusually cumbersome, we usually use a computer algebra system to assist in checking our calculations.

$$4y_p''' + 4y_p'' + 65y_p' = \left(-\frac{577}{27}A + \frac{286}{3}B\right)e^{-t/3}\cos 2t$$
$$+ \left(-\frac{286}{3}A - \frac{577}{27}B\right)e^{-t/3}\sin 2t$$
$$= e^{-t/3}\left(-\frac{286}{3}\cos 2t + \frac{577}{27}\sin 2t\right).$$

Equating coefficients gives us the system

$$-\frac{577}{27}A + \frac{286}{3}B = -\frac{286}{3}$$
$$-\frac{286}{3}A - \frac{577}{27}B = \frac{577}{27},$$

which has solution $A = 0$ and $B = -1$. Thus, $y_p = -e^{-t/3}\sin 2t$ and a general solution of the nonhomogeneous equation is $y = y_h + y_p = c_1 + e^{-t/2}(c_2\cos 4t + c_3\sin 4t - e^{-t/3}\sin 2t$ (see Figure 4.11(a)).

To solve the IVP, we first differentiate y twice resulting in

$$y' = \left(-\frac{1}{2}c_2 + 4c_3\right)e^{-t/2}\cos 4t$$
$$+ \left(4c_2 - \frac{1}{2}c_3\right)e^{-t/2}\sin 4t - 2e^{-t/3}\cos 2t$$
$$+ \frac{1}{3}e^{-t/3}\sin 2t$$

and

$$y'' = \left(-\frac{63}{4}c_2 + 4c_3\right)e^{-t/2}\cos 4t$$
$$+ \left(4c_2 - \frac{63}{4}c_3\right)e^{-t/2}\sin 4t + \frac{4}{3}e^{-t/3}\cos 2t$$
$$+ \frac{35}{9}e^{-t/3}\sin 2t.$$

Evaluating at $t = 0$ gives us the system of equations

$$
\begin{aligned}
y(0) = \quad c_1 \quad +c_2 \qquad\qquad &= \frac{4}{3} \\
y'(0) = -2 \quad -\frac{1}{2}c_2 \quad +4c_3 \quad &= -\frac{5}{2} \\
y''(0) = \frac{22}{3} \quad +\frac{191}{8}c_2 \quad -61c_3 \quad &= -\frac{173}{12},
\end{aligned}
$$

which has solution $c_1 = 1/3$, $c_2 = 1$, and $c_3 = 0$ so the solution to the IVP is $y = \frac{1}{3} + e^{-t/2}\cos 4t - e^{-t/3}\sin 2t$ (see Figure 4.11(b)).

What is the form of a particular solution of $4y''' + 4y'' + 65y' = t^2 + te^{-t/3}\cos 2t$?

Variation of Parameters

A particular solution of higher order non-homogeneous equations can be found through variation of parameters as well. Constructing a particular solution for higher order linear non-homogeneous equations is done in much the same way as for second-order equations.

We illustrate the concept for third and fourth-order equations. The third-order linear equation in standard form is

$$
\frac{d^3 y}{dt^3} + a_2(t)\frac{d^2 y}{dt^2} + a_1(t)\frac{dy}{dt} + a_0(t)y = f(t) \quad (4.24)
$$

with corresponding homogeneous equation

$$
\frac{d^3 y}{dt^3} + a_2(t)\frac{d^2 y}{dt^2} + a_1(t)\frac{dy}{dt} + a_0(t)y = 0. \quad (4.25)
$$

Let $S = \{y_1, y_2, y_3\}$ be a fundamental set for Equation (4.25) and let $y_h = c_1 y_1 + c_2 y_2 + c_3 y_3$ be a general solution of Equation (4.25). To find a particular solution we assume that $y_p = u_1 y_1 + u_2 y_2 + u_3 y_3$ where u_1, u_2 and u_3 are functions to be determined with the assumptions that $y_1 u_1' + y_2 u_2' + y_3 u_3' = 0$ and $y_1' u_1' + y_2' u_2' + y_3' u_3' = 0$, substituting y_p into the nonhomogeneous equations leads to $y_1'' u_1' + y_2'' u_2' + y_3'' u_3' = f(t)$ so we determine u_1', u_2' and u_3' by using Cramer's rule to solve the system

$$
\begin{aligned}
y_1 u_1' + y_2 u_2' + y_3 u_3' &= 0 \\
y_1' u_1' + y_2' u_2' + y_3' u_3' &= 0 \qquad (4.26) \\
y_1'' u_1' + y_2'' u_2' + y_3'' u_3' &= f(t).
\end{aligned}
$$

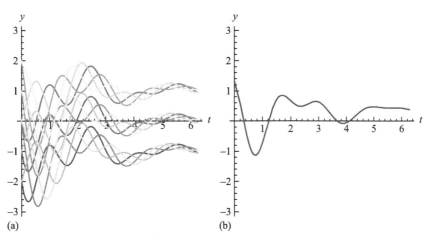

(a)   (b)

FIGURE 4.11 (a) Various solutions of the nonhomogeneous equation. (b) The solution of the nonhomogeneous equation that satisfies the initial conditions $y(0) = 4/3$, $y'(0) = -5/2$, and $y''(0) = -173/12$.

To use Cramer's rule, we first write system (4.26) in matrix form

$$\begin{pmatrix} y_1 & y_2 & y_3 \\ y_1' & y_2' & y_3' \\ y_1'' & y_2'' & y_3'' \end{pmatrix} \begin{pmatrix} u_1' \\ u_2' \\ u_3' \end{pmatrix} = \begin{pmatrix} 0 \\ 0 \\ f(t) \end{pmatrix}. \quad (4.27)$$

Observe that the coefficient matrix is the Wronskian matrix so it has nonzero determinant,

$$W(S) = \begin{vmatrix} y_1 & y_2 & y_3 \\ y_1' & y_2' & y_3' \\ y_1'' & y_2'' & y_3'' \end{vmatrix} \neq 0. \text{ Therefore, using}$$

Cramer's rule, the solution of system (4.27) is

$$u_1' = \frac{1}{W(S)} \begin{vmatrix} 0 & y_2 & y_3 \\ 0 & y_2' & y_3' \\ f(t) & y_2'' & y_3'' \end{vmatrix},$$

$$u_2' = \frac{1}{W(S)} \begin{vmatrix} y_1 & 0 & y_3 \\ y_1' & 0 & y_3' \\ y_1'' & f(t) & y_3'' \end{vmatrix},$$

$$\text{and} \quad u_3' = \frac{1}{W(S)} \begin{vmatrix} y_1 & y_2 & 0 \\ y_1' & y_2' & 0 \\ y_1'' & y_2'' & f(t) \end{vmatrix}.$$

For the fourth-order equation,

$$\frac{d^4y}{dt^4} + a_3(t)\frac{d^3y}{dt^3} + a_2(t)\frac{d^2y}{dt^2} + a_1(t)\frac{dy}{dt} + a_0(t)y = f(t) \quad (4.28)$$

we solve the linear system

$$\begin{aligned} y_1 u_1' + y_2 u_2' + y_3 u_3' + y_4 u_4' &= 0 \\ y_1' u_1' + y_2' u_2' + y_3' u_3' + y_4' u_4' &= 0 \\ y_1'' u_1' + y_2'' u_2' + y_3'' u_3' + y_4'' u_4' &= 0 \quad (4.29) \\ y_1''' u_1' + y_2''' u_2' + y_3''' u_3' + y_4''' u_4' &= f(t) \end{aligned}$$

for u_1', u_2', u_3', and u_4'.

The pattern solving for u_j' continues for the nth-order linear equation.

In general, if we are given the nonhomogeneous equation

$$y^{(n)} + a_{n-1}(t)y^{(n-1)} + \cdots + a_1(t)y' + a_0(t)y = f(t) \quad (4.30)$$

and fundamental set of solutions $S = \{y_1, y_2, \ldots, y_n\}$ of the corresponding homogeneous equation

$$y^{(n)} + a_{n-1}(t)y^{(n-1)} + \cdots + a_1(t)y' + a_0(t)y = 0, \quad (4.31)$$

we generalize the method for second-order equations (see Section 4.4) to find $u_1(t)$, $u_2(t)$, $\ldots, u_n(t)$ such that

$$y_p = u_1(t)y_1 + u_2(t)y_2 + \cdots + u_n(t)y_n \quad (4.32)$$

is a particular solution of the nonhomogeneous equation (4.30).

Assuming that a particular solution of the nonhomogeneous equation (4.30), has the form given by Equation (4.32) where $u_1(t)$, $u_2(t)$, $\ldots$, $u_n(t)$ are functions of t to be determined and that

$$u_1'(t)y_1^{(m-1)} + u_2'(t)y_2^{(m-1)} + \cdots + u_n'(t)y_n^{(m-1)} = 0 \quad (4.33)$$

for $m = 1, 2, \ldots, n - 1$, then

$$y_p^{(m)} = u_1(t)y_1^{(m)} + u_2(t)y_2^{(m)} + \cdots + u_n(t)y_n^{(m)} \quad (4.34)$$

for $m = 1, 2, \ldots, n - 1$.

Computing the nth derivative of y_p using the assumptions and consequences given in Equations (4.33) and (4.34) gives us

$$\begin{aligned} y_p^{(n)} &= u_1(t)y_1^{(n)} + u_2(t)y_2^{(n)} + \cdots + u_n(t)y_n^{(n)} \\ &\quad + u_1'(t)y_1^{(n)} + u_2'(t)y_2^{(n)} + \cdots + u_n'(t)y_n^{(n)}. \end{aligned} \quad (4.35)$$

With the assumptions given by Equation (4.33) and substitution of Equations (4.34) and (4.35) into the nonhomogeneous equation (4.30), and simplifying the results, we obtain the system of n equations

$$\begin{aligned} y_1 u_1'(t) &+ y_2 u_2'(t) &+ \cdots &+ y_n u_n'(t) &= 0 \\ y_1' u_1'(t) &+ y_2' u_2'(t) &+ \cdots &+ y_n' u_n'(t) &= 0 \\ y_1'' u_1'(t) &+ y_2'' u_2'(t) &+ \cdots &+ y_n'' u_n'(t) &= 0 \\ &\vdots && \vdots && \vdots \\ y_1^{(n-1)} u_1'(t) &+ y_2^{(n-1)} u_2'(t) &+ \cdots &+ y_n^{(n-1)} u_n'(t) &= f(t) \end{aligned}$$

written in matrix form as

$$
\begin{pmatrix}
y_1 & y_2 & y_3 & \cdots & y_n \\
y_1' & y_2' & y_3' & \cdots & y_n' \\
\vdots & \vdots & \vdots & \vdots & \vdots \\
y_1^{(n-1)} & y_2^{(n-1)} & y_3^{(n-1)} & \cdots & y_n^{(n-1)}
\end{pmatrix}
\begin{pmatrix}
u_1' \\ u_2' \\ \vdots \\ u_{n-1}' \\ u_n'
\end{pmatrix}
=
\begin{pmatrix}
0 \\ 0 \\ \vdots \\ 0 \\ f(t)
\end{pmatrix}
$$

(4.36)

that is linear in the variables u_1', u_2', ..., u_n' and can be solved for u_1', u_2', ..., u_n' using Cramer's rule.

Let $W(S)$ denote the Wronskian (i.e., the determinant of the Wronskian matrix) and $W_m(\{y_1, y_2, \ldots y_n\})$ denote the determinant of the matrix obtained by replacing the mth column of the Wronskian matrix, W,

$$
W = \begin{pmatrix}
y_1 & y_2 & y_3 & \cdots & y_n \\
y_1' & y_2' & y_3' & \cdots & y_n' \\
\vdots & \vdots & \vdots & \vdots & \vdots \\
y_1^{(n-1)} & y_2^{(n-1)} & y_3^{(n-1)} & \cdots & y_n^{(n-1)}
\end{pmatrix},
$$

by the column vector $\begin{pmatrix} 0 \\ 0 \\ \vdots \\ 0 \\ f(t) \end{pmatrix}$. Then, by Cramer's rule, the solutions of system (4.36) are given by

$$
u_i'(t) = \frac{1}{W(S)} f(t) W_i(\{y_1, y_2, \ldots y_n\}) \quad (4.37)
$$

for $i = 1, 2, \ldots, n$, and

$$
u_i(t) = \int \frac{1}{W(S)} f(t) W_i(\{y_1, y_2, \ldots y_n\}) \, dt,
$$

(4.38)

so $y_p = u_1(t)y_1 + u_2(t)y_2 + \cdots + u_n(t)y_n$ is a particular solution of the nonhomogeneous equation (4.30). A general solution of the nonhomogeneous equation is given by $y = y_h + y_p$, where y_h is a general solution of the corresponding homogeneous equation (4.31).

EXAMPLE 4.6.4

A fundamental set of solutions for $ty''' + 3y'' = 0, t > 0$, is $S = \{1, t, t^{-1}\}$. Solve $ty''' + 3y'' = -t^{-2}$, $t > 0$.

Solution

From the given information, a general solution of the corresponding homogeneous equation is $y_h = c_1 + c_2 t + c_3 t^{-1}$.

To use variation of parameters to find a particular solution of the nonhomogeneous equation, we first write the equation in standard form by dividing by t, resulting in $y''' + 3t^{-1}y'' = \underbrace{-t^{-3}}_{f(t)}$.

Then,

$$
W(S) = \begin{vmatrix}
1 & t & t^{-1} \\
0 & 1 & -t^{-2} \\
0 & 0 & 2t^{-3}
\end{vmatrix} = 2t^{-3}.
$$

Using Equation (4.37),

$$
u_1' = \frac{1}{W(S)} \begin{vmatrix}
0 & t & t^{-1} \\
0 & 1 & -t^{-2} \\
-t^{-3} & 0 & 2t^{-3}
\end{vmatrix} = \frac{1}{2}t^3 \times 2t^{-4} = t^{-1}
$$

so $u_1 = \int t^{-1} dt = \ln t$. Next,

$$
u_2' = \frac{1}{W(S)} \begin{vmatrix}
1 & 0 & t^{-1} \\
0 & 0 & -t^{-2} \\
0 & -t^{-3} & 2t^{-3}
\end{vmatrix} = \frac{1}{2}t^3 \times -t^{-5} = \frac{1}{2}t^{-2}
$$

so $u_2 = \frac{1}{2} \int t^{-2} dt = -\frac{1}{2}t^{-1}$. Finally,

$$
u_3' = \frac{1}{W(S)} \begin{vmatrix}
1 & t & 0 \\
0 & 1 & 0 \\
0 & 0 & -t^{-3}
\end{vmatrix} = \frac{1}{2}t^3 \times -t^{-3} = -\frac{1}{2}
$$

so $u_3 = -\frac{1}{2} \int dt = -\frac{1}{2}t$.

Then a particular solution of the nonhomogeneous equation is

$$
y_p = u_1 y_1 + u_2 y_2 + u_3 y_3
$$
$$
= \ln t \times 1 + \left(-\frac{1}{2}t^{-1}\right) \times t + \left(-\frac{1}{2}t\right) \times t^{-1}
$$
$$
= \ln t - 1 = \ln t.
$$

Note that term -1 is omitted (or dropped) from y_p because $y = -1$ is a solution of the corresponding homogeneous equation, represented by c_1, so does not contribute to the solution of the nonhomogeneous equation; it does not need to be included in y_p because it is included in y_h.

Thus, a general solution of the nonhomogeneous equation is $y = y_h + y_p = c_1 + c_2 t + c_3 t^{-1} + \ln t$.

To solve an IVP, first determine a general solution and then use the initial conditions to solve for the unknown constants in the general solution.

EXAMPLE 4.6.5

Solve $y^{(4)} + 9y'' = \sec^2 3t$, $-\pi/6 < t < \pi/6$, $y(0) = 0$, $y'(0) = 1$, $y''(0) = 0$, $y'''(0) = -3$.

Solution

The corresponding homogeneous equation is $y^{(4)} + 9y'' = 0$ with characteristic equation $r^4 + 9r^2 = r^2(r^2 + 9) = 0$ so $r_{1,2} = 0$ and $r_{3,4} = \pm 3i$. Thus, a fundamental set of solutions for the corresponding homogeneous equation is $S = \{1, t, \cos 3t, \sin 3t\}$; a general solution of the corresponding homogeneous equation is $y_h = c_1 + c_2 t + c_3 \cos 3t + c_4 \sin 3t$.

To find a particular solution of the nonhomogeneous equation, we first compute

$$W(S) = \begin{vmatrix} 1 & t & \cos 3t & \sin 3t \\ 0 & 1 & -3\sin 3t & 3\cos 3t \\ 0 & 0 & -9\cos 3t & -9\sin 3t \\ 0 & 0 & 27\sin 3t & -27\cos 3t \end{vmatrix}$$

$$= 243\cos^2 3t + 243\sin^2 3t = 243.$$

We now find u_i' using Equation (4.37). First,

$$u_1' = \frac{1}{243} \begin{vmatrix} 0 & t & \cos 3t & \sin 3t \\ 0 & 1 & -3\sin 3t & 3\cos 3t \\ 0 & 0 & -9\cos 3t & -9\sin 3t \\ \sec^2 3t & 0 & 27\sin 3t & -27\cos 3t \end{vmatrix}$$

$$= \frac{1}{243} \times -27t\sec^2 3t = -\frac{1}{9}t\sec^2 3t.$$

so $u_1 = -\frac{1}{9}\int t\sec^2 3t\,dt = -\frac{1}{81}\ln(\cos 3t) - \frac{1}{27}t\tan 3t$. Next,

$$u_2' = \frac{1}{243} \begin{vmatrix} 1 & 0 & \cos 3t & \sin 3t \\ 0 & 0 & -3\sin 3t & 3\cos 3t \\ 0 & 0 & -9\cos 3t & -9\sin 3t \\ 0 & \sec^2 3t & 27\sin 3t & -27\cos 3t \end{vmatrix}$$

$$= \frac{1}{243}\left(27 + 27\tan^2 3t\right) = \frac{1}{9}\sec^2 3t$$

so $u_2 = \frac{1}{9}\int \sec^2 3t\,dt = \frac{1}{27}\tan 3t$,

$$u_3' = \frac{1}{243} \begin{vmatrix} 1 & t & 0 & \sin 3t \\ 0 & 1 & 0 & 3\cos 3t \\ 0 & 0 & 0 & -9\sin 3t \\ 0 & 0 & \sec^2 3t & -27\cos 3t \end{vmatrix}$$

$$= \frac{1}{27}\sec 3t\tan 3t$$

so $u_3 = \frac{1}{27}\int \sec 3t\tan 3t\,dt = \frac{1}{81}\sec 3t$ and

$$u_4' = \frac{1}{243} \begin{vmatrix} 1 & t & \cos 3t & 0 \\ 0 & 1 & -3\sin 3t & 0 \\ 0 & 0 & -9\cos 3t & 0 \\ 0 & 0 & 27\sin 3t & \sec^2 3t \end{vmatrix} = -\frac{1}{27}\sec 3t$$

so $u_4 = -\frac{1}{81}\ln(\sec 3t + \tan 3t)$. Then, a particular solution of the nonhomogeneous equation is given by

$$y_p = u_1 y_1 + u_2 y_2 + u_3 y_3 + u_4 y_4$$

$$= \left(-\frac{1}{81}\ln(\cos 3t) - \frac{1}{27}t\tan 3t\right) \times 1$$

$$+ \frac{1}{27}\tan 3t \times t + \frac{1}{81}\sec 3t \times \cos 3t$$

$$- \frac{1}{81}\ln(\sec 3t + \tan 3t) \times \sin 3t$$

$$= \frac{1}{81}\left(1 + \ln(\sec 3t) - \ln(\sec 3t + \tan 3t)\sin 3t\right)$$

$$= \frac{1}{81}\left(\ln(\sec 3t) - \ln(\sec 3t + \tan 3t)\sin 3t\right),$$

where $1/81$ is discarded from the particular solution because constant functions are solutions of the corresponding homogeneous equation so need not be included in a particular solution of the nonhomogeneous equation.

Thus, a general solution of the nonhomogeneous equation is

$$y = y_h + y_p$$
$$= c_1 + c_2 t + c_3 \cos 3t + c_4 \sin 3t + \frac{1}{81} \left(\ln (\sec 3t) \right.$$
$$\left. - \ln(\sec 3t + \tan 3t) \sin 3t \right),$$

where we use properties of the natural logarithm: $\ln \cos t = -\ln \sec t$. Also, in the integrations we eliminated the absolute values in the logarithms because $\cos 3t$ and $\sec 3t$ are positive on the interval $-\pi/6 < t < \pi/6$.

To solve the IVP, we first differentiate three times,

$$y' = c_2 + 3c_4 \cos 3t - 3c_3 \sin 3t$$
$$- \frac{1}{27} \ln(\sec 3t + \tan 3t) \cos 3t$$

$$y'' = -9c_3 \cos 3t - 9c_4 \sin 3t$$
$$+ \frac{1}{9} \ln(\sec 3t + \tan 3t) \sin 3t - \frac{1}{9}$$

and

$$y''' = -27c_4 \cos 3t + 27 \sin 3t$$
$$+ \frac{1}{3} \ln(\sec 3t + \tan 3t) \cos 3t + \frac{1}{3} \tan 3t$$

and then apply the initial conditions

$$\begin{array}{llll} y(0) = & c_1 & +c_3 & = 0 \\ y'(0) = & c_2 & +3c_4 & = 1 \\ y''(0) = -\frac{1}{9} & & -9c_3 & = 0 \\ y'''(0) = & & -27c_4 & = -3 \end{array}$$

to see that $c_1 = 1/81$, $c_2 = 2/3$, $c_3 = -1/81$, and $c_4 = 1/9$. Thus, the solution to the IVP is

$$y = \frac{1}{81} + \frac{2}{3}t - \frac{1}{81}\cos 3t + \frac{1}{9}\sin 3t + \frac{1}{81} \left(\ln (\sec 3t) \right.$$
$$\left. - \ln(\sec 3t + \tan 3t) \sin 3t \right),$$

which is graphed in Figure 4.12(d) over the interval $(-\pi/6, \pi/6)$.

EXERCISES 4.6

In Exercises 1-10, use the method of undetermined coefficients to solve each nonhomogeneous equation.

1. $y''' + y'' = e^t$
2. $y^{(4)} - 16y = 1$
3. $y^{(5)} - y^{(4)} = 1$
4. $y^{(4)} + 9y'' = 1$
5. $y^{(4)} - 9y'' = 9e^{3t}$
6. $y''' + 10y'' + 34y' + 40y = te^{-4t} + 2e^{-3t} \cos t$
7. $y''' + 6y'' + 11y' + 6y = 2e^{-3t} - te^{-t}$
8. $y^{(4)} - 6y''' + 13y'' - 24y' + 36y = 108t$
9. $y''' + 6y'' - 14y' - 104y = -111e^t$
10. $y^{(4)} - 10y''' + 38y'' - 64y' + 40y = 153e^{-t}$

In Exercises 11-15, use the method of variation of parameters to solve each nonhomogeneous equation.

11. $y''' + 4y' = \tan 2t$, $-\pi/4 < t < \pi/4$
12. $y''' + 4y' = \sec 2t \tan 2t$, $-\pi/4 < t < \pi/4$
13. $y^{(4)} + 4y'' = \sec^2 2t$, $-\pi/4 < t < \pi/4$
14. $y^{(4)} + 4y'' = \tan^2 2t$, $-\pi/4 < t < \pi/4$
15. $y''' + 9y' = \sec^2 3t$, $-\pi/6 < t < \pi/6$

In Exercises 16-25, use any method to solve each nonhomogeneous equation.

16. $y''' + y' = -\sec t \tan t$, $-\pi/2 < t < \pi/2$
17. $y''' + 4y' = \sec 2t$, $-\pi/4 < t < \pi/4$
18. $y''' - 2y'' = -t^{-2} - 2t^{-1}$, $t > 0$
19. $y''' - 3y'' + 3y' - y = t^{-1}e^t$, $t > 0$
20. $y''' - 4y'' - 11y' + 30y = e^{4t}$
21. $y''' + 3y'' - 10y' - 24y = e^{-3t}$
22. $y''' - 13y' + 12y = \cos t$
23. $y''' + 3y'' + 2y' = \cos t$
24. $y^{(6)} + y^{(4)} = -24$
25. $y^{(4)} + y'' = \tan^2 t$, $-\pi/2 < t < \pi/2$

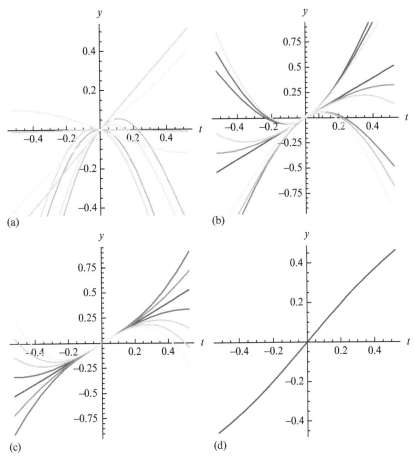

(a) (b) (c) (d)

FIGURE 4.12 (From left to right) (a) Various solutions of the nonhomogeneous equation that satisfy $y(0) = 0$. (b) Various solutions that satisfy $y(0) = 0$ and $y'(0) = 1$. (c) Various solutions that satisfy $y(0) = 0$, $y'(0) = 1$, and $y''(0) = 0$. (d) *The* solution that satisfies $y(0) = 0$, $y'(0) = 1$, $y''(0) = 0$, and $y'''(0) = -3$.

In Exercises 26-30, use any method to solve each IVP.

26. $y''' - y'' = 3t^2$, $y(0) = 0$, $y'(0) = 0$, $y''(0) = 0$

27. $y^{(4)} + y'' = \sec^2 t$, $-\pi/2 < t < \pi/2$,
    $y(0) = y''(0) = y'''(0) = 0$, $y'(0) = 1$

28. $y''' + y' = \sec t$, $-\pi/2 < t < \pi/2$,
    $y(0) = y'(0) = 0$, $y''(0) = 1$

29. $y'''' + y'' = \cos t$, $y(0) = 0$, $y'(0) = 0$,
    $y''(0) = 1$, $y'''(0) = 0$

30. $y'''' + y'' = t$, $y(0) = 0$, $y'(0) = 0$, $y''(0) = 1$,
    $y'''(0) = 0$

31. A fundamental set of solutions for
    $t^2 \ln t\, y''' - ty'' + y' = 0$, $t > 0$, is
    $S = \{1, t^2, t\ln t\}$. Use this information to
    solve $t^2 \ln t\, y'' - ty'' + y' = 1$.

32. A fundamental set of solutions for
    $(t^2 + t)y''' + (2 - t^2)y'' - (t + 2)y' = 0$, $t > 0$,
    is $S = \{1, e^t, \ln t\}$. Use this information to
    solve $(t^2 + t)y''' + (2 - t^2)y'' - (t + 2)y'$
    $= -2 - t$.

33. A fundamental set of solutions for
    $2t^3 y''' + t^2 y'' + ty' - y = 0$, $t > 0$, is
    $S = \{t, t\ln t, \sqrt{t}\}$. Use this information to

solve $2t^3y''' + t^2y'' + ty' - y = -3t^2, y(1) = 0,$ $y'(1) = 1, y''(1) = 0.$

34. A fundamental set of solutions for $ty^{(4)} + 2y''' = 0, t > 0,$ is $S = \{1, t, t\ln t, t^2\}.$ Use this information to solve $ty^{(4)} + 2y''' = \frac{45}{8}t^{-7/2}, y(1) = 0, y'(1) = 0,$ $y''(1) = 1, y'''(1) = 0.$

35. Show that if y_1 is a solution of $y^{(n)} + a_{n-1}(t)$ $y^{(n-1)} + \cdots + a_1(t)y' + a_0(t)y = f_1(t)$ and y_2 is a solution of $y^{(n)} + a_{n-1}(t)y^{(n-1)} + \cdots + a_1(t)y' + a_0(t)y = f_2(t),$ then $y_1 + y_2$ is a solution of $y^{(n)} + a_{n-1}(t)y^{(n-1)} + \cdots + a_1(t)y' + a_0(t)y = f_1(t) + f_2(t).$

36. If $\phi(t)$ is a solution of $a_ny^{(n)} + a_{n-1}y^{(n-1)} + \cdots + a_2y'' + a_1y' + a_0y = f(t),$ what is a solution of $a_ny^{(n+1)} + a_{n-1}y^{(n)} + \cdots + a_2y''' + a_1y'' + a_0y' = f'(t)?$

4.7 CAUCHY-EULER EQUATIONS

- SECOND-ORDER CAUCHY-EULER EQUATIONS
- NONHOMOGENEOUS CAUCHY-EULER EQUATIONS
- HIGHER ORDER CAUCHY-EULER EQUATIONS

In previous sections, we solved linear differential equations with constant coefficients. Generally, solving an arbitrary linear differential equation is a formidable task, particularly when the coefficients are not constants. However, we are able to solve certain equations with variable coefficients using techniques similar to those discussed previously. We begin by considering differential equations of the form

$$a_nx^ny^{(n)} + a_{n-1}x^{n-1}y^{(n-1)} + \cdots + a_1xy' + a_0y = g(x),$$

where $a_0, a_1, \ldots, a_n$ are constants, called *Cauchy-Euler equations*. (Notice that in Cauchy-Euler equations, the order of each derivative in

the equation equals the power of x in the corresponding coefficient.) Euler observed that equations of this type are reduced to linear equations with constant coefficients with the substitution $x = e^t.$[3]

Many mathematical terms are named after **Augustin Louis Cauchy** (1789-1857), including the Cauchy integral theorem and Cauchy-Rieman equations (complex analysis), Cauchy sequence (first year calculus), and the Cauchy-Kovalevskaya existence theorem (partial differential equations). Legend is that Cauchy was known for never admitting a mathematical or personal mistake, even when he had made one.

Definition 13 (Cauchy-Euler Equation). *A Cauchy-Euler differential equation is a differential equation of the form*

$$a_nx^ny^{(n)} + a_{n-1}x^{n-1}y^{(n-1)} + \cdots + a_1xy' + a_0y = g(x), \tag{4.39}$$

where $a_0, a_1, \ldots, a_n$ are constants.

Second-Order Cauchy-Euler Equations

Consider the second-order homogeneous Cauchy-Euler equation

$$3x^2y'' - 2xy' + 2y = 0, \quad x > 0.$$

Here, the solution y is a function of x. We assume that $x > 0$ because we are guaranteed solutions

[3]C.B. Boyer, A History of Mathematics. Princeton University Press, 1985, p. 496.

where the coefficient functions in $y'' - \frac{2}{3}x^{-1}y' + \frac{2}{3}x^{-2}y = 0$ are continuous. (Similarly, we can solve the equation for $x < 0$.)

Suppose that a solution of this differential equation is of the form $y = x^r$ for some constant(s) r. Substitution of $y = x^r$ with derivatives $y' = x^{r-1}$ and $y'' = r(r-1)x^{r-2}$ into this differential equation yields

$$3x^2 \times r(r-1)x^{r-2} - 2x \times rx^{r-1} + 2x^r = 0$$
$$3r(r-1)x^r - 2rx^r + 2x^r = 0$$
$$[3r(r-1) - 2r + 2]x^r = 0$$
$$\left(3r^2 - 4r + 2\right)x^r = 0.$$

We are seeking a nontrivial solution, so we solve the quadratic equation $3r^2 - 5r + 2 = 0$, the *auxiliary equation* for the Cauchy-Euler equation. Factoring gives us $(3r-2)(r-1) = 0$, so the roots are $r_1 = 2/3$ and $r_2 = 1$. Therefore, two solutions of this equation are $y_1 = x^{2/3}$ and $y_2 = x$. These two functions are linearly independent because they are not constant multiples of one another, so a general solution is $y = c_1 c^{2/3} + c_2 x$.

This approach works nicely if the roots of the auxiliary equation are real and distinct. However, if we obtain a repeated root or complex conjugate roots, additional work is required to determine a general solution (see Exercises 43-45). Therefore, we refer back to Euler's idea by using the substitution $x = e^t$.

In doing this, the chain rule must be used to transform the derivatives with respect to x into derivatives with respect to t. If $x = e^t$, $t = \ln x$, so $dt/dx = 1/x$,

$$\frac{dy}{dx} = \frac{dy}{dt}\frac{dt}{dx} = \frac{1}{x}\frac{dy}{dt} \qquad (4.40)$$

and

$$\frac{d^2y}{dx^2} = \frac{d}{dx}\left(\frac{1}{x}\frac{dy}{dt}\right) = \frac{1}{x}\frac{d}{dt}\left(\frac{dy}{dt}\right)\frac{dt}{dx} + \frac{d}{dx}\left(\frac{1}{x}\right)\frac{dy}{dt}$$
$$= \frac{1}{x^2}\frac{d^2y}{dt^2} - \frac{1}{x^2}\frac{dy}{dt} = \frac{1}{x^2}\left(\frac{d^2y}{dt^2} - \frac{dy}{dt}\right)$$

$$(4.41)$$

Substitution of these derivatives into a second-order Cauchy-Euler equation yields a second-order linear differential equation with constant coefficients.

EXAMPLE 4.7.1

Solve $x^2y'' - xy' + y = 0$, $x > 0$.

Solution

Substituting the derivatives given in Equations (4.40) and (4.41), we have

$$x^2 \times \frac{1}{x^2}\left(\frac{d^2y}{dt^2} - \frac{dy}{dt}\right) - x \times \frac{1}{x}\frac{dy}{dt} + y = 0$$

$$\frac{d^2y}{dt^2} - \frac{dy}{dt} - \frac{dy}{dt} + y = 0$$

$$\frac{d^2y}{dt^2} - 2\frac{dy}{dt} + y = 0.$$

 Graphs of the solutions corresponding to various values of c_1 and c_2 are shown in Figure 4.13. Regardless of the choice of c_1 and c_2, calculate $\lim_{x \to 0^+}(c_1 x + c_2 x \ln x)$. Why is a right-hand limit necessary? Are the solutions in Figure 4.13 bounded as $x \to \infty$?

This equation has constant coefficients, so solutions are of the form $y = e^{rt}$. In this case, the characteristic equation is $r^2 - 2r + 1 = (r-1)^2$ with repeated root $r_{1,2} = 1$. Therefore, a general solution of this equation is $y(t) = c_1 e^t + c_2 t e^t$. Returning to the original variable, $x = e^t$ (or $t = \ln x$), we have $y(x) = c_1 x + c_2 x \ln x$. In Figure 4.13, we graph the solution for several values of c_1 and c_2.

EXAMPLE 4.7.2

Solve $x^2y'' - 5xy' + 10y = 0$, $y(1) = 1$, $y'(1) = 0$.

Solution

After substituting the appropriate derivatives, we obtain the equation

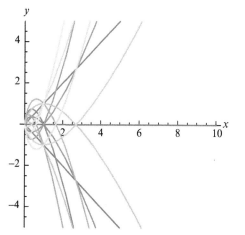

FIGURE 4.13   Plots of various solutions of $x^2 y'' - xy' + y = 0$, $x > 0$.

$$x^2 \times \frac{1}{x^2}\left(\frac{d^2 y}{dt^2} - \frac{dy}{dt}\right) - 5x \times \frac{1}{x}\frac{dy}{dt} + 10y = 0$$

or $\dfrac{d^2 y}{dt^2} - d\dfrac{dy}{dt} + 10y = 0.$

The characteristic equation is $r^2 - 6r + 10 = 0$ with complex conjugate roots $r_{1,2} = \frac{1}{2}\left(6 \pm \sqrt{36 - 40}\right) = 3 \pm i$. A general solution of the transformed differential equation is $y(t) = e^{3t}(c_1 \cos t + c_2 \sin t)$. We can transform the initial conditions as well. If $x = 1$ as indicated in the two initial conditions, then $t = \ln 1 = 0$. Therefore, in terms of t, the initial conditions are $y(0) = 1$ and $y'(0) = 0$, and the transformed IVP is

$$\frac{d^2 y}{dt^2} - 6\frac{dy}{dt} + 10y = 0, \quad y(0) = 1, \quad y'(0) = 0.$$

Then, $y(0) = c_1 = 1$. With

$$y'(t) = 3e^{3t}(c_1 \cos + c_2 \sin t) + e^{3t}(-c_1 \sin t + c_2 \cos t),$$

we find that $y'(0) = 3c_1 + c_2 = 0$, so $c_2 = -3c_1 = -3$. Therefore, $y(t) = e^{3t}(\cos t - 3\sin t)$. In terms of

x, this solution is $y(x) = x^3(\cos(\ln x) - 3\sin(\ln x))$ because $e^{3t} = (e^t)^3 = x^3$ and $t = \ln x$, which is graphed in Figure 4.14.

 What is $\lim_{x \to 0^+} y(x)$?

The examples illustrate the following theorem for solving second-order homogeneous Cauchy-Euler equations.

Theorem 27 (Solving Second-Order Homogeneous Cauchy-Euler Equations). *Consider the second-order homogeneous Cauchy-Euler equation $ax^2 y'' + bxy' + cy = 0$, $x > 0$. With the substitution $x = e^t$, this equation is transformed into the second order linear homogeneous differential equation with constant coefficients, $a\,d^2 y/dt^2 + (b-a)dy/dt + cy = 0$. Let r_1 and r_2 be solutions of the **auxiliary equation** $ar^2 + (b-a)r + c = 0$.*

1. *If $r_1 \neq r_2$ are real, a general solution of the Cauchy-Euler equation is $y(x) = c_1 x^{r_1} + c_2 x^{r_2}$.*
2. *If r_1, r_2 are real and $r_1 = r_2$, a general solution is $y(x) = c_1 x^{r_1} + c_2 x^{r_1}\ln x$.*
3. *If $r_1 = \bar{r}_2 = \alpha + \beta i$, $\beta > 0$, a general solution is $y(x) = x^\alpha(c_1 \cos(\beta \ln x) + c_2 \sin(\beta \ln x))$.*

Nonhomogeneous Cauchy-Euler Equations

Just as we encountered in nonhomogeneous linear differential equations with constant coefficients, the methods of variation of parameters and undetermined coefficients can be used to solve nonhomogeneous Cauchy-Euler equations. The following two examples illustrate two different approaches to solving nonhomogeneous Cauchy-Euler equations: (1) solving the transformed equation with the method of undetermined coefficients, and (2) solving the equation in its original form using variation of parameters.

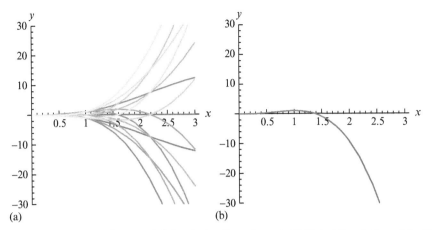

FIGURE 4.14   (a) Plots of various solutions of $x^2y'' - 5xy' + 10y = 0$, $x > 0$. (b) Plot of *the* solution of the IVP.

EXAMPLE 4.7.3

Solve $x^2y'' - 3xy' + 13y = 4 + 3x$, $x > 0$.

Solution

After substitution of the appropriate derivatives and functions of t, we obtain the differential equation

$$x^2 \times \frac{1}{x^2}\left(\frac{d^2y}{dt^2} - \frac{dy}{dt}\right) - 3x \times \frac{1}{x}\frac{dy}{dt} + 13y = 4 + 3e^t$$

$$\frac{d^2y}{dt^2} - 4\frac{dy}{dt} + 13y = 4 + 3e^t.$$

The characteristic equation of the corresponding homogeneous equation is $r^2 - 4r + 13 = 0$, which has complex conjugate roots $r_{1,2} = 2 \pm 3i$ and general solution $y_h(t) = e^{2t}(c_1 \cos 3t + c_2 \sin 3t)$. Using the method of undetermined coefficients to find a particular solution of the nonhomogeneous equation $y'' - 4y' + 13y = 4 + 3e^t$, we assume that a particular solution has the form $y_p = A + Be^t$. The derivatives of this function are $y_p' = Be^t$ and $y_p'' = Be^t$; thus, substitution into nonhomogeneous equation $y'' - 4y' + 13y = 4 + 3e^t$ yields

$$Be^t - 4Be^t + 13A + 13Be^t = 10Be^t + 13A = 4 + 3e^t.$$

Therefore, $B = 3/10$ and $A = 4/13$ so $y_p = 4/13 + 3e^t/10$. A general solution in the variable t is given by

$$y(t) = e^{2t}(c_1 \cos 3t + c_2 \sin 3t) + \frac{4}{13} + \frac{3}{10}e^t.$$

Returning to the original variable $x = e^t$ (or $t = \ln x$), we have

$$y(x) = x^2(c_1 \cos(3\ln x) + c_2 \sin(3\ln x)) + \frac{4}{13} + \frac{3}{10}x.$$

When solving nonhomogeneous Cauchy-Euler equations, the method of undetermined coefficients should be used *only* when the equation is transformed to a constant coefficient equation. On the other hand, the method of variation of parameters can be used with the original equation in standard form or with the transformed equation.

Remember that to use the variation of parameters formulas to find a particular solution of a nonhomogeneous linear equation, the equation must be in standard form.

EXAMPLE 4.7.4

Solve $x^2y'' + 4xy' + 2y = \ln x, y(1) = 2, y'(1) = 0$.

Solution

We begin by determining a solution of the corresponding homogeneous equation $x^2y'' + 4xy' + 2y = 0$, which is transformed into $d^2y/dt^2 + 3\,dy/dt + 2y = 0$ with characteristic equation $r^2 + 3r + 2 = (r+1)(r+2) = 0$. Therefore, $y_h(t) = c_1e^{-t} + c_2e^{-2t}$, so $y_h(x) = c_1x^{-1} + c_2x^{-2}$. Assuming that $y_p(x) = u_1(x)y_1(x) + u_2(x)y_2(x) = u_1x^{-1} + u_2x^{-2}$, we find using variation of parameters (see Section 4.4) with the equation $y'' + 4x^{-1}y' + 2x^{-2}y = x^{-2}\ln x$ that

$$u_1'(x) = -\frac{1}{-x^{-4}}x^{-4}\ln x = \ln x \quad \text{and}$$

$$u_2'(x) = \frac{1}{-x^{-4}}x^{-3}\ln x = -x\ln x.$$

Then $u_1(x) = \int \ln x\,dx = x\ln x - x$ and $u_2(x) = -\int x\ln x\,dx = \frac{1}{4}x^2 - \frac{1}{2}x^2\ln x$, so

$$y_p(x) = u_1y_1 + u_2y_2 = (x\ln x - x)x^{-1}$$
$$+ \left(\frac{1}{4}x^2 - \frac{1}{2}x^2\ln x\right)x^{-2}$$
$$= \frac{1}{2}\ln x - \frac{3}{4}.$$

We used integration by parts to evaluate these integrals with $dv = \ln x\,dx$.

A general solution of the nonhomogeneous equation is then

$$y(x) = y_h(x) + y_p(x) = c_1x^{-1} + c_2x^{-2} + \frac{1}{2}\ln x - \frac{3}{4},$$

where $y' = -c_1x^{-2} - 2c_2x^{-3} + \frac{1}{2}x^{-1}$.

Applying the initial conditions, we find that $y(1) = c_2 + c_2 - 3/4 = 2$ and $y'(1) = -c_1 - 2c_2 + 1/2 = 0$. Solving the, system for c_1 and c_2 shows that $c_1 = 5$ and $c_2 = -9/4$, so the solution to the IVP is

$$y(x) = y_h(x) + y_p(x) = 5x^{-1} - \frac{9}{4}x^{-2} + \frac{1}{2}\ln x - \frac{3}{4}.$$

We graph this function in Figure 4.15(b). Notice that the solution becomes unbounded as $x \to 0^+$.

Higher Order Cauchy-Euler Equations

Solutions of higher order homogeneous Cauchy-Euler equations are determined in the same manner as solutions of higher order homogeneous differential equations with constant coefficients.

EXAMPLE 4.7.5

Solve $2x^3y''' - 4x^2y'' - 20xy' = 0, x > 0$.

Solution

For this problem, if we assume that $y = x^r$ for $x > 0$, we have the derivatives $y' = rx^{r-1}$, $y'' = r(r-1)x^{r-2}$, and $y''' = r(r-1)(r-2)x^{r-3}$. Substitution into the differential equation and simplification yields $(2r^3 - 10r^2 - 12r)x^r = 0$. Because $x^r \neq 0$, we must solve

$$2r^3 - 10r^2 - 12r = 2r(r+1)(r-6) = 0$$

for r. Hence, $r_1 = 0, r_2 = -1$, and $r_3 = 6$, and a general solution is $y = c_1 + c_2x^{-1} + c_3x^6$.

Find and classify all relative extrema of $y(x)$ on the interval $(1,4)$ if y satisfies the boundary conditions $y(1) = y(4) = 0$ (see Figure 4.16(a)). Find the solution that satisfies $y(1) = y(4) = 0$ and $y'(1) = 2$ (see Figure 4.16(b)).

If a root r of the auxiliary equation is repeated m times, the m linearly independent solutions that correspond to r are $x^r, x^r\ln x, x^r(\ln x)^2, \ldots, x^r(\ln x)^{m-1}$ because in the transformed variable t, the corresponding solutions are $e^{rt}, te^{rt}, \ldots, t^{m-1}e^{rt}$.

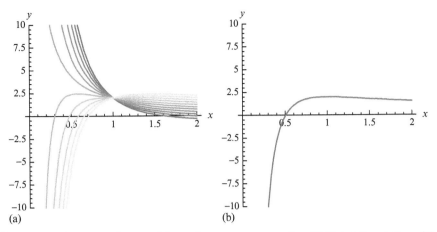

(a)   (b)

FIGURE 4.15 (a) Plots of various solutions of the nonhomogeneous equation. (b) Plot of *the* solution of the nonhomogeneous equation that satisfies the initial conditions.

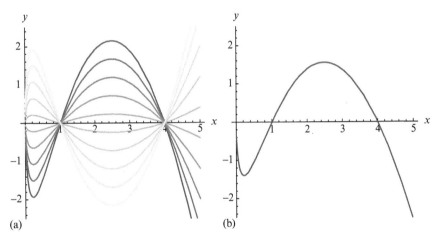

(a)   (b)

FIGURE 4.16 (a) Plots of various solutions of the differential equation that satisfy $y(1) = y(4) = 0$. (b) Plot of *the* solution of the differential equation that satisfies $y(1) = y(4) = 0$ and $y'(1) = 2$.

EXAMPLE 4.7.6

Solve $x^3 y''' + xy' - y = 0, x > 0$.

Solution

As in the previous example, we substitute $y = x^r$ into the equation. This gives us $x^r(r^3 -$

$3r^2 + 3r - 1) = 0$, where $r^3 - 3r^2 + 3r - 1 = (r - 1)^3 = 0$, so the roots of the auxiliary equation are $r_{1,2,3} = 1$. One solution to the differential equation is $y_1 = x^1 = x$, a second is $y_2 = x \ln x$, and a third is $y_3 = x(\ln x)^2$. Therefore, a general solution of the differential equation is $y = c_1 x + c_2 x \ln x + c_3 x (\ln x)^2$.

EXERCISES 4.7

In Exercises 1-20, find a general solution of the Cauchy-Euler equation. (Assume $x > 0$.)

1. $4x^2y'' - 8xy' + 5y = 0$
2. $3x^2y'' - 4xy' + 2y = 0$
3. $2x^2y'' - 8xy' + 8y = 0$
4. $2x^2y'' - 7xy' + 7y = 0$
5. $4x^2y'' + 17y = 0$
6. $9x^2y'' - 9xy' + 10y = 0$
7. $2x^2y'' - 2xy' + 20y = 0$
8. $x^2y'' - 5xy' + 10y = 0$
9. $4x^2y'' + 8xy' + y = 0$
10. $4x^2y'' + y = 0$
11. $x^2y'' - 5xy' + 9y = 0$
12. $x^2y'' + 7xy' + 9y = 0$
13. $x^3y''' + 22x^2y'' + 124xy' + 140y = 0$
14. $x^3y''' - 4x^2y'' - 46xy' + 100y = 0$
15. $x^3y''' + 2x^2y'' - 4xy' + 4y = 0$
16. $x^3y''' + 4x^2y'' + 6xy' + 4y = 0$
17. $x^3y''' + 2xy' - 2y = 0$
18. $x^3y''' + 3x^2y'' - 2xy' - 2y = 0$
19. $x^3y''' + 6x^2y'' + 7xy' + y = 0$
20. $x^3y^{(4)} + 6x^2y''' + 7xy'' + y' = 0$

In Exercises 21-30, solve the nonhomogeneous Cauchy-Euler equation. (Assume $x > 0$.)

21. $x^2y'' + 5xy' + 4y = x^{-5}$
22. $x^2y'' - 5xy' + 9y = x^3$
23. $x^2y'' + xy' + y = x^{-2}$
24. $x^2y'' + xy' + 4y = x^{-2}$
25. $x^2y'' + 2xy' - 6y = 2x$
26. $x^2y'' + xy' - 16y = \ln x$
27. $x^2y'' + xy' + 4y = 8$
28. $x^2y'' + xy' + 36y = x^2$
29. $x^3y''' + 3x^2y'' - 11xy' + 16y = x^{-3}$
30. $x^3y''' + 16x^2y'' + 70xy' + 80y = x^{-13}$

In Exercises 31-42, solve the IVP.

31. $3x^2y'' - 4xy' + 2y = 0$, $y(1) = 2$, $y'(1) = 1$
32. $2x^2y'' - 7xy' + 7y = 0$, $y(1) = -1$, $y'(1) = 1$
33. $x^2y'' + xy' + 4y = 0$, $y(1) = 1$, $y'(1) = 0$
34. $x^2y'' + xy' + 2y = 0$, $y(1) = 0$, $y'(1) = 2$

35. $x^3y''' + 10x^2y'' - 20xy' + 20y = 0$, $y(1) = 0$, $y'(1) = -1$, $y''(1) = 1$
36. $x^3y''' + 15x^2y'' + 54xy' + 42y = 0$, $y(1) = 5$, $y'(1) = 0$, $y''(1) = 0$
37. $x^3y''' - 2x^2y'' + 5xy' - 5y = 0$, $y(1) = 0$, $y'(1) = -1$, $y''(1) = 0$
38. $x^3y''' - 6x^2y'' + 17xy' - 17y = 0$, $y(1) = -2$, $y'(1) = 0$, $y''(1) = 0$
39. $2x^2y'' + 3xy' - y = x^{-2}$, $y(1) = 0$, $y'(1) = 2$
40. $x^2y'' + 4xy' + 2y = \ln x$, $y(1) = 2$, $y'(1) = 0$
41. $4x^2y'' + y = x^3$, $y(1) = 1$, $y'(1) = -1$
42. $9x^2y'' + 27xy' + 10y = x^{-1}$, $y(1) = 0$, $y'(1) = -1$

43. Suppose that a second-order Cauchy-Euler equation has solutions x^{r_1} and x^{r_2} where r_1 and r_2 are real and unequal. Calculate the Wronskian of these two solutions to show that they are linearly independent. Therefore, $y = c_1x^{r_1} + c_2x^{r_2}$ is a general solution.

44. (a) Show that the transformed equation for the Cauchy-Euler equation $ax^2y'' + bxy' + cy = 0$ is $a\,d^2y/dt^2 + (b-a)dy/dt + cy = 0$. (b) Show that if the characteristic equation for a transformed Cauchy-Euler equation has a repeated root, then this root is $r = -(b-a)/(2a)$. (c) Use reduction of order to show that a second solution to this equation is $x^r \ln x$. (d) Calculate the Wronskian of x^r and $x^r \ln x$ to show that these two solutions are linearly independent.

45. Suppose the roots of the characteristic equation for the transformed Cauchy-Euler equation are $r_{1,2} = \alpha \pm \beta i$, so that two solutions are $y_1 = x^{\alpha+\beta i}$ and $y_2 = x^{\alpha-\beta i}$. Use Euler's formula and the Principle of Superposition to show that $x^\alpha \cos(\beta \ln x)$ and $x^\alpha \sin(\beta \ln x)$ are solutions of the original equation. *Hint:*
$$x^{\beta i} = \left(e^{\ln x}\right)^{\beta i} = \cos(\beta \ln x) + i\sin(\beta \ln x).$$

46. Consider the equation $x^2y'' + Axy' + By = 0$, $x > 0$, where A and B are constants. Solve

the equation for each of the following. Investigate $\lim_{x\to 0^+} y(x)$ and $\lim_{x\to\infty} y(x)$.

(a) $A = -1$, $B = 2$
(b) $A = 4$, $B = 2$
(c) $A = 1$, $B = 1$

47. Use the results of Exercise 46 to show that solutions to $x^2 y'' + Axy' + By = 0$, $x > 0$, (a) approach zero as $x \to 0^+$ if $A < 1$ and $B > 0$; (b) approach zero as $x \to \infty$ if $A > 1$ and $B > 0$; (c) are bounded as $x \to 0^+$ and $x \to \infty$ if $A = 1$ and $B > 0$.

48. Show that if $x = e^t$ and $y = y(x)$, then

$$\frac{d^3 y}{dx^3} = \frac{1}{x^3}\left(\frac{d^3 y}{dt^3} - 3\frac{d^2 y}{dt^2} + 2\frac{dy}{dt}\right).$$

In Exercises 49-52, use the substitution in Exercise 48 to solve the following equations. (Assume $x > 0$.)

49. $x^3 y''' + 3x^2 y'' + 37xy' = 0$
50. $x^3 y''' + 3x^2 y'' - 3xy' = 0$
51. $x^3 y''' + xy' - y = 0$
52. $x^3 y''' + 3x^2 y'' - 3xy' = -8$

(*Other Substitutions*) Euler was lucky and smart to discover that the substitution $x = e^t$ transforms the Cauchy-Euler equation to an equation with constant coefficients. As you should suspect, other substitutions can also transform some equations that are difficult to solve in their original form to a form that we recognize and are able to solve based on what we have learned so far.

53. Let $x = \tan t$ so that $t = \tan^{-1} x$.
(a) Show that $\frac{dy}{dx} = \frac{1}{1+x^2}\frac{dy}{dt}$ and that

$$\frac{d^2 y}{dx^2} = \frac{1}{(1+x^2)^2}\left(-2x\frac{dy}{dt} + \frac{d^2 y}{dt^2}\right).$$

(b) Show that this substitution transforms equations of the form

$$a(1+x^2)^2 y'' + (2ax+b)(1+x^2)y' + cy = g(x)$$

to equations with constant coefficients.
(c) Solve $(1 + x^2)^2 y'' + 2x(1 + x^2)y' + 4y = 0$.
(d) Solve

$$(1+x^2)^2 y'' + 2x(1+x^2)y' + 4y = \tan^{-1} x.$$

(e) Solve $(1 + x^2)^2 y'' + 2x(1 + x^2)y' + 4y = 0$, $y(0) = 0$, $y'(0) = 1$.
(f) Solve

$$(1+x^2)^2 y'' + 2x(1+x^2)y' + 4y = \tan^{-1} x,$$

$y(0) = 0$, $y'(0) = 1$.

54. Let $x = \sin t$ so that $t = \sin^{-1} x$.
(a) Show that $\dfrac{dy}{dx} = \dfrac{1}{\sqrt{1-x^2}}\dfrac{dy}{dt}$ and that

$$\frac{d^2 y}{dx^2} = \frac{x}{(1-x^2)^{3/2}}\frac{dy}{dt} + \frac{1}{1-x^2}\frac{d^2 y}{dt^2}.$$

(b) Find conditions on $p_2(x)$, $p_1(x)$, and $p_0(x)$ so that the substitution $x = \sin t$ transforms the equation
$p_2(x)y'' + p_1(x)y' + p_0(x)y = g(x)$ to an equation with constant coefficients.
(c) Solve

$$(x^4 - 1)y'' + (x^3 - x)y' + (x^2 - 1)y = 0.$$

(d) Solve

$$(x^4 - 1)y'' + (x^3 - x)y' + (4x^2 - 4)y = 0.$$

(e) Solve

$$(x^4 - 1)y'' + (x^3 - x)y' + (x^2 - 1)y = 0,$$

$y(0) = 0$, $y'(0) = -1$.

55. Generalize the results from Exercises 53 and 54 for $x = \sec t$. Find an equation of the form $p_2(x)y'' + p_1(x)y' + p_0(x)y = g(x)$ with general solution $y = c_1\cos(\sec^{-1} x) + c_2\sin(\sec^{-1} x) = c_1 x + c_2\sqrt{1 - x^2}$. *Challenge:* Find an equation of the form $p_2(x)y'' + p_1(x)y' + p_0(x)y = g(x)$ *with general solution* $y = c_1\cos(2\sec^{-1} x) + c_2\sin(2\sec^{-1} x)$ *and* **simplify** *the result.*

56. Let $x = -e^t$. (a) Show that $\frac{dy}{dx} = \frac{dy}{dt}\frac{dt}{dx} = \frac{1}{x}\frac{dy}{dt}$ and $\frac{d^2 y}{dx^2} = \frac{1}{x^2}\left(\frac{d^2 y}{dt^2} - \frac{dy}{dt}\right)$. (b) Show that the differential equation $ax^2 y'' + bxy' + cy = f(x)$ is transformed into
$a\, d^2 y/dt^2 + (b - a)dy/dt + cy = f(-e^t)$.

In Exercises 57-60, use the substitution described in Exercise 56 to solve the indicated equation or IVP.

57. $x^2 y'' + 4xy' + 2y = 0$, $x < 0$
58. $x^2 y'' + xy' + y = x^2$, $x < 0$
59. $x^2 y'' + xy' + 4y = 0$, $y(-1) = 0$, $y'(-1) = 2$

60. $x^2 y'' - xy' + y = 0, y(-1) = 0, y'(-1) = 1$

61. Consider the second-order Cauchy-Euler equation $x^2 y'' + Bxy' + y = 0, x > 0$, where B is a constant.

    (a) Find $\lim_{x \to \infty} y(x)$ where $y(x)$ is a general solution of the equation using the given restriction on B.

    (b) Determine if the solution is bounded as $x \to \infty$ in each case: (i) $B = 1$; (ii) $B > 1$; (iii) $B < 1$.

62. Consider the second-order Cauchy-Euler equation stated in Exercise 61. Determine $\lim_{x \to 0^+} y(x)$ as well as if $y(x)$ is bounded as $x \to 0^+$ in each case: (a) $B = 1$; (b) $B > 1$; (c) $B < 1$.

63. Use variation of parameters to solve the nonhomogeneous differential equation that was solved in Example 4.7.3 with undetermined coefficients. Which method is easier?

64. Use a computer algebra system to assist in finding a general solution of the following Cauchy-Euler equations.

    (a) $x^3 y''' + 16x^2 y'' + 79xy' + 125y = 0$

    (b) $x^4 y^{(4)} + 5x^3 y''' - 12x^2 2y'' - 12xy' + 48y = 0$

    (c) $x^4 y^{(4)} + 14x^3 y''' + 55x^2 y'' + 65xy' + 15y = 0$

    (d) $x^4 y^{(4)} + 8x^3 y''' + 27x^2 y'' + 35xy' + 45y = 0$

    (e) $x^4 y^{(4)} + 10x^3 y''' + 27x^2 y'' + 21xy' + 4y = 0$

65. Solve the IVP
$x^3 y''' + 9x^2 y'' + 44xy' + 58y = 0, y(1) = 2,$
$y'(1) = 10, y''(1) = -2$. Graph the solution on the interval $[0.2, 1.8]$ and approximate all local minima and maxima of the solution on this interval.

66. (a) Solve the IVP $6x^2 y'' + 5xy' - y = 0$, $y(1) = a, y'(1) = b$.

    (b) Find conditions on a and b so that $\lim_{x \to 0^+} y(x) = 0$. Graph several solutions to confirm your results.

    (c) Find conditions on a and b so that $\lim_{x \to \infty} y(x) = 0$. Graph several solutions to confirm your results.

    (d) If both a and b are not zero, is it possible to find a and b so that both $\lim_{x \to 0^+} y(x) = 0$ and $\lim_{x \to \infty} y(x) = 0$? Explain.

67. An equation of the form $f(x, y, y')y'' + g(x, y, y') = 0$ that satisfies the system

$$f_{xx} + 2pf_{xy} + p^2 f_{yy} = g_{xp} + pg_{yp} - g_y$$
$$f_{xp} + pf_{yp} + 2f_y = g_{pp},$$

where $p = y'$ is called an *exact second-order differential equation.*

    (a) Show that the equation $x(y')^2 + yy' + xyy'' = 0$ is an exact second-order equation.

    (b) Show that $\phi(x, y, p) = h(x, y) + \int f(x, y, p)\, dp$ is a solution of the exact equation $f(x, y, y')y'' + g(x, y, y') = 0$ and that a general solution of $x(y')^2 + yy' + xyy'' = 0$ is $\frac{1}{2}y^2 = c_1 + c_2 \ln x$.

68. Find a general solution of the nonlinear equation $\frac{4}{5}(y' - ty'')^2 = 2yy'' - (y')^2$ by differentiating both sides of the equation and setting the result equal to zero, and factoring. (b) Is the Principle of Superposition valid for this nonlinear equation? Explain.

4.8 POWER SERIES SOLUTIONS OF ORDINARY DIFFERENTIAL EQUATIONS

- SERIES SOLUTIONS ABOUT ORDINARY POINTS
- LEGENDRE'S EQUATION

In calculus, we learned that Maclaurin and Taylor polynomials can be used to approximate *functions.* This idea can be extended to finding or approximating the *solution of a differential equation.*

Series Solutions About Ordinary Points

Consider the equation $a_2(x)y'' + a_1(x)y' + a_0(x)y = 0$ and let $p(x) = a_1(x)/a_2(x)$ and $q(x) = a_0(x)/a_2(x)$. Then, $a_2(x)y'' + a_1(x)y' + a_0(x)y = 0$ is equivalent to $y'' + p(x)y' + q(x)y = 0$, which is called the *standard form* of the equation. A number x_0 is an *ordinary point* if both $p(x)$ and $q(x)$ are *analytic* at x_0. If x_0 is not an ordinary point, x_0 is called a *singular point*. If x_0 is an ordinary point of the differential equation $y'' + p(x)y' + q(x)y = 0$, we can write $p(x) = \sum_{n=0}^{\infty} b_n(x - x_0)^n$, where $b_n = \frac{1}{n!}p^{(n)}(x_0)$, and $q(x) = \sum_{n=0}^{\infty} c_n(x - x_0)^n$, where $c_n = \frac{1}{n!}q^{(n)}(x_0)$. Substitution into the equation $y'' + p(x)y' + q(x)y = 0$ results in

$$y'' + y'\sum_{n=0}^{\infty} b_n(x - x_0)^n + y\sum_{n=0}^{\infty} c_n(x - x_0)^n = 0.$$

Note: A function is *analytic* at x_0 if its power series centered at x_0 has a positive radius of convergence.

If we assume that y is analytic at x_0, we can write $y(x) = \sum_{n=0}^{\infty} a_n(x - x_0)^n$. Because a power series can be differentiated term by term on its interval of convergence, we can compute the first and second derivatives of y,

$$y' = \sum_{n=1}^{\infty} na_n(x - x_0)^{n-1} = \sum_{n=0}^{\infty}(n+1)a_{n+1}(x - x_0)^n$$

and   (4.42)

$$y'' = \sum_{n=2}^{\infty} n(n-1)a_n(x - x_0)^{n-2}$$

$$= \sum_{n=0}^{\infty}(n+1)(n+2)a_{n+2}(x - x_0)^n,$$

and substitute back into the equation calculate the coefficients of a_n. Thus, we obtain a power series solution of the equation.

EXAMPLE 4.8.1

(a) Find a general solution of $(4 - x^2)y' + y = 0$ and (b) solve the IVP $(4 - x^2)y' + y = 0$, $y(0) = 1$.

Solution

Because $x_0 = 0$ is an ordinary point of the equation, we assume that $y = \sum_{n=0}^{\infty} a_n x^n$. Substitution of this function and its derivatives into the equation gives us

$$(4 - x^2)y' + y = (4 - x^2)\sum_{n=1}^{\infty} na_n x^{n-1} + \sum_{n=0}^{\infty} a_n x^n$$

$$= \sum_{n=1}^{\infty} 4na_n x^{n-1} - \sum_{n=1}^{\infty} na_n x^{n+1}$$

$$+ \sum_{n=0}^{\infty} a_n x^n = 0.$$

Note that the first term in these three series involves x^0, x^2, and x^0, respectively. Thus, if we pull off the first two terms in the first and third series, all three series will begin with an x^2 term. Doing so, we have

$$(4a_1 + a_0) + (8a_2 + a_1) + \sum_{n=3}^{\infty} 4na_n x^{n-1}$$

$$- \sum_{n=1}^{\infty} na_n x^{n+1} + \sum_{n=2}^{\infty} a_n x^n = 0.$$

Unfortunately, the indices of these three series do not match, so we must change two of the three to match the third. Substitution of $n + 1$ for n in $\sum_{n=3}^{\infty} 4na_n x^{n-1}$ yields

$$\sum_{n+1=3}^{\infty} 4(n+1)a_{n+1}x^{n+1-1} = \sum_{n=2}^{\infty} 4(n+1)a_{n+1}x^n.$$

Similarly, substitution of $n-1$ for n in $\sum_{n=1}^{\infty} na_n x^{n+1}$ yields

$$\sum_{n-1=1}^{\infty}(n-1)a_{n-1}x^{n-1+1} = \sum_{n=2}^{\infty}(n-1)a_{n-1}x^n.$$

TABLE 4.3 a_n Values for $n = 0, 1, 2, 3, \ldots, 10$

n	a_n	n	a_n	n	a_n	n	a_n
0	a_0	3	$-\frac{3}{128}a_0$	6	$\frac{69}{65536}a_0$	9	$-\frac{4859}{33554432}a_0$
1	$-\frac{1}{4}a_0$	4	$\frac{11}{2048}a_0$	7	$-\frac{187}{262144}a_0$	10	$\frac{12767}{268435456}a_0$
2	$\frac{1}{32}a_0$	5	$-\frac{31}{8192}a_0$	8	$\frac{1843}{8388608}a_0$		

Therefore after combining the three series, we have the equation

$$(4a_1 + a_0) + (8a_2 + a_1)$$
$$+ \sum_{n=2}^{\infty} [a_n + 4(n+1)a_{n+1} - (n-1)a_{n-1}] x^n = 0.$$

Because the sum of the terms on the left-hand side of the equation is zero, each coefficient must be zero. Equating the coefficients of x^0 and x to zero yields $a_1 = -\frac{1}{4}a_0$ and $a_2 = -\frac{1}{8}a_1 = \frac{1}{32}a_0$. When the series coefficient $a_n + 4(n+1)a_{n+1} - (n-1)a_{n-1}$ is set to zero, we obtain the recurrence relation $a_{n+1} = \frac{(n-1)a_{n-1} - a_n}{4(n+1)}$ for the indices in the series, $n \geq 2$. We use this formula to determine the values of a_n for $n = 2, 3, \ldots, 10$, and give these values in Table 4.3. Therefore,

$$y = a_0 - \frac{1}{4}a_0 x + \frac{1}{32}a_0 x^2 - \frac{3}{128}a_0 x^3 + \frac{11}{2048}a_0 x^4$$
$$- \frac{31}{8192}a_0 x^5 + \frac{69}{65536}a_0 x^6 - \frac{187}{262144}a_0 x^7 + \cdots$$

(b) When we apply the initial condition $y(0) = 1$, we substitute $x = 0$ into the general solution obtained in (a) and set the result equal to 1. Hence, $a_0 = 1$, so the series solution of the IVP is

$$y = 1 - \frac{1}{4}x + \frac{1}{32}x^2 - \frac{3}{128}x^3 + \frac{11}{2048}x^4 - \frac{31}{8192}x^5$$
$$+ \frac{69}{65536}x^6 - \frac{187}{262144}x^7 + \cdots$$

Notice that equation $(4 - x^2) y' + y = 0$ is separable, so we can compute the solution directly with separation of variables by rewriting the equation as $-\frac{1}{y}dy = \frac{1}{4-x^2}dx$. Integrating yields $\ln y =$

$\frac{1}{4}\left(\ln|x - 2| - \ln|x + 2|\right) = \ln\left|\frac{x-2}{x+2}\right|^{1/4} + C$. Applying the initial condition $y(0) = 1$ results in $y = \ln\left|\frac{x-2}{x+2}\right|^{1/4}$. We can approximate the solution of the problem by taking a finite number of terms of the series solution. The graph of the polynomial approximation of degree 10 is shown in Figure 4.17 along with the solution obtained through separation of variables. The graph shows that the accuracy of the approximation decreases near $x = \pm 2$, which are singular points of the differential equation. (Why?) The reason for the decrease in the accuracy of the approximation is discussed in the theorem following this example.

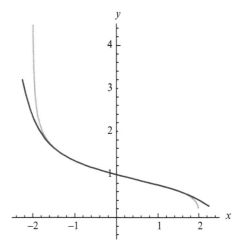

FIGURE 4.17 Comparison of exact (black) and approximate (gray) solutions to the initial-value problem $(4 - x^2)y' + y = 0$, $y(0) = 1$.

The following theorem explains where the approximation of the solution of the differential equation by the series is valid.

Theorem 28 (Convergence of a Power Series Solution). *Let* $x = x_0$ *be an ordinary point of the differential equation* $a_2(x)y'' + a_1(x)y' + a_0(x)y = 0$ *and suppose that* R *is the distance from* $x = x_0$ *to the closest singular point of the equation. Then the power series solution* $y = \sum_{n=0}^{\infty} a_n (x - x_0)^n$ *converges at least on the interval* $(x_0 - R, x_0 + R)$.

A proof of this theorem can be found in more advanced texts, such as Rabenstein's *Introduction to Ordinary Differential Equations* [25].

The theorem indicates that the approximation may not be as accurate near singular points of the equation. Hence, we understand why the approximation in Example 4.8.1 breaks down near $x = 2$, the closest singular point to the ordinary point $x = 0$.

Of course, $x = 0$ is not an ordinary point for every differential equation. However, because the series $y = \sum_{n=0}^{\infty} a_n (x - x_0)^n$ is easier to work with if $x_0 = 0$, we can always make a transformation so that we can use $y = \sum_{n=0}^{\infty} a_n x^n$ to solve any linear equation. For example, suppose that $x = x_0$ is an ordinary point of a linear equation. Then, if we make the change of variable $t = x - x_0$, then $t = 0$ corresponds to $x = x_0$, so $t = 0$ is an ordinary point of the transformed equation.

EXAMPLE 4.8.2

Solve $xy'' + y = 0$.

Solution

In standard form the equation becomes $y'' + x^{-1}y = 0$. Because x^{-1} is not analytic at $x = 0$ (Why?), this equation has a singular point at $x = 0$. All other values of x are classified as ordinary points, so we can select one to use in our power series solution. Choosing $x = 1$, we consider the

power series $y = \sum_{n=0}^{\infty} a_n(x-1)^n$. However, with the change of variable $t = x - 1$, we have that $t = 0$ corresponds to $x = 1$. Therefore, by changing variables, we can use the series $y = \sum_{n=0}^{\infty} a_n t^n$.

Notice that

$$\frac{dy}{dx} = \frac{dy}{dt}\frac{dt}{dx} = \frac{dy}{dt} \quad \text{and}$$

$$\frac{d^2y}{dx^2} = \frac{d}{dt}\left(\frac{dy}{dx}\right)\frac{dt}{dx} = \frac{d}{dt}\left(\frac{dy}{dt}\right) = \frac{d^2y}{dt^2},$$

so with these assumptions into $xy'' + y = 0$, we obtain $(t+1)\frac{d^2y}{dt^2} + y = 0$. Hence, we assume that $y = \sum_{n=0}^{\infty} a_n t^n$. Substitution into the transformed equation yields

$$(t+1)\sum_{n=2}^{\infty} n(n-1)a_n t^{n-2} + \sum_{n=0}^{\infty} a_n t^n = 0.$$

Simplification then gives us

$$\sum_{n=2}^{\infty} n(n-1)a_n t^{n-1}$$

$$+ \sum_{n=2}^{\infty} n(n-1)a_n t^{n-2} + \sum_{n=0}^{\infty} a_n t^n = 0$$

$$\sum_{n=2}^{\infty} n(n-1)a_n t^{n-1} + 2a_2 t^0$$

$$+ \sum_{n=3}^{\infty} n(n-1)a_n t^{n-2} + a_0 t^0 + \sum_{n=1}^{\infty} a_n t^n = 0.$$

In this case, we must change the index in two of the three series. If we substitute $n + 1$ for n in $\sum_{n=3}^{\infty} n(n-1)a_n t^{n-2}$ and $n - 1$ for n in $\sum_{n=1}^{\infty} a_n t^n$, we obtain

$$(2a_2 + a_0) + \sum_{n=2}^{\infty} n(n-1)a_n t^{n-1}$$

$$+ \sum_{n=2}^{\infty} (n+1)na_{n+1} t^{n-1} + \sum_{n=2}^{\infty} a_{n-1} t^{n-1} = 0$$

$$(2a_2 + a_0) + \sum_{n=2}^{\infty} [n(n-1)a_n + n(n+1)a_{n+1}$$

$$+ a_{n-1}] t^{n-1} = 0.$$

Equating the coefficients to zero, we determine that $a_2 = -a_0/2$ and

$$a_{n+1} = \frac{-a_{n-1} - n(n-1)a_n}{n(n+1)}$$

for the coefficients in the series, $n \geq 2$. We calculate the coefficients for $n = 2, \ldots, 7$ and display the results in Table 4.4.

Therefore,

$$y = a_0 + a_1 t - \frac{1}{2}a_0 t^2 + \frac{1}{6}(a_0 - a_1)t^3$$

$$+ \frac{1}{24}(2a_1 - a_0)t^4 + \frac{1}{120}(2a_0 - 5a_1)t^5$$

$$+ \frac{1}{720}(18a_1 - 7a_0)t^6 + \frac{1}{5040}(33a_0 - 85a_1)t^7 + \cdots$$

Returning to the original variable, we have

 What are two linearly independent solutions of $xy'' + y = 0$?

$$y = a_0 + a_1(x-1) - \frac{1}{2}a_0(x-1)^2$$

$$+ \frac{1}{6}(a_0 - a_1)(x-1)^3 + \frac{1}{24}(2a_1 - a_0)(x-1)^4$$

$$+ \frac{1}{120}(2a_0 - 5a_1)(x-1)^5$$

$$+ \frac{1}{720}(18a_1 - 7a_0)(x-1)^6$$

$$+ \frac{1}{5040}(33a_0 - 85a_1)(x-1)^7 + \cdots$$

Legendre's Equation

Legendre's equation is the equation

$$\left(1 - x^2\right)\frac{d^2y}{dx^2} - 2x\frac{dy}{dx} + k(k+1)y = 0, \quad (4.43)$$

where k is a constant, named after the French mathematician Adrien Marie Legendre (1752-1833). The *Legendre polynomials*, solutions of Legendre's equation, were introduced by Legendre in his three-volume work *Traité des fonctions elliptiques et des intégrales euleriennes* (1825-1832). Legendre encountered these polynomials while trying to determine the gravitational potential associated with a point mass.

Although most people have a life outside of their professional life, **Adrien Marie Legendre** (1752-1833) was one of those rare individual's whose professional life consumed their entire life. Because of politics, his pension was stopped in 1824 and he died a poor man in 1833.

EXAMPLE 4.8.3

Find a general solution of Legendre's equation.

Solution

In standard form, the equation is

$$\frac{d^2y}{dx^2} - \frac{2x}{1-x^2}\frac{dy}{dx} + \frac{k(k+1)}{1-x^2}y = 0.$$

TABLE 4.4   Coefficients for $n = 0, \ldots, 7$

n	a_n	n	a_n	n	a_n
0	a_0	3	$\frac{1}{6}(a_0 - a_1)$	6	$\frac{1}{720}(18a_1 - 7a_0)$
1	a_1	4	$\frac{1}{24}(2a_1 - a_0)$	7	$\frac{1}{5040}(33a_0 - 85a_1)$
2	$-\frac{1}{2}a_0$	5	$\frac{1}{120}(2a_0 - 5a_1)$		

There is a solution to the equation of the form $y = \sum_{n=0}^{\infty} a_n x^n$ because $x = 0$ is an ordinary point. This solution will converge at least on the interval $(-1,1)$ because the closest singular points to $x = 0$ are $x = \pm 1$.

Substitution of this function and its derivatives $y' = \sum_{n=0}^{\infty}(n+1)a_{n+1}x^n$ and $y'' = \sum_{n=0}^{\infty}(n+1)(n+2)a_{n+2}x^n$ into Legendre's equation (4.43) and simplifying the results yields

$$[2a_2 + k(k+1)a_0] + [-2a_1 + k(k+1)a_1 + 6a_3]x$$
$$+ \sum_{n=4}^{\infty} n(n-1)a_n x^{n-2} - \sum_{n=2}^{\infty} n(n-1)a_n x^n$$
$$- \sum_{n=2}^{\infty} 2na_n x^n + \sum_{n=2}^{\infty} k(k+1)a_n x^n = 0.$$

After substituting $n+2$ for each occurrence of n in the first series and simplifying, we have

$$[2a_2 + k(k+1)a_0] + [-2a_1 + k(k+1)a_1 + 6a_3]x$$
$$+ \sum_{n=2}^{\infty} \big\{(n+2)(n+1)a_{n+2} + [-n(n-1)$$
$$-2n + k(k+1)]a_n\big\} x^n = 0.$$

Equating the coefficients to zero, we find that

$$a_2 = -\frac{1}{2}k(k+1)a_0, \quad a_3 = -\frac{1}{6}[k(k+1) - 2]a_1$$
$$= -\frac{1}{6}(k-1)(k+2)a_1,$$

and

$$a_{n+2} = \frac{n(n-1) + 2n - k(k+1)}{(n+2)(n+1)}a_n$$
$$= \frac{(n-k)(n+k+1)}{(n+2)(n+1)}a_n, \quad n \geq 2.$$

We obtain a formula for a_n by replacing each occurrence of n in a_{n+2} by $n-2$,

$$a_n = \frac{(n-k-2)(n+k-1)}{n(n-1)}a_{n-2}.$$

Using these formula we find the following coefficients:

$$a_4 = \frac{(2-k)(3+k)}{4 \times 3}a_2 = -\frac{(2-k)(3+k)(k(k+1)}{4 \times 3 \times 2}a_0$$

$$a_5 = \frac{(3-k)(4+k)}{5 \times 4}a_3$$
$$= -\frac{(3-k)(4+k)(k-1)(k+2)}{5 \times 4 \times 3 \times 2}a_1$$

$$a_6 = \frac{(4-k)(5+k)}{6 \times 5}a_4$$
$$= -\frac{(4-k)(5+k)(2-k)(3+k)k(k+1)}{6 \times 5 \times 4 \times 3 \times 2}a_0$$

$$a_7 = \frac{(5-k)(6+k)}{7 \times 6}a_5$$
$$= \frac{(5-k)(6+k)(3-k)(4+k)(k-1)(k+2)}{7 \times 6 \times 5 \times 4 \times 3 \times 2}a_1$$

$$\vdots$$

Hence, we have the two linearly independent solutions

$$y_1 = a_0\Bigg(1 - \frac{k(k+1)}{2!}x^2 + \frac{(2-k)(3+k)k(k+1)}{4!}x^4$$
$$- \frac{(4-k)(5+k)(2-k)(3+k)k(k+1)}{6!}x^6 + \cdots\Bigg)$$

and

$$y_2 = a_1\Bigg(x - \frac{(k-1)(k+2)}{3!}x^3$$
$$+ \frac{(3-k)(4+k)(k-1)(k+2)}{5!}x^5$$
$$- \frac{(5-k)(6+k)(3-k)(4+k)(k-1)(k+2)}{7!}x^7$$
$$+ \cdots\Bigg)$$

so a general solution of Legendre's equation (4.43) is

$y = y_1 + y_2$

$$= a_1\left(x - \frac{(k-1)(k+2)}{3!}x^3\right.$$

$$+ \frac{(3-k)(4+k)(k-1)(k+2)}{5!}x^5$$

$$- \frac{(5-k)(6+k)(3-k)(4+k)(k-1)(k+2)}{7!}x^7$$

$$\left. + \cdots \right)$$

$$+ a_1\left(x - \frac{(k-1)(k+2)}{3!}x^3\right.$$

$$+ \frac{(3-k)(4+k)(k-1)(k+2)}{5!}x^5$$

$$- \frac{(5-k)(6+k)(3-k)(4+k)(k-1)(k+2)}{7!}x^7$$

$$\left. + \cdots \right).$$

An interesting observation from the general solution to Legendre's equation is that the series solutions terminate for integer values of k. If k is an even integer, the first series terminates while if k is an odd integer the second series terminates. Therefore, polynomial solutions are found for integer values of k. We list several of these polynomials for suitable choices of a_0 and a_1 in Table 4.5. Because these polynomials are useful

and are encountered in numerous applications, we have a special notation for them: $P_n(x)$ is called the *Legendre polynomial of degree n* and represents an nth degree polynomial solution to Legendre's equation.

Another interesting observation about the Legendre polynomials is that they satisfy the relationship $\int_{-1}^{1} P_m(x)P_n(x)\,dx = 0, m \neq n$, called an *orthogonality condition*.

 EXERCISES 4.8

In Exercises 1-3, determine the singular points of the equations. Use these points to find an upper bound on the radius of convergence of a series solution about x_0.

1. $x^2 y'' - 2xy' + 7y = 0, x_0 = 1$
2. $(x - 2)y'' + y' - y = 0, x_0 = -2$
3. $(x^2 - 4)y'' + 16(x + 2)y' - y = 0, x_0 = 1$

In Exercises 4-13, solve the differential equation with a power series expansion about $x = 0$. Write out at least the first five nonzero terms of each series.

4. $y'' + 3y' - 18y = 0$
5. $y'' - 11y' + 30y = 0$
6. $y'' + y = 0$
7. $y'' - y' - 2y = e^{-x}$
8. $(-2 - 2x)y'' + 2y' + 4y = 0$
9. $(2 + 3x)y'' + 3xy' = 0$
10. $(1 + 3x)y'' - 3y' - 2y = 0$
11. $(2 - x^2)y'' + 2(x - 1)y' + 4y = 0$
12. $y'' - xy' + 4y = 0$
13. $(2 + 2x^2)y'' + 2xy' - 3y = 0$
14. $(3 - 2x)y'' + 2y' - 2y = 0, y(0) = 3,$
    $y'(0) = -2$
15. $y'' - 4x^2 y = 0, y(0) = 1, y'(0) = 0$
16. $(2x^2 - 1)y'' + 2xy' - 3y = 0, y(0) = -2,$
    $y'(0) = 2$

In Exercises 14-16, determine at least the first five nonzero terms in a power series about $x = 0$ for the solution of each IVP.

TABLE 4.5 $P_n(x)$ for $n = 0,$ $1, \ldots, 5$

n	$P_n(x)$
0	$P_0(x) = 1$
1	$P_1(x) = x$
2	$P_2(x) = \frac{1}{2}(3x^2 - 1)$
3	$P_3(x) = \frac{1}{2}(5x^3 - 3x)$
4	$P_4(x) = \frac{1}{8}(35x^4 - 30x^2 + 3)$
5	$P_5(x) = \frac{1}{8}(63x^5 - 70x^3 + 15x)$

The English physicist **John William Strutt "Lord Rayleigh"** (1842-1919) won the Nobel prize in 1904 for his discovery of the inert gas argon in 1895. He donated the award money to the University of Cambridge to extend the Cavendish laboratories. His theory on traveling waves laid the foundations for the development of *soliton* theory.

In 1873, the French mathematician **Charles Hermite** (1822-1901) became the first person to prove that e is a transcendental number. "Analysis takes back with one hand what it gives with the other. I recoil in fear and loathing from that deplorable evil: continuous functions with no derivatives."

Interestingly, Hermite married Louise Bertrand, Joseph Bertrand's sister. One of Hermite's daughters married Emile Picard so Hermite was Picard's father-in-law.

In Exercises 17 and 18, solve the equation with a power series expansion about $t = 0$ by making the indicated change of variables.

17. $4xy'' + y' = 0, x = t + 1$
18. $4x^2y'' + (x + 1)y' = 0, x = t + 2$

In Exercises 19 and 20, determine at least the first five nonzero terms in a power series expansion about $x = 0$ of the solution to each nonhomogeneous IVP.

19. $y'' + xy' = \sin x, y(0) = 1, y'(0) = 0$
20. $y'' + y' + xy = \cos x, y(0) = 0, y'(0) = 1$
21. (a) If $y = a_0 + a_1x + a_2x^2 + a_3x^3 + \cdots$, what are the first three nonzero terms of the power series for y^2? (b) Use this series to find the first three nonzero terms in the power series solution about $x = 0$ to *Van-der-Pol's equation*,

$$y'' + (y^2 - 1)y' + y = 0$$

if $y(0) = 0$ and $y'(0) = 1$.
22. Use a method similar to that in Exercise 21 to find the first three nonzero terms of the power series solution about $x = 0$ to *Rayleigh's equation*,

$$y'' + \left(\frac{1}{3}(y')^2 - 1\right)y' + y = 0$$

if $y(0) = 1$ and $y'(0) = 0$, an equation that arises in the study of the motion of a violin string.
23. *Hermite's equation* is given by

$$y'' - 2xy' + 2ky, \quad k \geq 0.$$

Using a power series expansion about the ordinary point $x = 0$, obtain a general solution of this equation for (a) $k = 1$ and (b) $k = 3$. Show that if k is a nonnegative integer, then one of the solutions is a polynomial of degree k.
24. *Chebyshev's equation* is given by

$$(1 - x^2)y'' - xy' + k^2y = 0, \quad k \geq 0.$$

Using a power series expansion about the ordinary point $x = 0$, obtain a general solution of this equation for (a) $k = 1$ and (b) $k = 3$. Show that if k is a nonnegative integer, then one of the solutions is a polynomial of degree k.

In addition to his work with orthogonal poly-nomials, the Russian mathematician **Pafnuty Lvovich Chebyshev** (1821-1894) is also known for his contributions to number theory. Chebyshev was rich and never married but financially supported a daughter who he never recognized, although he did socialize with her relatively frequently, especially after she married.

25. **(a)** Show that Legendre's equation can be written as $\frac{d}{dx}\left[\left(1-x^2\right)y'\right]+k(k-1)y=0$.
   (b) Using the previous result, verify that the Legendre polynomial given in Table 4.5 satisfy the *orthogonality condition*

$$\int_{-1}^{1} P_m(x)P_n(x)dx = 0, \quad m \neq n.$$

   (c) Evaluate

$$\int_{-1}^{1} [P_n(x)]^2 \, dx$$

   for $n = 0, \ldots, 5$. How do these values compare to the value of $2/(2n+1)$, for $n = 0, \ldots, 5$? *Hint:* $P_m(x)$ and $P_n(x)$ satisfy the differential equations $\frac{d}{dx}\left[\left(1-x^2\right)P_n'(x)\right]+n(n+1)P_n(x)=0$ and $\frac{d}{dx}\left[\left(1-x^2\right)P_m'(x)\right]+m(m+1)P_m(x)=0$, respectively. Multiply the first equation by $P_m(x)$ and the second by $P_n(x)$ and subtract the results. Then integrate from -1 to 1.

26. Consider the IVP $y''+f(x)y'+y=0$, $y(0)=1, y'(0)=-1$ where
$$f(x) = \begin{cases} \frac{\sin x}{x}, & \text{if } x \neq 0 \\ 1, & \text{if } x = 0 \end{cases}$$
. **(a)** Show that $x=0$ is an ordinary point of the equation. **(b)** Find a power series solution of the equation and graph an approximation of the solution on an interval. **(c)** Generate a numerical solution of the equation. Explain any unexpected results.

27. **(a)** Use a power series to solve
$$y'' - y \cos x = \sin x, y(0) = 1, y'(0) = 0.$$
   (b) Compare the polynomial approximations of degree 4, 7, 10, and 13 to the numerical solution obtained with a computer algebra system.

Although the English scientist **George Biddell Airy** (1801-1892) made major contributions to mathematics and astronomy, by many he was considered to be a sarcastic snob. According to Eggen, "Airy was not a great scientist, but he made great science possible." Airy's son wrote "The life of Airy was essentially that of a hard-working business man, and differed from that of other hard-working people only in the quality and variety of his work. It was not an exciting life, but it was full of interest."

28. The differential equation $y'' - xy = 0$ is called *Airy's equation* and arises in electromagnetic theory and quantum mechanics. Two linearly independent solutions to Airy's equation, denoted by $Ai(x)$ and $Bi(x)$ are called the *Airy functions*. The function $Ai(x) \to 0$ as $x \to \infty$ while $Bi(x) \to \infty$ as $x \to \infty$. (Most computer algebra systems contain built-in definitions of the Airy functions.)

(a) If your computer algebra system contains built-in definitions of the Airy functions, graph each on the interval $[-15, 5]$. (b) Find a series solution of Airy's equation and obtain formulas for both $Ai(x)$ and $Bi(x)$. (c) Graph the polynomial approximation of degree n of $Ai(x)$ for $n = 6, 15, 30$, and 45 on the interval $[-15, 5]$. *Compare your results to (a) if applicable.* (d) Graph the polynomial approximation of degree n of $Bi(x)$ for $n = 6, 15, 30$, and 45 on the interval $[-15, 5]$. *Compare your results to (a) if applicable.*

4.9 SERIES SOLUTIONS OF ORDINARY DIFFERENTIAL EQUATIONS

- REGULAR AND IRREGULAR POINTS AND THE METHOD OF FROBENIUS
- GAMMA FUNCTION
- BESSEL'S EQUATION

Regular and Irregular Points and the Method of Frobenius

Let x_0 be a singular point of $y'' + p(x)y' + q(x)y = 0$. Then x_0 is a *regular singular point* of the equation if both $(x - x_0)p(x)$ and $(x - x_0)^2 q(x)$ are analytic at $x = x_0$. If x_0 is not an ordinary point or a regular singular point of the equation, then x_0 is called an *irregular singular point* of the equation.

Although the method of Frobenius was initiated by Euler, the method for finding a series expansion about a regular singular point was first published by Frobenius in 1873.

Theorem 29 (Method of Frobenius). *Let* $x = x_0$ *be a regular singular point of*

$$y'' + p(x)y' + q(x)y = 0. \qquad (4.44)$$

Then this differential equation has at least one solution of the form

$$y = \sum_{n=0}^{\infty} a_n (x - x_0)^{n+r},$$

where r is a constant that must be determined. This solution is convergent at least on some interval $|x - x_0| < R, R > 0$.

The German mathematician **Ferdinand Georg Frobenius** (1849-1917) preferred pure mathematics to applied mathematics. Although Frobenius may have been difficult to get along with, we quote Haubrich, "For Frobenius, conceptual argumentation played a somewhat secondary role. Although he argued in a comparatively abstract setting, abstraction was not an end in itself. Its advantages to him seemed to lie primarily in the fact that it can lead to much greater clearness and precision."

Suppose that $x = 0$ is a regular singular point of Equation (4.44). Then the functions $xp(x)$ and $x^2 q(x)$ are analytic at $x = 0$, which means that both of these functions have a power series in x centered at $x = 0$ with positive radius of convergence. Hence,

$$xp(x) = p_0 + p_1 x + p_2 x^2 + p_3 x^3 + \cdots \quad \text{and}$$

$$x^2 q(x) = q_0 + q_1 x + q_2 x^2 + q_3 x^3 + \cdots .$$

Therefore,

$$p(x) = p_0 x^{-1} + p_1 + p_2 x + p_3 x^2 + \cdots \quad \text{and}$$

$$q(x) = q_0 x^{-2} + q_1 x^{-1} + q_2 + q_3 x + q_4 x^2 + \cdots .$$

Substitution of these series into Equation (4.44) and multiplying through by the first term in the power series for $p(x)$ and $q(x)$, we see that the lowest term in the series involves x^{n+r-2}.

$$\sum_{n=0}^{\infty} a_n (n+r)(n+r-1) x^{n+r-2}$$

$$+ p_0 \sum_{n=0}^{\infty} a_n (n+r) x^{n+r-2}$$

$$+ \left(p_1 + p_2 x + p_3 x^2 + p_4 x^3 + \cdots \right)$$

$$\times \sum_{n=0}^{\infty} a_n (n+r) x^{n+r-1} + q_0 \sum_{n=0}^{\infty} a_n (n+r) x^{n+r-2}$$

$$+ \left(q_1 x^{-1} + q_2 + q_3 x + q_4 x^2 + q_5 x^3 + \cdots \right)$$

$$\times \sum_{n=0}^{\infty} a_n x^{n+r} = 0.$$

Then, with $n = 0$ we find the coefficient of x^{r-2} is

$$- r a_0 + r^2 a_0 + r a_0 p_0 + a_0 q_0$$

$$= a_0 \left(r^2 + (p_0 - 1) r + q_0 \right)$$

$$= a_0 \left[r(r-1) + p_0 r + q_0 \right].$$

Thus, for any equation of the form $y'' + p(x) y' + q(x) y = 0$ with regular singular point $x = 0$, we have the *indicial equation*

$$r(r-1) + p_0 r + q_0 = 0. \qquad (4.45)$$

The values of r that satisfy the indicial equation (4.45) are called the *exponents* or *indicial roots* and are

$$r_{1,2} = \frac{1}{2} \left(1 - p_0 \pm \sqrt{1 - 2p_0 + p_0^2 - 4q_0} \right).$$

$$(4.46)$$

Note that $r_1 \geq r_2$ and $r_1 - r_2 = \sqrt{1 - 2p_0 + p_0^2 - 4q_0}$.

Several situations can arise when finding the roots of the indicial equation.

1. If $r_1 \neq r_2$ and $r_1 - r_2 = \sqrt{1 - 2p_0 + p_0^2 - 4q_0}$ is *not* an integer, then there are two linearly independent solutions of the form $y_1 = x^{r_1} \sum_{n=0}^{\infty} a_n x^n$ and $y_2 = x^{r_2} \sum_{n=0}^{\infty} b_n x^n$.

2. If $r_1 \neq r_2$ and $r_1 - r_2 = \sqrt{1 - 2p_0 + p_0^2 - 4q_0}$ is an integer, then there are two linearly independent solutions of the form $y_1 = x^{r_1} \sum_{n=0}^{\infty} a_n x^n$ and $y_2 = c y_1 \ln x + x^{r_2} \sum_{n=0}^{\infty} b_n x^n$.

3. If $r_1 = r_2$, then there are two linearly independent solutions of the equation of the form $y_1 = x^{r_1} \sum_{n=0}^{\infty} a_n x^n$ and $y_2 = y_1 \ln x + x^{r_1} \sum_{n=0}^{\infty} b_n x^n$.

In any case, if $y_1(x)$ is a solution of the equation, a second linearly independent solution is given by

$$y_2(x) = y_1(x) \int \frac{1}{[y_1(x)]^2} e^{-\int p(x) dx} dx,$$

which can be obtained through reduction of order (see Equation 4.4).

When solving a differential equation in Case 2, first attempt to find a general solution using $y = x^{r_2} \sum_{n=0}^{\infty} a_n x^n = \sum_{n=0}^{\infty} a_n x^{n+r_2}$, where r_2 is the smaller of the two roots. A general solution can sometimes be found with this procedure. However, if the contradiction $a_0 = 0$ is reached, find solutions of the form $y_1 = x^{r_1} \sum_{n=0}^{\infty} a_n x^n$ and $y_2 = c y_1 \ln x + x^{r_2} \sum_{n=0}^{\infty} b_n x^n$.

EXAMPLE 4.9.1

Find a general solution of $xy'' + 3y' - y = 0$ using a series expansion about the regular singular point $x = 0$.

Solution

In standard form, this equation is $y'' + 3x^{-1}y' - x^{-1}y = 0$. Hence, $xp(x) = x \times 3x^{-1} = 3$ and $x^2q(x) = x^2 \times -x^{-1}$, so $p_0 = 3$ and $q_0 = 0$. Thus, the indicial equation is $r(r-1) + 3r = r^2 + 2r = r(r+2) = 0$ with roots $r_1 = 0$ and $r_2 = -2$. (Notice that we always use r_1 to denote the larger root of the indicial equation.) Therefore, we attempt to find a solution of the form $y = \sum_{n=0}^{\infty} a_n x^{n-2}$ with derivatives $y' = \sum_{n=0}^{\infty}(n-2)a_n x^{n-3}$ and $y'' = \sum_{n=0}^{\infty}(n-2)(n-3)a_n x^{n-4}$. Substitution into the differential equation, $xy'' + 3y' - y = 0$, yields

$$\sum_{n=0}^{\infty}(n-2)(n-3)a_n x^{n-3} +$$

$$\sum_{n=0}^{\infty} 3(n-2)a_n x^{n-3} - \sum_{n=0}^{\infty} a_n x^{n-2} = 0$$

$$(6a_0 - 6a_0)x^{-3} + \sum_{n=1}^{\infty}(n-2)(n-3)a_n x^{n-3}$$

$$+ \sum_{n=1}^{\infty} 3(n-2)a_n x^{n-3} - \sum_{n=0}^{\infty} a_n x^{n-2} = 0$$

$$\sum_{n=1}^{\infty}(n-2)(n-3)a_n x^{n-3} +$$

$$\sum_{n=1}^{\infty} 3(n-2)a_n x^{n-3} - \sum_{n=1}^{\infty} a_{n-1} x^{n-3} = 0$$

$$\sum_{n=1}^{\infty}\left\{[(n-2)(n-3)+3(n-2)]a_n - a_{n-1}\right\}x^{n-3} = 0.$$

Equating the coefficients to zero, we have

$$a_n = \frac{1}{(n-2)(n-3+3)}a_{n-1} = \frac{1}{n(n-2)}a_{n-1},$$

$$n \geq 1, \, n \neq 2.$$

Notice that from this formula, $a_1 = -a_0$. If $n = 2$, we refer to the recurrence relation $n(n-2)a_n - a_{n-1} = 0$ obtained from the coefficient in the

series solution. If $n = 2$, $2 \times 0 \times a_2 - a_1 = 0$, which indicates that $a_1 = 0$. Because $a_1 = -a_0$, we conclude that $a_0 = 0$. However, $a_0 \neq 0$ by assumption, so there is no solution of this form.

Because there is no series solution corresponding to $r_2 = -2$, we assume there is a solution of the form $y = \sum_{n=0}^{\infty} a_n x^n$ corresponding to $r_1 = 0$ with derivatives $y' = \sum_{n=1}^{\infty} n a_n x^{n-1}$ and $y'' = \sum_{n=2}^{\infty} n(n-1)a_n x^{n-2}$. Substitution into the differential equation, $xy'' + 3y' - y = 0$, yields

$$\sum_{n=2}^{\infty} n(n-1)a_n x^{n-1} + \sum_{n=1}^{\infty} 3n a_n x^{n-1} - \sum_{n=0}^{\infty} a_n x^n = 0$$

$$(3a_1 - a_0)x^0 + \sum_{n=2}^{\infty} n(n-1)a_n x^{n-1}$$

$$+ \sum_{n=2}^{\infty} 3n a_n x^{n-1} - \sum_{n=1}^{\infty} a_n x^n = 0$$

$$(3a_1 - a_0)x^0 + \sum_{n=2}^{\infty}\left\{[n(n-1)+3n]a_n - a_{n-1}\right\}x^{n-1} = 0.$$

Equating the coefficients to zero, we have $a_1 = a_0/3$ and

$$a_n = \frac{1}{n(n-1)+3n}a_{n-1} = \frac{1}{n^2+2n}a_{n-1}, \quad n \geq 2.$$

We use this formula to calculate a few more coefficients: $a_2 = a_0/24$ and $a_3 = a_0/360$ and use them to form one solution of the equation,

$$y_1 = a_0\left(1 + \frac{1}{3}x + \frac{1}{24}x^2 + \frac{1}{360}x^3 + \cdots\right).$$

To determine a second linearly independent solution, we assume that $y_2 = c y_1 \ln x + \sum_{n=0}^{\infty} b_n x^{n-2}$ and substitute this function into the differential equation to find the coefficients b_n. Because the derivatives of y_2 are

$$y_2' = cx^{-1}y_1 + c y_1' \ln x + \sum_{n=0}^{\infty}(n-2)b_n x^{n-3}$$

and

$$y_2'' = -cx^{-2}y_1 + 2x^{-1}c y_1' + c y_1'' \ln x$$

$$+ \sum_{n=0}^{\infty}(n-2)(n-3)b_n x^{n-4},$$

substitution into the differential equation yields

$$x\left[-cx^{-2}y_1 + 2x^{-1}cy_1' + cy_1'' \ln x\right.$$

$$\left. + \sum_{n=0}^{\infty}(n-2)(n-3)b_n x^{n-4}\right]$$

$$+ 3\left[cx^{-1}y_1 + cy_1' \ln x + \sum_{n=0}^{\infty}(n-2)b_n x^{n-3}\right]$$

$$-cy_1 \ln x - \sum_{n=0}^{\infty}b_n x^{n-2} = 0$$

$$-cx^{-1}y_1 + 2cy_1' + cxy_1'' \ln x$$

$$+ \sum_{n=0}^{\infty}(n-2)(n-3)b_n x^{n-3}$$

$$+ 3cx^{-1}y_1 + 3cy_1' \ln x + \sum_{n=0}^{\infty}3(n-2)b_n x^{n-3}$$

$$-cy_1 \ln x - \sum_{n=0}^{\infty}b_n x^{n-2} = 0$$

$$2cx^{-1}y_1 + 2cy_1' + \sum_{n=0}^{\infty}(n-2)(n-3)b_n x^{n-3}$$

$$+ \sum_{n=0}^{\infty}3(n-2)b_n x^{n-3} - \sum_{n=0}^{\infty}b_n x^{n-2}$$

$$+ \quad \underbrace{c(xy_1'' + 3y_1' - y_1)}_{\substack{= 0 \text{ because } y_1 \text{ is a solution} \\ \text{of the differential equation}}} \quad \ln x = 0.$$

Simplifying this expression gives us

$$2cx^{-1}y_1 + 2cy_1' + 6b_0 x^{-3} - 6b_0 x^{-3}$$

$$+ \sum_{n=1}^{\infty}(n-2)(n-3)b_n x^{n-3} + \sum_{n=1}^{\infty}3(n-2)b_n x^{n-3}$$

$$- \sum_{n=0}^{\infty}b_n x^{n-2} = 0$$

$$2cx^{-1}y_1 + 2cy_1' + \sum_{n=1}^{\infty}[(n-2)nb_n - b_{n-1}]x^{n-3} = 0.$$

Now we *choose* $a_0 = 1/c$, so

$$y_1 = \frac{1}{c}\left(1 + \frac{1}{3}x + \frac{1}{24}x^2 + \frac{1}{360}x^3 + \cdots\right) \quad \text{and}$$

$$y_1' = \frac{1}{c}\left(\frac{1}{3} + \frac{1}{12}x + \frac{1}{120}x^2 + \cdots\right).$$

Substitution into the previous equation the yields

$$2x^{-1}\left(1 + \frac{1}{3}x + \frac{1}{24}x^2 + \frac{1}{360}x^3 + \cdots\right)$$

$$+ 2\left(\frac{1}{3} + \frac{1}{12}x + \frac{1}{120}x^2 + \cdots\right)$$

$$+ \sum_{n=1}^{\infty}[(n-2)nb_n - b_{n-1}]x^{n-3} = 0$$

$$\left(2x^{-1} + \frac{4}{3} + \frac{1}{4}x + \frac{1}{45}x^2 + \cdots\right)$$

$$+ \sum_{n=1}^{\infty}[(n-2)nb_n - b_{n-1}]x^{n-3} = 0$$

$$\left(2x^{-1} + \frac{4}{3} + \frac{1}{4}x + \frac{1}{45}x^2 + \cdots\right)$$

$$+ (-b_1 - b_0)x^{-2} - b_1 x^{-1}$$

$$+ (3b_3 - b_2)x^0 + (8b_4 - b_3)x$$

$$+ (15b_5 - b_4)x^2 + \cdots = 0,$$

so we have the sequence of equations $-b_1 - b_0 = 0$, $-b_1 + 2 = 0$, $3b_3 - b_2 + \frac{4}{3} = 0$, $8b_4 - b_3 + \frac{1}{4} = 0$, $15b_5 - b_4 + \frac{1}{45} = 0$, ... Solving these equations, we see that $b_1 = 2$ and $b_0 = -2$. However, the other coefficients depend on the value of b_2. We compute the first few: $b_3 = \frac{1}{9}(3b_2 - 4)$, $b_4 = \frac{1}{288}(12b_2 - 25)$, and $b_5 = \frac{1}{21600}(60b_2 - 157)$. Hence a second linearly independent solution is given by

$$y_2 = cy_1 \ln x + x^{-2}\left(-2 + 2x + b_2 x^2\right.$$

$$\left. + \frac{1}{9}(3b_2 - 4)x^3 + \frac{1}{288}(12b_2 - 25)x^4 + \cdots\right)$$

$$= \left(1 + \frac{1}{3}x + \frac{1}{24}x^2 + \frac{1}{360}x^3 + \cdots\right)\ln x + x^{-2}$$

$$\times \left(-2 + 2x + b_2 x^2 + \frac{1}{9}(3b_2 - 4)x^3\right.$$

$$+\frac{1}{288}(12b_2-25)x^4$$

$$+\frac{1}{21600}(60b_2-157)x^5+\cdots\Bigg),$$

where b_2 is arbitrary. In particular, two linearly independent solutions of the equation ($c=1/a_0=1$)

$$y_1=1+\frac{1}{3}x+\frac{1}{24}x^2+\frac{1}{360}x^3+\cdots$$

and ($b_2=0$)

$$y_2=\left(1+\frac{1}{3}x+\frac{1}{24}x^2+\frac{1}{360}x^3+\cdots\right)\ln x$$

$$+x^{-2}\left(-2+2x-\frac{4}{9}x^3-\frac{25}{288}x^4-\frac{157}{21600}x^5+\cdots\right).$$

Explain why the choice $a_0=1/c$ does not affect the general solution obtained here. Hint: If $c=0$, $b_0=0$, which is impossible (why?), so c cannot be zero.

A general solution is then given by $y=c_1y_1+c_2y_2$, where c_1 and c_2 are arbitrary.

Several techniques can be used to solve a differential equation with equal indicial roots.

EXAMPLE 4.9.2

Find a general solution of $x^2y''+xy'+x^2y=0$, $x>0$, by using a series about the regular singular point $x=0$.

Solution

In standard form, the equation is $y''+x^{-1}y'+y=0$. Because $xp(x)=x\times x^{-1}=1$ and $x^2q(x)=x^2\times1=x^2$ are both analytic at $x=0$, $x=0$ is a regular singular point of the equation, $p_0=1$ and $q_0=0$. Therefore, the indicial equation is $r(r-1)+r=r^2-r+r=r^2=0$, so $r_{1,2}=0$. Hence, there is a solution of the form $y=\sum_{n=0}^{\infty}a_nx^n$. Substitution into $x^2y''+xy'+x^2y=0$ yields

$$x^2\sum_{n=2}^{\infty}n(n-1)a_nx^{n-2}+x\sum_{n=1}^{\infty}na_nx^{n-1}+x^2\sum_{n=0}^{\infty}a_nx^n=0$$

or

$$\sum_{n=2}^{\infty}n(n-1)a_nx^n+\sum_{n=1}^{\infty}na_nx^n+\sum_{n=0}^{\infty}a_nx^{n+2}=0.$$

After pulling off the first term of the second series and simplifying the expression, we have

$$a_1x+\sum_{n=2}^{\infty}\left\{[n(n-1)+n]a_n+a_{n-2}\right\}x^n=0.$$

Equating the coefficients to zero yields $a_1=0$ and $a_n=-n^{-2}a_{n-2}$, $n>3$. We use this formula to calculate several of these coefficients. $a_0\neq0$ is arbitrary, $a_1=0$, $a_2=-\frac{1}{4}a_0$, $a_3=0$, $a_4=\frac{1}{64}a_0$, ... Choosing $a_0=1$ we obtain that one solution to the equation is $y_1=1-\frac{1}{4}x^2+\frac{1}{64}x^4-\cdots$. We use the formula $y_2=y_1\int\frac{1}{(y_1)^2}e^{-\int p(x)dx}dx$ (Equation 4.4) to determine a second linearly independent solution to the differential equation as follows:

$$y_2=y_1\int\frac{1}{\left(1-\frac{1}{4}x^2+\frac{1}{64}x^4-\cdots\right)^2}e^{-\int(1/x)dx}dx$$

$$=y_1\int\frac{1}{1-\frac{1}{2}x^2+\frac{3}{32}x^4-\cdots}x^{-1}dx \quad\text{(squaring)}$$

$$=y_1\int\frac{1}{x}\frac{1}{1-\frac{1}{2}x^2+\frac{3}{32}x^4-\cdots}dx$$

$$=y_1\int\frac{1}{x}\left(1+\frac{1}{2}x^2+\frac{5}{32}x^4+\cdots\right)dx \quad\text{(long division)}$$

$$=y_1\int\left(\frac{1}{x}+\frac{1}{2}x+\frac{5}{32}x^3+\cdots\right)dx$$

$$=\left(\ln x+\frac{1}{4}x^2+\frac{5}{128}x^4+\cdots\right)y_1$$

$$=y_1(x)\ln x+y_1(x)\left(\frac{1}{4}x^2+\frac{5}{128}x^4+\cdots\right).$$

Notice that with this technique for finding a second linearly independent solution, we obtain a solution of the form that was stated in Case 3. A general solution of the equation is $y=c_1y_1+c_2y_2$, where c_1 and c_2 are arbitrary constants.

Note: We have not discussed the possibility of complex-valued roots of the indicial equation. When this occurs, the equation is solved using the procedures of Case 2. The solutions obtained are complex but they can be transformed into real solutions by taking the appropriate linear combinations, like those discussed for complex-valued roots of the characteristic equation.

Also, we have not mentioned if a solution can be found with a series expansion about an irregular singular point. If $x = x_0$ is an irregular singular point of $y'' + p(x)y' + q(x)y = 0$, then there may or may not be a solution of the form $y = \sum_{n=0}^{\infty} a_n(x - x_0)^{n+r}$.

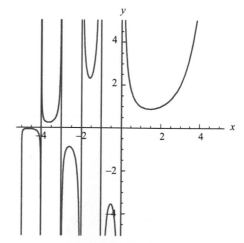

FIGURE 4.18 Although we have only defined $\Gamma(x)$ for $x > 0$, $\Gamma(x)$ can be defined for all real numbers *except* $x = 0$, $x = -1$, $x = -2$, ... (This topic is discussed in most complex analysis texts such as *Functions of One Complex Variable*, Second Edition, by John B. Conway, Springer-Verlag (1978), pp. 176-185.)

The Gamma Function

One of the more useful functions, which we will use shortly to solve *Bessel's equation*, is the *Gamma function*, first introduced by Euler in 1768, which is defined as follows.

Definition 14 (The Gamma Function). *The* **Gamma function**, *denoted* $\Gamma(x)$, *is given by*

$$\Gamma(x) = \int_0^{\infty} e^{-u}u^{x-1}du, \quad x > 0 \text{ (see Figure 4.18)}.$$

Notice that because integration is with respect to u, the result is a function of x.

A useful property associated with the Gamma function is

$$\Gamma(x + 1) = x\Gamma(x)$$

If x is a positive integer, using this property we have $\Gamma(2) = \Gamma(1 + 1) = 1 \times \Gamma(1) = \Gamma(1)$, $\Gamma(3) = \Gamma(2 + 1) = 2\Gamma(2) = 2$, $\Gamma(4) = \Gamma(3 + 1) = 3\Gamma(3) = 3 \times 2$, $\Gamma(5) = \Gamma(4+1) = 4\Gamma(4) = 4 \times 3 \times 2$, $\cdots$, and for the positive integer n, $\Gamma(n) = n!$. This property is used in solving the following equation.

EXAMPLE 4.9.3

Evaluate $\Gamma(1)$.

Solution

$$\Gamma(1) = \int_0^{\infty} e^{-u}u^{1-1}du = \int_0^{\infty} e^{-u}du$$

$$= -\lim_{b \to \infty} e^{-u}\Big|_0^b = -\lim_{b \to \infty}\left(e^{-b} - 1\right) = 1.$$

Bessel's Equation

Another important equation is *Bessel's equation (of order μ)*, named after the German astronomer Friedrich Wilhelm Bessel (1784-1846), who was a friend of Gauss. Bessel's equation is

$$x^2y'' + xy' + \left(x^2 - \mu^2\right)y = 0, \qquad (4.47)$$

where $\mu \geq 0$ is a constant. Bessel determined several representations of $J_\mu(x)$, the *Bessel* *function of the first kind of order μ, which is a solution of Bessel's equation, and noticed some of the important properties associated with the Bessel functions. The equation received its name due to Bessel's extensive work with $J_\mu(x)$, even though Euler solved the equation before Bessel.*

Friedrich Wilhelm Bessel (1784, Germany-1846, Russia) is probably more famous for his contributions to astronomy than to mathematics. The Bessel functions arise in problems from astronomy, applied mathematics, physics, and engineering. Interestingly, Bessel had no formal education past the age of 14.

To use a series method to solve Bessel's equation, first write the equation in standard form as

$$x^2 y'' + x^{-1} y + \left(x^2 - \mu^2\right) x^{-2} y = 0,$$

so $x = 0$ is a regular singular point. Using the method of Frobenius, we assume that there is a solution of the form $\sum_{n=0}^{\infty} a_n x^{n+r}$. We determine the value(s) of r with the indicial equation. Because $xp(x) = x \times x^{-1} = 1$ and $x^2 q(x) = x^2 \times \left(x^2 - \mu^2\right) x^{-2} = x^2 - \mu^2$, $p_0 = 1$ and $q_0 = -\mu^2$. Hence, the indicial equation is

$$r(r-1) + p_0 r + q_0 = r(r-1) + r - \mu^2 = r^2 - \mu^2$$
$$= (r - \mu)(r + \mu) = 0$$

with roots $r_1 = \mu$ and $r_2 = -\mu$. We assume that $y = \sum_{n=0}^{\infty} a_n x^{n+\mu}$ with derivatives $y' = \sum_{n=0}^{\infty} (n + \mu) a_n x^{n+\mu-1}$ and $y'' =$

$\sum_{n=0}^{\infty} (n + \mu)(n + \mu - 1) a_n x^{n+\mu-2}$. Substitution into Bessel's equation yields

$$x^2 \sum_{n=0}^{\infty} (n + \mu)(n + \mu - 1) a_n x^{n+\mu-2}$$

$$+ x \sum_{n=0}^{\infty} (n + \mu) a_n x^{n+\mu-1}$$

$$+ \left(x^2 - \mu^2\right) \sum_{n=0}^{\infty} a_n x^{n+\mu} = 0.$$

First we distribute,

$$\sum_{n=0}^{\infty} (n + \mu)(n + \mu - 1) a_n x^{n+\mu} + \sum_{n=0}^{\infty} (n + \mu) a_n x^{n+\mu}$$

$$+ \sum_{n=0}^{\infty} a_n x^{n+\mu+2} - \sum_{n=0}^{\infty} \mu^2 a_n x^{n+\mu} = 0.$$

Then we simplify,

$$\mu(\mu - 1) a_0 x^\mu + \mu(\mu + 1) a_1 x^{\mu+1}$$

$$+ \sum_{n=2}^{\infty} (n + \mu)(n + \mu - 1) a_n x^{n+\mu}$$

$$+ \mu a_0 x^\mu + (\mu + 1) a_1 x^{\mu+1} + \sum_{n=2}^{\infty} (n + \mu) a_n x^{n+\mu}$$

$$+ \sum_{n=0}^{\infty} a_n x^{n+\mu+2} - \mu^2 a_0 x^\mu - \mu^2 a_1 x^{\mu+1}$$

$$- \sum_{n=2}^{\infty} \mu^2 a_n x^{n+\mu} = 0$$

$$\underbrace{\left[\mu(\mu - 1) + \mu - \mu^2\right]}_{=0} a_0 x^\mu + \left[\mu(\mu + 1)\right.$$

$$\left. + (\mu + 1) - \mu^2\right] a_1 x^{\mu+1}$$

$$+ \sum_{n=2}^{\infty} \left\{\left[(n + \mu)(n + \mu - 1) + (n + \mu) - \mu^2\right]\right.$$

$$\left. a_n + a_{n-2}\right\} x^{n+\mu} = 0.$$

Notice that the coefficient of $a_0 x^{\mu}$ is zero because $r_1 = \mu$ is a root of the indicial equation. After simplifying the other coefficients and equating them to zero, we have $(2\mu + 1)a_1 = 0$ and

$$a_n = -\frac{1}{(n+\mu)(n+\mu-1)+(n+\mu)-\mu^2} a_{n-2}$$

$$= -\frac{1}{n(n+2\mu)} a_{n-2}, \quad n \geq 2.$$

From the equation $(2\mu + 1)a_1 = 0$, $a_1 = 0$. Therefore, from $a_n = -\frac{1}{n(n+2\mu)} a_{n-2}, n \geq 2, a_n = 0$ for all odd n. The coefficients that correspond to even indices are given by

$$a_2 = -\frac{1}{2(2\mu+2)} a_0 = -\frac{1}{2^2(\mu+1)} a_0,$$

$$a_4 = -\frac{1}{4(2\mu+4)} a_2 = \frac{1}{2^4 \times 2(\mu+2)(\mu+1)} a_0,$$

$$a_6 = -\frac{1}{6(2\mu+6)} a_4$$

$$= -\frac{1}{2^6 \times 3 \times 2(\mu+3)(\mu+2)(\mu+1)} a_0,$$

$$a_8 = -\frac{1}{8(2\mu+8)} a_6$$

$$= \frac{1}{2^8 \times 4 \times 3 \times 2(\mu+4)(\mu+3)(\mu+2)(\mu+1)} a_0$$

$$\vdots$$

A general formula for these coefficients is given by

$$a_{2n} = \frac{(-1)^n}{2^{2n} n!(\mu+n)(\mu+(n-1)) \cdots (\mu+2)(\mu+1)} a_0,$$

$$n \geq 2.$$

One solution of Bessel's equation can then be written as

$$y_1 = \sum_{n=0}^{\infty} a_{2n} x^{2n+\mu} = \sum_{n=0}^{\infty}$$

$$\times \frac{(-1)^n a_0}{2^{2n} n!(\mu+n)(\mu+(n-1)) \cdots (\mu+2)(\mu+1)} x^{2n+\mu}$$

$$= a_0 \sum_{n=0}^{\infty} \frac{(-1)^n 2^{\mu}}{n!(\mu+n)(\mu+(n-1)) \cdots (\mu+2)(\mu+1)}$$

$$\times \left(\frac{1}{2}x\right)^{2n+\mu}.$$

Using the Gamma function, $\Gamma(x)$, we write our first solution as

$$y_1 = \sum_{n=0}^{\infty} \frac{(-1)^n}{n!\Gamma(\mu+n+1)} \left(\frac{1}{2}x\right)^{2n+\mu},$$

$$\text{where } a_0 = \frac{1}{2^{\mu}\Gamma(\mu)}. \tag{4.48}$$

This function, denoted $J_{\mu}(x)$, is called the *Bessel function of the first kind of order* μ. For the other root, $r_2 = -\mu$, a similar derivation yields a second solution

$$y_2 = \sum_{n=0}^{\infty} \frac{(-1)^n}{n!\Gamma(-\mu+n+1)} \left(\frac{1}{2}x\right)^{2n-\mu},$$

which is the *Bessel function of the first kind of order* $-\mu$ and is denoted $J_{-\mu}(x)$.

Now, we must determine if the functions $J_{\mu}(x)$ and $J_{-\mu}(x)$ are linearly independent.

Notice that if $\mu = 0$, these two functions are the same. If $\mu > 0$, $r_1 - r_2 = \mu - (-\mu) = 2\mu$. If 2μ is not an integer, by the method of Frobenius the two solutions $J_{\mu}(x)$ and $J_{-\mu}(x)$ are linearly independent. Also, we can show that if 2μ is an odd integer, then $J_{\mu}(x)$ and $J_{-\mu}(x)$ are linearly independent. In both of these cases, a general solution is given by $y = c_1 J_{\mu}(x) + c_2 J_{-\mu}(x)$. The graphs of the functions $J_m(x)$ for $m = 0, 1, \ldots,$ 8 are shown in Figure 4.19. Notice that these functions have numerous zeros.

 What happens to the maximum value of $J_m(x)$ as m increases?

If μ is not an integer, we define the *Bessel function of the second kind of order* μ as the linear

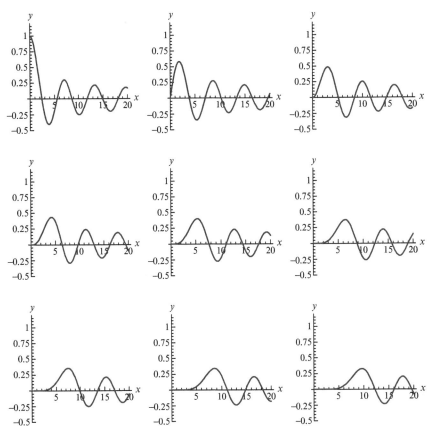

FIGURE 4.19 In the top row, from left to right, plots of $J_0(x)$, $J_1(x)$, and $J_2(x)$; in the second row, from left to right, plots of $J_3(x)$, $J_4(x)$, and $J_5(x)$; and in the third row, from left to right, plots of $J_6(x)$, $J_7(x)$, and $J_8(x)$.

combination of the functions $J_\mu(x)$ and $J_{-\mu}(x)$. This function, denoted by $Y_\mu(x)$, is given by

$$Y_\mu(x) = \frac{1}{\sin \mu\pi} \left(J_\mu(x) \cos \mu\pi - J_{-\mu}(x) \right). \quad (4.49)$$

We can show that $J_\mu(x)$ and $Y_\mu(x)$ are linearly independent solutions of Bessel's equation of order μ, so a general solution of the equation is $y = c_1 J_\mu(x) + c_2 Y_\mu(x)$. We show the graphs of the functions $Y_m(x)$ for $m = 0, 1, \ldots, 8$ in Figure 4.20. Notice that $\lim_{x \to 0^+} Y_\mu(x) = -\infty$. We can show that if m is an integer and if $Y_m(x) =$ $\lim_{\mu \to m} Y_\mu(x)$, then $J_m(x)$ and $Y_m(x)$ are linearly independent. Therefore, $y = c_1 J_\mu(x) + c_2 Y_\mu(x)$ is a general solution to Bessel's equation (4.47) for any value of μ.

A more general form of Bessel's equation is expressed in the form

$$x^2 y'' + xy' + \left(\lambda^2 x^2 - \mu^2 \right) y = 0. \quad (4.50)$$

Through a change of variables, we can show that a general solution of this parametric Bessel equation is $y = c_1 J_\mu(\lambda x) + c_2 Y_\mu(\lambda x)$.

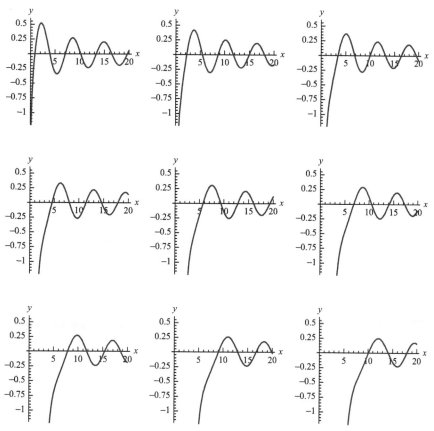

FIGURE 4.20 In the top row, from left to right, plots of $Y_0(x)$, $Y_1(x)$, and $Y_2(x)$; in the second row, from left to right, plots of $Y_3(x)$, $Y_4(x)$, and $Y_5(x)$; and in the third row, from left to right, plots of $Y_6(x)$, $Y_7(x)$, and $Y_8(x)$.

EXAMPLE 4.9.4

Find a general solution of each of the following equations: (a) $x^2y'' + xy' + (x^2 - 16) y = 0$; (b) $x^2y'' + xy' + \left(x^2 - \frac{1}{25}\right) y = 0$; (c) $x^2y'' + xy' + (9x^2 - 4) y = 0$.

Solution

(a) In this case, $\mu = 4$. Hence, $y = c_1J_4(x) + c_2Y_4(x)$. We graph this solution on $[0, 10]$ for various choices of the arbitrary constants in Figure 4.21(a). Notice that we must avoid graphing near $x = 0$ because of the behavior of $Y_4(x)$. (b) Because $\mu = 1/5$, $y = c_1J_{1/5}(x) + c_2Y_{1/5}(x)$. This solution is graphed for several

values of the arbitrary constants in Figure 4.21(b). (c) Using the parametric Bessel's equation with $\lambda = 3$ and $\mu = 2$, we have $y = c_1J_2(3x) + c_2Y_2(3x)$. We graph this solution for several choices of the arbitrary constants in Figure 4.21(c).

 **EXERCISES 4.9**

In Exercises 1-4, determine the singular points of each equation. For each equation, classify the singular point(s) as regular or irregular.

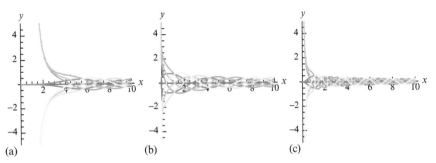

FIGURE 4.21   (a) $y = c_1 J_4(x) + c_2 Y_4(x)$, (b) $y = c_1 J_{1/5}(x) + c_2 Y_{1/5}(x)$, and (c) $y = c_1 J_2(3x) + c_2 Y_2(3x)$.

1. $x^2 y'' + 6y = 0$
2. $x(x+1) + x^{-2} y' + 5y = 0$
3. $(x^2 - 3x - 4)y'' - (x+1)y' + (x^2 - 1)y = 0$
4. $(x^2 - 25)^2 y'' - (x+5)y' + 10y = 0$

In Exercises 5-14, use the method of Frobenius to obtain two linearly independent solutions about the regular singular point $x = 0$.

5. $2xy'' - 5y' - 3y = 0$
6. $5xy'' + 8y' - xy = 0$
7. $9xy'' + 14y' + (x-1)y = 0$
8. $7xy'' + 10y' + (1 - x^2)y = 0$
9. $x^2 y'' + xy' + (x-1)y = 0$
10. $xy'' + 2xy' + y = 0$
11. $y'' + \frac{8}{3}x^{-1}y' - \left(\frac{2}{3}x^{-2} - 1\right)y = 0$
12. $y'' + \left(\frac{16}{3}x^{-1} - 1\right)y' - \frac{16}{3}x^{-2}y = 0$
13. $y'' + \left(\frac{1}{2}x^{-1} - 2\right)y' - \frac{35}{16}x^{-2}y = 0$
14. $y'' - \left(x^{-1} + 2\right)y' + \left(x^{-2} + x\right)y = 0$

In Exercises 15-17, solve the differential equation with a series expansion about $x = 0$. Compare these results with the solution obtained by solving the problem as a Cauchy-Euler equation.

15. $x^2 y'' + 7xy' - 7y = 0$
16. $x^2 y'' + 3xy' + y = 0$
17. $x^2 y'' - 3xy' + 4y = 0$
18. The differential equation
    $y'' + p(x)y' + q(x)y = 0$ has a *singular point at infinity* if after substitution of $w = 1/x$ the resulting equation has a singular point at

$w = 0$. Similarly, the equation has an *ordinary point at infinity* if the transformed equation has an ordinary point at $w = 0$. Use the chain rule and the substitution $w = 1/x$ to show that the differential equation $y'' + p(x)y' + q(x)y = 0$, where $'$ is with respect to x, is equivalent to

$$\frac{d^2 y}{dw^2} + \left(\frac{2}{w} + \frac{p(1/w)}{w^2}\right)\frac{dy}{dw} + \frac{q(1/w)}{w^4}y = 0.$$

19. Use the definition in Exercise 18 to determine if infinity is an ordinary point or a singular point of the given differential equation.
    (a) $y'' + xy = 0$
    (b) $x^2 y'' + xy' + (x^2 - k^2)y = 0$
    (c) $(1 - x^2)y'' - 2xy' + k(k+1)y = 0$
20. The *hypergeometric equation* is given by

$$x(1-x)y'' + [c - (a+b+1)x]y' - aby = 0,$$

where a, b, and c are constants. (a) Show that $x = 0$ and $x = 1$ are regular singular points. (b) Show that the roots of the indicial equation for the series $\sum_{n=0}^{\infty} a_n x^{n+r}$ are $r = 0$ and $r = 1 - c$. (c) Show that for $r = 0$, the solution obtained with the method of Frobenius is

$$y_1 = 1 + \frac{ab}{1!c}x + \frac{a(a+1)b(b+1)}{2!c(c+1)}x^2$$

$$+ \frac{a(a+1)(a+2)b(b+1)(b+2)}{3!c(c+1)(c+2)}x^3 + \cdots$$

where $c \neq 0, -1, -2, \ldots$. This series is called the *hypergeometric series*. Its sum, denoted $F(a,b,c;x)$, is called the *hypergeometric function*. (d) Show that $F(1,b,b;x) = 1/(1-x)$. (e) Find the solution of the equation that corresponds to $r = 1 - c$.

In Exercises 21-23, solve each hypergeometric equation. If necessary, express solutions in terms of the function $F(a,b,c;x)$. (*Note:* When either a or b is a negative integer, the solution is a polynomial.) (See Exercise 20.)

21. $x(1-x)y'' + \left(\frac{1}{2} - 3x\right)y' - y = 0$

22. $x(1-x)y'' + y' + 2y = 0$

23. $x(1-x)y'' + (1-2x)y' + 2y = 0$

24. *Laguerre's equation* is given by

$$xy'' + (1-x)y' + ky = 0,$$

The French mathematician **Edmond Nicolas Laguerre** (1834-1886) is well known for the special polynomials named after him that are discussed in this exercise. Although mathematics was important to Laguerre, so was his family. According to Bernkopf, "Laguerre was pictured by his contemporaries as a quiet, gentle man who was passionately devoted to his research, his teaching, and the education of his two daughters." He spent much time and energy to making sure his two daughters received the best education possible at the time.

where k is a constant (usually, it is assumed that $k > 0$). (a) Show that $x = 0$ is a regular singular point of Laguerre's equation. (b) Use the method of Frobenius to determine one solution of Laguerre's equation. (c) Show that if k is a positive integer, then the solution is a polynomial. This polynomial, denoted $L_k(x)$, is called the *Laguerre polynomial of order k*.

25. (*Relationships among Bessel functions*) (a) Using Equations (4.48) and $\Gamma(x+1) = x\Gamma(x)$, show that $\frac{d}{dx}\left[x^\mu J_\mu(x)\right] = x^\mu J_{\mu-1}(x)$. (b) Using Equation (4.48), show that $\frac{d}{dx}\left[x^{-\mu} J_\mu(x)\right] = -x^{-\mu} J_{\mu+1}(x)$. (c) Using the results of parts (a) and (b), show that $J_{\mu-1}(x) - J_{\mu+1}(x) = 2J'_\mu(x)$. (d) Evaluate $\int x^\mu J_{\mu-1}(x)dx$.

26. Show that $y = J_0(kx)$ where k is a constant is a solution of the parametric Bessel equation of order zero, $xy'' + y' + k^2xy = 0$.

27. Find a general solution of each equation.

(a) $x^2y'' + xy' + \left(x^2 - \frac{1}{4}\right)y = 0$

(b) $x^2y'' + xy' + \left(16x^2 - 25\right)y = 0$

28. Use the power series expansion of the Bessel function of the first kind of order n (n an integer), Equation (4.48) to verify that $J'_0(x) = -J_1(x)$.

29. Use the change of variables $y = v(x)x^{-1/2}$ to transform Bessel's equation $x^2y'' + xy' + \left(x^2 - k^2\right)y = 0$ into the equation $v'' + \left[1 + \left(\frac{1}{4} - k^2\right)x^{-2}\right]v = 0$. By substituting $k = 1/2$ into the transformed equation, derive the solution to Bessel's equation with $k = 1/2$.

30. Show that a solution of $xy'' + y' + v^{-1}y = 0$ is $y = J_0\left(2\sqrt{x/v}\right)$ where v is a constant.

31. Use integration by parts to verify that $\Gamma(p+1) = p\Gamma(p), p > 0$. *Note: p* is any real number.

32. (a) Show that $\Gamma(1/2) = 2\int_0^\infty e^{-x^2}dx$. (*Hint:* Let $u = x^2$.) (b) Use polar coordinates to

evaluate $\left(\int_0^\infty e^{-x^2}dx\right)\left(\int_0^\infty e^{-y^2}dy\right) =$
$\int_0^\infty \int_0^\infty e^{-(x^2+y^2)}dxdy$. (c) Use the results of (a) and (b) to evaluate $\Gamma(1/2)$ and then $\Gamma(3/2)$.

33. (a) Use a series solution to approximate the solution to the IVP
$xy'' + (2-x)y' + \frac{1}{4}x^{-1}y = 0, y(1) = 1,$
$y'(1) = -1$. (b) Use a computer algebra system to generate a numerical solution of the IVP and compare these results with those obtained in (a) by graphing the two approximations together.

34. *Laguerre's equation* (Refer back to Exercise 24 as needed.) (a) Show that the Laguerre polynomials satisfy the formula
$L_n(x) = \frac{1}{n!}e^x \frac{d^n}{dx^n}(x^n e^{-x})$. (b) Show that
$\int_0^\infty e^{-x} L_m(x) L_n(x)dx = 0$ for $n \neq m$, which indicates that the Laguerre polynomials are *orthogonal*. (c) Determine the value of
$\int_0^\infty e^{-x}[L_n(x)]^2 dx$ by experimenting with $n = 1, 2, \ldots$.

CHAPTER 4 SUMMARY: ESSENTIAL CONCEPTS AND FORMULAS

Second-order linear differential equation:
$a_2(t)y'' + a_1(t)y' + a_0(t)y = g(t)$
  Homogeneous: $a_2(t)y'' + a_1(t)y' + a_0(t)y = 0$
  Constant coefficients: $a_2y'' + a_1y' + a_0y = g(t)$
*n*th-order linear differential equation:
$a_n(t)y^{(n)} + \cdots + a_1(t)y' + a_0(t)y = g(t)$
  Homogeneous:
  $a_n(t)y^{(n)} + \cdots + a_1(t)y' + a_0(t)y = 0$
  Constant coefficients:
  $a_ny^{(n)} + \cdots + a_1y' + a_0y = g(t)$
Linearly dependent and independent set of functions: If $S = \{f_1, f_2, \ldots, f_m\}$, S is *linearly dependent* if there are constants $c_1, \ldots, c_n$ not all zero, so that $c_1f_1 + \cdots + c_nf_n = 0$. S is *linearly independent* if S is not linearly dependent.

Wronskian: Let $S = \{f_1, \ldots, f_n\}$ be a set of n functions for which each is differentiable at least $n-1$ times. The *Wronskian* of S, denoted by
$$W(S) = W(\{f_1, \ldots, f_n\}),$$
is the determinant
$$W(S) = \begin{vmatrix} f_1 & f_2 & \cdots & f_n \\ f_1' & f_2' & \cdots & f_n' \\ \vdots & \vdots & \vdots & \vdots \\ f_1^{(n-1)} & f_2^{(n-1)} & \cdots & f_n^{(n-1)} \end{vmatrix}.$$

Principle of Superposition: If $f_1, \ldots, f_m$ are solutions of the linear homogeneous differential equation $y^{(n)} + p_{n-1}(t)y^{(n-1)} + \cdots + p_1(t)y' + p_0(t)y = 0$ on the interval I, and if $c_1, c_2, \ldots, c_m$ are arbitrary constants, then $y = c_1f_1 + c_2f_2 + \cdots + c_mf_m$ is also a solution of this differential equation on I.

Fundamental set of solutions:
$S = \{f_1, \ldots, f_n\}$ is a *fundamental set* of solutions of the *n*th-order linear homogeneous equation if S is linearly independent and every function in S is a solution of the equation.

General solution of a homogeneous equation: y_h is the arbitrary linear combination of the functions in a fundamental set.

Reduction of order

Characteristic equation

Particular solution: y_p is a particular solution of an equation if it does not contain any arbitrary constants.

General solution of a nonhomogeneous equation: $y = y_h + y_p$
  Method of undetermined coefficients
  Variation of parameters

Cauchy-Euler equation

Ordinary and singular points

Power series solution method about an ordinary point: $y = \sum_{n=0}^\infty a_n(x - x_0)^n$

Regular and irregular singular points

Method of Frobenius:
$y = \sum_{n=0}^\infty a_n(x - x_0)^{n+r}$

CHAPTER 4 REVIEW EXERCISES

In Exercises 1-6, determine if the given set is linearly independent or linearly dependent.

1. $S = \{e^{5t}, 1\}$
2. $S = \{\cos^2 t, \sin^2 t\}$
3. $S = \{t, t \ln t\}$
4. $S = \{t, t - 1, 3t\}$
5. $S = \{t, \cos t, \sin t\}$
6. $S = \{e^t, e^{-2t}, e^{-t}\}$

In Exercises 7-9, verify that y is a general solution of the given differential equation.

7. $y'' - 7y' + 10y = 0; y = c_1 e^{5t} + c_2 e^{2t}$
8. $y'' - y' - 2y = 0; y = c_1 e^{2t} + c_2 e^{-t}$
9. $y'' - 2y' + 2y = 0; y = e^t(c_1 \cos t + c_2 \sin t)$

In Exercises 10 and 11, show that the function y_1 satisfies the differential equation and use reduction of order to find a second linearly independent solution.

10. $y_1 = t + 1; (t + 1)^2 y'' - 2(t + 1)y' + 2y = 0$
11. $y_1 = t^{-1} \sin t; ty'' + 2y' + ty = 0$

In Exercises 12-37, find a general solution for each equation.

12. $y'' + 7y' + 10y = 0$
13. $6y'' + 5y' - 4y = 0$
14. $y'' + 2y' + y = 0$
15. $y'' + 3y' + 2y = 0$
16. $y'' - 10y' + 34y = 0$
17. $2y'' - 5y' + 2y = 0$
18. $15y'' - 11y' + 2y = 0$
19. $20y'' + y' - y = 0$
20. $12y'' + 8y' + y = 0$
21. $2y''' + 3y'' + y' = 0$
22. $9y''' + 36y'' + 40y' = 0$
23. $9y''' + 12y'' + 13y' = 0$
24. $y'' - 2y' - 8y = -t$
25. $y'' + 5y' = 5t^2$
26. $y'' - 4y' = -3 \sin t$
27. $y'' + 2y' + 5y = 3 \sin 2t$
28. $y'' - 9y = 1/(1 + e^{3t})$
29. $y'' - 2y' = 1/(1 + e^{2t})$
30. $y'' - 3y' + 2y = -4e^{-2t}$

31. $y'' - 6y' + 13y = 3e^{-2t}$
32. $y'' + 9y' + 20y = -2te^t$
33. $y'' + 7y' + 12y = 3t^2 e^{-4t}$
34. $y''' + 3y'' - 9y' + 5y = e^t$
35. $y''' - 12y' - 16y = e^{4t} - e^{-2t}$
36. $y^{(4)} + 6y''' + 18y'' + 30y' + 25y = e^{-t} \cos 2t + e^{-2t} \sin t$
37. $y^{(4)} + 4y''' + 14y'' + 20y' + 25y = t^2$

In Exercises 38-45, solve the IVP.

38. $y'' + 5y' + 6y = 0, y(0) = 2, y'(0) = 0$
39. $y'' + 10y' + 16y = 0, y(0) = 0, y'(0) = 4$
40. $y'' + 16y = 0, y(0) = 0, y'(0) = -8$
41. $y'' + 25y = 0, y(0) = 1, y'(0) = 0$
42. $y'' - 4y = t, y(0) = 2, y'(0) = 0$
43. $y'' + 3y' - 4y = e^t, y(0) = 0, y'(0) = 4$
44. $y'' + 9y = \sin 3t, y(0) = 6, y'(0) = 0$
45. $y'' + y = \cos t, y(0) = 0, y'(0) = 0$

In Exercises 46-51, use variation of parameters to solve the indicated differential equations and IVPs.

46. $y'' + 4y = \tan 2t$
47. $y'' + y = \csc t$
48. $y'' - 8y' + 16y = t^{-3} e^{4t}$
49. $y'' - 8y' + 16y = t^{-3} e^{4t}, y(0) = 0, y'(0) = 1$
50. $y'' - 2y' + y = e^t \ln t$
51. $y'' - 2y' + y = e^t \ln t, y(1) = 1, y'(1) = 0$
52. Show that the substitution $u = y'/y$ converts the equation $p_2(t)y'' + p_1(t)y' + p_0(t)y = 0$ to $p_2(t)(u' + u^2) + p_1(t)u + p_0(t) = 0$.
53. Use the substitution in the previous problem to solve $y'' - 2ty' + t^2 y = 0$. *Hint:* You should obtain the differential equation $du/dt = -(u - t)^2$. Make the substitution $v = u - t$ so solve the equation.
54. Use Abel's formula to find the Wronskian (within a constant multiple) associated with the following differential equations. Also, obtain a fundamental set of solutions and compute the Wronskian directly. Compare the results (see Exercise 41 in Section 4.1).

(a) $y'' + 3y' - 4y = 0$
(b) $y'' + 4y' + 4y = 0$
(c) $t^2 y'' - 5ty' + 5y = 0$
(d) $t^2 y'' - ty' + 5y = 0$

55. Solve each of the following boundary-value problems.
   (a) $y'' - y = 0, y'(0) + 3y(0) = 0,$
       $y'(1) + y(1) = 1$
   (b) $y'' + \lambda y = 0, y(0) = 0, y(p) = 0 \ (p > 0)$
       Hint: Consider three cases: $\lambda = 0, \lambda < 0,$
       and $\lambda > 0$
   (c) $y'' + \lambda y = 0, y'(0) = 0, y'(p) = 0 \ (p > 0)$
       Hint: Consider three cases: $\lambda = 0, \lambda < 0,$
       and $\lambda > 0$
   a. $y'' + 2y' - (\lambda - 1)y = 0, y(0) = 0, y(2) = 0$
       Hint: Consider three cases: $\lambda = 0, \lambda < 0,$
       and $\lambda > 0$

In Exercises 56-62, solve the Cauchy-Euler equation.

56. $x^2 y'' + 7xy' + 8y = 0$
57. $x^2 y'' - 4xy' + 6y = 0$
58. $x^2 y'' + xy' + y = 0$
59. $2x^2 y'' + 5xy' + y = 0, y(1) = 1, y'(1) = 0$
60. $5x^2 y'' - xy' + 2y = 0$
61. $x^2 y'' - 7xy' + 25y = 0$
62. $x^2 y'' - 7xy' + 15y = 8x$

In Exercises 63-65, solve the differential equation with a series expansion about $x = 0$. Write out at least the first five terms of each series.

63. $y'' - 4y' + 4y = 0$
64. $y'' + 2y' - 3y = xe^x$
65. $(2x^2 - 1) y'' + 2xy' - 3y = 0$

In Exercises 66-68, use the method of Frobenius to obtain two linearly independent solutions about the regular singular point.

66. $3xy'' + 11y' - y = 0$
67. $2x^2 y'' + 5xy' - 2y = 0$
68. $x^2 y'' - 7xy' + (7 - 2x^2) y = 0$

In Exercises 69 and 70, find a solution to each hypergeometric equation. If necessary, express the solution in terms of the function $F(a, b, c; x)$ (see Exercise 20 in Section 4.9.)

69. $x(1 - x)y'' + (1 + 2x)y' + 10y = 0$
70. $x(1 + x)y'' + (1 - 2x) y' - 10y = 0$
71. *(Simple Modes of a Vibrating Chain)* The equation that describes the simple modes of a vibrating chain of length ℓ is

$$x\frac{d^2 y}{dx^2} + \frac{dy}{dx} + \frac{y}{v} = 0,$$

where y is the displacement and x is the distance from the bottom of the chain. The equation was studied extensively by Daniel Bernoulli around 1727.
   (a) If the chain is fixed at the top so that $y(\ell) = 0$ and $y(0) = 1$, show that a solution to this equation is $J_0(2\sqrt{x/v})$.
   (b) Convince yourself that for any value of ℓ, the equation $J_0(2\sqrt{\ell/v})$ has infinitely many solutions.
   (c) If $\ell = 1$, graph $J_0(2\sqrt{\ell/v})$ and approximate the last 10 solutions of $J_0(2\sqrt{\ell/v}) = 0$.
72. Complete the following table

Differential Equation	Characteristic Equation	Roots of Characteristic Equation	General Solution
$y^{(n)} = 0$			
	$(r - k)^n$		
		$r_{1,2} = \alpha \pm i\beta,$	
		$\beta \neq 0$	
			$y = e^{\alpha t}[(c_{1,1} + c_{1,2}t + \cdots c_{1,n-1}t^{n-1}) \cos \beta t$ $+ (c_{2,1} + c_{2,2}t + \cdots c_{2,n-1}t^{n-1}) \sin \beta t]$

73. Use the substitution $u = y'y$ to solve
$$t(y''y + y'^2) + y'y = 1, y(1) = 1, y'(1) = 1.$$
Hint: After solving for u, $y'y = \frac{1}{2}(y^2)$.

DIFFERENTIAL EQUATIONS AT WORK

A. Testing for Diabetes

Sources: D. N. Burghess and M. S. Borrie, *Modeling with Differential Equations*, Ellis Horwood Limited, pp. 113-116. Joyce M. Black and Esther Matassarin-Jacobs, *Luckman and Sorensen's Medical-Surgical Nursing: A Psychophysiologic Approach*, Fourth Edition, W. B. Saunders Company (1993), pp. 1775-1808.

Diabetes mellitus affects approximately 12 million Americans; approximately one-half of these people are unaware that they have diabetes. Diabetes is a serious disease: it is the leading cause of blindness in adults, the leading cause of renal failure, responsible for approximately one-half of all nontraumatic amputations in the United States. In addition, people with diabetes have an increased rate of coronary artery disease and strokes. People at risk for developing diabetes include those who are obese; those suffering from excessive thirst, hunger, urination, and weight loss; women who have given birth to a baby with weight greater than nine pounds; those with a family history of diabetes; those who are over 40 years of age.

People with diabetes cannot metabolize glucose because their pancreas produces an inadequate or ineffective supply of insulin. Subsequently, glucose levels rise. The body attempts to remove the excess glucose through the kidneys: the glucose acts as a diuretic, resulting in increased water consumption. Since some cells require energy, which is not being provided by glucose, fat and protein is broken down and ketone levels rise. Although there is no cure for diabetes at this time, many cases can be effectively managed by a balanced diet and insulin therapy in addition to maintaining an optimal weight.

Diabetes can be diagnosed by several tests. In the *fasting blood sugar test*, a patient fasts for at least four hours, and then the glucose level is measured. In a fasting state, the glucose level in normal adults ranges from 70 to 110 mg/ml. An adult in a fasting state with consistent readings of over 150 mg probably has diabetes. However, people with mild cases of diabetes might have fasting state glucose levels within the normal range because individuals vary greatly. In these cases, a highly accurate test which is frequently used to diagnose mild diabetes is the *glucose tolerance test* (GTT), which was developed by Drs. Rosevear and Molnar of the Mayo Clinic and Drs. Ackerman and Gatewood of the University of Minnesota. During the GTT, a blood and urine sample are taken from a patient in a fasting state to measure the glucose, G_0, hormone, H_0, and glycosuria levels, respectively. We assume that these values are equilibrium values. The patient is then given 100 g of glucose. Blood and urine samples are then taken at 1, 2, 3, and 4 h intervals. In a person without diabetes, glucose levels return to normal after two hours; in diabetics their blood sugar levels either take longer or never return to normal levels.

Let G denote the cumulative level of glucose in the blood, $g = G - G_0$, H the cumulative level of hormones that affect insulin production (like glucagon, epinephrine, cortisone and thyroxin), and $h = H - H_0$. Notice that g and h represent the fluctuation of the cumulative levels of glucose and hormones from their equilibrium values. The relationship between the rate of change of glucose in the blood and the rate of change of the cumulative levels of the hormones in the blood that affects insulin production is

$$\begin{cases} g' = f_1(g,h) + J(t) \\ h' = f_2(g,h) \end{cases},$$

where $J(t)$ represents the *external* rate at which the blood glucose concentration is being increased. If we assume that f_1 and f_2 are linear functions, then this system of equations becomes

$$\begin{cases} g' = -ag - bh + J(t) \\ h' = -ch + dg \end{cases},$$

where $a, b, c,$ and d represent positive numbers.

1. Show that if $g' = -ag - bh + J(t)$ then

$$h = \frac{1}{b}\left(-g' - ag + J\right) \quad \text{and}$$

$$h' = \frac{1}{b}\left(-g'' - ag' + J'\right). \quad (4.51)$$

2. Substitute Equations (4.51) into $h' = -ch + dg$ to obtain the second-order equation

$$\frac{1}{b}\left(-g'' - ag' + J'\right) = -\frac{c}{b}\left(-g' - ag + J\right) + dg$$

$$g'' + (a+c)g' + (ac+bd)g = J' + cJ. \quad (4.52)$$

Since the glucose solution is consumed at $t = 0$, for $t > 0$ we have that

$$g'' + (a+c)g' + (ac+bd)g = J' + cJ = 0. \quad (4.53)$$

3. Show that the solutions of the characteristic equation of the corresponding homogeneous equation (4.53) are

$$\frac{1}{2}\left(-a - c \pm \sqrt{(a-c)^2 - 4bd}\right).$$

4. Explain why it is reasonable to assume that glucose levels (in humans) are periodic and subsequently that $(a-c)^2 - 4bd < 0$.

5. If $(a-c)^2 - 4bd < 0$ and $t > 0$, show that a general solution of the corresponding homogeneous equation (4.53),

$$g'' + (a+c)g' + (ac+bd)g = 0, \text{ is}$$

$$g = e^{-(a+c)t/2}\left[c_1 \cos\left(\frac{1}{2}t\sqrt{4bd - (a-c)^2}\right)\right.$$

$$\left. + c_2 \sin\left(\frac{1}{2}t\sqrt{4bd - (a-c)^2}\right)\right]$$

and that

$$G = G_0 + e^{-(a+c)t/2}\left[c_1 \cos\left(\frac{1}{2}t\sqrt{4bd - (a-c)^2}\right)\right.$$

$$\left. + c_2 \sin\left(\frac{1}{2}t\sqrt{4bd - (a-c)^2}\right)\right].$$

Let $\alpha = \frac{1}{2}(a+c)$ and $\omega = \frac{1}{2}\sqrt{4bd - (a-c)^2}$. Then we can rewrite the general solution obtained here as

$$G(t) = G_0 + e^{-\alpha t}(c_1 \cos \omega t + c_2 \sin \omega t).$$

Research has shown that lab results of $2\pi/\omega > 4$ indicate a mild case of diabetes.

6. Suppose that you have given the GTT to four patients that you suspect of having a mild case of diabetes. The results for each patient are shown in the following table. Which patients, if any, probably have a mild case of diabetes?

	Patient 1	Patient 2	Patient 3	Patient 4
G_0	80.00	90.00	100.00	110.00
$t = 1$	85.32	91.77	103.35	114.64
$t = 2$	82.54	85.69	98.26	105.89
$t = 3$	78.25	92.39	96.59	108.14
$t = 4$	76.61	91.13	99.47	113.76

B. Modeling the Motion of a Skier

During a sporting event, an athlete loses strength because the athlete has to perform work against many physical forces. Some of the forces acting on an athlete include gravity (the athlete must usually work against gravity when moving body parts; friction is created between the ground and the athlete) and aerodynamic drag (a pressure gradient exists between the front and the back of the athlete; the athlete must overcome the force of air friction). In downhill skiing, friction and drag affect the skier most because the skier is not working against gravity (why?). The distance traveled by a skier moving down a slope is given by

$$m\frac{d^2s}{dt^2} = F_g - F_\mu - D,$$

where m is the mass of the skier, $s = s(t)$ is the distance traveled by the skier at time t, F_g is the gravitational force, F_μ is the friction force between ski and snow, and D is the aerodynamic drag.

If the slope has constant angle α, we can rewrite the equation as

$$\frac{d^2s}{dt^2} = g\left(\sin\alpha - \mu\cos\alpha\right) - \frac{C_D A\rho}{2m}\left(\frac{ds}{dt}\right)^2,$$

where μ is the coefficient of friction, C_D the drag coefficient, A the projected area of the skier, ρ the air density, and $g \approx 9.81$ m s^{-2} the gravitational constant.[4] Because g, α, μ, m, C_D, A, and ρ are constants, $g\left(\sin\alpha - \mu\cos\alpha\right)$ and $C_D A\rho/(2m)$ are constants. Thus, if we let $k^2 = (\sin\alpha - \mu\cos\alpha)$ (assuming that $(\sin\alpha - \mu\cos\alpha)$ is nonnegative) and $h^2 = C_D A\rho/(2m)$, we can rewrite this equation in the simpler form

$$\frac{d^2s}{dt^2} = k^2 - h^2\left(\frac{ds}{dt}\right)^2.$$

Remember that the relationship between displacement, s, and velocity, v, is $v = ds/dt$. Thus, we can find displacement by integrating velocity, if the velocity is known.

1. Use the substitution $v = ds/dt$ to rewrite $s'' = k^2 - h^2(s')^2$ as a first-order equation and find the solution that satisfies $v(0) = v_0$. *Hint:* Use the method of partial fractions.
2. Find $s(t)$ if $s(0) = 0$.

   Thus, in this case, we see that we are able to express both s and v as functions of t.

Typical values of the constants m, ρ, μ, and $C_D A$ are shown in the following table.

Constant	Typical Value
m	75 kg
ρ	1.29 kg m^{-3}
μ	0.06
$C_D A$	0.16 m^2

3. Use the values given in the previous table to complete the entries in the following table. For each value of α in the following table, graph $v(t)$ and $s(t)$ if $v(0) = 0, 5$, and 10 on the interval $[0, 40]$. In each case, calculate the maximum velocity achieved by the skier. What is the limit of the velocity achieved by the skier? How does changing the initial velocity affect the maximum velocity?

α	$k^2 = (\sin\alpha - \mu\cos\alpha)$	$h^2 = C_D A\rho/(2m)$
$\alpha = 30° = \pi/6$ rad		
$\alpha = 40° = 2\pi/9$ rad		
$\alpha = 45° = \pi/4$ rad		
$\alpha = 50° = 5\pi/18$ rad		

4. For each value of α in the previous table, graph $v(t)$ and $s(t)$ if $v(0) = 80$ and 100 on the interval $[0, 40]$, if possible. Interpret your results.

   Sometimes, it is more useful to express v (velocity) as a function of a different variable, like s (displacement). To do so, we write $dt/ds = 1/v$ because $ds/dt = v$. Multiplying $dv/dt = k^2 - h^2 v^2$ by $1/v$ and simplifying leads to

$$\frac{1}{v}\frac{dv}{dt} = \frac{1}{v}\left(k^2 - h^2 v^2\right)$$

$$\frac{dv}{dt}\frac{dt}{ds} = k^2 v^{-1} - h^2 v$$

$$\frac{dv}{ds} + h^2 v = k^2 v^{-1},$$

[4]S. Savolainen, R. Visuri, A Review of Athletic Energy Expenditure, Using Skiing as a Practical Example, *Journal of Applied Biomechanics*, **10** (3), 1994, 253-269.

which we recognize as a Bernoulli equation (see Section 2.5) and solve by letting $w = v^2$. Then,

$$\frac{dw}{ds} = 2v\frac{dv}{ds} \quad \text{so} \quad \frac{1}{2v}\frac{dw}{ds} = \frac{dv}{ds}$$

and the equation becomes $\frac{1}{2v}\frac{dw}{ds} + h^2 v = k^2 v^{-1}$. Multiplying this equation by $2v$ and applying the substitution $w = v^2$ leads us to the first-order linear differential equation $\frac{dw}{ds} + 2h^2 w = 2k^2$.

5. Show that a solution of this differential equation is given by $w(s) = (k/h)^2 + Ce^{-2h^2 s}$, where C is an arbitrary constant. Resubstitute $w = v^2$ to obtain $[v(s)]^2 = (k/h)^2 + Ce^{-2h^2 s}$.

6. Show that the solution that satisfies the initial condition $v(0) = v_0$ is

$$[v(s)]^2 = \frac{1}{h^2 e^{2hs}}\left(-k^2 + k^2 e^{2h^2 s} + h^2 v_0^2\right)$$

$$= \left(\frac{k}{h}\right)^2 \left(1 - e^{-2h^2 s}\right) + v_0^2 e^{-h^2 s}.$$

7. For each value of α in the previous table, graph $v(s)$ if $v(0) = 0, 5$, and 10 on the interval (0, 1500]. In each case, calculate the maximum velocity achieved by the skier. How does changing the initial velocity affect the maximum velocity? Compare your results to (3). Are your results consistent?

8. How do these results change if μ is increased or decreased? if $C_D A$ is increased or decreased? How much does a 10% increase or decrease in friction decrease or increase the skier's velocity?

9. The values of the constants considered here for three skiers are listed in the following table.

Constant	Skier 1	Skier 2	Skier 3
m	75 kg	75 kg	75 kg
ρ	1.29 kg m^{-3}	1.29 kg m^{-3}	1.29 kg m^{-3}
μ	0.05	0.06	0.07
$C_D A$	0.16 m^2	0.14 m^2	0.12 m^2

Suppose that these three skiers are racing down a slope with constant angle $\alpha = 30^\circ = \pi/6$ rad. Who wins the race if the length of the slope is 400 m? Who loses? Who wins the race if the length of the slope is 800 or 1200 m? Who loses? Does the length of the slope matter as to who wins or loses? What if the angle of the slope is increased or decreased? Explain.

10. Which of the variables considered here affects the velocity of the skier the most? Which variable do you think is the easiest to change? How would you advise a group of skiers who want to increase their maximum attainable velocity by 10%?

C. The Schrödinger Equation

The time independent *Schrödinger equation*, proposed by the Austrian physicist Erwin Schrödinger (1887-1961) in 1926, describes the relationship between particles and waves. The *Schrödinger equation in spherical coordinates is* given by[5]

Erwin Rudolf Josef Alexander Schrödinger (1887-1961): "Thus, the task is, not so much to see what no one has yet seen; but to think what nobody has yet thought, about that which everybody sees."

[5]S.T. Thornton, A. Rex, Modern Physics for Scientists and Engineers, Saunders College Publishing, 1993, pp. 248-253.

$$\frac{1}{r^2}\frac{\partial}{\partial r}\left(r^2\frac{\partial\psi}{\partial r}\right)+\frac{1}{r^2\sin\theta}\frac{\partial}{\partial\theta}\left(\sin\theta\frac{\partial\psi}{\partial\theta}\right)$$

$$+\frac{1}{r^2\sin^2\theta}\frac{\partial^2\psi}{\partial\phi^2}+\frac{2\mu}{h^2}(E-V)\psi=0.$$

If we assume that a solution to this partial differential equation of the form

$$\psi(r,\theta,\phi)=R(r)\Theta(\theta)g(\phi)$$

exists, then by a technique called *separation of variables*, the Schrödinger equation can be rewritten as three second-order ordinary differential equations:

Azimuthal equation: $\dfrac{d^2g}{d\phi^2}=-m_\ell^2 g$

Radial equation: $\dfrac{1}{r^2}\dfrac{d}{dr}\left(r^2\dfrac{dR}{dr}\right)+$

$$\frac{2\mu}{h^2}\left(E-V-\frac{h^2}{2\mu}\frac{\ell(\ell+1)}{r^2}\right)R=0$$

Angular equation: $\dfrac{1}{\sin\theta}\dfrac{d}{d\theta}\left(\sin\theta\dfrac{d\Theta}{d\theta}\right)+$

$$\left(\ell(\ell+1)-\frac{m_\ell^2}{\sin^2\theta}\right)\Theta=0.$$

The *orbital angular momentum quantum number* ℓ must be a nonnegative integer: $\ell = 0, \pm 1, \pm 2, \dots$ and the magnetic quantum number m_ℓ depends on ℓ: $m_\ell = -|\ell|, -|\ell|+1, \dots, -2, -1, 0, 1, 2, \dots, |\ell|, |\ell|+1$.

If $\ell = 0$, the radial equation is $\frac{1}{r^2}\frac{d}{dr}\left(r^2\frac{dR}{dr}\right)+\frac{2\mu}{h^2}(E-V)R=0$. Using $V(r)=-e^2/(4\pi\epsilon_0 r)$ (the constants μ, h, e, and ϵ_0 are described later), the equation becomes $\frac{1}{r^2}\frac{d}{dr}\left(r^2\frac{dR}{dr}\right)+\frac{2\mu}{h^2}\left(E+\frac{e^2}{4\pi\epsilon_0 r}\right)R=0$.

1. Show that if $R=Ae^{-r/a_0}$ is a solution of this equation, $a_0=4\pi\epsilon_0 h^2/(\mu e^2)$ and $E=-h^2/(2\mu a_0^2)$. The constants in the expression $4\pi\epsilon_0 h^2/(\mu e^2)$ are described as follows: $h\approx 1.054572\times 10^{-34}$ J s is the *Planck constant* (6.6260755×10^{-34} J s divided by 2π), $\epsilon_0\approx 8.854187817\times 10^{-12}$ F/m is the *permitivity of vacuum*, and $e\approx 1.60217733\times 10^{-19}$ C is the *elementary charge*. For the hydrogen atom, $\mu = 9.104434\times 10^{-31}$ kg is the reduced mass of the proton and electron.

2. Calculate a_0, $2a_0$, and E for the hydrogen atom.

3. The diameter of a hydrogen atom is approximately 10^{-10} m while the energy, E, is known to be approximately 13.6 eV How do the results you obtained in problem 2 compare to these?

For the hydrogen atom, the radial equation can be written in the form

$$\frac{d^2R}{dr^2}+\frac{2}{r}\frac{dR}{dr}+\left(2E+\frac{2Z}{r}-\frac{\ell(\ell+1)}{r^2}\right)R=0,$$

where E is the energy of the hydrogen atom and Z is a constant.[6]

When $E<0$ and we let $p=\sqrt{-2E}$, a solution of the radial equation for the hydrogen atom has the form $R(r)=e^{-pr}u(r)$. When this solution is substituted into the equation, we obtain the equation

$$\frac{d^2u}{dr^2}+2\left(\frac{1}{r}-p\right)\frac{du}{dr}$$

$$+\left[2\left(Z-\frac{p}{r}\right)-\frac{\ell(\ell+1)}{r^2}\right]u=0.$$

4. Show that the solutions of the indicial equation of this equation are ℓ and $-(\ell+1)$

5. Find a series solution to the equation of the form $u(r)=\sum_{n=0}^{\infty}a_n r^{n+\ell}$

6. Find a series solution of the radial equation for the hydrogen atom.

[6]Y.-H. Liu, W.-N. Mei, Solution of the Radial Equation for Hydrogen Atom: Series Solution or Laplace Transform? *International Journal of Mathematical Education in Science and Technology*, **21** (6), 1990, 913-918.

CHAPTER

5

Applications of Higher Order Differential Equations

In Chapter 4, we discussed several techniques for solving higher order differential equations. In this chapter, we illustrate how those methods can be used to solve some initial-value problems (IVPs) that model physical situations.

5.1 SIMPLE HARMONIC MOTION

Suppose that an object of mass m is attached to an elastic spring that is suspended from a rigid support such as a ceiling or a horizontal rod. The object causes the spring to stretch a distance s from its *natural length*. The position at which it comes to rest is called the *equilibrium position*. According to Hooke's law, the spring exerts a restoring force in the upward (opposite) direction that is proportional to the distance s that the spring is stretched. Mathematically, this is stated as $F = ks$ where $k > 0$ is the constant of proportionality or spring constant. In Figure 5.1 we see that the spring has natural length b. When the object is attached to the spring, it is stretched s units past its natural length to the equilibrium position $x = 0$. When the system is in motion, the displacement from $x = 0$ at time t is given by $x(t)$.

Although the English scientist **Robert Hooke** (1635-1703) discovered "Hooke's law" in 1660, he didn't formally state it until 1678. Hooke and Newton *did not* get along well and were *not* friends.

By Newton's second law of motion,

$$F = ma = m\frac{d^2x}{dt^2},$$

where m represents mass of the object and a represents acceleration. If we assume that there are no forces other than the force as a result of gravity acting on the mass, we determine the

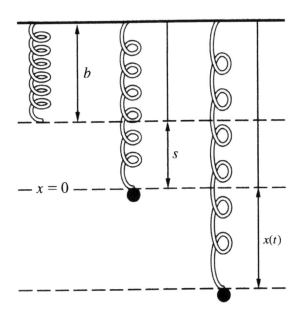

FIGURE 5.1   A spring-mass system.

differential equation that models this situation by summing the forces acting on the *spring-mass system* with

$$m\frac{d^2x}{dt^2} = \sum (\text{forces acting on the system})$$

$$= -k(s + x) + mg$$

$$= -ks - kx + mg.$$

At equilibrium $ks = mg$, so after simplification we obtain the differential equation

$$m\frac{d^2x}{dt^2} = -kx \quad \text{or} \quad m\frac{d^2x}{dt^2} + kx = 0.$$

The two initial conditions that are used with this problem are the *initial position* $x(0) = \alpha$ and the *initial velocity* $x'(0) = \beta$. The function $x(t)$ that describes the displacement of the object with respect to the equilibrium position at time t is found by solving the IVP

$$m\frac{d^2x}{dt^2} + kx = 0, \quad x(0) = \alpha, \quad x(0) = \beta$$

Based on the assumption made in deriving the differential equation, positive values of $x(t)$ indicate that the mass is below the equilibrium position and negative values of $x(t)$ indicate that the mass is above the equilibrium position. The units that are encountered in these problems are summarized in Table 5.1.

EXAMPLE 5.1.1

Determine the spring constant of the spring with natural length 10 in. that is stretched to a distance of 13 in. by an object weighing 5 lb.

Solution

Because the object weighs 5 lb, $F = 5$ lb, and because displacement from the equilibrium position is $13 - 10 = 3$ in., $s = 3$ in. $\times$ 1 ft./12 in. $= 1/4$ ft. Therefore,

$$F = ks \quad \text{indicates that} \quad 5 = \frac{1}{4}k \quad \text{so} \quad k = 20.$$

Notice that the spring constant is given in the units lb/ft. because $F = 5$ lb and $s = 1/4$ ft.

TABLE 5.1 Units Encountered When Solving Spring-Mass Systems Problems

System	Force	Mass	Length	k (Spring Constant)	Time
English	pounds (lb)	slugs (lbs-s^2/ft.)	feet (ft.)	lb/ft.	seconds (s)
Metric	Newton (N)	kilograms (kg)	meters (m)	N/m	seconds (s)

EXAMPLE 5.1.2

An object of weight 16 lb stretches a spring 3 in. Determine the IVP that models this situation if (a) the object is released from a point 4 in. below the equilibrium position with an upward initial velocity of 2 ft./s; (b) the object is released from rest 6 in. above the equilibrium position; (c) the object is released from the equilibrium position with a downward initial velocity of 8 ft./s.

Solution

We first determine the differential equation that models the spring mass system (we use the same equation for all three parts of the problem). The information is given in English units, so we use $g = 32$ ft./s^2. We must convert all measurements given in inches to feet. The object stretches the spring 3 in., so $s = 3$ in. $\times 1$ ft./12 in. $= 1/4$ ft. Also, because the mass weighs 16 lb, $F = 16$. According to Hooke's law, $16 = k \times 1/4$, so $k = 64$ lb/ft. We then find the mass m of the object with $F = mg$ to find that $16 = m \times 32$ or $m = 1/2$ slug. The differential equation used to find the displacement of the object at time t is

$$\frac{1}{2}\frac{d^2x}{dt^2} + 64x = 0.$$

The initial conditions for parts (a), (b), and (c) are then found.

(a) Because 4 in. $\times 1$ ft./12 in. $= 1/3$ ft. and down is the positive direction, $x(0) = 1/3$. Notice, however, that the initial velocity is 2 ft./s in the upward (negative) direction. Hence, $x'(0) = -2$.

(b) A position 6 in. (or 1/2 ft.) above the equilibrium position (in the negative direction) corresponds to initial position $x(0) = -1/2$. Being released from rest indicates that $dx/dt(0) = 0$.

(c) Because the mass is released from the equilibrium position, $x(0) = 0$. Also, the initial velocity is 8 ft./s in the downward (positive) direction, so $x'(0) = 8$.

EXAMPLE 5.1.3

An object weighing 60 lb stretches a spring 6 in. Determine the function $x(t)$ that describes the displacement of the object if it is released from rest 12 in. below the equilibrium position.

Solution

First, the spring constant k is determined from the supplied information. By Hooke's law, $F = ks$, so we have $60 = k \times 0.5$. Therefore, $k = 120$ lb/ft. Next, the mass m of the object is determined using $F = mg$. In this case, $60 = m \times 32$, so $m = 15/8$ slugs. Because $k/m = 64$, and 12 in. is equivalent to 1 ft., the IVP that models the situation is $x'' + 64x = 0$, $x(0) = 1$, $x'(0) = 0$.

The characteristic equation that corresponds to the differential equation is $r^2 + 64 = 0$. It has solutions $r_{1,2} = \pm 8i$, so a general solution of the equation is $x = c_1 \cos 8t + c_2 \sin 8t$.

To find the values of c_1 and c_2 that satisfy the initial conditions, we calculate $x' = -8c_1 \sin 8t + 8c_2 \cos 8t$. Then $x(0) = c_1 = 1$ and $x'(0) = 8c_2 = 0$ so $c_1 = 1$, $c_2 = 0$, and $x = \cos 8t$.

Notice that $x = \cos 8t$ indicates that the spring-mass system never comes to rest once it is set

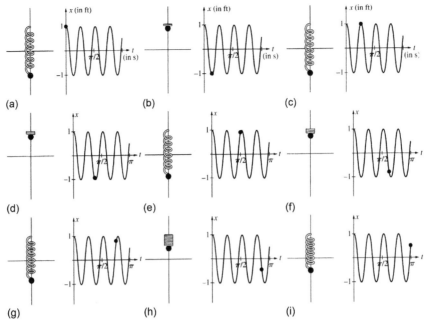

FIGURE 5.2   Simple harmonic motion.

into motion. The solution is periodic, so the mass moves vertically, retracing its motion, as shown in Figure 5.2. Motion of this type is called *simple harmonic motion*.

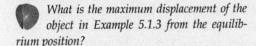
What is the maximum displacement of the object in Example 5.1.3 from the equilibrium position?

EXAMPLE 5.1.4

An objective weighing 2 lb stretches a spring 1.5 in. (a) Determine the function $x = x(t)$ that describes the displacement of the object if it is released with a downward initial velocity of 32 ft./s from 12 in. above the equilibrium position. (b) At what value of t does the object first pass through the equilibrium position?

Solution

(a) We begin by determining the spring constant. Because the force $F = 2$ lb stretches the spring $3/2$ in. × 1 ft./12 in. = $1/8$ ft., k is found by solving $2 = k \times 1/8$. Hence $k = 16$ lb/ft. With $F = mg$, we have $m = 2/32 = 1/16$ slug. The differential equation that models this situation is $\frac{1}{16}x'' + 16x = 0$ or $x'' + 256x = 0$. Because 12 in. is equivalent to 1 ft., the initial position above the equilibrium (in the negative direction) is $x(0) = -1$. The downward initial velocity (in the positive direction) is $x'(0) = 32$. Therefore, we must solve the IVP $x'' + 256x = 0$, $x(0) = -1$, $x'(0) = 32$. The characteristic equation associated with $x'' + 256x = 0$ is $r^2 + 256 = 0$ with roots $r_{1,2} = \pm 16i$ so a general solution of $x'' + 256x = 0$ is $x = c_1 \cos 16t + c_2 \sin 16t$ with derivative $x' = -16c_1 \sin 16t + 16c_2 \cos 16t$. Application of the initial conditions then yields $x(0) = c_1 \cos 0 + c_2 \sin 0 = c_1 = -1$ and $x'(0) = -16c_1 \sin 0 + 16c_2 \cos 0 = 16c_2 = 32$ so $c_2 = 2$. The position function is given by $x = -\cos 16t + 2\sin 16t$ (see Figure 5.3(a)).

(b) To determine when the object first passes through its equilibrium position, we solve the equation $x = -\cos 16t + 2\sin 16t = 0$ or $\tan 16t = 1/2$. Therefore, $t = \frac{1}{16}\tan^{-1}(1/2) \approx 0.03$ s, which appears to be a reasonable approximation based on the graph of $x = x(t)$.

Approximate the second time the mass considered in Example 5.1.4 passes through the equilibrium position.

A general formula for the solution of the IVP

$$m\frac{d^2x}{dt^2} + kx = 0, \quad x(0) = \alpha, \quad \frac{dx}{dt}(0) = \beta$$

is

$$x = \alpha\cos\omega t + \beta\omega^{-1}\sin\omega t, \quad \text{where } \omega = \sqrt{k/m}. \tag{5.1}$$

Through the use of the trigonometric identity $\cos(a+b) = \cos a\cos b - \sin a\sin b$, we can write $x = x(t)$ in terms of a cosine function with a phase shift. First, let $x = A\cos(\omega t - \phi)$. We then apply the identity to see that $x = A\cos\omega t\cos\phi + A\sin\omega t\sin\phi$. Comparing the functions

$$x = \alpha\cos\omega t + \beta\omega^{-1}\sin\omega t \quad \text{and}$$

$$x = A\cos\omega t\cos\phi + A\sin\omega t\sin\phi$$

indicates that

$$A\cos\phi = \alpha \quad \text{and} \quad A\sin\phi = \beta/\omega.$$

Thus $\cos\phi = \alpha/A$ and $\sin\phi = \beta/(A\omega)$. Because $\cos^2\phi + \sin^2\phi = 1$, $(\alpha/A)^2 + [\beta/(A\omega)]^2 = 1$. Therefore, the amplitude of the solution is $A = \sqrt{\alpha^2 + \beta^2/\omega^2}$ and $x = \sqrt{\alpha^2 + \beta^2/\omega^2}\cos(\omega t - \phi)$, where $\phi = \cos^{-1}\left(\frac{\alpha}{\sqrt{\alpha^2+\beta^2/\omega^2}}\right)$ and $\omega = \sqrt{k/m}$.

Note that the period of $x = x(t)$ is $T = 2\times\pi/\omega = 2\pi\sqrt{m/k}$. In many cases, questions about the displacement function are more easily answered if the solution is written in this form.

EXAMPLE 5.1.5

A 4-kg mass stretches a spring 0.392 m. (a) Determine the displacement function if the mass is released from 1 m below the equilibrium position with a downward initial velocity of 10 m/s. (b) What is the maximum displacement of the mass? (c) What is the approximate period of the displacement function?

How does an increase in the magnitude (absolute value) of the initial position and initial velocity affect the amplitude of the resulting motion of the spring-mass system? From experience, does this agree with the actual physical situation?

Solution

(a) Because the mass of the object (in metric units) is $m = 4$ kg, we use this with $F = mg$ to determine the force. We first compute $F = 4\times 9.8 = 39.2$ N. We then find the spring constant with $39.2 = k\times 0.392$, so $k = 100$ N/m. The differential equation that models this spring-mass system is $4x'' + 100x = 0$ or $x'' + 25x = 0$. The initial position is $x(0) = 1$, and the initial velocity is $x'(0) = 10$. We must solve the IVP $x'' + 25x = 0$, $x(0) = 1$, $x'(0) = 10$ either directly or with the general formula obtained in Equation (5.1). Using the general formula with $\alpha = 1$, $\beta = 10$, and $\omega = \sqrt{100/4} = 5$, we have

$$x = \sqrt{\alpha^2 + \beta^2/\omega^2}\cos(\omega t - \phi)$$

$$= \sqrt{1^2 + 10^2/5^2}\cos(5t - \phi)$$

$$= \sqrt{5}\cos(5t - \phi),$$

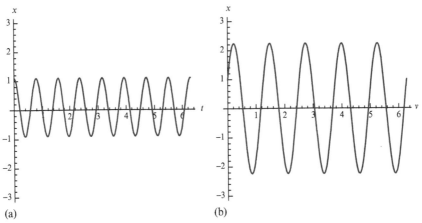

(a) (b)

FIGURE 5.3 Two plots of simple harmonic motion. (a) (on the left) is for Example 5.1.4 while (b) (on the right) is for Example 5.1.5.

where $\phi = \cos^{-1}\left(1/\sqrt{5}\right) \approx 1.11$ rad, which we graph in Figure 5.3(b).

(b) From our knowledge of trigonometric functions, we know that the maximum value of $x(t) = \sqrt{5}\cos(5t - \phi)$ is $x = \sqrt{5}$. Therefore, the maximum displacement of the mass from its equilibrium position is $\sqrt{5} \approx 2.236$ meters.

(c) The period of this trigonometric function is $T = 2\pi/\omega = 2\pi/5$. The mass returns to its initial position every $2\pi/5 \approx 1.257$ s.

 **EXERCISES 5.1**

In Exercises 1-4, determine the mass m and the spring constant k for the given spring-mass system,

$$m\frac{d^2x}{dt^2} + kx = 0, \quad x(0) = \alpha, \quad \frac{dx}{dt}(0) = \beta.$$

Interpret the initial conditions. (Assume the English system.)

1. $4x'' + 9x = 0, x(0) = -1, x'(0) = 0$
2. $9x'' + 4x = 0, x(0) = -1/2, x'(0) = 1$

3. $x'' + 64x = 0, x(0) = 3/4, x'(0) = -2$
4. $x'' + 100x = 0, x(0) = -1/4, x'(0) = 1$

In Exercises 5-10, express the solution of the IVP in the form $x = \sqrt{\alpha^2 + \beta^2/\omega^2}\cos(\omega t - \phi)$. What is the period and amplitude of the solution?

5. $x'' + x = 0, x(0) = 3, x'(0) = -4$
6. $x'' + 4x = 0, x(0) = 1, x'(0) = 1$
7. $x'' + 16x = 0, x(0) = -2, x'(0) = 1$
8. $x'' + 256x = 0, x(0) = 2, x'(0) = 4$
9. $x'' + 9x = 0, x(0) = 1/3, x'(0) = -1$
10. $10x'' + \frac{1}{10}x = 0, x(0) = -5, x'(0) = 1$
11. A 16-lb object stretches a spring 6 in. If the object is lowered 1 ft. below the equilibrium position and released, determine the displacement of the object. What is the maximum displacement of the object? When does it occur?
12. A 4-lb weight stretches a spring 1 ft. A 16-lb weight is then attached to the spring, and it comes to rest in its equilibrium position. If it is then put into motion with a downward velocity of 2 ft./s, determine the displacement of the mass. What is the maximum displacement of the object? When does it occur?

13. A 6-lb object stretches a spring 6 in. If the object is lifted 3 in. above the equilibrium position and released, determine the time required for the object to return to its equilibrium position. What is the displacement of the object at $t = 5\,\text{s}$? If the object is released from its equilibrium position with a downward initial velocity of $1\,\text{ft./s}$, determine the time required for the object to return to its equilibrium position.

14. A 16-lb weight stretches a spring 8 in. If the weight is lowered 4 in. below the equilibrium position and released, find the time required for the weight to return to the equilibrium position. What is the displacement of the weight at $t = 4\,\text{s}$? If the weight is released from its equilibrium position with an upward initial velocity of $2\,\text{ft./s}$, determine the time required for the weight to return to the equilibrium position.

15. Solve the IVP $x'' + kx = 0$, $x(0) = -1$, $x(0) = 0$ for values of $k = 1, 4,$ and 9. Comment on the effect that k has on the resulting motion.

16. Solve the IVP $x'' + 4m^{-1}x = 0$, $x(0) = -1$, $x'(0) = 0$ for values of $m = 1, 4,$ and 9. Comment on the effect that m has on the resulting motion.

17. Suppose that a 1-lb object stretches a spring $1/8\,\text{ft}$. The object is pulled downward and released from a position b feet beneath its equilibrium with an upward initial velocity of $1\,\text{ft./s}$. Determine the value of b so that the maximum displacement is 2 ft.

18. Suppose that a 70-g mass stretches a spring $5\,\text{cm}$ and that the mass is pulled downward and released from a position b units beneath its equilibrium with an upward initial velocity of $10\,\text{cm/s}$. Determine the value of b so that the mass first returns to its equilibrium at $t = 1\,\text{s}$ ($g \approx 980\,\text{cm/s}^2$).

19. The period of the motion of an undamped spring-mass system is $\pi/2$ seconds. Find the mass m if the spring constant is $k = 32\,\text{lb/ft}$.

20. Find the period of the motion of an undamped spring-mass system if the mass of the object is $4\,\text{kg}$ and the spring constant is $k = 0.25\,\text{N/m}$.

21. If the displacement, $x = x(t)$, of an object at time t satisfies the IVP $mx'' + kx = 0$, $x(0) = \alpha$, $x'(0) = \beta$, find the maximum velocity of the object.

22. Show that the solution to the IVP $mx'' + kx = 0$, $x(0) = \alpha$, $x'(0) = \beta$ can be written as $x(t) = u(t) + v(t)$, where u and v satisfy the same differential equation as x, u satisfies the initial conditions $u(0) = \alpha$, $u'(0) = 0$, and v satisfies the initial conditions $v(0) = 0$, $v'(0) = \beta$.

23. (Archimedes' principle) Suppose that an object of mass m is submerged (either partially or totally) in a liquid of density ρ. Archimedes' principle states that a body in liquid experiences a buoyant upward force equal to the weight of the liquid displaced by the body. The object is in equilibrium when the buoyant force of the displaced liquid equals the force of gravity on the object.

Consider the cylinder of radius r and height H of which h units of the height is submerged at equilibrium, as indicated in the following figure.

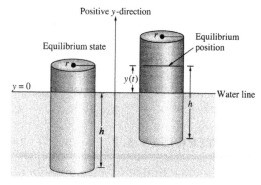

Positive y-direction

Equilibrium state

Equilibrium position

$y = 0$

$y(t)$

Water line

(a) Show that the weight of liquid displaced at equilibrium is $\pi r^2 h \rho$. Therefore, at equilibrium, $\pi r^2 h \rho = mg$.

(b) Let $y = y(t)$ represent the vertical displacement of the cylinder from equilibrium. Show that when the cylinder is raised out of the liquid, the downward force is $\pi r^2 (h - y)\rho$.

Use Newton's second law of motion to show that $my'' = \pi r^2 (h - y)\rho - mg$. Simplify this equation to obtain a second-order equation that models this situation.

24. Determine if the cylinder can float in a deep pool of water ($\rho \approx 62.5\,\text{lb/ft.}^2$) using the given radius r, height H, and weight W: (a) $r = 3\,\text{in.}, H = 12\,\text{in.}, W = 5\,\text{lb}$; (b) $r = 4\,\text{in.}$, $H = 8\,\text{in.}, W = 20\,\text{lb}$; (c) $r = 6\,\text{in.}, H = 9\,\text{in.}$, $W = 50\,\text{lb}$.

25. Determine the motion of the cylinder of weight $512\,\text{lb}$, radius $r = 1\,\text{ft.}$, and height $H = 4\,\text{ft.}$ if it is released with $3\,\text{ft.}$ of its height above the water ($\rho \approx 62.5\,\text{lb/ft}^2$) with a downward initial velocity of $3\,\text{ft./s}$. What is the maximum displacement of the cylinder from its equilibrium?

26. Consider the cylinder of radius $r = 3\,\text{in.}$, height $H = 12\,\text{in.}$, and weight $10\,\text{lb}$. Show that the portion of the cylinder submerged in water of density $\rho \approx 62.5\,\text{lb/ft}^2$ is $h \approx 0.815\,\text{ft} \approx 9.78\,\text{in.}$ Find the motion of the cylinder if it is released with $1.22\,\text{in.}$ of its height above the water with no initial velocity. (*Hint:* At $t = 0, h \approx 9.78\,\text{in.}$, so $0.22\,\text{in.}$ is above the water. Therefore, the initial position is $y(0) = 1$.)

27. An object with mass $m = 1$ slug is attached to a spring with spring constant $k = 4\,\text{lb/ft}$. (a) Determine the displacement function of the object if $x(0) = \alpha$ and $x'(0) = 0$. Graph the solution for $\alpha = 1, 4$, and -2. How does varying the value of α affect the solution? Does it change the values of t at which the mass passes through the equilibrium position? (b) Determine the displacement function of the object if $x(0) = 0$ and $x'(0) = \beta$. Graph the solution for $\beta = 1, 4$, and -2. How does varying the value of β affect the solution? Does it change the

values of t at which the mass passes through the equilibrium position?

28. An object of mass $m = 4$ slugs is attached to a spring with spring constant $k = 20\,\text{lb/ft}$. If the object is released from $7\,\text{in.}$ above its equilibrium with a downward initial velocity of $2.5\,\text{ft./s}$, find (a) the maximum displacement from the equilibrium position; (b) the time at which the object first passes through its equilibrium position; (c) the period of the motion.

29. If the spring in Problem 28 has the spring constant $k = 16\,\text{lb/ft}$, what is the maximum displacement from the equilibrium position? How does this compare to the result in the previous exercise? Determine the maximum displacement if $k = 24\,\text{lb/ft}$. Do these results agree with those that would be obtained with the general formula $A = \sqrt{\alpha^2 + \beta^2/\omega^2}$?

30. An object of mass $m = 3$ slugs is attached to a spring with spring constant $k = 15\,\text{lb/ft}$. If the object is released from $9\,\text{in.}$ below its equilibrium with a downward initial velocity of $1\,\text{ft./s}$, find (a) the maximum displacement from the equilibrium position; (b) the time at which the object first passes through its equilibriium position; (c) the period of the motion.

31. If the mass in Problem 30 is $m = 4$ slugs, find the maximum displacement from the equilibrium position. Compare this result to that obtained with $m = 3$ slugs.

5.2 DAMPED MOTION

Because the differential equation derived in Section 5.1 disregarded all retarding forces acting on the motion of the mass, a more realistic model is needed. Studies in mechanics reveal that the resistive force due to *damping* is a function of the velocity of the motion. For $c > 0$, functions such as

$$F_R = c\frac{dx}{dt}, \quad F_R = c\left(\frac{dx}{dt}\right)^3, \text{ and } F_R = c\,\text{sgn}\left(\frac{dx}{dt}\right),$$

$$\text{where sgn}\left(\frac{dx}{dt}\right) = \begin{cases} 1, & dx/dt > 0 \\ 0, & dx/dt = 0 \\ -1, & dx/dt < 0 \end{cases}$$

can be used to represent the damping force. We follow procedures similar to those used in Section 5.1 to model simple harmonic motion and to determine a differential equation that models the spring-mass system, which includes damping. Assuming that $F_R = c\,dx/dt$, after summing the forces acting on the spring-mass system, we have

$$m\frac{d^2x}{dt^2} = -c\frac{dx}{dt} - kx \quad \text{or} \quad m\frac{d^2x}{dt^2} + c\frac{dx}{dt} + kx = 0.$$

The displacement function is found by solving the IVP

$$m\frac{d^2x}{dt^2} + c\frac{dx}{dt} + kx = 0, \quad x(0) = \alpha, \quad \frac{dx}{dt}(0) = \beta. \tag{5.2}$$

From our experience with second-order ordinary differential equations (ODEs) with constant coefficients in Chapter 4, the solutions to IVPs of this type depend on the values of m, k, and c. Suppose we assume (as we did in Section 5.1) that solutions of the differential equation have the form $x = e^{rt}$ Then the characteristic equation is $mr^2 + cr + k = 0$ with solutions

$$r_{1,2} = \frac{1}{2m}\left(-c \pm \sqrt{c^2 - 4mk}\right).$$

The solution depends on the value of the quantity $c^2 - 4mk$. In fact, problems of this type are classified by the value of $c^2 - 4mk$ as follows.

Case 1: $c^2 - 4mk > 0$ This situation is said to be *overdamped*, because the damping coefficient c is large in comparison with the spring constant k.

Case 2: $c^2 - 4mk = 0$ This situation is said to be *critically damped* because the resulting

motion is oscillatory with a slight decrease in the damping coefficient c.

Case 3: $c^2 - 4mk < 0$ This situation is called *underdamped*, because the damping coefficient c is small in comparison with the spring constant k.

EXAMPLE 5.2.1

An 8 lb object is attached to a spring of length 4 ft. At equilibrium, the spring has length 6 ft. Determine the displacement function $x(t)$ if $F_R = 2\,dx/dt$ and (a) the object is released from the equilibrium position with a downward initial velocity of 1 ft./s; (b) the object is released 6 in. above the equilibrium position with an initial velocity of 5 ft./s in the downward direction.

Solution

Notice that $s = 6 - 4 = 2$ ft. and that $F = 8$ lb. We find the spring constant with $8 = k \times 2$, so $k = 4$ lb/ft. Also, the mass of the object is $m = 8/32 = 1/4$ slug. The differential equation that models this spring-mass system is

$$\frac{1}{4}\frac{d^2x}{dt^2} + 2\frac{dx}{dt} + 4x = 0 \quad \text{or} \quad \frac{d^2x}{dt^2} + 8\frac{dx}{dt} + 16x = 0.$$

The corresponding characteristic equation is

$$r^2 + 8r + 16 = (r+4)^2 = 0,$$

so $r_{1,2} = -4$ is a root of multiplicity two. A general solution is

$$x(t) = c_1 e^{-4t} + c_2 t e^{-4t}$$

Differentiating yields

$$x'(t) = (-4c_1 + c_2)e^{-4t} - 4c_2 t e^{-4t}.$$

(a) The initial conditions in this case are $x(0) = 0$ and $x'(0) = 1$, so $c_1 = 0$ and $-4c_1 + c_2 = 1$. Thus, $c_2 = 1$ and the solution is $x(t) = te^{-4t}$ which is shown in Figure 5.4(a). Notice that $x(t)$ is always positive, so the object is always below the equilibrium position and approaches zero (the equilibrium position) as

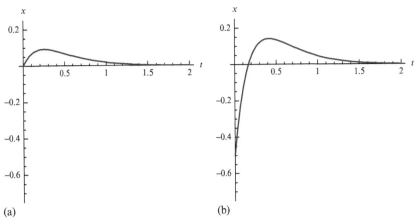

(a)   (b)

FIGURE 5.4   (a) On the left and (b) on the right illustrate critically damped motion.

t approaches infinity. Because of the resistive force due to damping, the object is not allowed to pass through its equilibrium position.

For Example 5.2.1(a), what is the maximum displacement from the equilibrium position? If you were looking at this spring, would you perceive its motion?

(b) In this case, $x(0) = -1/2$ and $x'(0) = 5$. When we apply these initial conditions, we find that $x(0) = c_1 = -1/2$ and $x'(0) = -4c_1 + c_2 = -4 \times -1/2 + c_2 = 5$. Hence $c_2 = 3$, and the solution is $x(t) = -\frac{1}{2}e^{-4t} + 3te^{-4t}$. This function, which is graphed in Figure 5.4(b), indicates the importance of the initial conditions on the resulting motion. In this case, the displacement is negative (above the equilibrium position) initially, but the positive initial velocity causes the function to become positive (below the equilibrium position) before approaching zero.

If the object in Example 5.2.1 is released from any point below its equilibrium position with an upward initial velocity of $1\,ft./s$, can it possibly pass through its equilibrium. If the object is released from below its equilibrium position with any upward initial velocity, can it possibly pass through the equilibrium position?

EXAMPLE 5.2.2

A 32 lb object stretches a spring 8 ft. If the resistive force due to damping is $F_R = 5\,dx/dt$, determine the displacement function if the object is released from 1 ft. below the equilibrium position with (a) an upward velocity of $1\,ft./s$; (b) an upward velocity of $6\,ft./s$.

Solution

(a) Because $F = 32$ lb, the spring constant is found with $32 = k \times 8$, so $k = 4$ lb/ft. Also, $m = 32/32 = 1$ slug. The differential equation that models this situation is

$4d^2x/dt^2 + 5\,dx/dt + 4x = 0$. The initial position is $x(0) = 1$, and the initial velocity in (a) is $x'(0) = 1$. The characteristic equation of the differential equation is $r^2 + 5r + 4 = (r+1)(r+4) = 0$ with roots $r_1 = -1$ and $r_2 = -4$. A general solution is

$$x(t) = c_1 e^{-t} + c_2 e^{-4t} \quad \text{and}$$

$$\frac{dx}{dt}(t) = -c_1 e^{-t} - 4c_2 e^{-4t}.$$

(Because $4c^2 - 4mk = 5^2 - 4 \times 1 \times 4 = 9 > 0$, the system is overdamped.) Application of the initial conditions yields the system of equations $c_1 + c_2 = 1$ and $-c_1 - 4c_2 = -1$ with solution $c_1 = 1$ and $c_2 = 0$, so the solution to the IVP is $x(t) = e^{-t}$. The graph of $x(t)$ is shown in Figure 5.5(a). Notice that this solution is always positive and, due to the damping, approaches zero as t approaches infinity. Therefore, the object is always below its equilibrium position.

(b) Using the initial velocity $x'(0) = -6$, we solve the system $c_1 + c_2 = 1$ and $-c_1 - 4c_2 = -6$, resulting in $c_1 = -2/3$ and $c_2 = 5/3$. The solution of the IVP is $x(t) = \frac{5}{3}e^{-4t} - \frac{2}{3}e^{-t}$. The graph of this function is shown in

Figure 5.5(b). As in Example 5.2.1, these results indicate the importance of the initial conditions on the resulting motion. In this case, the displacement is positive (below its equilibrium) initially, but the larger negative initial velocity causes the function to become negative (above its equilibrium) before approaching zero. Therefore, we see that the initial velocity in part (b) causes the object to pass through its equilibrium position.

Can you find a minimum value for the upward initial velocity v_0 in Example 5.2.2 so that the object passes through its equilibrium position for all values greater than v_0?

EXAMPLE 5.2.3

A 16 lb object stretches a spring 2 ft. Determine the displacement $x(t)$ if the resistive force due to damping is $F_R = 1/2\,dx/dt$ and the object is released from the equilibrium position with a downward velocity of 1 ft./s.

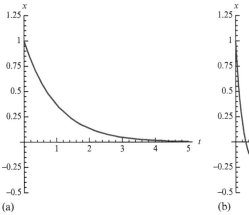

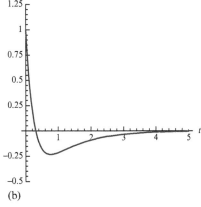

(a)   (b)

FIGURE 5.5 (a) On the left and (b) on the right illustrate overdamped motion.

Solution

Because $F = 16\,\text{lb}$, the spring constant is determined with $16 = k \times 2$. Hence, $k = 8\,\text{lb/ft}$. Also, $m = 16/32 = 1/2$ slug. Therefore, the differential equation is

$$\frac{1}{2}\frac{d^2x}{dt^2} + \frac{1}{2}\frac{dx}{dt} + 8x = 0 \quad \text{or} \quad \frac{d^2x}{dt^2} + \frac{dx}{dt} + 16x = 0.$$

The initial position is $x(0) = 0$ and the initial velocity is $x'(0) = 1$. We must solve the IVP

$$\frac{d^2x}{dt^2} + \frac{dx}{dt} + 16x = 0, \quad x(0) = 0, \quad \frac{dx}{dt}(0) = 1.$$

A general solution of $d^2x/dt^2 + dx/dt + 16x = 0$ is

$$x(t) = e^{-t/2}\left(c_1 \cos\left(\frac{3}{2}\sqrt{7}t\right) + c_2 \sin\left(\frac{3}{2}\sqrt{7}t\right)\right).$$

Notice that $c^2 - 4mk = (1/2)^2 - 4 \times 1/2 \times 8 = -63/4 < 0$, so the spring-mass system is underdamped. Because

$$x'(t) = -\frac{1}{2}e^{-t/2}\left(c_1 \cos\left(\frac{3}{2}\sqrt{7}t\right) + c_2 \sin\left(\frac{3}{2}\sqrt{7}t\right)\right)$$

$$+ \frac{3}{2}\sqrt{7}e^{-t/2}\left(-c_1 \sin\left(\frac{3}{2}\sqrt{7}t\right) + c_2 \cos\left(\frac{3}{2}\sqrt{7}t\right)\right),$$

application of the initial conditions yields $c_1 = 0$ and $c_2 = 2/(3\sqrt{7})$. Therefore, the solution is

$$x(t) = \frac{2}{3}\sqrt{7}e^{-t/2}\sin\left(\frac{3}{2}\sqrt{7}t\right).$$

Solutions of this type have several interesting properties. First, the trigonometric component of the solution causes the motion to *oscillate*. Also, the exponential portion forces the solution to approach zero as t approaches infinity These qualities are illustrated in the graph of $x(t)$ in Figure 5.6(a). Physically, the position of the mass in this case oscillates about the equilibrium position and eventually comes to rest in the equilibrium position. Of course, with our model the displacement function $x(t) \to 0$ as $t \to \infty$, but there are no values $t = T$ such that $x(t) = 0$ for $t > T$ as we might expect from the physical situation. Our model only approximates the behavior of the mass. Notice also that the solution is bounded above and below by the exponential term of the solution $e^{-t/2}$ and its reflection through the horizontal axis, $-e^{-t/2}$. This is illustrated with the simultaneous display of these functions in Figure 5.6(b); the motion of the spring is illustrated in Figure 5.7.

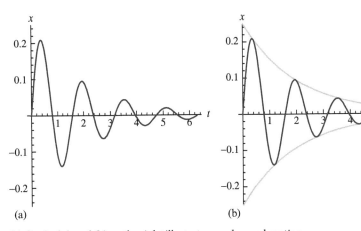

(a)   (b)

FIGURE 5.6   (a) On the left and (b) on the right illustrate overdamped motion.

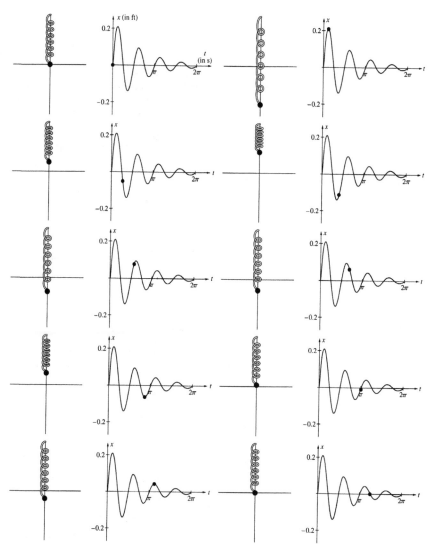

FIGURE 5.7 An underdamped spring.

Approximate the quasiperiod of the solution in Example 5.2.3.

Notice that when the system is underdamped as in Example 5.2.3, the amplitude (or *damped amplitude*) of the solution decreases as $t \to \infty$. The time interval between two successive local maxima (or minima) of $x(t)$ is called the *quasiperiod* of the solution.

In Section 5.1 we developed a general formula for the displacement function. We can do the same for systems that involve damping. Assuming that the spring-mass system is underdamped, the differential equation $m\,d^2x/dt^2 + c\,dx/dt + kx = 0$ has the characteristic equation $mr^2 + cr + k = 0$ with roots $r_{1,2} = \left(-c \pm i\sqrt{4mk - c^2}\right)/(2m)$. If we let $\rho = c/(2m)$ and $\mu = \sqrt{4mk - c^2}/(2m)$, then $r_{1,2} = -\rho \pm \mu i$ and a general solution is $x(t) = e^{-\rho t}(c_1 \cos \mu t + c_2 \sin \mu t)$. Applying the initial conditions $x(0) = \alpha$ and $x'(0) = \beta$ yields the solution $x(t) = e^{-\rho t}(\alpha \cos \mu t + (\beta + \alpha\rho)\mu^{-1} \sin \mu t)$, which can be written as

$$x(t) = Ae^{-\rho t} \cos(\mu t - \phi).$$

Then, $x(t) = e^{-\rho t}(A \cos \mu t \cos \phi + A \sin \mu t \sin \phi)$. Comparing the functions

$$x(t) = e^{-\rho t}\left(\alpha \cos \mu t + \frac{\beta + \alpha\rho}{\mu} \sin \mu t\right)$$

and

$$x(t) = e^{-\rho t}(A \cos \mu t \cos \phi + A \sin \mu t \sin \phi),$$

we have

$$A \cos \phi = \alpha \quad \text{and} \quad A \sin \phi = \frac{\beta + \alpha\rho}{\mu},$$

which indicates that

$$\cos \phi = \frac{\alpha}{A} \quad \text{and} \quad \sin \phi = \frac{\beta + \alpha\rho}{A\mu}.$$

Now, $\cos^2 \phi + \sin^2 \phi = 1$, so $(\alpha/A)^2 + [(\beta + \alpha\rho)/(A\mu)]^2 = 1$. Therefore, the decreasing amplitude of the solution is

$$A = \sqrt{\alpha^2 + \left(\frac{\beta + \alpha\rho}{\mu}\right)^2}$$

and

$$x(t) = e^{-\rho t}\sqrt{\alpha^2 + \left(\frac{\beta + \alpha\rho}{\mu}\right)^2} \cos(\mu t - \phi), \quad (5.3)$$

where $\phi = \cos^{-1}\left(\alpha/\sqrt{\alpha^2 + [(\beta + \alpha\rho)/\mu]^2}\right)$. The quantity

$$\frac{2\pi}{\mu} = \frac{4\pi m}{\sqrt{4km - c^2}}$$

is called the *quasiperiod* of the function. (Note that functions of this type are *not* periodic.) We can also determine the times at which the mass passes through the equilibrium position from the general formula given here. We do this by setting the argument equation to odd multiples of $\pi/2$, because the cosine function is zero at these values. The mass passes through the equilibrium position at

$$t = \frac{(2n + 1)\dfrac{\pi}{2} + \phi}{\mu} = \frac{m(2n + 1)\pi + 2m\phi}{\sqrt{4mk - c^2}},$$

$$n = 0, \pm 1, \pm 2, \ldots$$

 Use this formula to find the quasiperiod of the solution in Example 5.2.3.

EXAMPLE 5.2.4

An object of mass 1 slug is attached to a spring with spring constant $k = 13\,\text{lb/ft.}$ and is subject to a resistive force of $F_R = 4\,dx/dt$ due to damping. If the initial position is $x(0) = 1$ and the initial velocity is $x'(0) = 1$, determine the quasiperiod of the solution and find the values of t at which the object passes through the equilibrium position.

Solution

The IVP that models this situation is

$$\frac{d^2x}{dt^2} + 4\frac{dx}{dt} + 13x = 0, \quad x(0) = 1, \quad \frac{dx}{dt}(0) = 1.$$

The characteristic equation is $r^2 + 4r + 13 = 0$ with roots $r_{1,2} = -2 \pm 3i$. A general solution is $x(t) = e^{-2t}(c_1 \cos 3t + c_2 \sin 3t)$ with derivative

$$\frac{dx}{dt}(t) = -2e^{-2t}(c_1 \cos 3t + c_2 \sin 3t)$$
$$+ 3e^{-2t}(-c_1 \sin 3t + c_2 \cos 3t)$$

Application of the initial conditions yields the solution $x(t) = e^{-2t}(\cos 3t + \sin 3t)$, which can be written as

$$x(t) = e^{-2t}\sqrt{\alpha^2 + \left(\frac{\beta + \alpha\rho}{\mu}\right)^2}\cos(\mu t - \phi)$$

$$= e^{-2t}\sqrt{1^2 + \left(\frac{1+2}{3}\right)^2}\cos(3t - \phi)$$

$$= \sqrt{2}e^{-2t}\cos(3t - \phi).$$

The quasiperiod is

$$\frac{2\pi}{\mu} = \frac{4\pi m}{\sqrt{4km - c^2}} = \frac{4\pi}{\sqrt{4 \times 13 - 4^2}} = \frac{2\pi}{3}.$$

The values of t at which the object passes through the equilibrium are

$$t = \frac{m(2n+1)\pi + 2m\phi}{\sqrt{4mk - c^2}} = \frac{1}{6}[(2n+1)\pi + 2\phi],$$

$$n = 0, 1, 2, \ldots$$

where $\phi = \cos^{-1}\left(1/\sqrt{2}\right) = \pi/4$ (see Figure 5.8).

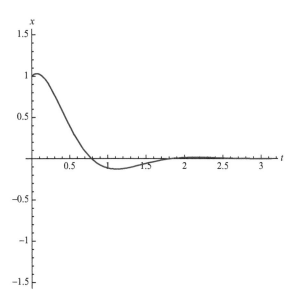

FIGURE 5.8   Plot of $x(t) = \sqrt{2}e^{-2t}\cos(3t - \phi)$, $\phi = \pi/4$.

EXERCISES 5.2

In Exercises 1-4, determine the mass m (slugs), spring constant k (lb./ft.), and damping coefficient c in $F_R = c\,dx/dt$ for the given spring-mass system $m\,d^2x/dt^2 + c\,dx/dt + kx = 0$, $x(0) = \alpha$, $dx/dt\,(0) = \beta$. Describe the initial conditions. (Assume the English system.)

1. $d^2x/dt^2 + 4\,dx/dt + 3x = 0$, $x(0) = 0$,
   $x'(0) = -4$
2. $1/32\,d^2x/dt^2 + 2\,dx/dt + x = 0$, $x(0) = 1$,
   $x'(0) = 0$
3. $1/4\,d^2x/dt^2 + 2\,dx/dt + x = 0$, $x(0) = -1/2$,
   $x'(0) = 1$
4. $4d^2x/dt^2 + 2/,dx/dt + 8x = 0$, $x(0) = 0$,
   $x'(0) = 2$

In Exercises 5-8, express the solution of the IVP in the form given by Equation (5.3). In each case, find the quasiperiod and the time at which the mass first passes through its equilibrium position.

5. $d^2x/dt^2 + 4\,dx/dt + 13x = 0$, $x(0) = 1$,
   $x'(0) = -1$
6. $d^2x/dt^2 + 4\,dx/dt + 20x = 0$, $x(0) = 1$,
   $x'(0) = 2$
7. $d^2x/dt^2 + 2\,dx/dt + 26x = 0$, $x(0) = 1$,
   $x'(0) = 1$
8. $d^2x/dt^2 + 10\,dx/dt + 41x = 0$, $x(0) = 3$,
   $x'(0) = -2$

In Exercises 9-16, solve the IVP. Classify each as overdamped or critically damped. Determine if the mass passes through its equilibrium position and, if so, when. Determine the maximum displacement of the object from the equilibrium position.

9. $d^2x/dt^2 + 8\,dx/dt + 15x = 0$, $x(0) = 0$,
   $x'(0) = 1$
10. $d^2x/dt^2 + 7\,dx/dt + 12x = 0$, $x(0) = -1$,
    $x'(0) = 4$
11. $2d^2x/dt^2 + 3\,dx/dt + x = 0$, $x(0) = -1$,
    $x'(0) = 2$

12. $d^2x/dt^2 + 5\,dx/dt + 4x = 0$, $x(0) = 0$,
    $x'(0) = 5$

13. $d^2x/dt^2 + 8\,dx/dt + 16x = 0$, $x(0) = 4$,
    $x'(0) = -2$

14. $d^2x/dt^2 + 6\,dx/dt + 9x = 0$, $x(0) = 3$,
    $x'(0) = -3$

15. $d^2x/dt^2 + 10\,dx/dt + 25x = 0$, $x(0) = -5$,
    $x'(0) = 1$

16. $2d^2x/dt^2 + 4\,dx/dt + x = 0$, $x(0) = -1$,
    $x'(0) = 2$

17. Suppose that an object with $m = 1$ is attached to the end of a spring with spring constant $k = 1$. After reaching its equilibrium position, the object is pulled one unit above the equilibrium and released with an initial velocity v_0. If the spring-mass system is critically damped, what is the value of c?

18. A weight having mass $m = 1$ is attached to the end of a spring with $k = 5/4$ and $F_R = 2\,dx/dt$. Determine the displacement of the object if the object is released from the equilibrium position with an initial velocity of 3 units/s in the downward direction.

19. A 32 lb weight is attached to the end of a spring with spring constant $k = 24$ lb/ft. If the resistive force is $F_R = 10\,dx/dt$, determine the displacement of the mass if it is released with no initial velocity from a position 6 in. above the equilibrium position. Determine if the mass passes through its equilibrium position and, if so, when. Determine the maximum displacement of the object from the equilibrium position.

20. An object weighing 8 lb stretches a spring 6 in. beyond its natural length. If the resistive force is $F_R = 4\,dx/dt$, find the displacement of the object if it is set into motion from its equilibrium position with an initial velocity of 1 ft./s in the downward direction.

21. An object of mass $m = 70$ kg is attached to the end of a spring and stretches the spring 0.25 m beyond its natural length. If the

restrictive force is $F_R = 280\,dx/dt$, find the displacement of the object if it is released from a position 3 m above its equilibrium position with no initial velocity. Does the object pass through its equilibrium position at any time?

22. Suppose that an object of mass $m = 1$ slug is attached to a spring with spring constant $k = 25$ lb/ft. If the resistive force is $F_R = 6\,dx/dt$, determine the displacement of the object if it is set into motion from its equilibrium position with an upward velocity of 2 ft./s. What is the quasiperiod of the motion?

23. An object of mass $m = 4$ slugs is attached to a spring with spring constant $k = 64$ lb/ft. If the resistive force is $F_R = c\,dx/dt$, find the value of c so that the motion is critically damped. For what values of c is the motion underdamped?

24. An object of mass $m = 2$ slugs is attached to a spring with spring constant k lb/ft. If the resistive force is $F_R = 8\,dx/dt$, find the value of k so that the motion is critically damped. For what values of k is the motion underdamped? For what values of k is the motion overdamped?

25. If the quasiperiod of the underdamped motion is $\pi/6$ seconds when a $1/13$ slug mass is attached to a spring with spring constant $k = 13$ lb/ft, find the damping constant c.

26. If a mass of 0.2 kg is attached to a spring with spring constant $k = 5$ N/m that undergoes damping equivalent to $6/5\,dx/dt$, find the quasiperiod of the resulting motion.

27. Show that the solution $x(t)$ of the IVP (5.2) can be written as $x(t) = u(t) + v(t)$, where u and v satisfy the same differential equation as x, u satisfies the initial conditions $u(0) = \alpha$, $u'(0) = 0$, and v satisfies the initial conditions $v(0) = 0$, $v'(0) = \beta$.

28. If the spring-mass system $m\,d^2x/dt^2 + c\,dx/dt + kx = 0$ is either

critically damped or overdamppd, show that the mass can pass through its equilibrium position at most one time (independent of the initial conditions).

29. Suppose that the spring-mass system (5.2) is critically damped. If $\beta = 0$, show that $\lim_{t\to\infty} x(t) = 0$, but that there is no value $t = t_0$ such that $x(t_0) = 0$. (The mass approaches but never reaches its equilibrium position.)

30. Suppose that the spring-mass system (5.2) is critically damped. If $\beta > 0$, find a condition on β so that the mass passes through its equilibrium position after it is released.

31. In the case of underdamped motion, show that the amount of time between two successive times at which the mass passes through its equilibrium is one-half of the quasiperiod.

32. In the case of underdamped motion, show that the amount of time between two successive maxima of the position function is $4\pi m/\sqrt{4km - c^2}$.

33. In the case of underdamped motion, show that the ratio between two consecutive maxima (or minima) is $e^{2c\pi/\sqrt{4mk-c^2}}$.

34. The natural logarithm of the ratio in Exercise 33, called the *logarithmic decrement*, is $d = 2c\pi/\sqrt{4mk - c^2}$ and indicates the rate at which the motion dies out because of damping. Notice that because m, k, and d are all quantities that can be measured in the spring-mass system, the value of d is useful in measuring the damping constant c. Compute the logarithmic decrement of the system in (a) Example 5.2.3 and (b) Example 5.2.4.

35. Determine how the value of c affects the solution of the IVP
$d^2x/dt^2 + c\,dx/dt + 6x = 0$, $x(0) = 0$,
$dx/dt(0) = 1$, where $c = 2\sqrt{6}, 4\sqrt{6}$, and $\sqrt{6}$.

36. In Problem 35, consider the solution that results using the damping coefficient that produces critical damping with $x'(0) = -1$, 0, 1, and 2. In which case does the object

pass through its equilibrium position? When does it pass through the equilibrium position?

37. In addition to using the values of $x'(0)$ and c in Problem 36, suppose that the equilibrium position is 1 unit above the floor. Does the object come into contact with the floor in any of the cases? If so, when?

38. Solve the IVP $4x'' + cx' + 5x = 0$, $x(0) = 3$, $x'(0) = 0$ using $c = 1, -4$, and 0. Plot the solutions and compare them.

39. Solve the IVP $x'' + cx' + x = 0$, $x(0) = -1$, $x'(0) = 3$ for values of $c = 2$ and $c = \sqrt{8}$. (Notice the effect of the coefficient of damping on the solution to the corresponding critically damped and overdamped equation.)

5.3 FORCED MOTION

In some cases, the motion of a spring is influenced by an external driving force, $f(t)$. Mathematically, this force is included in the differential equation that models the situation by

$$m\frac{d^2x}{dt^2} = -kx - c\frac{dx}{dt} + f(t).$$

The resulting IVP is

$$m\frac{d^2x}{dt^2} + c\frac{dx}{dt} + kx = f(t), \quad x(0) = \alpha, \quad \frac{dx}{dt}(0) = \beta.$$
$$(5.4)$$

Therefore, differential equations modeling forced motion are nonhomogeneous and require the method of undetermined coefficients or variation of parameters for solution. We first consider forced motion that is undamped.

EXAMPLE 5.3.1

An object of mass $m = 1$ slug is attached to a spring with spring constant $k = 4$ lb/ft. Assuming there is no damping and that the object begins from rest in the equilibrium position, determine the position function of the object if it is subjected to an external force of (a) $f(t) = 0$; (b) $f(t) = 1$; (c) $f(t) = \cos t$.

Solution

First, we note that we must solve the IVP

$$\frac{d^2x}{dt^2} + 4x = f(t), \quad x(0) = 0, \quad \frac{dx}{dt}(0) = 0.$$

for each of the forcing functions in (a), (b), and (c). A general solution of the corresponding homogeneous equation $d^2x/dt^2 + 4x = 0$ is $x_h(t) = c_1 \cos 2t + c_2 \sin 2t$.

(a) With $f(t) = 0$, the equation is homogeneous. so we apply the initial condition, to $x(t) = c_1 \cos 2t + c_2 \sin 2t$ with derivative $x'(t) = -2c_1 \sin 2t + 2c_2 \cos 2t$. Because $x(0) = c_1 = 0$ and $x'(0) = 2c_2 = 0$, $c_1 = c_2 = 0$, the solution is $x(t) = 0$. This solution indicates that the object does not move from the equilibrium position because there is no forcing function, no initial displacement from the equilibrium position, and no initial velocity,

(b) Using the method of undetermined coefficients with a particular solution of the form $x_p(t) = A$, substitution into the differential equation $d^2x/dt^2 + 4x = 1$ yields $4A = 1$, so $A = 1/4$. Hence,

$$x(t) = x_h(t) + x_p(t) = c_1 \cos 2t + c_2 \sin 2t + \frac{1}{4}$$

with derivative $x'(t) = -2c_1 \sin 2t + 2c_2 \cos 2t$. With $x(0) = c_1 + 1/4 = 0$ and $x'(0) = 2c_2 = 0$, we have $c_1 = -1/4$ and $c_2 = 0$ so

$$x(t) = -\frac{1}{4}\cos 2t + \frac{1}{4}.$$

Notice from the graph of this function in Figure 5.9(a) that the object never moves above the equilibrium position. (Positive values of x indicate that the mass is below the equilibrium position.)

(c) In this case, we assume that $x_p(t) = A \cos t + B \sin t$. Substitution into $d^2x/dt^2 + 4x = \cos t$ yields $3A \cos t + 3B \sin t = \cos t$, so $A = 1/3$ and $B = 0$. Therefore,

$$x(t) = x_h(t) + x_p(t) = c_1 \cos 2t$$
$$+ c_2 \sin 2t + \frac{1}{3}\cos t$$

with derivative

$$x'(t) = -2c_1 \sin 2t + 2c_2 \cos 2t - \frac{1}{3}\sin t.$$

Applying the initial conditions then gives us $x(0) = c_1 + 1/3 = 0$ and $x'(0) = 2c_2 = 0$, so $c_1 = -1/3$ and $c_2 = 0$. Thus,

$$x(t) = -\frac{1}{3}\cos 2t + \frac{1}{3}\cos t.$$

 How does changing the initial position to $x(0) = 1$ affect the solution in (c)? How does changing the initial velocity to $x'(0) = 1$ affect the solution? Is the maximum displacement affected in either case?

The graph of one period of this solution is shown in Figure 5.9(b). In this case, the mass passes through the equilibrium position twice (near $t = 2$ and $t = 4$) over the period, and it returns to the equilibrium position without passing through it at $t = 2\pi$. *(Can you predict other values of t where this occurs?)*

When we studied nonhomogeneous equations, we considered equations in which the nonhomogeneous (right-hand side) function was a solution of the corresponding homogeneous equation. We consider this type of situation with the IVP

$$\frac{d^2x}{dt^2} + \omega^2 x = F_1 \cos \omega t + F_2 \sin \omega t + G(t),$$

$$x(0) = \alpha, \quad x'(0) = \beta, \qquad (5.5)$$

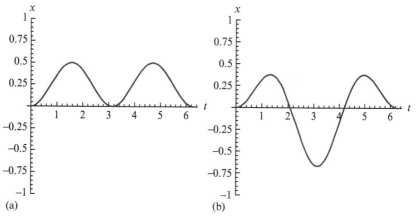

FIGURE 5.9   (a) Graph of $x(t) = \frac{1}{4}\cos t + \frac{1}{4}$. (b) Graph of $x(t) = -\frac{1}{3}\cos 2t + \frac{1}{3}\cos t$.

where F_1 and F_2 are constants and $G = G(t)$ is any function of t. (Note that one of the constants F_1 and F_2 can equal zero and G can be identically the zero function.) In this case, we say that ω is the *natural frequency of the system* because a general solution of the corresponding homogeneous equation is $x_h(t) = c_1 \cos \omega t + c_2 \sin \omega t$. In the case of this IVP, the *forced frequency*, the frequency of the trigonometric functions in $F_1 \cos \omega t + F_2 \sin \omega t + G(t)$, equals the natural frequency.

EXAMPLE 5.3.2

Investigate the effect that the forcing function $f(t) = \cos 2t$ has on the solution of the IVP

$$\frac{d^2x}{dt^2} + 4x = f(t), \quad x(0) = 0, \quad x'(0) = 0.$$

Solution

As we saw in Example 5.3.1, $x_h(t) = c_1 \cos 2t + c_2 \sin 2t$. Because $f(t) = \cos 2t$ is contained in this solution, we assume that $x_p(t) = At \cos 2t + Bt \sin 2t$, with first and second derivatives

$$x_p'(t) = A \cos 2t - 2At \sin 2t + B \sin 2t + 2Bt \cos 2t$$

and

$$x_p''(t) = -4A \sin 2t - 4At \cos 2t$$
$$+ 4B \cos 2t - 4Bt \sin 2t.$$

Substitution into $d^2x/dt^2 + 4x = \cos 2t$ yields $-4A \sin 2t + 4B \cos 2t = \cos 2t$. Thus $A = 0$ and $B = 1/4$, so $x_p(t) = \frac{1}{4}t \sin 2t$, and

$$x(t) = x_h(t) + x_p(t) = c_1 \cos 2t + c_2 \sin 2t + \frac{1}{4}t \sin 2t.$$

This function has derivative

$$x'(t) = -2c_1 \sin 2t + 2c_2 \cos 2t + \frac{1}{4}\sin 2t + \frac{1}{2}t \cos 2t,$$

so application of the initial conditions yields $x(0) = c_1 = 0$ and $x'(0) = 2c_2 = 0$. Therefore, $c_1 = c_2 = 0$, so

$$x(t) = \frac{1}{4}t \sin 2t.$$

The graph of this solution is shown in Figure 5.10. Notice that the amplitude increases as t increases. This indicates that the spring-mass system will encounter a serious problem; either the spring will break or the mass will eventually hit its support (like a ceiling or beam) or a lower boundary (like the ground or floor).

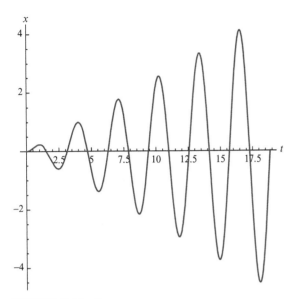

FIGURE 5.10   Resonance.

The phenomenon illustrated in Example 5.3.2 is called *resonance* and can be extended to other situations such as vibrations in an aircraft wing, a skyscraper, a glass, or a bridge. Some of the sources of excitation that lead to the vibration of these structures include unbalanced rotating devices, vortex shedding, strong winds, rough surfaces, and moving vehicles so we see that engineers must overcome many problems that are caused when structures and machines are subjected to forced vibrations.

> *Over a sufficient amount of time, do changes in the initial conditions affect the motion of a spring-mass system? Experiment by changing the initial conditions in the IVP in Example 5.3.2.*

Let us investigate in detail IVPs of the form

$$\frac{d^2x}{dt^2} + \omega^2 x = F\cos\beta t, \quad \omega \neq \beta, \quad x(0) = 0,$$

$$\frac{dx}{dt}(0) = 0. \tag{5.6}$$

A general solution of the corresponding homogeneous equation is $x_h(t) = c_1\cos\omega t + c_2\sin\omega t$. Using the method of undetermined coefficients, a particular solution is given by $x_p(t) = A\cos\beta t + B\sin\beta t$. The corresponding derivatives of this solution are

$$x_p'(t) = -A\beta\sin\beta t + B\beta\cos\beta t \quad \text{and}$$

$$x_p''(t) = -A\beta^2\cos\beta t - B\beta^2\sin\beta t.$$

Substituting into the nonhomogeneous equation $d^2x/dt^2 + \omega^2 x = F\cos\beta t$ and equating the corresponding coefficients yields

$$A = \frac{F}{\omega^2 - \beta^2} \quad \text{and} \quad B = 0.$$

Therefore, a general solution is

$$x(t) = x_h(t) + x_p(t) = c_1\cos\omega t + c_2\sin\omega t$$
$$+ \frac{F}{\omega^2 - \beta^2}\cos\beta t.$$

Application of the initial conditions yields the solution of IVP (5.6):

$$x(t) = \frac{F}{\omega^2 - \beta^2}(\cos\beta t - \cos\omega t).$$

Using the trigonometric identity $\frac{1}{2}[\cos(A - B) - \cos(A + B)] = \sin A \sin B$, we have

$$x(t) = \frac{2F}{\omega^2 - \beta^2}\sin\left(\frac{\omega + \beta}{2}t\right)\sin\left(\frac{\omega - \beta}{2}t\right).$$

Notice that the solution can be represented as

$$x(t) = A(t)\sin\left(\frac{\omega + \beta}{2}t\right), \quad \text{where}$$

$$A(t) = \frac{2F}{\omega^2 - \beta^2}\sin\left(\frac{\omega - \beta}{2}t\right).$$

Therefore, when the quantity $\omega - \beta$ is small, $\omega + \beta$ is relatively large in comparison. The function $\sin(\omega + \beta)t/2$ oscillates quite frequently because it has period $4\pi/(\omega+\beta)$. Meanwhile, the function $\sin(\omega-\beta)t/2$ oscillates relatively slowly because it has period $4\pi/(\omega-\beta)$. When we graph $x(t)$, we see that the functions $\pm\frac{2F}{\omega^2-\beta^2}\sin\left(\frac{\omega-\beta}{2}t\right)$ form an *envelope* for the solution.

EXAMPLE 5.3.3

 Solve the IVP

$$\frac{d^2x}{dt^2} + 4x = f(t), \quad x(0) = 0, \quad \frac{dx}{dt}(0) = 0,$$

with (a) $f(t) = \cos 3t$ and (b) $f(t) = \cos 5t$.

Solution

(a) A general solution of the corresponding homogeneous equation is $x_h(t) = c_1 \cos 2t + c_2 \sin 2t$. By the method of undetermined coefficients, we assume that $x_p(t) = A \cos 3t + B \sin 3t$. Substitution into $x'' + 4x = \cos 3t$ yields $-5A \cos 3t - 5B \sin 3t = \cos 3t$, so $A = -1/5$ and $B = 0$. Thus $x_p(t) = -\frac{1}{5} \cos 3t$ and

$$x(t) = x_t(t) + x_p(t) = c_1 \cos 2t$$
$$+ c_2 \sin 2t - \frac{1}{5} \cos 3t.$$

Because the derivative of this function is

$$x'(t) = -2c_1 \sin 2t + 2c_2 \cos 2t + \frac{3}{5} \sin 3t,$$

application of the initial conditions yields $x(0) = c_1 - 1/5 = 0$, and $x'(0) = 2c_2 = 0$. Therefore, $c_1 = 1/5$ and $c_2 = 0$, so

$$x(t) = \frac{1}{5} \cos 2t - \frac{1}{5} \cos 3t.$$

The graph of this function is shown in Figure 5.11(a) along with the envelope functions $x = \pm \frac{2}{5} \sin \left(\frac{1}{2} t \right)$.

(b) In a similar manner, the solution of the IVP if $f(t) = \cos 5t$ is found to be

$$x(t) = \frac{1}{21} \cos 2t - \frac{1}{21} \cos 5t.$$

The graph of the solution is shown in Figure 5.11(b) along with the envelope functions $x = \pm \frac{2}{21} \sin \left(\frac{3}{2} t \right)$.

Some computer algebra systems contain commands that allow you to "play" functions. If this is possible with your technology, see Exercises 27 and 28.

Oscillations like those illustrated in Example 5.3.3 are called *beats* because of the periodic variation of amplitude. This phenomenon is commonly encountered when two musicians try to

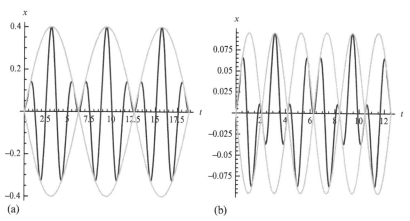

(a) (b)

FIGURE 5.11 (a) On the left and (b) on the right. Envelope functions help us see oscillations that are called *beats* because of the periodic variation of amplitude. The plot of the solution is darkest; the plots of the envelope solutions are lighter.

simultaneously tune their instruments or when two tuning forks with almost equivalent frequencies are played at the same time.

EXAMPLE 5.3.4

Investigate the effect that the forcing function $f(t) = e^{-t} \cos 2t$ has on the IVP

$$\frac{d^2x}{dt^2} + 4x = f(t), \quad x(0) = 0, \quad \frac{dx}{dt}(0) = 0,$$

Consider the problem $x'' + 4x = \cos \beta t$ for $\beta = 6, 8,$ and 10. What happens to the amplitude of the beats as β increases?

Solution

Using the method of undetermined coefficients, a general solution of the equation is

$$x(t) = x_h(t) + x_p(t) = c_1 \cos 2t + c_2 \sin 2t$$
$$+ \frac{1}{17}e^{-2t}(\cos 2t - 4\sin 2t).$$

Applying the initial conditions with $x(t)$ and

$$x'(t) = -2c_1 \sin 2t + 2c_2 \cos 2t$$
$$+ \frac{1}{17}e^{-2t}(-9\cos 2t + 2\sin 2t)$$

we have $x(0) = c_1 + 1/17 = 0$ and $x'(0) = 2c_2 - 9/17 = 0$. Therefore, $c_1 = -1/17$ and $c_2 = 9/34$ so

$$x(t) = -\frac{1}{17}\cos 2t + \frac{9}{34}\sin 2t$$
$$+ \frac{1}{17}e^{-2t}(\cos 2t - 4\sin 2t),$$

which is graphed in Figure 5.12. Notice that the effect of terms involving the exponential function diminishes as t increases. In this case, the forcing function $f(t) = e^{-t} \cos 2t$ approaches zero as t increases. Over time, the solution of the nonhomogeneous problem approaches that of the corresponding homogeneous problem, so we observe simple harmonic motion as $t \to \infty$.

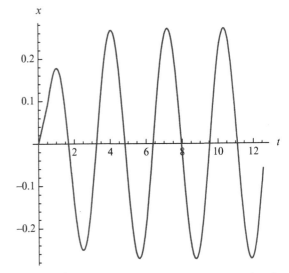

FIGURE 5.12   As t increases, the solution approaches that of the corresponding homogeneous problem.

We now consider spring problems that involve forces due to damping as well as external forces. In particular, consider the IVP

$$\frac{d^2x}{dt^2} + c\frac{dx}{dt} + kx = \rho \cos \lambda t, \quad x(0) = \alpha, \quad \frac{dx}{dt}(0) = \beta,$$
$$(5.7)$$

which has a solution of the form $x(t) = h(t) + s(t)$, where $\lim_{t\to\infty} h(t) = 0$ and $s(t) = c_1 \cos \lambda t + c_2 \sin \lambda t$.

The function $h(t)$ is called the *transient solution*, and $s(t)$ is known as the *steady-state solution*. Therefore, as t approaches infinity, the solution $x(t)$ approaches the steady-state solution. (Why?) Note that the steady-state solution corresponds to a particular solution obtained through the method of undetermined coefficients or variation of parameters.

EXAMPLE 5.3.5

Solve the IVP

$$\frac{d^2x}{dt^2} + 4\frac{dx}{dt} + 13x = \cos t, \quad x(0) = 0, \quad \frac{dx}{dt}(0) = 1$$

that models the motion of an object of mass $m = 1$ slug attached to a spring with spring constant

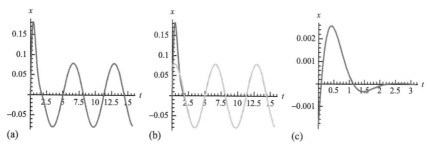

(a)   (b)   (c)

FIGURE 5.13   The transient solution (in (c)) approaches 0 as $t \to \infty$.

$k = 13$ lb/ft. that is subjected to a resistive force of $F_R = 4\,dx/dt$ and an external force of $f(t) = \cos t$. Identify the transient and steady-state solutions.

Solution

A general solution of $d^2x/dt^2 + 4\,dx/dt + 13x = 0$ is $x_h(t) = e^{-2t}(c_1 \cos 3t + c_2 \sin 3t)$. We assume that a particular solution has the form $x_p(t) = A \cos t + B \sin t$, with derivatives $x'_p(t) = -A \sin t + B \cos t$ and $x''_p(t) = -A \cos t - B \sin t$. After substitution into $d^2x/dt^2 + 4\,dx/dt + 13x = \cos t$, we see that the system $12A + 4B = 1$ and $-4A + 12B = 0$ must be satisfied. Hence $A = 3/40$ and $B = 1/40$, so

$$x_p(t) = \frac{3}{40}\cos t + \frac{1}{40}\sin t.$$

Therefore,

$$x(t) = x_h(t) + x_p(t) = e^{-2t}(c_1 \cos 3t + c_2 \sin 3t)$$
$$+ \frac{3}{40}\cos t + \frac{1}{40}\sin t,$$

with derivative

$$x'(t) = -2e^{-2t}(c_1 \cos 3t + c_2 \sin 3t)$$
$$+ 3e^{-2t}(-c_1 \sin 3t + c_2 \cos 3t)$$
$$- \frac{3}{40}\sin t + \frac{1}{40}\cos t.$$

Application of the initial conditions yields $x(0) = c_1 + 3/40 = 0$ and $x'(0) = -2c_1 + 3c_2 + 1/40 = 1$. Therefore, $c_1 = -3/40$ and $c_2 = 11/40$, so

$$x(t) = e^{-2t}\left(-\frac{3}{40}\cos 3t + \frac{11}{40}\sin 3t\right)$$
$$+ \frac{3}{40}\cos t + \frac{1}{40}\sin t.$$

Over a sufficient amount of time, do changes in the initial conditions in Example 5.3.5 affect the motion of the spring-mass system? Experiment by changing the initial conditions in the IVP in Example 5.3.5.

This indicates that the transient solution is $e^{-2t}\left(-\frac{3}{40}\cos 3t + \frac{11}{40}\sin 3t\right)$ and the steady-state solution is $\frac{3}{40}\cos t + \frac{1}{40}\sin t$. We graph this solution in Figure 5.13(a). The solution and the steady-state solution are graphed together in Figure 5.13(b). Notice that the two curves appear identical for $t > 2.5$. The reason for this is shown in the subsequent plot of the transient solution (in Figure 5.13(c)), which becomes quite small near $t = 2.5$.

 EXERCISES 5.3

1. An 8 lb weight stretches a spring 1 ft. If a 16 lb weight is attached to the spring, it comes to rest in its equilibrium position. If it is then put into motion with a downward initial velocity of 2 ft./s, determine the

displacement of the mass if there is no damping and an external force $f(t) = \cos 3t$. What is the natural frequency of the spring-mass system?

2. A 16 lb weight stretches a spring 6 in. If the mass is lowered 1 ft. below its equilibrium position and released, determine the displacement of the mass if there is no damping and an external force of $f(t) = 2\cos t$. What is the natural frequency of the spring-mass system?

3. A 16 lb weight stretches a spring 8 in. If the object is lowered 4 in. below its equilibrium position and released, determine the displacement of the object if there is no damping and an external force of $f(t) = 2\cos t$.

4. A 6 lb weight stretches a spring 6 in. The object is raised 3 in. above its equilibrium position and released. Determine the displacement of the object if there is no damping and an external force of $f(t) = 2\cos 5t$.

5. An object of mass $m = 1$ kg is attached to a spring with spring constant $k = 9$ kg/m. If there is no damping and the external force is $f(t) = 4\cos \omega t$, find the displacement of the object if $x(0) = 0$ and $x'(0) = 0$. What must be the value of ω for resonance to occur?

6. An object of mass $m = 2$ kg is attached to a spring with spring constant $k = 1$ kg/m. If the resistive force is $F_R = 3\,dx/dt$ and the external force is $f(t) = 2\cos \omega t$, find the displacement of the object if $x(0) = 0$ and $x'(0) = 0$. Will resonance occur for any values of ω?

7. An object of mass $m = 1$ slug is attached to a spring with spring constant $k = 25$ lb/ft. If the resistive force is $F_R = 8\,dx/dt$ and the external force is $f(t) = \cos t - \sin t$, find the displacement of the object if $x(0) = 0$ and $x'(0) = 0$.

8. An object of mass $m = 2$ slug is attached to a spring with spring constant $k = 6$ lb/ft. If

the resistive force is $F_R = 6\,dx/dt$ and the external force is $f(t) = 2\sin 2t + \cos t$, find the displacement of the object if $x(0) = 0$ and $x'(0) = 0$.

9. Suppose that an object of mass 1 slug is attached to a spring with spring constant $k = 4$ lb/ft. If the motion of the object is undamped and subjected to an external force of $f(t) = \cos t$, determine the displacement of the object if $x(0) = 0$ and $x'(0) = 0$. What functions envelope this displacement function? What is the maximum displacement of the object? If the external force is changed to $f(t) = \cos(t/2)$, does the maximum displacement increase or decrease?

10. An object of mass 1 slug is attached to a spring with spring constant $k = 25$ lb/ft. If the motion of the object is undamped and subjected to an external force of $f(t) = \cos \beta t$, $\beta \neq 5$, determine the displacement of the object if $x(0) = 0$ and $x'(0) = 0$. What is the value of β if the maximum displacement of the object is $2/11\sqrt{2}$ ft?

11. An object of mass 4 slugs is attached to a spring with spring constant $k = 26$ lb/ft. It is subjected to a resistive force of $F_R = 4\,dx/dt$ and an external force $f(t) = 250\sin t$. Determine the displacement of the object if $x(0) = 0$ and $x'(0) = 0$. What is the transient solution? What is the steady-state solution?

12. An object of mass 1 slug is attached to a spring with spring constant $k = 4000$ lb/ft. It is subjected to a resistive force of $F_R = 40\,dx/dt$ and an external force $f(t) = 600\sin t$. Determine the displacement of the object if $x(0) = 0$ and $x'(0) = 0$. What is the transient solution? What is the steady-state solution?

13. Find the solution of the differential equation $m\,d^2x/dt^2 + kx = F\sin \omega t$, $\omega \neq \sqrt{k/m}$ that satisfies the initial conditions: (a) $x(0) = \alpha$, $x'(0) = 0$; (b) $x(0) = 0$, $x'(0) = \beta$; (c) $x(0) = \alpha$, $x'(0) = \beta$.

14. Find the solution of the differential equation
$m\,d^2x/dt^2 + c\,dx/dt + kx = F\sin\omega t$,
$c^2 - 4mk < 0$ that satisfies the initial
conditions: (a) $x(0) = \alpha$, $x'(0) = 0$; (b)
$x(0) = 0$, $x'(0) = \beta$; (c) $x(0) = \alpha$, $x'(0) = \beta$.

15. Find the solution to the IVP
$$\frac{d^2x}{dt^2} + x = \begin{cases} 1, 0 \le t < \pi \\ 0, t \ge \pi \end{cases}, x(0) = 0,$$
$x'(0) = 0$. (*Hint:* Solve the IVP over each
interval. Choose constants appropriately so
that the functions x and x' are continuous.)

16. Find the solution to the IVP
$$\frac{d^2x}{dt^2} + x = \begin{cases} \cos t, 0 \le t < \pi \\ 0, t \ge \pi \end{cases}, x(0) = 0,$$
$x'(0) = 0$.

17. Find the solution to the IVP
$d^2x/dt^2 + x = f(t)$, $x(0) = 0$, $x'(0) = 0$, where
$$f(t) = \begin{cases} t, 0 \le t < 1 \\ 2 - t, 1 \le t < 2 \\ 0, t \ge 2 \end{cases}. \text{ (Hint: Solve the }$$
IVP over each interval. Choose constants
appropriately so that the functions x and x'
are continuous.)

18. Find the solution to the IVP
$d^2x/dt^2 + 4\,dx/dt + 13x = f(t)$, $x(0) = 0$,
$x'(0) = 0$, where
$$f(t) = \begin{cases} 1, 0 \le t < \pi \\ 1 - t, \pi \le t < 2\pi \\ 0, t \ge 2\pi \end{cases}.$$

19. Show that a general solution of
$d^2x/dt^2 + 2\lambda\,dx/dt + \omega^2 x = F\sin\gamma t$ $(\omega \ne \gamma$
if $\lambda = 0$) is
$$x(t) = Ae^{-\lambda t}\sin\left(\sqrt{\omega^2 - \lambda^2}t + \phi\right)$$
$$+ \frac{F}{\sqrt{(\omega^2 - \gamma^2)^2 + 4\lambda^2\gamma^2}}\sin(\gamma t + \theta)$$
where $A = \sqrt{c_1^2 + c_2^2}$, $\lambda < \omega$, and the phase
angles ϕ and θ are found with $\sin\phi = c_1/A$,
$\cos\phi = c_2/A$,

$$\sin\theta = \frac{-2\lambda\gamma}{\sqrt{(\omega^2 - \gamma^2)^2 + 4\lambda^2\gamma^2}} \text{ and}$$
$$\cos\theta = \frac{\omega^2 - \gamma^2}{\sqrt{(\omega^2 - \gamma^2)^2 + 4\lambda^2\gamma^2}}.$$

20. The steady-state solution (the
approximation of the solution for large
values of t) of the differential equation in
Exercise 19 is $x(t) = g(\gamma)\sin(\gamma t + \theta)$, where
$g(\gamma) = F/\sqrt{(\omega^2 - \gamma^2)^2 + 4\lambda^2\gamma^2}$. By
differentiating g with respect to γ, show
that the maximum value of $g(\gamma)$ occurs
when $\gamma = \sqrt{\omega^2 - 2\lambda^2}$. The quantity
$\sqrt{\omega^2 - 2\lambda^2}/(2\pi)$ is called the *resonance
frequency* for the system. Describe the
motion if the external force has frequency
$\sqrt{\omega^2 - 2\lambda^2}/(2\pi)$.

21. Solve the IVP $x'' + x = \cos t$, $x(0) = 0$,
$x'(0) = b$ using $b = 0$ and $b = 1$. Graph the
solutions simultaneously to determine the
effect that the nonhomogeneous initial
velocity has on the solution to the second
IVP as t increases.

22. Solve the IVP $x'' + x = \cos\omega t$, $x(0) = 0$,
$x'(0) = 1$ using values $\omega = 0.9$ and $\omega = 0.7$.
Graph the solutions simultaneously to
determine the effect that the value of ω has
on each solution.

23. Investigate the effect that the forcing
functions (a) $f(t) = \cos 1.9t$ and (b)
$f(t) = \cos 2.1t$ have on the IVP
$d^2x/dt^2 + 4x = f(t)$, $x(0) = 0$, $dx/dt(0) = 0$.
How do these results differ from those of
Example 5.3.3? Are there more or fewer
beats per time unit with these two
functions?

24. Solve the IVP $x'' + 0.1x' + x = 3\cos 2t$,
$x(0) = 0$, $x'(0) = 0$, using the method of

undetermined coefficients and compare the result with the forcing function $3 \cos 2t$. Determine the phase difference between these two functions.

25. Solve the following IVP involving a piecewise defined forcing function over

$$[0, 3]: \frac{d^2 x}{dt^2} + x = \begin{cases} t, & 0 \le t < 1 \\ 2 - t, & 1 \le t < 2 \\ 0, & t \ge 2 \end{cases},$$

$x(0) = a,$ $\dfrac{dx}{dt}(0) = b.$ Graph the solution using the initial conditions $a = 1, b = 1$; $a = 0, b = 1$; and $a = 1, b = 0$.

26. Consider the function
$g(\gamma) = F/\sqrt{\left(\omega^2 - \gamma^2\right)^2 + 4\lambda^2 \gamma^2}$ defined in Exercise 20. (a) Graph this function for $k = 4, m = 1, F = 2,$ and the damping coefficient: $c = 2\lambda = 2, 1, 0.75, 0.50,$ and 0.25. (b) Graph the function for $k = 49, m = 10,$ $F = 20,$ and the damping coefficient: $c = 2\lambda = 2, 1, 0.75, 0.50,$ and 0.25. In each case, describe what happens to the maximum magnitude of $g(\gamma)$ as $c \to 0$. Also, as $c \to 0$, how does the resonance frequency relate to the natural frequency of the corresponding undamped system?

27. To hear beats, solve the IVP
$d^2 x/dt^2 + \omega^2 x = F \cos \beta t, \omega \ne \beta, x(0) = 0,$
$dx/dt(0) = 0$ using $\omega^2 = 6000, \beta = 5991.62,$
and $F = 2$. In each case, plot and, if possible, play the solution. (*Note:* The purpose of the high frequencies is to assist in hearing the solutions when they are played.)

28. To hear resonance, solve the IVP
$d^2 x/dt^2 + \omega^2 x = F \cos \omega t, x(0) = 0,$
$dx/dt(0) = 0$ using $\omega^2 = 6000$ and $F = 2$. In each case, plot and, if possible, play the solution. (*Note:* The purpose of the high frequencies is to assist in hearing the solutions when they are played.)

5.4 OTHER APPLICATIONS

- L-R-C Circuits
- Deflection of a Beam

L-R-C Circuits

Second-order nonhomogeneous linear ODEs arise in the study of electrical circuits after the application of *Kirchhoff's law*. Suppose that $I(t)$ is the current in the L-R-C series electrical circuit (shown in Figure 5.14) where L, R, and C represent the inductance, resistance, and capacitance of the circuit, respectively.

The voltage drops across the circuit elements in Table 5.2 have been obtained from experimental data, where Q is the charge of the capacitor and $dQ/dt = I$.

Our goal is to model this physical situation with an IVP so that we can determine the current

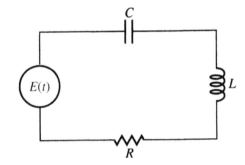

FIGURE 5.14   An L-R-C circuit.

TABLE 5.2   Voltage Drops Across an L-R-C Circuit

Circuit Element	Voltage Drop
Inductor	$L\dfrac{dI}{dt}$
Resistor	RI
Capacitor	$\dfrac{1}{C}Q$

TABLE 5.3 Terminology Used in Section 5.4

Electrical Quantities	Units
Inductance (L)	Henrys (H)
Resistance (R)	Ohms (Ω)
Capacitance (C)	Farads (F)
Charge (Q)	Coulombs (C)
Current (I)	Amperes (A)

and charge in the circuit. For convenience, the terminology used in this section is summarized in Table 5.3.

The physical principle needed to derive the differential equation that models the L-R-C series circuit is *Kirchhoff's law*.

The German physicist *Gustav Robert Kirchhoff* (1824-1887) worked in spectrum analysis, optics, and electricity.

Kirchhoff's law The sum of the voltage drops across the circuit elements is equivalent to the voltage $E(t)$ impressed on the circuit.

Applying Kirchhoff's law with the voltage drops in Table 5.2 yields the differential equation $L\frac{dI}{dt} + RI + C^{-1}Q = E(t)$. Using the fact that $dQ/dt = I$, we also have $d^2Q/dt^2 = dI/dt$. Therefore, the equation becomes $L\frac{d^2Q}{dt^2} + R\frac{dQ}{dt} + C^{-1}Q = E(t)$, which can be solved by the method

of undetermined coefficients or the method of variation of parameters. If the initial charge and current are $Q(0) = Q_0$ and $I(0) = Q'(0) = I_0$, we solve the IVP

$$L\frac{d^2Q}{dt^2} + R\frac{dQ}{dt} + \frac{1}{C}Q = E(t), \quad Q(0) = Q_0,$$

$$I(0) = \frac{dQ}{dt}(0) = I_0 \tag{5.8}$$

for the charge $Q(t)$. The solution is differentiated to find the current $I(t)$.

EXAMPLE 5.4.1

Consider the L-R-C circuit with $L = 1$ H, $R = 40\,\Omega$, $C = 1/4000$ F, and $E(t) = 24$ V. Determine the current in this circuit if there is zero initial current and zero initial charge.

Solution

Using the indicated values, the IVP that we must solve is

$$\frac{d^2Q}{dt^2} + 40\frac{dQ}{dt} + 4000Q = 24, \quad Q(0) = Q_0,$$

$$I(0) = \frac{dQ}{dt}(0) = I_0$$

The characteristic equation of the corresponding homogeneous equation is $r^2 + 40r + 4000 = 0$ with roots $r_{1,2} = -20 \pm 60i$, so a general solution of the corresponding homogeneous equation is $Q_h(t) = e^{-20t}(c_1 \cos 60t + c_2 \sin 60t)$. Because the voltage is the constant function $E(t) = 24$, we assume that the particular solution has the form $Q_p(t) = A$. Substitution into $d^2Q/dt^2 + 40\,dQ/dt + 4000Q = 24$ yields $4000A = 24$ or $A = 3/500$. Therefore, a general solution of the nonhomogeneous equation is

$$Q(t) = Q_h(t) + Q_p(t)$$

$$= e^{-20t}(c_1 \cos 60t + c_2 \sin 60t) + \frac{3}{500}$$

with derivative

$$\frac{dQ}{dt}(t) = -20e^{-20t}(c_1\cos 60t + c_2\sin 60t)$$
$$+ 60e^{-20t}(-c_1\sin 60t + c_2\cos 60t).$$

After application of the initial conditions, we have $Q(0) = c_1 + 3/500 = 0$ and $dQ/dt(0) = -20c_1 + 60c_2 = 0$. Therefore, $c_1 = -3/500$ and $c_2 = -1/500$, so the charge in the circuit is

$$Q(t) = e^{-20t}\left(-\frac{3}{500}\cos 60t - \frac{1}{500}\sin 60t\right) + \frac{3}{500},$$

and the current is given by

$$\frac{dQ}{dt}(t) = -20e^{-20t}\left(-\frac{3}{500}\cos 60t - \frac{1}{500}\sin 60t\right)$$
$$+ 60e^{-20t}\left(\frac{3}{500}\sin 60t - \frac{1}{500}\cos 60t\right)$$
$$= \frac{2}{5}e^{-20t}\sin 60t.$$

These results indicate that in time the charge approaches the constant value of 3/500, which is known as the *steady-state charge*. Also, because of the exponential term, the current approaches zero as t increases. We show the graphs of $Q(t)$ and $I(t)$ in Figure 5.15(a) and (b) to verify these observations.

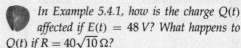

In Example 5.4.1, how is the charge $Q(t)$ affected if $E(t) = 48$ V? What happens to $Q(t)$ if $R = 40\sqrt{10}\ \Omega$?

Deflection of a Beam

An important mechanical model involves the deflection of a long beam that is supported at one or both ends, as shown in Figure 5.16. Assuming that in its undeflected form the beam is horizontal, then the deflection of the beam can be expressed as a function of x. Suppose that the shape of the beam when it is deflected is given by the graph of the function $y(x) = -s(x)$, where x is the distance from the left end of the beam and s the measure of the vertical deflection from the equilibrium position. The boundary value problem that models this situation is derived as follows.

Let $m(x)$ equal the turning moment of the force relative to the point x, and $w(x)$ represent the weight distribution of the beam. These two functions are related by the equation

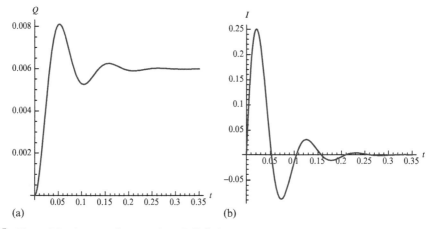

(a)   (b)

FIGURE 5.15   Plots of the charge and current in an *L-R-C* circuit.

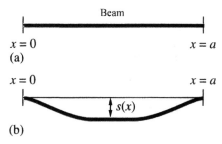

Beam

$x = 0$ $x = a$
(a)

$x = 0$ $x = a$

$s(x)$

(b)

FIGURE 5.16   Modeling the deflection of a beam.

$$\frac{d^2 m}{dx^2} = w(x).$$

Also, the turning moment is proportional to the curvature of the beam. Hence

$$m(x) = \frac{EI}{\left(\sqrt{1 + \left(\dfrac{ds}{dx}\right)^2}\right)^3} \frac{d^2 s}{dx^2},$$

where E and I are constants related to the composition of the beam and the shape and size of a cross section of the beam, respectively. Notice that this equation is nonlinear. This difficulty is overcome with an approximation. For small values of ds/dx, the denominator of the right-hand side of the equation can be approximated by the constant 1. Therefore, the equation is simplified to

$$m(x) = EI \frac{d^2 s}{dx^2}.$$

This equation is linear and can be differentiated twice to obtain

$$\frac{d^2 m}{dx^2} = EI \frac{d^4 s}{dx^4},$$

which is then used with the equation above relating $m(x)$ and $w(x)$ to obtain the single fourth order linear homogeneous differential equation

$$EI \frac{d^4 s}{dx^4} = w(x).$$

Boundary conditions for this problem may vary. In most cases, two conditions are given for each end of the beam. Some of these conditions, which are specified in pairs at $x = \rho$, where $\rho = 0$ or $\rho = a$, include $s(\rho) = 0$, $s'(\rho) = 0$ (fixed end); $s''(\rho) = 0$, $s'''(\rho) = 0$ (free end); $s(\rho) = 0$, $s''(\rho) = 0$ (simple support); and $s'(\rho) = 0$, $s(\rho) = 0$ (sliding clamped end).

EXAMPLE 5.4.2

Solve the beam equation over the interval $0 \le x \le 1$ if $E = I = 1$, $w(x) = 48$, and the following boundary conditions are used: $s(0) = 0$, $s'(0) = 0$ (fixed end at $x = 0$); and

(a) $s(1) = 0$, $s''(1) = 0$ (simple support at $x = 1$)
(b) $s''(1) = 0$, $s'''(1) = 0$ (free end at $x = 1$)
(c) $s'(1) = 0$, $s'''(1) = 0$ (sliding clamped end at $x = 1$)
(d) $s(1) = 0$, $s'(1) = 0$ (fixed end at $x = 1$).

Solution

We begin by noting that the differential equation is $d^4 s/dx^4 = 48$, which is a separable equation that can be solved by integrating each side four times to yield

$$s(x) = 2x^4 + c_1 x^3 + c_2 x^2 + c_3 x + c_4,$$

with derivatives $s'(x) = 8x^3 + 3c_1 x^2 + 2c_2 x + c_3$, $s''(x) = 24x^2 + 6c_1 x + 2c_2$, and $s'''(x) = 48x + 6c_1$ (Note that we could have used the method of undetermined coefficients or variation of parameters to find $s(x) = s_h(x) + s_p(x)$.) We next determine the arbitrary constants for the pair of boundary conditions at $x = 0$. Because $s(0) = 0$, $c_4 = 0$ and $s'(0) = c_3 = 0$,

$$s(x) = 2x^4 + c_1 x^3 + c_2 x^2.$$

(a) Because $s(1) = 2 + c_1 + c_2 = 0$ and $s''(1) = 24 + 6c_1 + 2c_2 = 0$, $c_1 = -5$ and $c_2 = 3$. Hence $s(x) = 2x^4 - 5x^3 + 3x^2$. We can visualize the shape of the beam by graphing $y = -s(x)$ as shown in Figure 5.17(a). (b) In this case, $s''(1) = 24 + 6c_1 +$

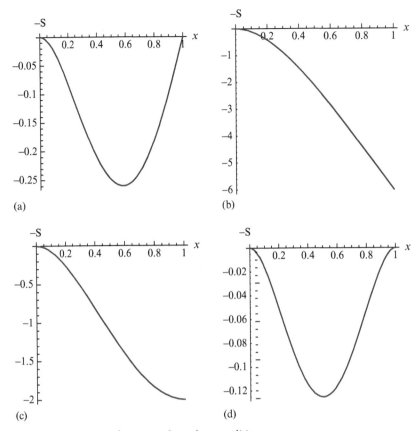

FIGURE 5.17   The shape of the beam for various boundary conditions.

$2c_2 = 0$ and $s'''(1) = 48 + 6c_1 = 0$, so $c_1 = -8$ and $c_2 = 12$. The deflection of the beam is given by $s(x) = 2x^4 - 8x^3 + 12x^2$. We graph $y = -s(x)$ in Figure 5.17(b). We see from the graph that the end is free at $x = 1$. (c) Because $s'(1) = 8 + 3c_1 + 2c_2 = 0$ and $s'''(1) = 48 + 6c_1 = 0$, $c_1 = -8$ and $c_2 = 8$. Therefore, $s(x) = 2x^4 - 8x^3 + 8x^2$. From the shape of the graph $y = -s(x)$ in Figure 5.17(c), we see that the end at $x = 1$ is clamped as compared to the free end in (b). (d) In this instance $s(1) = 2 + c_1 + c_2 = 0$ and $s'(1) = 8 + 3c_1 + 2c_2 = 0$, so $c_1 = -4$ and $c_2 = 2$. Thus, $s(x) = 2x^4 - 4x^3 + 2x^2$. The function

$y = -s(x)$ is graphed in Figure 5.17(d). Notice that both ends are fixed. Finally, all four graphs are shown together in Figure 5.18 to compare the different boundary conditions. (*Match each plot in Figure 5.17 with its plot in Figure 5.18.*)

If we had used free ends at both $x = 0$ and $x = 1$ in Example 5.4.2, what is the displacement? Is this what we should expect from the physical problem?

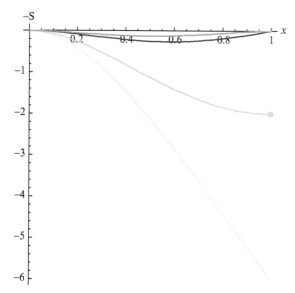

FIGURE 5.18 How the shape of a beam might vary based on different boundary conditions.

 EXERCISES 5.4

In Exercises 1-4, find the charge on the capacitor and the current in the L-C series circuit (in which $R = 0$) assuming that $Q(0) = 0$ and $I(0) = Q'(0) = 0$.

1. $L = 2\,\text{H}$, $C = 1/32\,\text{F}$, and $E(t) = 220\,\text{V}$.

2. $L = 2\,\text{H}$, $C = 1/50\,\text{F}$, and $E(t) = 220\,\text{V}$.

3. $L = 1/4\,\text{H}$, $C = 1/64\,\text{F}$, and $E(t) = 16t\,\text{V}$.

4. $L = 1/4\,\text{H}$, $C = 1/64\,\text{F}$, and $E(t) = 16\sin 4t\,\text{V}$.

5. Find the charge $Q(t)$ on the capacitor in an L-R-C series circuit if $L = 0.2\,\text{H}$, $R = 25\,\Omega$, $C = 0.001\,\text{F}$, $E(t) = 0$, $Q(0) = 0\,\text{C}$, and $I(0) = Q'(0) = 4\,\text{A}$. What is the maximum charge on the capacitor?

6. Consider the L-R-C circuit given in Exercise 5 with $E(t) = 1$. Determine the value of $Q(t)$ as t approaches infinity.

7. Consider the L-R-C circuit given in Exercise 5. In this exercise, let $E(t) = 126\cos t + 500$

$\sin t$. Determine the solution to this IVP. At what time does the charge first equal zero? What are the steady-state charge and current?

8. If the resistance, R, is changed in Exercise 5 to $R = 8\,\Omega$, what is the resulting charge on the capacitor? What is the maximum charge attained, and when does the charge first equal zero?

9. A beam of length 10 is fixed at both ends. Determine the shape of the beam if the weight distribution is the constant function $w(x) = 8$, with constants E and I such that $EI = 100$, $EI = 10$, and $EI = 1$. What is the displacement of the beam from $s = 0$ in each case? How does the value of EI affect the solution?

10. Suppose that the beam in Exercise 9 is fixed at $x = 0$ and has simple support at $x = 10$. Determine the maximum displacement using constants E and I such that $EI = 100$, $EI = 10$, and $EI = 1$.

11. Consider Exercise 9 with simple support at $x = 0$ and $x = 10$. How does the maximum displacement compare to that found in each case in Exercises 9 and 10?

12. Determine the shape of the beam of length 10 with constants E and I such that $EI = 1$, weight distribution $w(x) = x^2$, and boundary conditions:

    (a) $s(0) = 0$, $s'(0) = 0$ (fixed end at $x = 0$); $s(10) = 0$, $s''(10) = 0$ (simple support at $x = 10$)

    (b) $s(0) = 0$, $s'(0) = 0$ (fixed end at $x = 0$); $s''(10) = 0$, $s'''(10) = 0$ (free end at $x = 10$)

    (c) $s(0) = 0$, $s'(0) = 0$ (fixed end at $x = 0$); $s'(10) = 0$, $s'''(10) = 0$ (sliding clamped end at $x = 10$)

    (d) $s(0) = 0$, $s'(0) = 0$ (fixed end at $x = 0$); $s(10) = 0$, $s'(10) = 0$ (fixed end at $x = 10$)

    Discuss the differences brought about by these conditions (see Exercise 21).

13. Determine the shape of the beam of length 10 with constants E and I such that $EI = 1$,

weight distribution $w(x) = x^2$, and boundary conditions:

(a) $s(0) = 0$, $s''(0) = 0$ (simple support at $x = 0$); $s(10) = 0$, $s''(10) = 0$ (simple support at $x = 10$)

(b) $s(0) = 0$, $s''(0) = 0$ (simple support at $x = 0$); $s''(10) = 0$, $s'''(10) = 0$ (free end at $x = 10$)

(c) $s(0) = 0$, $s''(0) = 0$ (simple support at $x = 0$); $s'(10) = 0$, $s'''(10) = 0$ (sliding clamped end at $x = 10$)

(d) $s(0) = 0$, $s''(0) = 0$ (simple support at $x = 0$); $s(10) = 0$, $s'(10) = 0$ (fixed end at $x = 10$)

Discuss the differences brought about by these conditions (see Exercise 22).

14. Determine the shape of the beam of length 10 with constants E and I such that $EI = 1$, weight distribution $w(x) = 48 \sin(\pi x/10)$ and boundary conditions:

(a) $s(0) = 0$, $s''(0) = 0$ (simple support at $x = 0$); $s(10) = 0$, $s''(10) = 0$ (simple support at $x = 10$)

(b) $s(0) = 0$, $s''(0) = 0$ (simple support at $x = 0$); $s''(10) = 0$, $s'''(10) = 0$ (free end at $x = 10$)

(c) $s(0) = 0$, $s''(0) = 0$ (simple support at $x = 0$); $s'(10) = 0$, $s'''(10) = 0$ (sliding clamped end at $x = 10$)

(d) $s(0) = 0$, $s''(0) = 0$ (simple support at $x = 0$); $s(10) = 0$, $s'(10) = 0$ (fixed end at $x = 10$)

Discuss the differences brought about by these conditions (see Exercise 23).

15. Consider the L-C series circuit in which $E(t) = 0$ modeled by the IVP

$$L\frac{d^2Q}{dt^2} + \frac{1}{C}Q = 0, \quad Q(0) = Q_0,$$

$$I(0) = \frac{dQ}{dt}(0) = 0.$$

Find the charge Q and the current I. What is the maximum charge? What is the maximum current?

16. Consider the L-C series circuit in which $E(t) = 0$ modeled by the IVP

$$L\frac{d^2Q}{dt^2} + \frac{1}{C}Q = 0, \quad Q(0) = 0,$$

$$I(0) = \frac{dQ}{dt}(0) = I_0.$$

Find the charge Q and the current I. What is the maximum charge? What is the maximum current? How do these results compare to those of Exercise 15?

17. Consider the L-R-C circuit modeled by

$$L\frac{d^2Q}{dt^2} + R\frac{dQ}{dt} + \frac{1}{C}Q = E_0 \sin \omega t,$$

$$Q(0) = Q_0, \quad I(0) = \frac{dQ}{dt}(0) = I_0.$$

Find the *steady-state current*, $\lim_{t\to\infty} I(t)$, of this problem. Note that in this formula, $L\omega - 1/(C\omega)$ is called the *reactance* of the circuit, and $\sqrt{(L\omega - 1/(C\omega))^2 + R^2}$ is called the *impedance* of the circuit, where both of these quantities are measured in ohms.

18. Compute the reactance and the impedance of the circuit in Example 5.4.1 if $E(t) = 24 \sin 4t$.

19. Show that the maximum amplitude of the steady-state current found in Exercise 17 occurs when $\omega = 1/\sqrt{LC}$. (In this case, we say that *electrical resonance* occurs.)

20. (*Elastic shaft*) The differential equation that models the torsional motion of a weight suspended from an elastic shaft is $I\, d^2\theta/dt^2 + c\, d\theta/dt + k\theta = T(t)$, where θ represents the amount that the weight is twisted at time t, I is the moment of inertia, c is the damping constant, k is the elastic shaft constant (similar to the spring constant), and $T(t)$ is the applied torque. Consider the differential equation with $I = 1$, $c = 4$ and $k = 13$. Find $\theta(t)$ if (a) $T(t) = 0$, $\theta(0) = \theta_0$, and $\theta'(0) = 0$; (b) $T(t) = \sin \pi t$, $\theta(0) = \theta_0$, and $\theta'(0) = 0$. Describe the motion that results in each case.

To better understand the solutions to the elastic beam problem, we can use a graphics device to determine the shape of the beam under different boundary conditions.

21. Graph the solution of the beam equation if the beam has length 10, the constants E and I are such that $EI = 1$, the weight distribution is $w(x) = x^2$, and the boundary conditions are the same as in Problem 13.

22. Repeat Exercise 21 using the following boundary conditions in Problem 14.

23. Graph the solution of the beam equation if the beam has length 10, the constants E and I are such that $EI = 1$, the weight distribution is $w(x) = 48\sin(\pi x/10)$, and the boundary conditions are the same as in Problem 22.

24. Repeat Exercise 21 with $EI = 10$.

25. Repeat Exercise 22 with $EI = 100$. How do these solutions compare to those in Exercise 22?

26. Repeat Exercise 23 with $EI = 100$. How do these solutions compare to those in Exercise 23?

27. Attempt to find solutions of the beam equation in Exercises 21-23 using other combinations of boundary conditions. How do the differing boundary conditions affect the solution?

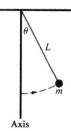

FIGURE 5.19 A swinging pendulum.

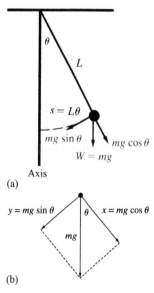

(a)

(b)

FIGURE 5.20 Two force diagrams for the swinging pendulum.

5.5 THE PENDULUM PROBLEM

Suppose that a mass m is attached to the end of a rod of length L, the weight of which is negligible (see Figure 5.19). We want to determine an equation that describes the motion of the mass in terms of the displacement $\theta(t)$, which is measured counterclockwise in radians from the vertical axis shown in Figure 5.19. This is possible if we are given an initial position and an initial velocity of the mass. A force diagram for this situation is shown in Figure 5.20(a).

Notice that the forces are determined with trigonometry using the diagram in Figure 5.20(a). In this instance, $\cos\theta = mg/x$ and $\sin\theta = mg/y$, so we obtain the forces

$$x = mg\cos\theta \quad \text{and} \quad y = mg\sin\theta,$$

which are indicated in Figure 5.20(b).

The momentum of the mass is given by $m\,ds/dt$, so the rate of change of the momentum is

$$\frac{d}{dt}\left(m\frac{ds}{dt}\right) = m\frac{d^2s}{dt^2},$$

where s represents the length of the arc formed by the motion of the mass. Then, because the force $mg \sin \theta$ acts in the opposite direction of the motion of the mass, we have the equation

$$m\frac{d^2s}{dt^2} = -mg \sin \theta.$$

(Notice that the force $mg \cos \theta$ is offset by the force of constraint in the rod, so mg and $mg \cos \theta$ cancel each other in the sum of the forces.) Using the relationship from geometry between the length of the arc, the length of the rod, and the angle θ, $s = L\theta$, we have the relationship

$$\frac{d^2s}{dt^2} = \frac{d^2}{dt^2}(L\theta) = L\frac{d^2\theta}{dt^2}.$$

The displacement $\theta(t)$ satisfies

$$mL\frac{d^2\theta}{dt^2} = -mg \sin \theta \text{ or } mL\frac{d^2\theta}{dt^2} + mg \sin \theta = 0,$$

which is a *nonlinear* equation. However, because we are only concerned with small displacements, we note from the Maclaurin series for $\sin \theta$,

$$\sin \theta = \theta - \frac{1}{3!}\theta^3 + \frac{1}{5!}\theta^5 - \frac{1}{7!}\theta^7 + \cdots,$$

that for small values of θ, $\sin \theta \approx \theta$. Therefore, we obtain the linear equation $mL\,d^2\theta/dt^2 + mg\theta = 0$ or $d^2\theta/dt^2 + (g/L)\theta = 0$, which approximates the original problem. If the initial displacement (position of the mass) is given by $\theta(0) = \theta_0$ and the initial velocity (the velocity with which the mass is set into motion) is given by $\theta'(0) = v_0$, we have the IVP

$$\frac{d^2\theta}{dt^2} + \frac{g}{L}\theta = 0, \quad \theta(0) = \theta_0, \quad \frac{d\theta}{dt}(0) = v_0 \quad (5.9)$$

to find the displacement function $\theta(t)$.

Suppose that $\omega^2 = g/L$ so that the differential equation becomes $d^2\theta/dt^2 + \omega^2\theta = 0$. Therefore, functions of the form

$$\theta(t) = c_1 \cos \omega t + c_2 \sin \omega t,$$

where $\omega = \sqrt{g/L}$, satisfy the equation $d^2\theta/dt^2 + g/L\,\theta = 0$. When we use the conditions $\theta(0) = \theta_0$ and $\theta'(0) = v_0$, we find that the function

$$\theta(t) = \theta_0 \cos \omega t + \frac{v_0}{\omega} \sin \omega t \quad (5.10)$$

satisfies the equation as well as the initial displacement and velocity conditions. As we did with the position function of spring-mass systems, we can write this function as a cosine function that includes a phase shift with

$$\theta(t) = \sqrt{\theta_0^2 + \frac{v_0^2}{\omega^2}} \cos(\omega t - \phi), \quad \text{where}$$

$$\phi = \cos^{-1}\left(\frac{\theta_0}{\sqrt{\theta_0^2 + v_0^2/\omega^2}}\right) \quad (5.11)$$

and $\omega = \sqrt{g/L}$.

Note that the period of $\theta(t)$ is $T = 2\pi/\omega = 2\pi\sqrt{L/g}$.

EXAMPLE 5.5.1

Determine the displacement of a pendulum of length $L = 8$ feet if $\theta(0) = 0$ and $\theta'(0) = 2$. What is the period? If the pendulum is part of a clock that ticks once for each time the pendulum makes a complete swing, how many ticks does the clock make in 1 min?

Solution

Because $g/L = 32/8 = 4$, the IVP that models this situation is

$$\frac{d^2\theta}{dt^2} + 4\theta = 0, \quad \theta(0) = 0, \quad \frac{d\theta}{dt}(0) = 2.$$

A general solution of the differential equation is $\theta(t) = c_1 \cos 2t + c_2 \sin 2t$, so application of the initial conditions yields the solution $\theta(t) = \sin 2t$. The period of this function is

$$T = 2\pi\sqrt{\frac{L}{g}} = 2\pi\sqrt{\frac{8\,\text{ft}}{32\,\text{ft/s}^2}} = \pi \text{ s}.$$

(Notice that we can use our knowledge of trigonometry to compare the period with $T = 2\pi/2 = \pi$.) Therefore, the number of ticks made by the clock per minute is calculated with the conversion

$$\frac{1\,\text{rev}}{\pi\,\text{s}} \times \frac{1\,\text{tick}}{1\,\text{rev}} \times \frac{60\,\text{s}}{1\,\text{min}} \approx 19.1\,\text{ticks/min}.$$

Hence the clock makes approximately 19 ticks in one minute.

 How is motion affected if the length of the pendulum in Example 5.5.1 is changed to L = 4?

If the pendulum undergoes a damping force that is proportional to the instantaneous velocity, the force due to damping is given by $F_R = b\,d\theta/dt$. Incorporating this force into the sum of the forces acting on the pendulum, we obtain the nonlinear equation $L\,d^2\theta/dt^2 + b\,d\theta/dt + g\sin\theta = 0$. Again, using the approximation $\sin\theta \approx \theta$ for small values of θ, we use the linear equation $L\,d^2\theta/dt^2 + b\,d\theta/dt + g\theta = 0$ to approximate the situation. Therefore, we solve the IVP

$$L\frac{d^2\theta}{dt^2} + b\frac{d\theta}{dt} + g\theta = 0, \quad \theta(0) = \theta_0, \quad \frac{d\theta}{dt}(0) = v_0$$

(5.12)

to find the displacement function $\theta(t)$.

EXAMPLE 5.5.2

A pendulum of length $L = 8/5$ ft. is subjected to the resistive force $F_R = 32/5\,d\theta/dt$ due to damping. Determine the displacement function if $\theta(0) = 1$ and $\theta'(0) = 2$.

Solution

The initial value problem that models this situation is

$$\frac{8}{5}\frac{d^2\theta}{dt^2} + \frac{32}{5}\frac{d\theta}{dt} + 32\theta = 0, \quad \theta(0) = 1, \quad \frac{d\theta}{dt}(0) = 2.$$

Simplifying the differential equation, we obtain $d^2\theta/dt^2 + 4\,d\theta/dt + 20\theta = 0$, which has characteristic equation $r^2 + 4r + 20 = 0$ with roots $r_{1,2} = -2 \pm 4i$. A general solution is $\theta(t) = e^{-2t}(c_1\cos 4t + c_2\sin 4t)$. Application of the initial conditions yields the solution $\theta(t) = e^{-2t}(\cos 4t + \sin 4t)$. We graph this solution in Figure 5.21. Notice that the damping causes the displacement of the pendulum to decrease over time.

 When does the object in Example 5.5.2 first pass through its equilibrium position? What is the maximum displacement from equilibrium?

In many cases, we can use computer algebra systems to obtain accurate approximations of nonlinear problems.

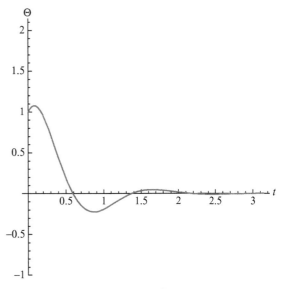

FIGURE 5.21 Plot of $\theta(t) = e^{-2t}(\cos 4t + \sin 4t)$.

EXAMPLE 5.5.3

Use a computer algebra system to approximate the solutions of the nonlinear problems (a) $\frac{d^2\theta}{dt^2} + \sin\theta = 0$, $\theta(0) = 0$, $\frac{d\theta}{dt}(0) = \frac{1}{2}$ and (b) $\frac{8}{5}\frac{d^2\theta}{dt^2} + \frac{32}{5}\frac{d\theta}{dt} + 32\sin\theta = 0$, $\theta(0) = 1$, $\frac{d\theta}{dt}(0) = 2$. Compare the results to the corresponding linear approximations.

Solution

(a) We show the results obtained with a typical computer algebra system in Figure 5.22. We see that as t increases, the approximate solution, $\theta = \frac{1}{2}\sin t$, becomes less accurate. However, for small values of t, the results are nearly identical.

(b) The exact solution to the corresponding linear approximation is obtained in Example 5.5.2. We show the results obtained with a typical computer algebra system in Figure 5.23. In this case, we see that the error diminishes as t increases. (Why?)

EXERCISES 5.5

1. Use the linear approximation of the model of the simple pendulum to determine the motion of a pendulum with rod length $L = 2$ ft. subject to the following sets of initial conditions:
   (a) $\theta(0) = 0.05$, $\theta'(0) = 0$
   (b) $\theta(0) = 0.05$, $\theta'(0) = 1$
   (c) $\theta(0) = 0.05$, $\theta'(0) = -1$
   In each case, determine the maximum displacement (in absolute value).

2. Consider the situation indicated in Exercise 1. However, use the initial velocity $\theta'(0) = 2$. How does the maximum displacement (in absolute value) differ from that in Exercise 1(b)-(c)?

3. Suppose that the pendulum in Exercise 1 is subjected to a resistive force with damping coefficient $b = 4\sqrt{7}$. Solve the IVPs given in Exercise 1, and compare the resulting motion to the undamped case.

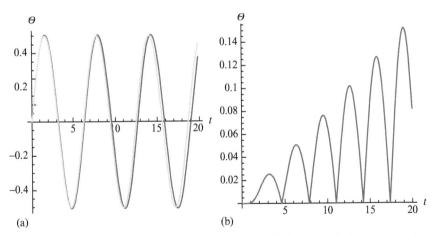

(a)   (b)

FIGURE 5.22 (a) The numerical solution to the nonlinear problem is in dark red (dark gray in print versions) and the exact solution to the linear approximation is in light red (light gray in print versions). (b) The absolute value of the difference between the two approximations.

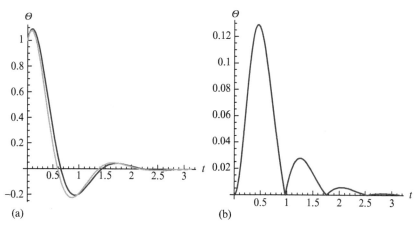

FIGURE 5.23   (a) The numerical solution to the nonlinear problem is in dark red (dark gray in print versions) and the exact solution to the linear approximation is in light red (light gray in print versions). (b) The absolute value of the difference between the two approximations.

4. Verify that $\theta(t) = c_1 \cos \omega t + c_2 \sin \omega t$, where $\omega = \sqrt{g/L}$ satisfies the equation $d^2\theta/dt^2 + g/L\,\theta = 0$.
5. Show that $\theta(t) = \theta_0 \cos \omega t + v_0/\omega \sin \omega t$, where $\omega = \sqrt{g/L}$ is the solution of IVP (5.9).
6. Let $\theta(t) = A \cos(\omega t - \phi)$. Use the identity $\cos(a+b) = \cos a \cos b - \sin a \sin b$ with the solution in Exercise 5 to find A so that $\theta(t)$ satisfies IVP (5.9).
7. Show that the phase angle in $\theta(t) = A \cos(\omega t - \phi)$ is
$$\phi = \cos^{-1}\left(\theta_0 / \sqrt{\theta_0^2 + v_0^2/\omega^2}\right).$$

In Exercises 8-11, approximate the period of the motion of the pendulum using the given length.

8. $L = 1\,\mathrm{m}$
9. $L = 2\,\mathrm{m}$
10. $L = 2\,\mathrm{ft}$
11. $L = 8\,\mathrm{ft}$
12. If $L = 1\,\mathrm{m}$, how many ticks does the clock make in 1 min if it ticks once for each time the pendulum makes a complete swing?
13. Assuming that a clock ticks once each time the pendulum makes a complete swing, how long (in meters) does the pendulum

need to be for the clock to tick once per second?
14. In Exercise 13, how long (in feet) does the pendulum need to be for the clock to tick once per second?
15. For the undamped problem, IVP (5.9) what is the maximum value of $\theta(t)$? For what values of t does the maximum occur?
16. For what values of t is the pendulum vertical in Exercise 15?
17. Solve IVP (5.12). Determine restrictions on the parameters L, b, and g that correspond to overdamping, critical damping, and underdamping. Describe the physical situation in each case.
18. Solve the IVP $L\frac{d^2\theta}{dt^2} + b\frac{d\theta}{dt} + g\theta = F \cos \gamma t$, $\theta(0) = 0$, $\frac{d\theta}{dt}(0) = 0$, assuming that $b^2 - 4gL < 0$. Describe the motion of the pendulum as $t \to \infty$.
19. Consider *Van-der-Pol's equation*, $\frac{d^2x}{dt^2} + \epsilon(x^2 - 1)\frac{dx}{dt} + x = 0$, where ϵ is a small positive number. Notice that this equation has a nonconstant damping coefficient. Because ϵ is small, we can approximate a solution of Van-der-Pol's equation with

$x(t) = A \cos \omega t$, a solution of $d^2x/dt^2 + x = 0$ (the equation obtained when $\epsilon = 0$) if $\omega = 1$. This method of approximation is called *harmonic balance*.

(a) Substitute $x(t) = A \cos \omega t$ into the nonlinear term in Van-der-Pol's equation to obtain

$$\epsilon(x^2 - 1)\frac{dx}{dt} = -\epsilon A\omega \left(\frac{1}{4}A^2 - 1\right)$$

$$\times \sin \omega t - \frac{1}{4}\epsilon A^4 \omega \sin 3\omega t.$$

(b) If we ignore the term involving the higher harmonic $\sin 3\omega t$, we have $\epsilon(x^2 - 1)dx/dt = -\epsilon A \omega \left(\frac{1}{4}A^2 - 1\right)\sin \omega t = \epsilon \left(\frac{1}{4}A^2 - 1\right) dx/dt$. Substitute this expression into Van-der-Pol's equation to obtain the linear equation

$$d^2x/dt^2 + \epsilon \left(\frac{1}{4}A^2 - 1\right) dx/dt + x = 0.$$

(c) If $A = 2$ in the linear equation in (b), is the approximate solution periodic?

(d) If $A \neq 2$ in the linear equation in (b), is the approximate solution periodic?

20. Comment on the behavior of solutions obtained in Exercise 19 if (a) $A < 2$ and (b) $A > 2$

21. Use a computer algebra system to solve the IVP $d^2\theta/dt^2 + \theta = 0$, $\theta(0) = \theta_0$, $d\theta/dt(0) = v_0$ subject to the following initial conditions:
(a) $\theta(0) = 0, \theta'(0) = 2$
(b) $\theta(0) = 2, \theta'(0) = 0$
(c) $\theta(0) = -2, \theta'(0) = 0$
(d) $\theta(0) = 0, \theta'(0) = -1$
(e) $\theta(0) = 0, \theta'(0) = -2$
(f) $\theta(0) = 1, \theta'(0) = -1$
(g) $\theta(0) = -1, \theta'(0) = 1$.

Plot each solution individually and plot the seven solutions simultaneously. Explain the physical interpretation of these solutions.

22. Solve the IVP $\frac{d^2\theta}{dt^2} + \frac{1}{2}\frac{d\theta}{dt} + \theta = 0$, $\theta(0) = \theta_0$, $\frac{d\theta}{dt}(0) = v_0$ subject to the initial conditions:
(a) $\theta(0) = 1, \theta'(0) = 0$
(b) $\theta(0) = -1, \theta'(0) = 0$
(c) $\theta(0) = 0, \theta'(0) = 1$
(d) $\theta(0) = 0, \theta'(0) = -1$
(e) $\theta(0) = 1, \theta'(0) = 1$
(f) $\theta(0) = 1, \theta'(0) = -1$
(g) $\theta(0) = -1, \theta'(0) = 1$
(h) $\theta(0) = -1, \theta'(0) = -1$
(i) $\theta(0) = 1, \theta'(0) = 2$
(j) $\theta(0) = 1, \theta'(0) = 3$
(k) $\theta(0) = -1, \theta'(0) = 2$
(l) $\theta(0) = -1, \theta'(0) = 3$.

23. Plot the solutions obtained in Exercise 22 individually and then plot them simultaneously. Give a physical interpretation of the results.

24. If the computer algebra system you are using has a built-in function that approximates the solution of nonlinear differential equations, solve the problem $\frac{d^2\theta}{dt^2} + \frac{1}{2}\frac{d\theta}{dt} + \sin\theta = 0$, $\theta(0) = \theta_0$, $\frac{d\theta}{dt}(0) = v_0$ with the initial conditions stated in Exercise 22. Compare the results you obtain with each.

25. Use a built-in computer algebra system function to approximate the solution of the pendulum problem with a variable damping coefficient $\frac{d^2\theta}{dt^2} + \frac{1}{2}(\theta^2 - 1)\frac{d\theta}{dt} + \theta = 0$, $\theta(0) = \theta_0$, $\frac{d\theta}{dt}(0) = v_0$ using the initial conditions stated in Exercise 22. Compare these results with those of Exercise 24.

26. Repeat Exercise 25 using $\frac{d^2\theta}{dt^2} + \frac{1}{2}(\theta^2 - 1)\frac{d\theta}{dt} + \sin\theta = 0$, $\theta(0) = \theta_0$, $\frac{d\theta}{dt}(0) = v_0$.

CHAPTER 5 SUMMARY: ESSENTIAL CONCEPTS AND FORMULAS

Hooke's Law: $F = ks$

Simple harmonic motion: The IVP
$mx'' + kx = 0$, $x(0) = \alpha$, $x'(0) = \beta$ has solution
$x = \alpha \cos \omega t + \beta/\omega \sin \omega t$, where $\omega = \sqrt{k/m}$;
the amplitude of the solution is
$A = \sqrt{\alpha^2 + \beta^2/\omega^2}$.

Damped motion: $mx'' + cx' + kx = 0$
  Overdamped: $c^2 - 4mk > 0$
  Critically damped: $c^2 - 4mk = 0$
  Underdamped: $c^2 - 4mk < 0$

Forced motion: $mx'' + cx' + kx = f(t)$

Kirchhoff's law: The sum of the voltage drops across the circuit elements is equivalent to the voltage $E(t)$ impressed on the circuit.

L-R-C Circuit: $LQ'' + RQ' + C^{-1}Q = E(t)$,
$Q(0) = Q_0$, $I(0) = Q'(0) = I_0$.

Deflection of a Beam: $EIs^{(4)} = w(x)$

Motion of a Pendulum: $\theta'' + g/L\,\theta = 0$,
$\theta(0) = \theta_0$, $\theta'(0) = v_0$ has solution
$\theta(t) = \theta_0 \cos \omega t + v_0/\omega \sin \omega t$.

CHAPTER 5 REVIEW EXERCISES

1. An object weighing 32 lb stretches a spring 6 in. If the object is lowered 4 in. below the equilibrium and released from rest, determine the displacement of the object, assuming there is no damping. What is the maximum displacement of the object from equilibrium? When does the object first pass through the equilibrium position? How often does the object return to the equilibrium position?

2. If the object in Exercise 1 is released from a point 3 in. above equilibrium with a downward initial velocity of 1 ft./s, determine the displacement of the object, assuming there is no damping. What is the maximum displacement of the object from

equilibrium? When does the object first pass through the equilibrium position? How often does the object return to the equilibrium position?

3. An object of mass 5 kg is attached to the end of a spring with spring constant $k = 65\,\text{N/m}$. If the object is released from the equilibrium position with an upward initial velocity of 1 m/s, determine the displacement of the object assuming the force due to damping is $F_R = 20\,dx/dt$. Find $\lim_{t\to\infty} x(t)$ and the quasiperiod. What is the maximum displacement of the object from equilibrium? When does the object first pass through the equilibrium position?

4. If the object in Exercise 3 is released from a point 1 in. below equilibrium with zero initial velocity, determine the displacement. Find $\lim_{t\to\infty} x(t)$ and the quasiperiod. What is the maximum displacement of the object from equilibrium? When does the object first pass through the equilibrium position?

5. An object of mass 4 slugs is attached to a spring with spring constant $k = 16\,\text{lb/ft}$. If there is no damping and the object is subjected to the forcing function $f(t) = 4$, determine the displacement function $x(t)$ if $x(0) = x'(0) = 0$. What is the maximum displacement of the object from equilibrium? When does the object first pass through the equilibrium position?

6. If the object in Exercise 5 is subjected to the forcing function $f(t) = 4 \cos 2t$, determine the displacement. Find $\lim_{t\to\infty} x(t)$ if it exists. Describe the physical phenomenon that occurs.

7. If the object in Exercise 5 is subjected to the forcing function $f(t) = 4 \cos t$, determine the displacement. Describe the physical phenomenon that occurs. Find the envelope functions.

8. An object of mass 2 slugs is attached to a spring with spring constant $k = 5\,\text{lb/ft}$. If the resistive force is $F_R = 6\,dx/dt$ and the

external force is $f(t) = 12 \cos 2t$, determine the displacement if $x(0) = x'(0) = 0$. What is the steady-state solution? What is the transient solution?

9. Find the charge and current in the L-R-C circuit if $L = 4$ H, $R = 80$ Ω, $C = 1/436$ F, and $E(t) = 100$ if $Q(0) = Q'(0) = 0$. Find $\lim_{t \to \infty} Q(t)$ and $\lim_{t \to \infty} I(t)$.

10. Find the charge and current in the L-R-C circuit in Exercise 9 if $E(t) = 100 \sin 2t$. Find $\lim_{t \to \infty} Q(t)$ and $\lim_{t \to \infty} I(t)$. How do these limits compare to those in Exercise 9?

11. Find the charge and current in the L-R-C circuit if $L = 1$ H, $R = 0$ Ω, $C = 10^{-4}$ F, and $E(t) = 220$ if $Q(0) = Q'(0) = 0$. Find $\lim_{t \to \infty} Q(t)$ and $\lim_{t \to \infty} I(t)$.

12. Find the charge and current in the L-R-C circuit in Exercise 11 if $E(t) = 100 \sin 10t$. Find $\lim_{t \to \infty} Q(t)$ and $\lim_{t \to \infty} I(t)$.

13. Determine the shape of the beam of length 10 with constants E and I such that $EI = 1$, weight distrbution $w(x) = x(10 - x)$, and fixed-end boundary conditions at $x = 0$ and $x = 10$.

14. Determime the shape of the beam in Exercise 13 if there are fixed-end boundary conditions at $x = 0$ and a sliding clamped end at $x = 10$.

15. Determine the shape of the beam in Exercise 13 if there are fixed-end boundary conditions at $x = 0$ and a free end at $x = 10$.

16. Determine the shape of the beam in Exercise 13 if there are fixed-end boundary conditions at $x = 0$ and simple support at $x = 10$.

17. Use the linear approximation of the model of the simple pendulum to determine the motion of a pendulum with rod length $L = 1/2$ ft. subject to the initial conditions $\theta(0) = 1$ and $d\theta/dt(0) = 0$. What is the maximum displacement of the pendulum from the vertical position? When does the pendulum first pass through the vertical position?

18. If the initial conditions in Exercise 17 are $\theta(0) = 0$ and $\theta'(0) = -1$, what is the maximum displacement of the pendulum from the vertical position? When does the pendulum first return to the vertical position?

19. How does the motion of the pendulum in Exercise 17 differ if it undergoes the damping force $F_R = 8\,d\theta/dt$?

20. Solve the model in Exercise 17 with the damping force $F_R = 8\sqrt{3}\,d\theta/dt$. How does the motion differ from that in Exercise 19?

21. Undamped torsional vibrations (rotations back and forth) of a wheel attached to a thin elastic rod or wire satisfy the differential equation $I_0\theta'' + k\theta = 0$, where θ is the angle measured from the state of equilibrium, I_0 is the polar moment of inertia of the wheel about its center, and k is the torsional stiffness of the rod. Solve this equation if $k/I_0 = 13.69\,\text{s}^{-2}$, the initial angle is $15° \approx 0.2168$ rad, and the initial angular velocity is $10°\,\text{s}^{-1} \approx 0.1745\,\text{rad s}^{-1}$.

22. Determine the displacement of the spring-mass system with mass 0.250 kg, spring constant $k = 2.25$ kg/s^2, and driving force $f(t) = \cos t - 4 \sin t$ if there is no damping, zero initial position, and zero initial velocity. For what frequency of the driving force would there be resonance?

23. The IVP

$$y'' + y = \begin{cases} 1 - t^2, & 0 \le t < 1 \\ 0, & t \ge 1 \end{cases}, \quad y(0) = y'(0) = 0$$

can be thought of as an undamped system in which a force F acts during the interval of time $0 \le t < 1$. This is the situation that occurs in a gun barrel when a shell is fired. The barrel is braked with heavy springs, as indicated in the following figure.

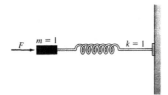

Solve this IVP.

24. Consider a buoy in the shape of a cylinder of radius r, height h, and density ρ where $\rho \leq 1/2 \, \text{g/cm}^3$ (the density of water is $1 \, \text{g/cm}^3$). Initially, the buoy sits with its base on the surface of the water. It is then released so that it is acted on by two forces: the force of gravity (in the downward direction) equal to the weight of the buoy, $F_1 = mg = \rho \pi r^2 h g$; and the force of buoyancy (in the upward direction) equal to the weight of the displaced water, $F_2 = \pi r^2 x g$, where $x = x(t)$ is the depth of the base of the cylinder from the surface of the water at time t. Using Newton's second law of motion with $m = \rho \pi r^2 h$, we have the differential equation $\rho \pi r^2 h x'' = -\pi r^2 x g$ or $\rho h x'' + g x = 0$. Find the displacement of the buoy if $x(0) = 1$ and $x'(0) = 0$. What is the period of the solution? What is the amplitude of the solution?

25. A cube-shaped buoy of side length ℓ and mass density ρ per unit volume is floating in a liquid of mass density ρ_0 per unit volume, where $\rho_0 > \rho$. If the buoy is slightly submerged into the liquid and released, it oscillates up and down. If there is no damping and no air resistance, the buoy is acted on by two forces: the force of gravity (in the downward direction), which is equal to the weight of the buoy, $F_1 = mg = \rho \ell^3 g$; and the force of buoyancy (in the upward direction), which is equal to

the weight of the displaced water, $F_2 = \ell^2 x g \rho_0$, where $x = x(t)$ is the depth of the base of the buoy from the surface of the water at time t. Then by Newton's second law with $m = \rho \ell^3$, we have the differential equation $\rho \ell^2 x'' + g \rho_0 x = 0$. Determine the amplitude of the motion if $\rho_0 = 1 \, \text{g/cm}^3$, $\rho = 1/4 \, \text{g/cm}^3$, $\ell = 100 \, \text{cm}$, $g = 980 \, \text{cm/s}^2$, $x(0) = 25 \, \text{cm}$ and $x'(0) = 0$.

26. (Pursuit Models) A rabbit starts at the origin and runs with speed a due north toward a hole in a fence located at the point $(0, d)$ on the y-axis. At the same time, a dog starts at the point $(c, 0)$ on the x-axis, running at speed b in pursuit of the rabbit. (Note: The dog runs directly toward the rabbit.) The slope of the tangent line to the dog's path is $dy/dx = -(at - y)/x$, which can be written as $xy' = y - at$. Differentiate both sides of this equation with respect to t to obtain $xy'' = -a \, dt/dx$. If s is the length of the arc from $(c, 0)$ along the dog's path, then $ds/dt = b$ is the dog's speed. Also,

$ds/dx = -\sqrt{1 + (y')^2}$, where the negative sign indicates that s increases as x decreases. By the chain rule,

$$\frac{dt}{dx} = \frac{dt}{ds}\frac{ds}{dx} = -\frac{1}{b}\sqrt{1 + (y')^2},$$

so substitution into the equation $xy'' = -a \, dt/dx$ yields the equation $xy'' = k\sqrt{1 + (y')^2}$, where $k = a/b$. Make the substitution $p = y'$ or $y'' = dp/dx$ to find the path of the dog with the initial condition $y'(c) = 0$. If $a < b$, when does the dog catch the rabbit? Does the problem make sense if $a = b$? (See the following figure.)

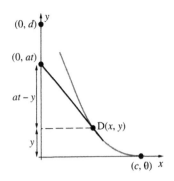

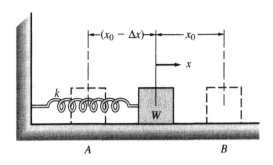

27. *(Coulomb damping)* Damping that results from dry friction (as when an object slides over a dry surface) is called *Coulomb damping* or *dry-friction damping*. On the other hand, *viscous damping* is damping that can be represented by a term proportional to the velocity (as when an object such as a vibrating spring vibrates in air or when an object slides over a lubricated surface).

Suppose that the kinetic coefficient of friction is μ. Friction forces oppose motion. Therefore, the force as a result of friction F is shown opposite the direction of motion in Figure 5.24. However, because F is a discontinuous function, we cannot use a single differential equation to model the motion as was done with previous damping problems. Instead, we have one equation for motion to the right and one equation for motion to the left:

    Motion to right: $x'' + \omega_n^2 x = -F/m$
    Motion to left: $x'' + \omega_n^2 x = F/m$,

where F represents the friction force, m the mass, and $\omega_n^2 = k/m$.[1]

(a) Find a general solution of
    $x'' + \omega_n^2 x = F/m$.
(b) Solve the IVP $x'' + \omega_n^2 x = F/m$, $x(0) = x_0$, $x'(0) = 0$.
(c) Find the values of t for which the solution in (b) is valid.

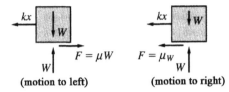

FIGURE 5.24 A system with coulomb (dry-friction) damping.

(d) At the right-hand endpoint of the interval obtained in (c), what is the displacement of the object?
(e) What must the force F be to guarantee that the object does not move under the given initial conditions?

28. *(Self-excited vibrations)* The differential equation that describes the motion of a spring-mass system with a single degree of freedom excited by the force Px' is $mx'' + cx' + kx = Px'$, which can be rewritten as $x'' + \frac{c-P}{m}x' + \frac{k}{m}x = 0$.[2]
(a) Show that the roots of the characteristic equation of this equation are

$$r_{1,2} = \frac{P-c}{2m} \pm \sqrt{\left(\frac{P-c}{2m}\right)^2 - \frac{k}{m}}.$$

(b) Show that if $P > c$ the motion of the system diverges *(dynamically unstable)*, if

[1] M. L. James, G. M. Smith, J. C. Wolford, and P. W. Whaley, *Vibration of Mechanical and Structural Systems with Microcomputer Applications*, Harper & Row, New York (1989), pp. 70-72.
[2] Robert K. Vierck, *Vibration Analysis*, Second Edition, HarperCollins, New York (1979), pp. 137-139.

$P = c$ the solution is the solution for a free undamped system, and if $P < c$ the solution is the solution for a free damped system.

DIFFERENTIAL EQUATIONS AT WORK

A. Rack-and-Gear Systems

Consider the rack-and-gear system shown in Figure 5.25. Let T represent the kinetic energy of the system and U the change in potential energy of the system from its potential energy in the static-equilibrium position. The kinetic energy of a system is a function of the velocities of the system masses. The potential energy of a system consists of the strain energy U, stored in elastic elements and the energy U_g, which is a function of the vertical distances between system masses.

The rack-and-gear system consists of two identical gears of pitch radius r and centroidal mass moment of inertia $\bar{I}$, a rack of weight W, and a linear spring of stiffness k, length l, and a mass of γ per unit length. To determine the differential equation to model the motion, we differentiate the law of conservation, $T + U =$ constant, to obtain

$$\frac{d}{dt}(T + U) = 0.$$

We then use $T_{max} = U_{max}$ to determine the natural circular frequency of the system, where T_{max} represents the maximum kinetic energy and U_{max} the maximum potential energy.

Because the static displacement x, of any point on the spring is proportional to its distance y from the spring support, we can write

$$x_s = \frac{yx}{l}.$$

When the rack is displaced a distance x below the equilibrium position, the gears have the angular displacements θ, which are related to the displacement of the rack by

$$\theta = \frac{x}{r}.$$

Then the angular velocity of each gear is given (through differentiation of θ with respect to t) by

$$\dot{\theta} = \frac{\dot{x}}{r}.$$

(Notice that we use a "dot" to indicate differentiation with respect to t. This is a common practice in physics and engineering.) Because $x = r\theta$, we have

$$x_s = \frac{yr\theta}{l},$$

so the velocity at any point on the spring is

$$\dot{x}_s = \frac{yr\dot{\theta}}{l}.$$

The kinetic energy of a differential element of length dy of the spring is

$$dT = \frac{1}{2}\gamma \, (\dot{x}_s)^2 \, dy.$$

The total kinetic energy of the system in terms of θ is

$$T = \underbrace{\bar{I}\dot{\theta}^2}_{\text{Gears}} + \underbrace{\frac{1}{2}\frac{W}{g}\left(r\dot{\theta}\right)^2}_{\text{Rack}} + \underbrace{\frac{\gamma}{2}\int_0^l \left(\frac{yr\dot{\theta}}{l}\right)^2 dy}_{\text{Spring}}$$

or

$$T = \left(\bar{I} + \frac{1}{2}\frac{W}{g}r^2 + \frac{1}{2\times 3}m_3 r^2\right) \times \left(\dot{\theta}\right)^2.$$

where $m_3 = \gamma l$ is the total mass of the spring.

The change in the strain energy U_e for a positive downward displacement x of the rack from the static-equilibrium position where the spring is already displaced Δ_s from its free length (see Figure 5.25(b)) is the area of the shaded region (see Figure 5.25(c)) given by $U_e = \frac{1}{2}kx^2 + Wx$. Similarly, the change in potential energy U_g as the rack moves a distance x below the static-equilibrium position is $U_g = -Wx$ and the total change in the potential energy is

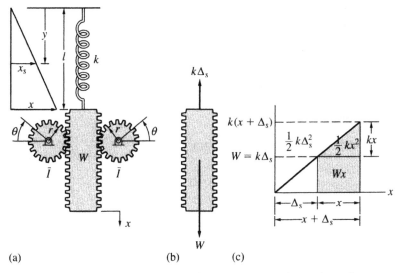

FIGURE 5.25 (a) A rack-and-gear system. (b) Static-equilibrium position. (c) Spring-force-diagram.

$$U = U_e + U_g = \left(\frac{1}{2}kx^2 + Wx\right) + (-Wx) = \frac{1}{2}kx^2$$

or $U = \frac{1}{2}k \times (r\theta)^2$. Substitution of these expressions into the equation $T + U = \text{constant}$ yields

$$\left(\bar{I} + \frac{1}{2}\frac{W}{g}r^2 + \frac{1}{2\times 3}m_3 r^2\right) \times \left(\dot{\theta}\right)^2$$

$$+ \frac{1}{2}k \times (r\theta)^2 = \text{constant}$$

Differentiating with respect to t then gives us the differential equation[3]

$$\ddot{\theta} + \frac{kr^2}{2\bar{I} + \frac{W}{g}r^2 + \frac{1}{3}m_3 r^2}\theta = 0.$$

1. Determine the natural frequency ω_n of the system.
2. Determine $\theta(t)$ if $\theta(0) = 0$ and $\dot{\theta}(0) = \theta_0 \omega_n$.

3. Find $T_{\max}$ and $U_{\max}$. Determine the natural frequency with these two quantities. Compare this value of ω_n with that obtained in Problem 1.
4. How does the natural frequency change as r increases?

B. Soft, Hard, and Aging Springs

In the case of a *soft spring*, the spring force weakens with compression or extension. For springs of this type, we model the physical system with the nonlinear IVP

$$m\frac{d^2x}{dt^2} + c\frac{dx}{dt} + kx - jx^3 = f(t),$$

$$x(0) = \alpha, \quad \frac{dx}{dt}(0) = \beta,$$

where j is a positive constant.

[3] M. L. James, G. M. Smith, I C. Wolford, and P. W Whaley, *Vibration of Mechanical and Structural Systems with Microcomputer Applications*, Harper & Row, New York (1989), pp. 82-86.

1. Approximate the solution to

$$\frac{d^2x}{dt^2} + 0.2\frac{dx}{dt} + 10x - 0.2x^3 = -9.8,$$

$$x(0) = \alpha, \quad \frac{dx}{dt}(0) = \beta,$$

for various values of α and β in the initial conditions.

2. Find conditions on α and β so that $\lim_{t\to\infty} x(t)$ (a) converges and (b) becomes unbounded.

3. If $\lim_{t\to\infty} x(t)$ exists, does the motion of the spring converge to its equilibrium position? Explain.

In the case of a *hard spring*, the spring force strengthens with compression or extension. For springs of this type, we model the physical system with

$$m\frac{d^2x}{dt^2} + c\frac{dx}{dt} + kx + jx^3 = f(t),$$

$$x(0) = \alpha, \quad \frac{dx}{dt}(0) = \beta,$$

where j is a positive constant.

4. Approximate the solution to

$$\frac{d^2x}{dt^2} + 0.3x + 0.04x^3 = 0,$$

$$x(0) = \alpha, \quad \frac{dx}{dt}(0) = \beta,$$

for various values of α and β in the initial conditions.

5. How does increasing the initial amplitude affect the period of the motion?

In the case of an *aging spring*, the spring constant weakens with time. For springs of this type, we model the physical system with

$$m\frac{d^2x}{dt^2} + c\frac{dx}{dt} + k(t)x = f(t),$$

$$x(0) = \alpha, \quad \frac{dx}{dt}(0) = \beta,$$

where $k(t) \to 0$ as $t \to \infty$.

6. Approximate the solution to

$$\frac{d^2x}{dt^2} + 4e^{-t/4}x = 0, \quad x(0) = \alpha, \quad \frac{dx}{dt}(0) = \beta,$$

for various values of α and β in the initial conditions.

7. What happens to the period of the oscillations as time increases? Is the motion oscillatory for large values of t? Why?

C. Bodé Plots

Consider the differential equation $d^2x/dt^2 + 2c\,dx/dt + k^2x = F_0 \sin \omega t$, where c and k are positive constants such that $c < k$. Therefore, the system is underdamped. To find a particular solution, we can consider the complex exponential form of the forcing function, $F_0 e^{i\omega t}$, which has imaginary part $F_0 \sin \omega t$. Assuming a solution of the form $z_p(t) = Ae^{i\omega t}$, substitution into the differential equation yields $A \times (-\omega^2 + 2ic\omega + k^2) = F_0 e^{i\omega t}$. Because $k^2 - \omega^2 + 2ic\omega = 0$ only when $k = \omega$ and $c = 0$, we find that $A = F_0/(k^2 - \omega^2 + 2ic\omega)$ or

$$A = \frac{F_0}{k^2 - \omega^2 + 2ic\omega} \times \frac{k^2 - \omega^2 - 2ic\omega}{k^2 - \omega^2 - 2ic\omega}$$

$$= \frac{k^2 - \omega^2 - 2ic\omega}{\left(k^2 - \omega^2\right)^2 + 4c^2\omega^2} F_0 = H(i\omega)F_0.$$

Therefore, a particular solution is $z_p(t) = H(i\omega)F_0 e^{i\omega t}$. Now we can write $H(i\omega)$ in polar form as $H(i\omega) = M(\omega)e^{i\phi(\omega)}$, where $M(\omega) = ((k^2 - \omega^2)^2 + 4c^2\omega^2)^{-1/2}$ and $\phi(\omega) = \cot^{-1}\left(\frac{\omega^2-k^2}{2c\omega}\right)$, $-\pi \le \phi \le 0$. A particular solution can then be written as $z_p(t) = M(\omega)F_0 e^{i\omega t} e^{i\phi(\omega)} = M(\omega)F_0 e^{i(\omega t + \phi(\omega))}$ with imaginary part $M(\omega)F_0 \sin(\omega t + \phi(\omega))$, so we take the particular solution to be $x_p(t) = M(\omega)F_0 \sin(\omega t + \phi(\omega))$ Comparing the forcing function to x_p, we see that the two functions have the same form but different amplitudes and phase shifts. The ratio of the amplitude of the particular solution (or steady state), $M(\omega)F_0$, to that of the forcing function, F_0, is $M(\omega)$ and is called the *gain*. Also, x_p is shifted in time by $|\phi(\omega)|/\omega$ radians to the right, so $\phi(\omega)$ is called the *phase shift*. When we graph the gain and the phase shift against ω (using a $\log_{10}$ on the ω-axis) we obtain the *Bodé plots*. Engineers refer to the value of $20\log_{10} M(\omega)$ as the gain in *decibels*.

1. Solve the IVP

$$\frac{d^2x}{dt^2} + 2\frac{dx}{dt} + 4x = \sin 2t, \quad x(0) = \frac{1}{2},$$

$$\frac{dx}{dt}(0) = 1.$$

2. Graph the solution simultaneously with the forcing function $f(t) = \sin 2t$.
3. Approximate $M(2)$ and $\phi(2)$ using the graph in Number 2.
4. Graph the corresponding Bodé plots.
5. Compare the values of $M(2)$ and $\phi(2)$ with those obtained in Number 2.

D. The Catenary

The solution of the second-order nonlinear IVP

$$\frac{d^2y}{dx^2} = \frac{1}{a}\sqrt{1 + \left(\frac{dy}{dx}\right)^2}, \quad y(0) = a, \quad \frac{dy}{dx}(0) = 0$$

is called a *catenary*.

1. Solve this IVP.
   A flexible wire or cable suspended between two poles of the same height takes the shape of the *catenary*,

$$y = c + a\cosh\left(\frac{x}{a}\right), \quad a > 0. \quad (5.13)$$

$y = \cosh x$ is defined by $\cosh x = \frac{1}{2}(e^x + e^{-x})$.

---

2. A flexible cable with length 150 ft. is to be suspended between two poles with height 100 ft. How far apart must the poles be spaced so that the bottom of the cable is 50 ft. off the ground?

3. According to our electric utility, *Excelsior Electric Membership Corp* (EMC), Metter, Georgia, due to terrain, easements, and so on, the average distance between utility poles ranges from 325 to 340 ft. Each pole is approximately 40 ft. long with 6 ft. buried so that the length of the pole from the ground to the top of the pole is 34 ft. The *Georgia Department of Transportation* states that the maximum height of a truck using interstates, national, and state routes is 13′6″. However, special permits may be granted by the *DOT* for heights up to 18′0″. With these restrictions in mind, EMC maintains a minimum clearance of 20′ under those lines it installs during cooler months because expansion causes lines to sag during warmer months. For the obvious reasons, EMC prefers that the distance from its lines to the ground is greater than 18′6″ at all times. Find c and a so that $f(x, c, a) = c + a\cosh\left(\frac{x}{a}\right)$ models this situation.

E. The Wave Equation on a Circular Plate

The vibrations of a circular plate satisfy the equation

$$D\nabla^4 w(r, \theta, t) + \rho h\frac{\partial^2 w(r, \theta, t)}{\partial t^2} = q(r, \theta, t), \quad (5.14)$$

For a classic approach to the subject see Graff's *Wave Motion in Elastic Solids*, [12].

---

where $\nabla^4 w = \nabla^2 \nabla^2 w$ and ∇^2 is the *Laplacian in polar coordinates*, which is defined by

$$\nabla^2 = \frac{1}{r}\frac{\partial}{\partial r}\left(r\frac{\partial}{\partial r}\right) + \frac{1}{r^2}\frac{\partial^2}{\partial\theta^2} = \frac{\partial^2}{\partial r^2} + \frac{1}{r}\frac{\partial}{\partial r} + \frac{1}{r^2}\frac{\partial^2}{\partial\theta^2}.$$

Assuming no forcing so that $q(r, \theta, t) = 0$ and $w(r, \theta, t) = W(r, \theta)e^{-i\omega t}$, Equation (5.14) can be written as

$$\nabla^4 W(r, \theta) - \beta^4 W(r, \theta) = 0, \quad \beta^4 = \omega^2\rho h/D.$$
$$(5.15)$$

For a clamped plate, the boundary conditions are $W(a, \theta) = \partial W(a, \theta)/\partial r = 0$ and after *much work* (see [12]) the *normal modes* are found to be

$$W_{nm}(r, \theta) = \left[J_n\left(\beta_{nm}r\right) - \frac{J_n\left(\beta_{nm}a\right)}{I_n\left(\beta_{nm}a\right)} I_n\left(\beta_{nm}r\right) \right]$$
$$\times \begin{pmatrix} \sin n\theta \\ \cos n\theta \end{pmatrix}. \qquad (5.16)$$

In Equation (5.16), $\beta_{nm} = \lambda_{nm}/a$ where λ_{nm} is the mth solution of

$$I_n(x)J_n'(x) - J_n(x)I_n'(x) = 0, \qquad (5.17)$$

where $J_n(x)$ is the Bessel function of the first kind of order n and $I_n(x)$ is the *modified Bessel function of the first kind* of order n, related to $J_n(x)$ by $i^n I_n(x) = J_n(ix)$.

Graph the first few normal modes of the clamped circular plate. (*Hint:* Graphing the sine and cosine part separately using n, $m = 1, \ldots, 4$ will satisfy most instructors.)

F. Duffing's Equation

Duffing's equation is the second-order nonlinear equation

$$\frac{d^2x}{dt^2} + k\frac{dx}{dt} - x + x^3 = \Gamma \cos \omega t, \qquad (5.18)$$

where k, Γ, and ω are positive constants. Depending upon the values of the parameters, solutions to Duffing's equation can exhibit *very interesting* behavior.

1. Graph the solution to Duffing's equation using 12 equally spaced values of Γ between 0 and 0.8 if $k = 0.3$, $\omega = 1.2$, $x_0 = 0$, and $y_0 = 1$ for $0 \le t \le 50$.

    The *Fourier transform*, X_k ($k = 1, 2, \ldots, N$) of N equally spaced values of a time series list $= \{x_1, x_2, \ldots, x_N\}$ is

$$X_k = \frac{1}{\sqrt{N}} \sum_{n=1}^{N} x_n e^{2\pi i(n-1)(k-1)/N}. \qquad (5.19)$$

The *power spectrum*, $P(\omega_k)$ ($k = 1, 2, \ldots, N$), of the list $\{X_1, X_2, \ldots, X_N\}$ is

$$P(\omega_k) = X_k \bar{X}_k = |x_k|^2. \qquad (5.20)$$

The power spectrum helps detect dominant frequencies. See Jordan and Smith [16].

2. Using a logarithmic scale, plot the power spectrum $P(\omega_{2000})$ of Duffing's equation if $k = 0.3$, $\Gamma = 0.5$, and $x_0 = y_0 = 0$ for 300 equally spaced values of ω between 0 and 3.

    For a second-order equation nonlinear equation like Duffing's equation, it is often desirable to generate a parametric plot of $x(t)$ versus $x'(t)$. To do so, we set $y = x'$. Then, $y' = x''$ and we see that Duffing's equation (5.18) can be rewritten as the nonlinear system

$$x' = y$$
$$y' + ky - x + x^3 = \Gamma \cos \omega t \qquad (5.21)$$

3. Plot $x(t)$ versus $x'(t)$ for 12 equally spaced values of ω betweeen 0 and 1.5 if $k = 0.3$, $\Gamma = 0.5$, $x(0) = 0$, and $y(0) = x'(0) = 1$ first for $0 \le t \le 50$ and then for $800 \le t \le 1000$. The *Poincaré plots* (or *returns*) are obtained by plotting

$$x = x\left(2n\pi/\omega\right)$$
$$x' = y = \left(2n\pi/\omega\right)$$

4. Generate a Poincaré plot for Duffing's equation if $k = 0.3$, $\Gamma = 0.4$, $\omega = 1.2$, $x(0) = 0$, and $y(0) = x'(0) = 1$.

G. Suspending an Object from a Cable

Suppose that a 1000 lb object is suspended from a steel cable with $k = 5000$ lb/in.

1. If the cable is being lowered at a rate of 25 ft./min and the cable suddenly stops, determine the frequency of the vibration produced.

2. What is the amplitude of the resulting vibration?
3. What is the maximum tension on the cable due to the vibration? (*Hint:* Consider the weight of the object added to the force due to the vibration at the point of maximum displacement.)
4. If the steel cable has a radius of 0.25 in., what is the maximum stress placed on the cable due to the vibration? (*Hint:* Stress is measured as a ratio of force and area.)
5. If the cable can withstand a stress of $8000 \, \text{lb/s}^2$, then will the cable break in this situation? If not, determine the approximate maximum weight of the object that could be supported.

H. Can Resonance Impact Machinery?

In order to apply a stamp to metal sheets, a company uses a hammering force by way of a plunger, which moves up and down in connection with a flywheel spinning at a constant speed. The metal sheet is positioned on a heavy base of mass $m = 4000 \, \text{kg}$ and the force acting on the base follows the motion of the forcing function $F(t) = 4000 \sin 10t$ where t is measured in seconds. The heavy base is also supported by an elastic pad with $k = 4 \times 10^5 \, \text{N/m}$.

1. If the initial displacement of the base from rest is 0.04 m, determine the IVP that models the displacement of the base, $x(t)$.
2. What is the homogeneous solution to the IVP in (1)?
3. What is a form for the particular solution of this IVP?
4. Solve the IVP. Does resonance occur? If so, approximate when the elastic support pad will break when it reaches an elongation of 0.4 m.

I. Inventory Management

Inventory management is important in order that companies not have too much of their product on hand to meet the demands of their customers. If they do, the company can lose money based of the costs of storing merchandise and have problems with the associated issues associated with space and safety. Let $P(t)$ represent the price at time t, which changes in proportion to the difference between the inventory level at time t, $I(t)$, and the optimal inventory level I_0. This yields the first-order ODE $\frac{dP}{dt} = -k(I - I_0)$ where k is a positive constant. Also assume that the sales can be determined by $S = 250 - 30P - 8\frac{dP}{dt}$ and that the production level is given by $Q = 120 - 4P$.[4]

1. If $\frac{dI}{dt} = Q(t) - S(t)$, differentiate $\frac{dP}{dt} = -k(I - I_0)$ to obtain the second-order ODE $\frac{d^2 P}{dt^2} + 8k\frac{dP}{dt} + 26kP = 130k$. Solve this ODE to obtain
$$P(t) = 5 + c_1 \exp(-4k - \sqrt{2}\sqrt{8k^2 - 13k}) + c_2 \exp\left(-4k + \sqrt{2}\sqrt{8k^2 - 13k}\right) \text{ if } k \neq 13/8 \text{ and}$$
$$P(t) = 5 + c_1 \exp\left(\frac{-13}{2}t\right) + c_2 t \exp\left(\frac{-13}{2}t\right) \text{ if } k = 13/8.$$
2. Investigate the behavior of $P(t)$ for the cases $k = 1, k = \frac{13}{8}$, and $k = 2$. Is the value of $P(t)$ stable in each case?

J. Heat Transfer

Suppose that two metal blocks with temperatures $T_1(t)$ and $T_2(t)$ are placed on a large plate kept at a temperature of $0\,^\circ\text{C}$.

1. Using the laws of heat conduction, derive two first-order ODEs that describe heat transfer in the two metal blocks.

---

[4] "Non-standard method for the solution of Ordinary Differential Equations arising from Pricing Policy for Optimal Inventory Leve" by Ibijola E.A., Obayomi, A.A., *Canadian Journal on Computing in Mathematics, Natural Sciences, Engineering and Medicine* Vol. 3 No. 7, December 2012.

2. Use elimination to rewrite this system of ODEs as a second- order ODE in terms of $T_1(t)$ and solve the ODE for $T_1(t)$.
3. Find $T_1(t)$ if $k=1$, $T_1(0)=10$ and $T_2(0)=40$. Use $\frac{dT_1}{dt} = -2kT_1 + kT_2$ with $k=1$ to find $T_2(t)$.

4. Use a computer algebra system (CAS) or graphing calculator to investigate the temperature in each block over the first few seconds after the blocks are placed on the plate.

6

Systems of Differential Equations

6.1 INTRODUCTION

Suppose that we consider the path of an object in the xy-plane. For example, if we were to launch a rocket or kick a ball, we would describe the object's position in the xy-plane with the vector-valued function $\mathbf{r}(t) = \langle x(t), y(t) \rangle = x(t)\mathbf{i} + y(t)\mathbf{j}$. In other words, the coordinates of the object at time $t = T$ are given at the terminal point of the vector $\mathbf{r}(T) = \langle x(T), y(T) \rangle$. Following

a method similar to that discussed in Chapter 3 concerning falling bodies, we will assume that the only forces acting on the object are caused by gravitational acceleration, $\mathbf{g} = \langle 0, -g \rangle = -g\mathbf{j}$. (Note that this force acts entirely in the vertical direction as expected.) Then if we let the velocity of the object be represented by the vector-valued function $\mathbf{v}(t) = \langle v_x(t), v_y(t) \rangle$, where $v_x(t)$ is the horizontal component of the velocity vector and $v_y(t)$ is the vertical component, and

Introductory Differential Equations
http://dx.doi.org/10.1016/B978-0-12-417219-7.00006-5

the acceleration function with $\mathbf{a}(t) = \mathbf{v}'(t) = \langle v'_x(t), v'_y(t) \rangle$, then we can use Newton's second law, $\mathbf{F} = m\mathbf{a}$, to model the situation. In this case, we have $m\mathbf{g} = m\mathbf{a}$ so that the acceleration must satisfy $\mathbf{a}(t) = \mathbf{g}$ or $\langle v'_x(t), v'_y(t) \rangle = \langle 0, -g \rangle$. Equating the corresponding components of these two vectors, we obtain the first-order system of differential equations

$$v'_x(t) = 0,$$
$$v'_y(t) = -g.$$

The equations in this system are *uncoupled* in that v_x does not appear in the equation involving v'_y and v_y does not appear in that describing v'_x. Therefore, we can solve this system by finding a solution of each individual first-order equation. If we assume that the object is launched with an initial speed v_0 making an angle α with the horizontal (Figure 6.1), the initial velocity vector is

$$\mathbf{v}(0) = \langle v_x(0), v_y(0) \rangle = \langle v_0 \cos \alpha, v_0 \sin \alpha \rangle.$$

This means that we determine the components of $\mathbf{v}(t) = \langle v_x(t), v_y(t) \rangle$ by solving the initial-value problem (IVP)

$$v'_x(t) = 0, \quad v_x(0) = v_0 \cos \alpha,$$
$$v'_y(t) = -g, \quad v_y(0) = v_0 \sin \alpha.$$

Through integration, we find that $v_x(t) = C_1$ and $v_y(t) = -gt + C_2$, so that application of the initial conditions $v_x(0) = v_0 \cos \alpha$ and $v_y(0) = v_0 \sin \alpha$ indicates that $C_1 = v_0 \cos \alpha$ and $C_2 = v_0 \sin \alpha$. Therefore, the solution to the IVP is

$$\mathbf{v}(t) = \langle v_0 \cos \alpha, -gt + v_0 \sin \alpha \rangle$$

with components $v_x(t) = v_0 \cos \alpha$ and $v_y(t) = -gt + v_0 \sin \alpha$. We determine the position function $\mathbf{r}(t)$ using the relationship $\mathbf{r}'(t) = \langle x'(t), y'(t) \rangle = \mathbf{v}(t)$. Equating the corresponding components as we did earlier, we obtain the system of first-order equations

$$x'(t) = v_0 \cos \alpha,$$
$$y'(t) = -gt + v_0 \sin \alpha.$$

If we also assume that the object is launched from the position $\mathbf{r}(0) = \langle 0, h \rangle$ (Figure 6.2), then we consider the IVP

$$x'(t) = v_0 \cos \alpha, \quad x(0) = 0,$$
$$y'(t) = -gt + v_0 \sin \alpha, \quad y(0) = h.$$

As in the previous system, we can solve the individual differential equations. Integration yields $x(t) = (v_0 \cos \alpha)t + C_3$ and $y(t) = -\frac{1}{2}gt^2 + (v_0 \sin \alpha)t + C_4$. Then, application of the initial conditions $x(0) = 0$ and $y(0) = h$ indicates that $C_3 = 0$ and $C_4 = h$, so

$$\mathbf{r}(t) = \left\langle (v_0 \cos \alpha)t, -\frac{1}{2}gt^2 + (v_0 \sin \alpha)t + h \right\rangle$$

For example, if $v_0 = 64$ ft./s, $\alpha = \pi/6$, $g \approx 32$ ft./s^2, and $h = 0$, then $\mathbf{r}(t) = \langle 32t\sqrt{3}, -16t^2 + 32t \rangle$. We graph this function (parametrically) in Figure 6.3 indicating the orientation of the curve. The object moves in the direction of the orientation as t increases.

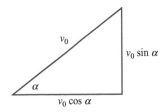

FIGURE 6.1   Sketch to determine components of $\mathbf{v}(0)$.

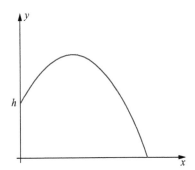

FIGURE 6.2   Sketch to determine $\mathbf{r}(0)$.

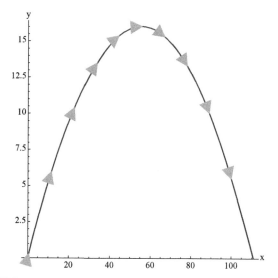

FIGURE 6.3 Path of projectile with orientation.

Of course, we may encounter systems of differential equations that are more difficult to solve than the systems discussed above. For example, when considering a population problem involving the interaction between two populations, such as a predator and prey relationship, we let $x(t)$ represent the size (or density) of the prey population at time t and $y(t)$ that of the predator population at time t. We can then model the situation with the IVP

$$dx/dt = ax - bxy, \quad x(0) = x_0,$$
$$dy/dt = -cy + dxy, \quad y(0) = y_0,$$

where the initial populations (or densities) are $x(0) = x_0$ and $y(0) = y_0$, and $a, b, c,$ and d are positive constants. Notice that when we have a system of two first-order equations, we must specify an initial condition for each of the unknowns, $x(t)$ and $y(t)$. The xy terms in each equation take into account interactions between the two populations. For example, if the two populations are foxes, denoted by y, and rabbits, denoted by x, then xy denotes a rabbit-fox interaction. This interaction hurts the rabbit population, as indicated by the negative coefficient of xy, $-b$, in

the dx/dt equation (the fox has eaten the rabbit), while it helps the fox population because the coefficient d of the xy term is positive in the dy/dt equation (the nutritious rabbit helps the fox stay healthy and reproduce).

We may also encounter systems that involve second-order differential equations. For example, we could consider a spring-mass system in which an object of mass m_1 is attached to the end of a spring with spring constant k_1 that is mounted on a support (as we did in Chapter 5). Then, if we attach a second spring with spring constant k_2 and mass m_2 to the end of the first spring, we can determine the displacement of the first and second springs, $x(t)$ and $y(t)$, respectively, by solving the system

$$m_1\, d^2x/dt^2 = -k_1 x + k_2(y - x),$$
$$m_2\, d^2y/dt^2 = -k_2(y - x).$$

Later, we discuss how this system is modeled using Hooke's law and Newton's second law. In the case of a system of two first-order equations, we must have two initial conditions for each unknown to define an initial-value problem. For example, we could have $x(0) = 0, x'(0) = -1,$ $y(0) = 1,$ and $y'(0) = 0.$

Although we can solve and approximate solutions of many different types of systems of differential equations, we will focus much of our attention on *systems of first-order linear differential equations*, which can be written in the form

$$x_1' = a_{11}(t)x_1 + a_{12}(t)x_2 + \cdots + a_{1n}(t)x_n + f_1(t),$$
$$x_2' = a_{21}(t)x_1 + a_{22}(t)x_2 + \cdots + a_{2n}(t)x_n + f_2(t),$$
$$\vdots$$
$$x_n' = a_{n1}(t)x_1 + a_{n2}(t)x_2 + \cdots + a_{nn}(t)x_n + f_n(t).$$
$$(6.1)$$

In system (6.1), there are n-dependent variables (unknown functions of t), $x_1 = x_1(t), x_2 = x_2(t),$ $\ldots, x_n = x_n(t),$ and n differential equations. If $f_j(t) \equiv 0, j = 1, 2, \ldots, n,$ then we say that the system is *homogeneous*. Otherwise, the system is *nonhomogeneous*. The system

$$x_1' = x_2 \quad \text{or} \quad \dot{x}_1 = x_2$$
$$x_2' = -x_1 \quad\quad\quad \dot{x}_2 = -x_1$$

Let $' = d/dt$ (differentiation with respect to t). Some prefer using an overdot to represent $'$. That is, $\dot{x} = x' = dx/dt$. Similarly, $\ddot{x} = x'' = d^2x/dt^2$.

---

is an example of a first-order linear homogeneous system, while

$$x_1' = x_2 \quad \text{or} \quad \dot{x}_1 = x_2$$
$$x_2' = -x_1 + \sin t \quad\quad \dot{x}_2 = -x_1 + \sin t$$

is a first-order linear nonhomogeneous system because $f_2(t) = \sin t$. A system such as

$$x_1' = x_1 - x_1^2 - x_1x_2 \quad (' \text{ denotes } d/dt)$$
$$x_2' = -x_2 + x_1x_2$$

is *nonlinear* because of the nonlinear terms x_1^2 and x_1x_2. Of course, dependent variable names other than $x_1(t), x_2(t), \ldots, x_n(t)$ can be used as well as independent variables names other than t. For example, the previous system can be written in terms of $x(s)$ and $y(s)$ with

$$dx/ds = x - x^2 - xy,$$
$$dy/ds = -y + xy.$$

We mentioned in Chapter 1 that there is a relationship between a single higher order equation and a system of first-order equations. We revisit that topic now by considering the second-order linear ordinary differential equation (ODE) with constant coefficients $x'' - 4x' + 13x = 2\sin t$. We can write this equation as a system of first-order equations by making the substitution $x' = y$ so that differentiation with respect to t yields $x'' = y'$. Solving $x'' - 4x' + 13x = 2\sin t$ for x'' gives us $x'' = 4x' - 13x + 2\sin t$. Substitution of $x'' = y'$ and $x' = y$ into this equation then gives us $y' = 4y - 13x + 2\sin t$. Therefore, the

second-order equation $x'' - 4x' + 13x = 2\sin t$ is equivalent to the system of first-order linear differential equations

$$x' = y$$
$$y' = 4y - 13x + 2\sin t \quad \text{or} \quad \begin{pmatrix} x' \\ y' \end{pmatrix}$$
$$= \begin{pmatrix} 0 & 1 \\ 4 & -13 \end{pmatrix}\begin{pmatrix} x \\ y \end{pmatrix} + \begin{pmatrix} 0 \\ 2\sin t \end{pmatrix}.$$

Matrices and their operations are reviewed in Section 6.2.

Sometimes, we can reverse this procedure as well. If we begin with a system of first-order linear equations and transform it into a higher order equation. For example, consider the system

$$dx/dt = x + 5y \quad \text{or} \quad \begin{pmatrix} x' \\ y' \end{pmatrix} = \begin{pmatrix} 1 & 5 \\ -1 & -1 \end{pmatrix}\begin{pmatrix} x \\ y \end{pmatrix}.$$
$$dy/dt = -x - y$$

If we differentiate $dx/dt = x + 5y$ with respect to t, we obtain $x'' = x' + 5y'$. Substituting $dy/dt = y' = -x - y$ into this equation then gives us

$$x'' = x' + 5(-x - y) = x' - 5x - 5y.$$

Solving $x' = x + 5y$ for y (or in this case, $5y$) yields $5y = x' - x$ so that substitution into $x'' = x' - 5x - 5y$ gives us the second-order equation $x'' = x' - (x' - x) = -4x$, or $x'' + 4x = 0$, which is a second-order equation that is equivalent to the original system. A general solution of this equation is $x(t) = c_1\cos 2t + c_2\sin 2t$. We can use $y = \frac{1}{5}(x' - x)$ to determine $y(t)$. This gives us

$$y(t) = \frac{1}{5}[(-2c_1\sin 2t + 2c_2\cos 2t)$$
$$- (c_1\cos 2t + c_2\sin 2t)]$$
$$= \frac{1}{5}(-c_1 + 2c_2)\cos 2t + \frac{1}{5}(-2c_1 - c_2)\sin 2t.$$

Therefore, we have solved the system of first-order linear equations by solving the equivalent second-order linear equation. Now, if we had considered the IVP

$$dx/dt = x + 5y, \quad x(0) = 0$$

$$dy/dt = -x - y, \quad y(0) = 1 \text{ or } \begin{pmatrix} x' \\ y' \end{pmatrix}$$

$$= \begin{pmatrix} 1 & 5 \\ -1 & -1 \end{pmatrix} \begin{pmatrix} x \\ y \end{pmatrix},$$

$$\begin{pmatrix} x(0) \\ y(0) \end{pmatrix} = \begin{pmatrix} 0 \\ 1 \end{pmatrix},$$

then we would apply the initial conditions to a general solution of the system. In our case, we have $x(0) = c_1 = 0$ so $c_1 = 0$. Then, $y(0) = 1/5 \times 2c_2$, so $c_2 = 5/2$, and the solution to the IVP is

$$x(t) = \frac{5}{2}\sin 2t, \quad y(t) = \cos 2t - \frac{1}{2}\sin 2t.$$

We graph these two functions of t in Figure 6.4(a). The graph of $x(t)$ is the solid curve, and that of $y(t)$ is the lighter curve. (We can recognize the functions by referring to the initial conditions.) We can also plot them parametrically in the xy-plane as done in Figure 6.4(b). *Note:* Observe that we include the orientation of this curve. The value of

$t = 0$ corresponds to the point $(0,1)$, while $t = \pi/4$ corresponds to $(5/2, -1/2)$, and $t = \pi/2$ corresponds to the point $(0, -1)$. Therefore, the orientation is clockwise because we move from $(0,1)$ to $(5/2, -1/2)$ as t increases.

Writing nonlinear differential equations as a system of first-order nonlinear equations can sometimes help us study the behavior of the nonlinear equation.

EXAMPLE 6.1.1

The *Van-der-Pol* equation is the nonlinear ODE

$$\frac{d^2w}{dt^2} - \mu\left(1 - w^2\right)\frac{dw}{dt} + w = 0. \quad (6.2)$$

Observe that the Van-der-Pol equation is similar to those encountered when studying spring-mass systems, except that the coefficient of w' that corresponds to damping is not constant. Therefore, we refer to this situation as *variable damping* and can observe its effect on the system.

Write the Van-der-Pol equation as a system.

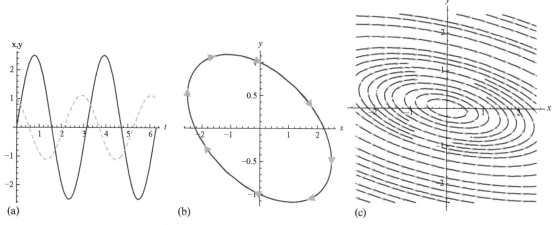

(a)   (b)   (c)

FIGURE 6.4 (a) Graphs of $x(t) = \frac{5}{2}\sin 2t$ and $y(t) = \cos 2t - \frac{1}{2}\sin 2t$. (b) Parametric plot of $\{x(t) = \frac{5}{2}\sin 2t, y(t) = \cos 2t - \frac{1}{2}\sin 2t\}$. (c) All solutions are periodic around the origin. We say that $(0,0)$ is a *stable center*.

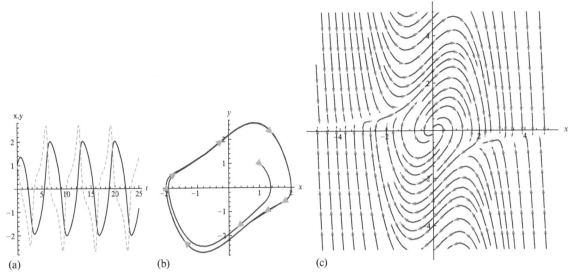

FIGURE 6.5 (a) Graph of $x(t)$ and $y(t)$ (dashed). (b) The graph of $\{x(t), y(t)\}$ indicates that the solution approaches an isolated periodic solution, which is called a *limit cycle*. (c) The phase portrait gives us a better understanding of the behavior of the solutions of the system.

Solution

Let $x = w$ and $y = w' = x'$. Then, $y' = w'' = \mu(1-w^2)w' - w = \mu(1-x^2)y - x$, so the Van-der-Pol equation is equivalent to the nonlinear system

$$dx/dt = y,$$

$$dy/dt = \mu\left(1-x^2\right)y - x. \qquad (6.3)$$

With $\mu = 1$, in Figure 6.5(a), we show graphs of the solution to the IVP

$$dx/dt = y, \quad x(0) = 1$$

$$dy/dt = \mu\left(1-x^2\right)y - x, \quad y(0) = 1$$

on the interval $[0, 25]$. Because we let $y = x'$, notice that $y(t) > 0$ when $x(t)$ is increasing and $y(t) < 0$ when $x(t)$ is decreasing. Note that these functions appear to become periodic as t increases.

Graph the solution to the IVP $w'' + (w^2 - 1)$ $w' + w = 0$, $w(0) = 1$, $w'(0) = 1$ on the interval $[0, 25]$.

The observation that these solutions become periodic is further confirmed by a graph of $x(t)$ (the horizontal axis) versus $y(t)$ (the vertical axis) shown in Figure 6.5(b), called the *phase plane*. We see that as t increases, the solution approaches a certain fixed path, called a *limit cycle*. We will find that many nonlinear equations are more easily studied when they are written as a system of first-order equations.

 **EXERCISES 6.1**

Solve each of the following uncoupled systems of equations:

1. $x' = 6, y' = \cos t$.
2. $x' = x, y' = 1$.
3. $x' = 0, y' = -2y$.
4. $x' = x^2, y' = e^t$.

Solve each of the following IVPs.

5. $\begin{cases} x_1' = -3x_1, x_1(0) = -1, \\ x_2' = 1, x_2(0) = 1, \end{cases}$

6. $\begin{cases} x_1' = -x_1 + 1, x_1(0) = 0, \\ x_2' = x_2, x_2(0) = 1. \end{cases}$

Solve each system by solving the equivalent second-order equation.

7. $x' = -3x + 6y, y' = 4x - y$.
8. $x' = 8x - y, y' = x + 6y$.
9. $x' = -x - 2y, y' = x + y$.
10. $x' = 4x + 2y, y' = -x + 2y$.
11. $x' = y, y' = -x + 1$.
12. $x' = y, y' = -x + \sin 2t$.

In Exercises 13-18, write each equation as an equivalent system of first-order equations.

13. $x'' - 3x' + 4x = 0$.
14. $x'' + 6x' + 9x = 0$.
15. $x'' + 16x = t \sin t$.
16. $x'' + x = e^t$.
17. $y''' + 3y'' + 6y' + 3y = t$.
18. $y^{(4)} + y'' = 0$.
19. *Rayleigh's equation* is

$$\frac{d^2x}{dt^2} + \mu \left[\frac{1}{3} \left(\frac{dx}{dt} \right)^2 \right] \frac{dx}{dt} + x = 0,$$

where μ is a constant.
  (a) Write Rayleigh's equation as a system.
  (b) Show that differentiating Rayleigh's equation and setting $x' = z$ reduces Rayleigh's equation to the Van-der-Pol equation.

20. (*Operator Notation*) Recall the differential operator $D = d/dt$, which was introduced in Exercises 4.2. Using this notation, we can write the system

$$y'' + 2y' - x' + 4x = \cos t$$
$$x'' + x + y'' - y' - 2y = e^t$$

as

$$(D^2 + 2D)y - (D - 4)x = \cos t$$
$$(D^2 + 1)x + (D^2 - D - 2)y = e^t.$$

Operator notation can be used to solve a system. For example, if we consider the system $Dx = y$, $Dy = -x$ or $Dx - y = 0, -Dx - D^2y = 0$, we can eliminate one of the variables by applying the operator $-D$ to the second equation to obtain $Dx - y = 0, -Dx - D^2y = 0$. When these two equations are added, we have the second-order ODE $-D^2y - y = 0$ or $D^2y + y = 0$. This equation has the characteristic equation $r^2 + 1 = 0$, with roots $r_{1,2} = \pm i$. Therefore, $y(t) = c_1 \cos t + c_2 \sin t$. At this point, we can repeat the elimination procedure to solve for x, or we can use the equation $Dy + x = 0$ or $x = -Dy$ to find x. Applying $-D$ to y yields

$$x(t) = -Dy(t) = -D(c_1 \cos t + c_2 \sin t)$$
$$= c_1 \sin t - c_2 \cos t.$$

Therefore, a general solution of the system is $x(t) = c_1 \sin t - c_2 \cos t, y(t) = c_1 \cos t + c_2 \sin t$.

In Exercises 21-27, solve the system using the operator method.

21. $x' = -2x - 2y + 4, y' = -5x + y$.
22. $x' = x + 2y + \cos t, y' = x$.
23. $x' = -3x + 2y - e^t, y' = -3y + 1$.
24. $x'' = 2y - 5x, y'' = 2x - 2y$.
25. $x'' + y'' + x + y = 0, y'' + 2x' - 2x - y = 0$.
26. $x' = -3x + 2y + 2z, y' = -2x + 3y + 2z$,
   $z' = x + y$.
27. $x' = x + 2y - 2z + \cos t, y' = x - y + 2z$,
   $z' = -x - y$.

In Exercises 28-30, solve the IVP. For each problem, graph $x(t)$, $y(t)$, and the parametric equations $\{x(t), y(t)\}$.

28. $x' - x - 2 = 0, y' - 2x - y = 0, x(0) = 0$,
   $y(0) = -1$.
29. $x' - 3x + 2y = 0, y' - 2x + y = 10, x(0) = 0$,
   $y(0) = 0$.
30. $x'' = y - x, y'' = y - x + \sin t, x(0) = 0$,
   $x'(0) = 0, y(0) = 0, y'(0) = 0$.

(*Degenerate Systems*) Suppose that the system of linear differential equations has the form $L_1x + L_2y = f_1(t), L_3x + L_4y = f_2(t)$ where L_1, L_2, L_3, and

L_4 are linear differential operators. For example, in the system

$$(D^2 + 2D)y - (D - 4)x = \cos t,$$

$$(D^2 + 1)x + (D^2 - D - 2)y = e^t,$$

$L_1 = -(D - 4)$, $L_2 = (D^2 + 2D)$, $L_3 = (D^2 + 1)$, and $L_4 = (D^2 - D - 2)$. If $\begin{vmatrix} L_1 & L_2 \\ L_3 & L_4 \end{vmatrix} = 0$, we say the system is *degenerate*, which means that the system has infinitely many solutions or no solutions. If $\begin{vmatrix} L_1 & L_2 \\ L_3 & L_4 \end{vmatrix} = \begin{vmatrix} f_1(t) & L_2 \\ f_2(t) & L_4 \end{vmatrix} = \begin{vmatrix} L_1 & f_1(t) \\ L_2 & f_2(t) \end{vmatrix} = 0$, then the system has *infinitely many solutions*. If at least one of the other two determinants is not zero, the system has *no solutions*.

In Exercises 31 and 32, show that the system has no solution.

31. $x' + x + y' = e^t, x' + x + y' = 2e^t.$
32. $x'' + y' = \cos t, 3x'' + 3y' = \sin t.$

In Exercises 33 and 34, show that the system has infinitely many solutions.

33. $x'' + y'' = t^2, 4x'' + 4y'' = 4t^2.$
34. $x'' - y'' = e^t, 2x'' - 2y'' = 2e^t.$

In Exercises 35 and 36, determine the value of k so that the system has infinitely many solutions.

35. $x' + x - y' = -t, 4y' - 4x' - 4x = kt.$
36. $y'' - kx = \cos t, 4y'' + kx = 4\cos t.$

In Exercises 37 and 38, determine restrictions on c so that the system has no solutions.

37. $x + x' - cy = e^{-t}, 2x' - 8y + 2x = 2e^{-t}.$
38. $x'' + 4y' = \cos t, 2x'' + 8y' = c\cos t.$
39. Suppose that we have a system of second-order equations of the form

$$x'' = F_1(t, x, y, x', y') \quad \text{and}$$
$$y'' = F_2(t, x, y, x', y').$$

If we let $x_1 = x$, $x_2 = x' = x_1'$, $y_1 = y$, and $y_2 = y' = y_1'$, then we can transform the original system to a system of first-order linear equations. For example, with these

substitutions, we transform the second-order linear system

$$\frac{d^2x}{dt^2} = -x + 2y \quad \text{and}$$

$$\frac{d^2y}{dt^2} = 3x - y + 5\sin t$$

into the equivalent first-order linear system

$$x_1' = x_2, \quad x_2' = -x_1 + 2y_1, \quad y_1' = y_2,$$
$$y_2' = 3x_1 - y_1 + 5\sin t.$$

Transform each of the following systems of second-order linear equations into a system of first-order linear equations:
(a) $d^2x/dt^2 = -3x + y, d^2y/dt^2 = -x - 2y.$
(b) $d^2x/dt^2 = -x + 2y + \cos 3t,$
      $d^2y/dt^2 = -2x + y.$
40. Consider the IVP

$$dx/dt = -2x - y, \quad x(0) = 1,$$
$$dy/dt = 5/4x, \quad y(0) = 0.$$

The solution obtained with one computer algebra system was

$$x(t) = \left(\frac{1}{2} - i\right) e^{(-1-i/2)t} + \left(\frac{1}{2} + i\right) e^{(-1+i/2)t}$$

and $y(t) = \frac{5}{4}ie^{(-1-i/2)t} - \frac{5}{4}e^{(-1+i/2)t},$

while that obtained with another was

$$x(t) = e^{-t}\left(\cos\left(\frac{1}{2}t\right) - 2\sin\left(\frac{1}{2}t\right)\right) \quad \text{and}$$

$$y(t) = \frac{5}{2}e^{-t}\sin\left(\frac{1}{2}t\right).$$

Show that these solutions are equivalent.

Often, we can use a computer algebra system to generate numerical solutions to a system of equations, which is particularly useful if the system under consideration is nonlinear. (Numerical techniques for systems are discussed in more detail in Section 6.7.)

6.2 REVIEW OF MATRIX ALGEBRA AND CALCULUS

- BASIC OPERATIONS
- DETERMINANTS AND INVERSES
- EIGENVALUES AND EIGENVECTORS
- MATRIX CALCULUS

When we encounter a system of linear first-order differential equations such as

$$dx/dt = -x + 2y,$$
$$dy/dt = 4x - 3y,$$

we will find that we often prefer to write the system in terms of *matrices*. Because of their importance in the study of systems of linear equations, we now briefly review matrices and the basic operations associated with them. Detailed discussions of the definitions and properties discussed here are found in introductory linear algebra texts.

Basic Operations

Definition 15 ($n \times m$ Matrix). *An $n \times m$ matrix is an array of the form*

$$\mathbf{A} = \begin{pmatrix} a_{11} & a_{12} & \dots & a_{1m} \\ a_{21} & a_{22} & \dots & a_{2m} \\ \vdots & \vdots & \ddots & \vdots \\ a_{n1} & a_{n2} & \dots & a_{nm} \end{pmatrix} \tag{6.4}$$

with n rows and m columns. This matrix can be denoted $\mathbf{A} = (a_{ij})$.

We generally call an $n \times 1$ matrix $\mathbf{v} = \begin{pmatrix} v_1 \\ v_2 \\ \vdots \\ v_n \end{pmatrix}$ a *column vector* and a $1 \times n$ matrix $\mathbf{v} = \begin{pmatrix} v_1 & v_2 & \cdots & v_n \end{pmatrix}$ a *row vector*.

Definition 16 (Transpose). *The **transpose** of the $n \times m$ matrix (6.4) is the $m \times n$ matrix*

$$\mathbf{A}^t = \begin{pmatrix} a_{11} & a_{21} & \cdots & a_{n1} \\ a_{12} & a_{22} & \cdots & a_{n2} \\ \vdots & \vdots & \ddots & \vdots \\ a_{1m} & a_{2m} & \cdots & a_{nm} \end{pmatrix} \tag{6.5}$$

Hence, $\mathbf{A}^t = (a_{ji})$

Definition 17 (Scalar Multiplication, Matrix Addition). *Let $\mathbf{A} = (a_{ij})$ be an $n \times m$ matrix and c a scalar. The **scalar multiple** of $\mathbf{A}$ by c is the $n \times m$ matrix given by $c\mathbf{A} = (ca_{ij})$.*

*If $\mathbf{B} = (b_{ij})$ is also an $n \times m$ matrix, then the **sum** of matrices $\mathbf{A}$ and $\mathbf{B}$ is the $n \times m$ matrix $\mathbf{A} + \mathbf{B} = (a_{ij}) + (b_{ij}) = (a_{ij} + b_{ij})$.*

Hence, $c\mathbf{A}$ is the matrix obtained by multiplying each element of $\mathbf{A}$ by c; $\mathbf{A} + \mathbf{B}$ is obtained by adding corresponding elements of the matrices $\mathbf{A}$ and $\mathbf{B}$ that have the same dimension.

EXAMPLE 6.2.1

Compute $3\mathbf{A} - 9\mathbf{B}$ if $\mathbf{A} = \begin{pmatrix} -1 & 4 & -2 \\ 6 & 2 & -10 \end{pmatrix}$ and $\mathbf{B} = \begin{pmatrix} 2 & -4 & 8 \\ 7 & 4 & 2 \end{pmatrix}$ What is $\mathbf{A}^t$?

Solution

Because $3\mathbf{A} = \begin{pmatrix} -3 & 12 & -6 \\ 18 & 6 & -30 \end{pmatrix}$ and $-9\mathbf{B} = \begin{pmatrix} -18 & 36 & -72 \\ -63 & -36 & -18 \end{pmatrix}$, $3\mathbf{A} - 9\mathbf{B} = 3\mathbf{A} + (-9\mathbf{B})$

$= \begin{pmatrix} -21 & 48 & -78 \\ -45 & -30 & -48 \end{pmatrix}$. $\mathbf{A}^t = \begin{pmatrix} -1 & 6 \\ 4 & 2 \\ -2 & -10 \end{pmatrix}$.

Definition 18 (Matrix Multiplication). *If $\mathbf{A} = (a_{ij})$ is an $n \times k$ matrix and $\mathbf{B} = (b_{ij})$ is an $k \times m$ matrix, $\mathbf{AB}$ is the unique $n \times m$ matrix $\mathbf{C} = (c_{ij})$ where*

$$c_{11} = a_{11}b_{11} + a_{12}b_{21} + \cdots + a_{1k}b_{k1} = \sum_{i=1}^{k} a_{1i}b_{i1},$$

$$c_{12} = a_{11}b_{12} + a_{12}b_{22} + \cdots + a_{1k}b_{k2} = \sum_{i=1}^{k} a_{1i}b_{i2},$$

and

$$c_{uv} = a_{u1}b_{1v} + a_{u2}b_{2v} + \cdots + a_{uk}b_{kv} = \sum_{i=1}^{k} a_{ui}b_{iv}.$$

In other words, the element c_{uv} is obtained by multiplying each member of the uth row of $\mathbf{A}$ by the corresponding entry in the vth column of $\mathbf{B}$ and adding the result.

EXAMPLE 6.2.2

Compute $\mathbf{AB}$ and $\mathbf{BA}$ if $\mathbf{A} = \begin{pmatrix} -1 & -5 & -5 & -4 \\ -3 & 5 & 3 & -2 \\ -4 & 4 & 2 & -3 \end{pmatrix}$

and $\mathbf{B} = \begin{pmatrix} 1 & -2 \\ -4 & 3 \\ 4 & -4 \\ -5 & -3 \end{pmatrix}.$

Solution

Because $\mathbf{A}$ is 3×4 and $\mathbf{B}$ is 4×2, $\mathbf{AB}$ is the 3×2 matrix

$$\mathbf{AB} = \begin{pmatrix} -1 \times 1 + -5 \times -4 + -5 \times 4 + -4 \times -5 \\ -3 \times 1 + 5 \times -4 + 3 \times 4 + -2 \times -5 \\ -4 \times 1 + 4 \times -4 + 2 \times 4 + -3 \times -5 \end{pmatrix.}$$

$$\begin{matrix} -1 \times -2 + -5 \times 3 + -5 \times -4 + -4 \times -3 \\ -3 \times -2 + 5 \times 3 + 3 \times -4 + -2 \times -3 \\ -4 \times -2 + 4 \times 3 + 2 \times -4 + -3 \times -3 \end{matrix}$$

$$= \begin{pmatrix} 19 & 19 \\ -1 & 15 \\ 3 & 21 \end{pmatrix}.$$

Since the number of columns of $\mathbf{B}$ is not the same as the number of rows of $\mathbf{A}$, $\mathbf{BA}$ is not defined.

Definition 19 (Identity Matrix). *The $n \times n$ matrix* $\begin{pmatrix} 1 & 0 & 0 & \dots & 0 \\ 0 & 1 & 0 & \dots & 0 \\ \vdots & \vdots & \vdots & \ddots & \vdots \\ 0 & 0 & 0 & \dots & 1 \end{pmatrix}$ *is called the $n \times n$ **identity** matrix, denoted by $\mathbf{I}$ or $\mathbf{I}_n$.*

If $\mathbf{A}$ is $n \times n$ (an $n \times n$ matrix is called a *square matrix*), then $\mathbf{IA} = \mathbf{AI} = \mathbf{A}$.

Determinants and Inverses

Definition 20 (Determinant). *If $\mathbf{A} = (a_{11})$, the **determinant** of $\mathbf{A}$, denoted by $\det(\mathbf{A})$ or $|\mathbf{A}|$, is* $\det(\mathbf{A}) = a_{11}$; *if $\mathbf{A} = \begin{pmatrix} a_{11} & a_{12} \\ a_{21} & a_{22} \end{pmatrix}$, then*

$$\det(\mathbf{A}) = |\mathbf{A}| = \begin{vmatrix} a_{11} & a_{12} \\ a_{21} & a_{22} \end{vmatrix} = a_{11}a_{22} - a_{12}a_{21}.$$

More generally, if $\mathbf{A} = (a_{ij})$ is an $n \times n$ matrix and $\mathbf{A}_{ij}$ is the $(n-1) \times (n-1)$ matrix obtained by deleting the ith row and jth column from $\mathbf{A}$, then

$$\det(\mathbf{A}) = |\mathbf{A}| = \sum_{j=1}^{n} (-1)^{i+j} a_{ij} \det(\mathbf{A}_{ij})$$

$$= \sum_{j=1}^{n} (-1)^{i+j} a_{ij} |\mathbf{A}_{ij}|.$$

*The number $(-1)^{i+j} \det(\mathbf{A}_{ij}) = \sum_{j=1}^{n} (-1)^{i+j} |\mathbf{A}_{ij}|$ is called the **cofactor** of a_{ij}. The **cofactor matrix**, $\mathbf{A}^c$, of $\mathbf{A}$ is the matrix obtained by replacing each element of $\mathbf{A}$ by its cofactor. Hence,*

$$\mathbf{A}^c = \begin{pmatrix} |\mathbf{A}_{11}| & -|\mathbf{A}_{12}| & \cdots & (-1)^{n+1}|\mathbf{A}_{1n}| \\ -|\mathbf{A}_{21}| & |\mathbf{A}_{22}| & \cdots & (-1)^{n}|\mathbf{A}_{2n}| \\ \vdots & \vdots & \ddots & \vdots \\ (-1)^{n+1}|\mathbf{A}_{n1}| & (-1)^{n}|\mathbf{A}_{n2}| & \cdots & |\mathbf{A}_{nn}| \end{pmatrix}.$$

EXAMPLE 6.2.3

Calculate $|\mathbf{A}|$ and $\mathbf{A}^c$ if $\mathbf{A} = \begin{pmatrix} -4 & -2 & -1 \\ 5 & -4 & -3 \\ 5 & 1 & -2 \end{pmatrix}.$

Solution

We illustrate that the determinant can be found by expanding along any row or column as stated in the theorem by computing the determinant in two ways. First, we choose to calculate $|\mathbf{A}|$ by expanding along the first row:

$$|\mathbf{A}| = (-1)^2 \times -4 \times \begin{vmatrix} -4 & -3 \\ 1 & -2 \end{vmatrix} + (-1)^3 \times -2$$

$$\times \begin{vmatrix} 5 & -3 \\ 5 & -2 \end{vmatrix} + (-1)^4 \times -1 \times \begin{vmatrix} 5 & -4 \\ 5 & 1 \end{vmatrix}$$

$$= -4(8+3) + 2(-10+15) - (5+20) = -59.$$

On the other hand, when we choose to expand along the second column we have:

$$|\mathbf{A}| = (-1)^3 \times -2 \times \begin{vmatrix} 5 & -3 \\ 5 & -2 \end{vmatrix} + (-1)^4 \times -4$$

$$\times \begin{vmatrix} -4 & -1 \\ 5 & -2 \end{vmatrix} + (-1)^5 \times 1 \times \begin{vmatrix} -4 & -1 \\ 5 & -3 \end{vmatrix}$$

$$= 2(-10+15) - 4(8+5) - (12+6) = -59.$$

The cofactor matrix is given by

$$\mathbf{A}^c = \begin{pmatrix} \begin{vmatrix} -4 & -3 \\ 1 & -2 \end{vmatrix} & -\begin{vmatrix} 5 & -3 \\ 5 & -2 \end{vmatrix} & \begin{vmatrix} 5 & -4 \\ 5 & 1 \end{vmatrix} \\ -\begin{vmatrix} -2 & -1 \\ 1 & -2 \end{vmatrix} & \begin{vmatrix} -4 & -1 \\ 5 & -2 \end{vmatrix} & -\begin{vmatrix} -4 & -2 \\ 5 & 1 \end{vmatrix} \\ \begin{vmatrix} -2 & -1 \\ -4 & -3 \end{vmatrix} & -\begin{vmatrix} -4 & -1 \\ 5 & -3 \end{vmatrix} & \begin{vmatrix} -4 & -2 \\ 5 & -4 \end{vmatrix} \end{pmatrix}$$

$$= \begin{pmatrix} 11 & -5 & 25 \\ -5 & 13 & -6 \\ 2 & -17 & 26 \end{pmatrix}.$$

Definition 21 (Adjoint and Inverse). **B** *is the inverse of the* $n \times n$ *matrix* **A** *means that* $\mathbf{AB} = \mathbf{BA} = \mathbf{I}$. *The adjoint,* $\mathbf{A}^a$, *of the* $n \times n$ *matrix* **A** *is the transpose of the cofactor matrix:* $\mathbf{A}^a = (\mathbf{A}^c)^t$. *If* $|\mathbf{A}| \neq 0$ *and* $\mathbf{B} = \dfrac{1}{|\mathbf{A}|}\mathbf{A}^a$, *then* $\mathbf{AB} = \mathbf{BA} = \mathbf{I}$. *Therefore, if* $|\mathbf{A}| \neq 0$, *the inverse of* **A** *is given by*

$$\mathbf{A}^{-1} = \frac{1}{|\mathbf{A}|}\mathbf{A}^a. \qquad (6.6)$$

If $|\mathbf{A}| \neq 0$ so that $\mathbf{A}^{-1}$ exists, as an alternative to using Equation (6.6) to find $\mathbf{A}^{-1}$, consider using row operations to reduce the matrix $(\mathbf{A}|\mathbf{I})$ to the form $(\mathbf{I}|\mathbf{B})$. When done correctly, $\mathbf{B} = \mathbf{A}^{-1}$. (Why?)

For a matrix **A**, the *elementary row operations* are

1. Multiply a row by a nonzero scalar.
2. Interchange two rows.
3. Add a scalar multiple of one row to another.

EXAMPLE 6.2.4

Find $\mathbf{A}^{-1}$ if $\mathbf{A} = \begin{pmatrix} -2 & -1 & 1 \\ 2 & 1 & 0 \\ 3 & 1 & -1 \end{pmatrix}$.

Solution

Using Equation (6.6), we begin by finding $|\mathbf{A}|$. Expanding along the third column, the determinant of **A**, $|\mathbf{A}|$, is given by

$$|\mathbf{A}| = 1 \times \begin{vmatrix} 2 & 1 \\ 3 & 1 \end{vmatrix} = -1 \times \begin{vmatrix} -2 & -1 \\ 2 & 1 \end{vmatrix}$$

$$= (2-3) - (-2+2) = -1.$$

We then calculate the cofactors

$$|\mathbf{A}_{11}| = \begin{vmatrix} 1 & 0 \\ 1 & -1 \end{vmatrix} = -1, \quad |\mathbf{A}_{12}| = -\begin{vmatrix} 2 & 0 \\ 3 & -1 \end{vmatrix} = 2,$$

$$|\mathbf{A}_{13}| = \begin{vmatrix} 2 & 1 \\ 3 & 1 \end{vmatrix} = -1, \quad |\mathbf{A}_{21}| = -\begin{vmatrix} -1 & 1 \\ 1 & -1 \end{vmatrix} = 0,$$

$$|\mathbf{A}_{22}| = \begin{vmatrix} -2 & 1 \\ 3 & -1 \end{vmatrix} = -1, \quad |\mathbf{A}_{23}| = -\begin{vmatrix} -2 & -1 \\ 3 & 1 \end{vmatrix} = -1,$$

$$|\mathbf{A}_{31}| = \begin{vmatrix} -1 & 1 \\ 1 & 0 \end{vmatrix} = -1, \quad |\mathbf{A}_{32}| = -\begin{vmatrix} -2 & 1 \\ 2 & 0 \end{vmatrix} = 2,$$

$$|\mathbf{A}_{33}| = \begin{vmatrix} -2 & -1 \\ 2 & 1 \end{vmatrix} = 0.$$

Then with Equation (6.6), we have

$$\mathbf{A}^{-1} = \frac{1}{|\mathbf{A}|}\mathbf{A}^a = \frac{1}{-1}\begin{pmatrix} -1 & 0 & -1 \\ 2 & -1 & 2 \\ -1 & -1 & 0 \end{pmatrix} = \begin{pmatrix} 1 & 0 & 1 \\ -2 & 1 & -2 \\ 1 & 1 & 0 \end{pmatrix}.$$

For convenience, we state the following theorem. The proof is left as an exercise.

Theorem 30 (Inverse of a 2 × 2 Matrix). *Let* $\mathbf{A} = \begin{pmatrix} a & b \\ c & d \end{pmatrix}$. *If* $|\mathbf{A}| = ad - bc \neq 0$, *Then*

$$\mathbf{A}^{-1} = \frac{1}{ad - bc}\begin{pmatrix} d & -b \\ -c & a \end{pmatrix}. \qquad (6.7)$$

EXAMPLE 6.2.5

Find $\mathbf{A}^{-1}$ if $\mathbf{A} = \begin{pmatrix} 5 & -1 \\ 2 & 3 \end{pmatrix}$.

Solution

Because $|\mathbf{A}| = 5 \times 3 - 2 \times -1 = 17$, applying formula (6.7) gives us

$$\mathbf{A}^{-1} = \frac{1}{17}\begin{pmatrix} 3 & 1 \\ -2 & 5 \end{pmatrix} = \begin{pmatrix} 3/17 & 1/17 \\ -2/17 & 5/17 \end{pmatrix}.$$

When performing row operations on matrices, we will use the convention $\mathbf{A} \xrightarrow{\alpha R_i + \beta R_j} \mathbf{B}$ to indicate that matrix $\mathbf{B}$ is obtained by replacing row j of matrix $\mathbf{A}$ by the sum of α times row i and β times row j of matrix $\mathbf{A}$.

We almost always take advantage of a computer algebra system to perform operations on higher dimension matrices. In addition, if you have taken linear algebra and are familiar with other techniques, you can use techniques such as row reduction to find the inverse of a matrix or solve a linear system of equations.

EXAMPLE 6.2.6

Find $\mathbf{A}^{-1}$ if $\mathbf{A} = \begin{pmatrix} 1 & \cos t & \sin t \\ 0 & -\sin t & \cos t \\ 0 & -\cos t & -\sin t \end{pmatrix}$.

Solution

You should verify that $|\mathbf{A}| = 1$ so $\mathbf{A}^{-1}$ exists. Rather than using formula (6.6), we illustrate how

to find $\mathbf{A}^{-1}$ by row reducing $(\mathbf{A}|\mathbf{I})$ to the form $(\mathbf{I}|\mathbf{A}^{-1})$ to find the inverse.

$$\begin{pmatrix} 1 & \cos t & \sin t & | & 1 & 0 & 0 \\ 0 & -\sin t & \cos t & | & 0 & 1 & 0 \\ 0 & -\cos t & -\sin t & | & 0 & 0 & 1 \end{pmatrix} \xrightarrow[\substack{-\csc t R_2 \\ -\cot t R_2 + R_3}]{}$$

$$\begin{pmatrix} 1 & \cos t & \sin t & | & 1 & 0 & 0 \\ 0 & 1 & -\cot t & | & 0 & -\csc t & 0 \\ 0 & 0 & -\csc t & | & 0 & -\cot t & 1 \end{pmatrix}.$$

Now multiply row three by $-\sin t$ and reduce back up the rows:

$$\begin{pmatrix} 1 & \cos t & \sin t & | & 1 & 0 & 0 \\ 0 & 1 & -\cot t & | & 0 & -\csc t & 0 \\ 0 & 0 & 1 & | & 0 & \cos t & -\sin t \end{pmatrix} \xrightarrow[\substack{-\cot R_3 + R_2 \\ -\sin t R_3 + R_1}]{}$$

$$\begin{pmatrix} 1 & \cos t & 0 & | & 1 & -\cos t \sin t & \sin^2 t \\ 0 & 1 & 0 & | & 0 & -\sin t & -\cos t \\ 0 & 0 & 1 & | & 0 & \cos t & -\sin t \end{pmatrix}.$$

Finally, multiply row two by $-\cos t$ and add to row one results in

$$\begin{pmatrix} 1 & 0 & 0 & | & 1 & 0 & 1 \\ 0 & 1 & 0 & | & 0 & -\sin t & -\cos t \\ 0 & 0 & 1 & | & 0 & \cos t & -\sin t \end{pmatrix}.$$

Thus, $\mathbf{A}^{-1} = \begin{pmatrix} 1 & 0 & 1 \\ 0 & -\sin t & -\cos t \\ 0 & \cos t & -\sin t \end{pmatrix}$.

If $\mathbf{A}^{-1}$ exists, it can be used to solve the linear system of equations $\mathbf{Ax} = \mathbf{b}$. For example, to solve $5x - y = -34$, $2x + 3y = 17$, we rewrite the system in matrix form, $\mathbf{Ax} = \mathbf{b}$, as $\underbrace{\begin{pmatrix} 5 & -1 \\ 2 & 3 \end{pmatrix}}_{\mathbf{A}} \underbrace{\begin{pmatrix} x \\ y \end{pmatrix}}_{\mathbf{x}} = \underbrace{\begin{pmatrix} -34 \\ 17 \end{pmatrix}}_{\mathbf{b}}$. As we saw previously, $\mathbf{A}^{-1} = \begin{pmatrix} 3/17 & 1/17 \\ -2/17 & 5/17 \end{pmatrix}$, so $\mathbf{x} = \mathbf{A}^{-1}\mathbf{b} = \begin{pmatrix} 3/17 & 1/17 \\ -2/17 & 5/17 \end{pmatrix}\begin{pmatrix} -34 \\ 17 \end{pmatrix} = \begin{pmatrix} -5 \\ 9 \end{pmatrix}$.

We will find several uses for the inverse in solving systems of differential equations as well.

Eigenvalues and Eigenvectors

Definition 22 (Eigenvalues and Eigenvectors). *A **nonzero** vector* $\mathbf{v}$ *is an **eigenvector** of the square matrix* $\mathbf{A}$ *if there is a number* λ, *called an **eigenvalue** of* $\mathbf{A}$, *so that*

$$\mathbf{Av} = \lambda\mathbf{v}.$$

Note: By definition, an eigenvector of a matrix is *never* the zero vector.

EXAMPLE 6.2.7

Show that $\mathbf{v}_1 = \begin{pmatrix} -1 \\ 2 \end{pmatrix}$ and $\mathbf{v}_2 = \begin{pmatrix} 1 \\ 1 \end{pmatrix}$ are eigenvectors of $\mathbf{A} = \begin{pmatrix} -1 & 2 \\ 4 & -3 \end{pmatrix}$ with eigenvalues $\lambda_1 = -5$ and $\lambda_1 = 1$, respectively.

Solution

Because $\begin{pmatrix} -1 & 2 \\ 4 & -3 \end{pmatrix} \begin{pmatrix} -1 \\ 2 \end{pmatrix} = \begin{pmatrix} 5 \\ -10 \end{pmatrix} = -5 \begin{pmatrix} -1 \\ 2 \end{pmatrix}$

and $\begin{pmatrix} -1 & 2 \\ 4 & -3 \end{pmatrix} \begin{pmatrix} 1 \\ 1 \end{pmatrix} = \begin{pmatrix} 1 \\ 1 \end{pmatrix} = 1 \begin{pmatrix} 1 \\ 1 \end{pmatrix}$, $\mathbf{v}_1$ and $\mathbf{v}_2$ are eigenvectors of $\mathbf{A}$ with corresponding eigenvalues λ_1 and λ_2, respectively.

If $\mathbf{v}$ is an eigenvector of $\mathbf{A}$ with corresponding eigenvalue λ, then $\mathbf{Av} = \lambda\mathbf{v}$. Because this equation is equivalent to the equation $(\mathbf{A} - \lambda\mathbf{I})\mathbf{v} = 0$, $\mathbf{v} \neq 0$, is an eigenvector if and only if $\det(\mathbf{A} - \lambda\mathbf{I}) = 0$.

If $\det(\mathbf{A} - \lambda\mathbf{I}) \neq 0$, *what is the solution of* $(\mathbf{A} - \lambda\mathbf{I})\mathbf{v} = 0$? *Can this solution vector be an eigenvector of* $\mathbf{A}$?

Definition 23 (Characteristic Polynomial). *If* $\mathbf{A}$ *is* $n \times n$, *equation* $(-1)^n \det(\mathbf{A} - \lambda\mathbf{I}) = 0$ *is called the **characteristic equation** of* $\mathbf{A}$; $(-1)^n \det(\mathbf{A} - \lambda\mathbf{I})$ *is called the **characteristic polynomial** of* $\mathbf{A}$.

Notice that the roots of the characteristic polynomial of $\mathbf{A}$ are the eigenvalues of $\mathbf{A}$.

EXAMPLE 6.2.8

Calculate the eigenvalues and corresponding eigenvectors of $\mathbf{A} = \begin{pmatrix} 4 & -6 \\ 3 & -7 \end{pmatrix}$.

Solution

The characteristic polynomial of $\mathbf{A}$ is

$$\begin{vmatrix} 4 - \lambda & -6 \\ 3 & -7 - \lambda \end{vmatrix} = \lambda^2 + 3\lambda - 10 = (\lambda + 5)(\lambda - 2),$$

so the eigenvalues are $\lambda_1 = -5$ and $\lambda_2 = 2$. Let $\mathbf{v}_1 = \begin{pmatrix} x_1 \\ y_1 \end{pmatrix}$ denote the eigenvectors corresponding to λ_1. Then,

$$(\mathbf{A} - \lambda_1\mathbf{I})\mathbf{v}_1 = \mathbf{0},$$

$$\left(\begin{pmatrix} 4 & -6 \\ 3 & -7 \end{pmatrix} - (-5) \begin{pmatrix} 1 & 0 \\ 0 & 1 \end{pmatrix} \right) \begin{pmatrix} x_1 \\ y_1 \end{pmatrix} = \begin{pmatrix} 0 \\ 0 \end{pmatrix},$$

$$\begin{pmatrix} 9 & -6 \\ 3 & -2 \end{pmatrix} \begin{pmatrix} x_1 \\ y_1 \end{pmatrix} = \begin{pmatrix} 0 \\ 0 \end{pmatrix}.$$

Note: We generally omit the column of zeros when forming the augmented matrix for a homogeneous linear system.

Row reducing the augmented matrix for this system,

$$\begin{pmatrix} 9 & -6 \\ 3 & -2 \end{pmatrix} \xrightarrow[\frac{1}{3}R_1]{-\frac{1}{3}R_1 + R_2} \begin{pmatrix} 3 & -2 \\ 0 & 0 \end{pmatrix}.$$

When finding an eigenvector $\mathbf{v}$ corresponding to the eigenvalue $\mathbf{A}$, we see that there is actually a collection (or family) of eigenvectors corresponding to $\mathbf{A}$. In the study of differential equations, we find that we only need to find linearly independent members of the collection of eigenvectors. Therefore, as we did in Example 6.2.8, we usually eliminate the arbitrary constants when we encounter them in eigenvectors by selecting particular values for the constants.

shows us that $3x_1 - 2y_1 = 0$. That is for any nonzero number t, $\mathbf{v}_1 = \begin{pmatrix} 2 \\ 3 \end{pmatrix} t$ is an eigenvector corresponding to λ_1. Because all these vectors are linearly dependent, we can write any one of them as a linear combination of particular one. So, we choose $\mathbf{v}_1 = \begin{pmatrix} 2 \\ 3 \end{pmatrix}$. Similarly, let $\mathbf{v}_2 = \begin{pmatrix} x_2 \\ y_2 \end{pmatrix}$ denote the eigenvectors corresponding to λ_2. Row reducing the augmented matrix for $(\mathbf{A} - \lambda_2\mathbf{I})\mathbf{v}_2 = \mathbf{0}$ gives us

$$\begin{pmatrix} 2 & -6 \\ 3 & -9 \end{pmatrix} \xrightarrow{\substack{\frac{-3}{2}R_1 + R_2 \\ \frac{1}{2}R_1}} \begin{pmatrix} 1 & -3 \\ 0 & 0 \end{pmatrix}$$

so $x_2 - 3y_2 = 0$. *Choosing* $y_2 = 1$ gives $x_2 = 3$ and $\mathbf{v}_2 = \begin{pmatrix} 3 \\ 1 \end{pmatrix}$.

EXAMPLE 6.2.9

Find the eigenvalues and corresponding eigenvectors of $\mathbf{A} = \begin{pmatrix} -4 & 5 \\ -1 & -2 \end{pmatrix}$.

Solution

In this case, the characteristic polynomial is $\begin{vmatrix} -4-\lambda & 5 \\ -1 & -2-\lambda \end{vmatrix} = \lambda^2 + 6\lambda + 13$ and solving $\lambda^2 + 6\lambda + 13 = 0$ gives us $\lambda_{1,2} = -3 \pm 2i$. Let $\mathbf{v}_1 = \begin{pmatrix} x_1 \\ y_1 \end{pmatrix}$ denote the eigenvectors corresponding to $\lambda_1 = -3 + 2i$. Then,

$$(\mathbf{A} - \lambda_1\mathbf{I})\mathbf{v}_1 = \mathbf{0},$$

$$\left(\begin{pmatrix} -4 & 5 \\ -1 & -2 \end{pmatrix} - (-3+2i)\begin{pmatrix} 1 & 0 \\ 0 & 1 \end{pmatrix}\right)\begin{pmatrix} x_1 \\ y_1 \end{pmatrix} = \begin{pmatrix} 0 \\ 0 \end{pmatrix},$$

$$\begin{pmatrix} -1-2i & 5 \\ -1 & 1-2i \end{pmatrix}\begin{pmatrix} x_1 \\ y_1 \end{pmatrix} = \begin{pmatrix} 0 \\ 0 \end{pmatrix}.$$

Row reducing the augmented matrix for this system gives us

$$\begin{pmatrix} -1-2i & 5 \\ -1 & 1-2i \end{pmatrix} \xrightarrow{\substack{-(1+2i)R_2 + R_1 \\ -R_2 \\ R_1 \leftrightarrow R_2}} \begin{pmatrix} 1 & -1+2i \\ 0 & 0 \end{pmatrix}$$

so $x_1 + (-1+2i)y_1 = 0$. Choosing $y_1 = 1$ gives us $x_1 = 1 - 2i$ and $\mathbf{v}_1 = \begin{pmatrix} 1-2i \\ 1 \end{pmatrix} = \underbrace{\begin{pmatrix} 1 \\ 1 \end{pmatrix}}_{\mathbf{a}} + \underbrace{\begin{pmatrix} -2 \\ 0 \end{pmatrix}}_{\mathbf{b}}i.$

In the same manner, we find $\mathbf{v}_2 = \begin{pmatrix} x_2 \\ y_2 \end{pmatrix}$ corresponding to $\lambda_2 = -3 - 2i$ by row reducing the augmented matrix for $(\mathbf{A} - \lambda_2\mathbf{I})\mathbf{v}_2 = \mathbf{0}$.

$$\begin{pmatrix} -1+2i & 5 \\ -1 & 1+2i \end{pmatrix} \xrightarrow{\substack{(-1+2i)R_2 + R_1 \\ -R_2 \\ R_1 \leftrightarrow R_2}} \begin{pmatrix} 1 & -1-2i \\ 0 & 0 \end{pmatrix}.$$

This indicates that $x_2 - (-1-2i)y_2 = 0$. Choosing $y_2 = 1$ results in $x_2 = 1 + 2i$ and $\mathbf{v}_2 = \begin{pmatrix} 1+2i \\ 1 \end{pmatrix} = \underbrace{\begin{pmatrix} 1 \\ 1 \end{pmatrix}}_{\mathbf{a}} - \underbrace{\begin{pmatrix} -2 \\ 0 \end{pmatrix}}_{\mathbf{b}}i.$

Recall that the complex conjugate of the complex number $z = \alpha + \beta i$ is $\bar{z} = \alpha - \beta i$. Similarly, the complex conjugate of the vector

$$\mathbf{v} = \begin{pmatrix} \alpha_1 + \beta_1 i \\ \alpha_2 + \beta_2 i \\ \vdots \\ \alpha_n + \beta_n i \end{pmatrix} = \underbrace{\begin{pmatrix} \alpha_1 \\ \alpha_2 \\ \vdots \\ \alpha_n \end{pmatrix}}_{\mathbf{a}} + \underbrace{\begin{pmatrix} \beta_1 \\ \beta_2 \\ \vdots \\ \beta_n \end{pmatrix}}_{\mathbf{b}} i = \mathbf{a} + \mathbf{b}i \text{ is}$$

the vector $\bar{\mathbf{v}} = \begin{pmatrix} \alpha_1 - \beta_1 i \\ \alpha_2 - \beta_2 i \\ \vdots \\ \alpha_n - \beta_n i \end{pmatrix} = \begin{pmatrix} \alpha_1 \\ \alpha_2 \\ \vdots \\ \alpha_n \end{pmatrix} - \begin{pmatrix} \beta_1 \\ \beta_2 \\ \vdots \\ \beta_n \end{pmatrix} i =$ $\mathbf{a} - \mathbf{b}i$.

In Example 6.2.9, the eigenvectors corresponding to the complex conjugate eigenvalues are complex conjugates. This is no coincidence. We can prove that the *eigenvectors that correspond to complex eigenvalues are themselves complex conjugates*. Moreover, if $\lambda_{1,2} = \alpha \pm \beta i$, $\beta \neq 0$, are complex conjugate eigenvalues of a matrix, our convention will be to call the eigenvector corresponding to $\lambda_1 = \alpha + \beta i$ $\mathbf{v}_1 = \mathbf{a} + \mathbf{b}i$ and the eigenvector corresponding to $\lambda_2 = \alpha - \beta i$ $\mathbf{v}_2 = \mathbf{a} - \mathbf{b}i$.

EXAMPLE 6.2.10

Calculate the eigenvalues and corresponding eigenvectors of the matrix $\mathbf{A} = \begin{pmatrix} -3 & 0 & -1 \\ -1 & -1 & -3 \\ 1 & 0 & -3 \end{pmatrix}$.

Solution

The eigenvalues are found by solving

$$\begin{vmatrix} -3-\lambda & 0 & -1 \\ -1 & -1-\lambda & -3 \\ 1 & 0 & -3-\lambda \end{vmatrix} = -10 - 16\lambda - 7\lambda^2 - \lambda^3$$

$$= -(\lambda+1)(\lambda^2 + 6\lambda + 10) = 0$$

resulting in $\lambda_1 = -1$ and $\lambda_{2,3} = -3 \pm i$.

Let $\mathbf{v}_i = \begin{pmatrix} x_i \\ y_i \\ z_i \end{pmatrix}$ denote an eigenvector corresponding to λ_i. For $\lambda_1 = -1$, $(\mathbf{A} - \lambda_1\mathbf{I})\mathbf{v}_1 = \mathbf{0}$ has augmented matrix $\begin{pmatrix} -2 & 0 & -1 \\ -1 & 0 & -3 \\ 1 & 0 & -2 \end{pmatrix}$, which reduces to $\begin{pmatrix} 1 & 0 & 0 \\ 0 & 0 & 1 \\ 0 & 0 & 0 \end{pmatrix}$ so $x_1 = y_1 = 0$ and z_1 is free. Choosing $z_1 = 1$ gives us $\mathbf{v}_1 = \begin{pmatrix} 0 \\ 0 \\ 1 \end{pmatrix}$. For $\lambda_2 = -3 + i$, $(\mathbf{A} - \lambda_2\mathbf{I})\mathbf{v}_2 = \mathbf{0}$ has augmented matrix $\begin{pmatrix} -2-3i & 0 & -1 \\ -1 & -3i & -3 \\ 1 & 0 & -2-3i \end{pmatrix}$, which reduces to

$\begin{pmatrix} 1 & 0 & -i \\ 0 & 1 & -1-i \\ 0 & 0 & 0 \end{pmatrix}$ so $x_2 = iz_2$, $y_2 = (1+i)z_2$, and z_2 is free. Choosing $z_2 = 1$ gives us $\mathbf{v}_2 = \begin{pmatrix} -i \\ 1+i \\ 1 \end{pmatrix} = \begin{pmatrix} 0 \\ 1 \\ 1 \end{pmatrix} + \begin{pmatrix} -1 \\ 1 \\ 0 \end{pmatrix} i$. Since eigenvectors of complex conjugate eigenvalues are also complex conjugates, $\mathbf{v}_3 = \begin{pmatrix} 0 \\ 1 \\ 1 \end{pmatrix} - \begin{pmatrix} -1 \\ 1 \\ 0 \end{pmatrix} i$.

Definition 24 (Eigenvalue of Multiplicity m). *Suppose that $(\lambda - \lambda_1)^m$ where m is a positive integer is a factor of the characteristic polynomial of the $n \times n$ matrix $\mathbf{A}$, while $(\lambda - \lambda_1)^{m+1}$ is not a factor of this polynomial. Then $\lambda = \lambda_1$ is an **eigenvalue of multiplicity** m.*

We often say that the eigenvalue of an $n \times n$ matrix $\mathbf{A}$ is repeated if it is of multiplicity m where $m \geq 2$ and $m \leq n$. When trying to find the eigenvector(s) corresponding to an eigenvalue of multiplicity m, two situations may be encountered: either m or fewer than m linearly independent eigenvectors can be found that correspond to the eigenvalue.

EXAMPLE 6.2.11

Find the eigenvalues and corresponding eigenvectors of $\mathbf{A} = \begin{pmatrix} 1 & -1 & 0 & 1 \\ 3 & -2 & -1 & 5 \\ -3 & 1 & 2 & -3 \\ 0 & 0 & 0 & 1 \end{pmatrix}$.

Solution

The eigenvalues are the roots of the characteristic polynomial

$$\begin{vmatrix} 1-\lambda & -1 & 0 & 1 \\ 3 & -\lambda-2 & -1 & 5 \\ -3 & 1 & 2-\lambda & -3 \\ 0 & 0 & 0 & 1-\lambda \end{vmatrix} = \lambda^4 - 2\lambda^3 + \lambda^2 = \lambda^2(\lambda-1)^2.$$

For this matrix, the eigenvalues $\lambda_{1,2} = 0$ and

$\lambda_{3,4} = 1$ each have multiplicity 2. Let $\mathbf{v}_i = \begin{pmatrix} x_i \\ y_i \\ z_i \\ w_i \end{pmatrix}$

denote the eigenvectors corresponding to λ_i. For λ_1, the augmented matrix of $(\mathbf{A} - \lambda_1\mathbf{I})\mathbf{v}_1 = \mathbf{0}$,

$$\begin{pmatrix} 1 & -1 & 0 & 1 \\ 3 & -2 & -1 & 5 \\ -3 & 1 & 2 & -3 \\ 0 & 0 & 0 & 1 \end{pmatrix}, \text{ row reduces to } \begin{pmatrix} 1 & 0 & -1 & 0 \\ 0 & 1 & -1 & 0 \\ 0 & 0 & 0 & 1 \\ 0 & 0 & 0 & 0 \end{pmatrix}.$$

This indicates that $w_1 = 0$ and z_1 is free: there is only one linearly independent eigenvector corresponding to $\lambda_{1,2} = 0$. Choosing $z_1 = 1$ yields

$x_1 = y_1 = 1$ and $\mathbf{v}_1 = \begin{pmatrix} 1 \\ 1 \\ 1 \\ 0 \end{pmatrix}$. For λ_3, the augmented

matrix of $(\mathbf{A} - \lambda_3\mathbf{I})\mathbf{v}_1 = \mathbf{0}$, $\begin{pmatrix} 0 & -1 & 0 & 1 \\ 3 & -3 & -1 & 5 \\ 1 & 1 & 1 & -3 \\ 0 & 0 & 0 & 0 \end{pmatrix}$, row

reduces to $\begin{pmatrix} 1 & 0 & -1/3 & 2/3 \\ 0 & 1 & 0 & -1 \\ 0 & 0 & 0 & 0 \\ 0 & 0 & 0 & 0 \end{pmatrix}$. This indicates that

z_3 and w_3 are free so we will be able to find two linearly independent eigenvectors corresponding to $\lambda_{3,4} = 1$. Setting $z_3 = s$ and $w_3 = t$, we find that $x_3 = \frac{1}{3}(s - 2t)$ and $y_3 = t$. Then,

$$\mathbf{v}_3 = \begin{pmatrix} \frac{1}{3}(s-2t) \\ t \\ s \\ t \end{pmatrix} = \begin{pmatrix} \frac{1}{3} \\ 0 \\ 1 \\ 0 \end{pmatrix} s + \begin{pmatrix} -\frac{2}{3} \\ 1 \\ 0 \\ 1 \end{pmatrix} t. \text{ Choosing}$$

$s = 3$ and $t = 0$ gives us $\mathbf{v}_3 = \begin{pmatrix} 1 \\ 0 \\ 3 \\ 0 \end{pmatrix}$ while choosing

$s = 0$ and $t = 3$ gives us $\mathbf{v}_4 = \begin{pmatrix} -2 \\ 3 \\ 0 \\ 3 \end{pmatrix}$. Observe that

$\mathbf{v}_3$ and $\mathbf{v}_4$ are linearly independent eigenvectors corresponding to the eigenvalue $\lambda_{3,4} = 1$. Moreover, every eigenvector corresponding to this repeated eigenvalue can be expressed as a linear combination of these two vectors.

Matrix Calculus

Definition 25 (Derivative and Integral of a Matrix). *The **derivative** of the $n \times m$ matrix $\mathbf{A}(t) =$*
$$\begin{pmatrix} a_{11}(t) & a_{12}(t) & \cdots & a_{1m}(t) \\ a_{21}(t) & a_{22}(t) & \cdots & a_{2m}(t) \\ \vdots & \vdots & \ddots & \vdots \\ a_{n1}(t) & a_{n2}(t) & \cdots & a_{nm}(t) \end{pmatrix}, \text{ where } a_{ij}(t) \text{ is differen-}$$
tiable for all values of i and j, is

$$\frac{d}{dt}\mathbf{A}(t) = \begin{pmatrix} \dfrac{d}{dt}a_{11}(t) & \dfrac{d}{dt}a_{12}(t) & \cdots & \dfrac{d}{dt}a_{1m}(t) \\ \dfrac{d}{dt}a_{21}(t) & \dfrac{d}{dt}a_{22}(t) & \cdots & \dfrac{d}{dt}a_{2m}(t) \\ \vdots & \vdots & \ddots & \vdots \\ \dfrac{d}{dt}a_{n1}(t) & \dfrac{d}{dt}a_{n2}(t) & \cdots & \dfrac{d}{dt}a_{nm}(t) \end{pmatrix}.$$

*The **integral** of $\mathbf{A}(t)$, where $a_{ij}(t)$ is integrable for all values of i and j, is*

$$\int \mathbf{A}(t)\, dt = \begin{pmatrix} \int a_{11}(t)\, dt & \int a_{12}(t)\, dt & \cdots & \int a_{1m}(t)\, dt \\ \int a_{21}(t)\, dt & \int a_{22}(t)\, dt & \cdots & \int a_{2m}(t)\, dt \\ \vdots & \vdots & \ddots & \vdots \\ \int a_{n1}(t)\, dt & \int a_{n2}(t)\, dt & \cdots & \int a_{nm}(t)\, dt \end{pmatrix}.$$

EXAMPLE 6.2.12

Find $(d/dt)\mathbf{A}(t)$ and $\int \mathbf{A}(t)\, dt$ if $\mathbf{A}(t) = \begin{pmatrix} \cos 3t & \sin 3t & e^{-t} \\ t & t\sin t^2 & \sec t \end{pmatrix}$

Solution

We find $(d/dt)\mathbf{A}(t)$ by differentiating each element of $\mathbf{A}(t)$. This yields

$$\frac{d}{dt}\mathbf{A}(t) = \begin{pmatrix} -3\sin 3t & 3\cos 3t & -e^{-t} \\ 1 & \sin t^2 + 2t^2\cos t^2 & \sec t\tan t \end{pmatrix}.$$

Similarly, we find $\int \mathbf{A}(t)\,dt$ by integrating each element of $\mathbf{A}(t)$.

$$\int \mathbf{A}(t)\,dt$$

$$= \begin{pmatrix} \frac{1}{3}\sin 3t + c_{11} & -\frac{1}{3}\cos 3t + c_{12} & -e^- + c_{13} \\ \frac{1}{2}t^2 + c_{21} & -\frac{1}{2}\cos t^2 + c_{22} & \ln|\sec t + \tan t| + c_{23} \end{pmatrix},$$

where each c_{ij} represents an arbitrary constant.

EXERCISES 6.2

In Exercises 1-4, perform the indicated calculation if $\mathbf{A} = \begin{pmatrix} 2 & -5 \\ 0 & 4 \end{pmatrix}$, $\mathbf{B} = \begin{pmatrix} -1 & 0 \\ 3 & 6 \end{pmatrix}$, and $\mathbf{C} = \begin{pmatrix} 4 & 1 \\ -2 & 7 \end{pmatrix}$.

1. $\mathbf{B} - \mathbf{A}$,
2. $2\mathbf{A} + \mathbf{C}$,
3. $(\mathbf{B} + \mathbf{C}) - 4\mathbf{A}$,
4. $(\mathbf{A} - 3\mathbf{C}) + (\mathbf{C} - 5\mathbf{B})$.

In Exercises 5-8, compute $\mathbf{AB}$ and $\mathbf{BA}$, when defined, using the given matrices.

5. $\mathbf{A} = \begin{pmatrix} 1 & 2 \\ 3 & 5 \end{pmatrix}$ and $\mathbf{B} = \begin{pmatrix} 1 & -1 \\ -1 & -5 \end{pmatrix}$.

6. $\mathbf{A} = \begin{pmatrix} -2 & 4 & 3 & 3 \\ -3 & 2 & 4 & -3 \end{pmatrix}$ and $\mathbf{B} = \begin{pmatrix} 0 & -2 & 2 \\ -2 & -4 & 1 \\ -2 & 0 & 4 \\ -1 & -4 & 5 \end{pmatrix}$.

7. $\mathbf{A} = \begin{pmatrix} 1 & -3 & -4 & 3 \\ -5 & -1 & -2 & -2 \\ 4 & -1 & 0 & -4 \end{pmatrix}$ and $\mathbf{B} = \begin{pmatrix} -5 & 2 & 4 \\ -5 & 0 & -5 \\ 0 & 4 & 5 \\ 4 & 3 & -4 \end{pmatrix}$.

8. $\mathbf{A} = \begin{pmatrix} 1 & t & \cos t & \sin t \\ 0 & 1 & -\sin t & \cos t \\ 0 & 0 & -\cos t & -\sin t \\ 0 & 0 & \sin t & -\cos t \end{pmatrix}$ and

$\mathbf{B} = \begin{pmatrix} 1 & -t & 1 & -t \\ 0 & 1 & 0 & 1 \\ 0 & 0 & -\cos t & \sin t \\ 0 & 0 & -\sin t & -\cos t \end{pmatrix}$.

In Exercises 9-13, find the determinant of the square matrix.

9. $\mathbf{A} = \begin{pmatrix} -1 & -4 \\ 5 & 3 \end{pmatrix}$.

10. $\mathbf{A} = \begin{pmatrix} 0 & 3 & 3 \\ 1 & 1 & -2 \\ -3 & 2 & -3 \end{pmatrix}$.

11. $\mathbf{A} = \begin{pmatrix} -1 & 2 & 0 & 0 \\ -1 & -1 & 0 & -1 \\ 0 & 0 & 0 & 1 \\ 0 & 1 & 1 & 1 \end{pmatrix}$.

12. $\mathbf{A} = \begin{pmatrix} 2 & 2 & 1 & 2 & 0 \\ 1 & 2 & -1 & -1 & 0 \\ 0 & -1 & -1 & 0 & 2 \\ 1 & 2 & -1 & -1 & 0 \\ 2 & 1 & 0 & 1 & 0 \end{pmatrix}$.

13. $\mathbf{A} = \begin{pmatrix} 1 & t & \cos t & \sin t \\ 0 & 1 & -\sin t & \cos t \\ 0 & 0 & -\cos t & -\sin t \\ 0 & 0 & \sin t & -\cos t \end{pmatrix}$.

In Exercises 14-18, find the inverse of each matrix.

14. $\mathbf{A} = \begin{pmatrix} 1 & -1 \\ 0 & 1 \end{pmatrix}$.

15. $\mathbf{A} = \begin{pmatrix} 0 & -1 \\ 2 & 2 \end{pmatrix}$.

16. $\mathbf{A} = \begin{pmatrix} -1 & -1 & -2 \\ 3 & 2 & 2 \\ 0 & 0 & -2 \end{pmatrix}$.

17. $\mathbf{A} = \begin{pmatrix} 0 & 2 & 2 & -1 \\ -1 & 1 & 0 & -1 \\ 0 & 1 & 0 & 0 \\ 1 & 0 & -1 & 2 \end{pmatrix}$.

18. $\mathbf{A} = \begin{pmatrix} 1 & e^t & e^{-t} \\ 0 & e^t & -e^{-t} \\ 0 & e^t & e^{-t} \end{pmatrix}$.

In Exercises 19-28, find the eigenvalues and corresponding eigenvectors of the matrix.

19. $\mathbf{A} = \begin{pmatrix} -6 & 1 \\ -2 & -3 \end{pmatrix}$.

20. $\mathbf{A} = \begin{pmatrix} -7 & 4 \\ -1 & -3 \end{pmatrix}$.

21. $\mathbf{A} = \begin{pmatrix} -1 & -1 \\ 5 & -3 \end{pmatrix}$.

22. $\mathbf{A} = \begin{pmatrix} -1 & -3 & 2 \\ -1 & 1 & 0 \\ 1 & 1 & -1 \end{pmatrix}$.

23. $\mathbf{A} = \begin{pmatrix} -1 & -1 & -2 \\ 2 & 2 & 2 \\ -2 & -1 & -1 \end{pmatrix}$.

24. $\mathbf{A} = \begin{pmatrix} 2 & 1 & -1 \\ 2 & 3 & 4 \\ -6 & -6 & -5 \end{pmatrix}$.

25. $\mathbf{A} = \begin{pmatrix} -3 & 0 & -1 \\ -1 & -1 & -3 \\ 1 & 0 & -3 \end{pmatrix}$.

26. $\mathbf{A} = \begin{pmatrix} -3 & -4 & 0 & 4 \\ -1 & -3 & 0 & 1 \\ 4 & 10 & -2 & -10 \\ -3 & -6 & 0 & 4 \end{pmatrix}$.

27. $\mathbf{A} = \begin{pmatrix} -2 & -1 & -3 & 1 \\ -2 & -3 & -2 & 2 \\ 1 & 1 & 0 & -1 \\ -1 & -1 & -3 & 0 \end{pmatrix}$.

28. $\mathbf{A} = \begin{pmatrix} -1 & -2 & 0 & 0 & 0 \\ 0 & 1 & 0 & 0 & 0 \\ 0 & 0 & -1 & 0 & 0 \\ 0 & 2 & 0 & -1 & 0 \\ 1 & 1 & -1 & 0 & 1 \end{pmatrix}$.

In Exercises 29-34, find $(d/dt)\mathbf{A}(t)$ and $\int \mathbf{A}(t)\,dt$.

29. $\mathbf{A} = \begin{pmatrix} e^{2t} \\ e^{-5t} \end{pmatrix}$.

30. $\mathbf{A} = \begin{pmatrix} \sin 3t \\ \cos 3t \end{pmatrix}$.

31. $\mathbf{A} = \begin{pmatrix} \cos t & t\cos t \\ \sin t & t\sin t \end{pmatrix}$.

32. $\mathbf{A} = \begin{pmatrix} e^{-t} & te^{-t} \\ t & t^2 \end{pmatrix}$.

33. $\mathbf{A} = \begin{pmatrix} e^{4t} \\ \cos 3t \\ \sin 3t \end{pmatrix}$.

34. $\mathbf{A} = \begin{pmatrix} t^{-1}\ln t \\ t\ln t \\ \ln t \end{pmatrix}$.

35. If $\mathbf{A} = \begin{pmatrix} a & b \\ c & d \end{pmatrix}$ and $|\mathbf{A}| = ad - bc \neq 0$, then

show that $\mathbf{A}^{-1} = \dfrac{1}{ad - bc} \begin{pmatrix} d & -b \\ -c & a \end{pmatrix}$.

Many computer software packages and calculators contain built-in functions for working with matrices. Use a computer or calculator to perform the following calculations to become familiar with these functions:

36. If $\mathbf{A} = \begin{pmatrix} 0 & 0 & -2 & 1 \\ 3 & -1 & 7 & 2 \\ -6 & 0 & 5 & -1 \\ -6 & 0 & 1 & -2 \end{pmatrix}$ and

$\mathbf{B} = \begin{pmatrix} -7 & -6 & -3 & -7 \\ 2 & -3 & 0 & 4 \\ 3 & 4 & 1 & 2 \\ 5 & 6 & 3 & 6 \end{pmatrix}$, compute (a)

$3\mathbf{A} - 2\mathbf{B}$; (b) $\mathbf{B}^t$; (c) $\mathbf{AB}$; (d) $|\mathbf{A}|$, $|\mathbf{B}|$, and $|\mathbf{AB}|$; and (e) $\mathbf{A}^{-1}$.

37. Find the eigenvalues and corresponding eigenvectors (numerically) of the matrices

$\mathbf{A} = \begin{pmatrix} -4 & 4 & -4 \\ 2 & 3 & -4 \\ 5 & 0 & -1 \end{pmatrix}$ and $\mathbf{B} = \begin{pmatrix} 1 & 3 & -5 \\ 5 & 5 & 0 \\ -5 & -2 & 3 \end{pmatrix}$.

38. Find $(d/dt)\mathbf{A}(t)$ and $\int \mathbf{A}(t)\,dt$ if $\mathbf{A}(t) = $
$\begin{pmatrix} te^{-t} & t^2\sin 2t & (9+4t^2)^{-1} \\ \cos^6 t & \sec^3 2t & t^{-1}(t-1)^{-1} \\ 4(1-t^2)^{-1/2} & \sin^5 t\cos t & t^3\sin^2 t \end{pmatrix}$.

39. Calculate the eigenvalues of $\begin{pmatrix} 0 & -1-k^2 \\ 1 & 2 \end{pmatrix}$

and $\begin{pmatrix} 0 & -1-k^2 \\ 1 & 2k \end{pmatrix}$. How do the eigenvalues change for $-\infty < k < \infty$?

40. Nearly every computer algebra system can compute exact values of the eigenvalues and corresponding eigenvectors of an $n \times n$ matrix for $n = 2, 3$, and 4. (a) Use a computer algebra system to compute the exact values of the eigenvalues and corresponding eigenvectors of

$$\mathbf{A} = \begin{pmatrix} 3 & 1 & 1 & 1 \\ -2 & -3 & 2 & -3 \\ -3 & 3 & -2 & 1 \\ 0 & 1 & 2 & 0 \end{pmatrix}. \text{ Write down a}$$

portion of the result (or print it and staple it to your homework). (b) Compute approximations of the eigenvalues and corresponding eigenvectors of $\mathbf{A}$. (c) In a sentence, explain which results are more meaningful to you.

6.3 AN INTRODUCTION TO LINEAR SYSTEMS

We first encounter systems of linear equations in elementary algebra courses. For example,

$$3x - 5y = -13$$
$$-3x + 6y = 15$$

written in matrix form as

$$\begin{pmatrix} 3 & -5 \\ -3 & 6 \end{pmatrix} \begin{pmatrix} x \\ y \end{pmatrix} = \begin{pmatrix} -13 \\ 15 \end{pmatrix}$$

is a system of two linear equations in two variables with solution $(x, y) = (-1, 2)$, which is written in matrix or vector notations using our convention as $\begin{pmatrix} x \\ y \end{pmatrix} = \begin{pmatrix} -1 \\ 2 \end{pmatrix}$. In the same manner, we can consider a system of linear differential equations.

We begin our study of systems of linear ODEs by introducing several definitions along with some convenient notation. Let

$$\mathbf{X}(t) = \begin{pmatrix} x_1(t) \\ x_2(t) \\ \vdots \\ x_n(t) \end{pmatrix},$$

$$\mathbf{A}(t) = \begin{pmatrix} a_{11}(t) & a_{12}(t) & \dots & a_{1n}(t) \\ a_{21}(t) & a_{22}(t) & \dots & a_{2n}(t) \\ \vdots & \vdots & \ddots & \vdots \\ a_{n1}(t) & a_{n2}(t) & \dots & a_{nn}(t) \end{pmatrix}, \text{ and}$$

$$\mathbf{F}(t) = \begin{pmatrix} f_1(t) \\ f_2(t) \\ \vdots \\ f_n(t) \end{pmatrix}.$$

Then, the homogeneous system of first-order linear differential equations

$$\begin{aligned} x_1' &= a_{11}(t)x_1 + a_{12}(t)x_2 + \dots + a_{1n}x_n(t), \\ x_2' &= a_{21}(t)x_1 + a_{22}(t)x_2 + \dots + a_{2n}x_n(t), \\ &\vdots \\ x_n' &= a_{n1}(t)x_1 + a_{n2}(t)x_2 + \dots + a_{nn}x_n(t) \end{aligned} \tag{6.8}$$

is equivalent to

$$\mathbf{X}'(t) = \mathbf{A}(t)\mathbf{X}(t) \tag{6.9}$$

and the nonhomogeneous system

$$\begin{aligned} x_1' &= a_{11}(t)x_1 + a_{12}(t)x_2 + \dots + a_{1n}x_n(t) + f_1(t), \\ x_2' &= a_{21}(t)x_1 + a_{22}(t)x_2 + \dots + a_{2n}x_n(t) + f_2(t), \\ &\vdots \\ x_n' &= a_{n1}(t)x_1 + a_{n2}(t)x_2 + \dots + a_{nn}x_n(t) + f_n(t) \end{aligned} \tag{6.10}$$

is equivalent to

$$\mathbf{X}'(t) = \mathbf{A}(t)\mathbf{X}(t) + \mathbf{F}(t). \tag{6.11}$$

For the nonhomogeneous system (6.11), the *corresponding homogeneous system* is system (6.9).

EXAMPLE 6.3.1

(a) Write the homogeneous system
$\begin{cases} x' = -5x + 5y \\ y' = -5x + y \end{cases}$ in matrix form. (b) Write the
nonhomogeneous system $\begin{cases} x' = x + 2y - \sin t \\ y' = 4x - 3y + t^2 \end{cases}$
in matrix form.

Solution

(a) The homogeneous system $\begin{cases} x' = -5x + 5y \\ y' = -5x + y \end{cases}$
is equivalent to the system $\begin{pmatrix} x \\ y \end{pmatrix}' = \begin{pmatrix} -5 & 5 \\ -5 & 1 \end{pmatrix} \begin{pmatrix} x \\ y \end{pmatrix}$.

With our notation, $\begin{pmatrix} x \\ y \end{pmatrix}' = \begin{pmatrix} x' \\ y' \end{pmatrix}$

(b) The nonhomogeneous system
$\begin{cases} x' = x + 2y - \sin t \\ y' = 4x - 3y + t^2 \end{cases}$ is equivalent to $\begin{pmatrix} x \\ y \end{pmatrix}' =$
$\begin{pmatrix} 1 & 2 \\ 4 & -3 \end{pmatrix} \begin{pmatrix} x \\ y \end{pmatrix} + \begin{pmatrix} -\sin t \\ t^2 \end{pmatrix}$.

The nth-order linear equation

$$y^{(n)}(t) + a_{n-1}(t)y^{(n-1)} + \cdots + a_2(t)y''$$
$$+ a_1(t)y' + a_0(t)y = f(t), \qquad (6.12)$$

The nth-order linear equation is discussed in Chapter 4.

---

discussed in previous chapters, can be written as a system of first-order equations as well. Let $x_1 = y$, $x_2 = dx_1/dt = y'$, $x_3 = dx_2/dt = y''$, $\ldots, x_{n-1} = dx_{n-2}/dt = y^{(n-2)}$, $x_n = dx_{n-1}/dt = y^{(n-1)}$. Then, Equation (6.12) is equivalent to the system

$$x_1' = x_2,$$
$$x_2' = x_3,$$
$$\vdots \qquad\qquad (6.13)$$
$$x_{n-1}' = x_n,$$
$$x_n' = -a_{n-1}x_n - \cdots - a_2x_3 - a_1x_2 - a_0x_1 + f(t),$$

which can be written in matrix form as

$$\begin{pmatrix} x_1 \\ x_2 \\ \vdots \\ x_{n-1} \\ x_n \end{pmatrix}' = \begin{pmatrix} 0 & 1 & 0 & \cdots & 0 \\ 0 & 0 & 1 & \cdots & 0 \\ \vdots & \vdots & \vdots & \cdots & \vdots \\ 0 & 0 & 0 & \cdots & 1 \\ -a_0 & -a_1 & -a_2 & \cdots & -a_n \end{pmatrix} \begin{pmatrix} x_1 \\ x_2 \\ \vdots \\ x_{n-1} \\ x_n \end{pmatrix}$$
$$+ \begin{pmatrix} 0 \\ 0 \\ \vdots \\ 0 \\ f(t) \end{pmatrix}. \qquad (6.14)$$

EXAMPLE 6.3.2

Write the equation $y'' + 5y' + 6y = \cos t$ as a system of first-order differential equations.

Solution

We let $x_1 = y$ and $x_2 = x_1' = y'$. Then,

$$x_2' = y'' = \cos t - 6y - 5y' = \cos t - 6x_1 - 5x_2$$

so the second-order equation $y'' + 5y' + 6y = \cos t$ is equivalent to the system

$$x_1' = x_2,$$
$$x_2' = \cos t - 6x_1 - 5x_2,$$

which can be written in matrix form as

$$\begin{pmatrix} x_1 \\ x_2 \end{pmatrix}' = \begin{pmatrix} 0 & 1 \\ -6 & -5 \end{pmatrix} \begin{pmatrix} x_1 \\ x_2 \end{pmatrix} + \begin{pmatrix} 0 \\ \cos t \end{pmatrix}.$$

At this point, given a system of ODEs, our goal is to construct either an explicit, numerical, or graphical solution of the system of equations.

We now state the theorems and terminology used in establishing the fundamentals of solving systems of differential equations. All proofs are omitted but can be found in advanced differential equations textbooks. In each case, we assume that the matrix $\mathbf{A} = \mathbf{A}(t)$ in the systems $\mathbf{X}'(t) = \mathbf{A}(t)\mathbf{X}(t)$ (Equation 6.9) and $\mathbf{X}'(t) = \mathbf{A}(t)\mathbf{X}(t) + \mathbf{F}(t)$ (Equation 6.11) is an $n \times n$ matrix.

Definition 26 (Solution Vector). *A **solution vector** (or solution) of the system $\mathbf{X}'(t) = \mathbf{A}(t)\mathbf{X}(t) + \mathbf{F}(t)$ (Equation 6.11) on the interval I is an $n \times 1$ matrix (or vector) of the form*

$$\Phi(t) = \begin{pmatrix} \phi_1(t) \\ \phi_2(t) \\ \vdots \\ \phi_n(t) \end{pmatrix},$$

where the $\phi_i(t)$ are differentiable functions, that satisfies $\mathbf{X}'(t) = \mathbf{A}(t)\mathbf{X}(t) + \mathbf{F}(t)$ on I.

EXAMPLE 6.3.3

Show that $\Phi_1(t) = \begin{pmatrix} -2e^{2t} \\ e^{2t} \end{pmatrix}$ is a solution of

$$\mathbf{X}' = \begin{pmatrix} 1 & -2 \\ 2 & 6 \end{pmatrix} \mathbf{X}.$$

Solution

Notice that $\Phi_1'(t) = \begin{pmatrix} -4e^{2t} \\ 2e^{2t} \end{pmatrix}$ and $\begin{pmatrix} 1 & -2 \\ 2 & 6 \end{pmatrix}\Phi_1 =$

$\begin{pmatrix} 1 & -2 \\ 2 & 6 \end{pmatrix}\begin{pmatrix} -2e^{2t} \\ e^{2t} \end{pmatrix} = \begin{pmatrix} -2e^{2t} - 2e^{2t} \\ -4e^{2t} + 6e^{2t} \end{pmatrix} = \begin{pmatrix} -4e^{2t} \\ 2e^{2t} \end{pmatrix} =$

Φ_1'. Then, because $\Phi_1' = \begin{pmatrix} 1 & -2 \\ 2 & 6 \end{pmatrix}\Phi_1$, Φ_1 is a solution of the system.

Theorem 31 (Principle of Superposition). *Suppose that $\Phi_1(t), \Phi_2(t), \ldots, \Phi_m(t)$ are m solutions of the linear homogeneous system $\mathbf{X}'(t) = \mathbf{A}(t)\mathbf{X}(t)$ (Equation 6.9) on the open interval I. Then, the linear combination*

$$\Phi(t) = c_1\Phi_1(t) + c_2\Phi_2(t) + \cdots + c_m\Phi_m(t),$$

where $c_1, c_2, \ldots, c_m$ are arbitrary constants, is also a solution of $\mathbf{X}'(t) = \mathbf{A}(t)\mathbf{X}(t)$.

EXAMPLE 6.3.4

Show that $\Phi_2(t) = \begin{pmatrix} -e^{5t} \\ 2e^{5t} \end{pmatrix}$ and $\Phi(t) =$

$$c_1\begin{pmatrix} -2e^{2t} \\ e^{2t} \end{pmatrix} + c_2\begin{pmatrix} -e^{5t} \\ 2e^{5t} \end{pmatrix} = \begin{pmatrix} -2e^{2t} & -e^{5t} \\ e^{2t} & 2e^{5t} \end{pmatrix}\begin{pmatrix} c_1 \\ c_2 \end{pmatrix}$$

are solutions of $\mathbf{X}' = \begin{pmatrix} 1 & -2 \\ 2 & 6 \end{pmatrix}\mathbf{X}$.

Solution

We let the reader follow the procedure used in Example 6.3.3 to show that Φ_2 satisfies $\mathbf{X}' = \begin{pmatrix} 1 & -2 \\ 2 & 6 \end{pmatrix}\mathbf{X}$. By the Principle of Superposition, the linear combination of the two solutions Φ_1 (from Example 6.3.3) and Φ_2 is also a solution. We verify this now by first writing $\Phi(t)$ as $\Phi(t) = \begin{pmatrix} -2c_1e^{2t} - c_2e^{5t} \\ c_1e^{2t} + 2c_2e^{5t} \end{pmatrix}$. Then, $\Phi'(t) = \begin{pmatrix} -4c_1e^{2t} - 5c_2e^{5t} \\ 2c_1e^{2t} + 10c_2e^{5t} \end{pmatrix}$ and

$$\begin{pmatrix} 1 & -2 \\ 2 & 6 \end{pmatrix}\begin{pmatrix} -4c_1e^{2t} - 5c_2e^{5t} \\ 2c_1e^{2t} + 10c_2e^{5t} \end{pmatrix}$$

$$= \begin{pmatrix} (-2c_1e^{2t} - c_2e^{5t}) - 2(c_1e^{2t} + 2c_2e^{5t}) \\ 2(-2c_1e^{2t} - c_2e^{5t}) + 6(c_1e^{2t} + 2c_2e^{5t}) \end{pmatrix}$$

$$= \begin{pmatrix} -4c_1e^{2t} - 5c_2e^{5t} \\ 2c_1e^{2t} + 10c_2e^{5t} \end{pmatrix}.$$

Therefore, the linear combination of the solutions is also a solution.

We define linear dependence and independence of a set of vector-valued functions $S = \{\Phi_1(t), \Phi_2(t), \ldots, \Phi_m(t)\}$ in a manner similar as to how we defined linear dependence and independence of sets of real-valued functions. The set $S = \{\Phi_1(t), \Phi_2(t), \ldots, \Phi_m(t)\}$ is *linearly dependent* on an interval I if there is a set of constants $c_1, c_2, \ldots, c_m$ not all zero such that

$$c_1 \Phi_1(t) + c_2 \Phi_2(t) + \cdots + c_m \Phi_m(t) = \mathbf{0};$$

otherwise, the set is *linearly independent*. (Note that $\mathbf{0}$ is the zero vector with the same dimensions as each of the $\Phi_i(t)$, $i = 1, 2, \ldots, m$.) As with two real-valued functions, two vector-valued functions are linearly dependent if they are scalar multiples of each other. Otherwise, the functions are linearly independent. For more than two vector-valued functions, we often use the Wronskian to determine if the functions are linearly independent or dependent.

Definition 27 (Wronskian of a Set of Vector-Valued Functions). *The **Wronskian** of $S = \{\Phi_1(t), \Phi_2(t), \ldots, \Phi_n(t)\}$ is the determinant of the*

$$\text{matrix with columns } \Phi_1(t) = \begin{pmatrix} \phi_{11}(t) \\ \phi_{21}(t) \\ \vdots \\ \phi_{n1}(t) \end{pmatrix}, \Phi_2(t) =$$

$$\begin{pmatrix} \phi_{12}(t) \\ \phi_{22}(t) \\ \vdots \\ \phi_{n2}(t) \end{pmatrix}, \ldots, \Phi_n(t) = \begin{pmatrix} \phi_{1n}(t) \\ \phi_{2n}(t) \\ \vdots \\ \phi_{nn}(t) \end{pmatrix},$$

$$W(S) = W\left(\{\Phi_1(t), \Phi_2(t), \ldots, \Phi_n(t)\}\right)$$

$$= \begin{vmatrix} \phi_{11}(t) & \phi_{12}(t) & \cdots & \phi_{1n}(t) \\ \phi_{21}(t) & \phi_{22}(t) & \cdots & \phi_{2n}(t) \\ \vdots & \vdots & \ddots & \vdots \\ \phi_{n1}(t) & \phi_{n2}(t) & \cdots & \phi_{nn}(t) \end{vmatrix}. \quad (6.15)$$

Theorem 32 (Wronskian of Solutions). *Suppose that*

$$S = \{\Phi_1(t), \Phi_2(t), \ldots, \Phi_n(t)\}$$

is a set of n solutions of the linear homogeneous system $\mathbf{X}'(t) = \mathbf{A}(t)\mathbf{X}(t)$ (Equation 6.9) on the open interval I, where each component of $\mathbf{A}(t)$ is continuous on I. If S is linearly dependent, then $W(S) = 0$ on I. If S is linearly independent, then $W(S) \neq 0$ for all values on I.

EXAMPLE 6.3.5

Verify that $\Phi_1(t) = \begin{pmatrix} -2e^{2t} \\ e^{2t} \end{pmatrix}$ and $\Phi_2(t) = \begin{pmatrix} -e^{5t} \\ 2e^{5t} \end{pmatrix}$ are linearly independent solutions of $\mathbf{X}' = \begin{pmatrix} 1 & -2 \\ 2 & 6 \end{pmatrix} \mathbf{X}.$

Solution

In Examples 6.3.3 and 6.3.4, we showed that $\Phi_1(t)$ and $\Phi_2(t)$ are solutions of the system $\mathbf{X}' = \begin{pmatrix} 1 & -2 \\ 2 & 6 \end{pmatrix} \mathbf{X}$. Therefore, we calculate

$$W(\{\Phi_1, \Phi_2\}) = \begin{vmatrix} -2e^{2t} & -e^{5t} \\ e^{2t} & 2e^{5t} \end{vmatrix} = -3e^{7t}. \text{ The}$$

vector-valued functions are linearly independent because $W(\{\Phi_1, \Phi_2\}) = -3e^{7t} \neq 0$ for all values of t.

Definition 28 (Fundamental Set of Solutions). *Any set*

$$S = \{\Phi_1(t), \Phi_2(t), \ldots, \Phi_n(t)\}$$

*of n linearly independent solutions of the linear homogeneous system $\mathbf{X}'(t) = \mathbf{A}(t)\mathbf{X}(t)$ (Equation 6.9) on an open interval I is called a **fundamental set of solutions** on I.*

EXAMPLE 6.3.6

Which of the following is a fundamental set of solutions for

$$\begin{pmatrix} x \\ y \end{pmatrix}' = \begin{pmatrix} -2 & -8 \\ 1 & 2 \end{pmatrix} \begin{pmatrix} x \\ y \end{pmatrix}?$$

(a) $S_1 = \left\{ \begin{pmatrix} \cos 2t \\ \sin 2t \end{pmatrix}, \begin{pmatrix} \sin 2t \\ \cos 2t \end{pmatrix} \right\}$ (b) $S_2 = \left\{ \begin{pmatrix} -2\sin 2t + 2\cos 2t \\ \sin 2t \end{pmatrix}, \begin{pmatrix} 4\cos 2t \\ \sin 2t - \cos 2t \end{pmatrix} \right\}$

Solution

We first remark that the equation $\begin{pmatrix} x \\ y \end{pmatrix}' = \begin{pmatrix} -2 & -8 \\ 1 & 2 \end{pmatrix}\begin{pmatrix} x \\ y \end{pmatrix}$ is equivalent to the system $\begin{cases} x' = -2x - 8y \\ y' = x + 2y \end{cases}$. (a) Differentiating we see that

$$\begin{pmatrix} \cos 2t \\ \sin 2t \end{pmatrix}' = \begin{pmatrix} -2\sin 2t \\ 2\cos 2t \end{pmatrix} \neq \begin{pmatrix} -2\cos 2t - 8\sin 2t \\ \cos 2t + 2\sin 2t \end{pmatrix},$$

which shows us that $\begin{pmatrix} \cos 2t \\ \sin 2t \end{pmatrix}$ is not a solution of the system. Therefore, S_1 is not a fundamental set of solutions. (b) You should verify that both $\begin{pmatrix} -2\sin 2t + 2\cos 2t \\ \sin 2t \end{pmatrix}$ and $\begin{pmatrix} 4\cos 2t \\ \sin 2t - \cos 2t \end{pmatrix}$ are solutions of the system. Computing the Wronskian we have

$$\begin{vmatrix} -2\sin 2t + 2\cos 2t & 4\cos 2t \\ \sin 2t & \sin 2t - \cos 2t \end{vmatrix}$$
$$= (-2\sin 2t + 2\cos 2t)(\sin 2t - \cos 2t)$$
$$- (4\cos 2t)(\sin 2t)$$
$$= -2\cos^2 2t - 2\sin^2 2t = -2.$$

Thus, the set S_2 is a set of two linearly independent solutions of the system and, consequently, a fundamental set of solutions.

Show that any linear combination of $\begin{pmatrix} -2\sin 2t + 2\cos 2t \\ \sin 2t \end{pmatrix}$ and $\begin{pmatrix} 4\cos 2t \\ \sin 2t - \cos 2t \end{pmatrix}$ is also a solution of the system.

The following theorem implies that a fundamental set of solutions cannot contain more than n vectors, because the solutions would not be linearly independent.

Theorem 33. *Any $n + 1$ nontrivial solutions of $\mathbf{X}'(t) = \mathbf{A}(t)\mathbf{X}(t)$ are linearly dependent.*

Finally, we state the following theorems, which state that a fundamental set of solutions

of $\mathbf{X}'(t) = \mathbf{A}(t)\mathbf{X}(t)$ exists and a general solution can (theoretically) be constructed.

Theorem 34. *There is a set of n nontrivial linearly independent solutions of $\mathbf{X}'(t) = \mathbf{A}(t)\mathbf{X}(t)$.*

Theorem 35 (General Solution). *Let $S =$*

$$\{\Phi_1(t), \Phi_2(t), \ldots, \Phi_n(t)\} = \left\{ \begin{pmatrix} \Phi_{1i} \\ \Phi_{2i} \\ \vdots \\ \Phi_{ni} \end{pmatrix} \right\}_{i=1}^n \text{ be a}$$

*fundamental set of solutions of $\mathbf{X}'(t) = \mathbf{A}(t)\mathbf{X}(t)$ on the open interval I, where each component of $\mathbf{A}(t)$ is continuous on I. Then every solution of $\mathbf{X}'(t) = \mathbf{A}(t)\mathbf{X}(t)$ is a linear combination of these solutions. Therefore, a **general solution** of $\mathbf{X}'(t) = \mathbf{A}(t)\mathbf{X}(t)$ is*

$$\mathbf{X}(t) = c_1\Phi_1(t) + c_2\Phi_2(t) + \cdots + c_n\Phi_n(t).$$

Thus, Examples 6.3.3, 6.3.4, and 6.3.5 imply that $\mathbf{X}(t) = c_1\begin{pmatrix} -2e^{2t} \\ e^{2t} \end{pmatrix} + c_2\begin{pmatrix} -e^{5t} \\ 2e^{5t} \end{pmatrix} = \begin{pmatrix} -2c_1e^{2t} - c_2e^{5t} \\ c_1e^{2t} + 2c_2e^{5t} \end{pmatrix} = \underbrace{\begin{pmatrix} -2e^{2t} & -e^{5t} \\ e^{2t} & 2e^{5t} \end{pmatrix}}_{\Phi \text{ "Fundamental Matrix"}} \underbrace{\begin{pmatrix} c_1 \\ c_2 \end{pmatrix}}_{\mathbf{C}}$ is

a general solution of $\mathbf{X}' = \begin{pmatrix} 1 & -2 \\ 2 & 6 \end{pmatrix}\mathbf{X}.$

EXAMPLE 6.3.7

Given that $\Phi_1(t) = \begin{pmatrix} t^{-1} \\ t \end{pmatrix}$ and

$\Phi_1(t) = \begin{pmatrix} 1 \\ t^{-1} \end{pmatrix}$ are solutions of

$$\mathbf{X}' = \begin{pmatrix} (t^4 - t)^{-1} & (t^3 - 1)^{-1} \\ 2t(1 - t^3)^{-1} & (1 + t^3)(t^4 - t)^{-1} \end{pmatrix}\mathbf{X},$$ find a

general solution of this equation.

Solution

To verify linear independence of these two solutions, we compute the Wronskian:

$$\begin{vmatrix} t^{-1} & 1 \\ t & t^{-1} \end{vmatrix} = t^{-2} - t \neq 0.$$

Hence, we have two linearly independent solutions of the equation, so a general solution is given by

$$\mathbf{X}(t) = c_1 \begin{pmatrix} t^{-1} \\ t \end{pmatrix} + c_2 \begin{pmatrix} 1 \\ t^{-1} \end{pmatrix}$$

$$= \underbrace{\begin{pmatrix} t^{-1} & 1 \\ t & t^{-1} \end{pmatrix}}_{\Phi \text{ "Fundamental Matrix"}} \underbrace{\begin{pmatrix} c_1 \\ c_2 \end{pmatrix}}_{\mathbf{C}}.$$

Definition 29 (Fundamental Matrix). *Suppose that* $S = \{\Phi_1(t), \Phi_2(t), \ldots, \Phi_n(t)\} = \left\{ \begin{pmatrix} \Phi_{1i} \\ \Phi_{2i} \\ \vdots \\ \Phi_{ni} \end{pmatrix} \right\}_{i=1}^{n}$ *is a fundamental set of solutions of* $\mathbf{X}'(t) = \mathbf{A}(t)\mathbf{X}(t)$ *on the open interval I, where each component of* $\mathbf{A}(t)$ *is continuous on I. The matrix*

$$\Phi(t) = \begin{pmatrix} \Phi_1(t) & \Phi_2(t) \ldots \Phi_n(t) \end{pmatrix}$$

is called a **fundamental matrix** *of the system* $\mathbf{X}'(t) = \mathbf{A}(t)\mathbf{X}(t)$ *on I.*

If $\Phi(t)$ is a fundamental matrix of the system $\mathbf{X}'(t) = \mathbf{A}(t)\mathbf{X}(t)$, a general solution can be written as $\mathbf{X}(t) = \Phi(t)\mathbf{C}$, where $\mathbf{C} = \begin{pmatrix} c_1 \\ c_2 \\ \vdots \\ c_n \end{pmatrix}$.

The theorems and definitions introduced in this section indicate that when solving an $n \times n$ homogeneous system of linear first-order equations, $\mathbf{X}'(t) = \mathbf{A}(t)\mathbf{X}(t)$, we find n linearly independent solutions. After finding these solutions, we form a fundamental matrix that can be used to form a general solution or solve an IVP.

Thus, $\Phi(t) = \begin{pmatrix} e^{-2t} & -3e^{5t} \\ 2e^{-2t} & e^{5t} \end{pmatrix}$ is a fundamental matrix for the system $\mathbf{X}' = \begin{pmatrix} 4 & -3 \\ -2 & -1 \end{pmatrix} \mathbf{X}$

because each column vector of Φ is a solution of the system:

$$\Phi'(t) = \begin{pmatrix} -2e^{-2t} & -15e^{5t} \\ -4e^{-2t} & 5e^{5t} \end{pmatrix}$$

$$= \begin{pmatrix} 4 & -3 \\ -2 & -1 \end{pmatrix} \begin{pmatrix} e^{-2t} & -3e^{5t} \\ 2e^{-2t} & e^{5t} \end{pmatrix}$$

and because these two vectors are linearly independent

$$\begin{vmatrix} e^{-2t} & -3e^{5t} \\ 2e^{-2t} & e^{5t} \end{vmatrix} = 7e^{3t} \neq 0.$$

EXAMPLE 6.3.8

Show that $\Phi = \begin{pmatrix} 0 & t^{-1} & 1 \\ t & 1 & 0 \\ 1 & 0 & t \end{pmatrix}, t > 0,$

is a fundamental matrix for the system $\mathbf{X}' = \begin{pmatrix} -(t+1)^{-1} & -t^{-2}(t+1)^{-1} & t^{-1}(t+1)^{-1} \\ -t(t+1)^{-1} & (t+1)^{-1} & (t+1)^{-1} \\ (t+1)^{-1} & -t^{-1}(t+1)^{-1} & (t+1)^{-1} \end{pmatrix} \mathbf{X}$. Use this matrix to find a general solution of $\mathbf{X}' = \begin{pmatrix} -(t+1)^{-1} & -t^{-2}(t+1)^{-1} & t^{-1}(t+1)^{-1} \\ -t(t+1)^{-1} & (t+1)^{-1} & (t+1)^{-1} \\ (t+1)^{-1} & -t^{-1}(t+1)^{-1} & (t+1)^{-1} \end{pmatrix} \mathbf{X}$.

Solution

Because

$$\begin{pmatrix} 0 \\ t \\ 1 \end{pmatrix}' = \begin{pmatrix} 0 \\ 1 \\ 0 \end{pmatrix}$$

$$= \begin{pmatrix} -(t+1)^{-1} & -t^{-2}(t+1)^{-1} & t^{-1}(t+1)^{-1} \\ -t(t+1)^{-1} & (t+1)^{-1} & (t+1)^{-1} \\ (t+1)^{-1} & -t^{-1}(t+1)^{-1} & (t+1)^{-1} \end{pmatrix}$$

$$\times \begin{pmatrix} 0 \\ t \\ 1 \end{pmatrix},$$

$$\begin{pmatrix} t^{-1} \\ 1 \\ 0 \end{pmatrix}' = \begin{pmatrix} -t^{-2} \\ 0 \\ 0 \end{pmatrix}$$

$$= \begin{pmatrix} -(t+1)^{-1} & -t^{-2}(t+1)^{-1} & t^{-1}(t+1)^{-1} \\ -t(t+1)^{-1} & (t+1)^{-1} & (t+1)^{-1} \\ (t+1)^{-1} & -t^{-1}(t+1)^{-1} & (t+1)^{-1} \end{pmatrix}$$

$$\times \begin{pmatrix} t^{-1} \\ 1 \\ 0 \end{pmatrix},$$

and

$$\begin{pmatrix} 1 \\ 0 \\ t \end{pmatrix}' = \begin{pmatrix} 0 \\ 0 \\ 1 \end{pmatrix}$$

$$= \begin{pmatrix} -(t+1)^{-1} & -t^{-2}(t+1)^{-1} & t^{-1}(t+1)^{-1} \\ -t(t+1)^{-1} & (t+1)^{-1} & (t+1)^{-1} \\ (t+1)^{-1} & -t^{-1}(t+1)^{-1} & (t+1)^{-1} \end{pmatrix}$$

$$\times \begin{pmatrix} 1 \\ 0 \\ t \end{pmatrix},$$

all three columns of Φ are solutions of the system. The solutions are linearly independent because the Wronskian of these three solutions is

$$\begin{vmatrix} 0 & t^{-1} & 1 \\ t & 1 & 0 \\ 1 & 0 & t \end{vmatrix} = -1 - t \neq 0 \quad \text{for } t > 0.$$

A general solution of the system is given by

$$\mathbf{X}(t) = \Phi(t)\mathbf{C} = \begin{pmatrix} 0 & t^{-1} & 1 \\ t & 1 & 0 \\ 1 & 0 & t \end{pmatrix} \begin{pmatrix} c_1 \\ c_2 \\ c_3 \end{pmatrix}$$

$$= c_1 \begin{pmatrix} 0 \\ t \\ 1 \end{pmatrix} + c_2 \begin{pmatrix} t^{-1} \\ 1 \\ 0 \end{pmatrix} + c_3 \begin{pmatrix} 1 \\ 0 \\ t \end{pmatrix}$$

$$= \begin{pmatrix} c_2 t^{-1} + c_3 \\ c_1 t + c_2 \\ c_1 + c_3 t \end{pmatrix}.$$

EXAMPLE 6.3.9

Solve
$$\begin{cases} t^2(t+1)x' = -t^2 x - y + tz \\ (t+1)y' = -tx + y + z \\ t(t+1)z' = tx - y + tz \\ x(1) = 0, y(1) = 1, z(1) = 2. \end{cases}$$

Solution

In matrix form, the system is equivalent to that in Example 6.3.8 so a general solution is $x = c_2 t^{-1} + c_3$, $y = c_1 t + c_2$, and $z = c_1 + c_3 t$. Application of the initial conditions results in $x(1) = c_2 + c_3 = 0$, $y(1) = c_1 + c_2 = 1$, and $z = c_1 + c_3 = 2$ so $c_1 = 3/2$, $c_2 = -1/2$ and $c_3 = 1/2$ and the solution to the IVP is $x = -\frac{1}{2}t^{-1} + \frac{1}{2}$, $y = \frac{3}{2}t - \frac{1}{2}$, and $z = \frac{3}{2} + \frac{1}{2}t$.

 EXERCISES 6.3

In Exercises 1-8, write the system of first-order equations in matrix form. (Throughout, ' denotes d/dt.)

1. $\begin{cases} x' = x + 4y \\ y' = 2x - y \end{cases}$.

2. $\begin{cases} x' = 2x + y \\ y' = x - 5y \end{cases}$.

3. $\begin{cases} x' - x = e^t \\ y' = x + y \end{cases}$.

4. $\begin{cases} x' - y = 0 \\ y' + x = 2\sin t \end{cases}$.

5. $\begin{cases} t(t^3 - 1)x' = x - ty \\ t(t^3 - 1)y' = -2t^2 x + (t^3 + 1)y \end{cases}$.

6. $\begin{cases} (\sin 2t - 2)x' = 2\sin^2 tx + 2\sin ty \\ \qquad\qquad + 2t\sin 2t - 4t \\ (\sin 2t - 2)y' = -2\cos tx + 2\cos^2 ty \end{cases}$.

7. $\begin{cases} x' = x + y + z + \cos t \\ y' = 2x + z \\ z' = -2y - z + \sin t \end{cases}$.

8. $\begin{cases} (2t^2 - t - 1)x' = 2tx - y - z \\ (2t^2 - t - 1)y' = -x + 2ty - z \\ \qquad\qquad + 2t^2 - t - 1 \\ (2t^2 - t - 1)z' = -x - y + 2tz \end{cases}$.

In Exercises 9-14, use the Wronskian to determine if the given set of vectors is linearly independent or dependent.

9. $\begin{pmatrix} e^t \\ 2e^t \end{pmatrix}, \begin{pmatrix} 3e^{-t} \\ e^{-t} \end{pmatrix}.$

10. $\begin{pmatrix} t \\ 1 \end{pmatrix}, \begin{pmatrix} -t \\ t \end{pmatrix}.$

11. $\begin{pmatrix} \cos 2t \\ -2\sin 2t \end{pmatrix}, \begin{pmatrix} \sin 2t \\ -2\cos 2t \end{pmatrix}.$

12. $\begin{pmatrix} \cos 2t \\ -2\sin 2t \end{pmatrix}, \begin{pmatrix} 1-2\sin^2 t \\ -4\sin t \cos t \end{pmatrix}.$

13. $\begin{pmatrix} e^{2t} \\ e^{2t} \\ 2e^{2t} \end{pmatrix}, \begin{pmatrix} e^t \\ -e^t \\ 3e^t \end{pmatrix}, \begin{pmatrix} e^{-t} \\ e^{-t} \\ e^{-t} \end{pmatrix}.$

14. $\begin{pmatrix} 6e^{4t} \\ 6e^{4t} \\ 3e^{4t} \end{pmatrix}, \begin{pmatrix} e^{-2t} \\ e^{-2t} \\ -e^{-2t} \end{pmatrix}, \begin{pmatrix} 2e^{4t} \\ 2e^{4t} \\ e^{4t} \end{pmatrix}$

In Exercises 15-22, verify that the given matrix is a fundamental matrix for the system. Use the fundamental matrix to obtain a general solution to the system of first-order homogeneous equations $\mathbf{X}' = \mathbf{AX}$.

15. $\Phi(t) = \begin{pmatrix} -2e^{-8t} & 5e^{-t} \\ e^{-8t} & e^{-t} \end{pmatrix}$ and

$\mathbf{X}' = \begin{pmatrix} -3 & 10 \\ 1 & -6 \end{pmatrix}\mathbf{X}.$

16. $\Phi(t) = \begin{pmatrix} 5\sin t & 5\cos t \\ -\cos t + 2\sin t & 2\cos t + \sin t \end{pmatrix}$ and

$\mathbf{X}' = \begin{pmatrix} 2 & -5 \\ 1 & -2 \end{pmatrix}\mathbf{X}.$

17. $\Phi(t) = \begin{pmatrix} e^{-t} & 2te^{-t} \\ -e^{-t} & -e^{-t}(2t+1) \end{pmatrix}$ and

$\mathbf{X}' = \begin{pmatrix} -3 & -2 \\ 2 & 1 \end{pmatrix}\mathbf{X}.$

18. $\Phi(t) = \begin{pmatrix} t^{-1} & 1 \\ t & t^{-1} \end{pmatrix}$ and $\mathbf{X}' =$

$\begin{pmatrix} t^{-1}(t^3-1)^{-1} & -(t^3-1)^{-1} \\ 2t(t^3-1)^{-1} & t(t^3+1)(t^3-1)^{-1} \end{pmatrix}\mathbf{X}.$

19. $\Phi(t) = \begin{pmatrix} 2 & e^t & 0 \\ 2 & e^t & e^{-t} \\ -1 & -e^t & e^{-t} \end{pmatrix}$ and

$\mathbf{X}' = \begin{pmatrix} -3 & 2 & -2 \\ -2 & 1 & -2 \\ 4 & -3 & 2 \end{pmatrix}\mathbf{X}.$

20. $\Phi(t) =$
$\begin{pmatrix} 0 & -2\cos 2t + \sin 2t & \cos 2t + 2\sin 2t \\ e^t & 3\cos 2t + \sin 2t & \cos 2t - 3\sin 2t \\ 0 & 3\cos 2t + \sin 2t & \cos 2t - 3\sin 2t \end{pmatrix}$ and

$\mathbf{X}' = \begin{pmatrix} 2 & 0 & 2 \\ -4 & 1 & -3 \\ -4 & 0 & -2 \end{pmatrix}\mathbf{X}.$

21. $\Phi(t) = \begin{pmatrix} 0 & t^{-1} & 1 \\ t & 1 & 0 \\ 1 & 0 & t \end{pmatrix}$ and

$\mathbf{X}' = \begin{pmatrix} 2t(2t^2-t-1)^{-1} & -(2t^2-t-1)^{-1} \\ -(2t^2-t-1)^{-1} & 2t(2t^2-t-1)^{-1} \\ -(2t^2-t-1)^{-1} & -(2t^2-t-1)^{-1} \end{pmatrix}$
$\begin{pmatrix} -(2t^2-t-1)^{-1} \\ -(2t^2-t-1)^{-1} \\ 2t(2t^2-t-1)^{-1} \end{pmatrix}\mathbf{X}.$

22. $\Phi(t) = \begin{pmatrix} 1 & e^t & e^{-t} & 0 \\ 2 & -e^t & e^{-t} & -2 \\ -1 & -e^t & -e^{-t} & 1 \\ 1 & -e^t & e^{-t} & 0 \end{pmatrix}$ and

$\mathbf{X}' = \begin{pmatrix} 2 & 1 & 2 & -2 \\ 1 & 1 & 2 & -1 \\ -2 & -1 & -2 & 2 \\ 1 & 1 & 2 & -1 \end{pmatrix}\mathbf{X}$

In Exercises 23-29, use the given fundamental matrix to obtain a general solution to the system of first-order homogeneous equations $\mathbf{X}' = \mathbf{AX}$. Find the solution that satisfies the given initial condition. *Graph the components individually as functions of t as well as parametrically (x vs. y as functions of t or x vs. y vs. z as functions of t).*

23. $\Phi(t) = \begin{pmatrix} 2e^{4t} & e^{-t} \\ 3e^{4t} & -e^{-t} \end{pmatrix}$, $\mathbf{A} = \begin{pmatrix} 1 & 2 \\ 3 & 2 \end{pmatrix}$,

$\begin{cases} x' = x + 2y \\ y' = 3x + 2y \\ x(0) = 3, y(0) = 2 \end{cases}$.

24. $\Phi(t) = \begin{pmatrix} 3e^{-2t} & e^{-t} \\ 2e^{-2t} & e^{-t} \end{pmatrix}$, $\mathbf{A} = \begin{pmatrix} -4 & 3 \\ -2 & 1 \end{pmatrix}$,

$\begin{cases} x' = -4x + 3y \\ y' = -2x + y \\ x(0) = 1, y(0) = 0 \end{cases}$.

25. $\Phi(t) = \begin{pmatrix} \sin t & \cos t \\ \cos t - \sin t & -\cos t - \sin t \end{pmatrix}$,

$\mathbf{A} = \begin{pmatrix} 1 & 1 \\ -2 & -1 \end{pmatrix}$, $\begin{cases} x' = x + y \\ y' = -2x - y \\ x(0) = 0, y(0) = 1 \end{cases}$.

26. $\Phi(t) =$
$\begin{pmatrix} e^{-t}(3\cos 3t + \sin 3t) & 5e^{-t}\cos 3t \\ 2e^{-t}\sin 3t & e^{-t}(\cos 3t + 3\sin 3t) \end{pmatrix}$,

$\mathbf{A} = \begin{pmatrix} 0 & -5 \\ 2 & -2 \end{pmatrix}$, $\begin{cases} x' = -5y \\ y' = 2x - 2y \\ x(0) = 0, y(0) = 1 \end{cases}$.

27. $\Phi(t) = \begin{pmatrix} t^{-1} & 1 \\ 1 & t^2 \end{pmatrix}$,

$\mathbf{A} = \begin{pmatrix} -(t-1)^{-1} & t^{-2}(t-1)^{-1} \\ -2t(t-1)^{-1} & -2(t-1)^{-1} \end{pmatrix}$,

$\begin{cases} (t-1)t^2x' = -t^2x + y \\ (t-1)y' = -2tx + 2y \\ x(2) = 1, y(2) = 0 \end{cases}$.

28. $\Phi(t) = \begin{pmatrix} -e^{-2t} & 0 & 1 \\ e^{-2t} & e^{-t} & 1 \\ 2e^{-2t} & 2e^{-2t} & 1 \end{pmatrix}$, $\mathbf{A} = \begin{pmatrix} 2 & 4 & -2 \\ 1 & -1 & 0 \\ 2 & -2 & 0 \end{pmatrix}$,

$\begin{cases} x' = 2x + 4y - 2z \\ y' = x - y \\ z' = 2x - 2y \\ x(0) = -1, y(0) = 0, z(0) = 1 \end{cases}$.

29. $\Phi(t) = \begin{pmatrix} t & 1 & 1 \\ \cos 3t & -\sin 3t & \sin 3t \\ \sin 3t & \cos 3t & -\cos 3t \end{pmatrix}$,

$\mathbf{A} = \begin{pmatrix} 0 & \cos 3t & \sin 3t \\ 0 & 0 & -3 \\ 0 & 3 & 0 \end{pmatrix}$,

$\begin{cases} x' = y\cos 3t + z\sin 3t \\ y' = -3z \\ z' = 3y \\ x(0) = 0, y(0) = 1, z(0) = -1 \end{cases}$.

30. (*Modeling the Motion of Spiked Volleyball*)[1] Under certain assumptions, the position of a spiked volleyball can be modeled by the system of two second-order nonlinear equations

$$X'' = \frac{1}{m}\left(C_M\omega^\alpha Y'((X')^2 + (Y')^2)^{(b-1)/2}\right.$$
$$\left. -\frac{1}{2}C_D\rho AX'\sqrt{(X')^2 + (Y')^2}\right)$$
$$Y'' = -g - \frac{1}{m}\left(C_M\omega^\alpha X'((X')^2 + (Y')^2)^{(b-1)/2}\right.$$
$$\left. +\frac{1}{2}C_D\rho AY'\sqrt{(X')^2 + (Y')^2}\right).$$

Write this system of two second-order nonlinear equations as a system of four first-order equations.

31. (*Principle of Superposition*) (a) Show that any linear combination of solutions of the homogeneous system $\mathbf{X}'(t) = \mathbf{A}(t)\mathbf{X}(t)$ is also a solution of the homogeneous system. (b) Is the Principle of Superposition ever valid for nonhomogeneous systems of equations? Explain.

32. Find two linearly independent solutions of $$\mathbf{X}' = \begin{pmatrix} -3 & -3 \\ 2 & 2 \end{pmatrix}\mathbf{X}$$ of the form $\mathbf{v}_1e^{\lambda_1 t}$ and $\mathbf{v}_2e^{\lambda_2 t}$. Use the two solutions to find a

---

[1] S.S. Kao, R.W. Sellens, J.M. Stevenson, A mathematical model for the trajectory of a spiked volleyball and its coaching applications, *Journal of Applied Biomechanics*, 1994, 95-109.

fundamental matrix for the system and a general solution of the system.

33. Find two linearly independent solutions of $X' = \begin{pmatrix} -3 & 2 \\ -4 & 3 \end{pmatrix} X$ of the form $v_1 e^{\lambda_1 t}$ and $v_2 e^{\lambda_2 t}$. Use the two solutions to find a fundamental matrix for the system and a general solution of the system.

34. Consider the second-order linear homogeneous equations with constant coefficients, $\frac{d^2 x}{dt^2} + b\frac{dx}{dt} + cx = 0$, where b and c are constants. Let $dx/dt = x' = y$ so that $d^2x/dy^2 = x'' = y'$. Show that $y' = -cx - by$ and that the second-order equation $(d^2x/dt^2) + b(dx/dt) + cx = 0$ can be written as the first-order system

$$\begin{pmatrix} x \\ y \end{pmatrix}' = \begin{pmatrix} 0 & 1 \\ -c & -b \end{pmatrix} \begin{pmatrix} x \\ y \end{pmatrix}.$$

(a) What is the characteristic equation for
$$\frac{d^2 x}{dt^2} + b\frac{dx}{dt} + cx = 0?$$

(b) What is the characteristic equation for
$$\begin{pmatrix} x \\ y \end{pmatrix}' = \begin{pmatrix} 0 & 1 \\ -c & -b \end{pmatrix} \begin{pmatrix} x \\ y \end{pmatrix}?$$

(c) Solve (a) and (b) for λ (or r). How do the characteristic equations for (a) and (b) differ?

(d) Generalize what you learned in (a)-(c) for higher order equations.

6.4 FIRST-ORDER LINEAR HOMOGENEOUS SYSTEMS WITH CONSTANT COEFFICIENTS

- DISTINCT REAL EIGENVALUES
- COMPLEX CONJUGATE EIGENVALUES
- ALTERNATE METHOD FOR SOLVING IVPS
- REPEATED EIGENVALUES

Now that we have covered the necessary terminology, we turn our attention to solving linear systems with constant coefficients. Let

$$A = \begin{pmatrix} a_{11} & a_{12} & \cdots & a_{1n} \\ a_{21} & a_{22} & \cdots & a_{2n} \\ \vdots & \vdots & \ddots & \vdots \\ a_{n1} & a_{n2} & \cdots & a_{nn} \end{pmatrix}$$ be an $n \times n$ matrix with real components. In this section, we will see that a general solution to the homogeneous system $X' = AX$ is determined by the eigenvalues and corresponding eigenvectors of A. We begin by considering the cases when the eigenvalues of A are distinct and real or the eigenvalues of A are distinct and complex, and then consider the case when A has repeated eigenvalues (eigenvalues of multiplicity greater than one).

Distinct Real Eigenvalues

In Section 6.3, we verified that a general solution of the 2×2 system, $X' = \begin{pmatrix} 4 & -3 \\ -2 & -1 \end{pmatrix} X$, where the eigenvalues of $\begin{pmatrix} 4 & -3 \\ -2 & -1 \end{pmatrix}$ are $\lambda_1 = -2$ and $\lambda_2 = 5$ and corresponding eigenvectors are $v_1 = \begin{pmatrix} 1 \\ 2 \end{pmatrix}$ and $v_2 = \begin{pmatrix} -3 \\ 1 \end{pmatrix}$, respectively, is

$$X(t) = \Phi(t)C = c_1 \begin{pmatrix} e^{-2t} \\ 2e^{-2t} \end{pmatrix} + c_2 \begin{pmatrix} -3e^{5t} \\ e^{5t} \end{pmatrix}$$

$$= c_1 \begin{pmatrix} 1 \\ 2 \end{pmatrix} e^{-2t} + c_2 \begin{pmatrix} -3 \\ 1 \end{pmatrix} e^{5t}.$$

More generally, if the eigenvalues of the $n \times n$ matrix A are distinct, we expect a general solution of the linear homogeneous system $X' = AX$ to have the form

$$X(t) = c_1 v_1 e^{\lambda_1 t} + c_2 v_2 e^{\lambda_2 t} + \cdots + c_n v_n e^{\lambda_n t},$$

where $\lambda_1, \lambda_2, \ldots, \lambda_n$ are the n distinct real eigenvalues of A with corresponding eigenvectors $v_1, v_2, \ldots, v_n$, respectively. We investigate this by assuming that $X = ve^{\lambda t}$ is a solution of $X' = AX$. Then, $X' = \lambda ve^{\lambda t}$ and substituting into the system of differential equations gives us

$$X' = AX,$$
$$\lambda ve^{\lambda t} = Ave^{\lambda t}.$$

Replacing $\mathbf{v}$ with $\mathbf{Iv} = \mathbf{v}$ on the left side of this equation yields

$$\lambda \mathbf{I v} e^{\lambda t} = \mathbf{A v} e^{\lambda t},$$

$$\mathbf{A v} e^{\lambda t} - \lambda \mathbf{I v} e^{\lambda t} = \mathbf{0},$$

$$(\mathbf{A} - \lambda \mathbf{I}) \mathbf{v} e^{\lambda t} = \mathbf{0}.$$

Then, because $e^{\lambda t} \neq 0$, we have $(\mathbf{A} - \lambda \mathbf{I}) \mathbf{v} = \mathbf{0}$. In order for this system of equations to have a solution other than $\mathbf{v} = \mathbf{0}$ (remember that eigenvectors are *never* the zero vector), we must have

$$|\mathbf{A} - \lambda \mathbf{I}| = 0.$$

A solution λ to this equation is an eigenvalue of $\mathbf{A}$, while a nonzero vector $\mathbf{v}$ satisfying $(\mathbf{A} - \lambda \mathbf{I}) \mathbf{v} = \mathbf{0}$ is an eigenvector that corresponds to λ. Hence, if $\mathbf{A}$ has n distinct eigenvalues $\lambda_1, \lambda_2, \ldots, \lambda_n$, we can find a set of n eigenvectors $\mathbf{v}_1, \mathbf{v}_2, \ldots, \mathbf{v}_n$. From these eigenvalues and corresponding eigenvectors, we form the n linearly independent solutions

$$\mathbf{X}_1 = \mathbf{v}_1 e^{\lambda_1 t}, \mathbf{X}_2 = \mathbf{v}_2 e^{\lambda_2 t}, \ldots, \mathbf{X}_n = \mathbf{v}_n e^{\lambda_n t}. \tag{6.16}$$

Therefore, if $\mathbf{A}$ is an $n \times n$ matrix with n distinct real eigenvalues $\{\lambda_k\}_{k=1}^{n}$, a general solution of $\mathbf{X}' = \mathbf{A X}$ is the linear combination of the set of solutions given in Equation (6.16),

$$\mathbf{X}(t) = c_1 \mathbf{v}_1 e^{\lambda_1 t} + c_2 \mathbf{v}_2 e^{\lambda_2 t} + \cdots + c_n \mathbf{v}_n e^{\lambda_n t}.$$

EXAMPLE 6.4.1

Find a general solution of $\begin{cases} x' = 2x + 3y \\ y' = -4x - 5y \end{cases}$.

Solution

In matrix form, this system is equivalent to $\mathbf{X}' = \mathbf{A X}$ with $\mathbf{A} = \begin{pmatrix} 2 & 3 \\ -4 & -5 \end{pmatrix}$. The eigenvalues of $\mathbf{A}$ are found with $\begin{vmatrix} 2 - \lambda & 3 \\ -4 & -5 - \lambda \end{vmatrix} = \lambda^2 + 3\lambda + 2 = (\lambda + 2)(\lambda + 1) = 0$. Thus, the eigenvalues of $\mathbf{A}$ are $\lambda_1 = -2$ and $\lambda_2 = -1$. An eigenvector

$\mathbf{v}_1 = \begin{pmatrix} x_1 \\ y_1 \end{pmatrix}$ corresponding to λ_1 satisfies $(\mathbf{A} - \lambda_1 \mathbf{I}) \mathbf{v}_1 = \mathbf{0}$. This system has augmented matrix $\begin{pmatrix} 4 & 3 \\ -4 & -3 \end{pmatrix}$ that reduces to $\begin{pmatrix} 4 & 3 \\ 0 & 0 \end{pmatrix}$, which indicates that $4x_1 + 3y_1 = 0$. Choosing $x_1 = 3$ and $y_1 = -4$ gives us $\mathbf{v}_1 = \begin{pmatrix} 3 \\ -4 \end{pmatrix}$. Similarly, an eigenvector

$\mathbf{v}_2 = \begin{pmatrix} x_2 \\ y_2 \end{pmatrix}$ corresponding to λ_2 satisfies $(\mathbf{A} - \lambda_2 \mathbf{I}) \mathbf{v}_2 = \mathbf{0}$. The augmented matrix for this system reduces to $\begin{pmatrix} 1 & 1 \\ 0 & 0 \end{pmatrix}$, which shows us that $x_2 + y_2 = 0$. Choosing $x_2 = 1$ and $y_2 = -1$ gives us $\mathbf{v}_2 = \begin{pmatrix} 1 \\ -1 \end{pmatrix}$. Thus, two linearly independent solutions of the system are

$$\mathbf{X}_1 = \begin{pmatrix} 3 \\ -4 \end{pmatrix} e^{-2t} \quad \text{and} \quad \mathbf{X}_2 = \begin{pmatrix} 1 \\ -1 \end{pmatrix} e^{-t};$$

a general solution of the system in matrix form is

$$\mathbf{X} = c_1 \begin{pmatrix} 3 \\ -4 \end{pmatrix} e^{-2t} + c_2 \begin{pmatrix} 1 \\ -1 \end{pmatrix} e^{-t}$$

$$= \begin{pmatrix} 3e^{-2t} & e^{-t} \\ -4e^{-2t} & -e^{-t} \end{pmatrix} \begin{pmatrix} c_1 \\ c_2 \end{pmatrix}.$$

 What is a fundamental matrix, Φ, for this system?

In terms of x and y, $\mathbf{X} = \begin{pmatrix} x \\ y \end{pmatrix}$, $x = 3c_1 e^{-2t} + c_2 e^{-t}$ and $y = -4c_1 e^{-2t} - c_2 e^{-t}$. Several solutions along with the direction field for the system are shown in Figure 6.6(a). (Notice that each curve corresponds to the parametric plot of the pair $\begin{cases} x = x(t) \\ y = y(t) \end{cases}$ for particular values of the constants c_1 and c_2.) Because both eigenvalues are negative, all nontrivial solutions move toward the origin as t increases. The arrows on the vectors in the direction field show this behavior.

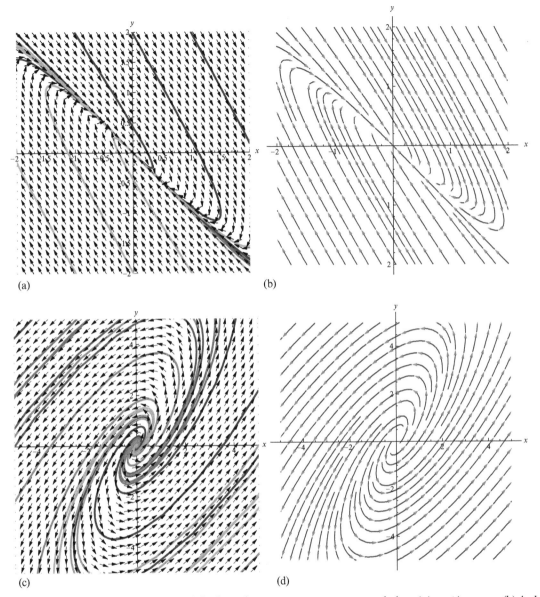

FIGURE 6.6   (a) All nontrivial solutions of this linear homogeneous system approach the origin as t increases. (b) A phase portrait confirms that all solutions tend to the origin as $t \to \infty$. (c) We see that all nontrivial solutions spiral away from the origin. (d) A phase portrait confirms that all solutions spiral away from the origin.

Complex Conjugate Eigenvalues

You may notice that general solutions you obtain when solving systems do not agree with those of your classmates or those given in Exercise 6.4 solutions. Before you become alarmed, realize that your solutions may be correct. You may have simply selected different values for the arbitrary constants in finding eigenvectors.

If $\mathbf{A}$ has complex conjugate eigenvalues $\lambda_{1,2} = \alpha \pm \beta i$ with corresponding eigenvectors $\mathbf{v}_{1,2} = \mathbf{a} \pm \mathbf{b}i$, respectively, one solution of $\mathbf{X}' = \mathbf{AX}$ is

$$\mathbf{X}(t) = \mathbf{v}_1 e^{\lambda_1 t} = (\mathbf{a} + \mathbf{b}i)e^{(\alpha + \beta i)t} = e^{\alpha t}(\mathbf{a} + \mathbf{b}i)e^{i\beta t}$$
$$= e^{\alpha t}(\mathbf{a} + \mathbf{b}i)(\cos \beta t + i \sin \beta t)$$
$$= \underbrace{e^{\alpha t}(\mathbf{a} \cos \beta t - \mathbf{b} \sin \beta t)}_{\text{named } \mathbf{X}_1(t)}$$
$$+ i \underbrace{e^{\alpha t}(\mathbf{b} \cos \beta t + \mathbf{a} \sin \beta t)}_{\text{named } \mathbf{X}_2(t)}$$
$$= \mathbf{X}_1(t) + i\mathbf{X}_2(t).$$

Because $\mathbf{X}$ is a solution to the system $\mathbf{X}' = \mathbf{AX}$, we have that $\mathbf{X}_1'(t) + i\mathbf{X}_2'(t) = \mathbf{AX}_1(t) + i\mathbf{AX}_2(t)$. Equating the real and imaginary parts of this equation yields $\mathbf{X}_1'(t) = \mathbf{AX}_1(t)$ and $\mathbf{X}_2'(t) = \mathbf{AX}_2(t)$. Therefore, $\mathbf{X}_1(t)$ and $\mathbf{X}_2(t)$ are solutions to $\mathbf{X}' = \mathbf{AX}$, and any linear combination of $\mathbf{X}_1(t)$ and $\mathbf{X}_2(t)$ is also a solution. We can show that $\mathbf{X}_1(t)$ and $\mathbf{X}_2(t)$ are linearly independent (see Exercise 45), so this linear combination forms a portion of a general solution of $\mathbf{X}' = \mathbf{AX}$ where $\mathbf{A}$ is a square matrix with real entries of *any size*.

Theorem 36. *Let* $\mathbf{A}$ *be a square matrix with real entries. If* $\mathbf{A}$ *has complex conjugate eigenvalues* $\lambda_{1,2} = \alpha \pm \beta i$, $\beta \neq 0$, *with corresponding eigenvectors* $\mathbf{v}_{1,2} = \mathbf{a} \pm \mathbf{b}i$, *respectively, two linearly independent solutions of* $\mathbf{X}' = \mathbf{AX}$ *are* $\mathbf{X}_1(t) = e^{\alpha t}(\mathbf{a} \cos \beta t - \mathbf{b} \sin \beta t)$ *and* $\mathbf{X}_2(t) = e^{\alpha t}(\mathbf{b} \cos \beta t + \mathbf{a} \sin \beta t)$.

Notice that in the case of complex conjugate eigenvalues, we are able to obtain two linearly independent solutions from know one of the eigenvalues and an eigenvector that corresponds to it.

EXAMPLE 6.4.2

Find a general solution of $\mathbf{X}' = \begin{pmatrix} 3 & -2 \\ 4 & -1 \end{pmatrix}\mathbf{X}$.

Solution

The eigenvalues of $\mathbf{A} = \begin{pmatrix} 3 & -2 \\ 4 & -1 \end{pmatrix}$ are $\lambda_{1,2} = 1 \pm 2i$. An eigenvector $\mathbf{v}_1 = \begin{pmatrix} x_1 \\ y_1 \end{pmatrix}$ corresponding to λ_1 satisfies $(\mathbf{A} - \lambda_1\mathbf{I})\mathbf{v}_1 = \mathbf{0}$. This system has augmented matrix $\begin{pmatrix} 2 - 2i & -2 \\ 4 & -2 - 2i \end{pmatrix}$ that reduces to $\begin{pmatrix} 1 - i & -1 \\ 0 & 0 \end{pmatrix}$, which indicates that $(1 - i)x_1 - y_1 = 0$. Choosing $x_1 = 1$, we obtain $\mathbf{v}_1 = \begin{pmatrix} x_1 \\ y_1 \end{pmatrix} = \begin{pmatrix} 1 \\ 1 - i \end{pmatrix} = \underbrace{\begin{pmatrix} 1 \\ 1 \end{pmatrix}}_{\mathbf{a}} + i \underbrace{\begin{pmatrix} 0 \\ -1 \end{pmatrix}}_{\mathbf{b}}$. Therefore, in the notation used in Theorem 36 with $\alpha = 1$ and $\beta = 2$, two linearly independent solutions of the system are

$$\mathbf{X}_1(t) = e^t\left[\begin{pmatrix} 1 \\ 1 \end{pmatrix} \cos 2t - \begin{pmatrix} 0 \\ -1 \end{pmatrix} \sin 2t\right]$$
$$= e^t\begin{pmatrix} \cos 2t \\ \cos 2t + \sin 2t \end{pmatrix}$$

and

$$\mathbf{X}_2(t) = e^t\left[\begin{pmatrix} 0 \\ -1 \end{pmatrix} \cos 2t + \begin{pmatrix} 1 \\ 1 \end{pmatrix} \sin 2t\right]$$
$$= e^t\begin{pmatrix} \sin 2t \\ -\cos 2t + \sin 2t \end{pmatrix}.$$

A general solution of the system is then

$$X(t) = c_1 X_1(t) + c_2 X_2(t)$$

$$= c_1 e^t \begin{pmatrix} \cos 2t \\ \cos 2t + \sin 2t \end{pmatrix} + c_2 e^t \begin{pmatrix} \sin 2t \\ -\cos 2t + \sin 2t \end{pmatrix}$$

$$= \underbrace{\begin{pmatrix} e^t \cos 2t & e^t \sin 2t \\ e^t(\cos 2t + \sin 2t) & e^t(-\cos 2t + \sin 2t) \end{pmatrix}}_{\text{"Fundamental Matrix"}} \begin{pmatrix} c_1 \\ c_2 \end{pmatrix}$$

$$= \begin{pmatrix} e^t(c_1 \cos 2t + c_2 \sin 2t) \\ e^t[(c_1 - c_2)\cos 2t + (c_1 + c_2)\sin 2t] \end{pmatrix}.$$

Figure 6.6(b) shows the graph of several solutions along with the direction field for the equation. Notice the spiraling motion of the vectors in the direction field. This is due to the product of exponential and trigonometric functions in the solution; the exponential functions cause $x(t)$ and $y(t)$ to increase, while the trigonometric functions lead to rotation about the origin.

Sketch the graphs of $x(t)$, $y(t)$, and $\{x(t), y(t)\}$ in Example 6.4.2 if $x(0) = 0$ and $y(0) = 1$. For these initial conditions, calculate $\lim_{t\to\infty} x(t)$, $\lim_{t\to\infty} y(t)$, $\lim_{t\to-\infty} x(t)$ and $\lim_{t\to-\infty} y(t)$.

Initial-value problems for linear homogeneous systems with constant coefficients can be solved through the use of eigenvalues and eigenvectors as well.

EXAMPLE 6.4.3

Solve $X' = \begin{pmatrix} 5 & 5 & 2 \\ -6 & -6 & -5 \\ 6 & 6 & 5 \end{pmatrix} X$ subject to

$X(0) = \begin{pmatrix} 0 \\ 0 \\ 2 \end{pmatrix}.$

Solution

The eigenvalues of $A = \begin{pmatrix} 5 & 5 & 2 \\ -6 & -6 & -5 \\ 6 & 6 & 5 \end{pmatrix}$ satisfy

$$\begin{vmatrix} 5-\lambda & 5 & 2 \\ -6 & -6-\lambda & -5 \\ 6 & 6 & 5-\lambda \end{vmatrix}$$

$$= \underbrace{(5-\lambda)\begin{vmatrix} -6-\lambda & -5 \\ 6 & 5-\lambda \end{vmatrix} - 5\begin{vmatrix} -6 & -5 \\ 6 & 5-\lambda \end{vmatrix} + 2\begin{vmatrix} -6 & -6-\lambda \\ 6 & 6 \end{vmatrix}}_{\text{Expand along the first row.}}$$

$$= (5-\lambda)(\lambda^2 + \lambda) - 5 \times 6\lambda + 2 \times 6\lambda = -\lambda(\lambda^2 - 4\lambda + 13) = 0.$$

Hence, $\lambda_1 = 0$ and $\lambda_{2,3} = (4\pm\sqrt{16-52})/2 = 2\pm 3i$.

An eigenvector $v_1 = \begin{pmatrix} x_1 \\ y_1 \\ z_1 \end{pmatrix}$ corresponding to λ_1 satisfies $(A - \lambda_1 I)v_1 = 0$. After row operations, the augmented matrix for this system reduces to $\begin{pmatrix} 1 & 1 & 0 \\ 0 & 0 & 1 \\ 0 & 0 & 0 \end{pmatrix}$, which indicates that $x_1 + y_1 = 0$ and $z_1 = 0$. If we choose $y_1 = 1$, $v_1 = \begin{pmatrix} -1 \\ 1 \\ 0 \end{pmatrix}$ and one solution of the system of differential equations is $X_1(t) = \begin{pmatrix} -1 \\ 1 \\ 0 \end{pmatrix} e^{0 \times t} = \begin{pmatrix} -1 \\ 1 \\ 0 \end{pmatrix}$. We find two solutions that correspond to the complex conjugate pair of eigenvalues by finding an eigenvector $v_2 = \begin{pmatrix} x_2 \\ y_2 \\ z_2 \end{pmatrix}$ corresponding to $\lambda_2 = 2+3i$. This vector satisfies the system $(A-\lambda_2 I)v_2 = 0$ whose augmented matrix *can* be reduced to $\begin{pmatrix} 1 & 0 & -\frac{1}{2}(1+i) \\ 0 & 1 & 1 \\ 0 & 0 & 0 \end{pmatrix}$. Therefore, the components of v_2 satisfy $x_2 = \frac{1}{2}(1+i)z_2$ and $y_2 = -z_2$. If we let $z_2 = 2$, then

$$\mathbf{v}_2 = \begin{pmatrix} x_2 \\ y_2 \\ z_2 \end{pmatrix} = \begin{pmatrix} 1+i \\ -2 \\ 2 \end{pmatrix} = \underbrace{\begin{pmatrix} 1 \\ -2 \\ 2 \end{pmatrix}}_{\mathbf{a}} + \underbrace{\begin{pmatrix} 1 \\ 0 \\ 0 \end{pmatrix}}_{\mathbf{b}} i = \mathbf{a} + \mathbf{b}i.$$

Thus, two linearly independent solutions of the system that correspond to the eigenvalues $\lambda_{2,3} = 2 \pm 3i$ are

$$\mathbf{X}_2(t) = e^{2t} \left[\begin{pmatrix} 1 \\ -2 \\ 2 \end{pmatrix} \cos 2t - \begin{pmatrix} 1 \\ 0 \\ 0 \end{pmatrix} \sin 2t \right]$$

$$= \begin{pmatrix} e^{2t}(\cos 3t - \sin 3t) \\ -2e^{2t}\cos 3t \\ 2e^{2t}\cos 3t \end{pmatrix}$$

and

$$\mathbf{X}_3(t) = e^{2t} \left[\begin{pmatrix} 1 \\ 0 \\ 0 \end{pmatrix} \cos 2t + \begin{pmatrix} 1 \\ -2 \\ 2 \end{pmatrix} \sin 2t \right]$$

$$= \begin{pmatrix} e^{2t}(\cos 3t + \sin 3t) \\ -2e^{2t}\sin 3t \\ 2e^{2t}\sin 3t \end{pmatrix}.$$

 It is a good exercise to verify this reduction by carefully carrying out the steps by hand.

Combining these three linearly independent solutions gives us a general solution of the system,

$$\mathbf{X}(t) = c_1 \mathbf{X}_1(t) + c_2 \mathbf{X}_2(t) + c_2 \mathbf{X}_3(t)$$

$$= c_1 \begin{pmatrix} -1 \\ 1 \\ 0 \end{pmatrix} + c_2 \begin{pmatrix} e^{2t}(\cos 3t - \sin 3t) \\ -2e^{2t}\cos 3t \\ 2e^{2t}\cos 3t \end{pmatrix}$$

$$+ c_3 \begin{pmatrix} e^{2t}(\cos 3t + \sin 3t) \\ -2e^{2t}\sin 3t \\ 2e^{2t}\sin 3t \end{pmatrix}$$

$$= \underbrace{\begin{pmatrix} -1 & e^{2t}(\cos 3t - \sin 3t) & e^{2t}(\cos 3t + \sin 3t) \\ 1 & -2e^{2t}\cos 3t & -2e^{2t}\sin 3t \\ 0 & 2e^{2t}\cos 3t & 2e^{2t}\sin 3t \end{pmatrix}}_{\text{"Fundamental Matrix"}} \begin{pmatrix} c_1 \\ c_2 \\ c_3 \end{pmatrix}$$

$$= \begin{pmatrix} -c_1 + e^{2t}[(c_2 + c_3)\cos 3t + (-c_2 + c_3)\sin 3t] \\ c_1 - 2c_2 e^{2t}\cos 3t - 2c_3 e^{2t}\sin 3t \\ 2c_2 e^{2t}\cos 3t + 2c_3 e^{2t}\sin 3t \end{pmatrix}.$$

Application of the initial condition yields $\mathbf{X}(0) = \begin{pmatrix} -c_1 + c_2 + c_3 \\ c_1 - 2c_2 \\ 2c_2 \end{pmatrix} = \begin{pmatrix} 0 \\ 0 \\ 2 \end{pmatrix}$, which gives us the system of equations $-c_1 + c_2 + c_3 = 0$, $c_1 - 2c_2 = 0$, $2c_2 = 2$ with solution $c_2 = 1$, $c_1 = 2$, and $c_3 = 1$. Therefore, the solution to the IVP is

$$\mathbf{X}(t) = \begin{pmatrix} -2 + e^{2t}\cos 3t \\ 2 - 2e^{2t}\cos 3t - 2e^{2t}\sin 3t \\ 2e^{2t}\cos 3t + 2e^{2t}\sin 3t \end{pmatrix}.$$

In Figure 6.7, we graph $x(t)$, $y(t)$, and $z(t)$ and show the parametric plot of $\{x(t), y(t), z(t)\}$ in three dimensions. Notice that $\lim_{t\to\infty} x(t)$ does not exist, $\lim_{t\to\infty} y(t)$ does not exist, and $\lim_{t\to\infty} z(t)$ does not exist, so over time the solution is directed away from the point $(0, 0, 2)$.

Alternate Method for Solving IVPs

We can also use a fundamental matrix to help us solve homogeneous IVPs. If $\Phi(t)$ is a fundamental matrix for the linear homogeneous system $\mathbf{X}' = \mathbf{AX}$, a general solution is $\mathbf{X}(t) = \Phi(t)\mathbf{C}$, where $\mathbf{C}$ is a constant vector. Given the initial condition $\mathbf{X}(0) = \mathbf{X}_0$, then through substitution into $\mathbf{X}(t) = \Phi(t)\mathbf{C}$,

$$\mathbf{X}(0) = \Phi(0)\mathbf{C} \quad \text{and} \quad \mathbf{X}_0 = \Phi(0)\mathbf{C} \quad \text{so}$$
$$\mathbf{C} = \Phi^{-1}(0)\mathbf{X}_0.$$

Therefore, the solution to the IVP $\mathbf{X}' = \mathbf{AX}$, $\mathbf{X}(0) = \mathbf{X}_0$ is $\mathbf{X}(t) = \Phi(t)\Phi^{-1}(0)\mathbf{X}_0$.

EXAMPLE 6.4.4

Use a fundamental matrix to solve the IVP $\mathbf{X}' = \begin{pmatrix} 1 & 1 \\ 4 & -2 \end{pmatrix} \mathbf{X}$, $\mathbf{X}(0) = \begin{pmatrix} 1 \\ -2 \end{pmatrix}$.

Solution

The eigenvalues of $\mathbf{A} = \begin{pmatrix} 1 & 1 \\ 4 & -2 \end{pmatrix}$ are $\lambda_1 = 2$ and $\lambda_2 = -3$ with corresponding eigenvectors

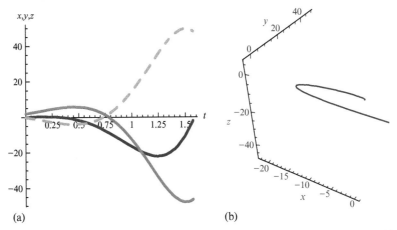

FIGURE 6.7   (a) Graphs of $x(t)$, $y(t)$ (dashed) and $z(t)$ (light pink; light gray in print versions). (b) Parametric plot of $\{x(t), y(t), z(t)\}$ in three dimensions for $0 \le t \le \pi/2$.

$\mathbf{v}_1 = \begin{pmatrix} 1 \\ 1 \end{pmatrix}$ and $\mathbf{v}_2 = \begin{pmatrix} 1 \\ -4 \end{pmatrix}$, respectively. A funda-

mental matrix is then given by $\Phi(t) = \begin{pmatrix} e^{2t} & e^{-3t} \\ e^{2t} & -4e^{-3t} \end{pmatrix}$.

We calculate $\Phi^{-1}(0)$ by observing that $\Phi(0) = \begin{pmatrix} 1 & 1 \\ 1 & -4 \end{pmatrix}$, so $\Phi^{-1}(0) = \dfrac{1}{-4-1}\begin{pmatrix} -4 & -1 \\ -1 & 1 \end{pmatrix} = \begin{pmatrix} 4/5 & 1/5 \\ 1/5 & -1/5 \end{pmatrix}$. Hence,

$$X(t) = \Phi(t)\Phi^{-1}(0)X_0 = \begin{pmatrix} e^{2t} & e^{-3t} \\ e^{2t} & -4e^{-3t} \end{pmatrix}\begin{pmatrix} 4/5 & 1/5 \\ 1/5 & -1/5 \end{pmatrix}\begin{pmatrix} 1 \\ -2 \end{pmatrix}$$

$$= \begin{pmatrix} e^{2t} & e^{-3t} \\ e^{2t} & -4e^{-3t} \end{pmatrix}\begin{pmatrix} 2/5 \\ 3/5 \end{pmatrix} = \begin{pmatrix} \frac{2}{5}e^{2t} + \frac{3}{5}e^{-3t} \\ \frac{2}{5}e^{2t} - \frac{12}{5}e^{-3t} \end{pmatrix}.$$

Figure 6.8 shows two graphs of the solution. Notice that as $t \to \infty$ the values of $x(t)$ and $y(t)$, where $\mathbf{X}(t) = \begin{pmatrix} x(t) \\ y(t) \end{pmatrix}$ are both close to $\frac{2}{5}e^{2t}$ because $\lim_{t\to\infty} e^{-3t} = 0$. This means that the solution approaches the line $y = x$ because for large values of t, $x(t)$ and $y(t)$ are approximately the same.

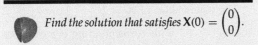

Find the solution that satisfies $\mathbf{X}(0) = \begin{pmatrix} 0 \\ 0 \end{pmatrix}$.

Repeated Eigenvalues

We now consider the case of repeated eigenvalues. This is more complicated than the other cases because two situations can arise. As we discovered in Section 6.2, an eigenvalue of multiplicity m can either have m corresponding linearly independent eigenvectors or have fewer than m corresponding linearly independent eigenvectors. In the case of m linearly independent eigenvectors, a general solution is found in the same manner as the case of m distinct eigenvalues.

EXAMPLE 6.4.5

Solve $\mathbf{X}' = \begin{pmatrix} 1 & -3 & 3 \\ 3 & -5 & 3 \\ 6 & -6 & 4 \end{pmatrix}\mathbf{X}$.

Solution

The eigenvalues of $\mathbf{A} = \begin{pmatrix} 1 & -3 & 3 \\ 3 & -5 & 3 \\ 6 & -6 & 4 \end{pmatrix}$ are $\lambda_1 = 4$ and $\lambda_{2,3} = -2$. An eigenvector corresponding to λ_1 is $\mathbf{v}_1 = \begin{pmatrix} 1 \\ 1 \\ 2 \end{pmatrix}$ so one solution of the system is

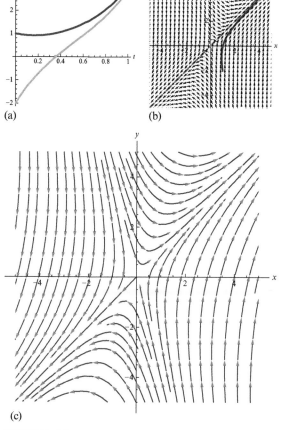

FIGURE 6.8 (a) Graph of $x(t)$ (dark red; dark black in print versions) and $y(t)$. (b) Graph of $\{x(t), y(t)\}$ along with the direction field associated with the system of equations. (c) Phase portrait.

$$\mathbf{X}_1(t) = \begin{pmatrix} 1 \\ 1 \\ 2 \end{pmatrix} e^{4t}. \text{ Let } \mathbf{v}_2 = \begin{pmatrix} x_2 \\ y_2 \\ z_2 \end{pmatrix} \text{ be an eigen-}$$

vector corresponding to $\lambda_{2,3}$. Then, $\mathbf{v}_2$ satisfies $(\mathbf{A} - \lambda_{2,3}\mathbf{I})\mathbf{v}_2 = \mathbf{0}$. The augmented matrix for this

system reduces to $\begin{pmatrix} 1 & -1 & 1 \\ 0 & 0 & 0 \\ 0 & 0 & 0 \end{pmatrix}$ which indicates that

$x_2 - y_2 + z_2 = 0$ and y_2 and z_2 are free. Hence, we can find two linearly independent eigenvectors corresponding to the repeated eigenvalue. In this

case choosing $y_2 = 1$ and then $z_2 = 1$ yields $\mathbf{v}_2 = \begin{pmatrix} 1 \\ 1 \\ 0 \end{pmatrix}$ and $\mathbf{v}_3 = \begin{pmatrix} -1 \\ 0 \\ 1 \end{pmatrix}$. Then, $\mathbf{X}_2(t) = \begin{pmatrix} 1 \\ 1 \\ 0 \end{pmatrix} e^{-2t}$

and $\mathbf{X}_3(t) = \begin{pmatrix} -1 \\ 0 \\ 1 \end{pmatrix} e^{-2t}$ are two additional lin-

early independent solutions of the system. Hence, $S = \{\mathbf{X}_1(t), \mathbf{X}_2(t), \mathbf{X}_3(t)\}$ is a fundamental set of solutions for the system. We verify linear independence by computing the Wronskian,

$$W(S) = \begin{vmatrix} e^{4t} & e^{-2t} & -e^{-2t} \\ e^{4t} & e^{-2t} & 0 \\ 2e^{4t} & 0 & e^{-2t} \end{vmatrix} = 2 \neq 0.$$

Thus, a general solution of the system is

$$\mathbf{X}(t) = c_1\mathbf{X}_1(t) + c_2\mathbf{X}_2(t) + c_3\mathbf{X}_3(t)$$

$$= \underbrace{\begin{pmatrix} e^{4t} & e^{-2t} & -e^{-2t} \\ e^{4t} & e^{-2t} & 0 \\ 2e^{4t} & 0 & e^{-2t} \end{pmatrix}}_{\text{Fundamental matrix}} \begin{pmatrix} c_1 \\ c_2 \\ c_3 \end{pmatrix}$$

$$= \begin{pmatrix} c_1e^{4t} + (c_2 - c_3)e^{-2t} \\ c_1e^{4t} + c_2e^{-2t} \\ 2c_1e^{4t} + c_3e^{-2t} \end{pmatrix}.$$

Find conditions on x_0, y_0, and z_0, if possible, so that the limit as $t \to \infty$ of the solution $\mathbf{X}(t)$ that satisfies the initial condition

$$\mathbf{X}(0) = \begin{pmatrix} x_0 \\ y_0 \\ z_0 \end{pmatrix} \text{ is } \mathbf{0}.$$

If an eigenvalue of multiplicity m has fewer than m linearly independent eigenvectors, we proceed in a manner that is similar to the situation that arose in Chapter 4 when we encountered repeated roots of characteristic equations. Consider a system with the repeated eigenvalue $\lambda_1 = \lambda_2$ and corresponding eigenvector $\mathbf{v}_1$. (Assume that there is not a second linearly independent eigenvector corresponding to

$\lambda_1 = \lambda_2$.) With the eigenvalue λ_1 and corresponding eigenvector $\mathbf{v}_1$, we obtain the solution to the system $\mathbf{X}_1 = \mathbf{v}_1 e^{\lambda_1 t}$. To find a second linearly independent solution corresponding to λ_1, instead of multiplying $\mathbf{X}_1$ by t as we did in Chapter 4, we suppose that a second linearly independent solution corresponding to λ_1 is of the form

$$\mathbf{X}_2 = (\mathbf{v}_2 t + \mathbf{w}_2) e^{\lambda_1 t}.$$

To find the vectors $\mathbf{v}_2$ and $\mathbf{w}_2$, we substitute $\mathbf{X}_2$ into $\mathbf{X}' = \mathbf{A}\mathbf{X}$. Because $\mathbf{X}_2' = \lambda_1(\mathbf{v}_2 t + \mathbf{w}_2) e^{\lambda_1 t} + \mathbf{v}_2 e^{\lambda_1 t}$, we have

$$\mathbf{X}_2' = \mathbf{A}\mathbf{X}_2,$$
$$\lambda_1(\mathbf{v}_2 t + \mathbf{w}_2) e^{\lambda_1 t} + \mathbf{v}_2 e^{\lambda_1 t} = \mathbf{A}(\mathbf{v}_2 t + \mathbf{w}_2) e^{\lambda_1 t},$$
$$\lambda_1 \mathbf{v}_2 t + (\lambda_1 \mathbf{w}_2 + \mathbf{v}_2) = \mathbf{A}\mathbf{v}_2 t + \mathbf{A}\mathbf{w}_2.$$

Equating coefficients yields $\lambda_1 \mathbf{v}_2 = \mathbf{A}\mathbf{v}_2$ and $\lambda_1 \mathbf{w}_2 + \mathbf{v}_2 = \mathbf{A}\mathbf{w}_2$. The equation $\lambda_1 \mathbf{v}_2 = \mathbf{A}\mathbf{v}_2$ indicates that $\mathbf{v}_2$ is an eigenvector that corresponds to λ_1, so we take $\mathbf{v}_2 = \mathbf{v}_1$. Simplifying $\lambda_1 \mathbf{w}_2 + \mathbf{v}_2 = \mathbf{A}\mathbf{w}_2$, we find that

$$\lambda_1 \mathbf{w}_2 + \mathbf{v}_2 = \mathbf{A}\mathbf{w}_2,$$
$$\mathbf{v}_2 = \mathbf{A}\mathbf{w}_2 - \lambda_1 \mathbf{w}_2,$$
$$\mathbf{v}_2 = (\mathbf{A} - \lambda_1 \mathbf{I})\mathbf{w}_2.$$

Hence, $\mathbf{w}_2$ satisfies the equation

$$(\mathbf{A} - \lambda_1 \mathbf{I})\mathbf{w}_2 = \mathbf{v}_1 \quad \text{(because } \mathbf{v}_2 = \mathbf{v}_1\text{).} \quad (6.17)$$

Therefore, a *second linearly independent solution* corresponding to the eigenvalue λ_1 has the form

$$\mathbf{X}_2 = (\mathbf{v}_1 t + \mathbf{w}_2) e^{\lambda_1 t}, \quad (6.18)$$

where $\mathbf{w}_2$ satisfies Equation (6.17).

Suppose that $\mathbf{A}$ has a repeated eigenvalue $\lambda_1 = \lambda_2$ and we can find only one corresponding (linearly independent) eigenvector $\mathbf{v}_1$. What happens if you try to find a second linearly independent solution of the form $\mathbf{v}_2 t e^{\lambda_1 t}$ as we did in solving higher order equations with repeated roots of the characteristic equation in Section 4.4?

Theorem 37 (Repeated Eigenvalues with One Eigenvector). *Let $\mathbf{A}$ be a square matrix with real, constant entries. If $\mathbf{A}$ has a repeated eigenvalue $\lambda_1 = \lambda_2$ with only one corresponding (linearly independent) eigenvector $\mathbf{v}_1$, two linearly independent solutions of $\mathbf{X}' = \mathbf{A}\mathbf{X}$ are $\mathbf{X}_1 = \mathbf{v}_1 e^{\lambda_1 t}$ and $\mathbf{X}_2 = (\mathbf{v}_1 t + \mathbf{w}_2) e^{\lambda_1 t}$, where $\mathbf{w}_2$ satisfies Equation (6.17).*

If $\mathbf{A}$ is a 2×2 matrix with the repeated eigenvalue $\lambda_1 = \lambda_2$ with only one corresponding (linearly independent) eigenvector $\mathbf{v}_1$ a general solution to $\mathbf{X}' = \mathbf{A}\mathbf{X}$ is

$$\mathbf{X}(t) = c_1 \mathbf{v}_1 e^{\lambda_1 t} + c_2(\mathbf{v}_1 t + \mathbf{w}_2) e^{\lambda_1 t}, \quad (6.19)$$

where $\mathbf{w}_2$ is found by solving Equation (6.17).

EXAMPLE 6.4.6

Find a general solution of $\mathbf{X}' = \begin{pmatrix} -8 & -1 \\ 16 & 0 \end{pmatrix} \mathbf{X}$.

Solution

The eigenvalues of $\mathbf{A} = \begin{pmatrix} -8 & -1 \\ 16 & 0 \end{pmatrix}$ are $\lambda_{1,2} = -4$. An eigenvector $\mathbf{v}_1$ that corresponds to $\lambda_{1,2}$ satisfies the system $(\mathbf{A} - \lambda_{1,2}\mathbf{I})\mathbf{v}_1 = \mathbf{0}$. After row operations, the augmented matrix for this system reduces to $\begin{pmatrix} 4 & 1 \\ 0 & 0 \end{pmatrix}$, which shows us that $4x_1 + y_1 = 0$ and that y_1 is free. There is *only one linearly independent eigenvector* corresponding to $\lambda_{1,2}$. Choosing $x_1 = 1$ yields $\mathbf{v}_1 = \begin{pmatrix} 1 \\ -4 \end{pmatrix}$ so one solution of the system is

$$\mathbf{X}_1(t) = \begin{pmatrix} 1 \\ -4 \end{pmatrix} e^{-4t}.$$

To find $\mathbf{w}_2 = \begin{pmatrix} x_2 \\ y_2 \end{pmatrix}$ in a second linearly independent solution of the form given by Equation (6.18), we solve Equation (6.17), which in this case is $\begin{pmatrix} -4 & -1 \\ 16 & 4 \end{pmatrix}\begin{pmatrix} x_2 \\ y_2 \end{pmatrix} = \begin{pmatrix} 1 \\ -4 \end{pmatrix}$. This system reduces to $\begin{pmatrix} 4 & 1 \\ 0 & 0 \end{pmatrix}\begin{pmatrix} x_2 \\ y_2 \end{pmatrix} = \begin{pmatrix} -1 \\ 0 \end{pmatrix}$, which indicates that $4x_2 + y_2 = -1$. If we let $x_2 = 0$, then $y_2 = -1$,

$\mathbf{w}_2 = \begin{pmatrix} 0 \\ -1 \end{pmatrix}$, and a second linearly independent solution of the system is

$$\mathbf{X}_2(t) = \left(\begin{pmatrix} 1 \\ -4 \end{pmatrix} t + \begin{pmatrix} 0 \\ -1 \end{pmatrix} \right) e^{-4t}.$$

Hence, a general solution is

$$\begin{aligned} \mathbf{X}(t) &= c_1 \begin{pmatrix} 1 \\ -4 \end{pmatrix} e^{-4t} + c_2 \left(\begin{pmatrix} 1 \\ -4 \end{pmatrix} t + \begin{pmatrix} 0 \\ -1 \end{pmatrix} \right) e^{-4t} \\ &= \underbrace{\begin{pmatrix} e^{-4t} & te^{-4t} \\ -4e^{-4t} & (-4t-1)e^{-4t} \end{pmatrix}}_{\text{"Fundamental Matrix"}} \begin{pmatrix} c_1 \\ c_2 \end{pmatrix} \\ &= \begin{pmatrix} c_1 + c_2 t \\ (-4c_1 - c_2) - 4c_2 t \end{pmatrix} e^{-4t}. \end{aligned}$$

Figure 6.9 shows the graph of several solutions to the system along with the direction field. Notice that the behavior of these solutions differs from those of the systems solved earlier in the section due to the repeated eigenvalues. We see from the formula for $\mathbf{X}(t)$ that $\lim_{t\to\infty} x(t) = 0$ and $\lim_{t\to\infty} y(t) = 0$ (why?). In addition, these solutions approach $(0,0)$ tangent to $y = -4x$, the line through the origin that is parallel to the vector $\mathbf{v}_1 = \begin{pmatrix} 1 \\ -4 \end{pmatrix}$. (We discuss why this occurs in Section 6.6.)

EXAMPLE 6.4.7

▼ Solve $\mathbf{X}' = \begin{pmatrix} 5 & 3 & -3 \\ 2 & 4 & -5 \\ -4 & 2 & -3 \end{pmatrix} \mathbf{X}$.

Solution

The eigenvalues of $\mathbf{A} = \begin{pmatrix} 5 & 3 & -3 \\ 2 & 4 & -5 \\ -4 & 2 & -3 \end{pmatrix}$ are determined with

$$\begin{vmatrix} 5-\lambda & 3 & -3 \\ 2 & 4-\lambda & -5 \\ -4 & 2 & -3-\lambda \end{vmatrix} = -\lambda^3 + 6\lambda^2 + 15\lambda + 8$$

$$= -(\lambda - 8)(\lambda + 1)^2 = 0$$

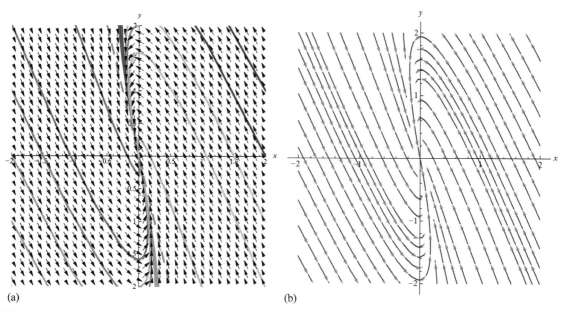

(a)   (b)

FIGURE 6.9 (a) All nontrivial solutions approach the origin. (b) A different view of the phase portrait emphasizes the behavior of the solutions.

to be $\lambda_{1,2} = -1$ and $\lambda_3 = 8$. An eigenvector $\mathbf{v}_1 = \begin{pmatrix} x_1 \\ y_1 \\ z_1 \end{pmatrix}$ that corresponds to $\lambda_{1,2} = -1$ satisfies $(\mathbf{A} - \lambda_{1,2}\mathbf{I})\mathbf{v}_1 = \mathbf{0}$. After row operations, the augmented matrix for this system reduces to $\begin{pmatrix} 1 & 0 & 0 \\ 0 & 1 & -1 \\ 0 & 0 & 0 \end{pmatrix}$. Hence, $x_1 = 0$ and $y_1 - z_1 = 0$, so there is only one linearly independent eigenvector corresponding to $\lambda_{1,2}$. If we let $y_1 = 1$, then $\mathbf{v}_1 = \begin{pmatrix} 0 \\ 1 \\ 1 \end{pmatrix}$. Therefore, one solution to the

system is $\mathbf{X}_1(t) = \begin{pmatrix} 0 \\ 1 \\ 1 \end{pmatrix} e^{-t}$. A second linearly

independent solution of the form $\mathbf{X}_2(t) = (\mathbf{v}_1 t + \mathbf{w}_2)e^{\lambda_1 t}$ is found by solving $(\mathbf{A} - \lambda_1\mathbf{I})\mathbf{w}_2 = \mathbf{v}_1$,

given by $\begin{pmatrix} 6 & 3 & -3 \\ 2 & 5 & -5 \\ -4 & 2 & -2 \end{pmatrix} \begin{pmatrix} x_2 \\ y_2 \\ z_2 \end{pmatrix} = \begin{pmatrix} 0 \\ 1 \\ 1 \end{pmatrix}$, for the vector

$\mathbf{w}_2 = \begin{pmatrix} x_2 \\ y_2 \\ z_2 \end{pmatrix}$. After row operations, the augmented

matrix for this system reduces to $\begin{pmatrix} 1 & 0 & 0 & | & -1/8 \\ 0 & 1 & -1 & | & 1/4 \\ 0 & 0 & 0 & | & 0 \end{pmatrix}$,

which indicates that $x_2 = -1/8$ and $y_2 - z_2 = 1/4$.

Hence, if $z_2 = 0$, $y_2 = 1/4$, so $\mathbf{w}_2 = \begin{pmatrix} -1/8 \\ 1/4 \\ 0 \end{pmatrix}$.

A second linearly independent solution of the

system is then $\mathbf{X}_2(t) = \left(\begin{pmatrix} 0 \\ 1 \\ 1 \end{pmatrix} t + \begin{pmatrix} -1/8 \\ 1/4 \\ 0 \end{pmatrix} \right) e^{-t}$.

A third linearly independent solution is found

using an eigenvector $\mathbf{v}_3 = \begin{pmatrix} x_3 \\ y_3 \\ z_3 \end{pmatrix}$ corresponding

to $\lambda_3 = 8$. After row operations, the augmented

matrix for $(\mathbf{A} - \lambda_3\mathbf{I})\mathbf{v}_3 = \mathbf{0}$ reduces to $\begin{pmatrix} 1 & 0 & 9/2 \\ 0 & 1 & 7/2 \\ 0 & 0 & 0 \end{pmatrix}$

so the components of $\mathbf{v}_3$ must satisfy $2x_3 + 9z_3 = 0$ and $2y_3 + 7z_3 = 0$. If we choose $z_3 = -2$, then

$x_3 = 9$ and $y_3 = 7$. Hence, $\mathbf{v}_3 = \begin{pmatrix} 9 \\ 7 \\ -2 \end{pmatrix}$ and a

third linearly independent solution of the system

is given by $\mathbf{X}_3(t) = \begin{pmatrix} 9 \\ 7 \\ -2 \end{pmatrix} e^{8t}$. A general solution

of the system is then given by

$\mathbf{X}(t) = c_1\mathbf{X}_1(t) + c_2\mathbf{X}_2(t) + c_3\mathbf{X}_3(t)$

$= c_1 \begin{pmatrix} 0 \\ 1 \\ 1 \end{pmatrix} e^{-t} + c_2 \left(\begin{pmatrix} 0 \\ 1 \\ 1 \end{pmatrix} t + \begin{pmatrix} -1/8 \\ 1/4 \\ 0 \end{pmatrix} \right) e^{-t}$

$+ c_3 \begin{pmatrix} 9 \\ 7 \\ -2 \end{pmatrix} e^{8t}$

$= \begin{pmatrix} 0 & \frac{-1}{8}e^{-t} & 9e^{8t} \\ e^{-t} & \left(t + \frac{1}{4} \right)e^{-t} & 7e^{8t} \\ e^{-t} & te^{-t} & -2e^{8t} \end{pmatrix} \begin{pmatrix} c_1 \\ c_2 \\ c_3 \end{pmatrix}$

$\underbrace{}_{\text{"Fundamental Matrix"}}$

$= \begin{pmatrix} -\frac{1}{8}c_2 e^{-t} + 9c_3 e^{8t} \\ c_1 e^{-t} + \left(t + \frac{1}{4} \right) c_2 e^{-t} + 7c_3 e^{8t} \\ c_1 e^{-t} + c_2 t e^{-t} - 2c_3 e^{8t} \end{pmatrix}$.

A similar method is carried out in the case of three equal eigenvalues $\lambda = \lambda_{1,2,3}$, where we can find only one (linearly independent) eigenvector $\mathbf{v}_1$. When we encounter this situation, we assume that

$\mathbf{X}_1(t) = \mathbf{v}_1 e^{\lambda_1 t}, \quad \mathbf{X}_2(t) = (\mathbf{v}_2 t + \mathbf{w}_2)e^{\lambda_1 t}, \quad$ and

$\mathbf{X}_3(t) = \left(\frac{1}{2}\mathbf{v}_3 t^3 + \mathbf{w}_3 t + \mathbf{u}_3 \right) e^{\lambda_1 t}$.

Substitution of these solutions into the system of differential equations yields the following system of equations, which is solved for the unknown vectors $\mathbf{v}_2$, $\mathbf{w}_2$, $\mathbf{v}_3$, $\mathbf{w}_3$, and $\mathbf{u}_3$:

$\lambda_1\mathbf{v}_2 = \mathbf{A}\mathbf{v}_2, \quad (\mathbf{A} - \lambda_1\mathbf{I})\mathbf{w}_2 = \mathbf{v}_2, \quad \lambda_1\mathbf{v}_3 = \mathbf{A}\mathbf{v}_3,$

$(\mathbf{A} - \lambda_1\mathbf{I})\mathbf{w}_3 = \mathbf{v}_3, \quad (\mathbf{A} - \lambda_1\mathbf{I})\mathbf{u}_3 = \mathbf{w}_3.$

Similar to the previous case, $\mathbf{v}_3 = \mathbf{v}_2 = \mathbf{v}_1$, $\mathbf{w}_2 = \mathbf{w}_3$, and the vector $\mathbf{u}_3$ is found by solving the system

$$(\mathbf{A} - \lambda_1 \mathbf{I})\mathbf{u}_3 = \mathbf{w}_2. \qquad (6.20)$$

The three linearly independent solutions have the form

$$\mathbf{X}_1(t) = \mathbf{v}_1 e^{\lambda_1 t}, \quad \mathbf{X}_2(t) = (\mathbf{v}_1 t + \mathbf{w}_2)e^{\lambda_1 t}, \quad \text{and}$$

$$\mathbf{X}_3(t) = \left(\frac{1}{2}\mathbf{v}_1 t^3 + \mathbf{w}_2 t + \mathbf{u}_3\right)e^{\lambda_1 t}.$$

Notice that this method is generalized for instances when the multiplicity of the repeated eigenvalue is greater than 3.

EXAMPLE 6.4.8

▼ Solve $\mathbf{X}' = \begin{pmatrix} 1 & 1 & 1 \\ 2 & 1 & -1 \\ -3 & 2 & 4 \end{pmatrix}\mathbf{X}$.

Solution

The eigenvalues are found by solving

$$\begin{vmatrix} 1-\lambda & 1 & 1 \\ 2 & 1-\lambda & -1 \\ -3 & 2 & 4-\lambda \end{vmatrix} = (1-\lambda)\begin{vmatrix} 1-\lambda & -1 \\ 2 & 4-\lambda \end{vmatrix}$$

$$-1 \times \begin{vmatrix} 2 & -1 \\ -3 & 4-\lambda \end{vmatrix} + 1 \times \begin{vmatrix} 2 & 1-\lambda \\ -3 & 2 \end{vmatrix}$$

$$= -\lambda^3 + 6\lambda^2 - 12\lambda + 8 = -(\lambda-2)(\lambda^2 - 4\lambda + 4)$$

$$= -(\lambda - 2)^3 = 0.$$

Hence, $\lambda_{1,2,3} = 2$. We find eigenvectors $\mathbf{v}_1 = \begin{pmatrix} x_1 \\ y_1 \\ z_1 \end{pmatrix}$ corresponding to λ_1 with the system $(\mathbf{A} - \lambda_1 \mathbf{I})\mathbf{v}_1 = 0$. With elementary row operations, the augmented matrix for this system reduces to $\begin{pmatrix} 1 & 0 & 0 \\ 0 & 1 & 1 \\ 0 & 0 & 0 \end{pmatrix}$. Thus, $x_1 = 0$ and $y_1 + z_1 = 0$. If we select

$z_1 = 1$, then $y_1 = -1$ and $\mathbf{v}_1 = \begin{pmatrix} 0 \\ -1 \\ 1 \end{pmatrix}$. Therefore,

we can find only one (linearly independent) eigenvector corresponding to $\lambda_{1,2,3} = 2$. Using this eigenvalue and eigenvector, we find that one solution to the system is $\mathbf{X}_1(t) = \mathbf{v}_1 e^{2t} \begin{pmatrix} 0 \\ -1 \\ 1 \end{pmatrix} e^{2t}$.

The vector $\mathbf{w}_2 = \begin{pmatrix} x_2 \\ y_2 \\ z_2 \end{pmatrix}$ in a second linearly independent solution of the form $\mathbf{X}_2(t) = (\mathbf{v}_1 t + \mathbf{w}_2)e^{2t}$ is found by solving the system $(\mathbf{A} - \lambda_1 \mathbf{I})\mathbf{w}_2 = \mathbf{v}_1$. The augmented matrix for this system is $\begin{pmatrix} -1 & 1 & 1 & 0 \\ 2 & -1 & -1 & -1 \\ -3 & 2 & 2 & 1 \end{pmatrix}$, which after

row operations reduces to $\begin{pmatrix} 1 & 0 & 0 & -1 \\ 0 & 1 & 1 & -1 \\ 0 & 0 & 0 & 0 \end{pmatrix}$. Hence,

$x_2 = -1$ and $y_2 + z_2 = -1$, so if we choose

$z_2 = 0$, then $y_2 = -1$. Therefore, $\mathbf{w}_2 = \begin{pmatrix} -1 \\ -1 \\ 0 \end{pmatrix}$,

so $\mathbf{X}_2(t) = \left(\begin{pmatrix} 0 \\ -1 \\ 1 \end{pmatrix} t + \begin{pmatrix} -1 \\ -1 \\ 0 \end{pmatrix}\right) e^{2t}$. Finally,

we must determine a vector $\mathbf{u}_3 = \begin{pmatrix} x_3 \\ y_3 \\ z_3 \end{pmatrix}$ in

a third linearly independent solution of the form $\mathbf{X}_3(t) = \left(\frac{1}{2}\mathbf{v}_1 t^2 + \mathbf{w}_2 t + \mathbf{u}_3\right)e^{\lambda_1 t}$ by solving system (6.20) for $\mathbf{u}_3$. The augmented matrix for this system is $\begin{pmatrix} -1 & 1 & 1 & -1 \\ 2 & -1 & -1 & -1 \\ -3 & 2 & 2 & 0 \end{pmatrix}$, which

after row operations reduces to $\begin{pmatrix} 1 & 0 & 0 & -2 \\ 0 & 1 & 1 & -3 \\ 0 & 0 & 0 & 0 \end{pmatrix}$.

Therefore, $x_3 = -2$ and $y_3 + z_3 = -3$. If we

select $z_3 = 0$, then $y_3 = -3$. Hence, $\mathbf{u}_3 = \begin{pmatrix} -2 \\ -3 \\ 0 \end{pmatrix}$,

so a third linearly independent solution is

$$X_3(t) = \left(\frac{1}{2}\begin{pmatrix}0\\-1\\1\end{pmatrix}t^2 + \begin{pmatrix}-1\\-1\\0\end{pmatrix}t + \begin{pmatrix}-2\\-3\\0\end{pmatrix}\right)e^{2t}. \quad A$$

general solution is then given by

$$X(t) = c_1X_1(t) + c_2X_2(t) + c_3X_3(t)$$

$$= c_1\begin{pmatrix}0\\-1\\1\end{pmatrix}e^{2t} + c_2\left(\begin{pmatrix}0\\-1\\1\end{pmatrix}t + \begin{pmatrix}-1\\-1\\0\end{pmatrix}\right)e^{2t}$$

$$+ c_3\left(\frac{1}{2}\begin{pmatrix}0\\-1\\1\end{pmatrix}t^2 + \begin{pmatrix}-1\\-1\\0\end{pmatrix}t + \begin{pmatrix}-2\\-3\\0\end{pmatrix}\right)e^{2t}$$

$$= \underbrace{\begin{pmatrix}0 & -e^{2t} & (-t-2)e^{2t}\\-e^{2t} & (-t-1)e^{2t} & (\frac{-1}{2}t^2-t-3)e^{2t}\\e^{2t} & te^{2t} & \frac{1}{2}t^2e^{2t}\end{pmatrix}}_{\text{"Fundamental Matrix"}}\begin{pmatrix}c_1\\c_2\\c_3\end{pmatrix}$$

$$= \begin{pmatrix}-c_2e^{2t} + c_3(-t-2)e^{2t}\\-c_1e^{2t} + c_2(-t-1)e^{2t} + c_3(\frac{-1}{2}t^2-t-3)e^{2t}\\c_1e^{2t} + c_2te^{2t} + \frac{1}{2}c_3t^2e^{2t}\end{pmatrix}.$$

EXERCISES 6.4

In Exercises 1-12, use the given eigenvalues and corresponding (linearly independent) eigenvectors of the matrix A to find a general solution of $X' = AX$.

1. $A = \begin{pmatrix}5 & -1\\3 & 1\end{pmatrix}$; $\lambda_1 = 2, v_1 = \begin{pmatrix}1\\3\end{pmatrix}$; $\lambda_2 = 4$, $v_2 = \begin{pmatrix}1\\1\end{pmatrix}$.

2. $A = \begin{pmatrix}4 & 0\\-1 & 0\end{pmatrix}$; $\lambda_1 = 0, v_1 = \begin{pmatrix}0\\1\end{pmatrix}$; $\lambda_2 = 4$, $v_2 = \begin{pmatrix}-4\\1\end{pmatrix}$.

3. $A = \begin{pmatrix}-4 & 0\\-1 & 2\end{pmatrix}$; $\lambda_1 = -4, v_1 = \begin{pmatrix}6\\1\end{pmatrix}$; $\lambda_2 = 2$, $v_2 = \begin{pmatrix}0\\1\end{pmatrix}$.

4. $A = \begin{pmatrix}5 & 0\\0 & 5\end{pmatrix}$; $\lambda_{1,2} = 5, v_1 = \begin{pmatrix}1\\0\end{pmatrix}, v_2 = \begin{pmatrix}0\\1\end{pmatrix}$.

5. $A = \begin{pmatrix}1 & 1\\-1 & 3\end{pmatrix}$; $\lambda_{1,2} = 2, v_1 = \begin{pmatrix}1\\1\end{pmatrix}$.

6. $A = \begin{pmatrix}0 & -6\\6 & 0\end{pmatrix}$; $\lambda_{1,2} = \pm6i, v_{1,2} = \begin{pmatrix}\pm i\\1\end{pmatrix}$.

7. $A = \begin{pmatrix}3 & 4\\-2 & -1\end{pmatrix}$; $\lambda_{1,2} = 1 \pm 2i$, $v_{1,2} = \begin{pmatrix}-1 \pm (-i)\\1\end{pmatrix}$.

8. $A = \begin{pmatrix}-1 & 4 & -2\\-3 & 4 & 0\\-3 & 1 & 3\end{pmatrix}$; $\lambda_1 = 1, v_1 = \begin{pmatrix}1\\1\\1\end{pmatrix}$; $\lambda_2 = 2, v_2 = \begin{pmatrix}2\\3\\3\end{pmatrix}$; $\lambda_3 = 3, v_3 = \begin{pmatrix}1\\3\\4\end{pmatrix}$.

9. $A = \begin{pmatrix}3 & 0 & 0\\0 & 2 & 1\\0 & 0 & 2\end{pmatrix}$; $\lambda_{1,2} = 2, v_1 = \begin{pmatrix}0\\1\\0\end{pmatrix}$; $\lambda_3 = 3$, $v_2 = \begin{pmatrix}1\\0\\0\end{pmatrix}$.

10. $A = \begin{pmatrix}2 & 1 & 0\\0 & 2 & 1\\0 & 0 & 2\end{pmatrix}$; $\lambda_{1,2,3} = 2, v_1 = \begin{pmatrix}1\\0\\0\end{pmatrix}$.

11. $A = \begin{pmatrix}0 & -1 & 0\\1 & 0 & 0\\0 & 0 & 2\end{pmatrix}$; $\lambda_{1,2} = \pm i, v_{1,2} = \begin{pmatrix}\pm i\\1\\0\end{pmatrix}$; $\lambda_3 = 2, v_3 = \begin{pmatrix}0\\0\\1\end{pmatrix}$.

12. $A = \begin{pmatrix}3 & 2 & -3\\1 & 1 & 1\\0 & -4 & -4\end{pmatrix}$; $\lambda_{1,2} = -2 \pm i$, $v_{1,2} = \begin{pmatrix}3 \pm i\\-2 \pm (-i)\\4\end{pmatrix}$; $\lambda_3 = 4, v_3 = \begin{pmatrix}-7\\-2\\1\end{pmatrix}$.

In Exercises 13-34, find a general solution of the system.

13. $X' = \begin{pmatrix}1 & -10\\-7 & 10\end{pmatrix}X$.

14. $\begin{cases} x' = x - 2y \\ y' = 2x + 6y \end{cases}$.

15. $\begin{cases} dx/dt = 6x - y \\ dy/dt = 5x \end{cases}$.

16. $X' = \begin{pmatrix} 4 & 3 \\ -5 & -4 \end{pmatrix} X$.

17. $\begin{cases} x' = 7x \\ y' = 5x - 8y \end{cases}$.

18. $\begin{cases} dx/dt = 8x + 9y \\ dy/dt = -2x - 3y \end{cases}$.

19. $X' = \begin{pmatrix} -5 & 3 \\ 2 & -10 \end{pmatrix} X$.

20. $\begin{cases} x' = 8x + 5y \\ y' = -10x - 6y \end{cases}$.

21. $\begin{cases} dx/dt = -6x - 4y \\ dy/dt = -3x - 10y \end{cases}$.

22. $X' = \begin{pmatrix} 1 & 8 \\ -2 & -7 \end{pmatrix} X$.

23. $\begin{cases} x' = -6x + 2y \\ y' = -2x - 10y \end{cases}$.

24. $X' = \begin{pmatrix} 0 & 8 \\ 2 & 0 \end{pmatrix} X$.

25. $X' = \begin{pmatrix} 0 & 8 \\ -2 & 0 \end{pmatrix} X$.

26. $\begin{cases} x' = y \\ y' = -13x - 4y \end{cases}$.

27. $X' = \begin{pmatrix} 4 & 0 & 1 \\ 0 & -2 & 0 \\ 0 & 0 & -1 \end{pmatrix} X$.

28. $\begin{cases} x' = -3x + 3y - 4z \\ y' = -3y \\ z' = -5y - 4z \end{cases}$.

29. $X' = \begin{pmatrix} 5 & -1 & 3 \\ -4 & -1 & -2 \\ -4 & 2 & -3 \end{pmatrix} X$.

30. $\begin{cases} dx/dt = -5x + 4y - 5z \\ dy/dt = -y \\ dz/dt = 5x + y + z \end{cases}$.

31. $\begin{cases} x' = x + 2y + 3z \\ y' = y + 2z \\ z' = -2y + z \end{cases}$.

32. $\begin{cases} x' = y \\ y' = z \\ z' = x - y + z \end{cases}$.

33. $\begin{cases} x' = -2x + 4y + 2z - 2w \\ y' = 6x - 10y - 7z + 4w \\ z' = -6x + 10y + 7z - 4w \\ w' = 9x - 16y - 10z + 7w \end{cases}$.

34. $X' = \begin{pmatrix} 3 & -4 & 4 & -2 \\ 7 & -9 & 9 & -4 \\ 3 & -4 & 4 & -2 \\ -2 & 2 & -2 & 0 \end{pmatrix} X$.

In Exercises 35-44, solve the IVP.

35. $X' = \begin{pmatrix} 1 & 1 \\ 0 & 2 \end{pmatrix} X, \, X(0) = \begin{pmatrix} 0 \\ 4 \end{pmatrix}$.

36. $X' = \begin{pmatrix} -4 & 0 \\ 2 & 4 \end{pmatrix} X, \, X(0) = \begin{pmatrix} 8 \\ 0 \end{pmatrix}$.

37. $X' = \begin{pmatrix} 4 & 0 \\ 2 & 4 \end{pmatrix} X, \, X(0) = \begin{pmatrix} 8 \\ 0 \end{pmatrix}$.

38. $X' = \begin{pmatrix} 4 & 8 \\ 0 & 4 \end{pmatrix} X, \, X(0) = \begin{pmatrix} 8 \\ 8 \end{pmatrix}$.

39. $X' = \begin{pmatrix} 0 & -4 \\ 4 & 0 \end{pmatrix} X, \, X(0) = \begin{pmatrix} 0 \\ 8 \end{pmatrix}$.

40. $X' = \begin{pmatrix} 0 & 13 \\ -1 & -4 \end{pmatrix} X, \, X(0) = \begin{pmatrix} 0 \\ 4 \end{pmatrix}$.

41. $X' = \begin{pmatrix} 4 & 0 & 1 \\ -2 & 1 & 0 \\ -2 & 0 & 1 \end{pmatrix} X, \, X(0) = \begin{pmatrix} -1 \\ 2 \\ 0 \end{pmatrix}$.

42. $X' = \begin{pmatrix} 1 & 0 & 2 \\ -1 & -1 & -1 \\ -1 & 0 & -2 \end{pmatrix} X, \, X(0) = \begin{pmatrix} 4 \\ 0 \\ 8 \end{pmatrix}$.

43. $X' = \begin{pmatrix} -4 & 1 & 8 & -3 \\ 2 & 0 & -2 & 0 \\ -1 & 0 & 1 & 0 \\ 2 & -1 & -6 & 3 \end{pmatrix} X, X(0) = \begin{pmatrix} 0 \\ 1 \\ 1 \\ 0 \end{pmatrix}.$

44. $X' = \begin{pmatrix} 2 & -2 & 0 & 4 \\ 1 & 1 & 0 & 0 \\ -1 & 2 & -1 & -2 \\ -1 & 2 & 0 & -3 \end{pmatrix} X, X(0) = \begin{pmatrix} 1 \\ 0 \\ 0 \\ 1 \end{pmatrix}.$

45. Show that if **a** and **b** are constant vectors, then $X_1(t) = e^{\alpha t}(\mathbf{a} \cos \beta t - \mathbf{b} \sin \beta t)$ and $X_2 = e^{\alpha t}(\mathbf{b} \cos \beta t + \mathbf{a} \sin \beta t)$ are linearly independent vector-valued functions.

46. Show that in the 3×3 system $X' = AX$ with the eigenvalue λ of multiplicity 3 with one corresponding (linearly independent) eigenvector $\mathbf{v_1}$, three linearly independent solutions of the system are $X_1 = \mathbf{v_1} e^{\lambda t}$, $X_2 = (\mathbf{v_1} t + \mathbf{w_2}) e^{\lambda t}$, and $X_3 = \left(\frac{1}{2} \mathbf{v_1} t^2 + \mathbf{w_2} t + \mathbf{u_3} \right) e^{\lambda t}$, where $\mathbf{u_3}$ satisfies $(A - \lambda I)\mathbf{u_3} = \mathbf{w_2}$ and $\mathbf{w_2}$ satisfies $(A - \lambda I)\mathbf{w_2} = \mathbf{v_1}$.

47. Show that both
$$\begin{cases} x = e^{2t}(c_1 \cos t - c_2 \sin t) \\ y = e^{2t}(c_2 \cos t + c_1 \sin t) \end{cases} \text{ and }$$

$$\begin{cases} x = \frac{1}{2} e^{(2-i)t} \left[c_1(1 + e^{2it}) + c_2 i(e^{2it} - 1) \right] \\ y = \frac{1}{2} e^{(2-i)t} \left[c_2(1 + e^{2it}) + c_1 i(1 - e^{2it}) \right] \end{cases}$$

are general solutions of the system
$$\begin{cases} x' = 2x - y \\ y' = x + 2y \end{cases}.$$

48. Solve the systems (a) $\begin{cases} x' = 2x - y \\ y' = -x + 3y \end{cases}$;

(b) $\begin{cases} x' = 2x \\ y' = 3x + 2y \end{cases}$; (c) $\begin{cases} x' = x + 4y \\ y' = -2x - y \end{cases}$

subject to $x(0) = 1$ and $y(0) = 1$. In each case, graph the solution parametrically and individually.

49. Find a general solution of the system
$$\begin{cases} x' = 3z \\ y' = x - 4y + 2z \\ z' = -4y + z \end{cases}.$$ Graph the solution for various values of the constant.

50. How do the general solutions and direction field of the system $X' = \begin{pmatrix} 2 & \lambda \\ 1 & 0 \end{pmatrix} X$ change as λ ranges from -2 to 0? Solve the system for 25 equally spaced values of λ between -2 and 0 and note how the solution changes.

51. Consider the IVP $X' = \begin{pmatrix} 1 & -7/3 \\ -2 & -8/3 \end{pmatrix} X$, $x(0) = x_0, y(0) = y_0$. (a) Graph the direction field associated with the system $X' = \begin{pmatrix} 1 & -7/3 \\ -2 & -8/3 \end{pmatrix} X$. (b) Find conditions on x_0 and y_0 so that at least one of $x(t)$ or $y(t)$ approaches zero as t approaches infinity. (c) Is it possible to choose x_0 and y_0 so that both $x(t)$ and $y(t)$ approach zero as t approaches infinity?

52. (a) Find a general solution of $X' = AX$ if
$$X(t) = \begin{pmatrix} x_1 \\ x_2 \\ x_3 \\ x_4 \end{pmatrix} \text{ and (i) } A = \begin{pmatrix} \lambda & 0 & 0 & 0 \\ 0 & \lambda & 0 & 0 \\ 0 & 0 & \lambda & 0 \\ 0 & 0 & 0 & \lambda \end{pmatrix}, \text{ (ii)}$$

$$A = \begin{pmatrix} \lambda & 1 & 0 & 0 \\ 0 & \lambda & 0 & 0 \\ 0 & 0 & \lambda & 0 \\ 0 & 0 & 0 & \lambda \end{pmatrix}, \text{ (iii) } A = \begin{pmatrix} \lambda & 1 & 0 & 0 \\ 0 & \lambda & 1 & 0 \\ 0 & 0 & \lambda & 0 \\ 0 & 0 & 0 & \lambda \end{pmatrix}, \text{ and}$$

(i) $A = \begin{pmatrix} \lambda & 1 & 0 & 0 \\ 0 & \lambda & 1 & 0 \\ 0 & 0 & \lambda & 1 \\ 0 & 0 & 0 & \lambda \end{pmatrix}$. (b) For each system in (a), find the solution that satisfies the initial condition $X(0) = \begin{pmatrix} -1 \\ 0 \\ 1 \\ 2 \end{pmatrix}$ if $\lambda = -1/2$ and then graph $x_1, x_2, x_3,$ and x_4 for $0 \le t \le 10$. How are the solutions similar? How are they different? (c) Indicate how to generalize the results obtained in (a). How would you would find a general solution of $X' = AX$ for the 5×5 matrix

$$A = \begin{pmatrix} \lambda & 1 & 0 & 0 & 0 \\ 0 & \lambda & 1 & 0 & 0 \\ 0 & 0 & \lambda & 1 & 0 \\ 0 & 0 & 0 & \lambda & 1 \\ 0 & 0 & 0 & 0 & \lambda \end{pmatrix}?$$ How would you find a

general solution of $X' = AX$ for the $n \times n$

$$\text{matrix } A = \begin{pmatrix} \lambda & 1 & 0 & \cdots & 0 & 0 \\ 0 & \lambda & 1 & \cdots & 0 & 0 \\ \vdots & \vdots & \ddots & \ddots & \vdots & \vdots \\ \vdots & \vdots & \vdots & \ddots & \ddots & \vdots \\ 0 & 0 & 0 & \cdots & \lambda & 1 \\ 0 & 0 & 0 & \cdots & 0 & \lambda \end{pmatrix}?$$

6.5 FIRST-ORDER LINEAR NONHOMOGENEOUS SYSTEMS: UNDETERMINED COEFFICIENTS AND VARIATION OF PARAMETERS

- UNDETERMINED COEFFICIENTS
- VARIATION OF PARAMETERS

In Chapter 4, we learned how to find a particular solution of a nonhomogeneous differential equation through the use of undetermined coefficients or variation of parameters. Here, we approach the solution of systems of nonhomogeneous equations using these methods.

$$\text{Let } X = X(t) = \begin{pmatrix} x_1(t) \\ x_2(t) \\ \vdots \\ x_n(t) \end{pmatrix}, A = A(t) =$$

$$\begin{pmatrix} a_{11}(t) & a_{12}(t) & \cdots & a_{1n}(t) \\ a_{21}(t) & a_{22}(t) & \cdots & a_{2n}(t) \\ \vdots & \vdots & \ddots & \vdots \\ a_{n1}(t) & a_{n2}(t) & \cdots & a_{nn}(t) \end{pmatrix}, F(t) = \begin{pmatrix} f_1(t) \\ f_2(t) \\ \vdots \\ f_n(t) \end{pmatrix}, \text{ and}$$

$\Phi(t)$ be a fundamental matrix of the corresponding homogeneous system $X' = AX$. Then a general solution to the corresponding homogeneous

system $X' = AX$ is $X_h = \Phi(t)C$ where $C = \begin{pmatrix} c_1 \\ c_2 \\ \vdots \\ c_n \end{pmatrix}$

is an $n \times 1$ constant matrix.

To find a general solution to the linear nonhomogeneous system $X' = AX + F(t)$, we proceed in the same way we did with linear nonhomogeneous equations in Chapter 4. If $X_p(t)$ is a particular solution of the nonhomogeneous system, then all other solutions X of the system can be written in the form $X(t) = X_h(t) + X_p(t) = \Phi(t)C + X_p(t)$ (see Exercise 36).

Undetermined Coefficients

We use the method of undetermined coefficients to find a particular solution X_p to a nonhomogeneous linear system with constant coefficient matrix in much the same way as we approached nonhomogeneous higher order linear equations with constant coefficients in Chapter 4. The main difference is that the coefficients are *constant vectors* when we work with systems. For example, if we consider $X' = AX + F(t)$ where the entries of A are constant and $F(t) = \begin{pmatrix} e^{-2t} \\ 4 \end{pmatrix} = \begin{pmatrix} 1 \\ 0 \end{pmatrix} e^{-2t} + \begin{pmatrix} 0 \\ 4 \end{pmatrix}$ and none of the terms in $F(t)$ satisfy the corresponding homogeneous system $X' = AX$, we assume that a particular solution has the form $X_p(t) = ae^{-2t} + b$ where a and b are constant vectors. On the other hand if $\lambda = 2$ is an eigenvalue of A (with multiplicity one), we assume that $X_p(t) = ate^{-2t} + be^{-2t} + c$. (Note that even in this situation, a could be the zero vector.)

For constant vectors, a, b, c, ..., our convention will be that $a = \begin{pmatrix} a_1 \\ a_2 \\ \vdots \\ a_n \end{pmatrix}$, $b = \begin{pmatrix} b_1 \\ b_2 \\ \vdots \\ b_n \end{pmatrix}$, $c = \begin{pmatrix} c_1 \\ c_2 \\ \vdots \\ c_n \end{pmatrix}$, ...

EXAMPLE 6.5.1

Solve $X' = \begin{pmatrix} 0 & 8 \\ 2 & 0 \end{pmatrix} X + \begin{pmatrix} e^{3t} \\ t \end{pmatrix}$.

Solution

In this case, $F(t) = \begin{pmatrix} e^{3t} \\ t \end{pmatrix} = \begin{pmatrix} 1 \\ 0 \end{pmatrix} e^{3t} + \begin{pmatrix} 0 \\ 1 \end{pmatrix} t$ and a general solution of the corresponding homogeneous system is $X_h(t) = c_1 \begin{pmatrix} 2 \\ 1 \end{pmatrix} e^{4t} + c_2 \begin{pmatrix} -2 \\ 1 \end{pmatrix} e^{4t}$.

Notice that none of the components of $F(t)$ are in $X_h(t)$, so we assume that there is a particular solution of the form $X_p(t) = ae^{3t} + bt + c$. Then, $X'_p = 3ae^{3t} + b$ and substitution into the nonhomogeneous system $X' = AX + F(t)$, where $A = \begin{pmatrix} 0 & 8 \\ 2 & 0 \end{pmatrix}$, yields

$$3ae^{3t} + b = Aae^{3t} + Abt + Ac + \begin{pmatrix} 1 \\ 0 \end{pmatrix} e^{3t} + \begin{pmatrix} 0 \\ 1 \end{pmatrix} t.$$

Collecting like terms, we obtain the system of equations

$$3a = Aa + \begin{pmatrix} 1 \\ 0 \end{pmatrix} \quad \text{(Coefficient of } e^{3t}\text{)},$$

$$b = Ac \quad \text{(Constant terms)},$$

$$Ab + \begin{pmatrix} 0 \\ 1 \end{pmatrix} = 0 \quad \text{(Coefficients of } t\text{)}.$$

From the coefficients of e^{3t}, we find that $(A - 3I)a = \begin{pmatrix} -1 \\ 0 \end{pmatrix}$ or $\begin{pmatrix} -3 & 8 \\ 2 & -3 \end{pmatrix} \begin{pmatrix} a_1 \\ a_2 \end{pmatrix} = \begin{pmatrix} -1 \\ 0 \end{pmatrix}$. This system has the unique solution $a = \begin{pmatrix} -3/7 \\ -2/7 \end{pmatrix}$.

Next, we solve the system $Ab + \begin{pmatrix} 0 \\ 1 \end{pmatrix} = 0$ or $\begin{pmatrix} 0 & 8 \\ 2 & 0 \end{pmatrix} \begin{pmatrix} b_1 \\ b_2 \end{pmatrix} = \begin{pmatrix} 0 \\ -1 \end{pmatrix}$ for b. This yields the unique solution $b = \begin{pmatrix} -1/2 \\ 0 \end{pmatrix}$. Finally, we solve $b = Ac$ or $\begin{pmatrix} 0 & 8 \\ 2 & 0 \end{pmatrix} \begin{pmatrix} c_1 \\ c_2 \end{pmatrix} = \begin{pmatrix} -1/2 \\ 0 \end{pmatrix}$ for c, which gives

us $c = \begin{pmatrix} 0 \\ -1/16 \end{pmatrix}$. A particular solution to the nonhomogeneous system is then

$$X_p(t) = ae^{3t} + bt + c = \begin{pmatrix} -3/7 \\ -2/7 \end{pmatrix} e^{3t} + \begin{pmatrix} -1/2 \\ 0 \end{pmatrix} t$$

$$+ \begin{pmatrix} 0 \\ -1/16 \end{pmatrix},$$

so a general solution to the nonhomogeneous system is

$$X(t) = X_h(t) + X_p(t) = \begin{pmatrix} 2c_1 e^{4t} - 2c_2 e^{-4t} - \frac{3}{7} e^{3t} - \frac{1}{2} t \\ c_1 e^{4t} + c_2 e^{-4t} - \frac{2}{7} e^{3t} - \frac{1}{16} \end{pmatrix}.$$

To give another illustration of how the form of a particular solution is selected, suppose that $F(t) = \begin{pmatrix} 4 \sin 2t \\ e^{-t} \end{pmatrix} = \begin{pmatrix} 4 \\ 0 \end{pmatrix} \sin 2t + \begin{pmatrix} 0 \\ 1 \end{pmatrix} e^{-t}$ in Example 6.5.1. In this case, we assume that $X_p(t) = a \cos 2t + b \sin 2t + c e^{-t}$ and find the vectors a, b, and c through substitution into the nonhomogeneous system.

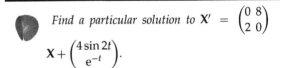

Find a particular solution to $X' = \begin{pmatrix} 0 & 8 \\ 2 & 0 \end{pmatrix} X + \begin{pmatrix} 4 \sin 2t \\ e^{-t} \end{pmatrix}$.

EXAMPLE 6.5.2

Solve $X' = AX + F(t)$ if $A = \begin{pmatrix} -3 & 4 \\ -2 & 3 \end{pmatrix}$ and (a) $F(t) = \begin{pmatrix} -3 \\ -1 \end{pmatrix} e^t$ and (b) $F(t) = \begin{pmatrix} -4 \\ -2 \end{pmatrix} e^t$.

Solution

A general solution of the corresponding homogeneous system $X' = \begin{pmatrix} -3 & 4 \\ -2 & 3 \end{pmatrix} X$ is

$X_h(t) = c_1 \begin{pmatrix} 2 \\ 1 \end{pmatrix} e^{-t} + c_2 \begin{pmatrix} 1 \\ 1 \end{pmatrix} e^t$. (a) Because an e^t term is a solution to the corresponding homogeneous system, we search for a particular solution of the nonhomogeneous system of the form $X_p(t) = a t e^t + b e^t$. Differentiating gives us $X'_p = a t e^t + (a + b)e^t$ and substituting into the nonhomogeneous system results in

$$a t e^t + (a+b)e^t = A\left(a t e^t + b e^t\right) + \begin{pmatrix} -3 \\ -1 \end{pmatrix} e^t$$

$$a t e^t + (a+b)e^t = A a t e^t + \left(A b + \begin{pmatrix} -3 \\ -1 \end{pmatrix}\right) e^t.$$

Equating coefficients gives us the system

$$a = A a \qquad\qquad (A - I)a = 0$$
$$a + b = A B + \begin{pmatrix} -3 \\ -1 \end{pmatrix} \quad \text{or} \quad a - (A - I)b = \begin{pmatrix} -3 \\ -1 \end{pmatrix}.$$

The equation $(A - I)a = 0$ has solution $a = cv$ where v is an eigenvector corresponding to the eigenvalue $\lambda = 1$. Choosing $v = \begin{pmatrix} 1 \\ 1 \end{pmatrix}$ gives us $a = c \begin{pmatrix} 1 \\ 1 \end{pmatrix}$. The equation $(A - I)b = \begin{pmatrix} 3 \\ 1 \end{pmatrix} + a$ or $\begin{pmatrix} -4 & 4 \\ -2 & 2 \end{pmatrix} \begin{pmatrix} b_1 \\ b_2 \end{pmatrix} = \begin{pmatrix} 3 \\ 1 \end{pmatrix} + c \begin{pmatrix} 1 \\ 1 \end{pmatrix}$ has zero solutions unless $3 + c = 2(1 + c)$ which yields $c = 1$. With $c = 1$, $\begin{pmatrix} -4 & 4 \\ -2 & 2 \end{pmatrix} \begin{pmatrix} b_1 \\ b_2 \end{pmatrix} = \begin{pmatrix} 4 \\ 2 \end{pmatrix}$ so $-b_1 + b_2 = 1$. Choosing $b_1 = 0$ and $b_2 = 1$ gives us $b = \begin{pmatrix} 0 \\ 1 \end{pmatrix}$. Then a particular solution of the nonhomogeneous system is

$$X_p(t) = a t e^t + b e^t = \begin{pmatrix} 1 \\ 1 \end{pmatrix} t e^t + \begin{pmatrix} 0 \\ 1 \end{pmatrix} e^t$$

and a general solution is

$$X(t) = X_h(t) + X_p(t) = c_1 \begin{pmatrix} 2 \\ 1 \end{pmatrix} e^{-t} + c_2 \begin{pmatrix} 1 \\ 1 \end{pmatrix} e^t$$
$$+ \begin{pmatrix} 1 \\ 1 \end{pmatrix} t e^t + \begin{pmatrix} 0 \\ 1 \end{pmatrix} e^t.$$

(b) Proceeding in the same manner as in (a), we assume that a particular solution of the nonhomogeneous system has the form $X_p(t) = a t e^t + b e^t$. Differentiating gives us $X'_p = a t e^t + (a + b)e^t$ and substituting into the nonhomogeneous system results in

$$a t e^t + (a+b)e^t = A\left(a t e^t + b e^t\right) + \begin{pmatrix} -4 \\ -2 \end{pmatrix} e^t,$$

$$a t e^t + (a+b)e^t = A a t e^t + \left(A b + \begin{pmatrix} -4 \\ -2 \end{pmatrix}\right) e^t.$$

Equating coefficients gives us the system

$$a = A a \qquad\qquad (A - I)a = 0$$
$$a + b = A B + \begin{pmatrix} -4 \\ -2 \end{pmatrix} \quad \text{or} \quad a - (A - I)b = \begin{pmatrix} -4 \\ -2 \end{pmatrix}.$$

The equation $(A - I)a = 0$ has solution $a = cv$ where v is an eigenvector corresponding to the eigenvalue $\lambda = 1$. Choosing $v = \begin{pmatrix} 1 \\ 1 \end{pmatrix}$ gives us $a = c \begin{pmatrix} 1 \\ 1 \end{pmatrix}$. The equation $(A - I)b = \begin{pmatrix} 3 \\ 1 \end{pmatrix} + a$ or $\begin{pmatrix} -4 & 4 \\ -2 & 2 \end{pmatrix} \begin{pmatrix} b_1 \\ b_2 \end{pmatrix} = \begin{pmatrix} 4 \\ 2 \end{pmatrix} + c \begin{pmatrix} 1 \\ 1 \end{pmatrix}$ has zero solutions unless $4 + c = 2(2 + c)$ which yields $c = 0$. With $c = 0$, $\begin{pmatrix} -4 & 4 \\ -2 & 2 \end{pmatrix} \begin{pmatrix} b_1 \\ b_2 \end{pmatrix} = \begin{pmatrix} 4 \\ 2 \end{pmatrix}$ so $-b_1 + b_2 = 1$. Choosing $b_1 = 0$ and $b_2 = 1$ gives us $b = \begin{pmatrix} 0 \\ 1 \end{pmatrix}$. Then a particular solution of the nonhomogeneous system is

$$X_p(t) = a t e^t + b e^t = \begin{pmatrix} 0 \\ 1 \end{pmatrix} e^t$$

and a general solution is

$$X(t) = X_h(t) + X_p(t) = c_1 \begin{pmatrix} 2 \\ 1 \end{pmatrix} e^{-t}$$
$$+ c_2 \begin{pmatrix} 1 \\ 1 \end{pmatrix} e^t + \begin{pmatrix} 0 \\ 1 \end{pmatrix} e^t.$$

 Find a particular solution to $X' = AX + F(t)$ *if* $A = \begin{pmatrix} -3 & 4 \\ -2 & 3 \end{pmatrix}$ *and* $F(t) = \begin{pmatrix} -3 \\ -3 \end{pmatrix} e^t$.

Variation of Parameters

Variation of parameters can be used to solve linear nonhomogeneous systems as well. In much the same way that we derived the method of variation of parameters for solving higher order differential equations.

Let $\Phi(t)$ be a fundamental matrix for the corresponding homogeneous system of the linear system $X' = AX + F(t)$. We assume that a particular solution has the form $X_p(t) = \Phi(t)U(t)$, where

$$U(t) = \begin{pmatrix} u_1(t) \\ u_2(t) \\ \vdots \\ u_n(t) \end{pmatrix}. \text{ Differentiating } X_p(t) \text{ gives us}$$

$$X_p'(t) = \Phi'(t)U(t) + \Phi(t)U'(t).$$

Substituting into equation $X' = AX + F(t)$ results in

$$\Phi'(t)U(t) + \Phi(t)U'(t) = A\Phi(t)U(t) + F(t),$$
$$\Phi(t)U'(t) = F(t),$$
$$U'(t) = \Phi^{-1}(t)F(t),$$
$$U(t) = \int \Phi^{-1}(t)F(t)\,dt,$$

where we have used the fact that $\Phi'(t)U(t) - A\Phi(t)U(t) = (\Phi' - A\Phi(t))U(t) = 0$ because the fundamental matrix $\Phi(t)$ satisfies the corresponding homogeneous system $X' = AX$. It follows that

$$X_p(t) = \Phi(t)\int \Phi^{-1}(t)F(t)\,dt. \qquad (6.21)$$

A general solution is then

$$X(t) = X_h(t) + X_p(t)$$
$$= \Phi(t)C + \Phi(t)\int \Phi^{-1}(t)F(t)\,dt$$
$$= \Phi(t)\left(C + \int \Phi^{-1}(t)F(t)\,dt\right)$$
$$= \Phi(t)\int \Phi^{-1}(t)F(t)\,dt,$$

where we have incorporated the constant vector C into the indefinite integral $\int \Phi^{-1}(t)F(t)\,dt$.

Remember that if $ad - bc \neq 0$, the inverse of the 2×2 matrix $\begin{pmatrix} a & b \\ c & d \end{pmatrix}$ is $\dfrac{1}{ad - bc}\begin{pmatrix} d & -b \\ -c & a \end{pmatrix}$.

EXAMPLE 6.5.3

Solve $\begin{cases} dx/dt = -3x + 9y \\ dy/dt = -2x + 3y + \tan 3t \\ x(0) = 1, y(0) = -1, -\pi/6 < t < \pi/6 \end{cases}$.

Solution

In matrix form, the system can be written as

$$X' = AX + F(t), \text{ where } X = \begin{pmatrix} x \\ y \end{pmatrix}, A = \begin{pmatrix} -3 & 9 \\ -2 & 3 \end{pmatrix},$$

and $F(t) = \begin{pmatrix} 0 \\ \tan 3t \end{pmatrix}$. A has eigenvalues $\lambda_{1,2} = \pm 3i$ with corresponding eigenvectors $v_{1,2} = \begin{pmatrix} 3 \\ 2 \end{pmatrix} \pm \begin{pmatrix} -3 \\ 0 \end{pmatrix} i$. Then, two linearly independent solutions of the corresponding homogeneous system, $X' = AX$, are

$$X_1(t) = \begin{pmatrix} 3 \\ 2 \end{pmatrix}\cos 3t - \begin{pmatrix} -3 \\ 0 \end{pmatrix}\sin 3t = \begin{pmatrix} 3\cos 3t + 3\sin 3t \\ 2\cos 3t \end{pmatrix}$$

and

$$X_2(t) = \begin{pmatrix} 3 \\ 2 \end{pmatrix}\sin 3t + \begin{pmatrix} -3 \\ 0 \end{pmatrix}\cos 3t = \begin{pmatrix} -3\cos 3t + 3\sin 3t \\ 2\sin 3t \end{pmatrix},$$

so a fundamental matrix for the corresponding homogeneous system is

$$\Phi(t) = \begin{pmatrix} 3\cos 3t + 3\sin 3t & -3\cos 3t + 3\sin 3t \\ 2\cos 3t & 2\sin 3t \end{pmatrix}$$

Remember that a general solution of the corresponding homogeneous system is $X_h = \Phi(t)C$, where $C = \begin{pmatrix} c_1 \\ c_2 \end{pmatrix}$.

with inverse

$$\Phi^{-1}(t) = \begin{pmatrix} \dfrac{1}{2}\sin 3t & \dfrac{1}{2}(\cos 3t - \sin 3t) \\ -\dfrac{1}{3}\cos 3t & \dfrac{1}{2}(\cos 3t + \sin 3t) \end{pmatrix}.$$

We now compute:

$$U(t) = \int \Phi^{-1}(t) F(t)\, dt$$

$$= \int \begin{pmatrix} \frac{1}{2}\sin 3t & \frac{1}{2}(\cos 3t - \sin 3t) \\ -\frac{1}{3}\cos 3t & \frac{1}{2}(\cos 3t + \sin 3t) \end{pmatrix} \begin{pmatrix} 0 \\ \tan 3t \end{pmatrix} dt$$

$$= \int \begin{pmatrix} \frac{1}{2}\sin 3t - \frac{1}{2}\sin 3t \tan 3t \\ \frac{1}{2}\sin 3t + \frac{1}{2}\sin 3t \tan 3t \end{pmatrix} dt$$

$$= \int \begin{pmatrix} \frac{1}{2}\sin 3t - \frac{1}{2}(\sec 3t - \cos 3t) \\ \frac{1}{2}\sin 3t + \frac{1}{2}(\sec 3t - \cos 3t) \end{pmatrix} dt$$

$$= \frac{1}{6} \begin{pmatrix} -\cos 3t - \ln(\sec 3t + \tan 3t) + \sin 3t \\ -\cos 3t + \ln(\sec 3t + \tan 3t) - \sin 3t \end{pmatrix}.$$

Then, a particular solution to the nonhomogeneous system is

$$X_p(t) = \Phi(t) \int \Phi^{-1}(t) F(t)\, dt$$

$$= \begin{pmatrix} -\cos 3t \ln \sec 3t + \tan 3t \\ \frac{1}{3}[-1 - \cos 3t \ln \sec 3t + \tan 3t + \ln(\sec 3t + \tan 3t)]\sin 3t \end{pmatrix}$$

and a general solution is given by

$$x(t) = c_1(3\cos 3t + 3\sin 3t) + c_2(-3\cos 3t + 3\sin 3t)$$
$$\qquad - \cos 3t \ln \sec 3t + \tan 3t$$

and

$$y(t) = 2c_1 \cos 3t + 2c_2 \sin 3t$$
$$\qquad + \frac{1}{3}[-1 - \cos 3t \ln \sec 3t + \tan 3t$$
$$\qquad + \ln(\sec 3t + \tan 3t)]\sin 3t.$$

Applying the initial conditions gives us that

$$3c_1 - 3c_2 = 1,$$
$$2c_1 = -1,$$

so $c_1 = -1/2$ and $c_2 = -5/6$ so the solution to the IVP is

$$x(t) = -\frac{1}{2}(3\cos 3t + 3\sin 3t) - \frac{5}{6}(-3\cos 3t + 3\sin 3t)$$
$$\qquad - \cos 3t \ln \sec 3t + \tan 3t$$

and

$$y(t) = -\cos 3t + \frac{5}{3}\sin 3t$$
$$\qquad + \frac{1}{3}[-1 - \cos 3t \ln \sec 3t + \tan 3t$$
$$\qquad + \ln(\sec 3t + \tan 3t)]\sin 3t,$$

which are graphed individually ($x = x(t)$ and $y = y(t)$ as functions of t) in Figure 6.10(a) and parametrically (x vs. y) in Figure 6.10(b).

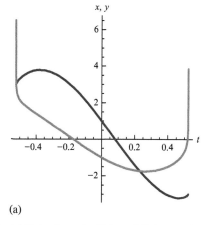

(a)

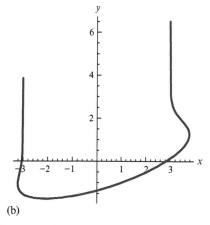

(b)

FIGURE 6.10   (a) Plots of x and y (light red; light gray in print versions) as functions of t. (b) Parametric plot of x versus y.

EXAMPLE 6.5.4

▼ Solve $\begin{cases} dx/dt = -4x + y - 6z + \sec t \\ dy/dt = -x - 2z \\ dz/dt = 3x - y + 4z - \tan t \\ x(0) = 0, y(0) = -1, z(0) = 1 \end{cases}$,

$-\pi/2 < t < \pi/2$.

Solution

In matrix form, this system is equivalent to

$$\mathbf{X}' = \underbrace{\begin{pmatrix} -4 & 1 & -6 \\ -1 & 0 & -2 \\ 3 & -1 & 4 \end{pmatrix}}_{\mathbf{A}} \mathbf{X} + \mathbf{F}(t), \text{ where } \mathbf{F}(t) =$$

$\begin{pmatrix} \sec t \\ 0 \\ -\tan t \end{pmatrix}$. $\mathbf{A}$ has eigenvalues $\lambda_1 = 0$ and $\lambda_{2,3} = \pm i$

with corresponding eigenvectors $\mathbf{v}_1 = \begin{pmatrix} -2 \\ -2 \\ 1 \end{pmatrix}$ and

$\mathbf{v}_{2,3} = \underbrace{\begin{pmatrix} -3 \\ -1 \\ 2 \end{pmatrix}}_{\mathbf{a}} + \underbrace{\begin{pmatrix} 1 \\ 1 \\ 0 \end{pmatrix}}_{\mathbf{b}} i$. The corresponding homo-

geneous system, $\mathbf{X}' = \mathbf{AX}$, has fundamental ma-

trix $\Phi(t) = \begin{pmatrix} -2 & -3\cos t - \sin t & \cos t - 3\sin t \\ -2 & -\cos t - \sin t & \cos t - \sin t \\ 1 & 2\cos t & 2\sin t \end{pmatrix}$

with inverse $\Phi^{-1}(t) =$

$\begin{pmatrix} 1 & -1 & 1 \\ \frac{1}{2}(-\cos t - 3\sin t) & \frac{1}{2}(\cos t + \sin t) & -2\sin t \\ \frac{1}{2}(3\cos t - \sin t) & \frac{1}{2}(-\cos t + \sin t) & 2\cos t \end{pmatrix}$.

Thus, a general solution of the corresponding homogeneous system is $\mathbf{X}_h(t) = \Phi(t)\mathbf{C} = \begin{pmatrix} x_h(t) \\ y_h(t) \\ z_h(t) \end{pmatrix}$, where

$x_h(t) = -2c_1 + (-3c_2 + c_3)\cos t - (c_2 + 3c_3)\sin t$

$y_h(t) = -2c_1 + (-c_2 + c_3)\cos t - (c_2 + c_3)\sin t$

and

$$z_h(t) = c_1 + 2c_2 \cos t + 2c_3 \sin t. \qquad (6.22)$$

We now use Equation (6.21) to find a particular solution of the nonhomogeneous system. First,

$$\mathbf{U}(t) = \int \Phi^{-1}(t)\mathbf{F}(t)\,dt$$

$$= \int \begin{pmatrix} \sec t + \tan t \\ \frac{1}{2}[-1 - (3 + 4\sin t)\tan t] \\ \frac{1}{2}(3 + 4\sin t - \tan t) \end{pmatrix} dt$$

$$= \begin{pmatrix} \ln(\sec t + \tan t) + \ln \sec t \\ -\frac{1}{2}t - \frac{3}{2}\ln\sec t + 2\sin t - 2\ln(\sec t + \tan t) \\ \frac{3}{2}t - 2\cos t - \frac{1}{2}\ln\sec t \end{pmatrix}.$$

Then, a particular solution of the nonhomogeneous system is $\mathbf{X}_p(t) = \Phi(t)\mathbf{U}(t) = \begin{pmatrix} x_p(t) \\ y_p(t) \\ z_p(t) \end{pmatrix}$,

where

$x_p(t) = -2\left(1 + \ln(\sec t) + \ln(\sec t + \tan t)\right)$
$\qquad + \cos t\left(3t + 4\ln(\sec t) + 6\ln(\sec t + \tan t)\right)$
$\qquad + \left(-4t + 3\ln(\sec t) + 2\ln(\sec t + \tan t)\right)\sin t,$

$y_p(t) = -2\left(1 + \ln(\sec t) + \ln(\sec t + \tan t)\right)$
$\qquad + \cos t\left(\ln(\sec t) + 2\left(t + \ln(\sec t + \tan t)\right)\right)$
$\qquad + \left(-t + 2\ln(\sec t) + 2\ln(\sec t + \tan t)\right)\sin t,$

and

$z_p(t) = \ln(\sec t) + \ln(\sec t + \tan t)$
$\qquad - \cos t\left(t + 3\ln(\sec t) + 4\ln(\sec t + \tan t)\right)$
$\qquad + \left(3t - \ln(\sec t)\right)\sin t. \qquad (6.23)$

Adding $\mathbf{X}_h$ and $\mathbf{X}_p$ gives us a general solution of the nonhomogeneous system,

$$\mathbf{X}(t) = \mathbf{X_h}(t) + \mathbf{X_p}(t) = \begin{pmatrix} x_h(t) + x_p(t) \\ y_h(t) + y_p(t) \\ z_h(t) + z_p(t) \end{pmatrix}, \quad (6.24)$$

where x_h, y_h, and z_h are given by Equation (6.22) and x_p, y_p, and z_p are given by Equation (6.23).

To solve the IVP, we apply the initial conditions, which results in the system of equations

$$x(0) = -2 - 2c_1 - 3c_2 + c_3 = 0,$$

$$y(0) = -2 - 2c_1 - c_2 + c_3 = -1,$$

$$z(0) = c_1 + 2c_2 = 1$$

that has solution $c_1 = 2$, $c_2 = -1/2$, and $c_3 = 9/2$. We substitute these values into Equation (6.24) and show the graphs of $x = x(t)$, $y = y(t)$, and $z = z(t)$ in Figure 6.11.

 EXERCISES 6.5

In Exercises 1-26, solve the system by undetermined coefficients or variation of parameters.

1. $\mathbf{X}' = \begin{pmatrix} -2 & 2 \\ -1 & 1 \end{pmatrix} \mathbf{X} + \begin{pmatrix} -5e^t \\ -e^t \end{pmatrix}$.

2. $\mathbf{X}' = \begin{pmatrix} 3 & 4 \\ -2 & -3 \end{pmatrix} \mathbf{X} + \begin{pmatrix} -3t \\ 2t+1 \end{pmatrix}$.

3. $\begin{cases} dx/dt = -5x - 4y - 13t + 4 \\ dy/dt = 6x + 5y + 16t - 8 \end{cases}$.

4. $\begin{cases} dx/dt = 2x - y + 2e^{2t} \\ dy/dt = 3x - 2y + 11e^{2t} \end{cases}$.

5. $\mathbf{X}' = \begin{pmatrix} 1 & -3 \\ 2 & -4 \end{pmatrix} \mathbf{X} + \begin{pmatrix} -2\cos t + 4\sin t \\ -\cos t + 5\sin t \end{pmatrix}$.

6. $\mathbf{X}' = \begin{pmatrix} 2 & 2 \\ -6 & -5 \end{pmatrix} \mathbf{X} + \begin{pmatrix} -4e^{-2t} \\ 6e^{-2t} \end{pmatrix}$.

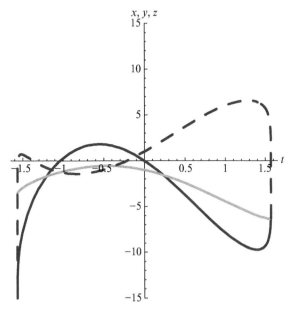

FIGURE 6.11   Plots of $x = x(t)$, $y = y(t)$ (in pink; light gray in print versions), and $z = z(t)$ (dashed).

7. $X' = \begin{pmatrix} 2 & 2 \\ -6 & -5 \end{pmatrix} X + \begin{pmatrix} e^{-2t} \\ -e^{-2t} \end{pmatrix}$.

8. $X' = \begin{pmatrix} -4 & 3 \\ -2 & 1 \end{pmatrix} X + \begin{pmatrix} 2e^{-t} \\ e^{-t} \end{pmatrix}$.

9. $X' = \begin{pmatrix} -4 & 3 \\ -2 & 1 \end{pmatrix} X + \begin{pmatrix} 6e^{-t} \\ 4e^{-t} \end{pmatrix}$.

10. $X' = \begin{pmatrix} 1 & -2 \\ 2 & -3 \end{pmatrix} X + \begin{pmatrix} -e^{-t} \\ -e^{-t} \end{pmatrix}$.

11. $X' = \begin{pmatrix} 1 & -2 \\ 2 & -3 \end{pmatrix} X + \begin{pmatrix} -2e^{-t} \\ -2e^{-t} \end{pmatrix}$.

12. $X' = \begin{pmatrix} 1 & -2 \\ 2 & -3 \end{pmatrix} X + \begin{pmatrix} -4\cos t + 2\sin t \\ -4\cos t + 6\sin t \end{pmatrix}$.

13. $\begin{cases} x' = x - 6y + f_1(t) \\ y' = x - 4y + f_2(t) \end{cases}$ where
   (a) $f_1(t) = 0, f_2(t) = 0$.
   (b) $f_1(t) = 0, f_2(t) = -2e^t$.
   (c) $f_1(t) = -2e^{-t}, f_2(t) = -e^{-t}$.

14. $X' = \begin{pmatrix} 2 & -5 \\ 1 & -2 \end{pmatrix} X + F(t)$ for
   (a) $F(t) = 0$.
   (b) $F(t) = \begin{pmatrix} -2\cos t + 4\sin t \\ 2\sin t \end{pmatrix}$.

15. $X' = \begin{pmatrix} 2 & 3 \\ -3 & -4 \end{pmatrix} X + F(t)$ for
   (a) $F(t) = 0$.
   (b) $F(t) = \begin{pmatrix} e^{-t} \\ -e^{-t} \end{pmatrix}$.
   (c) $F(t) = \begin{pmatrix} 2e^{-t} \\ -e^{-t} \end{pmatrix}$.

16. $\begin{cases} x' = -y + 4\tan t \\ y' = x \end{cases}$, $-\pi/2 < t < \pi/2$.

17. $\begin{cases} x' = -y + \sec t \\ y' = x \end{cases}$, $-\pi/2 < t < \pi/2$.

18. $\begin{cases} x' = -y \\ y' = x - 2\csc t \end{cases}$, $0 < t < \pi$.

19. $\begin{cases} x' = y \\ y' = -x - 2\cot t \end{cases}$, $0 < t < \pi$.

20. $\begin{cases} dx/dt = -2x - 2y - 2z + 1 \\ dy/dt = -2y + z + t^2 \\ dz/dt = -2y - 5z + t \end{cases}$.

21. $X' = \begin{pmatrix} 0 & 0 & 4 \\ 2 & -2 & 5 \\ -3 & 4 & -1 \end{pmatrix} X + \begin{pmatrix} t \\ 1 \\ t^2 \end{pmatrix}$.

22. $X' = \begin{pmatrix} 1 & -1 & -2 \\ -4 & 1 & 4 \\ 2 & -1 & -3 \end{pmatrix} + F(t)$ if
   (a) $F(t) = 0$.
   (b) $F(t) = \begin{pmatrix} 2e^{-t} \\ -4e^{-t} \\ 2e^{-t} \end{pmatrix}$.

23. $\begin{cases} x' = -2x + y - z + f_1(t) \\ y' = 5x - 2y + 5z + f_2(t) \\ z' = 2x - 2y + z + f_3(t) \end{cases}$, where
   (a) $f_i(t) = 0, i = 1, 2, 3$.
   (b) $f_1(t) = 3e^t - \cos 3t + \sin 2t$,
      $f_2(t) = -5e^t + 2\cos 2t - 7\sin 2t$,
      $f_3(t) = -2e^t + 4\cos 2t - \sin 2t$.

24. $X' = \begin{pmatrix} 3 & 3 & 1 \\ -3 & -3 & -1 \\ -8 & -4 & 0 \end{pmatrix} + \begin{pmatrix} 0 \\ 0 \\ \sec 2t \end{pmatrix}$.

25. $X' = \begin{pmatrix} 3 & 3 & 1 \\ -3 & -3 & -1 \\ -8 & -4 & 0 \end{pmatrix} + \begin{pmatrix} 0 \\ 0 \\ \cos 2t \end{pmatrix}$.

26. $\begin{cases} dx/dt = -x - 3y - 2z + \tan 3t \\ dy/dt = -6x - 3y - 12z \\ dz/dt = 2x + 3y + 4z \end{cases}$.

In Exercises 27-35, solve the IVP.

27. $X' = \begin{pmatrix} -1 & 3 \\ 0 & 2 \end{pmatrix} X + \begin{pmatrix} t \\ 0 \end{pmatrix}, X(0) = \begin{pmatrix} 0 \\ 0 \end{pmatrix}$.

28. $X' = \begin{pmatrix} 0 & 1 \\ -3 & -4 \end{pmatrix} X + \begin{pmatrix} 0 \\ \sin t \end{pmatrix}, X(0) = \begin{pmatrix} 0 \\ 1 \end{pmatrix}$.

29. $\begin{cases} x' = 3x + 4y \\ y' = 2x + y + e^t \\ x(0) = -1, y(0) = 1 \end{cases}$.

30. $\begin{cases} x' = 6x - 3y + t \\ y' = 2x + y + t^2 - 1 \\ x(0) = 0, y(0) = 1 \end{cases}$.

31. $\mathbf{X}' = \begin{pmatrix} 3 & 2 \\ -5 & 1 \end{pmatrix} \mathbf{X} + \begin{pmatrix} 0 \\ 10 \end{pmatrix}, \mathbf{X}(0) = \begin{pmatrix} 1 \\ -2 \end{pmatrix}$.

32. $\mathbf{X}' = \begin{pmatrix} 3 & -1 \\ 4 & -1 \end{pmatrix} \mathbf{X} + \begin{pmatrix} \cos t \\ \sin t \end{pmatrix}, \mathbf{X}(0) = \begin{pmatrix} 0 \\ 0 \end{pmatrix}$.

33. $\begin{cases} x' = -x + 5y - \tan 2t \\ y' = -x + y + \sec 2t \qquad , -\pi/4 < t < \pi/4. \\ x(0) = 1, y(0) = 0 \end{cases}$

34. $\begin{cases} dx/dt = x - y + e^{-t} \sec 3t \\ dy/dt = 13x - 3y \qquad , \\ x(0) = 0, y(0) = 1 \\ -\pi/6 < t < \pi/6. \end{cases}$

35. $\begin{cases} dx/dt = -4x + 5y \\ dy/dt = -4x + 4y + \tan 2t \quad , \\ x(0) = 1, y(0) = 0 \\ -\pi/2 < t < \pi/2. \end{cases}$

36. Let $\mathbf{X} = \mathbf{X}(t) = \begin{pmatrix} x_1(t) \\ x_2(t) \\ \vdots \\ x_n(t) \end{pmatrix}, \mathbf{A} = \mathbf{A}(t) =$

$\begin{pmatrix} a_{11}(t) & a_{12}(t) & \cdots & a_{1n}(t) \\ a_{21}(t) & a_{22}(t) & \cdots & a_{2n}(t) \\ \vdots & \vdots & \ddots & \vdots \\ a_{n1}(t) & a_{n2}(t) & \cdots & a_{nn}(t) \end{pmatrix}, \mathbf{F}(t) = \begin{pmatrix} f_1(t) \\ f_2(t) \\ \vdots \\ f_n(t) \end{pmatrix},$

and $\Phi(t)$ be a fundamental matrix for the linear homogeneous system $\mathbf{X}' = \mathbf{AX}$. (a) Show that if $\mathbf{X}_1(t)$ and $\mathbf{X}_2(t)$ are any two solutions of $\mathbf{X}' = \mathbf{AX} + \mathbf{F}(t)$, then $\mathbf{X}_1(t) - \mathbf{X}_2(t)$ is a solution to $\mathbf{X}' = \mathbf{AX}$. (b) Show that if $\mathbf{X}_p(t)$ is a particular solution of $\mathbf{X}' = \mathbf{AX} + \mathbf{F}(t)$ and $\mathbf{X}_{any}(t)$ is *any* solution of $\mathbf{X}' = \mathbf{AX} + \mathbf{F}(t)$, then there is an $n \times 1$ constant vector $\mathbf{C} = \begin{pmatrix} c_1 \\ c_2 \\ \vdots \\ c_n \end{pmatrix}$ so that

$\mathbf{X}_{any}(t) = \Phi(t)\mathbf{C} + \mathbf{X}_p(t)$.

37. Given that $\Phi(t) = \begin{pmatrix} 1 & \sqrt{t} \\ t & t \end{pmatrix}, t > 0$, is a fundamental matrix for

$\mathbf{X}' = \begin{pmatrix} \dfrac{1}{\left(-2 + 2\sqrt{t}\right)\sqrt{t}} & \dfrac{1}{\left(2 - 2\sqrt{t}\right)t^{\frac{3}{2}}} \\ 0 & t^{-1} \end{pmatrix} \mathbf{X},$

solve

$\mathbf{X}' = \begin{pmatrix} \dfrac{1}{\left(-2 + 2\sqrt{t}\right)\sqrt{t}} & \dfrac{1}{\left(2 - 2\sqrt{t}\right)t^{\frac{3}{2}}} \\ 0 & t^{-1} \end{pmatrix} \mathbf{X}$

$+ \begin{pmatrix} \dfrac{1}{2}t^{-3/2}\left(-1 - \sqrt{t} - t + t^{3/2}\right) \\ -t^{-1} \end{pmatrix}$.

38. Given that $\Phi(t) = \begin{pmatrix} 1 & \ln t \\ \ln t & -1 \end{pmatrix}, t > 0$, is a fundamental matrix for

$\mathbf{X}' = \begin{pmatrix} \dfrac{\ln(t)}{t + t\ln(t)^2} & \dfrac{1}{-t - t\ln(t)^2} \\ \dfrac{1}{t + t\ln(t)^2} & \dfrac{\ln(t)}{t + t\ln(t)^2} \end{pmatrix} \mathbf{X},$ solve

$\mathbf{X}' = \begin{pmatrix} \dfrac{\ln(t)}{t + t\ln(t)^2} & \dfrac{1}{-t - t\ln(t)^2} \\ \dfrac{1}{t + t\ln(t)^2} & \dfrac{\ln(t)}{t + t\ln(t)^2} \end{pmatrix} \mathbf{X}$

$+ \begin{pmatrix} \dfrac{1}{t + t\ln(t)^2} \\ -\dfrac{\ln(t)}{t + t\ln(t)^2} \end{pmatrix}$.

39. Assume that $f(t)$ is a differentiable function of t. Given that $\Phi(t) = \begin{pmatrix} 1 & f(t) \\ f(t) & 1 \end{pmatrix}$ is a fundamental matrix for

$\mathbf{X}' = \begin{pmatrix} \dfrac{f(t)f'(t)}{-1 + f(t)^2} & \dfrac{f'(t)}{1 - f(t)^2} \\ \dfrac{f'(t)}{1 - f(t)^2} & \dfrac{f(t)f'(t)}{-1 + f(t)^2} \end{pmatrix} \mathbf{X},$ solve

$$X' = \begin{pmatrix} \dfrac{f(t)f'(t)}{-1+f(t)^2} & \dfrac{f'(t)}{1-f(t)^2} \\ \dfrac{f'(t)}{1-f(t)^2} & \dfrac{f(t)f'(t)}{-1+f(t)^2} \end{pmatrix} X + \begin{pmatrix} \dfrac{f'(t)}{1+f(t)} \\ \dfrac{f'(t)}{1+f(t)} \end{pmatrix}.$$

40. Assume that $f(t)$ is a twice differentiable function of t. Given that $\Phi(t) = \begin{pmatrix} f(t) & -f'(t) \\ f'(t) & f(t) \end{pmatrix}$ is a fundamental matrix for

$$X' = \begin{pmatrix} \dfrac{f'(t)\left(f(t)+f''(t)\right)}{f(t)^2+f'(t)^2} & \dfrac{-f'(t)^2+f(t)f''(t)}{f(t)^2+f'(t)^2} \\ \dfrac{f'(t)^2-f(t)f''(t)}{f(t)^2+f'(t)^2} & \dfrac{f'(t)\left(f(t)+f''(t)\right)}{f(t)^2+f'(t)^2} \end{pmatrix} X,$$

solve

$$X' = \begin{pmatrix} \dfrac{f'(t)\left(f(t)+f''(t)\right)}{f(t)^2+f'(t)^2} & \dfrac{-f'(t)^2+f(t)f''(t)}{f(t)^2+f'(t)^2} \\ \dfrac{f'(t)^2-f(t)f''(t)}{f(t)^2+f'(t)^2} & \dfrac{f'(t)\left(f(t)+f''(t)\right)}{f(t)^2+f'(t)^2} \end{pmatrix} X$$

$$+ \begin{pmatrix} \dfrac{f'(t)^3-f(t)f'(t)f''(t)}{f(t)^2+f'(t)^2} \\ \dfrac{f(t)\left(-f'(t)^2+f(t)f''(t)\right)}{f(t)^2+f'(t)^2} \end{pmatrix}.$$

41. Use a computer algebra system to solve each of the following IVPs. In each case, graph $x(t)$, $y(t)$, and the parametric equations $\{x(t), y(t)\}$ for the indicated values of t.

(a) $\begin{cases} x' = -7x - 3y + 1 \\ y' = -2x - 2y + te^{-t} \\ x(0) = 0, y(0) = 1 \end{cases}$; $0 \le t \le 10$.

(b) $\begin{cases} x' = 8x + 10y + t^2 e^{-2t} \\ y' = -7x - 9y - te^{t} \\ x(0) = 1, y(0) = 0 \end{cases}$; $0 \le t \le 10$.

(c) $\begin{cases} x' = 2x - 5y + \sin 4t \\ y' = 4x - 2y - te^{-t} \\ x(0) = 1, y(0) = 1 \end{cases}$; $0 \le t \le 2\pi$.

42. (a) Show that

$$x = -\frac{6}{13}e^{-4t} + \frac{6}{13}e^{9t} - \frac{3}{13}e^{9t}\int_0^t e^{-9s}f(s)\,ds$$

$$+ \frac{16}{13}e^{-4t}\int_0^t e^{4s}f(s)\,ds$$

$$y = -\frac{3}{13}e^{-4t} + \frac{16}{13}e^{9t} - \frac{8}{13}e^{9t}\int_0^t e^{-9s}f(s)\,ds$$

$$+ \frac{8}{13}e^{-4t}\int_0^t e^{4s}f(s)\,ds$$

is the solution to the IVP

$$\begin{cases} dx/dt = -7x + 6y + f(t) \\ dy/dt = -8x + 12y \\ x(0) = 0, y(0) = 1 \end{cases}.$$

(b) Is it possible to choose $f(t)$ so that $\lim_{t\to\infty} x(t) = 0$? So that $\lim_{t\to\infty} y(t) = 0$? So that both $\lim_{t\to\infty} x(t) = 0$ and $\lim_{t\to\infty} y(t) = 0$? Why or why not? Provide evidence of your results by graphing various solutions.

In Exercises 43-45, graphically compare solutions to each nonhomogeneous system and the corresponding homogeneous system.

43. $\begin{cases} dx/dt = -5x + 6y + 1 \\ dy/dt = -7y + t \\ x(0) = 1, y(0) = -1 \end{cases}.$

44. $\begin{cases} dx/dt = -6x + t \\ dy/dt = -x - 6y + t^2 \\ x(0) = 0, y(0) = 1 \end{cases}.$

45. $\begin{cases} dx/dt = 9x + 5y + e^{3t}\sin 2t \\ dy/dt = -8x - 3y + e^{3t} \\ x(0) = 1, y(0) = 0 \end{cases}.$

6.6 PHASE PORTRAITS

- REAL DISTINCT EIGENVALUES
- REPEATED EIGENVALUES
- COMPLEX CONJUGATE EIGENVALUES
- STABILITY

In this section, we consider solutions of

$$dx/dt = ax + by,$$
$$dy/dt = cx + dy \qquad (6.25)$$

from a geometric point of view. We have solved many systems of this type using the eigenvalues and corresponding eigenvectors of $\mathbf{A} = \begin{pmatrix} a & b \\ c & d \end{pmatrix}$ and have seen that the solutions vary greatly.

Notice that $(0,0)$ is a solution of the system of equations obtained by setting the right sides of the differential equations equal to zero,

$$ax + by = 0$$
$$cx + dy = 0 \qquad (6.26)$$

for all values of a, b, c, and d. In fact, if $|\mathbf{A}| = \begin{vmatrix} a & b \\ c & d \end{vmatrix} \neq 0$, then the only solution of this system is the origin. We call $(0,0)$ an *equilibrium point* (or *rest point*) of the system of ODEs (6.25) because $(0,0)$ satisfies the corresponding system of algebraic equations (6.26). That is, $\{x = 0, y = 0\}$ is a *constant solution* of the system. In this section, we investigate the properties of the equilibrium point based on the eigenvalues and corresponding eigenvectors of the coefficient matrix A by viewing the *phase portrait*, a picture of a set of solutions (trajectories) and equilibrium point(s) of the system.

Note: We make the assumption that $|\mathbf{A}| \neq 0$ throughout this section.

In fact, $(0,0)$ is the only constant solution of system (6.25).

The behavior of the solutions of system (6.25) and the classification of the equilibrium point depend on the eigenvalues and corresponding eigenvectors of the system. We more thoroughly investigate the cases that can arise in solving this system by considering the classification of the equilibrium point $(0,0)$ based on the eigenvalues and corresponding eigenvectors of $\mathbf{A}$.

Real Distinct Eigenvalues

Suppose that λ_1 and λ_2 are real eigenvalues of $\mathbf{A}$ where $\lambda_2 < \lambda_1$ with corresponding eigenvectors $\mathbf{v}_1$ and $\mathbf{v}_2$, respectively. Then, a general solution of $\mathbf{X}' = \mathbf{A}\mathbf{X}$ is

$$\mathbf{X}(t) = \begin{pmatrix} x(t) \\ y(t) \end{pmatrix} = c_1\mathbf{v}_1 e^{\lambda_1 t} + c_2\mathbf{v}_2 e^{\lambda_2 t}$$
$$= e^{\lambda_1 t}(c_1\mathbf{v}_1 + c_2\mathbf{v}_2 e^{(\lambda_2 - \lambda_1)t}).$$

1. If both eigenvalues are negative, with $\lambda_2 < \lambda_1 < 0$ then $\lambda_2 - \lambda_1 < 0$. This means that $e^{(\lambda_2 - \lambda_1)t}$ is very small for large values of t, so $\mathbf{X}(t) \approx c_1\mathbf{v}_1 e^{\lambda_1 t}$ is small for large values of t. If $c_1 \neq 0$, then $\lim_{t \to \infty} \mathbf{X}(t) = \mathbf{0}$ along the line through $(0,0)$ in the direction of $\mathbf{v}_1$. If $c_1 = 0$, then $\mathbf{X}(t) = c_2\mathbf{v}_2 e^{\lambda_2 t}$. Again, because $\lambda_2 < 0$, $\lim_{t \to \infty} \mathbf{X}(t) = \mathbf{0}$ along the line through $(0,0)$ in the direction of $\mathbf{v}_2$. In this case, $(0,0)$ is a *stable improper node*.

2. If both eigenvalues are positive, with $0 < \lambda_2 < \lambda_1$. Then $e^{\lambda_1 t}$ and $e^{\lambda_2 t}$ both become unbounded as t increases. If $c_1 \neq 0$, then $\mathbf{X}(t)$ becomes unbounded along the line through $(0,0)$ in the direction of $\mathbf{v}_1$. If $c_1 = 0$, then $\mathbf{X}(t)$ becomes unbounded along the line through $(0,0)$ in the direction given by $\mathbf{v}_2$. In this case, $(0,0)$ is an *unstable improper node*.

3. If the eigenvalues have opposite signs with $\lambda_2 < 0 < \lambda_1$ and $c_1 \neq 0$. Then, $\mathbf{X}(t)$ becomes unbounded along the line through $(0,0)$ in the direction of $\mathbf{v}_1$ as it did in (2). However, if $c_1 = 0$, then due to the fact that $\lambda_2 < 0$, $\lim_{t \to \infty} \mathbf{X}(t) = \mathbf{0}$ along the line through $(0,0)$ determined by $\mathbf{v}_2$. If the initial point $\mathbf{X}(0)$ is not on the line through $(0,0)$ determined by $\mathbf{v}_2$, then the line given by $\mathbf{v}_1$ is an asymptote for the solution. We say that $(0,0)$ is a *saddle point* in this case.

EXAMPLE 6.6.1

Classify the equilibrium point $(0,0)$ in the systems (a) $\begin{cases} x' = x \\ y' = 2y \end{cases}$ and (b) $\begin{cases} x' = y \\ y' = x \end{cases}$.

Solution

(a) The coefficient matrix in this case is $\mathbf{A} = \begin{pmatrix} 1 & 0 \\ 0 & 2 \end{pmatrix}$, so we identify the eigenvalues of $\mathbf{A}$ by solving $\begin{vmatrix} 1-\lambda & 0 \\ 0 & 2-\lambda \end{vmatrix} = (1-\lambda)(2-\lambda) = 0$. The eigenvalues $\lambda_1 = 1$ and $\lambda_2 = 2$ are real and unequal, so we classify $(0,0)$ as an unstable improper node. Substitution of $\lambda_1 = 1$ into $(\mathbf{A} - \lambda_1\mathbf{I})\mathbf{v}_1 = \mathbf{0}$ indicates that $\mathbf{v}_1 = \begin{pmatrix} 1 \\ 0 \end{pmatrix}$ is an eigenvector associated with $\lambda_1 = 1$. Similarly, we find that $\mathbf{v}_2 = \begin{pmatrix} 0 \\ 1 \end{pmatrix}$ is an eigenvector corresponding to $\lambda_2 = 2$. Therefore, a general solution of the system is

$$\mathbf{X}(t) = c_1 \begin{pmatrix} 1 \\ 0 \end{pmatrix} e^t + c_2 \begin{pmatrix} 0 \\ 1 \end{pmatrix} e^{2t} = \begin{pmatrix} c_1 e^t \\ c_2 e^{2t} \end{pmatrix}$$

or $\begin{cases} x = c_1 e^t \\ y = c_2 e^{2t} \end{cases}$.

Notice that both components of the solution move away from zero ($\lim_{t \to \infty} x(t) = \infty$ and $\lim_{t \to \infty} y(t) = \infty$) as $t \to \infty$ while both approach zero as $t \to -\infty$. (Explain?)

There is a relationship between solutions of the system and the eigenlines, lines through the equilibrium point in the direction of the eigenvectors. In this case, the eigenlines corresponding to $\mathbf{v}_1$ and $\mathbf{v}_2$ are the y-axis and the x-axis, respectively. To see the relationship between solutions and the eigenlines, we find trajectories of the system by writing the system as the first-order equation

$$\frac{dy}{dx} = \frac{dy/dt}{dx/dt} = \frac{2y}{x}.$$

Separating variables, integrating, and simplifying give us $y = Cx^2$, where C is an arbitrary constant, so solution curves are parabolas with

vertices at the origin. (We can obtain the same result by eliminating the parameter from the solution for $\mathbf{X}(t)$ found earlier.) To sketch the phase portrait, we begin by graphing the eigenlines. Because each is associated with a positive eigenvalue, we place arrows directed away from the origin on each eigenline. Next, we graph several parabolas (Figure 6.12(a)). These trajectories are also directed away from the origin because of the positive eigenvalues. We can also graph several members of this family along with the direction field of the system, as done in Figure 6.12(b), to view the orientation of trajectories. Based on our earlier observation, solutions become tangent to one eigenline, the x-axis, as $t \to \infty$ (as we move against the direction of the arrows) and become parallel to the other eigenline, the y-axis, as $t \to -\infty$. (We will find that this behavior holds in other systems that include a node.) Notice that the trajectories are tangent to the eigenline associated with the smaller eigenvalue (in absolute value). They become parallel to the eigenline associated with the larger eigenvalue (in absolute value).

(b) By solving $\begin{vmatrix} -\lambda & 1 \\ 1 & -\lambda \end{vmatrix} = \lambda^2 - 1 = 0$, we find that the eigenvalues are $\lambda_1 = 1$ and $\lambda_2 = -1$. In this case, one eigenvalue is positive and one is negative, so we classify $(0,0)$ as a saddle point. Note that the eigenvectors corresponding to $\lambda_{1,2} = \pm 1$ are $\mathbf{v}_{1,2} = \begin{pmatrix} 1 \\ \pm 1 \end{pmatrix}$, respectively, so the eigenlines associated with $\mathbf{v}_1$ and $\mathbf{v}_2$ are $y = x$ and $y = -x$, respectively. A general solution is

$$\mathbf{X}(t) = c_1 \begin{pmatrix} 1 \\ 1 \end{pmatrix} e^t + c_2 \begin{pmatrix} 1 \\ -1 \end{pmatrix} e^{-t} = \begin{pmatrix} c_1 e^t + c_2 e^{-t} \\ c_1 e^t - c_2 e^{-t} \end{pmatrix}$$

or $\begin{cases} x = c_1 e^t + c_2 e^{-t} \\ y = c_1 e^t - c_2 e^{-t} \end{cases}$.

To determine the behavior of solutions as $t \to \infty$, notice that if $c_1 \ne 0$, then $\mathbf{X}(t)$ becomes like $c_1 \begin{pmatrix} 1 \\ 1 \end{pmatrix} e^t$ because $\lim_{t \to \infty} e^{-t} = 0$. In other words, solutions become asymptotic to the eigenline

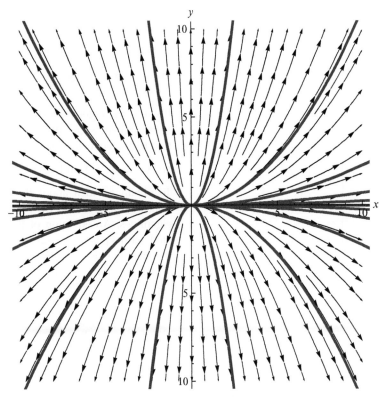

FIGURE 6.12 (a) Phase portrait of $\{x' = x, y' = y\}$. (b) Solutions with direction field of $\{x' = x, y' = y\}$.

associated with the positive eigenvalue, $y = x$, as $t \to \infty$. In a similar manner, we determine the behavior of solutions as $t \to -\infty$. In this case, solutions are dominated by the term $c_2 \begin{pmatrix} 1 \\ -1 \end{pmatrix} e^{-t}$ (if $c_2 \neq 0$) because $\lim_{t\to-\infty} e^t = 0$. Therefore, solutions approach the eigenline associated with the negative eigenvalue, $y = -x$, as $t \to -\infty$. Also, if $c_2 = 0$ and $c_1 \neq 0$, then the solution approaches the origin along the eigenline $y = x$ as $t \to \infty$. When we solve

$$\frac{dy}{dx} = \frac{dy/dt}{dx/dt} = \frac{x}{y},$$

we obtain $x^2 - y^2 = C$, which represents a family of hyperbolas (for $C \neq 0$) and the lines $y = \pm x$

(for $C = 0$). When we sketch the phase portrait in Figure 6.13, we begin by drawing the eigenlines. The line $y = x$ is associated with a positive eigenvalue so we place arrows directed away from the origin on this line. In contrast, $y = -x$ corresponds to a negative eigenvalue, so the arrows are directed toward the origin on this line. Next, we sketch hyperbolas centered at the origin and assign orientation based on that of the eigenlines (asymptotes of the hyperbolas). Observe that all solutions approach the eigenlines. They become asymptotic to the eigenline $y = x$ associated with $\lambda_1 = 1$ (the positive eigenvalue) as $t \to \infty$ and become asymptotic to the eigenline $y = -x$ associated with $\lambda_2 = -1$ (the negative eigenvalue) as $t \to -\infty$. This is a property linked to other saddle points we encounter.

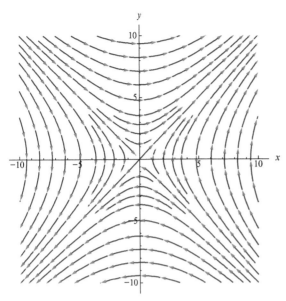

FIGURE 6.13 Phase portrait of $\{x' = y, y' = x\}$.

Note: Eigenlines are trajectories of the system because each can be obtained by setting one of the arbitrary constants c_1 or c_2 in the general solution equal to zero. Therefore, other trajectories cannot intersect the eigenlines because doing so would contradict the uniqueness of solutions of IVPs involving a first-order linear homogeneous system of ODEs.

EXAMPLE 6.6.2

Classify the equilibrium point $(0,0)$ of the systems and sketch the phase portrait:

(a) $\begin{cases} x' = 5x + 3y \\ y' = -4x - 3y \end{cases}$; (b) $\begin{cases} x' = x - 2y \\ y' = 3x - 4y \end{cases}$; and

(c) $\begin{cases} x' = -x - 2y \\ y' = 3x + 4y \end{cases}$.

Solution

(a) The eigenvalues are found by solving

$$\begin{vmatrix} 5 - \lambda & 3 \\ -4 & -3 - \lambda \end{vmatrix} = \lambda^2 - 2\lambda - 3 = (\lambda - 3)(\lambda + 1) = 0.$$

Because the eigenvalues $\lambda_1 = 3$ and $\lambda_2 = -1$ have opposite signs, $(0,0)$ is a saddle point. Note that the eigenvectors corresponding to λ_1 and λ_2 are $\mathbf{v}_1 = \begin{pmatrix} -3 \\ 2 \end{pmatrix}$ and $\mathbf{v}_2 = \begin{pmatrix} 1 \\ -2 \end{pmatrix}$, respectively. To obtain an idea about the behavior of solutions of this system, we find equations of the eigenlines. An equation of the eigenline associated with $\mathbf{v}_1$ is $y = -2x/3$ and an equation of the eigenline associated with $\mathbf{v}_2$ is $y = -2x$. We sketch the phase portrait in Figure 6.14(a). First, we draw the eigenlines, $y = -2x/3$ and $y = -2x$, with arrows directed toward and away from the origin, respectively. Then, we draw curves using the eigenlines as asymptotes and follow the directions associated with the eigenlines to assign the proper orientation to the trajectories. As mentioned in a previous example involving a saddle point, solutions become asymptotic to the eigenline associated with the positive eigenvalue as $t \to \infty$ and they become asymptotic to that associated with the negative eigenvalue as $t \to -\infty$.

Note: To find an equation of the eigenline associated with the eigenvector $\mathbf{v} = \begin{pmatrix} a \\ b \end{pmatrix}$, eliminate the parameter from $x = at, y = bt$: $t = x/a$ so $y = bt = bx/a$.

For another interpretation of the behavior of solutions as $t \to \infty$, notice that in a general solution of this system, $\mathbf{X}(t) = c_1 \begin{pmatrix} -3 \\ 2 \end{pmatrix} e^{3t} +$ $c_2 \begin{pmatrix} 1 \\ -2 \end{pmatrix} e^{-t}$, if $c_1 \neq 0$, then as $t \to \infty, \mathbf{X}(t)$ behaves

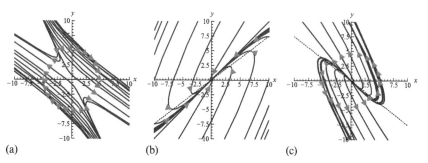

(a)   (b)   (c)

FIGURE 6.14 (a) Phase portrait for Example 6.6.2, solution (a). (b) Phase portrait for Example 6.6.2, solution (b). (c) Phase portrait for Example 6.6.2, solution (c).

more like $c_1 \begin{pmatrix} -3 \\ 2 \end{pmatrix} e^{3t}$ because $\lim_{t\to\infty} e^{-t} = 0$. In other words, solutions become asymptotic to the eigenline associated with $\mathbf{v}_1$ as $t \to \infty$. In a similar manner, we can determine the behavior of solutions as $t \to -\infty$. In this case, solutions are dominated by the term $c_2 \begin{pmatrix} 1 \\ -2 \end{pmatrix} e^{-t}$ (if $c_2 \neq 0$) because $\lim_{t\to-\infty} e^{3t} = 0$. Therefore, solutions approach the eigenline associated with $\mathbf{v}_2$ as $t \to -\infty$. Notice that the behavior of the solutions agrees with what we observed in our earlier discussion.

(b) Because the characteristic equation is

$$\begin{vmatrix} 1-\lambda & -2 \\ 3 & -4-\lambda \end{vmatrix} = \lambda^2 + 3\lambda + 2 = (\lambda+1)(\lambda+2) = 0,$$

the eigenvalues $\lambda_1 = -1$ and $\lambda_2 = -2$ are distinct and both negative; $(0,0)$ is a stable improper node. In this case, corresponding eigenvectors are $\mathbf{v}_1 = \begin{pmatrix} 1 \\ 1 \end{pmatrix}$ and $\mathbf{v}_2 = \begin{pmatrix} 2 \\ 3 \end{pmatrix}$, so the solutions approach $(0,0)$ along the lines through $(0,0)$ in the direction of these vectors, $y = x$ and $y = 3x/2$. Based on our earlier findings concerning nodes, we know that trajectories are tangent to one eigenline and become parallel to the other. To better understand the shape of the phase portrait, we convert the system to the first-order equation

$$\frac{dy}{dx} = \frac{dy/dt}{dx/dt} = \frac{3x-4y}{x-2y},$$

and we observe that trajectories have horizontal tangent lines, where $dy/dx = 0$ (when $3x-4y = 0$). In other words, trajectories cross the line $y = 3x/4$ horizontally. This fact and the vectors in the direction field indicate that trajectories approach the eigenline $y = x$ as $t \to \infty$ and are parallel to $y = 3x/2$ as $t \to -\infty$. (Remember that trajectories cannot cross eigenlines.) We also recall from Example 6.6.1 that trajectories are tangent to the eigenline associated with the smaller eigenvalue (in absolute value), $y = x$, and becomes parallel (as $t \to \pm\infty$) to the eigenline associated with the larger eigenvalue (in absolute value), $y = 3x/2$. We graph the phase portrait in Figure 6.14(b). The line $y = 3x/4$ is dashed. Both eigenlines have associated with them arrows directed toward the origin because both eigenvalues are negative.

(c) Because

$$\begin{vmatrix} -1-\lambda & -2 \\ 3 & 4-\lambda \end{vmatrix} = \lambda^2 - 3\lambda + 2 = (\lambda-2)(\lambda-1) = 0,$$

the eigenvalues $\lambda_1 = 2$ and $\lambda_2 = 1$ are real and unequal. Therefore, $(0,0)$ is an improper node. Note that corresponding eigenvectors are $\mathbf{v}_1 = \begin{pmatrix} 2 \\ -3 \end{pmatrix}$ and $\mathbf{v}_2 = \begin{pmatrix} 1 \\ -1 \end{pmatrix}$, respectively. The solutions become unbounded along the lines through $(0,0)$ determined by these vectors, the eigenlines $y = -3x/2$ and $y = -x$. We graph these eigenlines in Figure 6.14(c) and place arrows on them directed away from the origin due to the positive eigenvalues. Based on our earlier findings, trajectories are tangent to $y = -x$ as $t \to \infty$ and become parallel to $y = -3x/2$ as $t \to -\infty$. Following a method similar to that in (b), we see that trajectories cross the line $y = -3x/4$ horizontally (this line is dashed in Figure 6.14(c)).

Repeated Eigenvalues

We recall from our previous experience with repeated eigenvalues of a 2×2 system that the eigenvalue can have two linearly independent eigenvectors associated with it or only one (linearly independent) eigenvector associated with it. We investigate the behavior of solutions in the case of repeated eigenvalues by considering both of these possibilities.

1. If the eigenvalue $\lambda = \lambda_{1,2}$ has two corresponding linearly independent eigenvectors $\mathbf{v}_1$ and $\mathbf{v}_2$, a general solution is

$$\mathbf{X}(t) = c_1\mathbf{v_1}e^{\lambda t} + c_2\mathbf{v_2}e^{\lambda t} = (c_1\mathbf{v_1} + c_2\mathbf{v_2})e^{\lambda t}.$$

If $\lambda > 0$, then $\mathbf{X}(t)$ becomes unbounded along the lines through $(0,0)$ determined by the vectors $c_1\mathbf{v}_1 + c_2\mathbf{v}_2$, where c_1 and c_2 are arbitrary constants. In this case, we call the equilibrium point an *unstable star node*. However, if $\lambda < 0$, then $\mathbf{X}(t)$ approaches $(0,0)$ along these lines, and we call $(0,0)$ a *stable star node*.

Note: The name "star" was selected due to the shape of the solutions.

---

2. If the eigenvalue $\lambda = \lambda_{1,2}$ has only one corresponding (linearly independent) eigenvector $\mathbf{v} = \mathbf{v}_1$, a general solution is

$$\mathbf{X}(t) = c_1\mathbf{v}e^{\lambda t} + c_2(\mathbf{v}t + \mathbf{w})e^{\lambda t}$$
$$= (c_1\mathbf{v} + c_2\mathbf{w})e^{\lambda t} + c_2\mathbf{v}te^{\lambda t}$$

where $\mathbf{w}$ satisfies $(\mathbf{A} - \lambda\mathbf{I})\mathbf{w} = \mathbf{v}$. If we write this solution as

$$\mathbf{X}(t) = te^{\lambda t}\left[(c_1\mathbf{v} + c_2\mathbf{w})\frac{1}{t} + c_2\mathbf{v}\right],$$

we can more easily investigate the behavior of this solution. If $\lambda < 0$, then $\lim_{t \to \infty} te^{\lambda t} = 0$ and $\lim_{t \to \infty}[(c_1\mathbf{v} + c_2\mathbf{w})(1/t) + c_2\mathbf{v}] = c_2\mathbf{v}$. The solutions approach $(0,0)$ along the line through $(0,0)$ determined by $\mathbf{v}$, and we call $(0,0)$ a *stable deficient node*. If $\lambda > 0$, the solutions become unbounded along this line, and we say that $(0,0)$ is an *unstable deficient node*.

EXAMPLE 6.6.3

Classify the equilibrium point $(0,0)$ in the systems: (a) $\begin{cases} x' = x + 9y \\ y' = -x - 5y \end{cases}$ and (b)

$$\begin{cases} x' = 2x \\ y' = 2y \end{cases}.$$

Solution

(a) The eigenvalues are found by solving

$$\begin{vmatrix} 1-\lambda & 9 \\ -1 & -5-\lambda \end{vmatrix} = \lambda^2 + 4\lambda + 4 = (\lambda+2)^2 = 0.$$

Hence, $\lambda_{1,2} = -2$. In this case, an eigenvector $\mathbf{v}_1 = \begin{pmatrix} x_1 \\ y_1 \end{pmatrix}$ satisfies $\begin{pmatrix} 3 & 9 \\ -1 & -3 \end{pmatrix}\begin{pmatrix} x_1 \\ y_1 \end{pmatrix} = \begin{pmatrix} 0 \\ 0 \end{pmatrix}$, which is equivalent to $\begin{pmatrix} 1 & 3 \\ 0 & 0 \end{pmatrix}\begin{pmatrix} x_1 \\ y_1 \end{pmatrix} = \begin{pmatrix} 0 \\ 0 \end{pmatrix}$, so there is only one corresponding (linearly independent) eigenvector $\mathbf{v}_1 = \begin{pmatrix} -3y_1 \\ y_1 \end{pmatrix} = \begin{pmatrix} -3 \\ 1 \end{pmatrix}y_1$. Because $\lambda = -2 < 0$, $(0,0)$ is a degenerate stable node. In this case, the eigenline is $y = -x/3$. We graph this line in Figure 6.15(a) and direct the arrows toward the origin because of the negative eigenvalue. Next, we sketch trajectories that become tangent to the eigenline as $t \to \infty$ and associate with each arrows directed toward the origin.

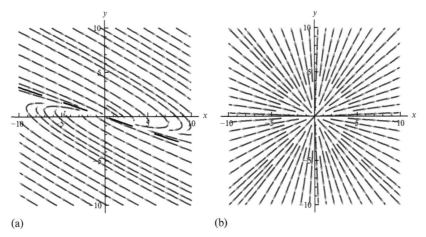

FIGURE 6.15 (a) Phase portrait for Example 6.6.3, solution (a). (b) Phase portrait for Example 6.6.3, solution (b).

(b) Solving the characteristic equation

$$\begin{vmatrix} 2-\lambda & 0 \\ 0 & 2-\lambda \end{vmatrix} = (2-\lambda)^2 = 0,$$

we have $\lambda = \lambda_{1,2} = 2$. However, because an eigenvector $\mathbf{v}_1 = \begin{pmatrix} x_1 \\ y_1 \end{pmatrix}$ satisfies the system $\begin{pmatrix} 0 & 0 \\ 0 & 0 \end{pmatrix}\begin{pmatrix} x_1 \\ y_1 \end{pmatrix} = \begin{pmatrix} 0 \\ 0 \end{pmatrix}$, any nonzero choice of $\mathbf{v}_1$ is an eigenvector. If we select two linearly independent vectors such as $\mathbf{v}_1 = \begin{pmatrix} 1 \\ 0 \end{pmatrix}$ and $\mathbf{v}_2 = \begin{pmatrix} 0 \\ 1 \end{pmatrix}$, we obtain two linearly independent eigenvectors corresponding to $\lambda_{1,2} = 2$. (*Note:* The choice of these two vectors does not change the value of the solution, because of the form of the general solution in this case.) Because $\lambda = 2 > 0$, we classify $(0,0)$ as a degenerate unstable star node. A general solution of the system is $\mathbf{X}(t) = c_1 \begin{pmatrix} 1 \\ 0 \end{pmatrix} e^{2t} + c_2 \begin{pmatrix} 0 \\ 1 \end{pmatrix} e^{2t}$, so when we eliminate the parameter, we obtain $y = c_2 x / c_1$. Therefore, the trajectories of this system are lines passing through the origin. In Figure 6.15(b), we graph several trajectories. Because of the positive eigenvalue, we associate with each an arrow directed away from the origin.

Complex Conjugate Eigenvalues

We have seen that if $\mathbf{A}$ is a matrix with real entries two eigenvalues of $\mathbf{A}$ are $\lambda_{1,2} = \alpha \pm \beta i$, $\beta \neq 0$, with corresponding eigenvectors $\mathbf{v}_{1,2} = \mathbf{a} \pm \mathbf{b}i$, then two linearly independent solutions of the system $\mathbf{X}' = \mathbf{AX}$ are

$$\mathbf{X}_1(t) = e^{\alpha t}(\mathbf{a}\cos\beta t - \mathbf{b}\sin\beta t) \quad \text{and}$$
$$\mathbf{X}_2(t) = e^{\alpha t}(\mathbf{b}\cos\beta t + \mathbf{a}\sin\beta t)$$

If $\mathbf{A}$ is 2×2, a general solution of the system is $\mathbf{X}(t) = c_1\mathbf{X}_1(t) + c_2\mathbf{X}_2(t)$, so there are constants $A_1, A_2, B_1,$ and B_2 such that $x(t)$ and $y(t)$, where $\mathbf{X}(t) = \begin{pmatrix} x(t) \\ y(t) \end{pmatrix}$, are given by

$$\mathbf{X}(t) = \begin{pmatrix} x(t) \\ y(t) \end{pmatrix} = \begin{pmatrix} A_1 e^{\alpha t}\cos\beta t + A_2 e^{\alpha t}\sin\beta t \\ B_1 e^{\alpha t}\cos\beta t + B_2 e^{\alpha t}\sin\beta t \end{pmatrix}.$$

1. If $\alpha = 0$, a general solution is $\mathbf{X}(t) = \begin{pmatrix} x(t) \\ y(t) \end{pmatrix} = \begin{pmatrix} A_1\cos\beta t + A_2\sin\beta t \\ B_1\cos\beta t + B_2\sin\beta t \end{pmatrix}$. Both x and y are periodic. In fact, if $A_2 = B_1 = 0$, then $\mathbf{X}(t) = \begin{pmatrix} x(t) \\ y(t) \end{pmatrix} = \begin{pmatrix} A_1\cos\beta t \\ B_2\sin\beta t \end{pmatrix}$. In rectangular coordinates, this solution is $(1/A_1^2)x^2 + (1/B_2^2)y^2 = 1$ where the graph is either a circle or an ellipse centered at $(0,0)$

depending on the value of A_1 and B_2. Hence, $(0,0)$ is classified as a *center*. Note that the motion around these circles or ellipses is either clockwise or counterclockwise for all solutions.

2. If $\alpha \neq 0$, the $e^{\alpha t}$ term is present in the solution. The $e^{\alpha t}$ term causes the solution to spiral or to away from the equilibrium point. If $\alpha > 0$, the solution spirals away from $(0,0)$, so we classify $(0,0)$ as an *unstable spiral point*. If $\alpha < 0$, the solution spirals toward $(0,0)$, so we say that $(0,0)$ is a *stable spiral point*.

EXAMPLE 6.6.4

Classify the equilibrium point $(0,0)$ in the systems: (a) $\begin{cases} x' = -y \\ y' = x \end{cases}$ and (b) $\begin{cases} x' = x - 5y \\ y' = x - 3y. \end{cases}$

Solution

(a) The eigenvalues are found by solving

$$\begin{vmatrix} -\lambda & -1 \\ 1 & -\lambda \end{vmatrix} = \lambda^2 + 1 = 0.$$

Hence, $\lambda_{1,2} = \pm i$. Because these eigenvalues have a zero real part (i.e., they are purely imaginary), $(0,0)$ is a center. We can view solutions of this system by writing the system as a first-order equation

with $\dfrac{dy}{dx} = \dfrac{dy/dt}{dx/dt} = \dfrac{-x}{y}$. Separating variables, we have $y\,dy = -x\,dx$, so that integration yields $\frac{1}{2}y^2 = -\frac{1}{2}x^2 + C$, or $x^2 + y^2 = K$ (where $K = 2C$). Therefore, solutions of the system are circles (if $K > 0$) centered at the origin. The solution is $(0,0)$ if $K = 0$. Several solutions are graphed in Figure 6.16(a). The arrows indicate that the solutions move counterclockwise around $(0,0)$. We see this by observing the equations in the system, $x' = -y$ and $y' = x$. In the first quadrant where $x > 0$ and $y > 0$, these equations indicate that $x' < 0$ and $y' > 0$. Therefore, on solutions in the first quadrant, x decreases while y increases. Similarly, in the second quadrant where $x < 0$ and $y > 0$, $x' < 0$ and $y' < 0$, so x and y both decrease in this quadrant. (Similar observations are made in the other two quadrants.)

(b) Because the characteristic equation is

$$\begin{vmatrix} 1 - \lambda & -5 \\ 1 & -3 - \lambda \end{vmatrix} = \lambda^2 + 2\lambda + 2 = 0,$$

the eigenvalues are $\lambda_{1,2} = \frac{1}{2}\left(-2 \pm \sqrt{4 - 8}\right) = -1 \pm i$. Thus, $(0,0)$ is a stable spiral point because $\alpha = -1 < 0$. To better understand the behavior of solutions, consider a general solution of this system,

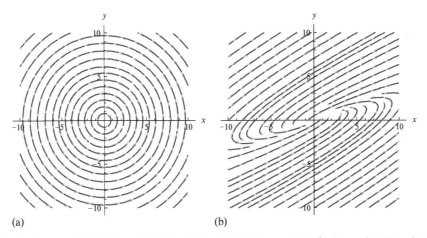

(a)   (b)

FIGURE 6.16   (a) Phase portrait for Example 6.6.4, solution (a). (b) Phase portrait for Example 6.6.4, solution (b).

$$\mathbf{X}(t) = e^{-t} \begin{pmatrix} -c_1 \cos t + (-2c_1 + 5c_2) \sin t \\ c_2 \cos t + (c_1 - 2c_2) \sin t \end{pmatrix}.$$

The trigonometric functions cause solutions to rotate about the origin, while e^{-t} forces both components to zero as $t \to \infty$ In Figure 6.15(b), we graph the phase portrait. Solutions spiral in toward the origin.

Note: In the case of centers and spiral points, the direction field may be particularly useful in determining the shape of solutions as well as their orientation.

Stability

We now understand the behavior of solutions of systems of linear equations based on the eigenvalues of the coefficient matrix. In addition, we would like to comment on the stability of the equilibrium point of the system using the same information. Looking back, we classified equilibrium points as saddle points, nodes,

spirals, or centers. In the case of saddle points, some nodes, and some spiral points in which at least one eigenvalue is positive or has a positive real part, trajectories move away from the equilibrium point as $t \to \infty$. Each of these classes of equilibrium points is *unstable*. Then, when we consider nodes and spiral points in which both eigenvalues are negative or have a negative real part, solutions converge to the equilibrium point as $t \to \infty$. These equilibrium points are *asymptotically stable* because of this convergence. Finally, centers are *stable* if a solution that starts sufficiently close to the equilibrium point remains close to the equilibrium point. An equilibrium point that is not stable is considered *unstable*. We summarize our findings in Table 6.1.

Note: In the case of a saddle point, all solutions except the eigenline associated with the negative eigenvalue move away from the equilibrium point as $t \to \infty$.

To better understand the meaning of the terms *unstable, asymptotically stable,* and *stable,* we take another look at graphs of solutions to several

TABLE 6.1   Classification of Equilibrium Point in Linear System

Eigenvalues	Geometry	Stability
λ_1, λ_2 real; $\lambda_1 > \lambda_2 > 0$	Improper node	Unstable
λ_1, λ_2 real; $\lambda_1 = \lambda_2 > 0$; 1 eigenvector	Deficient node	Unstable
λ_1, λ_2 real; $\lambda_1 = \lambda_2 > 0$; 2 eigenvectors	Star node	Unstable
λ_1, λ_2 real; $\lambda_2 < \lambda_1 < 0$;	Improper node	Asymptotically stable
λ_1, λ_2 real; $\lambda_1 = \lambda_2 < 0$; 1 eigenvector	Deficient note	Asymptotically stable
λ_1, λ_2 real; $\lambda_1 = \lambda_2 < 0$; 2 eigenvectors	Star node	Asymptotically stable
λ_1, λ_2 real; $\lambda_2 < 0 < \lambda_1$	Saddle point	Unstable
$\lambda_1 = \alpha + \beta i, \lambda_2 = \alpha - \beta i, \beta \neq 0, \alpha > 0$	Spiral point	Unstable
$\lambda_1 = \alpha + \beta i, \lambda_2 = \alpha - \beta i, \beta \neq 0, \alpha < 0$	Spiral point	Asymptotically stable
$\lambda_1 = \beta i, \lambda_2 = -\beta i, \beta \neq 0$	Center	Stable

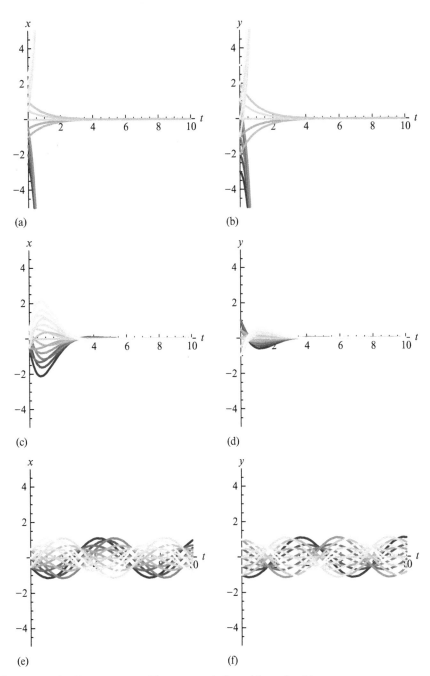

FIGURE 6.17 Graphically illustrating unstable, asymptotically stable, and stable.

systems in the example problems. Instead of graphing these solutions in the phase plane, however, we graph them in the tx- and ty-planes. Consider the system in Example 6.6.2 that included a saddle point with general solution $\mathbf{X}(t) = c_1 \begin{pmatrix} -3 \\ 2 \end{pmatrix} e^{3t} + c_2 \begin{pmatrix} 1 \\ -2 \end{pmatrix} e^{-t}$. When we graph $x(t) = -3c_1 e^{3t} + c_2 e^{-t}$ and $y(t) = 2c_1 e^{3t} - 2c_2 e^{-t}$ for various values of c_1 and c_2 in Figure 6.17(a) and (b), we see that some solutions diverge from $x(t) = 0$ and $y(t) = 0$, which are the coordinates of the equilibrium point of this system. We therefore call the equilibrium point in this case *unstable*. In contrast, in Figure 6.17(c) and (d), when we graph $x(t) = e^{-t}[-c_1 \cos t + (-2c_1 + 5c_2) \sin t]$ and $y(t) = e^{-t}[c_2 \cos t + (c_1 - 2c_2 \sin t)]$, solutions in Example 6.6.4(b) that involved a stable spiral point, we see that all solutions asymptotically approach $x(t) = 0$ and $y(t) = 0$. Therefore, we classify the equilibrium point in a system possessing this property as *asymptotically stable*. Finally, when we graph solutions to the system in Example 6.6.4(a) containing the center $x(t) = c_1 \cos t + c_2 \sin t$ and $y(t) = c_2 \cos t - c_1 \sin t$ in Figure 6.17(e) and (f), we notice that solutions neither converge to nor diverge from $x(t) = 0$ and $y(t) = 0$. Hence, the equilibrium point is *stable*.

EXERCISES 6.6

In Exercises 1-12, classify the equilibrium point $(0,0)$ of the system.

1. $x' = -5x - 4y, y' = -9x - 5y$.
2. $x' = -x - 7y, y' = -3x + 3y$.
3. $x' = 8x - 4y, y' = 9x - 4y$.
4. $x' = 4x - 2y, y' = -6x + 8y$.
5. $x' = -10x - 2y, y' = -2x - 10y$.
6. $x' = 6y, y' = -3x - 9y$.
7. $x' = -x, y' = -y$.
8. $x' = 3x, y' = 3y$.
9. $x' = -9y, y' = 6x + 4y$.

10. $x' = -8x + 5y, y' = -2x - 2y$.
11. $x' = -3x + 5y, y' = -10x + 3y$.
12. $x' = -4x - 4y, y' = 8x$.
13. (*Linear Systems with Zero Eigenvalues*)
 (a) Show that the eigenvalues of the system $x' = -2x, y' = -4x$ are $\lambda_1 = 0$ and $\lambda_2 = -2$.
 (b) Show that all points on the y-axis are equilibrium points.
 (c) By solving $dy/dx = (-4x)/(-2x) = 2$, show that the trajectories are the lines $y = 2x + C$, where C is an arbitrary constant.
 (d) Show that a general, solution of the system is $\mathbf{X}(t) = \begin{pmatrix} c_2 e^{-2t} \\ c_1 + 2c_2 e^{-2t} \end{pmatrix}$.
 (e) Find $\lim_{t\to\infty} \mathbf{X}(t)$.
 (f) Use the information in (c) and (e) to sketch the phase portrait of the system.

Follow steps similar to those in Exercise 13 to sketch the phase portrait of each of the following systems.

14. $x' = 2x - y, y' = -4x - 2y$.
15. $x' = 2x, y' = -4x$.
16. $x' = 12x - 6y, y' = 18x - 9y$.

In Exercises 17-20, use the given eigenvalues to (a) develop a corresponding second-order linear equation; (b) develop a corresponding system of first-order linear equations; (e) classify the stability of the equilibrium point $(0,0)$.

17. $\lambda_1 = 2, \lambda_2 = -1$.
18. $\lambda_1 = -3, \lambda_2 = -1$.
19. $\lambda_1 = -3, \lambda_2 = -3$.
20. $\lambda_1 = 3i, \lambda_2 = -3i$.
21. (*Linear Nonhomogeneous Systems*) Consider the system $\begin{cases} x' = x - 1 \\ y' = x + y \end{cases}$
 (a) Show that $(1, -1)$ is an equilibrium point.
 (b) Show that with the change of variables $u = x - 1$ and $v = y + 1$, we can

transform the system to $\begin{cases} u' = u \\ v' = u + v \end{cases}$.

(Notice that if the equilibrium point is (x_0, y_0), then the change of variables is $u = x - x_0$ and $v = y - y_0$.)

(c) Classify the equilibrium point $(1, -1)$ of the original system by classifying the equilibrium point $(0, 0)$ of the transformed system.

(d) Sketch the phase portrait of the original system by translating that of the transformed system.

For each of the following systems, follow steps similar to those in Exercise 21 to locate the equilibrium point of the system, classify, the equilibrium, and sketch the phase portrait.

22. $x' = 2x + y$, $y' = -y + 6$.
23. $x' = -2x + y - 6$, $y' = x - 2y$.
24. $x' = x - 2y - 1$, $y' = 5x - y - 5$.

25. Consider the system $\begin{cases} x' = ax + by \\ y' = cx + dy \end{cases}$ where

a, b, c, and d are real constants. Show that the eigenvalues of this system are found by solving $\lambda^2 - (a + d)\lambda + (ad - bc) = 0$. If we let $p = a + d$ and $q = ad - bc$, show that the eigenvalues are $\lambda_{1,2} = \frac{1}{2}\left(P \pm \sqrt{\Delta}\right)$ where $\Delta = p^2 - 4q$. Show that each of the following is true of the equilibrium point.

(a) $(0, 0)$ is a node if $q > 0$ and $\Delta \geq 0$. It is asymptotically stable if $p < 0$ and unstable if $p > 0$.

(b) $(0, 0)$ is a saddle point if $q < 0$.

(c) $(0, 0)$ is a spiral point if $p \neq 0$ and $\Delta < 0$. It is asymptotically stable if $p < 0$ and unstable if $p > 0$.

(d) $(0, 0)$ is a center if $p = 0$ and $q > 0$.

(e) Place the classifications in parts (a)-(d) in the regions shown in Figure 6.18. The parabola is the graph of $\Delta = p^2 - 4q = 0$.

26. Suppose that substance X decays into substance Y at rate $k_1 > 0$ which in turn decays into another substance at rate $k_2 > 0$. If $x(t)$ and $y(t)$ represent the amount of X

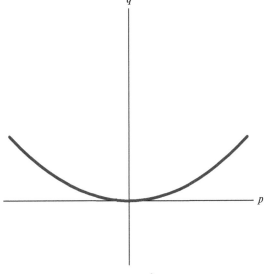

FIGURE 6.18   Graph of $\Delta = p^2 - 4q = 0$.

and Y, respectively, then the system

$\begin{cases} x' = -k_1 x \\ y' = k_1 x - k_2 y \end{cases}$ is solved to determine

$x(t)$ and $y(t)$. Show that $(0, 0)$ is the equilibrium solution of this system. Find the eigenvalues of the system and classify the equilibrium solution. Also, solve the system. Find $\lim_{t \to \infty} x(t)$ and $\lim_{t \to \infty} y(t)$. Do these limits correspond to the physical situation?

27. In the circular chemical reaction, compounds A_1 and A_2 combine at rate k_A to produce compound B_1. This compound reacts with compound B_2 at rate k_B to form C_1, which in turn combines with compound C_2 at rate k_C to form A. Therefore, the system of equations

$$dA_1/dt = k_C C_1 C_2 - k_A A_1 A_2,$$
$$dB_1/dt = k_A A_1 A_2 - k_B B_1 B_2,$$
$$dC_1/dt = k_B B_1 B_2 - k_C C_1 C_2,$$

models this situation. Let $K_1 = k_A A_2$, $K_2 = k_B B_2$, and $K_3 = k_C C_2$, and determine the equilibrium solutions of the system.

6.7 NONLINEAR SYSTEMS

We now turn our attention to systems in which at least one of the equations is not linear. These nonlinear systems possess many of the same properties as linear systems. In fact, we rely on our study of the classification of equilibrium points of linear systems to assist us in our understanding of the behavior of nonlinear systems. For example, consider the system

$$dx/dt = y,$$
$$dy/dt = x + x^2, \qquad (6.27)$$

which is nonlinear because of the x^2 term in the second equation. If we remove this nonlinear term, we obtain the linear system

$$dx/dt = y,$$
$$dy/dt = x. \qquad (6.28)$$

This system has an equilibrium point at $(0,0)$, which is also an equilibrium point of the system of nonlinear equations. Using the techniques discussed in Section 6.6, we can quickly show that the linear system has a saddle point at $(0,0)$. We show several trajectories of this system together with its direction field in Figure 6.19(a). Next, in Figure 6.19(b), we graph several trajectories of the nonlinear system along with its direction field. We see that the nonlinear term

affects the behavior of the trajectories. However, the, behavior near $(0,0)$ is "saddle-like" in that solutions appear to approach the origin but eventually move away from it. When we zoom in on the origin in Figure 6.19(c), we see how much it resembles a saddle point. Notice that the curves that separate the trajectories resemble the lines $y = x$ and $y = -x$, the asymptotes of trajectories in the linear system, near the origin. These curves, along with the origin, form the *separatrix*, because solutions within the portion of the separatrix to the left of the y-axis behave differently than solutions in other regions of the xy-plane. In this section, we discuss how we classify equilibrium points of nonlinear systems by using an associated linear system in much the same way as we analyzed systems (6.27) and (6.28).

The equilibrium solutions to a system of differential equations in which each differential equation does not explicitly depend on the independent variable (typically, t) are the constants solutions of the system. Thus, the equilibrium solutions of $x_1' = f_1(x_1, x_2, \ldots, x_n)$, $x_2' = f_2(x_1, x_2, \ldots, x_n)$, ..., $x_n' = f_n(x_1, x_2, \ldots, x_n)$ are found by solving the system $f_1(x_1, x_2, \ldots, x_n) = 0, f_2(x_1, x_2, \ldots, x_n) = 0$, ..., $f_n(x_1, x_2, \ldots, x_n) = 0$ for $x_1, x_2, \ldots, x_n$.

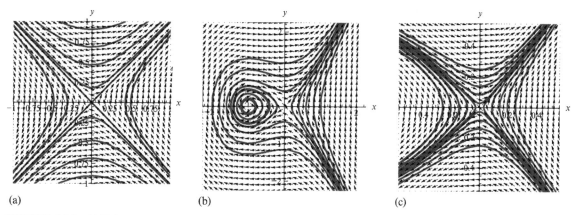

(a) (b) (c)

FIGURE 6.19 (a) Trajectories of corresponding linear system with direction field. (b) Trajectories of nonlinear system with direction field. (c) Trajectories of nonlinear system with direction field near the origin.

When working with nonlinear systems, we can often gain a great deal of information concerning the system by making a *linear approximation* near each equilibrium point of the nonlinear system and solving the linear system. Although the solution to the linearized system only approximates the solution to the nonlinear system, the general behavior of solutions to the nonlinear system near each equilibrium point is the same as that of the corresponding linear system *in many cases*. The first step toward approximating a nonlinear system near each equilibrium point is to find the equilibrium points of the system and to linearize the system at each of these points.

Recall from multivariable calculus that if $z = F(x, y)$ is a differentiable function, the tangent plane to the surface S given by the graph of $z = F(x, y)$ at the point (x_0, y_0)

$$z = F_x(x_0, y_0)(x-x_0)+F_y(x_0, y_0)(y-y_0)+F(x_0, y_0).$$
$$(6.29)$$

Near each equilibrium point (x_0, y_0) of the (nonlinear) autonomous system

$$dx/dt = f(x, y),$$
$$dy/dt = g(x, y), \qquad (6.30)$$

System (6.30) is *autonomous* because $f(x, y)$ and $g(x, y)$ do not explicitly depend on the independent variable, which is t in this situation.

---

under certain conditions the system's solution(s) can be *approximated* with the system

$$dx/dt = f_x(x_0, y_0)(x - x_0) + f_y(x_0, y_0)(y - y_0)$$
$$+ f(x_0, y_0),$$
$$dy/dt = g_x(x_0, y_0)(x - x_0) + g_y(x_0, y_0)(y - y_0)$$
$$+ g(x_0, y_0), \qquad (6.31)$$

where we have used the tangent plane to approximate $f(x, y)$ and $g(x, y)$ in the two dimensional autonomous system (6.30). Because $f(x_0, y_0) = 0$ and $g(x_0, y_0) = 0$ (Why?), the *approximate* system is

$$dx/dt = f_x(x_0, y_0)(x - x_0) + f_y(x_0, y_0)(y - y_0),$$

$$dy/dt = g_x(x_0, y_0)(x - x_0) + g_y(x_0, y_0)(y - y_0),$$
$$(6.32)$$

which can be written in matrix form as

$$\begin{pmatrix} x \\ y \end{pmatrix}' = \begin{pmatrix} f_x(x_0, y_0) & f_y(x_0, y_0) \\ g_x(x_0, y_0) & g_y(x_0, y_0) \end{pmatrix} \begin{pmatrix} x - x_0 \\ y - y_0 \end{pmatrix}. \quad (6.33)$$

Note that we often call system (6.33) the *linearized system corresponding to the (nonlinear) system* (6.30) or the *associated linearized system* due to the fact that we have removed the nonlinear terms from the original system.

If $f(x_0, y_0) = 0$ and $g(x_0, y_0) = 0$, the equilibrium point (x_0, y_0) of the system $\begin{cases} dx/dt = f(x, y) \\ dy/dt = g(x, y) \end{cases}$ is classified by the eigenvalues of the matrix

$$J(x_0, y_0) = \begin{pmatrix} f_x(x_0, y_0) & f_y(x_0, y_0) \\ g_x(x_0, y_0) & g_y(x_0, y_0) \end{pmatrix}, \quad (6.34)$$

which is called the *Jacobian matrix*. After determining the Jacobian matrix for each equilibrium point, we find the eigenvalues of the matrix in order to classify the corresponding equilibrium point according to the following criteria.

Notice that the linearization must be carried out for *each* equilibrium point.

Classification of Equilibrium Points of a Nonlinear System

Let (x_0, y_0) be an equilibrium point of system (6.30) and let λ_1 and λ_2 be eigenvalues of the Jacobian matrix (6.34) of the associated linearized system about the equilibrium point (x_0, y_0).

1. If (x_0, y_0) is classified as an asymptotically stable or unstable improper node (because the eigenvalues of $J(x_0, y_0)$ are real and distinct), a saddle point, or an asymptotically stable or unstable spiral in the associated linear system, (x_0, y_0) has the same classification in the, nonlinear system.
2. If (x_0, y_0) is classified as a center in the associated linear system, (x_0, y_0) may be a center, unstable spiral point, or

TABLE 6.2 Classification of Equilibrium Point in Nonlinear System

Eigenvalues of $J(x_0, y_0)$	Geometry	Stability
λ_1, λ_1 real; $\lambda_1 > \lambda_2 > 0$	Improper node	Unstable
λ_1, λ_2 real; $\lambda_1 = \lambda_2 > 0$	Node or spiral point	Unstable
λ_1, λ_2 real; $\lambda_2 < \lambda_1 < 0$	Improper node	Asymptotically stable
λ_1, λ_2 real; $\lambda_1 = \lambda_2 < 0$	Node or spiral point	Asymptotically stable
λ_1, λ_2 real; $\lambda_2 < 0 < \lambda_1$	Saddle point	Unstable
$\lambda_1 = \alpha + \beta i, \lambda_2 = \alpha - \beta i, \beta \neq 0, \alpha > 0$	Spiral point	Unstable
$\lambda_1 = \alpha + \beta i, \lambda_2 = \alpha - \beta i, \beta \neq 0, \alpha < 0$	Spiral point	Asymptotically stable
$\lambda_1 = \beta i, \lambda_2 = -\beta i, \beta \neq 0$	Center or spiral point	Inconclusive

asymptotically stable spiral point in the nonlinear system, so we cannot classify (x_0, y_0) in this situation (see Exercise 28).

3. If the eigenvalues of $J(x_0, y_0)$ are real and equal, then (x_0, y_0) may be a node or a spiral point in the nonlinear system. If $\lambda_1 \leq \lambda_2 < 0$, then (x_0, y_0) is asymptotically stable. If $\lambda_1 \leq \lambda_2 > 0$, then (x_0, y_0) is unstable.

These findings are summarized in Table 6.2.

EXAMPLE 6.7.1

Find and classify the equilibrium points of
$$\begin{cases} x' = 1 - y \\ y' = x^2 - y^2 \end{cases}.$$

Solution

We begin by finding the equilibrium points of this nonlinear system by solving $\begin{cases} 1 - y = 0 \\ x^2 - y^2 = 0 \end{cases}$. Because $y = 1$ from the first equation, substitution into the second equation yields $x^2 - 1 = 0$. Therefore, $x = \pm 1$, so the two equilibrium points are $(1,1)$ and $(-1,1)$. Because $f(x,y) = 1 - y$ and $g(x,y) = x^2 - y^2$, $f_x(x,y) = 0$, $f_y(x,y) = -1$, $g_x(x,y) = 2x$, and $g_y(x,y) = -2y$, so the Jacobian matrix is $J(x,y) = \begin{pmatrix} 0 & -1 \\ 2x & -2y \end{pmatrix}$. Next, we classify

each equilibrium point by finding the eigenvalues of the Jacobian matrix of each linearized system.

For $(1,1)$, we obtain the Jacobian matrix $J(1,1) = \begin{pmatrix} 0 & -1 \\ 2 & -2 \end{pmatrix}$ with eigenvalues that satisfy $\begin{vmatrix} -\lambda & -1 \\ 2 & -2-\lambda \end{vmatrix} = \lambda^2 + 2\lambda + 2 = 0$. Hence, $\lambda_{1,2} = -1 \pm i$. Because these eigenvalues are complex-valued with negative real part, we classify $(1,1)$ as an asymptotically stable spiral in the associated linearized system. Therefore, $(1,1)$ is an asymptotically stable spiral in the nonlinear system.

For $(-1,1)$, we obtain $J(-1,1) = \begin{pmatrix} 0 & -1 \\ -2 & -2 \end{pmatrix}$. In this case, the eigenvalues are solutions of $\begin{vmatrix} -\lambda & -1 \\ -2 & -2-\lambda \end{vmatrix} = \lambda^2 + 2\lambda - 2 = 0$. Thus, $\lambda_1 = \frac{1}{2}(-2 + 2\sqrt{3}) = -1 + \sqrt{3} > 0$ and $\lambda_2 = \frac{1}{2}(-2 - 2\sqrt{3}) = -1 - \sqrt{3} > 0$ so $(-1,1)$ is a saddle point in the associated linearized system and this classification carries over to the nonlinear system. In Figure 6.20(a), we graph solutions to this nonlinear system approximated with the use of a computer algebra system. We can see how the solutions move toward and away from the equilibrium points by observing the arrows on the vectors in the direction field.

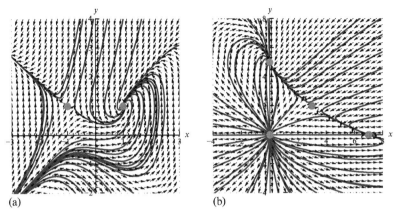

(a)   (b)

FIGURE 6.20 (a) $(1,1)$ is a stable spiral and $(-1,1)$ is a saddle. (b) $(0,0)$ is an unstable node, $(0,5)$ is an asymptotically stable improper node, $(7,0)$ is an asymptotically stable improper node, and $(3,2)$ is a saddle point.

Note: In Example 6.7.1, the linear approximation about $(1,1)$ is $\begin{pmatrix} x \\ y \end{pmatrix}' = \begin{pmatrix} 0 & -1 \\ 2 & -2 \end{pmatrix} \begin{pmatrix} x-1 \\ y-1 \end{pmatrix} = \begin{pmatrix} -y+1 \\ 2x-2y \end{pmatrix}$ or $\begin{cases} x' = -y+1 \\ y' = 2x-2y \end{cases}$. Notice that this (nonhomogeneous) linear system has an equilibrium point at $(1,1)$. With the change of variable $u = x-1$ and $v = y-1$, where the change of variable for the equilibrium point (x_0, y_0) is $u = x - x_0$ and $v = y - y_0$, we can transform this system to $\begin{cases} u' = -v \\ v' = 2u - 2v \end{cases}$ with equilibrium point $(0,0)$. Finding the eigenvalues of the matrix of coefficients indicates that $(0,0)$ is a saddle point, and we can sketch the phase portrait of the linearized system by translating the axes back to the original variables.

EXAMPLE 6.7.2

Find and classify the equilibrium points of $\begin{cases} x' = x(7 - x - 2y) \\ y' = y(5 - x - y) \end{cases}$.

Solution

The equilibrium points of this system satisfy $\begin{cases} x(7 - x - 2y) = 0 \\ y(5 - x - y) = 0. \end{cases}$ Thus, $\{x = 0 \text{ or } 7 - x - 2y = 0\}$ and $\{y = 0 \text{ or } 5 - x - y = 0\}$. If $x = 0$, then $y(5 - y) = 0$ so $y = 0$ or $y = 5$, and we obtain the equilibrium points $(0,0)$ and $(0,5)$. If $y = 0$, then $x(7 - x) = 0$, which indicates that $x = 0$ or $x = 7$. The corresponding equilibrium points are $(0,0)$ (which we found earlier) and $(7,0)$. The other possibility that leads to an equilibrium point is the solution to $\begin{cases} 7 - x - 2y = 0 \\ 5 - x - y = 0 \end{cases}$, which is $x = 3$ and $y = 2$, resulting in the equilibrium point $(3,2)$.

The Jacobian matrix is $J(x,y) = \begin{pmatrix} 7 - 2x - 2y & -2x \\ -y & 5 - x - 2y \end{pmatrix}$. We classify each of the equilibrium points (x_0, y_0) of the associated linearized system using the eigenvalues of $J(x_0, y_0)$:

$J(0,0) = \begin{pmatrix} 7 & 0 \\ 0 & 5 \end{pmatrix}$; $\lambda_1 = 7, \lambda_2 = 5$; $(0,0)$ is an unstable node.

$J(0,5) = \begin{pmatrix} -3 & 0 \\ -5 & -5 \end{pmatrix}$; $\lambda_1 = -3, \lambda_2 = -5$; $(0,5)$ is an asymptotically stable improper node.

$J(7,0) = \begin{pmatrix} -7 & -14 \\ 0 & -2 \end{pmatrix}$; $\lambda_1 = -2, \lambda_2 = -7$; $(7,0)$
is an asymptotically stable improper node.

$J(3,2) = \begin{pmatrix} -3 & -6 \\ -2 & -2 \end{pmatrix}$; $\lambda_1 = 1, \lambda_2 = -6$; $(3,2)$ is
a saddle point.

In each case, the classification carries over to the nonlinear system. In Figure 6.20(b), we graph several approximate solutions and the direction field to this nonlinear system through the use of a computer algebra system. Notice the behavior near each equilibrium point.

EXAMPLE 6.7.3

Investigate the stability of the equilibrium point $(0,0)$ of the nonlinear system

$$\frac{dx}{dt} = y + \frac{1}{2}x\left(x^2 + y^2\right),$$

$$\frac{dy}{dt} = -x + \frac{1}{2}y\left(x^2 + y^2\right).$$

Solution

First, we find the Jacobian matrix, $J(x,y) = \begin{pmatrix} \frac{3}{2}x^2 + \frac{1}{2}y^2 & 1 + xy \\ -1 + xy & \frac{1}{2}x^2 + \frac{3}{2}y^2 \end{pmatrix}$. Then, at the equilibrium

point $(0,0)$, we have $J(0,0) = \begin{pmatrix} 0 & 1 \\ -1 & 0 \end{pmatrix}$, so the

linear approximation is

$$dx/dt = y \quad dy/dt = -x$$

with eigenvalues $\lambda_{1,2} = \pm i$. Therefore, $(0,0)$ is a (stable) center in the linearized system. However, when we graph the direction field for the original (nonlinear) system in Figure 6.21(a), we observe that $(0,0)$ is not a center. Instead, trajectories appear to spiral away from $(0,0)$ (see Figure 6.21(b)), so $(0,0)$ is an unstable spiral point of the nonlinear system. The nonlinear terms not only affect the classification of the equilibrium point but also change the stability. *Note:* This is the only case in which we cannot assign the same classification to the equilibrium point in the nonlinear system as we do in the associated linear system. When an equilibrium point is a center in the associated linear system, then we cannot draw any conclusions concerning its classification in the nonlinear system.

EXERCISES 6.7

In Exercises 1-20, classify the equilibrium points of the nonlinear system. *If you have access to software that lets you plot direction fields for two-dimensional systems, verify your results by plotting the direction field for each system.*

1. $x' = y, y' = x - x^2$.
2. $x' = x + y, y' = x + xy$.
3. $x' = y + y^2, y' = x + y$.
4. $x' = 2y, y' = 2x + y^2$.
5. $x' = y, y' = -x + xy$.
6. $x' = 4y, y' = -4x + y^3$.
7. $x' = x + y + y^2, y' = y$.
8. $x' = y + xy, y' = x + y$.
9. $x' = -2x + x^2 - xy, y' = -4y + y^2 + xy$.
10. $x' = 10x - 2x^2 - 2xy, y' = -4y + 2xy$.
11. $x' = -3y - xy - 4, y' = y^2 - x^2$.
12. $x' = y, y' = \sin x$.
13. $x' = x^2 - 10x + 16, y' = y + 1$.
14. $x' = x(3 - y), y' = x + y + 1$.
15. $x' = y - x, y' = x + y - 2xy$.
16. $x' = 1 - x^2, y' = y$.
17. $x' = xy, y' = x^2 + y^2 - 4$.
18. $x' = 2x + y^2, y' = -2x + 4y$.
19. $x' = 1 - xy, y' = y - x^3$.
20. $x' = y - \frac{1}{2}x, y' = 1 - xy$.
21. (*Population*) Suppose that we consider the relationship between a host population and a parasite population that is modeled by the nonlinear system

$$dx/dt = (a_1 - b_1x - c_1y)x,$$
$$dx/dt = (-a_2 + c_2x)y,$$

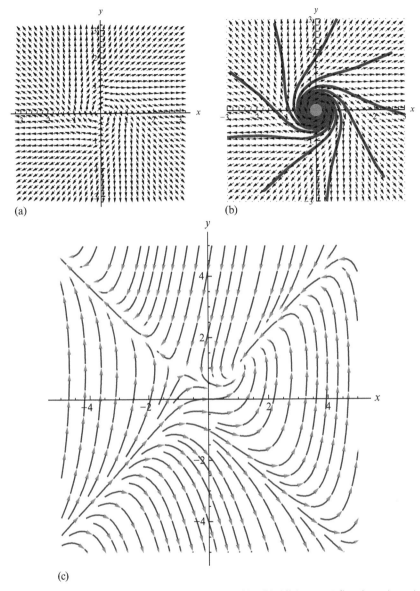

FIGURE 6.21   (a) The direction field indicates that $(0,0)$ is unstable. (b) All (nontrivial) trajectories spiral away from the origin. (c) Phase portrait.

where $x(t) \geq 0$ represents the number (or density) in the host population at time t, $y(t) \geq 0$ the number (or density) in the parasite population at time t, $a_1 > 0$ the growth rate of the host population, $a_2 > 0$ the death rate of the parasite population,

and $b_1 > 0$ the death rate of the host population. The constants c_1 and c_2 relate the interactions between the two populations leading to decay in the host population and growth in the parasite population, respectively. Find and classify the equilibrium point(s) of the system.

22. (*Economics*) Let $x(t)$ represent the income of a company and $y(t)$ the amount of consumer spending. Also suppose that z represents the rate of company expenditures. A system of nonlinear equations that models this situation is $\begin{cases} dx/dt = x - ay \\ dy/dt = b(x - y - z) \end{cases}$, where $a > 1$ and $b \geq 1$.

(a) If $z = z_0$ is constant, find and classify the equilibrium point of the system. Also consider the special case if $b = 1$.

(b) If the expenditure depends on income according to the relationship $z = z_0 + cx$ ($c > 0$), find and classify the equilibrium points (if they exist) of the system.

Compare Hamiltonian systems to first-order exact equations, which we discussed in Section 2.4.

---

23. (*Hamiltonian Systems*) As we saw in Chapter 2, solving nonlinear equations is usually difficult with the exception of some special situations, such as with first-order exact equations. The same situation is encountered with systems of differential equations. Systems of differential equations that correspond to first-order exact equations are called *Hamiltonian systems*. For the two-dimensional autonomous system $\begin{cases} dx/dt = f(x, y) \\ dy/dt = g(x, y) \end{cases}$ assume that f, g, $\partial f/\partial x$, $\partial f/\partial y$, $\partial g/\partial x$, and $\partial g/\partial y$ are continuous on a region R in the xy-plane. The system is a *Hamiltonian system* if and only if $(\partial f/\partial x)(x, y) = -(\partial g/\partial y)(x, y)$. If a system is Hamiltonian, then there is a function $H(x, y)$ (called a *Hamiltonian for the system*) so that $\partial H(x, y)/\partial x = -g(x, y)$ and $\partial H(x, y)/\partial y = f(x, y)$. Level curves of $z = H(x, y)$ correspond to trajectories of the system. By displaying the trajectories

together with the direction field or using the linearization techniques discussed in this section, we can (usually) classify the equilibrium points.

(a) Verify that the system
$$\begin{cases} dx/dt = -2xy \cos(x^2 y) + y^2 \sin(xy^2) \\ dy/dt = x^2 \cos(x^2 y) - 2xy \sin(xy^2) \end{cases}$$
is Hamiltonian.

(b) Integrate either $\partial H(x, y)/\partial x = x^2 \cos(x^2 y) - 2xy \sin(xy^2)$ or $\partial H(x, y)/\partial y = -2xy \cos(x^2 y) + y^2 \sin(xy^2)$ to obtain a formula for $H(x, y)$.

(c) Differentiate the result obtained in (b) with respect to the *other* variable and set the result equal to the *other* unknown partial (be sure to take sign into account). Integrate to obtain your unknown function and then form H.

(d) Verify that your result is equivalent to the Hamiltonian $H(x, y) = \cos(xy^2) + \sin(x^2 y)$ (see Figure 6.22).

In Exercises 24-27, verify that the system is Hamiltonian and find a Hamiltonian for the system. *If you have access to software that lets you plot direction fields and level curves, confirm your results by graphing both and displaying the results together on R.*

24. $x' = 2y$, $y' = 2 \sin 2x$; $R = [-\pi, \pi] \times [-\pi, \pi]$.
25. $x' = 2y + x^{-1}y^{-2}$, $y' = -2x - x^{-2}y^{-1}$; $R = [0.01, 5] \times [0.01, 5]$.
26. $x' = -2y + x \cos xy$, $y' = 2x - y \cos xy$; $R = [-\pi, \pi] \times [-\pi, \pi]$.
27. $x' = x \cos xy - \sin 2y$, $y' = -y \cos xy + \sin 2x$; $R = [-\pi, \pi] \times [-\pi, \pi]$.
28. Consider the nonlinear system
$$dx/dt = y + \mu x \left(x^2 + y^2 \right)$$
$$dy/dt = -x + \mu y \left(x^2 + y^2 \right)$$

(a) Show that this system has the equilibrium point $(0, 0)$ and the linearized system about this point is

FIGURE 6.22 Direction field for $dx/dt = -2xy\cos(x^2y) + y^2\sin(xy^2)$, $dy/dt = x^2\cos(x^2y) - 2xy\sin(xy^2)$ along with plots of $H(x,y) = C$ for various values of C.

$$\begin{cases} dx/dt = y \\ dy/dt = -x \end{cases}$$. Also show that $(0,0)$ is classified as a center of the linearized system.

(b) Consider the change of variables to polar coordinates $x = r\cos\theta$ and $y = r\sin\theta$. Use these equations with the chain rule to show that $dx/dt = \cos\theta(dr/dt) - r\sin\theta d\theta/dt$ and $dy/dt = \sin\theta(dr/dt)$ $+r\cos\theta d\theta/dt$. Use elimination with these equations to show that $dr/dt = \cos\theta(dx/dt) + \sin\theta dy/dt$ and $rd\theta dt = -\sin\theta(dx/dt) + \cos\theta dy/dt$. Transform the original system to polar coordinates and substitute the equations that result in the equations for dr/dt and

$rd\theta/dt$ to obtain the system in polar coordinates $$\begin{cases} dr/dt = \mu r^3 \\ d\theta/dt = -1 \end{cases}.$$

(c) What does the equation $d\theta/dt = -1$ indicate about the rotation of solutions to the system? According to $dr/dt = \mu r^3$, does r increase or decrease for $r > 0$? Using these observations, is the equilibrium point $(0,0)$, which was classified as a center for the linearized system, also classified as a center for the nonlinear system? If not, how would you classify it?

29. (a) Find a general solution of the system $$\begin{cases} x' = -x \\ y' = y + x^2 \end{cases}$$ by solving the first equation and by substituting the result into the second. (b) Sketch the phase plane and determine if the origin is a stable equilibrium point. (c) Solve the linearized system about $(0,0)$ and compare the result to that found in (b). Is $(0,0)$ assigned the same classification as in (b)?

30. Consider the *Lorenz system*

$$dx/dt = -\sigma x + \sigma y,$$
$$dy/dt = rx - y - xz,$$
$$dz/dt = -bz + xy,$$

which was developed in the 1960s by Edward Lorenz to study meteorology. (a) Show that $(0,0,0)$ is an equilibrium point. (b) If $0 < r < 1$, $\sigma > 0$, and $b > 0$, show that the system is asymptotically stable at $(0,0,0)$. (c) Graph the solution to the Lorenz equations if $\sigma = 7$, $r = 27.2$, and $b = 3$ if the initial conditions are $x(0) = 3$, $y(0) = 4$ and $z(0) = 2$. *Suggestion:* Generate two-dimensional plots of (t,x), (t,y), (t,z), (x,y), (y,z), and (x,z) for $0 \le t \le 25$ and then for $950 \le t \le 1000$. Then, generate a three-dimensional plot of (x,y,z) for the same interval. (d) How do the results change if you change r from 27.2 to 28.

31. (*The Chen System:*[2] *Anti-Control of the Lorenz System*) In some cases, scientists and engineers try to create chaos through the use of anti-control techniques. In this problem, we try to destabilize existing stable equilibria in the Lorenz system. Consider the controlled Lorenz system

$$\begin{cases} \dfrac{dx}{dt} = a\,(y - x) \\[2mm] \dfrac{dy}{dt} = cx - xz - y + u \\[2mm] \dfrac{dz}{dt} = xy - bz \end{cases} \quad \text{where the}$$

parameters $a, b, c > 0$ are not in the range to produce chaos and u, the linear feedback controller, is given by $u = k_1 x + k_2 y + k_3 z$. ($k_1$, k_2 and k_3 are gains that must be determined.)

(a) Show that the Jacobian of this controlled system evaluated at the equilibrium point (x^*, y^*, z^*) is given by

$$J = \begin{pmatrix} -a & a & 0 \\ c + k_1 - z^* & k_2 - 1 & k_3 - x^* \\ y^* & x^* & -b \end{pmatrix}.$$

(b) Show that one of the equilibrium solutions is $(0,0,0)$ and that nonzero equilibria satisfy

$$x = y = \tfrac{1}{2}k_3 \pm \tfrac{1}{2}\sqrt{k_3^2 + 4b\,(c + k_1 + k_2 - 1)}.$$

(c) Evaluate the Jacobian at $(0,0,0)$ and show that the corresponding eigenvalues do not depend on k_3.

(d) For simplicity, let $k_3 = 0$. Also consider $k_1 = -a$ and $k_2 = 1 + c$. Show that the resulting system is

$$\begin{cases} \dfrac{dx}{dt} = a\,(y - x) \\[2mm] \dfrac{dy}{dt} = (c - a)\,x - xz + cy \\[2mm] \dfrac{dz}{dt} = xy - bz \end{cases} \quad \text{, which is}$$

known as the Chen system.

(e) Let $a = 35$, $b = 3$, and $c = 28$, Show that the resulting nonzero equilibrium points are unstable.

32. Consider the nonlinear system

$$\begin{cases} dx/dt = 2x - xy \\ dy/dt = -3y + xy \end{cases} \quad \text{. (a) Graph the}$$

direction field associated with the system for $0 \le x \le 6$ and $0 \le y \le 6$. Use the direction field to sketch the graph of the solutions that satisfy the initial conditions (b) $x(0) = 2$ and $y(0) = 3$; and (c) $x(0) = 3$ and $y(0) = 2$. (d) How are the solutions alike? How are they different?

---

[2] L. Jinhu, C. Guanrong, A new chaotic attractor coined, *International Journal of Bifurcation and Chaos*, **12**(3), 2002, 659-661.

33. An exciting system to explore is the system

$$dx/dt = \mu x + y - x\left(x^2 + y^2\right),$$
$$dy/dt = \mu y - x - y\left(x^2 + y^2\right),$$

for fixed values of μ.

(a) Show that the only equilibrium point of the system is $(0,0)$.

(b) Show that the eigenvalues of the Jacobian matrix evaluated at $(0,0)$ are $\lambda_{1,2} = \mu \pm i$ and classify the equilibrium point $(0,0)$.

(c) How do you *think* the direction field of the system changes as μ ranges from -1 to 1

(d) Graph the direction field and various solutions of the system for several values of μ between -1 and 1.

(e) How do your results in (d) differ from your predictions in (c)?

6.8 NUMERICAL METHODS

- EULER'S METHOD
- RUNGE-KUTTA METHOD
- COMPUTER ALGEBRA SYSTEMS AND OTHER SOFTWARE

Because it may be difficult or even impossible to construct an explicit solution to some systems of differential equations, we now turn our attention to some numerical methods that are used to construct solutions to systems of differential equations.

Euler's Method

Euler's method for approximation, which was discussed for first-order equations, can be extended to include systems of first-order equations. The IVP

$$dx/dt = f(t, x, y),$$
$$dy/dt = g(t, x, y),$$
$$x(t_0) = x_0, \quad y(t_0) = y_0$$

As you have seen throughout this text **Leonhard Euler** (1707-1783) made such significant contributions to mathematics that he may have been the best and most prolific mathematician to have ever lived, yet.

is approximated at each step by the recursive relationship based on the Taylor series expansion of x and y with stepsize h,

$$x_{n+1} = x_n + hf(t_n, x_n, y_n),$$
$$y_{n+1} = y_n + hg(t_n, x_n, y_n),$$

where $t_n = t_0 + nh, n = 0, 1, 2, \ldots$.

EXAMPLE 6.8.1

Use Euler's method with $h = 0.1$ and $h = 0.05$ to approximate the solution to the IVP:

$$dx/dt = x - y + 1,$$
$$dy/dt = x + 3y + e^{-t},$$
$$x(0) = 0, y(0) = 1.$$

Compare these results with the exact solution to the system of equations.

Solution

In this case, $f(x, y) = x - y + 1, g(x, y) = x + 3y + e^{-t}, t_0 = 0, x_0 = 0$, and $y_0 = 1$, so we use the formulas

$$x_{n+1} = x_n + h \times \left(x_n - y_n + 1\right),$$
$$y_{n+1} = y_n + h \times \left(x_n + 3y_n + e^{-t_n}\right),$$

where $t_n = 0.1 \times n, n = 0, 1, 2, \ldots$.

If $n = 0$, then

$$x_1 = x_0 + h \times (x_0 - y_0 + 1) = 0,$$
$$y_1 = y_0 + h \times (x_0 + 3y_0 + e^{-t_0}) = 1.4.$$

The exact solution of this problem, which can be determined using the method of undetermined coefficients or variation of parameters, is

$$x(t) = -\frac{3}{4} - \frac{1}{9}e^{-t} + \frac{31}{36}e^{2t} - \frac{11}{6}te^{2t},$$
$$y(t) = \frac{1}{4} - \frac{2}{9}e^{-t} + \frac{35}{36}e^{2t} + \frac{11}{6}te^{2t}.$$

In Table 6.3, we display the results obtained with this method and compare them to rounded values of the exact results.

Because the accuracy of this approximation diminishes as t increases, we attempt to improve the approximation by decreasing the increment size. We do this by entering the value $h = 0.05$ and repeating the above procedure. We show the results of this method in Table 6.4. Notice that the approximations are more accurate with the smaller value of h.

Runge-Kutta Method

Because we would like to be able to improve the approximation without using such a small value for h, we seek to improve the method. As with first-order equations, the Runge-Kutta method can be extended to systems. In this case, the recursive formula at each step is

$$\begin{cases} x_{n+1} = x_n + \frac{1}{6}h\,(k_1 + 2k_2 + 2k_3 + k_4) \\ y_{n+1} = y_n + \frac{1}{6}h\,(m_1 + 2m_2 + 2m_3 + m_4) \end{cases},$$
$$(6.35)$$

TABLE 6.3 Numerical Results

t_n	x_n (approx)	x_n (exact)	y_n (approx)	y_n (exact)
0.0	0.0	0.0	1.0	1.0
0.1	0.0	−0.02270	1.4	1.46032
0.2	−0.04	−0.10335	1.91048	2.06545
0.3	−0.13505	−0.26543	−2.5615	2.85904
0.4	−0.30470	−0.54011	13.39053	3.89682
0.5	−0.57423	−0.96841	4.44423	5.24975
0.6	−0.97607	−1.60412	5.78076	7.00806
0.7	−1.55176	−2.51737	7.47226	9.28638
0.8	−2.35416	−3.79926	9.60842	12.23
0.9	−3.45042	−5.56767	12.3005	16.0232
1.0	−4.9255	−7.97468	15.6862	20.8987

TABLE 6.4 Numerical Results

t_n	x_n (approx)	x_n (exact)	y_n (approx)	y_n (exact)
0.0	0.0	0.0	1.0	1.0
0.05	0.0	−0.00532	1.2	1.21439
0.10	−0.01	−0.02270	1.42756	1.46032
0.15	−0.03188	−0.05447	1.68644	1.74321
0.20	−0.06779	−0.10335	1.98084	2.06545
0.25	−0.12023	−0.17247	2.31552	2.43552
0.30	−0.192013	−0.26543	2.69577	2.85904
0.35	−0.28640	−0.38639	3.12758	3.34338
0.40	−0.40710	−0.54011	3.61763	3.89682
0.45	−0.55834	−0.73203	4.17344	4.52876
0.50	−0.74493	−0.96841	4.80342	5.24975
0.55	−0.97234	−1.25639	5.51701	6.07171
0.60	−1.24681	−1.60412	6.32479	7.00806
0.65	−1.57529	−2.02091	7.23861	8.07394
0.70	−1.96609	−2.51737	8.27174	9.28638
0.75	−2.42798	−3.10558	9.43902	10.6645
0.80	−2.97133	−3.79926	10.7571	12.23
0.85	−3.60776	−4.61405	12.2446	14.0071
0.90	−4.35037	−5.56767	13.9222	16.0232
0.95	−5.214	−6.68027	15.8134	18.3088
1.00	−6.21537	−7.97468	17.944	20.8987

Carle David Runge (1856-1927) and famous physicist Max Planck were good friends. In 1877, Runge's interests turned from physics to mathematics.

where

$$k_1 = f\left(t_n, x_n, y_n\right) \, m_1 = g\left(t_n, x_n, y_n\right), \quad (6.36)$$

$$k_2 = f\left(t_n + \frac{1}{2}h, x_n + \frac{1}{2}hk_1, y_n + \frac{1}{2}hm_1\right),$$

$$m_2 = g\left(t_n + \frac{1}{2}h, x_n + \frac{1}{2}hk_1, y_n + \frac{1}{2}hm_1\right),$$

$$(6.37)$$

$$k_3 = f\left(t_n + \frac{1}{2}h, x_n + \frac{1}{2}hk_2, y_n + \frac{1}{2}hm_2\right),$$

$$m_3 = g\left(t_n + \frac{1}{2}h, x_n + \frac{1}{2}hk_2, y_n + \frac{1}{2}hm_2\right),$$

$$(6.38)$$

$$k_4 = f\left(t_n + h, x_n + hk_3, y_n + hm_3\right),$$

$$m_4 = g\left(t_n + h, x_n + hk_3, y_n + hm_3\right). \quad (6.39)$$

EXAMPLE 6.8.2

Use the Runge-Kutta method to approximate the solution of the IVP from Example 6.8.1 using $h = 0.1$. Compare these results to those of the exact solution of the system of equations as well as those obtained with Euler's method.

Solution

Because $f(x, y) = x - y + 1, g(x, y) = x + 3y + e^{-t}$, $t_0 = 0, x_0 = 0$, and $y_0 = 1$, we use formulas (6.35) where

$$k_1 = f(t_n, x_n, y_n) = x_n - y_n + 1$$

$$m_1 = g(t_n, x_n, y_n) = x_n + 3y_n + e^{-t_n},$$

$$k_2 = \left(x_n + \frac{1}{2}hk_1\right) - \left(y_n + \frac{1}{2}hm_1\right) + 1,$$

$$m_2 = \left(x_n + \frac{1}{2}hk_1\right) + 3\left(y_n + \frac{1}{2}hm_1\right) + e^{-(t_n + h/2)},$$

$$k_3 = \left(x_n + \frac{1}{2}hk_2\right) - \left(y_n + \frac{1}{2}hm_2\right) + 1,$$

$$m_3 = \left(x_n + \frac{1}{2}hk_2\right) + 3\left(y_n + \frac{1}{2}hm_2\right) + e^{-(t_n + h/2)},$$

$$k_4 = (x_n + hk_3) - (y_n + hm_3) + 1$$

$$m_4 = (x_n + hk_3) + 3(y_n + hm_3) + e^{-(t_n + h)}.$$

For example, if $n = 0$, then

$$k_1 = x_0 - y_0 + 1 = 0 - 1 + 1 = 0,$$

$$m_1 = x_0 + 3y_0 + e^{-t_0} = 0 + 3 + 1 = 4,$$

$$k_2 = \left(x_n + \frac{1}{2}hk_1\right) - \left(y_n + \frac{1}{2}hm_1\right) + 1$$

$$= -1 - \frac{1}{2} \times 4 \times 0.1 + 1 = -0.2,$$

$$m_2 = \left(x_0 + \frac{1}{2}hk_1\right) + 3\left(y_0 + \frac{1}{2}hm_1\right) + e^{-(t_0 + h/2)}$$

$$= 3\left(1 + \frac{1}{2} \times 4 \times 0.1\right) + e^{-0.05}$$

$$\approx 4.55123,$$

$$k_3 = \left(x_0 + \frac{1}{2}hk_2\right) - \left(y_0 + \frac{1}{2}hm_2\right) + 1$$

$$= \frac{1}{2} \times 0.1 \times 0.2 - 1 - \frac{1}{2} \times 0.1 \times 4.55123 + 1$$

$$\approx -0.23756$$

$$m_3 = \left(x_0 + \frac{1}{2}hk_2\right) + 3\left(y_0 + \frac{1}{2}hm_2\right) + e^{-(t_0 + h/2)}$$

$$= \frac{1}{2} \times 0.1 \times 0.2 + 3\left(1 + \frac{1}{2} \times 0.1 \times 4.55123\right)$$

$$+ e^{-0.05} \approx 4.62391,$$

$$k_4 = (x_0 + hk_3) - (y_0 + hm_3) + 1$$

$$= 0.1 \times -0.23756 - 1 + +0.1 \times 4.62391 + 1$$

$$\approx -0.48615,$$

and

$$m_4 = (x_0 + hk_3) + 3(y_0 + hm_3) + e^{-(t_0+h)}$$

$$= 0.1 \times -0.23756 + 3 (1 + 0.1 \times 4.62391)$$

$$+ e^{-0.1} \approx 5.26826.$$

Therefore,

$$x_1 = x_0 + \frac{1}{6} \times 0.1 \times (k_1 + 2k_2 + 2k_3 + k_4)$$

$$= 0 + \frac{1}{6} \times 0.1 \times (0 + 2 \times -0.2$$

$$+ 2 \times -0.23756 + -0.48615) \approx -0.226878$$

and

$$y_1 = y_0 + \frac{1}{6} \times 0.1 \times (m_1 + 2m_2 + 2m_3 + m_4)$$

$$= 1 + \frac{1}{2} \times 0.1 \times (4 + 2 \times 4.55123$$

$$+ 2 \times 4.62391 + 5.26826) \approx 1.46031.$$

In Table 6.5, we show the results obtained with this method and compare them to rounded values of the exact results. Notice that the Runge-Kutta method is much more accurate than Euler's method. In fact, the Runge-Kutta method with $h = 0.1$ is more accurate than Euler's method with $h = 0.05$ for this IVP. (Compare the results here to those given in Table 6.4.)

Detailed discussions regarding the error involved in using Euler's method or the Runge-Kutta method to approximate solutions of systems of differential equations can be found in advanced numerical analysis texts.

The Runge-Kutta method can be extended to systems of first-order equations so it can be used to approximate solutions of higher order equations that can be written as systems of first-order equations. This is accomplished by

TABLE 6.5 Numerical Results

t_n	x_n (approx)	x_n (exact)	y_n (approx)	y_n (exact)
0.0	0.0	0.0	1.0	1.0
0.1	−0.02269	−0.02270	1.46031	1.46032
0.2	−0.10332	−0.10335	2.06541	2.06545
0.3	−0.26538	−0.26543	2.85897	2.85904
0.4	−0.54002	−0.54011	3.8967	3.89682
0.5	−0.96827	−0.96841	5.24956	5.24975
0.6	−1.60391	−1.60412	7.00778	7.00806
0.7	−2.51707	−2.51737	9.28596	9.28638
0.8	−3.79882	−3.79926	12.2294	12.23
0.9	−5.56704	−5.56767	16.0223	16.0232
1.0	−7.97379	−7.97468	20.8975	20.8987

transforming the higher order equation into a system of first-order equations. We illustrate this with the pendulum equation that we have approximately solved in several situations by using the approximation $\sin x \approx x$ for small values of x.

EXAMPLE 6.8.3

Use the Runge-Kutta method to approximate the solution of the IVP $x'' + \sin x = 0$, $x(0) = 0, x'(0) = 1$.

Solution

We begin by transforming the second-order equation into a system of first-order equations. We this by letting $x' = y$, so $y' = x'' = -\sin x$. Hence, $f(t, x, y) = y$ and $g(t, x, y) = \sin x$. With the Runge-Kutta method, we obtain the approximate values given in Table 6.6 under the heading "R-K." Also in Table 6.6 under the heading "linear," we give the corresponding values of the solution of the IVP $x'' + x = 0$, $x(0) = 0, x'(0) = 1$, which is $x = \sin t$ with $y = x' = \cos t$. We approximate the

TABLE 6.6   Numerical Results

t_n	x_n (R-K)	x_n (linear)	y_n (R-K)	y_n (linear)
0.0	0.0	0.0	1.0	1.0
0.1	0.09983	0.09983	0.99500	0.99500
0.2	0.19867	0.198669	0.98013	0.98007
0.3	0.29553	0.29552	0.95566	0.95534
0.4	0.38950	0.389418	0.922061	0.92106
0.5	0.47966	0.47943	0.87994	0.87758
0.6	0.56523	0.56464	0.83002	0.82534
0.7	0.64544	0.64422	0.77309	0.764842
0.8	0.71964	0.71736	0.70999	0.69671
0.9	0.78726	0.78333	0.641545	0.62161
1.0	0.84780	0.84147	0.568569	0.54030

nonlinear equations $x'' + \sin x = 0$ with the linear equation $x'' + x = 0$ because $\sin x \approx x$ for small values of x. Because the use of the approximation $\sin x \approx x$ is linear, we expect the approximations of the values of the solution to the nonlinear problem obtained with the Runge-Kutta method to be more accurate than the approximations of the values of the solution to the nonlinear problems obtained with the linear IVP.

Computer Algebra Systems and Other Software

Numerical and graphical solutions generated by computer algebra systems and other software packages are useful in helping us observe and understand behavior of the solution(s) to a differential equation, especially when we do not wish to carry out the implementation of numerical methods like Euler's method or the Runge-Kutta method.

EXAMPLE 6.8.4

Rayleigh's equation is the nonlinear equation $\left(d^2x/dt^2\right) + \left(\frac{1}{3}(dx/dt)^2 - 1\right)(dx/dt) + x = 0$ and arises in the study of the motion of a violin string. We write Rayleigh's equation as a system by letting $y = dx/dt$. Then,

$$\frac{dy}{dt} = \frac{d^2x}{dt^2} = -\left(\frac{1}{3}\left(\frac{dx}{dt}\right)^2 - 1\right)\frac{dx}{dt} - x$$

$$= -\left(\frac{1}{3}y^2 - 1\right)y - x$$

so Rayleigh's equation is equivalent to the system

$$\frac{dx}{dt} = y \quad \text{and} \quad \frac{dy}{dt} = -\left(\frac{1}{3}y^2 - 1\right)y - x.$$

By examination, we see that the only equilibrium point of this system is $(0,0)$. (a) Classify the equilibrium point $(0,0)$. (b) Is it possible to find $y_0 \neq 0$ so that the solution of the IVP $x' = y$, $y' = -\left(\frac{1}{3}y^2 - 1\right)y - x$, $x(0) = 0$, $y(0) = y_0$ is periodic?

Solution

(a) The associated linearized system about the point $(0,0)$ is $\mathbf{X}' = \begin{pmatrix} 0 & 1 \\ -1 & 1 \end{pmatrix}\mathbf{X}$ and the eigenvalues of $\begin{pmatrix} 0 & 1 \\ -1 & 1 \end{pmatrix}$ are $\lambda_{1,2} = \frac{1}{2}\left(1 \pm \sqrt{3}i\right)$ so $(0,0)$ is an unstable spiral point. This result is confirmed by the graph of the direction field for $-5 \leq x \leq 5$ and $-5 \leq y \leq 5$, shown in Figure 6.23(a) and (b).

In the direction field (Figure 6.23(b)), we see that all solutions tend to a closed curve, L. Choosing initial conditions for $x(0)$ and $y(0)$ inside and outside of L have the same result: the solution tends to L.

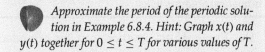

Approximate the period of the periodic solution in Example 6.8.4. Hint: Graph $x(t)$ and $y(t)$ together for $0 \leq t \leq T$ for various values of T.

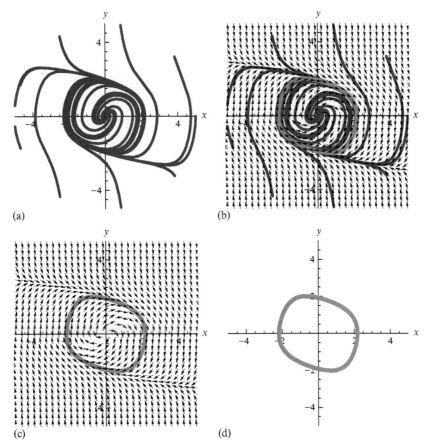

(a)   (b)

(c)   (d)

FIGURE 6.23 (a) If $x(0)$ and $y(0)$ are both close to 0, the solutions spiral outward while if $x(0)$ and $y(0)$ are both sufficiently large, the solutions spiral inward. (b) All nontrivial solutions tend to a closed curve, L. (c) An isolated period solution like this is called a *limit cycle*. This limit cycle is stable because all nontrivial solutions spiral into it. (d) Finding L.

(b) Can we find L? We use Figure 6.23(c) to approximate a point at which L intersects the y-axis. We obtain $y = 1.9$. We find a numerical solution that satisfies the initial conditions $x(0) = 0$ and $y(0) = 1.9$ and graph the result in Figure 6.23(d). The solution does appear to be periodic.[1]

___

[1] See D.W. Jordan, P. Smith (Eds.), *Nonlinear ODEs: An Introduction to Dynamical Systems*, third ed., Oxford University Press, 1999, for detailed discussions about limit cycles and their significance.

 EXERCISES 6.8

In Exercises 1-6, use Euler's method with $h = 0.1$ to approximate the solution of the IVP at the given value of t.

1. $x' = -x + 2y + 5, y' = 2x - y + 4, x(0) = 1, y(0) = 0, t = 1$.

2. $x' = x + y + t, y' = x - y, x(0) = 1, y(0) = -1, t = 1$.

3. $x' = 3x - 5y, y' = x - 2y + t^2, x(0) = -1,$
   $y(0) = 0, t = 1.$
4. $x' = x + 2y + \cos t, y' = 5x - 2y, x(0) = 0,$
   $y(0) = 1, t = 1.$
5. $x' = x^2 - y, y' = x + y, x(0) = -1, y(0) = 0,$
   $t = 1.$
6. $x' = xy, y' = x - y, x(0) = 1, y(0) = 1, t = 1.$

In Exercises 7-12, use the Runge-Kutta method with $h = 0.1$ to approximate the solution of the IVP at the given value of t.

7. $x' = x - 3y + e^t, y' = -x + 6y, x(0) = 0,$
   $y(0) = 1, t = 1.$
8. $x' = 4x - y, y' = -x + 5y + 6\sin t, x() = -1,$
   $y(0) = 0, y = 1.$
9. $x' = x - 8y, y' = 3x - y + e^t \cos 2t, x(0) = 0,$
   $y(0) = 0, t = 1.$
10. $x' = 5x + y + \sqrt{t+1}, y' = x - 2y, x(0) = 0,$
    $y(0) = 0, t = 2.$
11. $x' = 2y, y' = -xy, x(0) = 0, y(0) = 1, t = 1.$
12. $x' = x\sqrt{y}, y' = x - y, x(1) = 1, y(1) = 1,$
    $t = 2.$

In Exercises 13-18, use Euler's method with $h = 0.1$ to approximate the solution of the IVP by transforming the second-order equation to a system of first-order equations. Compare the approximations with the exact solution at the given value of t.

13. $x'' + 3x' + 2x = 0, x(0) = 0, x'(0) = -3, t = 1.$
14. $x'' + 4x' + 4x = 0, x(0) = 4, x'(0) = 0, t = 1.$
15. $x'' + 9x = 0, x(0) = 0, x'(0) = 3, t = 1.$
16. $x'' + 4x' + 13x = 0, x(0) = 0, x'(0) = 12, t = 1.$
17. $t^2 x'' + tx' + 16x = 0, x(1) = 0, x'(1) = 4,$
    $t = 2.$
18. $t^2 x'' + 3tx' + x = 0, x(1) = 0, x'(1) = 2, t = 2.$

In Exercises 19-24, use the Runge-Kutta method with $h = 0.1$ to approximate the solution of the IVP in the earlier exercise given. Compare the results obtained to those in Exercises 13-18.

19. Exercise 13
20. Exercise 14
21. Exercise 15
22. Exercise 16
23. Exercise 17
24. Exercise 18

In Exercise 25-27, find the exact solution of the IVP and then approximate the solution with the Runge-Kutta method using $h = 0.1$. Compare the results by graphing the two solutions together.

25. $x' = y, y' = 2x - y, x(0) = 0, y(0) = 1,$
    $0 \le t \le 5.$
26. $x' = y, y' = -13x - 4y, x(0) = -1, y(0) = 1,$
    $0 \le t \le 10.$
27. $x' = y, y' = -x - 2y, x(0) = 2, y(0) = 0,$
    $0 \le t \le 2.$
28. (a) Graph the direction field associated with the nonlinear system $dx/dt = y,$
    $dy/dt = -\sin x$ for $-7 \le x \le 7$ and $-4 \le y \le 4.$
    (b) (i) Approximate the solution to the IVP $x_1' = y_1, y_1' = -\sin x_1,$
    $x_1(0) = 0, y_1(0) = 1.$
    (ii) Graph $\{x_1(t), y_1(t)\}$ for $0 \le t \le 7$ and display the graph together with the direction field. Does it appear as though the vectors in the vector field are tangent to the solution curve?
    (iii) Approximate the solution to the IVP $x_2' = y_2, y_2' = -\sin x_2,$
    $x_2(0) = 0, y_2(0) = 2.$
    (iv) Graph $\{x_2(t), y_2(t)\}$ for $0 \le t \le 7$ and display the graph together with the direction field. Does it appear as though the vectors in the vector field are tangent to the solution curve?
    (v) Graph $\{x_1(t) + x_2(t), y_1(t) + y_2(t)\}$ for $0 \le t \le 7$ and display the graph together with the direction field. Does it appear as though the vectors in the vector field are tangent to the solution curve?
    (c) Approximate the solution to the IVP $x' = y, y' = -\sin x, x(0) = 0, y(0) = 3$

and graph the solution parametrically for $0 \leq t \leq 7$. Is this the graph of $\{x_1(t) + x_2(t), y_1(t) + y_2(t)\}$ found in (b) (v)?

(d) Is the Principle of Superposition valid for nonlinear systems? Explain.

CHAPTER 6 SUMMARY: ESSENTIAL CONCEPTS AND FORMULAS

System of ODEs: A *system* of ODEs is a simultaneous set of equations that involves two or more dependent variables that depend on one independent variable. A *solution* to the system is a set of functions that satisfies each equation on a common interval I.

Matrices and their Operations: Matrix, transpose of a matrix, scalar multiplication, matrix addition, matrix multiplication, identity matrix, determinant, eigenvalues and eigenvectors, characteristic polynomial, eigenvalue of multiplicity m, derivative and integral of a matrix.

Solution Vector: A *solution vector* of the system $\mathbf{X}' = \mathbf{A}(t)\mathbf{X}(t) + \mathbf{F}(t)$ on the interval I is an $n \times 1$ matrix of the form $\mathbf{X}(t) = \begin{pmatrix} x_1(t) \\ x_2(t) \\ \vdots \\ x_n(t) \end{pmatrix}$,

where the $x_i(t)$ are differentiable functions, that satisfies $\mathbf{X}' = \mathbf{A}(t)\mathbf{X}(t) + \mathbf{F}(t)$ on I.

Fundamental Set of Solutions: A set $\{\Phi_i(t)\}_{i=1}^{n}$ of n linear independent solution vectors of $\mathbf{X}' = \mathbf{A}(t)\mathbf{X}(t)$.

Wronskian: $|\Phi(t)| = \begin{vmatrix} \Phi_1 & \Phi_2 & \dots & \Phi_n \end{vmatrix}$

General Solution: $\Phi(t) = \sum_{i=1}^{n} c_1 \Phi_i(t)$

Fundamental Matrix: $\Phi(t) = \begin{pmatrix} \Phi_1 & \Phi_2 & \dots & \Phi_n \end{pmatrix}$

General Solution of $\mathbf{X}' = \mathbf{A}\mathbf{X}(t)$: Distinct real eigenvalues, complex conjugate eigenvalues, repeated eigenvalues

General Solution of $\mathbf{X}' = \mathbf{A}(t)\mathbf{X}(t) + \mathbf{F}(t)$: Undetermined coefficients, variation of parameters:

$$\mathbf{X}(t) = \Phi(t)\mathbf{C} + \mathbf{X}_\mathrm{p}(t)$$

$$= \Phi(t)\mathbf{C} + \Phi(t) \int \Phi^{-1}(t)\mathbf{F}(t) \, dt.$$

Stability Analysis: Improper node, deficient node, star node, saddle point, spiral point, center.

Numerical Methods: Euler's method, Runge-Kutta method

CHAPTER 6 REVIEW EXERCISES

In Exercises 1-7, find the eigenvalues and corresponding eigenvectors of $\mathbf{A}$.

1. $\mathbf{A} = \begin{pmatrix} -1 & 6 \\ 6 & 8 \end{pmatrix}$.

2. $\mathbf{A} = \begin{pmatrix} 2 & 13 \\ -1 & -2 \end{pmatrix}$.

3. $\mathbf{A} = \begin{pmatrix} -1 & -3 \\ 3 & -7 \end{pmatrix}$.

4. $\mathbf{A} = \begin{pmatrix} 3 & 1 & 1 \\ -1 & 3 & 0 \\ 2 & 0 & 3 \end{pmatrix}$.

5. $\mathbf{A} = \begin{pmatrix} 6 & -13 & 1 \\ 3 & -8 & -1 \\ -6 & 14 & 1 \end{pmatrix}$.

6. $\mathbf{A} = \begin{pmatrix} 0 & -3 & -3 \\ 0 & -3 & -3 \\ 0 & 2 & 2 \end{pmatrix}$.

7. $\mathbf{A} = \begin{pmatrix} 1 & -1 & 1 \\ 0 & 3 & -3 \\ -1 & 4 & -4 \end{pmatrix}$.

In Exercises 8-22, find a general solution of each homogeneous system or solve the IVP.

8. $\mathbf{X}' = \begin{pmatrix} -1 & -5 \\ 2 & 6 \end{pmatrix} \mathbf{X}$.

9. $X' = \begin{pmatrix} 2 & -4 \\ -1 & 5 \end{pmatrix} X.$

10. $x' = -4x + 2y, y' = -3x + y, x(0) = 0,$
$y(0) = 1.$

11. $x' = 4x + 2y, y' = -9x - 5y.$

12. $x' = -4x + 5y, y' = -5x + 4y.$

13. $X' = \begin{pmatrix} -4 & 8 \\ -4 & 4 \end{pmatrix} X, X(0) = \begin{pmatrix} 0 \\ 1 \end{pmatrix}.$

14. $x' = -x - 13y, y' = x + 5y.$

15. $x' = -x + 17y, y' = -x - 3y, x(0) = 0,$
$y(0) = 1.$

16. $x' = -x - 3y, y' = 3x - y, x(0) = 1, y(0) = 0.$

17. $X' = \begin{pmatrix} 1 & 2 \\ -2 & -3 \end{pmatrix} X.$

18. $X' = \begin{pmatrix} -1 & 1 \\ -1 & -3 \end{pmatrix} X, X(0) = \begin{pmatrix} 1 \\ 1 \end{pmatrix}.$

19. $x' = -5x + 6y - 4z, y' = -4x + 4y - 4z,$
$z' = -x - 2z.$

20. $x' = x + 2z, y' = 2x - y + 2z, z' = -3x - 4z,$
$x(0) = 0, y(0) = -1, z(0) = 1.$

21. $X' = \begin{pmatrix} 6 & 0 & -5 \\ 0 & 0 & -1 \\ 9 & 0 & -6 \end{pmatrix} X, X(0) = \begin{pmatrix} 1 \\ 0 \\ -1 \end{pmatrix}.$

22. $X' = \begin{pmatrix} -3 & 1 & -2 \\ -5 & 2 & -3 \\ 1 & 0 & 1 \end{pmatrix} X.$

In Exercises 23-31, find a particular solution and general solution of each nonhomogeneous system.

23. $X' = \begin{pmatrix} 8 & -9 \\ 1 & -2 \end{pmatrix} X - 8 \begin{pmatrix} 1 \\ -1 \end{pmatrix} e^{7t}.$

24. $X' = \begin{pmatrix} 1 & -3 \\ 1 & 5 \end{pmatrix} X + \begin{pmatrix} -2 \\ 2 \end{pmatrix} e^{4t}.$

25. $x' = 3x + 4y + 1, y' = 2x + y - 1.$

26. $x' = 3x + 3y - 1, y' = -4x - 10y + 1.$

27. $x' = -y - \sin t, y' = x + \cos t.$

28. $x' = -3x - \frac{5}{2}y + \frac{5}{2}\sin 2t,$
$y' = 4x + 3y + 2\cos 2t - 3\sin 2t.$

29. $x' = -6x + 9y - \ln t, y' = -4x + 6y + \ln t.$

30. $X' = \begin{pmatrix} -4 & 5 \\ -5 & 4 \end{pmatrix} X + \begin{pmatrix} \tan 3t \\ -\sec 3t \end{pmatrix}.$

31. $x' = -x + z + \sqrt{t}, y' = -9x + 3y + 9z - \sqrt{t},$
$z' = 2x - y - 2z + \sqrt{t}.$

32. Find and classify the equilibrium points for the system
$$dx/dt = x(x - 1)(y + 1)$$
$$dy/dt = (x + 1)y(y - 1).$$
Check your result by graphing the direction associated with this nonlinear system on an appropriate region.

33. Find and classify the equilibrium points for the system
$$dx/dt = x(1 - x)(y + 1), \quad dy/dt = y(x + 1).$$
Check your result by graphing the direction associated with this nonlinear system on an appropriate region.

34. (*Modeling Testosterone Production*) The level of testosterone in men can be modeled by the system of delay equations[3]
$$dR/dt = f(T) - b_1 R,$$
$$dL/dt = g_1 R - b_2 L,$$
$$dT/dt = g_2 L(t - \tau) - b_3 T.$$
Show that if $f(0) > 0$ and $f(T)$ is a one-to-one decreasing functions, then the equilibrium point of this system is (R_0, L_0, T_0), where
$$L_0 = \frac{b_3 T_0}{g_2}, \quad R_0 = \frac{b_3 b_2 T_0}{g_1 g_2}, \quad \text{and}$$
$$f(T_0) - \frac{b_1 b_2 b_3 T_0}{g_1 g_2} = 0$$
and that the associated linearized system is
$$dx/dt = f'(T_0)z - b_1 x,$$
$$dy/dt = g_1 x - b_2 y,$$
$$dz/dt = g_2 y(t - \tau) - b_3 z.$$

[3]J.D. Murray, *Mathematical Biology*, Springer-Verlag, 1990, pp. 166-175.

DIFFERENTIAL EQUATIONS AT WORK

A. Modeling a Fox Population in Which Rabies Is Present

Under various assumptions, the nonlinear system of differential equations

$$dX/dt = rX - \gamma XN - \beta XY,$$
$$dI/dt = \beta XY - (\sigma + b + \gamma N) I,$$
$$dY/dt = \sigma I - (\alpha + \beta + \gamma N) Y, \quad (6.40)$$
$$dN/dt = aX - (b + \gamma N) N - \alpha Y$$

has been successfully used to model a fox population in which rabies is present.[4] Here, $X(t)$ represents the population of foxes susceptible to rabies at time t, $I(t)$ the population that has contracted the rabies virus but is not yet ill, $Y(t)$ the population that has developed rabies, and $N(t)$ the total population of the foxes. The symbols a, b, r, γ, σ, α, and β represent constants and are described in the following table:

1. Generate a numerical solution to the system that satisfies the initial conditions $X(0) = 0.93$, $I(0) = 0.035$, $Y(0) = 0.035$, and $N(0) = 1.0$ valid for $0 \le t \le 40$ using the values given in the previous table if $K = 1, 2, 3, 4$, and 8. In each case, graph $X(t)$, $I(t)$, $Y(t)$, and $N(t)$ for $0 \le t \le 40$.

2. Repeat (1) using the initial conditions $X(0) = 0.93$, $I(0) = 0.02$, $Y(0) = 0.05$, and $N(0) = 2.0$.

3. For both (1) and (2), estimate the smallest value of K, say K_T, so that $Y(t)$ is a periodic function.

4. What happens to $Y(t)$ for $K < K_T$? Explain why this result does or does not make sense.

5. Define the *basic reproductive rate* R to be

$$R = \frac{\sigma \beta K}{(\sigma + a)(\alpha + a)} \text{ and } K_T = \frac{(\sigma + a)(\alpha + a)}{\sigma b}.$$

Show that if $R > 1$ then $K > K_T$ and if $R < 1$ then $K < K_T$.

Constant	Description	Typical Value(s)
a	a represents the *average per capita birth rate* of foxes.	1
b	$1/b$ denotes *fox life expectancy* (without resource limitations), which is typically in the range of 1.5-2.7 years.	0.5
r	$r = a - b$ represents the *intrinsic per capita population growth rate*.	0.5
γ	$K = r/\gamma$ represents the *fox-carrying capacity* of the defined area, which is typically in the range of 0.1-4 foxes per km^2. We will compute K and r and then approximate γ.	Varies
σ	$1/\sigma$ represents the *average latent period*. This represents the average time (in years) that a fox can carry the rabies virus but not actually be ill with rabies. Typically, $1/\sigma^2$ is between 28 and 30 days.	12.1667
α	α represents the *death rate* of foxes with rabies. $1/\alpha$ is the expectancy (in years) of a fox with rabies and is typically between 3 and 10 days.	73
β	β represents a *transmission coefficient*. Typically, $1/\beta$ is between 4 and 6 days.	80

[4] R.M. Anderson, H.C. Jackson, R.N. May, A.M. Smith, Population dynamics of fox rabies in Europe, *Nature* **289**, 1981, 765-771.

6. Use the values in the table to calculate K_T. Compare the result to your approximations in (3).

7. Predict how the solution would change if the transmission coefficient β were decreased or the death rate α were increased. What if the average latent period σ were increased? Experiment with different conditions to see if you are correct.

B. Controlling the Spread of a Disease

See the subsection *Modeling the Spread of a Disease* at the end of Chapter 2 for an introduction to the terminology used in this section.

If a person becomes immune to a disease after recovering from it, and births and deaths in the population are not taken into account, then the percentage of persons susceptible to becoming infected with the disease $S(t)$, the percentage of people in the population infected with the disease $I(t)$, and the percentage of people in the population recovered and immune to the disease $R(t)$ can be modeled by the system

$$dS/dt = -\lambda SI,$$
$$dI/dt = \lambda SI - \gamma I,$$
$$dR/dt = \gamma I,$$
$$S(0) = S_0, I(0) = I_0, R(0) = 0.$$

Because $S(t) + I(t) + R(t) = 1$, once we know S and I, we can compute R with $R(t) = 1 - S(t) - I(t)$. This model is called an *SIR model without vital dynamics* because once a person has had the disease, the person becomes immune to the disease, and because births and deaths are

not taken into consideration. This model might be used to model diseases that are *epidemic* to a population: those diseases that persist in a population for short periods of time (less than 1 year). Such diseases typically include influenza, measles, rubella, and chicken pox.

1. Show that if $S_0 < \gamma/\lambda$, the disease dies out, while an epidemic results if $S_0 > \gamma/\lambda$.

2. Show that $dI/dS = -((\lambda S - \gamma)I)/\lambda SI = -1 + \rho/S$, where $\rho = \gamma/\lambda$, has solution $I + S - \rho \ln S = I_0 + S_0 - \rho \ln S_0$.

3. What is the maximum value of I? When diseases persist in a population for long periods of time, births and deaths must be taken into consideration. If a person becomes immune to a disease after recovering from it and births and deaths in the population are taken into account, then the percentage of persons susceptible to becoming infected with the disease $S(t)$ and the percentage of people in the population infected with the disease $I(t)$ can be modeled by the system

$$dS/dt = -\lambda SI + \mu - \mu S,$$
$$dI/dt = \lambda SI - \gamma I - \mu I, \qquad (6.41)$$
$$S(0) = S_0, I(0) = I_0.$$

This model is called an *SIR model with vital dynamics*[5] because once a person has had the disease, the person becomes immune to it, and because births and deaths are taken into consideration. This model might be used to model diseases that are *endemic* to a population: those diseases that persist in a population for long periods of time (10 or 20 years). Smallpox is an example of a disease that was endemic until it was eliminated in 1977.

[5] H.W. Hethcote, Three Basic Epidemiological Models, in: *Applied Mathematical Ecology*, S.A. Levin, T.G. Hallan, L.J. Gross (Eds.), Springer-Verlag, 1989, pp. 119-143. R.M. Anderson, R.M. May, Directly transmitted infectious diseases: control by vaccination, *Science*, **215**, 1982, 1053-1060. J.D. Murray, *Mathematical Biology*, Springer-Verlag, 1990, pp. 611-618.

4. Show that the equilibrium points of system (6.41) are $(S_0, I_0) = (1, 0)$ and $(S_A, I_A) = ((\gamma+\mu)/\lambda, \mu[\lambda-(\gamma+\mu)]/[\lambda(\gamma+\mu)])$. Because $S(t) + I(t) + R(t) = 1$, it follows that $S(t) + I(t) \leq 1$.

5. Use the fact that $S(t) + I(t) \leq 1$ to determine conditions on γ, μ, and λ so that system (6.41) has the equilibrium point (S_A, I_A). In this case, classify the equilibrium point.

6. Use the fact that $S(t) + I(t) \leq 1$ to determine conditions on γ, μ, and λ so that system (6.41) does *not* have the equilibrium point (S_A, I_A).

The following table shows the average infectious period and typical contact numbers for several diseases during certain epidemics:

Disease	Infectious) Period (Average) $1/\gamma$	γ	Typical Contact Number σ
Measles	6.5	0.153846	14.9667
Chicken pox	10.5	0.0952381	11.3
Mumps	19	0.0526316	8.1
Scarlet fever	17.5	0.0571429	8.5

Let's assume that the average lifetime is $1/\mu$ is 70 years so that $\mu = 0.0142857$.

7. For each of the diseases listed in the following table, use the formula $\sigma = \lambda/(\gamma + \mu)$ to calculate the daily contact rate λ.

Disease	λ
Measles	
Chicken pox	
Mumps	
Scarlet fever	

Diseases such as those listed above can be controlled once an effective and inexpensive vaccine has been developed. It is virtually impossible to vaccinate everybody against a disease; we would like to determine the percentage of a population that needs to be vaccinated to eliminate a disease from the population under consideration. A population of people has *herd immunity* to a disease when enough people are immune to the disease so that if it is introduced into the population, it will not spread throughout the population. To have herd immunity, an infected person must infect less than one uninfected person during the time the infected person is infectious. Thus, we must have $\sigma S < 1$. Because $I + S + R = 1$, when $I = 0$ we have that $S = 1 - R$; consequently, herd immunity is achieved when

$$\sigma(1 - r) < 1 \text{ or } \sigma - \sigma R < 1,$$
$$-\sigma R < 1 - \sigma,$$
$$R > \frac{\sigma - 1}{\sigma} = 1 - \frac{1}{\sigma}.$$

8. For each of the diseases listed in the following table, estimate the minimum percentage of a population that needs to be vaccinated to achieve herd immunity.

Disease	Minimum Value of R to Achieve Herd Immunity
Measles	
Chicken pox	
Mumps	
Scarlet fever	

9. Using the values obtained in the previous exercises, for each disease in the tables, graph the direction field and several solutions ($I(t)$, $S(t)$, $R(t)$, and $\{S(t), I(t)\}$ parametrically as S vs. I) using both models. Discuss scenarios in which each model is valid and note any significant differences between the two models.

10. What are some possible ways that an epidemic can be controlled?

C. FitzHugh-Nagumo Model

Under certain assumptions, the *FitzHugh-Nagumo equation*, which arises in the study of the impulses in a nerve fiber, can be written as system of ODEs

$$dV/d\xi = W,$$
$$dW/d\xi = F(V) + R - uW,$$
$$dR/d\xi = \frac{\epsilon}{u}(bR - V - a),$$

were $F(V) = \frac{1}{3}V^3 - V$.[6]

1. Graph the solution to system (6.42) that satisfies the initial conditions $V(0) = 1$, $W(0) = 0$, $R(0) = 1$, if $\epsilon = 0.08$, $a = 0.7$, $b = 0$, and $u = 1$.
2. Graph the solution to system (6.42) that satisfies the initial conditions $V(0) = 1$, $W(0) = 0.5$, $R(0) = 0.5$, if $\epsilon = 0.08$, $a = 0.7$, $b = 0.8$, and $u = 0.6$.
3. Approximate the maximum and minimum values, if they exist, of V, W, and R in (1) and (2).

D. An Agricultural Model

Consider the following model of an insect population that is hunted by two spider populations, which are distinguishable because they live in two different habitats,[7]

$$\begin{cases} \dfrac{dp}{dT} = rp\left(1 - \dfrac{p}{K}\right) - aps_\omega - bps_c, \\ \dfrac{ds_\omega}{dT} = s_\omega\left(eap - \mu_\omega - \dfrac{s_\omega}{W}\right), \\ \dfrac{ds_c}{dT} = s_c\left(ebp - \mu_c - \dfrac{s_c}{W}\right). \end{cases}$$

In this model, p represents the size of the insect population, while s_ω and s_c represent the sizes of the webbuilder spider populations living

in two separate habitats. The positive constants a and b represent the *per capita* prey capturing rates for the spiders in the different habitats, and e $(0 < e < 1)$ stands for the conversion coefficient of turning the prey into new biomass. The spiders compete with each other intra-specifically at rates $1/W$ and $1/V$, respectively, and their mortality rates are μ_ω and μ_c, respectively.

1. A common approach to working with applied models is to make the system dimensionless. Use the dimensionless variables $P = p/K$, $S_\omega = as_\omega/r$, $S_c = bs_c/r$ and $t = rT$ to obtain the dimensionless system

$$\begin{cases} \dfrac{dP}{dt} = P(1 - P) - PS_\omega - PS_c \\ \dfrac{dS_\omega}{dt} = \alpha S_\omega P - \delta_\omega S_\omega - \dfrac{S_\omega^2}{W_1} \\ \dfrac{dS_c}{dt} = \beta S_c P - \delta_c S_c - \dfrac{S_c^2}{V1} \end{cases}$$

where $\alpha = eaK/r$, $\beta = ebK/r$, $\delta_\omega = \mu_\omega/r$, $\delta_c = \mu_c/r$, $W_1 = aW$, and $V_1 = bV$.
2. Show that the point $(P, S_\omega, S_c) = (0,0,0)$ is an equilibrium point. If $P = 0$, show that the resulting coordinates for S_ω and S_c are less than or equal to zero so that $(0,0,0)$ is the only equilibrium point that corresponds to $P = 0$.
3. Show that if $S_\omega = 0$, then the coordinates of the resulting equilibrium point are
$P = \dfrac{1 + \delta_c V_1}{1 + \beta V_1}$ and $S_c = \dfrac{(\beta - \delta_c)V_1}{1 + \beta V_1}$. Is there a restriction on the parameters so that the solution is feasible?
4. Show that if $S_c = 0$, then the coordinates of the resulting equilibrium point are $P = (1 + \delta_\omega W_1)/(1 + \alpha W_1)$ and $S_\omega = ((\alpha - \delta_\omega)W_1)/(1 + \alpha W_1)$.

[6] J.D. Murray, *Mathematical Biology*, Springer-Verlag (1990), pp. 161-166. A.C. Scott, The electrophysics of a nerve fiber, *Reviews of Modern Physics*, **47**(2), 1975, 487-533.

[7] M. Sen, M. Banerjee, E. Venturino, A model for biological control in agriculture, *Mathematics and Computers in Simulation* **87**, 2013, 30-44.

5. Show that the only interior equilibrium point has coordinates $P = (1 + \delta_\omega W_1 + \delta_c V_1)/(1 + \alpha W_1 + \beta V_1)$, $S_\omega = ([\alpha(1 + \delta_c V_1) - \delta_\omega(1 + \beta V_1)]W_1)/(1 + \alpha W_1 + \beta V_1)$, and $S_c = ([\beta(1 + \delta_\omega W_1) - \delta_c(\alpha W_1 + 1)]V_1)/(1 + \alpha W_1 + \beta V_1)$. Are there restrictions on the parameters so that the solution is feasible?

6. Use the parameter values $\alpha = 2$, $\beta = 3$, $\delta_\omega = 0.1$, $\delta_c = 0.2$, $W_1 = 2$, and $V_1 = 2.5$ and initial conditions $P(0) = 0.6$, $S_\omega(0) = 0.4$ and $S_c(0) = 0.2$ with a CAS to investigate the stability of the resulting solutions.

E. Modeling the Spread of Dengue in Indonesia

Dengue is a virus spread in Indonesia through mosquito bites. We can present a simple model for this situation with the system of equations[8]

$$\begin{cases} \dfrac{dH_S}{dt} = Q_S - \delta_S H_S M_{DS} - \mu_H H_S \\[2mm] \dfrac{dH_D}{dt} = \delta_S M_{DS} H_S - \gamma H_D - \mu_H H_D \\[2mm] \dfrac{dH_R}{dt} = \gamma H_D - \mu_H H_R \\[2mm] \dfrac{dM_S}{dt} = R_S - \beta_S M_S H_D - v_S M_S \\[2mm] \dfrac{dM_{DS}}{dt} = \beta_S M_S H_D - v_S M_{DS} \end{cases}$$

where H_S, H_D, and H_R represent the susceptible, dengue-infested, and recovered human populations, respectively; and M_S and M_{DS} represent the susceptible mosquito and dengue-infected mosquito populations, respectively. The parameters Q_S and R_S represent the susceptible human and mosquito recruitment rates while β_S represents the dengue infection rates from human to M_{DS} mosquitos and δ_S represents the dengue infection rates to human from M_{DS} mosquitos. In

addition, the parameter γ represents the recovery rate of humans from dengue, μ_H represents the human mortality rate, and v_S represents the natural mortality rate of the M_S mosquitos.

Since there is no commercially available vaccine for dengue, a prevention approach is to reduce the mosquito population. In one such treatment, a wolbachia (bacteria) infection is introduced to the mosquito population. This leads to the more extensive mathematical model

$$\begin{cases} \dfrac{dH_S}{dt} = Q_S - \delta_S H_S M_{DS} - \delta_W H_S M_{DW} - \mu_H H_S \\[2mm] \dfrac{dH_D}{dt} = (\delta_S M_{DS} + \delta_W M_{DW}) H_S - \gamma H_D - \mu_H H_D \\[2mm] \dfrac{dH_R}{dt} = \gamma H_D - \mu_H H_D \\[2mm] \dfrac{dM_S}{dt} = R_S - \alpha M_S M_W - \beta_S M_S H_D - v_S M_S \\[2mm] \dfrac{dM_W}{dt} = R_W + \alpha M_S M_W - \beta_S (1 - \zeta M_W) \\ \qquad\qquad \times M_W H_D - v_S (1 + \xi M_W) M_W \\[2mm] \dfrac{dM_{DS}}{dt} = \beta_S M_S H_D - v_{DS} M_{DS} \\[2mm] \dfrac{dM_{DW}}{dt} = \beta_S (1 - \zeta M_W) M_W H_D - v_{DW} M_{DW} \end{cases}$$

where in addition to the variables used in the simple model, M_W and M_{DW} represent the wolbachia-infected mosquito and dengue-infected mosquito from the wolbachia-infected population, respectively. The parameter β_W represents the dengue infection rates from human to M_{DW} mosquitos. Similarly, δ_W represents the dengue infection rates to human from M_{DW} mosquitos. In addition, the parameters v_{DS}, v_W, and v_{DW} represent the natural mortality rates of the M_{DS} mosquitos and the wolbachia-induced mortality rates of the M_W and M_{DW} mosquitos. Also important to the model are the parameters α (the wolbachia infection rate

[8] A.K., Supriatna, Anggriani, N., System dynamics model of wolbachia in dengue transmission, *Procedia Engineering* **50**, 2012, 12-18.

among mosquitos) and R_W (the wolbachia infected recruitment rate). In this case, the wolbachia-induced mortality rate is assumed to be $v_S(1 + \xi M_W)$ while the dengue infection rate from human to M_{DW} mosquitos is assumed to be $\beta_S(1 - \zeta M_W)$ where ξ and ζ are small positive constants.

1. Show that the simple model has two equilibrium solutions,
$E_0^* = (H_{0S}^*, H_{0D}^*, H_{0R}^*, M_{0S}^*, M_{0DS}^*)$ where $H_{0S}^* = Q_S/\mu_H$, $M_{0S}^* = R_S/v_S$ and all other components are zero; and $E_e^* = (H_{eS}^*, H_{eD}^*, H_{eR}^*, M_{eS}^*, M_{eDS}^*)$ where $H_{eS}^* = (v_S Q_S \beta_S + (\gamma + \mu_H)v_S^2)/(\beta_S(R_S\delta_S + \mu_H v_S))$,
$H_{eD}^* = (Q_S R_S \beta_S \delta_S - \mu_H(\gamma + \mu_H)v_S^2)/(\beta_S(\gamma + \mu_H)(R_S\delta_S + \mu_H v_S))$,
$H_{eR}^* = (\gamma Q_S R_S \beta_S \delta_S - \gamma \mu_H(\gamma + \mu_H)v_S^2)/(\beta_S \mu_H(\gamma + \mu_H)(R_S\delta_S + \mu_H v_S))$, $M_{eS}^* = ((\gamma + \mu_H)(R_S\delta_S + \mu_H v_S))/(\delta_S(Q_S\beta_S + (\gamma + \mu_H)v_S))$ and $M_{eDS}^* = (Q_S R_S \beta_S \delta_S - \mu_H(\gamma + \mu_H)v_S^2)/(\delta_S v_S(Q_S\beta_S + (\gamma + \mu_H)v_S))$.
(Use of a computer algebra system is suggested.) Note that E_0^* is the disease-free equilibrium while E_e^* is the endemic equilibrium.

2. Find the Jacobian for the simple system.

3. In the simple model, consider the parameter values $Q_S = 50$, $\delta_S = 0.0001$, $\mu_H = 0.1$, $\gamma = 0.35$, $R_S = 1000$, $\beta_S = 0.0005$, and $v_S = 0.25$. Use these values to determine the stability of the system at E_0^*. Reinvestigate the stability of E_0^* if γ is changed to $\gamma = 0.25$. Research shows that the disease-free equilibrium is stable if the system's reproduction number R_0 is less than zero and unstable otherwise. If $R_0 = (Q_S R_S \beta_S \delta_S)/(\mu_H(\gamma + \mu_H)v_S^2)$, find R_0 for the given parameter values above. How does the stability compare to the findings using the eigenvalues of the system?

4. In the wolbachia model, consider the parameter values above (with $\gamma = 0.35$) and these additional values: $\delta_W = 0.0001$. $R_W = 1000$, $v_{DS} = 0.25$, $\alpha = 0.025$, $\xi = 0.025$, and $\zeta = 0.00025$. If the initial populations are $H_S(0) = Q_S/\mu_H$, $H_D(0) = 10$, $H_R(0) = 0$, $M_S(0) = R_S/v_H$, $M_W(0) = R_W/v_W$ and $M_{DS}(0) = M_{DW}(0) = 0$, use a computer algebra system to investigate the long term behavior of the dengue-infected human and mosquito populations, H_D and $M_{DS} + M_{DW}$. Is the same true in the case of the simple model?

Applications of Systems of Ordinary Differential Equations

7.1 MECHANICAL AND ELECTRICAL PROBLEMS WITH FIRST-ORDER LINEAR SYSTEMS

- *L-R-C* Circuits with Loops
  - *L-R-C* Circuits with One Loop
  - *L-R-C* Circuits with Two Loops
- Spring-Mass Systems

L-R-C Circuits with Loops

As indicated in Chapter 5, an electrical circuit can be modeled with a linear ordinary differential equation (ODE) with constant coefficients. In this section, we illustrate how a circuit involving loops can be described as a system of linear

ODEs with constant coefficients. This derivation is based on the following principles.

> *Kirchhoff's Current Law*: The current entering a point of the circuit equals the current leaving the point.
> *Kirchhoff's Voltage Law*: The sum of the changes in voltage around each loop in the circuit is zero.

As was the case in Chapter 5, we use the following standard symbols for the components of the circuit:

$$I(t) = \text{current}$$

where $I(t) = dQ(t)/dt$, $Q(t)$ is the charge, R is the resistance, C is the capacitance, E is the voltage, and L is the inductance.

TABLE 7.1   Circuit Elements and Corresponding Voltage Drops

Circuit Element	Voltage Drop
Inductor	$L\dfrac{dI}{dt}$
Resistor	RI
Capacitor	$\dfrac{1}{C}Q$
Voltage source	$-E(t)$

The relationships corresponding to the drops in voltage in the various components of the circuit are restated in Table 7.1.

L-R-C Circuit with One Loop

In determining the drops in voltage around the circuit, we consistently add the voltages in the clockwise direction. The positive direction is from the negative symbol toward the positive symbol associated with the voltage source. In summing the voltage drops encountered in the circuit, a drop across a component is added to the sum if the positive direction through the component agrees with the clockwise direction. Otherwise, this drop is subtracted. In the case of the L-R-C circuit with one loop involving each type of component shown in Figure 7.1, the current is equal around the circuit by Kirchhoff's current law.

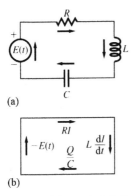

(a)

(b)

FIGURE 7.1   (a) and (b) A simple L-R-C circuit.

Also, by Kirchhoff's voltage law, we have

$$RI + L\frac{dI}{dt} + \frac{1}{C}Q - E(t) = 0.$$

Solving this equation for dI/dt and using the relationship $dQ/dt = I$, so $dI/dt = d^2Q/dt^2$, we have the system of differential equations

$$\frac{dQ}{dt} = I, \quad \frac{dI}{dt} = -\frac{1}{LC}Q - \frac{R}{L}I + \frac{1}{L}E(t) \quad (7.1)$$

with initial conditions $Q(0) = Q_0$ and $I(0) = I_0$ on charge and current, respectively. The method of variation of parameters (for systems) can be used to solve problems of this type.

EXAMPLE 7.1.1

Determine the charge and current in the L-R-C circuit with $L = 1$ H, $R = 2\,\Omega$, $C = 4/3$ F and $E(t) = e^{-t}$ V if $Q(0) = 1$ C and $I(0) = 1$ A.

Solution

We begin by modeling the circuit with a system of differential equations. In this case, we have

$$\frac{dQ}{dt} = I, \quad \frac{dI}{dt} = -\frac{3}{4}Q - 2I + e^{-t}$$

with initial conditions $Q(0) = 1$ and $I(0) = 1$. We can write this nonhomogeneous system in matrix form as $\begin{pmatrix} dQ/dt \\ dI/dt \end{pmatrix} = \begin{pmatrix} 0 & 1 \\ -3/4 & -2 \end{pmatrix} \begin{pmatrix} Q \\ I \end{pmatrix} + \begin{pmatrix} 0 \\ e^{-t} \end{pmatrix}$.

The eigenvalues of the corresponding homogeneous system are $\lambda_1 = -1/2$ and $\lambda_2 = -3/2$, with corresponding eigenvectors $\mathbf{v}_1 = \begin{pmatrix} -2 \\ 1 \end{pmatrix}$ and $\mathbf{v}_2 = \begin{pmatrix} 2 \\ -3 \end{pmatrix}$, respectively. Thus a fundamental matrix is

$$\Phi(t) = \begin{pmatrix} -2e^{-t/2} & 2e^{-3t/2} \\ e^{-t/2} & -3e^{-3t/2} \end{pmatrix} \quad \text{with}$$

$$\Phi^{-1}(t) = \begin{pmatrix} -\frac{3}{4}e^{t/2} & -\frac{1}{2}e^{t/2} \\ -\frac{1}{4}e^{3t/2} & -\frac{1}{2}e^{3t/2} \end{pmatrix}.$$

Then, by variation of parameters, if we let $\mathbf{X}(t) = \begin{pmatrix} Q(t) \\ I(t) \end{pmatrix}$, we have

$$\mathbf{X}(t) = \Phi(t)\Phi^{-1}(0)\mathbf{X}(0) + \Phi(t)\int_0^t \Phi^{-1}(u)\mathbf{F}(u)du$$

$$= \begin{pmatrix} -2e^{-t/2} & 2e^{-3t/2} \\ e^{-t/2} & -3e^{-3t/2} \end{pmatrix}\begin{pmatrix} -\frac{3}{4} & -\frac{1}{2} \\ -\frac{1}{4} & -\frac{1}{2} \end{pmatrix}\begin{pmatrix} 1 \\ 1 \end{pmatrix}$$

$$+ \begin{pmatrix} -2e^{-t/2} & 2e^{-3t/2} \\ e^{-t/2} & -3e^{-3t/2} \end{pmatrix}\int_0^t \begin{pmatrix} -\frac{3}{4}e^{u/2} & -\frac{1}{2}e^{u/2} \\ -\frac{1}{4}e^{3u/2} & -\frac{1}{2}e^{3u/2} \end{pmatrix}\begin{pmatrix} 0 \\ e^{-u} \end{pmatrix}du$$

$$= \begin{pmatrix} -2e^{-t/2} & 2e^{-3t/2} \\ e^{-t/2} & -3e^{-3t/2} \end{pmatrix}\begin{pmatrix} -\frac{5}{4} \\ -\frac{3}{4} \end{pmatrix}$$

$$+ \begin{pmatrix} -2e^{-t/2} & 2e^{-3t/2} \\ e^{-t/2} & -3e^{-3t/2} \end{pmatrix}\int_0^t \begin{pmatrix} -\frac{1}{2}e^{-u/2} \\ -\frac{1}{2}e^{u/2} \end{pmatrix}du$$

$$= \begin{pmatrix} \frac{5}{2}e^{-t/2} - \frac{3}{2}e^{-3t/2} \\ -\frac{5}{4}e^{-t/2} + \frac{9}{4}e^{-3t/2} \end{pmatrix} + \begin{pmatrix} -4e^{-t} + 2e^{-t/2} + 2e^{-3t/2} \\ 4e^{-t} - e^{-t/2} - 3e^{-3t/2} \end{pmatrix}$$

$$= \begin{pmatrix} \frac{9}{2}e^{-t/2} + \frac{1}{2}e^{-3t/2} - 4e^{-t} \\ -\frac{9}{4}e^{-t/2} - \frac{3}{4}e^{-3t/2} + 4e^{-t} \end{pmatrix}.$$

In Example 7.1.1, do the limits $\lim_{t\to\infty} Q(t) = \lim_{t\to\infty} I(t) = 0$ hold for all choices of the initial conditions?

We plot the solution $\mathbf{X}(t) = \begin{pmatrix} Q(t) \\ I(t) \end{pmatrix}$ parametrically in Figure 7.2(a). Notice that $\lim_{t\to\infty} Q(t) = \lim_{t\to\infty} I(t) = 0$, so the solution approaches $(0,0)$ as t increases. We also plot $Q(t)$ (dark red; dark

gray in print versions) and $I(t)$ simultaneously in Figure 7.2(b). Finally, in Figure 7.2(c), we graph $\mathbf{X}(t) = \begin{pmatrix} Q(t) \\ I(t) \end{pmatrix}$ for other initial conditions.

L-R-C Circuit with Two Loops

The differential equations that model the circuit become more difficult to derive as the number of loops in the circuit increases. For example, consider the circuit that contains two loops, as shown in Figure 7.3.

In this case, the current through the capacitor is equivalent to $I_1 - I_2$. Summing the voltage drops around each loop, we obtain the system of equations

$$R_1 I_1 + \frac{1}{C}Q - E(t) = 0, \quad L\frac{dI_2}{dt} + R_2 I_2 - \frac{1}{C}Q = 0. \tag{7.2}$$

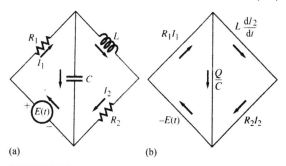

(a)   (b)

FIGURE 7.3 (a) and (b) A two-loop circuit.

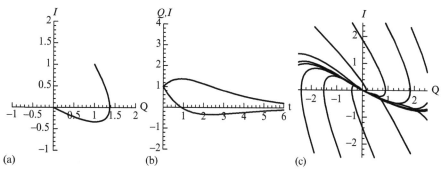

(a)   (b)   (c)

FIGURE 7.2 From left to right, (a)-(c).

Solving the first equation for I_1 yields $I_1 = (1/R_1)E(t) - (1/R_1 C)Q$. Using the relationship $dQ/dt = I = I_1 - I_2$ gives us the system

$$\frac{dQ}{dt} = -\frac{1}{R_1 C}Q - I_2 + \frac{1}{R_1}E(t), \qquad \frac{dI_2}{dt} = \frac{1}{LC}Q - \frac{R_2}{L}I_2.$$
$$(7.3)$$

EXAMPLE 7.1.2

Find $Q(t)$, $I(t)$, $I_1(t)$, and $I_2(t)$ in the L-R-C circuit with two loops given that $R_1 = R_2 = 1\,\Omega$, $C = 1\,F$, $L = 1\,H$, and $E(t) = e^{-t}\,V$ if $Q(0) = 1\,C$ and $I_2(0) = 3\,A$.

Solution

The nonhomogeneous system that models this circuit is

$$\frac{dQ}{dt} = -Q - I_2 + e^{-t}, \qquad \frac{dI_2}{dt} = Q - I_2$$

with initial conditions $Q(0) = 1$ and $I_2(0) = 3$. As in Example 7.1.1, we use the method of variation of parameters to solve the problem. In matrix form this system is $\begin{pmatrix} dQ/dt \\ dI_2/dt \end{pmatrix} = \begin{pmatrix} -1 & -1 \\ 1 & -1 \end{pmatrix} \begin{pmatrix} Q \\ I_2 \end{pmatrix} + \begin{pmatrix} e^{-t} \\ 0 \end{pmatrix}$. The eigenvalues of the corresponding homogeneous system are $\lambda_{1,2} = -1 \pm i$, and an eigenvector corresponding to λ_1 is $\mathbf{v}_1 = \begin{pmatrix} i \\ 1 \end{pmatrix} = \begin{pmatrix} 0 \\ 1 \end{pmatrix} + \begin{pmatrix} 1 \\ 0 \end{pmatrix}i$. Two linearly independent solutions of the corresponding homogeneous system are

$$\mathbf{X}_1(t) = \begin{pmatrix} Q(t) \\ I_2(t) \end{pmatrix} = e^{-t}\cos t \begin{pmatrix} 0 \\ 1 \end{pmatrix} - e^{-t}\sin t \begin{pmatrix} 1 \\ 0 \end{pmatrix}$$

$$= \begin{pmatrix} -e^{-t}\sin t \\ e^{-t}\cos t \end{pmatrix}$$

and

$$\mathbf{X}_2(t) = \begin{pmatrix} Q(t) \\ I_2(t) \end{pmatrix} = e^{-t}\cos t \begin{pmatrix} 1 \\ 0 \end{pmatrix} + e^{-t}\sin t \begin{pmatrix} 0 \\ 1 \end{pmatrix}$$

$$= \begin{pmatrix} e^{-t}\cos t \\ e^{-t}\sin t \end{pmatrix},$$

so a fundamental matrix is

$$\Phi(t) = \begin{pmatrix} -e^{-t}\sin t & e^{-t}\cos t \\ e^{-t}\cos t & e^{-t}\sin t \end{pmatrix} \quad \text{with}$$

$$\Phi^{-1}(t) = \begin{pmatrix} -e^{t}\sin t & e^{t}\cos t \\ e^{t}\cos t & e^{t}\sin t \end{pmatrix}.$$

Therefore,

$$\mathbf{X}(t) = \Phi(t)\Phi^{-1}(0)\mathbf{X}(0) + \Phi(t)\int_0^t \Phi^{-1}(u)\mathbf{F}(u)\,du$$

$$= \begin{pmatrix} -e^{-t}\sin t & e^{-t}\cos t \\ e^{-t}\cos t & e^{-t}\sin t \end{pmatrix}\begin{pmatrix} 0 & 1 \\ 1 & 0 \end{pmatrix}\begin{pmatrix} 1 \\ 3 \end{pmatrix}$$

$$+ \begin{pmatrix} -e^{-t}\sin t & e^{-t}\cos t \\ e^{-t}\cos t & e^{-t}\sin t \end{pmatrix}\int_0^t \begin{pmatrix} -e^{u}\sin u & e^{u}\cos u \\ e^{u}\cos u & e^{u}\sin u \end{pmatrix}\begin{pmatrix} e^{-u} \\ 0 \end{pmatrix}\,du$$

$$= \begin{pmatrix} -e^{-t}\sin t & e^{-t}\cos t \\ e^{-t}\cos t & e^{-t}\sin t \end{pmatrix}\begin{pmatrix} 3 \\ 1 \end{pmatrix}$$

$$+ \begin{pmatrix} -e^{-t}\sin t & e^{-t}\cos t \\ e^{-t}\cos t & e^{-t}\sin t \end{pmatrix}\int_0^t \begin{pmatrix} -\sin u \\ \cos u \end{pmatrix}\,du$$

$$= \begin{pmatrix} e^{-t}\cos t - 3e^{-t}\sin t \\ 3e^{-t}\cos t + e^{-t}\sin t \end{pmatrix} + \begin{pmatrix} -e^{-t}\sin t & e^{-t}\cos t \\ e^{-t}\cos t & e^{-t}\sin t \end{pmatrix}\begin{pmatrix} \cos t - 1 \\ \sin t \end{pmatrix}$$

$$= \begin{pmatrix} e^{-t}\cos t - 3e^{-t}\sin t \\ 3e^{-t}\cos t + e^{-t}\sin t \end{pmatrix} + \begin{pmatrix} e^{-t}\sin t \\ e^{-t} - e^{-t}\cos t \end{pmatrix}$$

$$= \begin{pmatrix} e^{-t}\cos t - 2e^{-t}\sin t \\ 2e^{-t}\cos t + e^{-t} + e^{-t}\sin t \end{pmatrix}.$$

Because $dQ/dt = I$ and $Q(t) = e^{-t}\cos t - 2e^{-t}\sin t$, differentiation yields $I(t) = -3e^{-t}\cos t + e^{-t}\sin t$. Also, because $I_1(t) = I(t) + I_2(t)$, $I_1(t) = -e^{-t}\cos t + 2e^{-t}\sin t + e^{-t}$. We graph $Q(t)$ and $I_2(t)$ in Figure 7.4(a) and $I(t)$ and $I_1(t)$ in Figure 7.4(b). In Figure 7.4(c), we graph $\{Q(t), I_2(t)\}$ parametrically to show the phase plane for the system of nonhomogeneous equations using several different initial conditions. Notice that some of the graphs overlap, which does not occur if a system is homogeneous.

Find the limits of $Q(t)$, $I_1(t)$, $I_2(t)$, and $I(t)$ as $t \to \infty$. Does a change in initial conditions affect these limits?

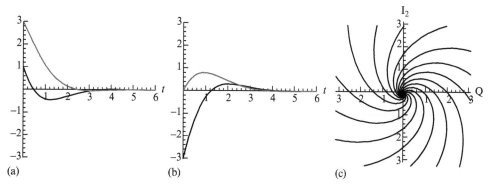

FIGURE 7.4   (a) $Q(t)$ (dark red; dark gray in print versions) and $I_2(t)$. (b) $I(t)$ (dark red; dark gray in print versions) and $I_1(t)$. (c) Parametric plots of solutions that satisfy other initial conditions.

Spring-Mass Systems

The displacement of a mass attached to the end of a spring was modeled with a second-order linear differential equation with constant coefficients in Chapter 5. This situation can be expressed as a system of first-order ODEs as well. Recall that if there is no external forcing function, the second-order differential equation that models the situation is

$$m\frac{d^2x}{dt^2} + c\frac{dx}{dt} + kx = 0,$$

where m is the mass of the object attached to the end of the spring, c is the damping coefficient, and k is the spring constant found with Hooke's law. This equation is transformed into a system of equations with the substitution $dx/dt = y$. Then, solving the differential equation for d^2x/dt^2, we have $dy/dt = d^2x/dt^2 = -k/mx - c/m\,dx/dt = -k/mx - c/m\,y$, which yields the system

$$\frac{dx}{dt} = y, \quad \frac{dy}{dt} = -\frac{k}{m}x - \frac{c}{m}y. \quad (7.4)$$

In previous chapters, the displacement of the spring was illustrated as a function of time. Problems of this type may also be investigated using the phase plane. In the following example, the phase plane corresponding to the various

situations encountered by spring-mass systems discussed in previous sections (undamped, damped, overdamped, and critically damped) are determined.

EXAMPLE 7.1.3

Solve the system of differential equations to find the displacement of the mass if $m = 1, c = 0$, and $k = 1$.

Solution

In this case, the system is

$$\frac{dx}{dt} = y, \quad \frac{dy}{dt} = -x.$$

The eigenvalues are solutions of $\begin{vmatrix} -\lambda & 1 \\ -1 & -\lambda \end{vmatrix} = \lambda^2 + 1 = 0$, so $\lambda_{1,2} = \pm i$. An eigenvector corresponding to λ_1 is $\mathbf{v}_1 = \begin{pmatrix} 1 \\ i \end{pmatrix} = \begin{pmatrix} 1 \\ 0 \end{pmatrix} + \begin{pmatrix} 0 \\ 1 \end{pmatrix}i$, so two linearly independent solutions are $\mathbf{X}_1(t) = \begin{pmatrix} 1 \\ 0 \end{pmatrix}\cos t - \begin{pmatrix} 0 \\ 1 \end{pmatrix}\sin t = \begin{pmatrix} \cos t \\ -\sin t \end{pmatrix}$ and $\mathbf{X}_2(t) = \begin{pmatrix} 0 \\ 1 \end{pmatrix}\cos t + \begin{pmatrix} 1 \\ 0 \end{pmatrix}\sin t = \begin{pmatrix} \sin t \\ \cos t \end{pmatrix}$. A general solution is

$$\mathbf{X}(t) = \begin{pmatrix} x(t) \\ y(t) \end{pmatrix} = c_1 \mathbf{X}_1(t) + c_2 \mathbf{X}_2(t)$$

$$= \begin{pmatrix} c_1 \cos t + c_2 \sin t \\ c_2 \cos t - c_1 \sin t \end{pmatrix}.$$

By observing the phase plane in Figure 7.5(c) and the corresponding system of differential equations in Example 7.1.3, describe the motion of the object in each quadrant and determine the sign on the velocity dx/dt of the object in each quadrant.

Notice that this system is equivalent to the second-order differential equation $x'' + x = 0$, which we solved in Chapters 4 and 5. At that time, we found a general solution to be $x(t) = c_1 \cos t + c_2 \sin t$, which is the same as the first component of $\mathbf{X}(t) = \begin{pmatrix} x(t) \\ y(t) \end{pmatrix}$ obtained in this instance. We graph $x(t)$ and $y(t)$ for several values of the arbitrary constants in Figures 7.5(a) and (b), respectively, to illustrate the periodic motion, of the mass. Also notice that $(0,0)$ is the equilibrium point of the system. Because the eigenvalues are $\lambda_{1,2} = \pm i$, we classify the origin as a center. We graph the phase plane of this system in Figure 7.5(c).

 EXERCISES 7.1

Solve each of the following systems for charge and current using the procedures discussed in the example problems.

1. Solve the one-loop L-R-C circuit with $L = 3\,\text{H}$, $R = 10\,\Omega$, $C = 1/10\,\text{F}$, and
   (a) $E(t) = 0\,\text{V}$; (b) $E(t) = e^{-t}\,\text{V}$. ($Q(0) = 0\,\text{C}$, $I(0) = 1\,\text{A}$.)

2. Solve the one-loop L-R-C circuit with $L = 1\,\text{H}$, $R = 20\,\Omega$, $C = 1/100\,\text{F}$, and
   (a) $E(t) = 0\,\text{V}$; (b) $E(t) = 1200\,\text{V}$. ($Q(0) = 0\,\text{C}$, $I(0) = 1\,\text{A}$.)

3. Find $Q(t)$, $I(t)$, $I_1(t)$, and $I_2(t)$ in the two-loop L-R-C circuit with $L = 1\,\text{H}$, $R_1 = 2\,\Omega$, $R_2 = 1\,\Omega$, $C = 1/2\,\text{F}$, and
   (a) $E(t) = 0\,\text{V}$; (b)$E(t) = 2e^{-t/2}\,\text{V}$. ($Q(0) = 10^{-6}\,\text{C}$, $I_2(0) = 0\,\text{A}$.)

4. Find $Q(t)$, $I(t)$, $I_1(t)$, and $I_2(t)$ in the two-loop L-R-C circuit with $L = 1\,\text{H}$, $R_1 = 2\,\Omega$, $R_2 = 1\,\Omega$, $C = 1/2\,\text{F}$, and
   (a) $E(t) = 90\,\text{V}$; (b)$E(t) = 90 \sin t\,\text{V}$. ($Q(0) = 0\,\text{C}$, $I_2(0) = 0\,\text{A}$.)

5. Find $Q(t)$, $I(t)$, $I_1(t)$, and $I_2(t)$ in the two-loop L-R-C circuit with $L = 1\,\text{H}$, $R_1 = 1\,\Omega$, $R_2 = 3\,\Omega$, $C = 1\,\text{F}$, and
   (a) $E(t) = 0\,\text{V}$, $Q(0) = 10^{-6}$, and $I_2(0) = 0$; (b)$E(t) = 90\,\text{V}$, $Q(0) = 0\,\text{C}$, $I_2(0) = 0\,\text{A}$.

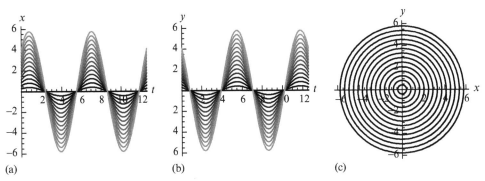

(a)   (b)   (c)

FIGURE 7.5   Graphs associated with Example 7.1.3 (a)-(c).

EXERCISES 7.1

6. Consider the circuit made up of three loops illustrated in the following figure.

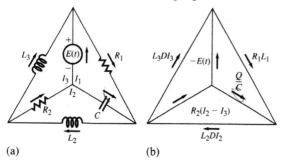

(a)   (b)

In this circuit, the current through the resistor R_2 is $I_2 - I_3$, and the current through the capacitor is $I_1 - I_2$. Using these quantities in the voltage-drop sum equations, model this circuit with the three-dimensional system:

$$-E(t) + R_1 I_1 + \frac{1}{C}Q = 0,$$

$$-\frac{1}{C}Q + L_2\frac{dI_2}{dt} + R_2(I_2 - I_3) = 0,$$

$$E(t) - R_2(I_2 - I_1) + L_3\frac{dI_3}{dt} = 0.$$

Using the relationship $dQ/dt = I_1 - I_2$ and solving the first equation for I_1, show that we obtain the system

$$\frac{dQ}{dt} = -\frac{1}{R_1 C}Q - I_2 + \frac{1}{R_1}E(t),$$

$$\frac{dI_2}{dt} = \frac{1}{L_2 C}Q - \frac{R_2}{L_2}I_2 + \frac{R_2}{L_2}I_3,$$

$$\frac{dI_3}{dt} = \frac{R_2}{L_3}I_2 - \frac{R_2}{L_3}I_3 - \frac{1}{L_3}E(t).$$

In Exercises 7-11, solve the three-loop circuit using the given values and initial conditions.

7. $L_2 = L_3 = 1\,\text{H}$, $R_1 = R_2 = 1\,\Omega$, $C = 1\,\text{F}$, and (a) $E(t) = 0\,\text{V}$; (b) $E(t) = e^{-t}\,\text{V}$. ($Q(0) = 10^{-6}\,\text{C}$, $I_2(0) = I_3(0) = 0\,\text{A}$.)

8. $L_2 = L_3 = 1\,\text{H}$, $R_1 = R_2 = 1\,\Omega$, $C = 1\,\text{F}$, and (a) $E(t) = 90\,\text{V}$; (b) $E(t) = 90\sin t\,\text{V}$. ($Q(0) = 0\,\text{C}$, $I_2(0) = I_3(0) = 0\,\text{A}$.)

9. $L_2 = L_3 = 1\,\text{H}$, $R_1 = R_2 = 1\,\Omega$, $C = 1\,\text{F}$, and (a) $E(t) = 90\,\text{V}$; (b) $E(t) = 90\sin t\,\text{V}$. ($Q(0) = 0\,\text{C}$, $I_2(0) = 1\,\text{A}$, $I_3(0) = 0\,\text{A}$.)

10. $L_2 = 3\,\text{H}$, $L_3 = 1\,\text{H}$, $R_1 = R_2 = 1\,\Omega$, $C = 1\,\text{F}$, and (a) $E(t) = 0\,\text{V}$; (b) $E(t) = 90\sin t\,\text{V}$. ($Q(0) = 0\,\text{C}$, $I_2(0) = 1\,\text{A}$, $I_3(0) = 0\,\text{A}$.)

11. $L_2 = 4\,\text{H}$, $L_3 = 1\,\text{H}$, $R_1 = R_2 = 1\,\Omega$, $C = 1\,\text{F}$, and (a) $E(t) = 90\,\text{V}$; (b) $E(t) = 90\sin t\,\text{V}$. ($Q(0) = 0\,\text{C}$, $I_2(0) = 1\,\text{A}$, $I_3(0) = 0\,\text{A}$.)

Show that the system of differential equations that models the four-loop circuit shown in the following figure

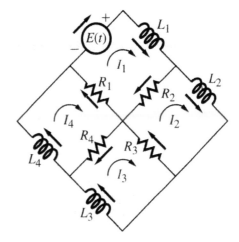

is

$$L_1\frac{dI_1}{dt} = -(R_1 + R_2)I_1 + R_2 I_2 + R_1 I_4 + E(t),$$

$$L_2\frac{dI_2}{dt} = R_2 I_1 - (R_2 + R_3)I_2 + R_3 I_3,$$

$$L_3\frac{dI_3}{dt} = R_3 I_2 - (R_3 + R_4)I_3 + R_4 I_4,$$

$$L_4\frac{dI_4}{dt} = R_1 I_1 + R_4 I_3 - (R_1 + R_4)I_4.$$

In Exercises 13-20, transform the second-order equation to a system of first-order equations and classify the system as unoverdamped, underdamped, or critically damped by finding the eigenvalues of the corresponding system. Also classify the equilibrium point $(0,0)$.

13. $x'' + 9x = 0$.

14. $x'' + 16x = 0$.

15. $x'' + 10x' + 9x = 0$.

16. $x'' + 4x' + 5x = 0$.

17. $x'' + 10x' + 50x = 0$.

18. $x'' + 4x' + 13x = 0$.

19. $x'' + 10x' + 25x = 0$.

20. $x'' + 6x' + 9x = 0$.

21. Solve Exercises 13, 15, and 17 with the initial conditions $x(0) = 1$, $x'(0) = y(0) = 0$.

22. Solve Exercises 17-20 with the initial conditions $x(0) = 0$, $x'(0) = y(0) = 1$.

23. Find the eigenvalues for the spring-mass system (7.4). How do these values relate to overdamping, critical damping, and underdamping?

24. (a) Find the equilibrium point of the spring-mass system (7.4). (b) Find restrictions on m, c, and k to classify this point as a center, stable node, or stable spiral. (c) Can the equilibrium point be unstable for any choice of the positive constants m, c, and k? Is a saddle possible?

Use a graphing device to graph the solutions to Exercises 25-28 simultaneously and parametrically. Also determine the limit of these solutions as $t \to \infty$.

25. Solve the one-loop L-R-C circuit with $L = 1\,\mathrm{H}$, $R = 40\,\Omega$, $C = 1/250\,\mathrm{F}$, and $E(t) = 120 \sin t\,\mathrm{V}$. ($Q(0) = 0\,\mathrm{C}$, $I(0) = 0\,\mathrm{A}$.)

26. Solve the one-loop L-R-C circuit with $L = 4\,\mathrm{H}$, $R = 80\,\Omega$, $C = 2/25\,\mathrm{F}$, and $E(t) = 120e^{-t} \sin t\,\mathrm{V}$. ($Q(0) = 10^{-6}\,\mathrm{C}$, $I(0) = 0\,\mathrm{A}$.)

27. Solve the two-loop L-R-C circuit with $L = 1\,\mathrm{H}$, $R_1 = R_2 = 40\,\Omega$, $C = 1/250\,\mathrm{F}$, and $E(t) = 220 \cos t\,\mathrm{V}$. ($Q(0) = 0\,\mathrm{C}$, $I_2(0) = 0\,\mathrm{A}$.)

28. Solve the two-loop L-R-C circuit with $L = 1\,\mathrm{H}$, $R_1 = 40\,\Omega$, $R_2 = 80\,\Omega$, $C = 1/250\,\mathrm{F}$, and $E(t) = 150e^{-t} \cos t\,\mathrm{V}$. ($Q(0) = 10^{-6}\,\mathrm{C}$, $I_2(0) = 0\,\mathrm{A}$.)

29. Use the system derived in Exercise 6 to solve the three-loop circuit shown in the exercise if $R_1 = R_2 = 2\,\Omega$, $L_1 = L_2 = L_3 = 1\,\mathrm{H}$,

$C = 1\,\mathrm{F}$, $E(t) = 90\,\mathrm{V}$, and the initial conditions are $Q(0) = I_2(0) = I_3(0) = 0$. Find $I_1(t)$ and determine $\lim_{t \to \infty} Q(t)$, $\lim_{t \to \infty} I_1(t)$, $\lim_{t \to \infty} I_2(t)$, and $\lim_{t \to \infty} I_3(t)$.

30. Solve the system of differential equations to find the displacement of the spring-mass system if $m = 1$, $c = 1$, and $k = 1/2$. Graph several solutions in the phase plane for this system. How is the equilibrium point $(0,0)$ classified?

31. Solve the system of differential equations to find the displacement of the spring-mass system given that $m = 1$, $c = 2$, and $k = 3/4$. Graph several solutions in the phase plane for this system. How is the equilibrium point $(0,0)$ classified?

7.2 DIFFUSION AND POPULATION PROBLEMS WITH FIRST-ORDER LINEAR SYSTEMS

- DIFFUSION THROUGH A MEMBRANE
- MIXTURE PROBLEMS
- POPULATION PROBLEMS

Diffusion Through a Membrane

Solving problems to determine the diffusion of a substance (such as glucose or salt) in a medium (such as a blood cell) also lead to first-order systems of linear ODEs. For example, consider the situation shown in Figure 7.6 in which two solutions of a substance are separated by a membrane. The amount of substance that passes through the membrane at any particular time is proportional to the difference in the concentrations of the two solutions. The constant of proportionality, P, is called the *permeability* of the membrane and describes the ability of the substance to permeate the membrane (where $P > 0$). If we let $x(t)$ and $y(t)$ represent the amount of substance at time t on each side of the membrane,

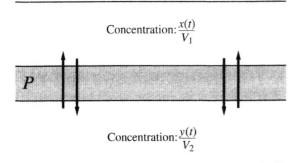

Concentration: $\dfrac{x(t)}{V_1}$

P

Concentration: $\dfrac{y(t)}{V_2}$

FIGURE 7.6 Two solutions separated by a permeable membrane.

and V_1 and V_2 represent the (constant) volume of each solution, respectively, then the system of differential equations is given by

$$\frac{dx}{dt} = \left(\frac{1}{V_2} y - \frac{1}{V_1} x \right)$$

$$\frac{dy}{dt} = \left(\frac{1}{V_1} x - \frac{1}{V_2} y \right), \qquad (7.5)$$

> In this system, if $y(t)/V_2 > x(t)/V_1$, is $dx/dt > 0$ or is $dx/dt < 0$? Also, is $dy/dt > 0$ or is $dy/dt < 0$? Using these results, does the substance move from the side with a lower concentration to that with a higher concentration, or is the opposite true?

where the initial amounts of x and y are given with the initial conditions $x(0) = x_0$ and $y(0) = y_0$. (Notice that the amount of the substance divided by the volume is the concentration of the solution.)

EXAMPLE 7.2.1

Suppose that two salt concentrations of equal volume V are separated by a membrane of permeability P. Given that $P = V$, determine the amount of salt in each concentration at time t if $x(0) = 2$ and $y(0) = 10$.

Solution

In this case, the initial-value problem (IVP) that models the situation is

$$\frac{dx}{dt} = y - x, \quad \frac{dy}{dt} = x - y, \quad x(0) = 2, y(0) = 10.$$

The eigenvalues of $\mathbf{A} = \begin{pmatrix} -1 & 1 \\ 1 & -1 \end{pmatrix}$ are $\lambda_1 = 0$ and $\lambda_2 = -2$. Corresponding eigenvectors are found to be $\mathbf{v}_1 = \begin{pmatrix} 1 \\ 1 \end{pmatrix}$ and $\mathbf{v}_2 = \begin{pmatrix} -1 \\ 1 \end{pmatrix}$, respectively, so a general solution is

$$\mathbf{X}(t) = \begin{pmatrix} x(t) \\ y(t) \end{pmatrix} = c_1 \begin{pmatrix} 1 \\ 1 \end{pmatrix} + c_2 \begin{pmatrix} -1 \\ 1 \end{pmatrix} e^{-2t}$$

$$= \begin{pmatrix} c_1 - c_2 e^{-2t} \\ c_1 + c_2 e^{-2t} \end{pmatrix}.$$

> In Example 7.2.1, if $x(0) = x_0$ and $y(0) = y_0$, does $\lim_{t \to \infty} x(t) = \lim_{t \to \infty} y(t) = \frac{1}{2}(x_0 + y_0)$?

Because $\mathbf{X}(0) = \begin{pmatrix} c_1 - c_2 \\ c_1 + c_2 \end{pmatrix} = \begin{pmatrix} 2 \\ 10 \end{pmatrix}$, we have that $c_1 = 6$ and $c_2 = 4$, so the solution to the IVP is $\mathbf{X}(t) = \begin{pmatrix} x(t) \\ y(t) \end{pmatrix} = \begin{pmatrix} 6 - 4e^{-2t} \\ 6 + 4e^{-2t} \end{pmatrix}$. Notice that $\lim_{t \to \infty} x(t) = \lim_{t \to \infty}(6 - 4e^{-2t}) = 6$ and $\lim_{t \to \infty} y(t) = \lim_{t \to \infty}(6 + 4e^{-2t}) = 6$, which is the average value of the two initial amounts.

Mixture Problems

Consider the interconnected tanks that are shown in Figure 7.7 in which a solution with a given concentration of some substance (like salt) is allowed to flow according to the given information. Let $x(t)$ and $y(t)$ represent the amount of the substance in Tank 1 and Tank 2, respectively. Using this information, we set up two differential equations to describe the rate at which x and

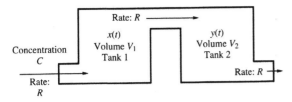

FIGURE 7.7 Illustrating a mixture problem for two interconnected tanks.

y change with respect to time. Notice that the rate at which liquid flows into each tank equals the rate at which it flows out, so the volume of liquid in each tank remains constant. If we consider Tank 1, we can determine a first-order differential equation for dx/dt with

$$\frac{dx}{dt} = (\text{Rate at which substance enters Tank 1})$$
$$- (\text{Rate at which substance leaves Tank 1}),$$

where the rate at which the substance enters Tank I is $R\,\text{gal/min} \times C\,\text{lb/gal} = RC\,\text{lb/min}$, and the rate at which it leaves is $R\,\text{gal/min} \times x/V_1\,\text{lb/gal} = Rx/V_1\,\text{lb/min}$, where x/V_1 is the substance concentration in Tank 1. Therefore, $dx/dt = RC - (R/V_1)x$. Similarly, we find dy/dt to be $\frac{dy}{dt} = (R/V_1)x - (R/V_2)y$. We use the initial conditions $x(0) = x_0$ and $y(0) = y_0$ to solve the nonhomogeneous system

$$\frac{dx}{dt} = RC - \frac{R}{V_1}x, \qquad (7.6)$$
$$\frac{dy}{dt} = \frac{R}{V_1}x - \frac{R}{V_2}y,$$

for $x(t)$ and $y(t)$.

In deriving this system of equations, we used rates in gal/min and concentrations in lb/gal. In general, rates are given by (volume of liquid)/(time) and (concentration of substance)/(volume of liquid).

EXAMPLE 7.2.2

Determine the amount of salt in each tank in Figure 7.7 if $V_1 = V_2 = 500\,\text{gal}$, $R = 5\,\text{gal/min}$, $C = 3\,\text{lb/gal}$, $x_0 = 50\,\text{lb}$ and $y_0 = 100\,\text{lb}$.

Solution

In this case, the IVP is

$$\frac{dx}{dt} = 5 \times 3 - \frac{5}{500}x = 15 - \frac{1}{100}x$$
$$\frac{dy}{dt} = \frac{5}{500}x - \frac{5}{500}y = \frac{1}{100}x - \frac{1}{100}y,$$
$$x(0) = 50, y(0) = 100,$$

which in matrix form is $\mathbf{X}' = \begin{pmatrix} -1/100 & 0 \\ 1/100 & -1/100 \end{pmatrix}\mathbf{X} + \begin{pmatrix} 15 \\ 0 \end{pmatrix} = \mathbf{AX} + \mathbf{F}(t)$, $\mathbf{X}(0) = \begin{pmatrix} 50 \\ 100 \end{pmatrix}$. The matrix $\mathbf{A}$ has the repeated eigenvalue $\lambda_{1,2} = -1/100$, for which we can find only one (linearly independent) eigenvector, $\mathbf{v}_1 = \begin{pmatrix} 0 \\ 1 \end{pmatrix}$. Therefore, one solution of the corresponding homogenous system is $\mathbf{X}_1(t) = \begin{pmatrix} 0 \\ 1 \end{pmatrix}e^{-t/100}$, and after some work a second linearly independent solution is found to be $\mathbf{X}_2(t) = \left[\begin{pmatrix} 0 \\ 1 \end{pmatrix}t + \begin{pmatrix} 100 \\ 0 \end{pmatrix}\right]e^{-t/100}$, so a general solution of the corresponding homogeneous system is

$$\mathbf{X}_h(t) = c_1\begin{pmatrix} 0 \\ 1 \end{pmatrix}e^{-t/100} + c_2\left[\begin{pmatrix} 0 \\ 1 \end{pmatrix}t + \begin{pmatrix} 100 \\ 0 \end{pmatrix}\right]e^{-t/100}$$
$$= \begin{pmatrix} 100c_2e^{-t/100} \\ c_1e^{-t/100} + c_2te^{-t/100} \end{pmatrix}.$$

Notice that $\mathbf{F}(t) = \begin{pmatrix} 15 \\ 0 \end{pmatrix}$ is not a solution to the corresponding homogeneous system, so with the method of undetermined coefficients we assume a particular solution of the nonhomogeneous system has the form $\mathbf{X}_p(t) = \mathbf{a} = \begin{pmatrix} a_1 \\ a_2 \end{pmatrix}$ and substitute this vector-valued function into the nonhomogeneous equation $\mathbf{X}' = \mathbf{AX} + \mathbf{F}(t)$. This yields

$$\begin{pmatrix} 0 \\ 0 \end{pmatrix} = \begin{pmatrix} -1/100 & 0 \\ 1/100 & -1/100 \end{pmatrix} \begin{pmatrix} a_1 \\ a_2 \end{pmatrix} + \begin{pmatrix} 15 \\ 0 \end{pmatrix}$$

$$= \begin{pmatrix} -a_1/100 + 15 \\ a_1/100 - a_2/100 \end{pmatrix}$$

with solution $a_1 = 1500$ and $a_2 = 1500$. Therefore,
$\mathbf{X}_p(t) = \begin{pmatrix} 1500 \\ 1500 \end{pmatrix}$ and

$$\mathbf{X}(t) = \mathbf{X}_h(t) + \mathbf{X}_p(t)$$

$$= \begin{pmatrix} 100c_2 e^{-t/100} + 1500 \\ c_1 e^{-t/100} + c_2 t e^{-t/100} + 1500 \end{pmatrix}.$$

Application of the initial conditions then gives us

$$\mathbf{X}(0) = \begin{pmatrix} 100c_2 + 1500 \\ c_1 + 1500 \end{pmatrix} = \begin{pmatrix} 50 \\ 100 \end{pmatrix}, \text{ so } c_1 = -1400$$

and $c_2 = -1450/100 = -29/2$. The solution to the IVP is

$$\mathbf{X}(t) = \begin{pmatrix} x(t) \\ y(t) \end{pmatrix}$$

$$= \begin{pmatrix} -1450 e^{-t/100} + 1500 \\ -1400 e^{-t/100} - \frac{29}{2} t e^{-t/100} + 1500 \end{pmatrix}.$$

Notice that $\lim_{t \to \infty} x(t) = \lim_{t \to \infty} y(t) = 1500$, which means that the amount of salt in each tank tends toward a value of 1500 lb.

In Example 7.2.2, is there a value of t for which $x(t) = y(t)$? If so, what is this value? Which function increases most rapidly for smaller values of t?

Population Problems

In Chapter 3, we discussed population problems that were based on the simple principle that the rate at which a population grows (or decays) is proportional to the number present in the population at any time t. This idea can be extended to problems involving more than one

population and leads to systems of ODEs. We illustrate several situations through the following examples. Note that in each problem we determine the rate at which a population P changes with the equation

$$\frac{dP}{dt} = (\text{rate entering}) - (\text{rate leaving}).$$

We begin by determining the population in two neighboring territories where the populations x and y of the territories depend on several factors. The birth rate of x is a_1, and that of y is b_1. The rate at which citizens of x move to y is a_2, and that at which citizens move from y to x is b_2. After assuming that the mortality rate of each territory is disregarded, we determine the respective populations of these two territories for any time t.

Using the simple principles of previous examples, the rate at which population $x = x(t)$ changes is

$$\frac{dx}{dt} = a_1 x - a_2 x + b_2 y = (a_1 - a_2)x + b_2 y,$$

and the rate at which the population $y = y(t)$ changes is

$$\frac{dy}{dt} = b_1 y - b_2 y + a_2 x = (b_1 - b_2)y + a_2 x.$$

Therefore, the system of equations that is solved is

$$\frac{dx}{dt} = (a_1 - a_2)x + b_2 y,$$

$$\frac{dy}{dt} = a_2 x + (b_1 - b_2)y, \qquad (7.7)$$

where the initial populations of the two territories are $x(0) = x_0$ and $y(0) = y_0$.

EXAMPLE 7.2.3

Determine the populations $x(t)$ and $y(t)$ in each territory if $a_1 = 5, a_2 = 4, b_1 = 5$, and $b_2 = 3$, given that $x(0) = 14$ and $y(0) = 7$.

Solution

In this case, the IVP that models the situation is

$$\begin{array}{l} dx/dt = x + 3y \\ dy/dt = 4x + 2y \end{array}, \quad x(0) = 14 \quad \text{and} \quad y(0) = 7.$$

The eigenvalues of $\mathbf{A} = \begin{pmatrix} 1 & 3 \\ 4 & 2 \end{pmatrix}$ are $\lambda_1 = -2$ and $\lambda_2 = 5$. Corresponding eigenvectors are found to be $\mathbf{v}_1 = \begin{pmatrix} 1 \\ -1 \end{pmatrix}$ and $\mathbf{v}_2 = \begin{pmatrix} 3 \\ 4 \end{pmatrix}$, so a general solution is

$$\mathbf{X}(t) = \begin{pmatrix} x(t) \\ y(t) \end{pmatrix} = c_1 \begin{pmatrix} 1 \\ -1 \end{pmatrix} e^{-2t} + c_2 \begin{pmatrix} 3 \\ 4 \end{pmatrix} e^{5t}$$

$$= \begin{pmatrix} c_1 e^{-2t} + 3c_2 e^{5t} \\ -c_1 e^{-2t} + 4c_2 e^{5t} \end{pmatrix}.$$

Application of the initial condition $\mathbf{X}(0) = \begin{pmatrix} x(0) \\ y(0) \end{pmatrix} = \begin{pmatrix} 14 \\ 7 \end{pmatrix}$ yields the system $\begin{cases} c_1 + 3c_2 = 14 \\ -c_1 + 4c_2 = 7 \end{cases}$, so $c_1 = 5$ and $c_2 = 3$. Therefore, the solution is

$$\mathbf{X}(t) = \begin{pmatrix} x(t) \\ y(t) \end{pmatrix} = \begin{pmatrix} 5e^{-2t} + 9e^{5t} \\ -5e^{-2t} + 12e^{5t} \end{pmatrix}.$$ We graph these two population functions in Figure 7.8(a). In Figure 7.8(b), we graph several solutions to the system of differential equations for various initial conditions in the phase plane. As we can see, all solutions move away from the origin in the direction of the eigenvector $\mathbf{v}_2 = \begin{pmatrix} 3 \\ 4 \end{pmatrix}$, which corresponds to the positive eigenvalue $\lambda_2 = 5$.

Population problems that involve more than two neighboring populations can be solved with a system of differential equations as well. Suppose that the population of three neighboring territories, x, y, and z, depends on several factors. The birth rates of x, y, and z are a_1, b_1, and c_1, respectively. The rate at which citizens of x move to y is a_2, and that at which citizens move from x to z is a_3. Similarly, the rate at which citizens of y move to x is b_2, and that at which citizens move from y to z is b_3. Also, the rate at which citizens of z move to x is c_2, and that at which citizens move from z to y is c_3:

From	To			Birth Rate
	x	y	z	
x		a_2	a_3	a_1
y	b_2		b_3	b_1
z	c_2	c_3		c_1

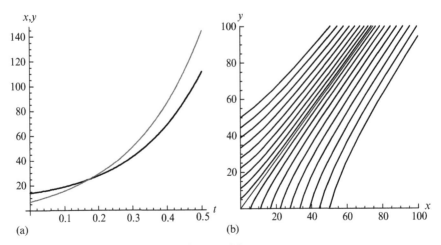

(a)   (b)

FIGURE 7.8   (a) Identify $x(t)$ and $y(t)$. (b) Various solutions of the system.

If the mortality rate of each territory is ignored in the model, we can determine the respective populations of the three territories for any time t.

The system of equations to determine $x(t)$, $y(t)$, and $z(t)$ is similar to that derived in the previous example. The rate at which population x changes is

$$\frac{dx}{dt} = a_1 x - a_2 x - a_3 x + b_2 y + c_2 z$$
$$= (a_1 - a_2 - a_3)x + b_2 y + c_2 z,$$

and the rate at which population y changes is

$$\frac{dy}{dt} = b_1 y - b_2 y - b_3 y + a_2 x + c_3 z$$
$$= a_2 x + (b_1 - b_2 - b_3)y + c_3 z,$$

and that of z is

$$\frac{dz}{dt} = c_1 z - c_2 z - c_3 z + a_3 x + b_3 y$$
$$= a_3 x + b_3 y + (c_1 - c_2 - c_3)z.$$

We must solve the 3×3 system

$$\frac{dx}{dt} = (a_1 - a_2 - a_3)x + b_2 y + c_2 z,$$
$$\frac{dy}{dt} = a_2 x + (b_1 - b_2 - b_3)y + c_3 z, \qquad (7.8)$$
$$\frac{dz}{dt} = a_3 x + b_3 y + (c_1 - c_2 - c_3)z,$$

where the initial populations $x(0) = x_0$, $y(0) = y_0$, and $z(0) = z_0$ are given.

EXAMPLE 7.2.4

Determine the population of the three territories if $a_1 = 3$, $a_2 = 0$, $a_3 = 2$, $b_1 = 4$, $b_2 = 2$, $b_3 = 1$, $c_1 = 5$, $c_2 = 3$, and $c_3 = 0$ if $x(0) = 50$, $y(0) = 60$, and $z(0) = 25$.

Solution

In this case, the system of differential equations is

$$\begin{aligned} dx/dt &= x + 2y + 3z \\ dy/dt &= y \\ dz/dt &= 2x + y + 2z \end{aligned} \qquad \text{or} \qquad \begin{pmatrix} x \\ y \\ z \end{pmatrix}' = \begin{pmatrix} 1 & 2 & 3 \\ 0 & 1 & 0 \\ 2 & 1 & 2 \end{pmatrix} \begin{pmatrix} x \\ y \\ z \end{pmatrix},$$

where $' = d/dt$. Because the characteristic polynomial for the coefficient matrix of this system is

$$(-1)^3 \begin{vmatrix} 1-\lambda & 2 & 3 \\ 0 & 1-\lambda & 0 \\ 2 & 1 & 2-\lambda \end{vmatrix} = -(1-\lambda)^2(2-\lambda) - 3 \times (-2) \times (1-\lambda)$$

$$= (\lambda - 1)(\lambda + 1)(\lambda - 4) = 0,$$

the eigenvalues are $\lambda_1 = 4$, $\lambda_2 = 1$, and $\lambda_3 = -1$, with corresponding eigenvectors $\mathbf{v}_1 = \begin{pmatrix} 1 \\ 0 \\ 1 \end{pmatrix}$, $\mathbf{v}_2 = \begin{pmatrix} 1 \\ -6 \\ 4 \end{pmatrix}$, and $\mathbf{v}_3 = \begin{pmatrix} -3 \\ 0 \\ 2 \end{pmatrix}$, respectively. A general solution is

$$\begin{pmatrix} x(t) \\ y(t) \\ z(t) \end{pmatrix} = c_1 \begin{pmatrix} 1 \\ 0 \\ 1 \end{pmatrix} e^{4t} + c_2 \begin{pmatrix} 1 \\ -6 \\ 4 \end{pmatrix} e^t + c_3 \begin{pmatrix} -3 \\ 0 \\ 2 \end{pmatrix} e^{-t}$$

$$= \begin{pmatrix} c_1 e^{4t} + c_2 e^t - 3c_3 e^{-t} \\ -6c_2 e^t \\ c_1 e^{4t} + 4c_2 e^t + 2c_3 e^{-t} \end{pmatrix}.$$

Using the initial conditions, we find that

$$\begin{cases} c_1 + c_2 - 3c_3 = 50 \\ -6c_2 = 60 \\ c_1 + 4c_2 + 2c_3 = 25 \end{cases}, \text{ so } c_1 = 63, c_2 = -10,$$

and $c_3 = 1$. Therefore, the solution to the IVP is

$$\begin{pmatrix} x(t) \\ y(t) \\ z(t) \end{pmatrix} = \begin{pmatrix} 63e^{4t} - 10e^t - 3e^{-t} \\ 60e^t \\ 63e^{4t} - 40e^t + 2e^{-t} \end{pmatrix}. \text{ We graph these}$$

three population functions in Figure 7.9. Notice that although population y is initially greater than populations x and z, these populations increase at a much higher rate than does y.

In Example 7.2.4, does population y approach a limit or do all three populations increase exponentially?

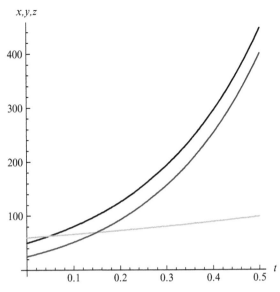

FIGURE 7.9   Identify $x(t)$, $y(t)$, and $z(t)$.

EXERCISES 7.2

In Exercises 1-6, solve the diffusion problem with one permeable membrane with the indicated initial conditions and parameter values. Find the limiting concentration of each solution.

1. $P = 0.5$, $V_1 = V_2 = 1$, $x(0) = 1$, $y(0) = 2$.
2. $P = 0.5$, $V_1 = V_2 = 1$, $x(0) = 2$, $y(0) = 1$.
3. $P = 2$, $V_1 = 4$, $V_2 = 2$, $x(0) = 0$, $y(0) = 4$.
4. $P = 2$, $V_1 = 1/2$, $V_2 = 1/4$, $x(0) = 8$,
$\quad y(0) = 0$.
5. $P = 6$, $V_1 = 2$, $V_2 = 8$, $x(0) = 4$, $y(0) = 1$.
6. $P = 6$, $V_1 = 2$, $V_2 = 8$, $x(0) = 1$, $y(0) = 4$.

In Exercises 7-8, use the tanks shown in Figure 7.7 with $R = 4\,\text{gal/min}$, $C = 1/2$ lb/gal, $V_1 = V_2 = 20\,\text{gal}$, and the given initial conditions. (a) Determine $x(t)$ and $y(t)$. (b) Find $\lim_{t\to\infty} x(t)$ and $\lim_{t\to\infty} y(t)$. (c) Does one of the tanks contain more of the substance (like salt) than the other tank for all values of t?

7. $x(0) = y(0) = 0$.
8. $x(0) = 0$, $y(0) = 2$.

9. Suppose that pure water is pumped into Tank 1 (in Figure 7.7) at a rate of 4 gal/min (i.e., $R = 4$, $C = 0$) and that $V_1 = V_2 = 20$ gal. Determine the amount of the substance (like salt) in each tank at time t if $x(0) = y(0) = 4$. Calculate $\lim_{t\to\infty} x(t)$ and $\lim_{t\to\infty} y(t)$. Which function decreases more rapidly?

10. If $x(0) = y(0) = 0$ in Exercise 9, how much of the substance (like salt) is in each tank at any time t?

11. Use the tanks in Figure 7.7 with $R = 5\,\text{gal/min}$, $C = 3\,\text{lb/gal}$, $V_1 = 100\,\text{gal}$, $V_2 = 50\,\text{gal}$, and the initial conditions $x(0) = y(0) = 0$. (a) Determine $x(t)$ and $y(t)$. (b) Find $\lim_{t\to\infty} x(t)$ and $\lim_{t\to\infty} y(t)$.

12. Solve Exercise 11 with $V_1 = 50$ and $V_2 = 100$. How do the limiting values of $x(t)$ and $y(t)$ compare to those in Exercise 11?

13. Consider the tanks shown in the following figure, where $R_1 = 3\,\text{L/min}$, $R_2 = 4\,\text{L/min}$, $R_3 = 1\,\text{L/min}$, $C = 1\,\text{kg/L}$, and $V_1 = V_2 = 50$ L.

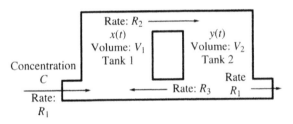

If $x(0) = y(0) = 0$, determine the amount of the substance (like salt) in each tank at time t. Find $\lim_{t\to\infty} x(t)$ and $\lim_{t\to\infty} y(t)$. Is there a time (other than $t = 0$) at which each tank contains the same amount of salt?

14. Solve the tank problem described in Exercise 13 using the initial conditions $x(0) = 0$ and $y(0) = 6$. For how many values of t do the functions $x(t)$ and $y(t)$ agree?

15. Find the amount of the substance (like salt) in each tank in the following figure if $R = 1\,\text{gal/min}$, $C = 1\,\text{lb/gal}$, $V_1 = 1\,\text{gal}$, $V_2 = 1/2\,\text{gal}$, $x(0) = 2$, and $y(0) = 4$. What is the maximum amount of salt (at any value of t) in each tank? Find $\lim_{t\to\infty} x(t)$ and $\lim_{t\to\infty} y(t)$.

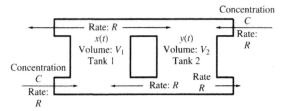

16. Solve the tank problem described in Exercise 15 using the initial conditions $x(0) = 4$ and $y(0) = 0$. How do these functions differ from those in Exercise 15?

17. Consider the three tanks in the following figure in which the amount of salt in Tanks 1, 2, and 3 is given by $x(t)$, $y(t)$, and $z(t)$, respectively. Find $x(t)$, $y(t)$, and $z(t)$ if $R = 5\,\text{gal}/\text{min}$, $C = 2\,\text{lb}/\text{gal}$, $V_1 = V_2 = V_3 = 50\,\text{gal}$, and $x(0) = y(0) = z(0) = 0$. Find $\lim_{t\to\infty} x(t)$, $\lim_{t\to\infty} y(t)$, and $\lim_{t\to\infty} z(t)$.

18. Solve the problem described in Exercise 17 using the unequal volumes $V_1 = 100$, $V_2 = 50$, and $V_3 = 25$. How do the limiting values of $x(t)$, $y(t)$, and $z(t)$ differ from those in Exercise 17?

In Exercises 19-22, solve the IVP using the given parameters to find the population in two neighboring territories. Do either of the populations approach a finite limit? If so, what is the limit?

19. $a_1 = 10, a_2 = 9, b_1 = 2, b_2 = 1; x(0) = 10, y(0) = 20.$

20. $a_1 = 4, a_2 = 4, b_1 = 1, b_2 = 1; x(0) = 4, y(0) = 4.$

21. $a_1 = 2, a_2 = 0, b_1 = 2, b_2 = 3; x(0) = 5, y(0) = 10.$

22. $a_1 = 1, a_2 = 1, b_1 = 1, b_2 = 1; x(0) = 5, y(0) = 10.$

In Exercises 23-26, solve the IVP using the given parameters to find the population in three neighboring territories. Which population is largest at $t = 1$?

23. $a_1 = 10, a_2 = 6, a_3 = 7, b_1 = 6, b_2 = 3, b_3 = 3, c_1 = 5, c_2 = 7, c_3 = 1; x(0) = 17, y(0) = 0, z(0) = 34.$

24. $a_1 = 2, a_2 = 1, a_3 = 4, b_1 = 6, b_2 = 4, b_3 = 5, c_1 = 2, c_2 = 8, c_3 = 4; x(0) = 0, y(0) = 4, z(0) = 2.$

25. $a_1 = 7, a_2 = 2, a_3 = 4, b_1 = 7, b_2 = 5, b_3 = 8, c_1 = 7, c_2 = 1, c_3 = 2; x(0) = 8, y(0) = 2, z(0) = 0.$

26. $a_1 = 7, a_2 = 2, a_3 = 4, b_1 = 7, b_2 = 5, b_3 = 8, c_1 = 7, c_2 = 1, c_3 = 2; x(0) = 0, y(0) = 0, z(0) = 16.$

27. Suppose that a radioactive substance X decays into another unstable substance Y, which in turn decays into a stable substance Z. Show that we can model this situation through the system of differential equations
$$\begin{cases} dx/dt = -ax \\ dy/dt = ax - by \\ dz/dt = by \end{cases}$$, where a and b are positive constants. (Assume that one unit of X decomposes into one unit of Y, and one unit of Y decomposes into one unit of Z.)

28. Solve the system of differential equations in Exercise 27 if $a \neq b$, $x(0) = x_0$, $y(0) = y_0$, and $z(0) = z_0$. Find $\lim_{t\to\infty} x(t)$, $\lim_{t\to\infty} y(t)$ and $\lim_{t\to\infty} z(t)$.

29. Solve the system of differential equations in Exercise 27 if $a = b$, $x(0) = x_0$, $y(0) = y_0$, and $z(0) = z_0$. Find $\lim_{t\to\infty} x(t)$, $\lim_{t\to\infty} y(t)$ and $\lim_{t\to\infty} z(t)$.

30. In Exercise 27, what is the half-life of substance X?

31. In the reaction described in Exercise 27, show that if k units of X are added per year and h units of Z are removed, then the situation is described with the system

$$\begin{cases} dx/dt = -ax + k \\ dy/dt = ax - by \\ dz/dt = by - h \end{cases}, \text{ where } a, b, k, \text{ and } h \text{ are}$$

positive constants.

32. Solve the system described in Exercise 31 if $a \neq b$, $x(0) = x_0$, $y(0) = y_0$, and $z(0) = z_0$. Find $\lim_{t \to \infty} x(t)$, $\lim_{t \to \infty} y(t)$ and $\lim_{t \to \infty} z(t)$. How do these limits differ from those in Exercise 27? Find $\lim_{t \to \infty} z(t)$ if $k = h$, then find $\lim_{t \to \infty} z(t)$ if $k > h$. Describe the corresponding physical situation.

In Exercises 33-36, solve the radioactive decay model in Exercise 27 with the given parameter values and initial conditions. If $z(t)$ represents the amount (grams) of substance Z after t hours, how many grams of Z are eventually produced?

33. $a = 6$, $b = 1$, $x_0 = 7$, $y_0 = 1$, $z_0 = 8$.
34. $a = 4$, $b = 2$, $x_0 = 10$, $y_0 = 2$, $z_0 = 4$.
35. $a = 4$, $b = 2$, $x_0 = 1$, $y_0 = 1$, $z_0 = 1$.
36. $a = 1$, $b = 4$, $x_0 = 2$, $y_0 = 2$, $z_0 = 2$.

In Exercises 37-40, solve the radioactive decay model in Exercise 31 with $a = 1$, $b = 1$, and the given parameter values and initial conditions. Describe what eventually happens to the amount of each substance.

37. $k = 2$, $h = 1$, $x_0 = 2$, $y_0 = 1$, $z_0 = 2$.
38. $k = 0$, $h = 10$, $x_0 = 4$, $y_0 = 2$, $z_0 = 1$.
39. $k = 10$, $h = 0$, $x_0 = 8$, $y_0 = 2$, $z_0 = 2$.
40. $k = 0$, $h = 5$, $x_0 = 1$, $y_0 = 10$, $z_0 = 5$.
41. Solve the IVP to find the concentration of a substance on each side of a permeable membrane modeled by the system

$$\begin{aligned} \frac{dx}{dt} &= P\left(\frac{1}{V_2}y - \frac{1}{V_1}x\right) \\ \frac{dy}{dt} &= P\left(\frac{1}{V_1}x - \frac{1}{V_2}y\right) \end{aligned}, \quad x(0) = a, y(0) = b.$$

(a) Find $\lim_{t \to \infty} x(t)$ and $\lim_{t \to \infty} y(t)$.
(b) Determine a condition so that $x(t) > y(t)$ as $t \to \infty$. When does $\lim_{t \to \infty} x(t) = \lim_{t \to \infty} y(t)$?

(c) Find a condition so that $x(t)$ is an increasing function. Find a condition so that $y(t)$ is an increasing function. Can these functions increase simultaneously?
(d) If $V_1 = V_2$ and $a = b$, describe what eventually happens to $x(t)$ and $y(t)$.

42. Investigate the effect that the membrane permeability has on the diffusion of a substance. Suppose that $V_1 = V_2 = 1$, $x(0) = 1$, $y(0) = 2$, and (a) $P = 0.25$, (b) $P = 0.5$, (c) $P = 1.0$, and (d) $P = 2.0$. Graph the solution in each case both parametrically and simultaneously. Describe the effect that the value of P has on the corresponding solution.

43. Solve the IVP

$$\begin{aligned} dx/dt &= (a_1 - a_2)x + b_1 y \\ dy/dt &= a_2 x + (b_1 - b_2)y \end{aligned}, \quad x(0) = x_0, y(0) = y_0$$

Are there possible parameter values so that the functions $x(t)$ and $y(t)$ are periodic? Are there possible parameter values so that the functions $x(t)$ and $y(t)$ experience exponential decay? For what parameter values do the two populations experience exponential growth?

7.3 NONLINEAR SYSTEMS OF EQUATIONS

- BIOLOGICAL SYSTEMS: PREDATOR-PREY INTERACTION
- PHYSICAL SYSTEMS: VARIABLE DAMPING

Several special equations and systems that arise in the study of many areas of applied mathematics can be solved using the techniques of Chapter 6. These include the predator-prey population dynamics problem, the Van-der-Pol equation that models variable damping in a spring-mass system, and the Bonhoeffer-Van-der-Pol (BVP) oscillator.

Alfred James Lotka (1880-1949) American biophysicist, born in Ukraine; wrote first text on mathematical biology.

Biological Systems: Predator-Prey Interaction

Let $x(t)$ and $y(t)$ represent the number of members at time t of the prey and predator populations, respectively. (Examples of such populations include fox-rabbit and shark-seal.) Suppose that the positive constant a is the birth rate of $x(t)$, so that in the absence of the predator $dx/dt = ax$ and that $c > 0$ is the death rate of y, which indicates that $dy/dt = -cy$ in the absence of the prey-population. In addition to these factors, the number of interactions between predator and prey affects the number of members in the two populations. Note that an interaction increases the growth of the predator population and decreases the growth of the prey population, because an interaction between the two populations indicates that a predator overtakes a member of the prey population. To include these interactions in the model, we assume that the number of interactions is directly proportional to the product of $x(t)$ and $y(t)$. Therefore, the rate at which $x(t)$ changes with respect to time is

$$dx/dt = ax - bxy,$$

where $b > 0$. Similarly, the rate at which $y(t)$ changes with respect to time is

$$dy/dt = -cy + dxy,$$

where $d > 0$. These equations and initial conditions form the *Lotka-Volterra problem*

$$\begin{matrix} dx/dt = ax - bxy \\ dy/dt = -cy + dxy \end{matrix}, \quad x(0) = x_0, y(0) = y_0. \quad (7.9)$$

Vito Volterra (May 3, 1860, Ancona, Italy-October 11, 1940, Rome, Italy) During World War I, Volterra was a member of the Italian Air Force. After the War, he returned to the University of Rome and his research interests moved from mathematical physics to mathematical biology.

EXAMPLE 7.3.1

Find and classify the equilibrium points of the Lotka-Volterra system.

Solution

Solving $\begin{cases} ax - bxy = x(a - by) = 0 \\ -cy + dxy = y(-c + dx) = 0 \end{cases}$ for x

and y we have $x = 0$ or $y = a/b$ from the first equation and $y = 0$ or $x = c/d$ from the second equation. Thus, the equilibrium points are $E_0 = (0,0)$ and $E_A = (c/d, a/b)$. The Jacobian matrix of the

nonlinear system is $J(x,y) = \begin{pmatrix} a - by & -bx \\ dy & -c + dx \end{pmatrix}$.
Evaluated at $E_0 = (0,0)$, the Jacobian is $J(E_0) = \begin{pmatrix} a & 0 \\ 0 & -c \end{pmatrix}$ with eigenvalues $\lambda_1 = -c$ and $\lambda_2 = a$.
Because these eigenvalues are real with opposite
signs, we classify $E_0 = (0,0)$ as a saddle. Similarly, evaluated at $E_A = (c/d, a/b)$, the Jacobian is
$J(E_A) = \begin{pmatrix} 0 & -bc/d \\ ad/b & 0 \end{pmatrix}$, with eigenvalues $\lambda_{1,2} = \pm i\sqrt{ac}$, so the equilibrium point $E_A = (c/d, a/b)$ is
classified as a center.

In Figure 7.10(a), we show several solutions
(which were found using different initial populations) parametrically in the phase plane of this
system with $a = 2, b = 1, c = 3$, and $d = 1$. Notice
that all of the solutions oscillate about the center.
These solutions reveal the relationship between
the two populations: prey, $x(t)$, and predator, $y(t)$.
As we follow one cycle counterclockwise beginning on the left extreme of the cycle, we notice
that as the prey population, $x(t)$, increases, the
predator population, $y(t)$, first slightly decreases
(is that really possible?), and then increases until the predator becomes overpopulated. Then,

because the prey population is too small to supply the predator population, the predator population decreases, which leads to an increase in
the population of the prey. At this point, because
the number of predators becomes too small to
control the prey population, the number in the
prey population becomes overpopulated and the
cycle repeats itself, as illustrated in Figure 7.10(b).

 *How does the period of a solution with
an initial point near the equilibrium point
$(c/d, a/b)$ compare to that of a solution with an
initial point that is not located near this point?*

Physical Systems: Variable Damping

In some physical systems, energy is fed into
the system when there are small oscillations,
and energy is taken from the system when there
are large oscillations. This indicates that the system undergoes "negative damping" for small
oscillations and "positive damping" for large

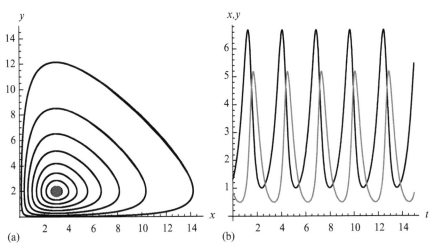

FIGURE 7.10 (a) Typical solutions of the Lotka-Volterra system-x versus y. (b) A typical solution to the Lotka-Volterra system, x (in dark red; dark gray in print versions) and y as functions of t.

oscillations. A differential equation that models this situation is *Van-der-Pol's equation*

$$x'' + \mu\left(x^2 - 1\right) x' + x = 0, \qquad (7.10)$$

Balthasar Van-der-Pol (1889-1959) Dutch applied mathematician and engineer.

where μ is a positive constant. We can transform this second-order differential equation into a system of first-order differential equations with the substitution $x' = y$. Hence $y' = x'' = -x - \mu\left(x^2 - 1\right) x' = -x - \mu\left(x^2 - 1\right) y$, so the corresponding system of equations is

$$x' = y,$$
$$y' = -x - \mu\left(x^2 - 1\right) y, \qquad (7.11)$$

which is solved using an initial position $x(0) = x_0$ and an initial velocity $y(0) = x'(0) = y_0$. Notice that $\mu\left(x^2 - 1\right)$ represents the damping coefficient. This system models variable damping because $\mu\left(x^2 - 1\right) < 0$ when $-1 < x < 1$ and $\mu\left(x^2 - 1\right) > 0$ when $|x| > 1$. Therefore, damping is negative for the small oscillations, $-1 < x < 1$, and positive for the large oscillations, $|x| > 1$.

EXAMPLE 7.3.2

Find and classify the equilibrium points of the system of differential equations that is equivalent to Van-der-Pol's equation.

Solution

We find the equilibrium points by solving

$$y = 0,$$
$$-x - \mu\left(x^2 - 1\right) y = 0.$$

From the first equation, we see that $y = 0$. Substitution of $y = 0$ into the second equation yields $x = 0$ as well. Therefore, the (only) equilibrium point is $(0,0)$.

The Jacobian matrix for this system is

$$\mathbf{J}(0,0) = \begin{pmatrix} 0 & 1 \\ -1 - 2\mu xy & -\mu\left(x^2 - 1\right) \end{pmatrix}.$$

At $(0,0)$, we have the matrix $\mathbf{J}(0,0) = \begin{pmatrix} 0 & 1 \\ -1 & \mu \end{pmatrix}$. We find the eigenvalues of $\mathbf{J}(0,0)$ by solving

$$\begin{vmatrix} -\lambda & 1 \\ -1 & -\mu - \lambda \end{vmatrix} = \lambda^2 - \mu\lambda + 1 = 0,$$

> Graph several solution curves in the phase plane for the Van-der-Pol equation if $\mu = 1/10$ and $\mu = 1/1000$. Compare these graphs to the corresponding solution curves for the equation $x'' + x = 0$. Are they similar? Why?

which has roots $\lambda_{1,2} = \frac{1}{2}\left(\mu \pm \sqrt{\mu^2 - 4}\right)$. Notice that if $\mu > 2$, then both eigenvalues are positive and real, so we classify $(0,0)$ as an unstable node. On the other hand, if $0 < \lambda < 2$, the eigenvalues are a complex conjugate pair with a positive real part. Hence $(0,0)$ is an unstable spiral point. (We omit the case when $\mu = 2$ because the eigenvalues are repeated.) In Figure 7.11, we show several curves in the phase plane that begin at different points for various values of μ. In each figure, we see that all of the curves approach a curve called a *limit cycle*. Physically, the fact that the system has a limit cycle indicates that for all oscillations

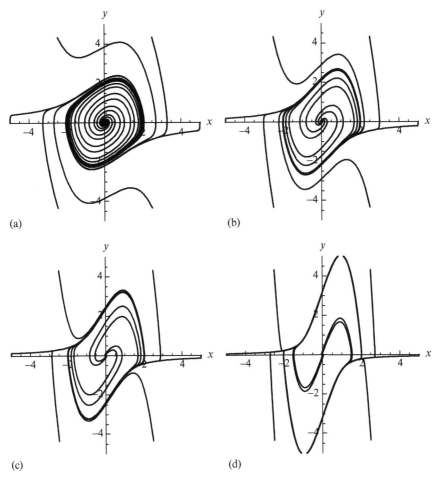

(a)   (b)

(c)   (d)

FIGURE 7.11   From left to right, (a) $\mu = 1/2$, (b) $\mu = 1$, (c) $\mu = 3/2$, (d) $\mu = 3$.

the motion eventually becomes periodic, which is represented by a closed curve in the phase plane.

On the other hand, in Figure 7.12 we graph the solution that satisfies the initial conditions $x(0) = 1$ and $y(0) = 0$ parametrically and individually for various values of μ. Notice that for small values of μ the system more closely approximates that of the harmonic oscillator because the damping coefficient is small. The curves are more circular than for those for larger values of μ.

 EXERCISES 7.3

1. The phase paths of the Lotka-Volterra model (7.9) are given by
$\frac{dy}{dx} = \frac{dy/dt}{dx/dt} = \frac{-cy+dxy}{ax-bxy}$. Solve this separable equation to find an implicit equation of the phase paths.

2. Consider the predator-prey model
$$dx/dt = (a_1 - b_1 x - c_1 y)\,x,$$
$$dy/dt = (-a_2 + c_2 x)\,y,$$

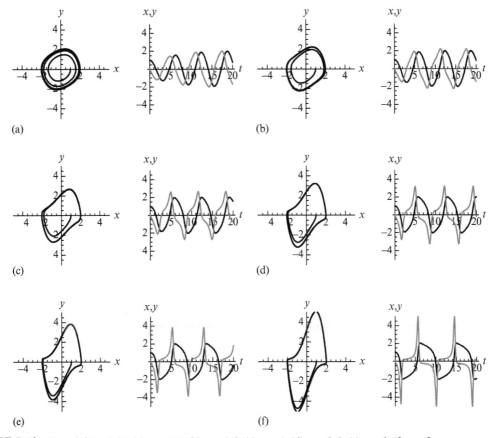

(a)

(b)

(c)

(d)

(e)

(f)

FIGURE 7.12   From left to right, (a) $\mu = 1/4$, (b) $\mu = 1/2$, (c) $\mu = 1$, (d) $\mu = 3/2$, (e) $\mu = 2$, (f) $\mu = 3$.

where $x(t)$ represents the prey population, $y(t)$ the predator population, and the constants a_1, a_2, b_1, c_1, c_2 are all positive. In this model, the term $-b_1 x^2$ represents the interference that occurs when the prey population becomes too large. (a) If $a_1 = 2$, $a_2 = b_1 = c_1 = c_2 = 1$, locate and classify the three equilibrium points. (b) If $a_2 = 2$, $a_1 = b_1 = c_1 = c_2 = 1$, locate and classify the three equilibrium points.

3. The national economy can be modeled with a nonlinear system of differential equations. If I represents the national income, C the rate of consumer spending, and G the rate of government spending, a simple model of the economy is

$$dI/dt = I - aC,$$
$$dC/dt = b(I - C - G),$$

where a and b are constants. (a) Suppose that $a = b = 2$ and $G = k$ (k = constant). Find and classify the equilibrium point. (b) If $a = 2$, $b = 1$, and $G = k$, find and classify the equilibrium point.

4. If $a = b = 2$ and $G = k_1 + k_2 I$ ($k_2 > 0$), show that there is no equilibrium point if $k_2 \geq 1/2$ for the national economy model in Exercise 3. Describe the economy under these conditions.

5. The differential equation $x'' + \sin x = 0$ can be used to describe the motion of a pendulum. In this case, $x(t)$ represents the

displacement from the position $x = 0$. Represent this second-order equation as a system of first-order equations. Find and classify the equilibrium points of this system. Describe the physical significance of these points. Graph several paths in the phase plane of the system.

6. Show that the paths in the phase plane of $x'' + \sin x = 0$ satisfy the first-order equation $dy/dx = -(\sin x)y$ (see Exercise 5). Use separation of variables to show that paths are $y^2 - 2\cos x = C$, where C is a constant. Graph several paths. Do these graphs agree with your result in Exercise 5?

7. (a) Write the equation $x'' + k^2 x = 0$, which models the *simple harmonic oscillator*, as a system of first-order equations. (b) Show that paths in the phase plane satisfy $dy/dx = -k^2 x/y$ (see Exercise 6). (c) Solve the equation in (b) to find the paths. (d) What is the equilibrium point of this system and how is it classified? How do the paths in (c) compare with what you expect to see in the phase plane?

8. Repeat Exercise 7 for the equation $x'' - k^2 x = 0$.

9. Suppose that a satellite is in flight on the line between a planet of mass M_1 and its moon of mass M_2, which are a constant distance R apart. The distance x between the satellite and the planet satisfies the nonlinear second-order equation $x'' = -gM_1 x^{-2} + gM_2(R - x)^{-2}$ where g is the gravitational constant. Transform this equation into a system of first-order equations. Find and classify the equilibrium point of the linearized system.

10. Consider the nonlinear autonomous system

$$dx/dt = -y + x\left(1 - x^2 - y^2\right),$$
$$dy/dt = x + y\left(1 - x^2 - y^2\right).$$

(a) Show that $(0,0)$ is an equilibrium point of this system. Classify $(0,0)$.

(b) Show that $x = r\cos\theta, y = r\sin\theta$ transforms the system to the (uncoupled) system

$$dr/dt = r\left(1 - r^2\right),$$
$$d\theta/dt = 1.$$

(c) Show that this system has solution $r(t) = \left(1 + ae^{-2t}\right)^{-1/2}, \theta(t) = t + b$ so that $x(t) = \left(1 + ae^{-2t}\right)^{-1/2}\cos(t + b)$, $y(t) = \left(1 + ae^{-2t}\right)^{-1/2}\sin(t + b)$. (d) Calculate $\lim_{t\to\infty} x(t)$ and $\lim_{t\to\infty} y(t)$ to show that all solutions approach the circle $x^2 + y^2 = 1$ as $t \to \infty$. Thus $x^2 + y^2 = 1$ is a limit cycle.

11. Consider Liénard's equation $x'' + f(x)x' + g(x) = 0$, where $' = d/dt$ and $f(x)$ and $g(x)$ are continuous. Show that this equation can be written as the system

$$dx/dt = y - F(x), \quad \text{where } F(x) = \int_0^x f(u)du.$$
$$dy/dt = -g(x)$$

12. (See Exercise 11.) *Liénard's theorem* states that if (i) $F(x)$ is an odd function, (ii) $F(x)$ is zero only at $x = 0, x = a, x = -a$ (for some $a > 0$), (iii) $F(x) \to \infty$ monotonically for $x > a$, and (iv) $g(x)$ is an odd function where $g(x) > 0$ for all $x > 0$; then Liénard's equation has a unique limit cycle. Use Liénard's theorem to determine which of the following equations has a unique limit cycle:

$$x'' + \epsilon\left(1 - x^2\right)x' + x = 0, \quad \epsilon > 0,$$
$$x'' + 3x^2 x' + x^3 = 0.$$

13. Consider the system of autonomous equations $\{dx/dt = f(x,y), dy/dt = g(x,y)\}$. *Bendixson's theorem* (or *negative criterion*) states that if $f_x(x,y) + g_y(x,y)$ is a continuous function that is either always positive or always negative in a particular region R of the phase plane, then the system has no limit cycle in R. Use this theorem to determine if the given system has no limit cycle in the phase plane.

(a) $x' = x^3 + x + 7y, y' = x^2y.$
(b) $x' = y - x, y' = 3x - y.$
(c) $x' = xy^2, y' = x^2 + 8y.$

14. Let E and s be positive constants and suppose that $f(x)$ is a continuous odd function that approaches a finite limit as $x \to \infty$, is increasing, and is concave down for $x > 0$. The voltages over the deflection plate in a sweeping circuit for an oscilloscope are determined by solving

$$dV_1/dt = -sV_1 + f(E - V_2),$$
$$dV_2/dt = -sV_2 + f(E - V_1).$$

Use Bendixson's theorem (see Exercise 13) to show that this system has no limit cycles.

15. Consider the relativistic equation for the central orbit of a planet, $d^2u/d\theta^2 + u - ku^2 = \alpha$, where k and α are positive constants, (k is very small), $u = 1/r$, and r and θ are polar coordinates. (a) Write this second-order equation as a system of first-order equations. (b) Show that $((1 - \sqrt{1 - 4k\alpha})/(2k), 0)$ is a center in the linearized system.

16. Consider the system of equations $\{dx/dt = P(x, y), dy/dt = Q(x, y)\}$ where P and Q are polynomials of degree n. Finding the maximum number of limit cycles of this system, called the *Hilbert number H_n* has been investigated as part of Hilbert's 16th problem. Several of these numbers are known: $H_0 = 0, H_1 = 0, H_2 \geq 4$, $H_n \geq (n - 1)/2$ (if n is odd), and $H_n < \infty$. Determine the Hilbert number (or a restriction on the Hilbert number) for each system:

(a) $dx/dt = 10, dy/dt = -5.$
(b) $dx/dt = x - y, dy/dt = y + 1.$
(c) $dx/dt = y^2 - 2xy, dy/dt = x^2 + y^2.$
(d) $dx/dt = x + 5y^7, dy/dt = 10x^7 + y^4.$

17. In a mechanical system, suppose that $x(t)$ represents position at time t, K kinetic energy, and V potential energy, where $K = 1/2m(x)(dx/dt)^2$ ($m(x)$ is a positive function) and $V = V(x)$. If the system is

conservative, then the total energy E of the system remains constant during motion, which indicates that $1/2m(x)(dx/dt)^2 = E$. Show that with the change of variable $u = \int \sqrt{m(x)}\,dx$ this equation becomes $d^2u/dt^2 + Q'(u) = 0$, where $Q'(u) = V'(x)/\sqrt{m(x)}$. This means that an equation of the form $d^2x/dt^2 = f(x)$ is a *conservative system*, where f is a function representing force per unit mass that does not depend on dx/dt.

18. Consider the conservative system $d^2x/dt^2 = f(x)$ in which f is a continuous function. Suppose that $V(x) = -\int f(x)\,dx$ so that $V'(x) = -f(x)$. (a) Use the substitution $dx/dt = y$ to transform $d^2x/dt^2 = f(x)$ into the system of first-order equations $\{dx/dt = y, dy/dt = -V'(x)\}$. (b) Find the equilibrium points of the system in (a). What is the physical significance of these points? (c) Use the chain rule and $dx/dt = y$ to show that $dy/dt = y\,dy/dx$. (d) Show that the paths in the phase plane are $y^2 + 2V(x) = C.$

19. (See Exercise 18.) What are the paths in the phase plane if $V(x) = \frac{1}{2}x^2$? How does this compare with the classification of the equilibrium point of the corresponding system of first-order equations? Notice that if V has a local minimum at $x = a$, the system has a center at the corresponding point in the phase plane.

20. (See Exercise 18.) What are the paths in the phase plane if $V(x) = -\frac{1}{2}x^2$? How does this compare with the classification of the equilibrium point of the corresponding system of first-order equations? Notice that if V has a local maximum at $x = a$, the system has a saddle at the corresponding point in the phase plane.

21. Use the observations made in Exercises 19-20 concerning the relationship between the local extrema of V and the classification of equilibrium points in the phase plane to

classify the equilibrium points of a conservative system with (a) $V'(x) = x^2 - 1$; (b) $V'(x) = x - x^3$.

22. Which of the following physical systems are conservative? (a) The motion of a pendulum modeled by $d^2x/dt^2 + \sin x = 0$. (b) A spring-mass system that disregards damping and external forces. (c) A spring-mass system that includes damping.

23. Consider a brake that acts on a wheel. Assuming that the force due to friction depends only on the angular velocity of the wheel, $d\theta/dt$, we have $Id^2\theta/dt^2 = -FR \, \text{sgn} \, (d\theta/dt)$ to describe the spinning motion of the wheel, where R is the radius of the brake drum, F is the frictional force, I is the moment of inertia of the wheel, and $\text{sgn}(x) = \begin{cases} 1, x > 0 \\ 0, x = 0 \\ -1, x < 0 \end{cases}$.

(a) Let $d\theta/dt = y$ and transform this second-order equation into a system of first-order equations. (b) What are the equilibrium points of this system? (c) Show that $d^2\theta/dt = (d\theta/dt)(d/d\theta)(d\theta/dt)$. (d) Use the relationship in (c) to show that the paths in the phase plane are $I(d\theta/dt)^2 = -2FR\theta + C$, $d\theta/dt > 0$ and $I(d\theta/dt)^2 = 2FR\theta + C$, $d\theta/dt < 0$. What are these paths?

24. The equation $d^2x/dt^2 + k^2 \sin x = 0$ that models the motion of a pendulum is approximated with $d^2x/dt^2 + k^2 \left(x - \frac{1}{6}x^3 \right) = 0$. Why? Write this equation as a system, then locate and classify its equilibrium points. How do these findings differ from those of $d^2x/dt^2 + k^2 \sin x = 0$?

25. Consider the system

$x' = y - x \left(1 - x^2 - y^2 - z^2 \right)$

$y' = -x + y \left(1 - x^2 - y^2 - z^2 \right)$ (′ denotes d/dt).

$z' = 0$

(a) Show that $(x^2 + y^2 + z^2)' = 0$ if $x^2 + y^2 + z^2 = 1$ so $x^2 + y^2 + z^2 = 1$ is called an *invariant set*. (b) Show that a solution with initial point (x_0, y_0, z_0) remains in the plane $z = z_0$. Therefore, $z = z_0$ is an invariant set. (c) Show that the limit cycle for the solution with initial point $(1/2, 1/2, 1/2)$ is the circle $x^2 + y^2 = 3/4$ in the plane $z = 1/2$.

26. The *BVP oscillator* is the system of ODEs

$$\frac{dx}{dt} = x - \frac{1}{3}x^3 - y + I(t), \qquad (7.12)$$

$$\frac{dy}{dt} = c(x + a - by).$$

(a) Find and classify the equilibrium points of this system if $I(t) = 0$, and $a = b = c = 1$. (b) Graph the direction field associated with the system and then approximate the phase plane by graphing several solutions near each equilibrium point.

27. Find $x_0 > 0$ so that the solution to the initial-value problem $x' = y$, $y' = -x - (x^2 - 1) y$, $x(0) = x_0, y(0) = y_0$ is periodic. Confirm your result graphically.

28. We saw that solutions to the Lotka-Volterra problem, system (7.9), oscillate periodically, where the period T and amplitude of $x(t)$ and $y(t)$ depend on the initial conditions. *Volterra's principle* states that the time averages of $x(t)$ and $y(t)$ remain constant and equal the corresponding equilibrium values, $\bar{x} = c/d$ and $\bar{y} = a/b$. In symbols, this statement is represented with

$$\bar{x} = \frac{1}{T} \int_0^T x(t)dt \quad \text{and} \quad \bar{y} = \frac{1}{T} \int_0^T y(t)dt.$$

Use the following steps to prove that $\bar{y} = \frac{1}{T} \int_0^T y(t)dt$.

(a) Show that $(d/dt)(\ln x) = a - by$. *Hint:* $(d/dt)(\ln x) = (1/x)dx/dt$.

(b) Integrate each side of the equation obtained in (a) from 0 to T to obtain
$\ln x(T) - \ln x(0) = aT - b \int_0^T y(t)dt.$

(c) Follow steps similar to those outlined in (a) and (b) to prove that $\bar{x} = \dfrac{1}{T}\int_0^T x(t)dt.$

CHAPTER 7 SUMMARY: ESSENTIAL CONCEPTS AND FORMULAS

L-R-C **Circuits:** *L-R-C* Circuit with One or Two loops
Spring-Mass Systems
Diffusion through a Membrane
Mixture Problem with Two Tanks
Population Problems
Predator-Prey (Lotka-Volterra System
Van-der-Pol's Equation (System)

CHAPTER 7 REVIEW EXERCISES

1. Solve the one-loop *L-R-C* circuit with $L = 1\,$H, $R = 5/3\,\Omega$, $C = 3/2\,$F, $Q(0) = 10^{-6}\,$C, $I(0) = 0\,$A, and (a) $E(t) = 0\,$V; (b) $E(t) = e^{-t}\,$V.

2. Solve the one-loop *L-R-C* circuit with $L = 3\,$H, $R = 10\,\Omega$, $C = 1/10\,$F, $Q(0) = 0\,$C, $I(0) = 0\,$A, and (a) $E(t) = 120\,$V; (b) $E(t) = 120\sin t\,$V.

3. Solve the two-loop *L-R-C* circuit with $L = 1\,$H, $R_1 = R_2 = 1\,\Omega$, $C = 1\,$F, $Q(0) = 10^{-6}\,$C; $I_2(0) = 0\,$A, and (a) $E(t) = 0\,$V; (b) $E(t) = 120\,$V.

4. Solve the two-loop *L-R-C* circuit with $L = 1\,$H, $R_1 = R_2 = 1\,\Omega$, $C = 1\,$F, $Q(0) = 0\,$C, $I_2(0) = 0\,$A, and (a) $E(t) = 120e^{-t/2}\,$V; (b) $E(t) = 120\cos t\,$V.

5. Transform the second-order equation to a system of first-order equations and classify the systems as undamped, overdamped,

underdamped, or critically damped by finding the eigenvalues of the corresponding system. Solve with the initial conditions $x(0) = 1$, $dx(0)/dt = y(0) = 0$. (a) $x'' + 2x' + x = 0$; (b) $x'' + 4x = 0$.

6. Transform the second-order equation to a system of first-order equations and classify the system as undamped, overdamped, underdamped, or critically damped by finding the eigenvalues of the corresponding system. Solve with the initial conditions $x(0) = 0$, $dx(0)/dt = y(0) = 1$. (a) $x'' + 4x = 0$; (b) $4x'' + 16x' + 7x = 0$.

7. Solve the diffusion problem with one permeable membrane with $P = 1/2$, $V_1 = V_2 = 1$, $x(0) = 5$, and $y(0) = 10$.

8. Solve the mixture problem with two tanks with $R = 4\,$gal/min, $V_1 = V_2 = 80\,$gal, $C = 3\,$lb/gal, $x(0) = 10$, and $y(0) = 20$.

9. Investigate the behavior of solutions of the two-dimensional population problem with $a_1 = 3$, $a_2 = 1$, $b_1 = 3$, $b_2 = 1$, $x(0) = 20$, and $y(0) = 10$.

10. Investigate the behavior of solutions of the two-dimensional population problem with $a_1 = 1$, $a_2 = 2$, $b_1 = 1$, $b_2 = 1$, $x(0) = 4$, and $y(0) = 8$.

11. Solve the three-dimensional population problem with $a_1 = 2$, $a_2 = 1$, $a_3 = 4$, $b_1 = 6$, $b_2 = 4$, $b_3 = 5$, $c_1 = 2$, $c_2 = 8$, $c_3 = 4$, $x(0) = 4$, $y(0) = 4$, and $z(0) = 8$.

12. The nonlinear second-order equation that describes the motion of a damped pendulum is $d^2x/dt^2 + b\,dx/dt + g/L\sin x = 0$. (a) Make the substitution $dx/dt = y$ to write the equation as the system

$$\frac{dx}{dt} = y,$$

$$\frac{dy}{dt} = -by - \frac{g}{L}\sin x.$$

(b) Show that the equilibrium points of the system are $(n\pi, 0)$, where n is an integer.
(c) Show that if n is odd, then $(n\pi, 0)$ is a

saddle. (d) Show that if n is even, then $(n\pi, 0)$ is either a stable spiral point or a stable node.

13. Suppose that the predator-prey model is altered so that the prey population $x(t)$ follows the logistic equation when the predator population y(t) is not present. With this assumption, the system is

$$dx/dt = ax - kx^2 - bxy,$$
$$dy/dt = -cy + dxy,$$

where a, b, c, d, and k are positive constants, $a \neq k$, and the ratio a/k is much larger than the ratio c/d. Find and classify the equilibrium points of this system.

14. (Production of Monochlorobenzene) The Ajax Pharmaceutical Company has often thought of making its own monochlorobenzene C_6H_5Cl from benzene C_6H_6 and chloride Cl_2 (with a small amount of ferric chloride as a catalyst) instead of purchasing it from another company. The chemical reactions are

$$C_6H_6 + Cl_2 \longrightarrow C_6H_5Cl + HCl,$$
$$C_6H_5Cl + Cl_2 \longrightarrow C_6H_4Cl_2 + HCl,$$
$$C_6H_4Cl_2 + HCl_2 \longrightarrow C_6H_4Cl_3 + HCl,$$

where HCl is hydrogen chloride, $C_6H_4Cl_2$ is dichlorobenzene, and $C_6H_4Cl_3$ is trichlorobenzene. Experimental data indicate that the formation of trichlorobenzene is small. Therefore, we will neglect it, so the rate equations are given by

$$-dx_A/dt = k_1 x_A,$$
$$dx_B/dt = k_1 x_A - k_2 x_B,$$

where x_A is the mole fraction (dimensionless) of benzene and x_B is the mole fraction of monochlorobenzene. The formation of trichlorobenzene is neglected, so we assume that x_C, the mole fraction of

dichlorobenzene, is found with $x_C = 1 - x_A - x_B$.

The rate constants k_1 and k_2 have been determined experimentally. We give these constants in the following table.[1]

	40 °C	55 °C	70 °C
k_1 (h^{-1})	0.0965	0.412	1.55
k_2 (h^{-1})	0.0045	0.055	0.45

If $x_A(0) = 1$ and $x_B(0) = 0$, solve the system for each of these three temperatures. Graph $x_A(t)$ and $x_B(t)$ simultaneously for each temperature. Is there a relationship between the temperature and the time at which $x_B(t) > x_A(t)$?

15. (The Rössler Attractor) The Rössler attractor is the system

$$dx/dt = -y - z,$$
$$dy/dt = x + ay, \qquad (7.13)$$
$$dz/dt = bx - cz + xz.$$

Observe that system (7.13) is nonlinear because of the product of the x and z terms in the z' equation.

See texts like Jordan and Smith's *Nonlinear ODEs* [16], for discussions of ways to analyze systems like the Rössler attractor and the Lorenz equations.

If $a = 0.4, b = 0.3, x_0 = 1, y_0 = 0.4$, and $z(0) = 0.7$, how does the value of c affect solutions to the initial-value problem

$$dx/dt = -y - z$$
$$dy/dt = x + ay \qquad ?$$
$$dz/dt = bx - cz + xz \qquad (7.14)$$
$$x(0) = x_0, y(0) = y_0, z(0) = z_0$$

---

[1] S.W. Bodman, *The Industrial Practice of Chemical Process Engineering*, MIT Press, Cambridge, MA, 1968, pp. 18-24.

Suggestion: Numerically solve the system for $c = 1.4, 2.4, 2.6, 3.4$, and 3.44. Then generate two dimensional plots of $(t, x(t))$, $(t, y(t))$, $(t, z(t))$, $(x(t), y(t))$, $(x(t), z(t))$, and $(y(t), z(t))$ and three dimensional plots of $(t, x(t), y(t))$, $(t, x(t), z(t))$, $(t, y(t), z(t))$, and $(x(t), y(t), z(t))$. Explain why the result for $c = 1.4$ is called a *limit cycle*, for $c = 2.4$ is called a *two-cycle*, for $c = 2.6$ is called a *four-cycle*, and for $c = 3.4$ and for 3.44 is called *chaos*.

16. An interesting variation of the Lotka-Volterra equations is to assume that a depends strongly on environmental factors and might be given by the differential equation

$$\frac{da}{dt} = -ax + \bar{a} + k\sin(\omega t + \phi), \qquad (7.15)$$

where the term $-ax$ represents the loss of nutrients due to species x; $\bar{a}$, k, ω, and ϕ are constants. Observe that incorporating Equation (7.15) into system (7.9) results in a nonautonomous system.

Suppose that $x(0) = y(0) = a(0) = 0.5$, $b = d = 1, c = 0.5, \bar{a} = 0.25, k = 0.125$ and $\phi = 0$. Plot $x(t)$ and $y(t)$ if $\omega = 0.1, 0.25, 0.5, 0.75, 1, 1.25, 1.5$, and 2.5.

Refer to Gray's outstanding text, *Modern Differential Geometry of Curves and Surfaces*, [13]

---

17. *Curvature* Let C be a piecewise-smooth curve with parametrization $\mathbf{r}(t) = \langle x(t), y(t) \rangle$, $a \le t \le b$. The *unit tangent vector* to C at t is

$$\mathbf{T} = \frac{\mathbf{r}'(t)}{\|\mathbf{r}'(t)\|}. \qquad (7.16)$$

The *arc length function*, $s = s(t)$, is defined by

$$s(t) = \int_a^t \|\mathbf{r}'(u)\| \, du$$

$$= \int_a^t \sqrt{\left(\frac{dx}{du}\right)^2 + \left(\frac{dy}{du}\right)^2} \, du. \qquad (7.17)$$

Solving Equation (7.17) for t, we have $t = t(s)$ and the *parametrization of C with respect to arc length* is $\mathbf{r}(s) = \langle x(t(s)), y(t(s)) \rangle$. When C is parametrized by arc length, $\|\mathbf{r}'(s)\| = 1$ so the unit tangent vector (7.16) is given by $\mathbf{T}(s) = \mathbf{r}'(s)$. The *curvature of C*, $\kappa(s)$, is

$$\kappa(s) = \left\| \frac{d\mathbf{T}}{ds} \right\|. \qquad (7.18)$$

Thus, for the curve C parametrized by arc length, $\kappa(s) = \|\mathbf{r}''(s)\|$.

Conversely, a given curvature function determines a plane curve: the curve C parametrized by arc length with curvature $\kappa(s)$ has parametrization $\mathbf{r}(s) = \langle x(s), y(s) \rangle$ where

$$
\begin{aligned}
dx/ds &= \cos\theta, \\
dy/ds &= \sin\theta, \\
d\theta/ds &= \kappa, \qquad (7.19) \\
x(a) &= c, \quad y(a) = d, \quad \theta(0) = \theta_0.
\end{aligned}
$$

You can generate stunning curves by specifying a curve function. For each of the following curvature functions, $\kappa(s)$, plot the curve C, $(x(s), y(s))$ for $-40 \le s \le 40$ (or other appropriate s values) if $x(0) = y(0) = \theta(0) = 0$.
(a) $\kappa(s) = e^{-s} + e^s$.
(b) $\kappa(s) = s + \sin s$.
(c) $\kappa(s) = sJ_1(s)$ ($J_1(x)$ denotes the Bessel function of order 1).
(d) $\kappa(s) = sJ_2(s)$ ($J_2(x)$ denotes the Bessel function of order 2).
(e) $\kappa(s) = s\sin(\sin s)$.
(f) $\kappa(s) = |s\sin(\sin s)|$.

18. (*Tumor and Organism Growth*) When considering the growth of tumors, researchers have discovered that cells on the outer surface of the tumor are more likely to grow because they have better access to nutrients and oxygen external to the tumor. Therefore, if the tumor is assumed to be shaped like a sphere, then the ratio of the

surface area to the volume is given by $4\pi r^2/\frac{4}{3}\pi r^3 = \frac{3}{r}$, where r is the radius of the tumor. This indicates that the growth rate decreases as r increases. Assuming that the volume of the tumor increases at a rate proportional to its total volume while the growth rate decreases as the radius increases, we can model the growth of the tumor with the IVP $\{dV/dt = r(t)V,$ $dr/dt = -kr, V(0) = V_0, r(0) = r_0\}$. Show that the solution to this system is $V(t) = \exp((r_0/k) - (r_0/k)e^{-kt} + \ln V_0)$.

DIFFERENTIAL EQUATIONS AT WORK

A. Competing Species

The system of equations

$$dx/dt = x\left(a - b_1 x - b_2 y\right),$$
$$dy/dt = y\left(c - d_1 x - d_2 y\right), \quad (7.20)$$

where a, b_1, b_2, c, d_1, and d_2 represent positive constants, can be used to model the population of two species, represented by $x(t)$ and $y(t)$, competing for a common food supply.

1. (a) Find and classify the equilibrium points of the system if $a = 1, b_1 = 2, b_2 = 1, c = 1,$ $d_1 = 0.75$, and $d_2 = 2$. (b) Graph several solutions by using different initial populations parametrically in the phase plane. (c) Find $\lim_{t\to\infty} x(t)$ and $\lim_{t\to\infty} y(t)$ if both $x(0)$ and $y(0)$ are not zero. Compare your result to (b).
2. (a) Find and classify the equilibrium points of the system if $a = 1, b_1 = 1, b_2 = 1, c = 0.67,$ $d_1 = 0.75$, and $d_2 = 1$. (b) Graph several solutions by using different initial populations parametrically in the phase plane. (c) Determine the fate of the species with population $y(t)$. What happens to the

species with population $x(t)$? What happens to the species with population $y(t)$ if the species with population $x(t)$ is suddenly removed? (*Hint:* How does the term $-d_1 xy$ affect the equation $dy/dt = y(c - d_1 x - d_2 y)$)
3. Find conditions on the positive constants a, b_1, b_2, c, d_1, and d_2 so that (a) the system has exactly one equilibrium point in the first quadrant and (b) the system has no equilibrium point in the first quadrant. (c) Is it possible to choose positive constants $a, b_1,$ b_2, c, d_1, and d_2 so that each population grows without bound? If so, illustrate graphically.

B. Food Chains

The food chain in the ocean can be divided into five trophic levels, where the base level is made up primarily of seaweeds and phytoplankton. Suppose that $x_1(t)$ represents the size of this base level population and let $x_2(t)$ represent that of the second level in the food chain, which feeds on the base level.[2] In the absence of the second level, assume that $x_1(t)$ follows the logistic equation. However, an interaction between members of the first two levels of the food chain prohibits growth of $x_1(t)$, and it enhances that of $x_2(t)$. In addition, assume that $x_2(t)$ decays exponentially without the food supply of the base level. Therefore, we model this situation with

$$dx_1/dt = ax_1 - bx_1^2 - cx_1 x_2, \quad (7.21)$$
$$dx_2/dt = -dx_2 + ex_1 x_2.$$

1. Show that the equilibrium solution of this system is $(d/e, (ae - bd)/(ce))$.
2. Suppose that $a = 36, b = 4, c = 2,$ and $d = 1$. What is the equilibrium solution in this case?
3. Numerically solve the system using the initial populations $x_1(0) = 3$ and $x_2(0) = 2$

[2] J. Raloff, How Long Will We Go Fishing for Dinner? *Science News*, **153**, 1998, 86.

and graph these functions on the same set of axes. What is $\lim_{t\to\infty} x_1(t)$? What is $\lim_{t\to\infty} x_2(t)$? Describe the corresponding physical situation. At how many values of t does $x_1(t) = x_2(t)$?

4. Repeat part 3 using $x_1(0) = 1$ and $x_2(0) = 2$.

Suppose that we harvest the base level at the rate h. The system becomes

$$dx_1/dt = ax_1 - bx_1^2 - cx_1x_2 - h,$$
$$dx_2/dt = -dx_2 + ex_1x_2. \qquad (7.22)$$

5. Solve system (7.22) if $a = 36, b = 4, c = 2$, $d = 1, h = 10, x_1(0) = 1$ and $x_2(0) = 2$. Compare the solution to that obtained in (4).

6. How does harvesting affect the populations?

7. Repeat (5) with $h = 20$. Compare the results with those of (5).

8. Repeat (5) with $h = 25$. Compare the results with those of (5) and (7).

9. Repeat (5) with $h = 26$. Compare the results with those of (5), (7), and (8).

Suppose that we harvest the second level at the rate h. Then system (7.20) becomes

$$dx_1/dt = ax_1 - bx_1^2 - cx_1x_2,$$
$$dx_2/dt = -dx_2 + ex_1x_2 - h. \qquad (7.23)$$

10. Solve system (7.23) if $a = 36, b = 4, c = 2$, $d = 1, h = 1, x_1(0) = 1$, and $x_2(0) = 2$. Compare the solution to that obtained in (6). How does harvesting affect the populations?

11. Solve system (7.23) if $a = 36, b = 4, c = 2$, $d = 1, h = 4, x_1(0) = 1$, and $x_2(0) = 2$. Compare the solution to that obtained in (10).

12. Solve system (7.23) if $a = 36, b = 4, c = 2$, $d = 1, h = 4.5, x_1(0) = 1$, and $x_2(0) = 2$. Compare the solution to that obtained in (10) and (11).

Suppose that we consider all five trophic levels in the food chain, where we let $x_3(t)$ represent the size of the population at the third level, $x_4(t)$ that at the fourth level, and

$x_5(t)$ that of the fifth (or top) level. This situation is modeled with the system

$$dx_1/dt = a_1x_1 - a_2x_1^2 - a_3x_1x_2,$$
$$dx_2/dt = -b_1x_2 + b_2x_1x_2 - b_3x_2x_3,$$
$$dx_3/dt = -c_1x_3 + c_2x_2x_3 - c_3x_3x_4, \quad (7.24)$$
$$dx_4/dt = -d_1x_4 + d_2x_3x_4 - d_3x_4x_5,$$
$$dx_5/dt = -e_1x_5 + e_2x_4x_5.$$

Notice that we assume each level benefits from an encounter with a member of the next lower level.

13. Suppose that $a_1 = a_2 = a_3 = 1$, $b_1 = b_2 = b_3 = 1, c_1 = c_2 = c_3 = 1$, $d_1 = d_2 = d_3 = 1, e_1 = e_2 = e_3 = 1, x_1(0) = 2$, $x_2(0) = 3, x_3(0) = 8, x_4(0) = 10$, and $x_5(0) = 12$. Numerically solve system (7.24) using the initial populations and graph these functions on the same set of axes. Determine $\lim_{t\to\infty} x_1(t), \lim_{t\to\infty} x_2(t)$, $\lim_{t\to\infty} x_3(t), \lim_{t\to\infty} x_4(t)$, and $\lim_{t\to\infty} x_5(t)$? Describe the corresponding physical situation.

14. Repeat (13) with $a_3 = 2$.

15. Repeat (13) with $a_3 = 10$.

If we fish at the base level at rate h, we obtain the system

$$dx_1/dt = a_1x_1 - a_2x_1^2 - a_3x_1x_2 - h,$$
$$dx_2/dt = -b_1x_2 + b_2x_1x_2 - b_3x_2x_3,$$
$$dx_3/dt = -c_1x_3 + c_2x_2x_3 - c_3x_3x_4, \quad (7.25)$$
$$dx_4/dt = -d_1x_4 + d_2x_3x_4 - d_3x_4x_5,$$
$$dx_5/dt = -e_1x_5 + e_2x_4x_5.$$

16. Suppose that $h = 1, a_1 = 10, a_2 = a_3 = 1$, $b_1 = b_2 = b_3 = 1, c_1 = c_2 = c_3 = 1$, $d_1 = d_2 = d_3 = 1, e_1 = e_2 = e_3 = 1, x_1(0) = 2$, $x_2(0) = 3, x_3(0) = 8, x_4(0) = 10$, and $x_5(0) = 12$. Numerically solve system (7.25) using the initial populations and graph these functions on the same set of axes. Determine $\lim_{t\to\infty} x_1(t), \lim_{t\to\infty} x_2(t)$, $\lim_{t\to\infty} x_3(t), \lim_{t\to\infty} x_4(t)$, and $\lim_{t\to\infty} x_5(t)$? Describe the corresponding physical situation.

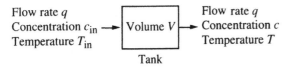

Flow rate q
Concentration c_{in} → | Volume V | →   Flow rate q
Temperature T_{in} Concentration c
       Temperature T

Tank

FIGURE 7.13   Continuous-flow stirred tank reactor.

17. Suppose that we fish at the third level at rate $h = 8$ and the other values are the same as in (16). How does this form of harvesting affect the population sizes?

C. Chemical Reactor

Consider the continuous-flow stirred-tank reactor shown in Figure 7.13. In this reaction, a stream of chemical C flows into the tank of volume V, and products as well as residue flow out of the tank at a constant rate q. Because the reaction is continuous, the composition and temperature of the contents of the tank are constant and are the same as the composition and temperature of the stream flowing out of the tank. Suppose that a concentration c_{in}, of the chemical C flows into the tank and a concentration c flows out of the tank. In addition, suppose that the chemical C changes into products at a rate proportional to the concentration c of C, where the constant of proportionality $k(T)$ depends on the temperature T, $k(T) = Ae^{-B/T}$. Because

$$\begin{pmatrix} \text{Rate of change} \\ \text{of amount of } C \end{pmatrix} = (\text{Rate in of } C) - $$

$$(\text{Rate out of } C) - \begin{pmatrix} \text{Rate that} \\ C \text{ disappears} \\ \text{by the reaction} \end{pmatrix},$$

we have the differential equation[3]

$$\frac{d}{dt}(Vc) = qc_{in} - qc - Vk(T)c.$$

1. Show that if V is constant, then this equation is equivalent to
$$dc/dt = qV^{-1}(c_{in} - c) - k(T)c.$$

2. In a similar manner, we balance the heat of the reaction with

$$\begin{pmatrix} \text{Rate of change} \\ \text{of heat content} \end{pmatrix} = (\text{Rate in of heat})$$

$$- (\text{Rate out of heat}) - \begin{pmatrix} \text{Heat removed} \\ \text{by cooling} \end{pmatrix}$$

$$+ \begin{pmatrix} \text{Heat produced} \\ \text{by reaction} \end{pmatrix}.$$

Let C_p be the specific heat so that the heat content per unit volume of the reaction mixture at temperature T is $C_p T$. If H is the rate at which heat is generated by the reaction and $VS(T)$ is the rate at which heat is removed from the system by a cooling system, then we have the differential equation

$$VC_p \frac{dT}{dt} = qC_p T_{in} - qC_p T - VS(t) + HVk(T)c,$$

where T_{in} is the temperature at which the chemical flows into the tank. Solve this equation for dT/dt.

3. The equations in (1) and (2) form a system of nonlinear ODEs. Show that the equilibrium point of this system satisfies the equations

$$c_{in} - c = \frac{V}{q}ck(T) \quad \text{and} \quad T - T_{in} = \frac{HV}{qC_p}ck(T),$$

if we assume there is no cooling system. Solve the first equation for c and substitute into the second equation to find that

$$T - T_{in} = \frac{HVc_{in}}{qC_p + VC_p k(T)}k(T). \quad (7.26)$$

We call values of T that satisfy this equation *steady state temperatures*.

---

[3]*Applications in Undergraduate Mathematics in Engineering*, Mathematical Association of America, Macmillan Company, New York, 1967, pp. 122-125. W. Fred Ramirez, *Computational Methods for Process Simulation*, Butterworths Series in Chemical Engineering, Boston, 1989, pp. 175-182.

4. If $T_{in} = 1$, $HVC_{in} = VC_p = qCp = 1$, and $k(T) = e^{-1/T}$ graph $y = T - T_{in}$ and $y = HVC_{in}/(qC_p + VC_p k(T))k(T)$ to determine the number of roots of Equation (7.26).

5. If $T_{in} = 0.15$, $HVc_{in} = 1$, $VC_p = 0.25$, $qC_p = 0.015$, and $k(T) = e^{-3/T}$ graph $y = T - T_{in}$ and $y = HVC_{in}/(qC_p + VC_p k(T))k(T)$ to determine the number of roots of Equation (7.26). How does this situation differ from that in (4)?

6. Notice that $y = T - T_{in}$ describes heat removal and $y = HVC_{in}/(qC_p + VC_p k(T))k(T)$ describes heat production. Therefore, if the slope of the heat production curve is greater than that of the heat removal curve, the steady state is unstable. On the other hand, if the slope of the heat production curve is less than or equal to that of the heat removal curve, then the steady state is stable. Use this information to determine which of the temperatures found in (4) and (5) are stable and which are unstable. What do the slopes of these curves represent?

7. Approximate the solution to the system of differential equations using the parameter values given in (4) and (5). Does the stability of the system correspond to those found in (6)?

D. Food Chains in a Chemostat

Simple Food Chain in a Chemostat

The equations that describe a simple food chain in a chemostat are

$$\frac{dS}{dt} = 1 - S - \frac{m_1 x S}{a_1 + S},$$

$$\frac{dx}{dt} = \frac{m_1 x S}{a_1 + S} - x - \frac{m_2 x y}{a_2 + x}, \qquad (7.27)$$

$$\frac{dy}{dt} = \frac{m_2 x y}{a_2 + x} - y,$$

$$S(0) = S_0, x(0) = x_0, y(0) = y_0.$$

In system (7.27), y (the predator) consumes x (the prey) and x consumes the nutrient S.

See Smith and Waltman's, *The Theory of the Chemostat: Dynamics of Microbial Competition* [27], for a detailed discussion of various chemostat models.

Now let $\Sigma = 1 - S - x - y$. Then $\Sigma' = -S' - x' - y' = -(1 + S + x + y) = -\Sigma$ so $\Sigma = \Sigma_0 e^{-t}$ and $\lim_{t \to \infty} \Sigma = 0$. In the limit as $t \to \infty$, $\Sigma = 0 = 1 - S - x - y$ so $S = 1 - x - y$ and system (7.27) becomes

$$\frac{dx}{dt} = \frac{m_1 x (1 - x - y)}{1 + a_1 - x - y} - x - \frac{m_2 x y}{a_2 + x},$$

$$\frac{dy}{dt} = \frac{m_2 x y}{a_2 + x} - y, \qquad (7.28)$$

$$x(0) = x_0, y(0) = y_0.$$

The analysis of system (7.27) is quite technical but we can use a CAS to assist us carrying out a few of the computations needed when analyzing system (7.27).

See Chapter 3 of Smith and Waltman's *The Theory of the Chemostat: Dynamics of Microbial Competition* [27], for a detailed analysis of system (7.27).

---

1. Find the equilibrium points (or rest points) of system (7.27) by solving

$$\frac{m_1 x (1 - x - y)}{1 + a_1 - x - y} - x - \frac{m_2 x y}{a_2 + x} = 0,$$

$$\frac{m_2 x y}{a_2 + x} - y = 0.$$

For convenience we will refer to $(0, 0)$ as E_0, $\left(1 - \dfrac{a_1}{m_1 - 1}, 0\right)$ as E_1 and the other rest point as E_A.

2. Compute the Jacobian, **J**, of system (7.27).

3. Find conditions so that E_0 is unstable.

4. Let $\lambda_i = a_i/(m_i - 1)$. Show that E_1 is stable if $\lambda_1 + \lambda_2 > 1$ and an unstable saddle if $\lambda_1 + \lambda_2 < 1$.

Incorporating a second predator, z, of x into system (7.29) results in

$$\frac{dS}{dt} = 1 - S - \frac{m_1 x S}{a_1 + S},$$

$$\frac{dx}{dt} = \frac{m_1 x S}{a_1 + S} - x - \frac{m_2 x y}{a_2 + x} - \frac{m_3 x z}{a_3 + x},$$

$$\frac{dy}{dt} = \frac{m_2 x y}{a_2 + x} - y, \qquad (7.29)$$

$$\frac{dz}{dt} = \frac{m_3 x z}{a_3 + x} - z,$$

$$S(0) = S_0, \quad x(0) = x_0, \quad y(0) = y_0, \quad z(0) = z_0.$$

In the same way as with system (7.29), we let $\Sigma = 1 - S - x - y - z$. Then, $\Sigma' = -\Sigma$ so $\lim_{t \to \infty} \Sigma = 0$. Substitution of Σ into system (7.29) and taking the limit $t \to \infty$ results in

$$\frac{dx}{dt} = \frac{m_1 x \left(1 - x - y - z\right)}{1 + a_1 - x - y - z} - x - \frac{m_2 x y}{a_2 + x} - \frac{m_3 x z}{a_3 + x},$$

$$\frac{dy}{dt} = \frac{m_2 x y}{a_2 + x} - y, \qquad (7.30)$$

$$\frac{dz}{dt} = \frac{m_3 x z}{a_3 + x} - z,$$

$$S(0) = S_0, \quad x(0) = x_0, \quad y(0) = y_0, \quad z(0) = z_0.$$

System (7.30) can exhibit *very* interesting behavior.

5. Let $a_1 = 0.3$, $a_2 = 0.4$, $m_1 = 8$, $m_2 = 4.5$, and $m_3 = 5.0$. If $x(0) = 0.1$, $y(0) = 0.1$, and $z(0) = 0.3$, how does varying a_3 affect the solutions of system (7.30)? *Suggestion:* Let a_3 range for 30 equally spaced values from 0.35 to 0.55 For each value of a_3, solve the system and plot (x, y, z) for $0 \le t \le 60$ and display the results together.

In system (7.27), y preys on x. Incorporating a predator z of y into system (7.27) results in

$$\frac{dS}{dt} = 1 - S - \frac{m_1 x S}{a_1 + S},$$

$$\frac{dx}{dt} = \frac{m_1 x S}{a_1 + S} - x - \frac{m_2 x y}{a_2 + x},$$

$$\frac{dy}{dt} = \frac{m_2 x y}{a_2 + x} - y - \frac{m_3 y z}{a_3 + y}, \qquad (7.31)$$

$$\frac{dz}{dt} = \frac{m_3 y z}{a_3 + y} - z.$$

As with system (7.27), in the limit as $t \to \infty$, $S = 1 - x - y - z$ so system (7.31) can be rewritten as

$$\frac{dx}{dt} = x \left[f_1(1 - x - y - z) - 1 \right] - y f_2(x),$$

$$\frac{dy}{dt} = y \left[f_2(x) - 1 \right] - z f_3(x), \qquad (7.32)$$

$$\frac{dz}{dt} = z \left[f_3(y) - 1 \right],$$

where

$$f_i(u) = \frac{m_i u}{a_i + u}. \qquad (7.33)$$

Of course, rigorous analysis of system (7.32) is even more complicated than the analysis of system (7.27).

6. Let $a_1 = 0.08$, $a_2 = 0.23$, $m_1 = 10$, $m_2 = 4.0$, and $m_3 = 3.5$. If $x(0) = 0.3$, $y(0) = 0.1$, and $z(0) = 0.3$, how does varying a_3 affect the solutions of system (7.32)? *Suggestion:* Follow the same approach as in (5). Use $a_3 = 0.3$, 0.26, 0.24, 0.22, and 0.2. Plot for $50 \le t \le 60$ and then for $975 \le t \le 1000$. Explain why we say that the solution looks *chaotic* if $a_3 = 0.2$.

E. The Rössler System and Attractor

The Rössler system[4]
$$\begin{cases} \dfrac{dx}{dt} = -y - z \\ \dfrac{dy}{dt} = x + ay \\ \dfrac{dz}{dt} = b + z \left(x - c \right) \end{cases}$$
is

a non-linear system of ODE's that was studied

[4] B. Abdelkrim, M. Naim, Generalized chaos control and synchronization by nonlinear high-order approach, *Mathematics and Computers in Simulation* **82**, 2012, 2268-32281.

by Otto Rössler, a German biochemist known for his work in chaos theory. The parameters a, b and c are positive.

1. Analyze the system in the xy-plane by assuming that $z = 0$. Show that the origin $(0,0)$ is an equilibrium solution and that the Jacobian $\begin{pmatrix} 0 & -1 \\ 1 & a \end{pmatrix}$ has eigenvalues $\lambda_{1,2} = \left(a \pm \sqrt{a^2 - 4} \right) / 2$. For what values of a are the eigenvalues complex? Do the eigenvalues indicate that the origin is stable or unstable?

2. Find the equilibrium solutions of the original system and find the eigenvalues associated with each equilibrium solution. One of the equilibrium points resides in the center of the attractor while the other is more remote. Use a computer algebra system to determine which is more central to the attractor. Assume the values of $a = 0.2$, $b = 0.2$, and $c = 5.7$ with initial conditions $x(0) = 2$, $y(0) = 2$ and $z(0) = 3$.

F. Cell Dynamics in Colon Cancer

In order to model colon cancer, we must understand how tumor growth occurs in the colon, which has 10^7 crypts (or invaginations in the lining of the colon) to help prevent uncontrolled growth of cancer cells. Each crypt has stem cells at its base, which produce a variety of cell types needed for tissue renewal and regeneration after injury. Stem cells produce semi-differentiated cells, which migrate up the crypt wall to the surface; along the way, these cells differentiate into several types of cells. Once they reach the

surface, these cells either die or are transported away. To build a basic model, we divide the cells into three groups, stem cells, semi-differentiated cells and fully differentiated cells, denoted by N_0, N_1 and N_2, respectively.

We assume that a fraction α_0 of the stem cells die, a fraction α_1 differentiate and a fraction α_2 renew. This yields the ODE $dN_0/dt = (\alpha_3 - \alpha_1 - \alpha_2) N_0$. The renewed stem cells, $\alpha_3 N_0$, then contribute the N_1 population as does the renewed N_1 by a fraction β_3. In addition, a fraction β_1 and β_2 die and differentiate, respectively, which yields $dN_1/dt = (\beta_3 - \beta_1 - \beta_2) N_1 + \alpha_2 N_0$. Similarly, the renewed cells, $\beta_2 N_1$, contribute to the N_2 population, whereas the fraction γ of the N_2 population dies, which gives us $dN_2/dt = \beta_2 N_1 - \gamma N_2$.[5]

1. Let $\alpha = \alpha_3 - \alpha_1 - \alpha_2$ and show that the solution to the IVP $dN_0/dt = \alpha N_0$, $N_0(0) = N_{00}$ is $N_0(t) = N_{00}e^{\alpha t}$.

2. Use the solution in (1) to solve the IVP for dN_1/dt with $\beta = \beta_3 - \beta_1 - \beta_2$ and $N_1(0) = N_{10}$.

3. Using the solutions in (1) and (2), solve $dN_2/dt = \beta_2 N_1 - \gamma N_2$.

4. Notice that the stem cell population $N_0(t)$ grows exponentially if $\alpha > 0$ decays exponentially if $\alpha < 0$ or remains constant if $\alpha = 0$. Use the fact that $\alpha_1 + \alpha_2 + \alpha_3 = 1$ to show that the stem cell population size remains constant if $\alpha_3 = \frac{1}{2}$.

5. Show that if the stem cell population size is constant (i.e., $\alpha = 0$) then the limiting population sizes for N_1 and N_2 when $\beta < 0$ and $\gamma > 0$ are $N_{00}/(\beta_1 + \beta_2 - \beta_3)$ and $\beta_2 N_{00}/((\alpha - \beta)(\alpha + \gamma))$, respectively.

[5] M.D. Johnston, C.M. Edwards, A.F. Bodmer, P.K. Malni, J. Chapman, Mathematical modeling of cell population dynamics in the colonic crypt and in colorectal cancer, *Proceedings of the National Academy of Sciences of the U.S.A.*

8

Introduction to the Laplace Transform

In previous chapters we investigated solving the equation

$$a_n(t)y^{(n)} + a_{n-1}(t)y^{(n-1)} + \cdots + a_1(t)y' + a_0(t)y = f(t) \tag{8.1}$$

for y. We saw that if the coefficients $a_n, a_{n-1}, \ldots, a_0$ are constants, a general solution of the corresponding homogeneous equation

$$a_n(t)y^{(n)} + a_{n-1}(t)y^{(n-1)} + \cdots + a_1(t)y' + a_0(t)y = 0 \tag{8.2}$$

is determined by the solutions of an algebraic equation, the characteristic equation,

$$a_n r^n + a_{n-1}r^{n-1} + \cdots + a_1 r + a_0 = 0. \tag{8.3}$$

If the coefficient functions of Equation (8.1) or (8.3) are not constants, the situation is more difficult. In some cases, as when Equation (8.1), is a Cauchy-Fuler equation, similar techniques can be used. In other cases, we might be able to use series to find solutions. In all of these cases, however, the function $f(t)$ has typically been a *smooth* function. In cases when $f(t)$ is not a smooth function, such as when $f(t)$ is a piecewise-defined function, implementing solution techniques like undetermined coefficients, variation of parameters, or series methods to solve Equation (8.1) are usually more difficult.

In this chapter, we discuss a technique that transforms the *differential* equation (8.1) into an *algebraic* equation that can often be solved to obtain a solution of the differential equation, *even if $f(t)$ it not a smooth function*.

8.1 THE LAPLACE TRANSFORM: PRELIMINARY DEFINITIONS AND NOTATION

- DEFINITION OF THE LAPLACE TRANSFORM
- PROPERTIES OF THE LAPLACE TRANSFORM

We are already familiar with several operations (or transforms) on functions. In previous courses, we learned to add, subtract, multiply, divide, and compose functions. Another operation on functions is differentiation, which transforms the differentiable function $f(t)$ to its derivative. Similarly, the operation of integration transforms the integrable function $f(t)$ to its integral.

In this section, we introduce another operation on functions, the *Laplace transform*, and discuss several of its properties.

Definition of the Laplace Transform

Definition 30 (Laplace Transform). *Let $f(t)$ be a function defined on the interval $(0, \infty)$. The **Laplace transform** of $f(t)$ is the function (of s)*

$$F(s) = \mathcal{L}\{f(t)\} = \int_0^\infty e^{-st}f(t)\,dt, \qquad (8.4)$$

provided that the improper integral exists.

Because the Laplace transform yields a function of s, we often use the notation $\mathcal{L}\{f(t)\} = F(s)$ to denote the Laplace transform of $f(t)$.

We use the *capital* letter ($F(s)$) to denote the Laplace transform of the function named with the corresponding *small* letter ($f(t)$).

Pierre Simon de Laplace (1749-1827), French mathematician and astronomer, introduced this integral transform in his work *Théorie analytique des probabilités*, published in 1812. However, Laplace is probably most famous for his contributions to astronomy and probability.

The Laplace transform is defined as an improper integral. Recall that we evaluate improper integrals of this form by taking the limit of a definite integral:

$$\int_0^\infty f(t)\,dt = \lim_{M \to \infty} \int_0^M f(t)\,dt \quad \text{or}$$

$$\int_0^\infty f(t)\,dt = \lim_{\epsilon \to 0^+, M \to \infty} \int_\epsilon^M f(t)\,dt,$$

if $f(t)$ is discontinuous at $t = 0$.

EXAMPLE 8.1.1

Compute $F(s) = \mathcal{L}\{f(t)\}$ if $f(t) = 1$.

Solution

Using Equation (8.4) and a u-substitution with $u = -st$ so that $-\dfrac{1}{s}du = dt$, we have

$$F(s) = \mathcal{L}\{1\} = \int_0^\infty e^{-st} \times 1\,dt = \lim_{M \to \infty} \int_0^M e^{-st} \times 1\,dt$$

$$= \lim_{M \to \infty} \left[-\frac{1}{s}e^{-st} \right]_0^M = -\frac{1}{s}\lim_{M \to \infty}(e^{-sM} - 1).$$

If $s > 0$, $\lim_{M\to\infty} e^{-sM} = 0$. (Otherwise, the limit does not exist.) Therefore,

$$F(s) = \mathcal{L}\{1\} = -\frac{1}{s} \lim_{M\to\infty} (e^{-sM} - 1) = -\frac{1}{s}(0 - 1)$$

$$= \frac{1}{s}, \quad s > 0.$$

Notice that the limit in Example 8.1.1 does not exist if $s < 0$. This means that $\mathcal{L}\{1\}$ is defined only for $s > 0$, so the domain of $\mathcal{L}\{1\}$ is $s > 0$. We will find that the domain of the Laplace transform of many functions ($F(s)$) is restricted by the given function ($f(t)$).

EXAMPLE 8.1.2

Compute $\mathcal{L}\{f(t)\}$ if $f(t) = e^{at}$.

Solution

As before, we have

$$\mathcal{L}\{f(t)\} = \int_0^\infty e^{-st} f(t)\, dt = \int_0^\infty e^{-st} e^{at}\, dt$$

$$= \int_0^\infty e^{-(s-a)t}\, dt = \lim_{M\to\infty} \left[-\frac{1}{s-a} e^{-(s-a)t} \right]_{t=0}^{t=M}$$

$$= -\frac{1}{s-a} \lim_{M\to\infty} (e^{-(s-a)M} - 1)$$

If $s - a > 0$, then $\lim_{M\to\infty} e^{-(s-a)M} = 0$. Therefore,

$$\mathcal{L}\{f(t)\} = -\frac{1}{s-a} \lim_{M\to\infty} (e^{-(s-a)M} - 1) = \frac{1}{s-a}, s > a.$$

The formula $\mathcal{L}\{e^{at}\} = \dfrac{1}{s-a}$ can now be used to avoid using the definition.

EXAMPLE 8.1.3

Compute (a) $\mathcal{L}\{e^{-3t}\}$ and (b) $\mathcal{L}\{e^{5t}\}$.

Solution

We have that (a) $\mathcal{L}\{e^{-3t}\} = \dfrac{1}{s - (-3)} = \dfrac{1}{s+3}$, $s > -3$, and (b) $\mathcal{L}\{e^{5t}\} = \dfrac{1}{s-5}$, $s > 5$.

EXAMPLE 8.1.4

Compute $F(s) = \mathcal{L}\{f(t)\}$ if $f(t) = t$.

Solution

To compute $F(s) = \mathcal{L}\{f(t)\} = \int_0^\infty t e^{-st}\, dt$ we use integration by parts with $u = t$ and $dv = e^{-st} dt$. Then, $du = dt$ and $v = -s^{-1} e^{-st}$, so

$$F(s) = \mathcal{L}\{f(t)\} = \int_0^\infty t e^{-st}\, dt = \lim_{M\to\infty} \int_0^M t e^{-st}\, dt$$

$$= \lim_{M\to\infty} \left[-\frac{1}{s} t e^{-st} \right]_{t=0}^{t=M} + \lim_{M\to\infty} \frac{1}{s} \int_0^M e^{-st}\, dt$$

$$= 0 - \frac{1}{s^2} \lim_{M\to\infty} [e^{-st}]_{t=0}^{t=M}.$$

If $s > 0$, $\lim_{M\to\infty} e^{-sM} = 0$. Therefore,

$$F(s) = \mathcal{L}\{f(t)\} = -\frac{1}{s^2} \lim_{M\to\infty} (e^{-sM} - 1)$$

$$= \frac{1}{s^2}, \quad s > 0.$$

EXAMPLE 8.1.5

Compute $\mathcal{L}\{\sin t\}$.

Solution

You should verify that evaluating the improper integral that results requires integration by parts twice.

$$\mathcal{L}\{\sin t\} = \int_0^\infty e^{-st} \sin t\, dt = \lim_{M\to\infty} \int_0^M e^{-st} \sin t\, dt$$

$$= \lim_{M\to\infty} \left[-\frac{1}{s^2+1} e^{-st} (s \sin t + \cos t) \right]_{t=0}^{t=M}$$

$$= -\frac{1}{s^2+1} \lim_{M\to\infty} [e^{-sM} (s \sin M + \cos M) - 1]$$

If $s > 0$, $\lim_{M\to\infty} e^{-sM}(s \sin M + \cos M) = 0$. (Why?). Therefore,

$$\mathcal{L}\{\sin t\} = -\frac{1}{s^2+1} \lim_{M\to\infty} [e^{-sM} (s \sin M + \cos M) - 1]$$

$$= \frac{1}{s^2+1}, \quad s > 0.$$

As we can see from these examples, the definition of the Laplace transform can be difficult to apply. For example, if we wanted to calculate $\mathcal{L}\{t^n\}$ with the definition, we would have to integrate by parts n times; a time-consuming task if done with pencil and paper. We now investigate other properties of the Laplace transform so that we can determine the Laplace transform of many functions more easily. We begin by discussing the *linearity property*, which enables us to use the transforms that we have already found to find the Laplace transforms of other functions.

Theorem 38 (Linearity Property of the Laplace Transform). *Let a and b be constants, and suppose that the Laplace transform of the functions f(t) and g(t) exists. Then,*

$$\mathcal{L}\{af(t) + bg(t)\} = a\mathcal{L}\{f(t)\} + b\mathcal{L}\{g(t)\}. \quad (8.5)$$

Proof.

$$\mathcal{L}\{af(t) + bg(t)\} = \int_0^\infty e^{-st}\left(af(t) + bg(t)\right)\,dt$$
$$= a\int_0^\infty e^{-st}f(t)\,dt + b\int_0^\infty e^{-st}g(t)\,dt$$
$$= a\mathcal{L}\{f(t)\} + b\mathcal{L}\{g(t)\}.$$

EXAMPLE 8.1.6

Calculate (a) $\mathcal{L}\{6\}$ and (b) $\mathcal{L}\{5 - 2e^{-t}\}$.

Solution

(a) Using the result from Example 8.1.1 and the linearity property, $\mathcal{L}\{6\} = 6\mathcal{L}\{6\} = 6 \times s^{-1}$.
(b) $\mathcal{L}\{5 - 2e^{-t}\} = 5\mathcal{L}\{1\} - 2\mathcal{L}\{e^{-t}\} = 5 \times \frac{1}{s} - 2 \times \frac{1}{s-(-1)} = \frac{5}{s} - \frac{2}{s+1}.$

Exponential Order, Jump Discontinuities, and Piecewise-Continuous Functions

In calculus, we learn that some improper integrals diverge, which indicates that the Laplace transform may not exist for some functions. For example, $f(t) = t^{-1}$ grows too rapidly near $t = 0$ for the improper integral $\int_0^\infty e^{-st}f(t)\,dt$ to exist

and $f(t) = e^{t^2}$ grows too rapidly as $t \to \infty$ for the improper integral $\int_0^\infty e^{-st}f(t)\,dt$ to exist. Therefore, we present the following definitions and theorems so that we can better understand the types of functions for which the Laplace transform exists.

Definition 31 (Exponential Order). *A function $y = f(t)$ is of **exponential order** b if there are numbers b, C > 0, and T > 0 such that*

$$|f(t)| \le Ce^{bt}$$

for $t > T$.

In the following sections, we will see that the Laplace transform is particularly useful in solving equations involving piecewise or recursively defined functions.

Definition 32 (Jump Discontinuity). *A function $y = f(t)$ has a **jump discontinuity** at $t = c$ on the closed interval $[a, b]$ if the one-sided limits $\lim_{t \to c^+} f(t)$ and $\lim_{t \to c^-} f(t)$ are finite, but unequal, values. $y = f(t)$ has a **jump discontinuity** at $t = a$ if $\lim_{t \to a^+} f(t)$ is a finite value different from $f(a)$. $y = f(t)$ has a **jump discontinuity** at $t = b$ if $\lim_{t \to b^-} f(t)$ is a finite value different from $f(b)$.*

Definition 33 (Piecewise Continuous). *A function $y = f(t)$ is **piecewise continuous** on the finite interval $[a, b]$ if $y = f(t)$ is continuous at every point in $[a, b]$ except at finitely many points at which $y = f(t)$ has a jump discontinuity.*

*A function $y = f(t)$ is **piecewise continuous** on $[0, \infty)$ if $y = f(t)$ is piecewise continuous on $[0, N]$ for all N.*

Theorem 39 (Sufficient Condition for Existence of $\mathcal{L}\{f(t)\}$). *Suppose that $y = f(t)$ is a piecewise-continuous function on the interval $[0, \infty)$ and that it is of exponential order b for $t > T$. Then, $\mathcal{L}\{f(t)\}$ exists for s > b.*

Proof. We need to show that the integral $\int_0^\infty e^{-st}f(t)\,dt$ converges for $s > b$, assuming that $f(t)$ is a piecewise-continuous function on the interval $[0, \infty)$ and that it is of exponential order b for $t > T$. First, we write the integral as

$$\int_0^\infty e^{-st}f(t)\,dt = \int_0^T e^{-st}f(t)\,dt + \int_T^\infty e^{-st}f(t)\,dt,$$

where T is selected so that $\left|f(t)\right| \leq Ce^{bt}$ for the constants b and C, $C > 0$.

Note: In this textbook, typically we work with functions that are piecewise continuous and of exponential order. However, in the exercises, we explore functions that may or may not have these properties.

Notice that because $f(t)$ is a piecewise-continuous function, so is $e^{-st}f(t)$. The first of these integrals, $\int_0^T e^{-st}f(t)\,dt$, exists because it can be written as the sum of integrals over which $e^{-st}f(t)$ is continuous. The fact that $e^{-st}f(t)$ is piecewise continuous on $[T, \infty)$ is also used to show that the second integral, $\int_T^\infty e^{-st}f(t)\,dt$, converges. Because there are constants C and b such that $\left|f(t)\right| \leq Ce^{bt}$, we have

$$\left| \int_T^\infty e^{-st}f(t)\,dt \right| \leq \int_T^\infty |e^{-st}f(t)|\,dt$$

$$\leq C \int_T^\infty e^{-st}e^{bt}\,dt = C \int_T^\infty e^{-(s-b)t}\,dt$$

$$= C \lim_{M\to\infty} \int_T^M e^{-(s-b)t}\,dt$$

$$= C \lim_{M\to\infty} \left[-\frac{1}{s-b}e^{-(s-b)t} \right]_{t=T}^{t=M}$$

$$= -\frac{C}{s-b} \lim_{M\to\infty} (e^{-(s-b)M} - e^{-(s-b)T}).$$

Then, if $s - b > 0$, $\lim_{M\to\infty} e^{-(s-b)M} = 0$, so

$$\left| \int_T^\infty e^{-st}f(t)\,dt \right| \leq \frac{C}{s-b}e^{-(s-b)T}, \quad s > b.$$

Because both of the integrals $\int_0^T e^{-st}f(t)\,dt$ and $\int_T^\infty e^{-st}f(t)\,dt$ exist, $\int_0^\infty e^{-st}f(t)\,dt$ also exists for $s > b$.

EXAMPLE 8.1.7

Find the Laplace transform of $f(t) = \begin{cases} -1, & 0 \leq t < 4 \\ 1, & t \geq 4 \end{cases}$.

Solution

Because $y = f(t)$ is a piecewise-continuous function on $[0, \infty)$ and of exponential order, $\mathcal{L}\{f(t)\}$ exists. We use the definition and evaluate the integral using the sum of two integrals. We assume that $s > 0$:

$$\mathcal{L}\{f(t)\} = \int_0^\infty f(t)e^{-st}\,dt = \int_0^4 -1 \times e^{-st}\,dt + \int_4^\infty e^{-st}\,dt$$

$$= \left[\frac{1}{s}e^{-st} \right]_{t=0}^{t=4} + \lim_{M\to\infty} \left[-\frac{1}{s}e^{-st} \right]_{t=4}^{t=M}$$

$$= \frac{1}{s}(e^{-4s} - 1) - \frac{1}{s}\lim_{M\to\infty}(e^{-Ms} - e^{-4s})$$

$$= \frac{1}{s}(2e^{-4s} - 1).$$

Theorem 39 gives a sufficient condition and not a necessary condition for the existence of the Laplace transform. In other words, there are functions such as $f(t) = t^{-1/2}$ that do not satisfy the hypotheses of the theorem for which the Laplace transform exists (see Exercises 62 and 63).

Properties of the Laplace Transform

The definition of the Laplace transform is not easy to apply to most functions. Therefore, we now discuss several properties of the Laplace transform so that numerous transformations can be made without having to repeatedly reuse the definition. Most of the properties discussed here follow directly from our knowledge of integrals so the proofs are omitted.

Theorem 40 (Shifting Property). *If $\mathcal{L}\{f(t)\} = F(s)$ exists for $s > b$, then*

$$\mathcal{L}\{e^{at}f(t)\} = F(s - a). \qquad (8.6)$$

EXAMPLE 8.1.8

Find the Laplace transform of (a) $f(t) = e^{-2t}\cos t$ and (b) $f(t) = 4te^{3t}$.

Solution

(a) In this case, $f(t) = \cos t$ and $a = -2$. Using $F(s) = \mathcal{L}\{\cos t\} = \dfrac{s}{s^2+1}$, we replace each s with $s - a = s + 2$. Therefore,

$$\mathcal{L}\left\{e^{-2t}\cos t\right\} = \frac{s+2}{(s+2)^2 + 1} = \frac{s+2}{s^2+4s+5}.$$

(b) Using the linearity property, we have $\mathcal{L}\left\{4te^{3t}\right\} = 4\mathcal{L}\left\{te^{3t}\right\}$. To apply the shifting property we have $f(t) = t$ and $a = 3$, so we replace s in $F(s) = \mathcal{L}\{t\} = s^{-2}$ by $s - a = s - 3$. Therefore,

$$\mathcal{L}\left\{4te^{3t}\right\} = \frac{4}{(s-3)^2}.$$

Multiplying $f(t)$ by t^n, $n \geq 0$ an integer, and taking the Laplace transform corresponds to taking the nth derivative of the Laplace transform of $f(t)$ and multiplying by a power of -1.

Theorem 41 (Derivatives of the Laplace Transform). *Suppose that $F(s) = \mathcal{L}\left\{f(t)\right\}$ where $y = f(t)$ is a piecewise-continuous function on $[0, \infty)$ and of exponential order b. Then, for $s > b$,*

$$\mathcal{L}\left\{t^n f(t)\right\} = (-1)^n \frac{d^n F}{ds^n}(s). \qquad (8.7)$$

EXAMPLE 8.1.9

Find the Laplace transform of (a) $f(t) = t\cos 2t$ and (b) $f(t) = t^2 e^{-3t}$.

Solution

(a) In this case, $n = 1$ and $F(s) = \mathcal{L}\{\cos 2t\} = \dfrac{s}{s^2+4}$. Then,

$$\mathcal{L}\{t\cos 2t\} = (-1)\frac{d}{ds}\left(\frac{s}{s^2+4}\right)$$

$$= -\frac{(s^2+4) - s \times 2s}{(s^2+4)^2} = \frac{s^2-4}{(s^2+4)^2}.$$

(b) Because $n = 2$ and $F(s) = \mathcal{L}\left\{e^{-3t}\right\} = \dfrac{1}{s+3}$, we have

$$\mathcal{L}\left\{t^2 e^{-3t}\right\} = (-1)^2 \frac{d^2}{ds^2}\left(\frac{1}{s+3}\right) = \frac{2}{(s+3)^2}.$$

EXAMPLE 8.1.10

Find $\mathcal{L}\{t^n\}$.

Solution

Using the theorem with $\mathcal{L}\{t^n\} = \mathcal{L}\{t^n \times 1\}$, we have $f(t) = 1$. Then, $F(s) = \mathcal{L}\{1\} = s^{-1}$. Calculating the derivatives of F, we obtain

$$\frac{dF}{ds}(s) = -\frac{1}{s^2}$$

$$\frac{d^2 F}{ds^2}(s) = \frac{2}{s^3}$$

$$\frac{d^3 F}{ds^3}(s) = -\frac{3 \times 2}{s^4}$$

$$\vdots$$

$$\frac{d^n F}{ds^n}(s) = (-1)^n \frac{n!}{s^{n+1}}$$

Recall that for nonnegative integers n, $\Gamma(n+1) = n!$.

Therefore,

$$\mathcal{L}\{t^n\} = \mathcal{L}\{t^n \times 1\} = (-1)^n(-1)^n \frac{n!}{s^{n+1}}$$

$$= (-1)^{2n}\frac{n!}{s^{n+1}} = \frac{n!}{s^{n+1}}.$$

EXAMPLE 8.1.11

Compute the Laplace transform of $f(t)$, $f'(t)$, and $f''(t)$ if $f(t) = (3t-1)^3$.

TABLE 8.1 Laplace Transforms of Frequently Encountered Functions

$f(t)$	$F(s) = \mathcal{L}(f(t))$	$f(t)$	$F(s) = \mathcal{L}(f(t))$
1	$\dfrac{1}{s}, s > 0$	$t^n, n = 1, 2, \ldots$	$\dfrac{n!}{s^{n+1}}, s > 0$
e^{at}	$\dfrac{1}{s-a}, s > a$	$t^n e^{at}, n = 1, 2, \ldots$	$\dfrac{n!}{(s-a)^{n+1}}$
$\sin kt$	$\dfrac{k}{s^2 + k^2}$	$e^{at} \sin kt$	$\dfrac{k}{(s-a)^2 + k^2}$
$\cos kt$	$\dfrac{s}{s^2 + k^2}$	$e^{at} \cos kt$	$\dfrac{s-a}{(s-a)^2 + k^2}$
$\sinh kt$	$\dfrac{k}{s^2 - k^2}$	$e^{at} \sinh kt$	$\dfrac{k}{(s-a)^2 - k^2}$
$\cosh kt$	$\dfrac{s}{s^2 - k^2}$	$e^{at} \cosh kt$	$\dfrac{s-a}{(s-a)^2 - k^2}$
$t^n f(t), n = 1, 2, \ldots$	$(-1)^n \dfrac{d^n}{ds^n} \mathcal{L}\{f(t)\} = (-1)^n \dfrac{d^n F}{ds^n}(s)$		

Solution

First, $f(t) = (3t-1)^3 = 27t^3 - 27t^2 + 9t - 1$ and $\mathcal{L}\{t^n\} = \dfrac{n!}{s^{n+1}}$ so

$$\mathcal{L}\{f(t)\} = 27\frac{3!}{s^4} + 27\frac{2!}{s^3} + 9\frac{1}{s^2} - \frac{1}{s}$$

$$= \frac{1}{s^4}\left(162 - 54s + 9s^2 - s^3\right).$$

By the previous theorem, $\mathcal{L}\{f'(t)\} = s\mathcal{L}\{f(t)\} - f(0)$. Hence,

$$\mathcal{L}\{f'(t)\} = s \times \frac{1}{s^4}\left(162 - 54s + 9s^2 - s^3\right) - f(0)$$

$$= \frac{1}{s^3}\left(162 - 54s + 9s^2 - s^3\right) + 1$$

$$= \frac{9}{s^3}\left(18 - 6s + s^2\right).$$

Similarly, $\mathcal{L}\{f''(t)\} = s^2 \mathcal{L}\{f(t)\} - sf(0) - f'(0)$:

$$\mathcal{L}\{f''(t)\} = s^2 \frac{9}{s^3}\left(18 - 6s + s^2\right) - sf(0) - f'(0)$$

$$= \frac{54}{s^2}(3 - s).$$

Using the properties of the Laplace transform discussed here and in the exercises, we can compute the Laplace transforms of a large number of frequently encountered functions. See Table 8.1. A more comprehensive table of Laplace transforms of many frequently encountered functions is found on the inside cover of this textbook.

 EXERCISES 8.1

In Exercises 1-16, use the definition of the Laplace transform to compute the Laplace transform of each function.

1. $f(t) = 21t$
2. $f(t) = 7e^{-t}$
3. $f(t) = 2e^t$
4. $f(t) = -8\cos 3t$
5. $f(t) = 2\sin 2t$
6. $f(t) = \begin{cases} 1, & \text{if } 0 \le t \le 2 \\ 0, & \text{if } t > 2 \end{cases}$
7. $f(t) = \begin{cases} 0, & \text{if } 0 \le t \le 1 \\ 1, & \text{if } t > 1 \end{cases}$
8. $f(t) = \begin{cases} 1 - t, & \text{if } 0 \le t \le 1 \\ 0, & \text{if } t > 1 \end{cases}$
9. $f(t) = \begin{cases} \cos t, & \text{if } 0 \le t \le \pi/2 \\ 0, & \text{if } t > \pi/2 \end{cases}$

10. $f(t) = \begin{cases} \sin t, & \text{if } 0 \le t \le 2\pi \\ 0, & \text{if } t > 2\pi \end{cases}$

11. $f(t) = \begin{cases} 3 - t, & \text{if } 0 \le t \le 3 \\ 0, & \text{if } t > 3 \end{cases}$

12. $f(t) = \begin{cases} 3t + 2, & \text{if } 0 \le t \le 5 \\ 0, & \text{if } t > 5 \end{cases}$

13. $f(t) = \begin{cases} 1, & \text{if } 0 \le t < 10 \\ -1, & \text{if } t \ge 10 \end{cases}$

14. $f(t) = \begin{cases} t, & \text{if } 0 \le t < 2 \\ 2, & \text{if } t \ge 2 \end{cases}$

15. $f(t) = \sin kt$

16. $f(t) = \cos kt$

17. Use the definitions, $\cosh kt = \frac{1}{2}(e^{kt} + e^{-kt})$ and $\sinh kt = \frac{1}{2}(e^{kt} - e^{-kt})$ to verify each of the following.
(a) $\mathcal{L}(\cosh kt) = s/(s^2 - k^2)$
(b) $\mathcal{L}(\sinh kt) = k/(s^2 - k^2)$
(c) $\mathcal{L}(e^{at}\cosh kt) = (s - a)/[(s - a)^2 - k^2]$
(d) $\mathcal{L}(e^{at}\sinh kt) = k/[(s - a)^2 - k^2]$

18. Use the definitions $\cos kt = \frac{1}{2}(e^{ikt} + e^{-ikt})$ and $\sin kt = \frac{1}{2}(e^{ikt} - e^{-ikt})$ to verify each of the following. (Note that $i = \sqrt{-1}$.)
(a) $\mathcal{L}(e^{at}\cos kt) = (s - a)/[(s - a)^2 + k^2]$
(b) $\mathcal{L}(e^{at}\sin kt) = k/[(s - a)^2 + k^2]$

19. *(Shifting property)* Suppose that $\mathcal{L}(f(t)) = F(s)$ exists for $s > a$. Use the definition of $\mathcal{L}\{e^{at}f(t)\}$ to show that $\mathcal{L}\{e^{at}f(t)\} = F(s - a)$.

In Exercises 20-25, use the result of Exercise 19 to compute the following. Compare your results with those found using Table 8.1.

20. $\mathcal{L}\{e^t \sin 2t\}$
21. $\mathcal{L}\{e^t \cos 3t\}$
22. $\mathcal{L}\{e^{-2t} \cos 4t\}$
23. $\mathcal{L}\{e^{-t} \sin 5t\}$
24. $\mathcal{L}\{te^{3t}\}$
25. $\mathcal{L}\{t^2 e^{-t/2}\}$

In Exercises 26-51, use properties of the Laplace transform and Table 8.1 or the table on the inside cover of this textbook to compute the Laplace transform of each function.

26. $f(t) = 28e^t$
27. $f(t) = -18e^{3t}$
28. $f(t) = \frac{1}{4}t \sin 2t$
29. $f(t) = t \cos 5t$
30. $f(t) = 1 + \cos 2t$
31. $f(t) = 1 + \sin 5t$
32. $f(t) = te^{-2t} \sin t$
33. $f(t) = -te^{-t} \cos 2t$
34. $f(t) = -6t^4 e^{-7t}$
35. $f(t) = \frac{1}{5040}t^7 e^{-t}$
36. $f(t) = 1 + t + t^2$
37. $f(t) = t - \frac{1}{6}t^3 + \frac{1}{120}t^5$
38. $f(t) = t^4 \cos t$
39. $f(t) = t^4 \sin t$
40. $f(t) = t \cosh 4t$
41. $f(t) = t \sinh 3t$
42. $f(t) = -2t \cos 3t$
43. $f(t) = t \sin 7t$
44. $f(t) = t^2 \sinh 6t$
45. $f(t) = t^2 \cosh 3t$
46. $f(t) = e^t \cosh t$
47. $f(t) = e^{-t} \sinh 2t$
48. $f(t) = -7e^{6t} \sin 2t$
49. $f(t) = e^{-2t} \cos 4t$
50. $f(t) = e^{2t} \cos 7t$
51. $f(t) = e^{-5t} \sin 4t$

52. Recall that the geometric series $\sum_{n=0}^{\infty} ar^n = a + ar + ar^2 + \cdots = a/(1 - r)$ if $|r| < 1$. (The series diverges if $|r| \ge 1$.) Also, recall that the Maclaurin series for $f(t) = e^t$ is $e^t = \sum_{n=0}^{\infty} \frac{1}{n!}t^n = 1 + t + \frac{1}{2!}t^2 + \frac{1}{3!}t^3 + \cdots$. Take the Laplace transform of each term in the Maclaurin series and show that the sum of the resulting series converges to $\mathcal{L}\{e^t\} = 1/(s - a)$.

53. Use the Maclaurin series
$$\sin t = \sum_{n=0}^{\infty}\frac{(-1)^n}{(2n+1)!}t^{2n+1}$$ to verify that
$\mathcal{L}\{\sin t\} = 1/(s^2+1)$.

54. Use the Maclaurin series
$$\cos t = \sum_{n=0}^{\infty}\frac{(-1)^n}{(2n)!}t^{2n}$$ to verify that
$\mathcal{L}\{\cos t\} = s/(s^2+1)$.

55. Find (a) $\mathcal{L}\{\cos^2 kt\}$ and (b) $\mathcal{L}\{\sin^2 kt\}$.

56. Use the identities
$\sin(A+B) = \sin A \cos B + \cos A \sin B$ and
$\cos(A+B) = \cos A \cos B - \sin A \sin B$ to
assist in finding the Laplace transform of
the following functions.
(a) $f(t) = \sin(t+\pi/4)$
(b) $f(t) = \cos(t+\pi/4)$
(c) $f(t) = \cos(t+\pi/6)$
(d) $f(t) = \sin(t+\pi/6)$

57. The *Gamma function*, $\Gamma(x)$, is defined by
$\Gamma(x) = \int_0^{\infty} t^{x-1}e^{-t}dt$, $x > 0$. Show that
$\mathcal{L}\{t^{\alpha}\} = \Gamma(\alpha+1)s^{-(\alpha+1)}$, $\alpha > -1$.

58. Use the result of Exercise 57 to find the
Laplace transform of (a) $f(t) = 1/\sqrt{t}$ and
(b) $f(t) = \sqrt{t}$.

59. (a) Use the definition of *exponential order* to
show that if $\lim_{t\to\infty} f(t)e^{-bt}$ exists and is
finite, $f(t)$ is of exponential order and if
$\lim_{t\to\infty} f(t)e^{-bt} = \infty$ for every value of b,
$f(t)$ is not of exponential order. (b) Give an
example of a function $f(t)$ of exponential
order b for which $\lim_{t\to\infty} f(t)e^{-bt}$ does not
exist.

60. Let $f(t)$ be a piecewise-continuous function
on the interval $[0,\infty)$ that is of exponential
order b for $t > T$. Show that $h(t) = \int_0^t f(u)\,du$
is also of exponential order.

61. Determine if $F(s)$ is the Laplace transform of
a piecewise-continuous function of
exponential order. (a) $F(s) = \dfrac{s}{4-s}$;
(b) $F(s) = \dfrac{3s}{s+1}$; (c) $F(s) = \dfrac{s^2}{4s+10}$;
(d) $F(s) = \dfrac{5s^3}{s^2+1}$.

62. (a) Show that $f(t) = t^3$ is of exponential
order. (b) Show that $f(t) = e^{t^2}$ is not of
exponential order. (c) Is $f(t) = t^n$, where n is
a positive integer, of exponential
order? Why?

63. Determine values of C and T that show that
$f(t) = (\sin t \cos 2t)/t$ is of exponential
order 1/2.

64. Let $f_n(t) = \begin{cases} 1/n, & \text{if } 0 \le t \le n \\ 0, & \text{if } t > n \end{cases}$. (a) Graph
$f_n(t)$ for $n = 10, 1, 1/10$, and $1/100$. Is $f_n(t)$
of exponential order? Why? (b) Evaluate
$\int_0^{\infty} f_n(t)\,dt$ and $\lim_{n\to 0^+}\int_0^{\infty} f_n(t)\,dt$.
(c) Describe $\lim_{n\to 0^+} f_n(t)$. Is this limit a
function? (d) Calculate $F_n(s) = \mathcal{L}\{f_n(t)\}$
and then $\lim_{n\to 0^+} F_n(s)$. Are you surprised
by the result? Explain. Is every function $F(s)$
the Laplace transform of some function
$f(t)$? Why?

8.2 THE INVERSE LAPLACE TRANSFORM

- **DEFINITION OF THE INVERSE LAPLACE TRANSFORM**
- **PROPERTIES OF THE INVERSE LAPLACE TRANSFORM**

Up to now, we were concerned with finding
the Laplace transform of a given function. Now,
we reverse the process: If possible, given a function $F(s)$ we want to find a function $f(t)$ such that
$\mathcal{L}(f(t)) = F(s)$.

Definition 34 (Inverse Laplace Transform).
*The **inverse Laplace transform** of the function $F(s)$
is the function $f(t)$, if such a function exists, that
satisfies $\mathcal{L}\{f(t)\} = F(s)$. If the inverse of $F(s)$ exists,
we denote the inverse Laplace transform of $F(s)$ with*

$$f(t) = \mathcal{L}^{-1}\{F(s)\}.$$

As we mentioned, there is a one-to-one cor-
respondence between continuous functions and

their Laplace transforms. With this in mind, we can compile a list of functions with the corresponding Laplace transforms. Such a table of Laplace transforms, given on the inside cover of this textbook as well as in Table 8.1 at the end of this section, is useful in finding the inverse Laplace transform of a given function. In the table, we look for $F(s)$ in the right-hand column to find $f(t) = \mathcal{L}^{-1}\{F(s)\}$ in the corresponding left-hand column. For example, the second row of the table indicates that $\mathcal{L}^{-1}\{4!/s^5\} = t^4$ and the ninth row shows that $\mathcal{L}^{-1}\{(s-1)/[(s-1)^2 + 16]\} = e^t \cos 4t$. This example illustrates shifting in the variable s.

Laplace was known for his overuse of the phrase "It is therefore obvious that..." in his *Celestial Mechanics*. In several works, Laplace indicated the existence of galaxies besides the Milky Way and what we now call "black holes"—stars so dense that not even light can escape their gravity. His observations were made over 100 years before these discoveries were formally made!

Theorem 42 (Shifting in s). *If* $\mathcal{L}\{f(t)\} = F(s)$ *exists, then* $\mathcal{L}^{-1}\{F(s-a)\} = e^{at}f(t)$.

EXAMPLE 8.2.1

Find the inverse Laplace transform of (a) $F(s) = 1/(s-6)$; (b) $F(s) = 2/(s^2 + 4)$; (c) $F(s) = 6/s^4$; and (d) $F(s) = 6/(s+2)^4$.

Solution

(a) Because $\mathcal{L}\{e^{6t}\} = 1/(s-6)$, $\mathcal{L}^{-1}\{1/(s-6)\} = e^{6t}$. (b) Note that $\mathcal{L}\{\sin 2t\} = 2/(s^2+2^2) = 2/(s^2+4)$, so $\mathcal{L}^{-1}\{2/(s^2+4)\} = \sin 2t$. (c) Because $\mathcal{L}\{t^3\} = 3!/s^4$, $\mathcal{L}^{-1}\{6/s^4\} = t^3$. (d) Notice that $F(s) = 6/(s+2)^4$ is obtained from $6/s^4$ by substituting $(s+2)$ for s. Therefore, by the shifting property,

$$\mathcal{L}\{e^{-2t}t^3\} = \frac{6}{(s+2)^4}, \quad \text{so } \mathcal{L}^{-1}\left\{\frac{6}{(s+2)^4}\right\} = e^{-2t}t^3.$$

Find $\mathcal{L}^{-1}\left\{\dfrac{1}{s-4}\right\}$ *and* $\mathcal{L}^{-1}\left\{\dfrac{s}{s^2+4}\right\}$.

Just as we use the linearity property of the Laplace transform to find $\mathcal{L}\{af(t) + bg(t)\} = a\mathcal{L}\{f(t)\} + b\mathcal{L}\{g(t)\}$ for functions $f(t)$ and $g(t)$ and constants a and b, the inverse Laplace transform is linear as well.

Theorem 43 (Linearity Property of the Inverse Laplace Transform). *Suppose that* $\mathcal{L}^{-1}\{f(s)\}$ *and* $\mathcal{L}^{-1}\{G(s)\}$ *exist and are continuous on* $[0,\infty)$ *and that a and b are constants. Then,*

$$\mathcal{L}^{-1}\{aF(s) + bG(s)\} = a\mathcal{L}^{-1}\{f(s)\} + b\mathcal{L}^{-1}\{G(s)\}.$$

If the functions are not in the forms presented in the table on the inside cover of the book or in Table 8.1, we can make use of the linearity property to determine the inverse Laplace transform.

EXAMPLE 8.2.2

Find the inverse Laplace transform of (a) $F(s) = 1/s^3$; (b) $F(s) = -7/(s^2 + 16)$; (c) $F(s) = 1/s^3 - 7/(s^2 + 16)$; and (d) $F(s) = 5/s - 2/(s - 10)$.

Solution

Refer to the formulas given in Table 8.1. (a) $\mathcal{L}^{-1}\left\{\dfrac{1}{s^3}\right\} = \mathcal{L}^{-1}\left\{\dfrac{1}{2} \times \dfrac{2}{s^3}\right\} = \dfrac{1}{2}\mathcal{L}^{-1}\left\{\dfrac{2}{s^3}\right\} = \dfrac{1}{2}t^2.$

(b)

$$\mathcal{L}^{-1}\left\{-\frac{7}{s^2+16}\right\} = -7\mathcal{L}^{-1}\left\{\frac{1}{s^2+16}\right\}$$

$$= -7\mathcal{L}^{-1}\left\{\frac{1}{4} \times \frac{4}{s^2+4^2}\right\}$$

$$= \frac{-7}{4}\mathcal{L}^{-1}\left\{\frac{4}{s^2+4^2}\right\} = -\frac{7}{4}\sin 4t.$$

(c) Combining the results from (a) and (b),

$$\mathcal{L}^{-1}\left\{\frac{1}{s^3} - \frac{7}{s^2+16}\right\} = \mathcal{L}^{-1}\left\{\frac{1}{s^3}\right\} + \mathcal{L}^{-1}\left\{-\frac{7}{s^2+16}\right\}$$

$$= \frac{1}{2}t^2 - \frac{7}{4}\sin 4t.$$

Thus, for (d), $\mathcal{L}^{-1}\left\{\dfrac{5}{s} - \dfrac{2}{s-10}\right\} = 5\mathcal{L}^{-1}\left\{\dfrac{1}{s}\right\} - 2\mathcal{L}^{-1}\left\{\dfrac{1}{s-10}\right\} = 5 - 2e^{10t}.$

EXAMPLE 8.2.3

Find the inverse Laplace transform of $F(s) = \dfrac{2s-9}{s^2+25}$.

Solution

Remember to divide each term of the numerator by the denominator.

$$\mathcal{L}^{-1}\left\{\frac{2s-9}{s^2+25}\right\} = 2\mathcal{L}^{-1}\left\{\frac{s}{s^2+25}\right\} - 9\mathcal{L}^{-1}\left\{\frac{1}{s^2+25}\right\}$$

$$= 2\mathcal{L}^{-1}\left\{\frac{s}{s^2+25}\right\} - \frac{9}{5}\mathcal{L}^{-1}\left\{\frac{5}{s^2+25}\right\}$$

$$= 2\cos 5t - \frac{9}{5}\sin 5t.$$

Find $\mathcal{L}^{-1}\left\{\dfrac{3s+1}{s^2+16}\right\}$.

Sometimes we must complete the square in the denominator of $F(s)$ before finding $\mathcal{L}^{-1}\{F(s)\}$.

EXAMPLE 8.2.4

Determine $\mathcal{L}^{-1}\left\{\dfrac{s}{s^2+2s+5}\right\}$.

Solution

Notice that many of the forms of $F(s)$ in Table 8.1 and the table of Laplace transforms on the inside cover of this book involve a term of the form $s^2 + k^2$ in the denominator. Through shifting, this term is replaced by $(s-a)^2 + k^2$. We obtain a term of this form in the denominator by completing the square. This yields

$$\frac{s}{s^2+2s+5} = \frac{s}{(s^2+2s+1)+4} = \frac{s}{(s+1)^2+4}.$$

Because the variable s appears in the numerator, we must write it in the form $(s+1)$ to find the inverse Laplace transform. Doing so, we find that

$$\frac{s}{s^2+2s+5} = \frac{s}{(s+1)^2+4} = \frac{(s+1)-1}{(s+1)^2+4}.$$

Therefore,

$$\mathcal{L}^{-1}\left\{\frac{s}{s^2+2s+5}\right\} = \mathcal{L}^{-1}\left\{\frac{(s+1)-1}{(s+1)^2+4}\right\}$$

$$= \mathcal{L}^{-1}\left\{\frac{s+1}{(s+1)^2+4}\right\} - \frac{1}{2}\mathcal{L}^{-1}\left\{\frac{2}{(s+1)^2+4}\right\}$$

$$= e^{-t}\cos 2t - \frac{1}{2}e^{-t}\sin 2t.$$

Find $\mathcal{L}^{-1}\left\{\dfrac{2s}{s^2-4s+8}\right\}$.

When applicable, the following theorem is useful in showing that the inverse Laplace transform of a function $F(s)$ does not exist.

Theorem 44. *Suppose that $f(t)$ is a piecewise-continuous function on $[0,\infty)$ and of exponential order b Then,*

$$\lim_{s\to\infty} F(s) = \lim_{s\to\infty} \mathcal{L}\{f(t)\} = 0.$$

Thus, if $\lim_{s\to\infty} F(s) \neq 0$, it follows that $\mathcal{L}^{-1}\{F(s)\}$ does not exist.

EXAMPLE 8.2.5

For the following functions, determine if $F(s)$ is the Laplace transform of a piecewise-continuous function of exponential order. (a) $F(s) = 2s/(s-6)$ and (b) $F(s) = s^3/(s^2+16)$.

Solution

(a) $\lim_{s\to\infty} F(s) = \lim_{s\to\infty} 2s/(s-6) = 2 \neq 0$ and (b) $\lim_{s\to\infty} F(s) = \lim_{s\to\infty} s^3/(s^2+16) = \infty \neq 0$. In each case, $\lim_{s\to\infty} F(s) \neq 0$, so $F(s)$ is not the Laplace transform of a piecewise-continuous function of exponential order.

 Does $\mathcal{L}^{-1}\left\{\dfrac{s^2}{s^2+8s+16}\right\}$ exist?

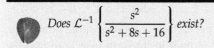 **EXERCISES 8.2**

In Exercises 1-25, compute the inverse Laplace transform of the given function.

1. $F(s) = 1/s^8$
2. $F(s) = 1/s^2$
3. $F(s) = 1/(s+5)$
4. $F(s) = 1/(s-3)$
5. $F(s) = 1/(s+6)^3$
6. $F(s) = 4/(s-2)^2$
7. $F(s) = 1/(s^2+9)$
8. $F(s) = s/(s^2+16)$
9. $F(s) = 1/(s^2+15s+56)$
10. $F(s) = 1/(s^2+12s+61)$
11. $F(s) = s/(s^2-5s-14)$
12. $F(s) = (s+3)/(s^2+6s+5)$
13. $F(s) = (s-2)/(s^2-4s)$
14. $F(s) = 1/(s^2+2s-24)$
15. $F(s) = 1/(s^2-4s-12)$
16. $F(s) = \dfrac{2s^2+1}{s(s^2+1)}$
17. $F(s) = \dfrac{2s^2-s+2}{s(s^2+1)}$
18. $F(s) = \dfrac{s^2+s+1}{s^2(s+1)}$
19. $F(s) = \dfrac{3s-4}{s(s-2)}$
20. $F(s) = \dfrac{1-s}{1+s^2}$
21. $F(s) = \dfrac{s^3+s^2+4}{s^2(s^2+4)}$
22. $F(s) = \dfrac{s+1}{(s-1)(s^2+1)}$
23. $F(s) = (s^2+s+2)/s^3$
24. $F(s) = (s^4-s^2+1)/s^6$
25. $F(s) = (s^3-s^2+2s-6)/s^5$
26. (a) Compute $f(t) = \mathcal{L}^{-1}\{\tan^{-1}(1/s)\}$ with $\mathcal{L}\{t^n f(t)\} = (-1)^n d^n/ds^n F(s)$ in the form $(n=1)$

$$f(t) = -\frac{1}{t}\mathcal{L}^{-1}\left\{\frac{d}{ds}F(s)\right\}$$
$$= -\frac{1}{t}\mathcal{L}^{-1}\left\{\frac{d}{ds}\tan^{-1}\left(\frac{1}{s}\right)\right\}.$$

(b) Show that $\mathcal{L}^{-1}\{\tan^{-1}(a/s)\} = (\sin at)/t$.
(c) Show that $\mathcal{L}^{-1}\{1/2 \tan^{-1}((a+b)/s) + 1/2 \tan^{-1}((a-b)/s)\} = (\sin at \cos bt)/t$.
27. (a) Use the linearity of the Laplace transform to compute $\mathcal{L}\{a\sin bt - b\sin at\}$ and $\mathcal{L}\{\cos bt - \cos at\}$. (b) Use these results to find $\mathcal{L}^{-1}\{1/[(s^2+a^2)(s^2+b^2)]\}$ and $\mathcal{L}^{-1}\{s/[(s^2+a^2)(s^2+b^2)]\}$.
28. Use the results of Exercise 27 to find (a) $\mathcal{L}^{-1}\left\{\dfrac{10s}{(s^2+1)(s^2+16)}\right\}$ and (b) $\mathcal{L}^{-1}\left\{\dfrac{7}{(s^2+100)(s^2+1)}\right\}$.
29. (a) Assume that $\mathcal{L}\{f(t)\} = F(s)$. Verify that $\mathcal{L}^{-1}\left\{\dfrac{1}{s}F(s)\right\} = \int_0^t f(\tau)d\tau$ and $\mathcal{L}^{-1}\left\{\dfrac{1}{s^2}F(s)\right\} = \int_0^t\int_0^\tau f(\lambda)d\lambda\,d\tau$. (b) Use (a) to find $\mathcal{L}^{-1}\left\{\dfrac{k}{s(s^2+k^2)}\right\}$ and $\mathcal{L}^{-1}\left\{\dfrac{k}{s^2(s^2+k^2)}\right\}$.

8.3 SOLVING INITIAL-VALUE PROBLEMS WITH THE LAPLACE TRANSFORM

In this section, we show how the Laplace transform is used to solve initial-value problems. To do this, we first need to understand how the Laplace transform of the derivatives of a function relates to the function itself. We begin with the first derivative.

Theorem 45 (Laplace Transform of the First Derivative). *Suppose that $f(t)$ is continuous for all $t \geq 0$ and is of exponential order b for $t > T$. Also, suppose that $f'(t)$ is piecewise continuous on any closed subinterval of $[0, \infty)$. Then, for $s > b$*

$$\mathcal{L}\{f'(t)\} = s\,\mathcal{L}\{f(t)\} - f(0).$$

Proof. Using integration by parts with $u = e^{-st}$ and $dv = f'(t)\,dt$, we have

$$\mathcal{L}\{f'(t)\} = \int_0^\infty e^{-st}f'(t)\,dt$$

$$= \lim_{M \to \infty} \left(e^{-st}f(t)\Big|_{t=0}^{t=M} + s\int_0^\infty e^{-st}f(t)\,dt \right)$$

$$= -f(0) + s\mathcal{L}\{f(t)\} = s\mathcal{L}\{f(t)\} - f(0).$$

Proof of Theorem 45 assumes that f' is a continuous function. If we use the assumption that f' is continuous on $0 < t_1 < t_2 < \cdots t_n < \infty$, we complete the proof by using

$$\mathcal{L}\{f'(t)\} = \int_0^{t_1} e^{-st}f'(t)\,dt + \int_{t_1}^{t_2} e^{-st}f'(t)\,dt$$

$$+ \cdots + \int_{t_n}^\infty e^{-st}f'(t)\,dt.$$

This is the same integration by parts formula shown in the proof of Theorem 45 for each integral. Now we make the same assumptions of f' and f'' as we did of f and f', respectively, in the statement of Theorem 45 and use Theorem 45 to develop an expression for $\mathcal{L}\{f''(t)\}$:

$$\mathcal{L}\{f''(t)\} = s\mathcal{L}\{f'(t)\} - f'(0)$$

$$= s[s\mathcal{L}\{f(t)\} - f(0)] - f'(0)$$

$$= s^2\mathcal{L}\{f(t)\} - sf(0) - f'(0).$$

Continuing this process, we can construct similar expressions for the Laplace transform of higher order derivatives, which leads to the following theorem.

Theorem 46 (Laplace Transform of Higher Derivatives). *More generally, if $f^{(i)}(t)$ is a continuous function of exponential order b on $[0, \infty)$ for $i = 0, 1, \ldots, n-1$ and $f^{(n)}(t)$ is piecewise continuous on any closed subinterval of $[0, \infty)$, then for $s > b$*

$$\mathcal{L}\{f^{(n)}(t)\} = s^n\mathcal{L}\{f(t)\} - s^{n-1}f(0) - \cdots -$$

$$sf^{(n-2)}(0) - f^{(n-1)}(0).$$

We will use this theorem and corollary in solving initial-value problems. However, we can also use them to find the Laplace transform of a function when we know the Laplace transform of the derivative of the function.

EXAMPLE 8.3.1

Find $\mathcal{L}\left\{\sin^2 kt\right\}$.

Solution

We can use the theorem to find the Laplace transform of $f(t) = \sin^2 kt$. Notice that $f'(t) = 2k\sin kt\cos kt = k\sin 2kt$. Then, because $\mathcal{L}\left\{f'(t)\right\} = s\mathcal{L}\left\{f(t)\right\} - f(0)$ and

$$\mathcal{L}\left\{f'(t)\right\} = \mathcal{L}\{k\sin 2kt\} = k\frac{2k}{s^2 + (2k)^2} = \frac{2k^2}{s^2 + 4k^2},$$

we have $\dfrac{2k^2}{s^2 + 4k^2} = s\mathcal{L}\left\{f(t)\right\} - 0$. Therefore,

$$\mathcal{L}\left\{f(t)\right\} = \frac{2k^2}{s\left(s^2 + 4k^2\right)}.$$

We now show how the Laplace transform can be used to solve initial-value problems. Typically, when we solve an initial-value problem that involves $y(t)$, we use the following steps:

1. compute the Laplace transform of each term in the differential equation;
2. solve the resulting equation for $\mathcal{L}\{y(t)\} = Y(s)$; and
3. determine $y(t)$ by computing the inverse Laplace transform of $Y(s)$.

The advantage of this method is that through the use of the property

$$\mathcal{L}\{f^{(n)}(t)\} = s^n\mathcal{L}\{f(t)\} - s^{n-1}f(0) - \cdots - sf^{(n-2)}(0) - f^{(n-1)}(0) \qquad (8.8)$$

we change the *differential* equation to an *algebraic* equation that can be solved for $\mathcal{L}\{f(t)\}$.

EXAMPLE 8.3.2

Solve the initial-value problem $y' - 4y = e^{4t}$, $y(0) = 0$.

 How does the solution change if $y(0) = 1$?

Solution

We begin by taking the Laplace transform of both sides of the differential equation. Because $\mathcal{L}\{y'\} = sY(s) - y(0) = sY(s)$, we have

$$\mathcal{L}\{y' - 4y\} = \mathcal{L}\{e^{4t}\}$$
$$\mathcal{L}\{y'\} - 4\mathcal{L}\{y\} = \frac{1}{s-4}$$
$$\underbrace{sY(s) - y(0)}_{\mathcal{L}\{y'\}} - 4\underbrace{Y(s)}_{\mathcal{L}\{y\}} = \frac{1}{s-4}$$
$$(s-4)Y(s) = \frac{1}{s-4}.$$

Solving this equation for $Y(s)$ yields $Y(s) = \frac{1}{(s-4)^2}$. By using the shifting property with $\mathcal{L}\{t\} = 1/s^2$, we have $y(t) = \mathcal{L}^{-1}\left\{\frac{1}{(s-4)^2}\right\} = te^{4t}$.

In many cases, we must determine a partial fraction decomposition of $Y(s)$ to obtain terms for which the inverse Laplace transform can be found.

EXAMPLE 8.3.3

Solve the initial-value problem $y'' - 4y' = 0$, $y(0) = 3$, $y'(0) = 8$.

Solution

Let $Y(s) = \mathcal{L}\{y(t)\}$. Then, applying the Laplace transform to the equation gives us $\mathcal{L}\{y'' - 4y'\} = \mathcal{L}\{0\}$. Because

$$\mathcal{L}\{y'' - 4y'\} = \mathcal{L}\{y''\} - 4\mathcal{L}\{y'\}$$
$$= \underbrace{s^2Y(s) - sy(0) - y'(0)}_{\mathcal{L}\{y''\}} - 4\underbrace{(sY(s) - y(0))}_{\mathcal{L}\{y'\}}$$

and $\mathcal{L}\{0\} = 0$ (why?), this equation becomes

$$s^2Y(s) - sy(0) - y'(0) - 4sY(s) + 4Y(0) = 0.$$

Applying the initial conditions $y(0) = 3$ and $y'(0) = 8$ to this results in

$$s^2Y(s) - 3s - 8 - 4sY(s) + 4 \times 3 = 0 \quad \text{or}$$
$$s(s-4)Y(s) = 3s - 4.$$

Solving for $Y(s)$, we find that $Y(s) = \frac{3s-4}{s(s-4)}$. If we expand the right-hand side of this equation in partial fractions based on the two linear factors in the denominator of $Y(s)$, we obtain $\frac{3s-4}{s(s-4)} = \frac{A}{s} + \frac{B}{s-4}$. Multiplying both sides of the equation by the denominator $s(s-4)$, we have $3s-4 = A(s-4)+Bs$. If we substitute $s = 0$ into this equation, we have $3 \times 0 - 4 = A \cdot (0 - 4) + B \times 0$ or $-4 = -4A$ so that $A = 1$. Similarly, is we substitute $s = 4$, we find that $8 = 4B$ so that $B = 2$. Therefore, $\frac{3s-4}{s(s-4)} = \frac{1}{s} + \frac{2}{s-4}$ so

$$y(t) = \mathcal{L}^{-1}\left\{\frac{1}{s} + \frac{2}{s-4}\right\} = 1 + 2e^{4t}.$$

A partial fraction decomposition involving a repeated linear factor is illustrated in the following example.

EXAMPLE 8.3.4

Solve the initial-value problem $y'' + 2y' + y = 6$, $y(0) = 5$, $y'(0) = 10$.

Solution

Let $\mathcal{L}\{y(t)\} = Y(s)$. Then,

$$\mathcal{L}\{y'' + 2y' + y\} = \mathcal{L}\{6\}$$
$$\mathcal{L}\{y''\} + 2\mathcal{L}\{y'\} + \mathcal{L}\{y\} = 6/s$$
$$\underbrace{s^2 Y(s) - s y(0) - y'(0)}_{\mathcal{L}\{y''\}}$$
$$+ 2\underbrace{(s Y(s) - y(0))}_{\mathcal{L}\{y'\}} + Y(s) = 6/s$$
$$(s^2 + 2s + 1)Y(s) = 6/s + 5s + 20$$
$$(s^2 + 2s + 1)Y(s) = (6 + 5s^2 + 20s)/s.$$

Solving for $Y(s)$ and factoring the denominator yields $Y(s) = \frac{5s^2 + 20s + 6}{s(s+1)^2}$. In this case, the denominator contains the nonrepeated linear factor s and the repeated linear factor $(s + 1)$. The partial fraction decomposition of $Y(s)$ is

$$\frac{5s^2 + 20s + 6}{s(s+1)^2} = \frac{A}{s} + \frac{B}{s+1} + \frac{C}{(s+1)^2},$$

so multiplication on each side of this equation by $s(s + 1)^2$ results in the equation

$$5s^2 + 20s + 6 = A(s+1)^2 + Bs(s+1) + Cs$$

or

$$5s^2 + 20s + 6 = (A + B)s^2 + (2A + B + C)s + A.$$

Equating the coefficients, we obtain the system

$$A + B = 5$$
$$2A + B + C = 20$$
$$A = 6,$$

which has solution $A = 6$, $B = -1$, and $C = 9$ so

$$\frac{5s^2 + 20s + 6}{s(s+1)^2} = \frac{6}{s} - \frac{1}{s+1} + \frac{9}{(s+1)^2}.$$

 Use Laplace transforms to solve $y' - y = 0$.

Therefore,

$$y(t) = \mathcal{L}^{-1}\left\{\frac{6}{s} - \frac{1}{s+1} + \frac{9}{(s+1)^2}\right\} = 6 - e^{-t} + 9te^{-t}.$$

In some cases, $F(s)$ involves irreducible quadratic factors as we see in the next example.

EXAMPLE 8.3.5

Solve the initial-value problem $y''' + 4y' = -10e^t$, $y(0) = 2$, $y'(0) = 2$, $y''(0) = -10$.

Solution

Let $\mathcal{L}\{y(t)\} = Y(s)$. Taking the Laplace transform of the equation and solving for $Y(s)$ gives us,

$$\mathcal{L}\{y'''\} + 4\mathcal{L}\{y'\} = \mathcal{L}\{-10e^t\}$$
$$\underbrace{s^3 Y(s) - s^2 y(0) - s y'(0) - y''(0)}_{\mathcal{L}\{y'''\}}$$
$$+ 4\underbrace{(s Y(s) - y(0))}_{\mathcal{L}\{y'\}} = -10/(s-1)$$
$$(s^3 + 4s)Y(s) = -10/(s-1)$$
$$+ 2s^2 + 2s - 2$$
$$Y(s) = -\frac{10}{(s-1)(s^3 + 4s)}$$
$$+ \frac{2s^2 + 2s - 2}{s^3 + 4s}$$
$$Y(s) = \frac{2s^3 - 4s - 8}{s(s-1)(s^2 + 4)}.$$

Finding the partial fraction decomposition of the right-hand side of this equation, we obtain $\frac{2s^3 - 4s - 8}{s(s-1)(s^2+4)} = \frac{A}{s} + \frac{B}{s-1} + \frac{Cs+D}{s^2+4}$. Multiplying each side of this equation by $s(s-1)(s^2 + 4)$ gives us

$$2s^3 - 4s - 8 = A(s-1)(s^2 + 4)$$
$$+ Bs(s^2 + 4) + (Cs + D)s(s - 1)$$
$$2s^3 - 4s - 8 = (A + B + C)s^3 + (-A - C + D)s^2$$
$$+ (4A + 4B - D)s - 4A.$$

Equating coefficients yields the system of equations

$$A + B + C = 2$$

$$-A - C + D = 0$$

$$4A + 4B - D = -4$$

$$-4A = -8,$$

with solution $A = 2, B = -2, C = 2,$ and $D = 4$, so

$$\frac{2s^3 - 4s - 8}{s(s-1)(s^2+4)} = \frac{A}{s} + \frac{B}{s-1} + \frac{Cs+D}{s^2+4}$$

$$= \frac{2}{s} - \frac{2}{s-1} + \frac{2s+4}{s^2+4}.$$

Therefore,

$$y(t) = \mathcal{L}^{-1}\left\{\frac{2}{s} - \frac{2}{s-1} + \frac{2s+4}{s^2+4}\right\}$$

$$= 2\mathcal{L}^{-1}\left\{\frac{1}{s}\right\} - 2\mathcal{L}^{-1}\left\{\frac{1}{s-1}\right\} + 2\mathcal{L}^{-1}\left\{\frac{s}{s^2+4}\right\}$$

$$+ 2\mathcal{L}^{-1}\left\{\frac{2}{s^2+4}\right\}$$

$$= 2 - 2e^t + 2\cos 2t + 2\sin 2t.$$

 EXERCISES 8.3

In Exercises 1-15, use Laplace transforms to solve the initial-value problems. Briefly describe how each equation could be solved using other methods such as undetermined coefficients or variation of parameters.

1. $y'' + 11y' + 24y = 0, y(0) = -1, y'(0) = 0$
2. $y'' + 8y' + 7y = 0, y(0) = 0, y'(0) = 1$
3. $y'' + 3y' - 10y = 0, y(0) = -1, y'(0) = 1$
4. $y'' - 13y' + 40y = 0, y(0) = 0, y'(0) = -2$
5. $y'' - y' = e^t, y(0) = 0, y'(0) = 1$
6. $y'' + 4y = \cos 2t, y(0) = y'(0) = 0$
7. $y''' + y' = e^t, y(0) = y'(0) = y''(0) = 0$
8. $y^{(4)} + y'' = 0, y(0) = y'(0) = y''(0) = 0, y'''(0) = 1$

9. $y^{(4)} + y'' = 1, y(0) = y'(0) = y''(0) = y'''(0) = 0$
10. $y''' + y' = e^{-t}, y(0) = y'(0) = y''(0) = 0$
11. $y'' + 2y' + y = e^{-t}, y(0) = y'(0) = 0$
12. $y''' + y' = 0, y(0) = y'(0) = 0, y''(0) = 1$
13. $y''' - y' = e^t + e^{-t}, y(0) = y'(0) = y''(0) = 0$
14. $18y''' + 27y'' + 13y' + 2y = 0,$ $y(0) = y'(0) = 0, y''(0) = 1$
15. $18y''' + 27y'' + 13y' + 2y = 1,$ $y(0) = y'(0) = y''(0) = 0$
16. (*Nonconstant coefficients*) Use the property that $\mathcal{L}\{tf(t)\} = -F'(s)$ to solve the initial-value problem $y'' + 4ty' - 8y = 4$, $y(0) = y'(0) = 0$. (*Hint*: Applying the Laplace transform to each term in the equation yields $Y'(s) + (3/s - s/4)Y(s) = -1/s^2$. Use the integrating factor $\mu(s) = e^{\int p(s)\,ds}$, where $p(s) = 3/s - s/4$, to find that the solution of this linear first-order equation is $Y(s) = s^{-3}(4 + Ce^{s^3/8})$. For what value of C does $\mathcal{L}^{-1}\{Y(s)\}$ exist? Substitute this value and find $y(t) = \mathcal{L}^{-1}\{Y(s)\} = 2t^2$.)

In Exercises 17-22, solve the initial-problem involving nonconstant coefficients.

17. $y'' + 3ty' - 6y = 3, y(0) = y'(0) = 0$
18. $y'' + 3ty' - 6y = 0, y(0) = 1, y'(0) = 0$
19. $y'' + ty' - 2y = 0, y(0) = 1, y'(0) = 0$
20. $y'' + ty' - 2y = 4, y(0) = y'(0) = 0$
21. $y'' + 2ty' - 4y = 2, y(0) = y'(0) = 0$
22. $y'' + 2ty' - 4y = 2, y(0) = 1, y'(0) = 0$
23. Find the differential equation satisfied by $Y(s)$ for (a) $y'' - ty = 0, y(0) = 1, y'(0) = 0$ (*Airy's equation*) and for (b) $(1 - t^2)y'' - 2ty' + n(n+1)y = 0, y(0) = 0,$ $y'(0) = 1$ (*Legendre's equation*). What is the order of each differential equation involving Y? Is there a relationship between the power of the independent variable in the original equation and the order of the differential equation involving Y?
24. Show that application of the Laplace transform to the initial-value problem

(involving a Cauchy-Euler equation)
$at^2y'' + bty' + cy = 0, y(0) = \alpha, y'(0) = \beta$
yields $as^2Y'' + (4a - b)sY' + (2a - b + c)Y = 0$,
where $Y(s) = \mathcal{L}\{y(t)\}$. Is the Laplace
transform method worthwhile in the case of
Cauchy-Euler equations?

25. Consider the initial-value problem
(involving *Bessel's equation* of order 0)
$ty'' + y' + ty = 0, y(0) = 1, y'(0) = 0$, with
solution $J_0(t)$, the Bessel function of order 0.
Use Laplace transforms to show that
$\mathcal{L}\{y(t)\} = \mathcal{L}\{J_0(t)\} = k/\sqrt{s^2 + 1}$, where k is a
constant. (*Hint:* $\mathcal{L}\{ty''\} = -d/ds\,(s^2Y(s) - s)$
and $\mathcal{L}\{ty\} = -d/ds\,Y(s)$.)

26. (See Exercise 25.) Use the binomial series to
expand $\mathcal{L}\{J_0(t)\} = \dfrac{k}{\sqrt{s^2+1}} = \dfrac{k}{s}\left(1 + \dfrac{1}{s^2}\right)^{-1/2}$.
Take the inverse transform of each term in
the expansion to show that
$J_0(t) = k\sum_{n=0}^{\infty} \dfrac{(-1)^n}{2^{2n}(n!)^2}t^{2n}$. Use the initial
condition $J_0(0) = 1$ to show that $k = 1$.
Therefore, $\mathcal{L}\{J_0(t)\} = 1/\sqrt{s^2 + 1}$.

Use the method of Laplace transforms to
solve the following initial-value problems.
In each case, use a computer algebra system
to assist in computing the inverse Laplace
transform either by using a system
command for performing the task or by
determining the partial fraction
decomposition. Graph the solution of each
problem.

27. $y'' + 2y' + 4y = t - e^{-t}, y(0) = 1, y'(0) = -1$
28. $y'' + 4y' + 13y = e^{-2t}\cos 3t + 1$,
    $y(0) = y'(0) = 1$
29. $y'' + 2y' + y = 2te^{-t} - e^{-t}, y(0) = 1$,
    $y'(0) = -1$
30. *Bessel's equation* is the equation
$$x^2\frac{d^2y}{dx^2} + x\frac{dy}{dx} + (x^2 - \mu^2)y = 0,$$
where $\mu \geq 0$ is a constant, and a general
solution is given by $y = c_1J_\mu(x) + c_2Y_\mu(x)$.
(a) Find α and β so that the solution of the
equation that satisfies the conditions

$y(0) = \alpha$ and $y'(0) = \beta$ is $y = J_\mu(x)$. (b)
Compute the Laplace transform of each side
of the equation $x^2y'' + xy' + (x^2 - \mu^2)x = 0$,
substitute $\mathcal{L}\{y\} = Y(s)$, and solve the
resulting second-order equation for $\mu = 1$,
2, 3, 4, 5. (c) Use the results obtained in (b)
to determine the Laplace transform of $J_\mu(x)$
for $\mu = 1, 2, 3, 4, 5$. (d) Can you generalize
the result you obtained in (c)?

31. Consider the definition of the Laplace
transform, $F(s) = \mathcal{L}\{f(t)\} = \int_0^\infty e^{-st}f(t)\,dt$. If
f satisfies the conditions given in
Theorem 10 (Laplace Transform of the First
Derivative), we can differentiate within the
integral with respect to s if $s > a$. Show that
$F'(s) = \mathcal{L}\{-tf(t)\}$ and $F''(s) = \mathcal{L}\{t^2f(t)\}$. Use
the result to generalize
$F^{(n)}(s) = \mathcal{L}\{(-1)^nt^nf(t)\}$ so that
$\mathcal{L}\{t^nf(t)\} = (-1)^nF^{(n)}(s)$.

8.4 LAPLACE TRANSFORMS OF SEVERAL IMPORTANT FUNCTIONS

- Piecewise Defined Functions: The Unit Step Function
- Periodic Functions
- Impulse Functions: The Delta Function

Piecewise Defined Functions: The Unit Step Function

An important function in modeling many
physical situations is the unit step function $\mathcal{U}$, shown in Figure 8.1 and defined as
follows.

Definition 35 (Unit Step Function). *The **unit
step function** $\mathcal{U}(t - a)$, where a is a given number, is
defined by*

$$\mathcal{U}(t - a) = \begin{cases} 0, & \text{if } t < a \\ 1, & \text{if } t \geq a \end{cases}.$$

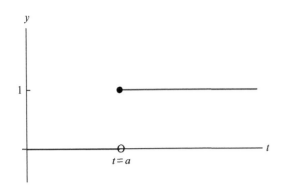

FIGURE 8.1   Graph of $\mathcal{U}(t - a)$.

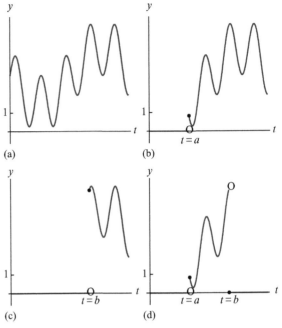

FIGURE 8.3   (a) Graph of a function $h(t)$. (b) $h(t)\mathcal{U}(t - a)$. (c) $h(t)\mathcal{U}(t - b)$. (d) $g(t) = h(t)\mathcal{U}(t - a) - h(t)\mathcal{U}(t - b) = h(t)\mathcal{U}(t - a) - \mathcal{U}(t - b)$. The graph of the function obtained by subtracting $h(t)\mathcal{U}(t - b)$ from $h(t)\mathcal{U}(t - a)$ is the graph of the function $h(t)$ between a and b.

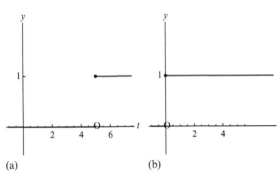

FIGURE 8.2   (a) Graph of $\mathcal{U}(t - 5)$. (b) Graph of $\mathcal{U}(t)$.

EXAMPLE 8.4.1

Graph (a) $\mathcal{U}(t - 5)$ and (b) $\mathcal{U}(t)$.

Solution

(a) In this case, $\mathcal{U}(t - 5) = \begin{cases} 0, & t < 5 \\ 1, & t \geq 5 \end{cases}$, so the jump occurs at $t = 5$. We graph $\mathcal{U}(t - 5)$ in Figure 8.2(a). (b) Here $\mathcal{U}(t) = \mathcal{U}(t - 0)$, so $\mathcal{U}(t) = 1$ for $t \geq 0$. We graph this function in Figure 8.2(b).

The unit step function is useful in defining functions that are piecewise continuous. Consider the function $f(t) = \mathcal{U}(t - a) - \mathcal{U}(t - b)$. If $t < a$, then $f(t) = 0 - 0 = 0$. If $a \leq t < b$,

then $f(t) = 1 - 0 = 1$. Finally, if $t \geq b$, then $f(t) = 1 - 1 = 0$. Hence, $\mathcal{U}(t - a) - \mathcal{U}(t - b) = \begin{cases} 0, & \text{if } t < a \text{ or } t \geq b \\ 1, & \text{if } a \leq t < b \end{cases}$. Thus, we can define the

function $g(t) = \begin{cases} 0, & \text{if } t < a \\ h(t), & \text{if } a \leq t < b \\ 0, & \text{if } t \geq b \end{cases}$ using the

unit step function as $g(t) = h(t)(\mathcal{U}(t - a) - \mathcal{U}(t - b))$, which is illustrated in Figure 8.3. Similarly, a function such as $f(t) = \begin{cases} g(t), & \text{if } t < a \\ h(t), & \text{if } t \geq a \end{cases}$

can be written in terms of the unit step function as

$$f(t) = g(t)[\mathcal{U}(t - 0) - \mathcal{U}(t - a)] + h(t)\mathcal{U}(t - a)$$
$$= g(t)[1 - \mathcal{U}(t - a)] + h(t)\mathcal{U}(t - a).$$

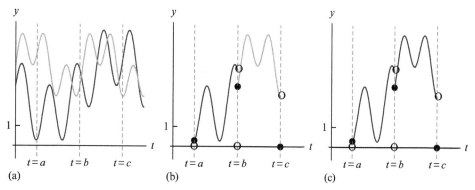

FIGURE 8.4   (a) Graph of $g(t)$ in black and $h(t)$ in gray. (b) Graph of $g(t)[\mathcal{U}(t-a)-\mathcal{U}(t-b)]$ in black and $h(t)[\mathcal{U}(t-a)-\mathcal{U}(t-b)]$ in gray. (c) Graph of $g(t)[\mathcal{U}(t-a)-\mathcal{U}(t-b)]+h(t)[\mathcal{U}(t-a)-\mathcal{U}(t-b)]$.

Thus, in terms of the unit step function, a function such as $f(t) = \begin{cases} 0, & \text{if } t < a \text{ or } t \geq c \\ g(t), & \text{if } a \leq t < b \\ h(t), & \text{if } b \leq t < c \end{cases}$ is

expressed as $f(t) = g(t)[\mathcal{U}(t-a) - \mathcal{U}(t-b)] + h(t)[\mathcal{U}(t-b) - \mathcal{U}(t-c)]$. See Figure 8.4.

Write $f(t) = \begin{cases} 1, & \text{if } t < 1 \\ -1, & \text{if } 1 \leq t < 2 \end{cases}$ *in terms of the unit step function.*

The reason for writing piecewise-continuous functions in terms of the unit step function is because we encounter functions of this type when solving initial-value problems. Using the methods in previous chapters, we solve the problem over each subinterval on which the function was continuous (i.e., "each piece of the function"). However, the method of Laplace transforms can be used to avoid the repeated calculations encountered then. We start to explain the process with the following theorem.

Theorem 47. *Suppose that* $F(s) = \mathcal{L}\{f(t)\}$ *exists for* $s > b \geq 0$. *If a is a positive constant, then*

$$\mathcal{L}\{f(t-a)\mathcal{U}(t-a)\} = e^{-as}F(s).$$

Proof. Using the definition of the Laplace transform, we obtain

$$\mathcal{L}\{f(t-a)\mathcal{U}(t-a)\} = \int_0^\infty e^{-st}f(t-a)\mathcal{U}(t-a)\,dt$$

$$= \int_0^a e^{-st}f(t-a)\underbrace{\mathcal{U}(t-a)}_{=0}\,dt$$

$$+ \int_a^\infty e^{-st}f(t-a)\underbrace{\mathcal{U}(t-a)}_{=1}\,dt$$

$$= \int_a^\infty e^{-st}f(t-a)\,dt.$$

Changing the variables with $u = t - a$ (where $du = dt$ and $t = u + a$) and changing the limits of integration, we have

$$\int_0^\infty e^{-s(u+a)}f(u)\,du = e^{-as}\int_0^\infty e^{-su}f(u)\,du$$

$$= e^{-as}\mathcal{L}\{f(t)\} = e^{-as}F(s).$$

EXAMPLE 8.4.2

Find (a) $\mathcal{L}\{\mathcal{U}(t-a)\}, a > 0$; (b) $\mathcal{L}\{(t-3)^5\mathcal{U}(t-s)\}$; and (c) $\mathcal{L}\{\sin(t - \pi/6)\mathcal{U}(t - \pi/6)\}$

Solution

(a) Because $\mathcal{L}\{\mathcal{U}(t-a)\} = \mathcal{L}\{1 \times \mathcal{U}(t-a)\}$, $f(t) = 1$. Thus, $f(t - a) = 1$ and

$$\mathcal{L}\{\mathcal{U}(t-a)\} = \mathcal{L}\{1 \times \mathcal{U}(t-a)\}$$

$$= e^{-as}\mathcal{L}\{1\} = e^{-as} \times \frac{1}{s} = \frac{e^{-as}}{s}.$$

(b) In this case $a = 3$ and $f(t) = t^5$. Thus,

$$\mathcal{L}\{(t-3)^5\mathcal{U}(t-s)\} = e^{-3s}\mathcal{L}\{t^5\}$$

$$= e^{-3s} \times \frac{5!}{s^6} = 120s^{-6}e^{-3s}.$$

(c) Here $a = \pi/6$ and $f(t) = \sin t$. Therefore,

$$\mathcal{L}\left\{\sin\left(t - \frac{\pi}{6}\right)\mathcal{U}\left(t - \frac{\pi}{6}\right)\right\} = e^{-\pi s/6}\mathcal{L}\{\sin t\}$$

$$= e^{-\pi s/6} \times \frac{1}{s^2 + 1} = \frac{e^{-\pi s/6}}{s^2 + 1}.$$

 Find $\mathcal{L}\{\cos(t - \pi/6)\mathcal{U}(t - \pi/6)\}$.

In most cases, we must calculate $\mathcal{L}\{g(t)\mathcal{U}(t-a)\}$ rather than $\mathcal{L}\{f(t-a)\mathcal{U}(t-a)\}$. To solve this problem, we let $g(t) = f(t-a)$, so $f(t) = g(t+a)$. Therefore,

$$\mathcal{L}\{g(t)\mathcal{U}(t-a)\} = e^{-as}\mathcal{L}\{g(t+a)\}.$$

EXAMPLE 8.4.3

Calculate (a) $\mathcal{L}\{t^2\mathcal{U}(t-1)\}$ and (b) $\mathcal{L}\{\sin t\,\mathcal{U}(t-\pi)\}$.

Solution

(a) Because $g(t) = t^2$ and $a = 1$,

$$\mathcal{L}\{t^2\mathcal{U}(t-1)\} = e^{-s}\mathcal{L}\{(t+1)^2\}$$

$$= e^{-s}\mathcal{L}\{t^2 + 2t + 1\} = e^{-s}\left(\frac{2}{s^3} + \frac{2}{s^2} + \frac{1}{s}\right)$$

(b) In this case, $g(t) = \sin t$ and $a = \pi$. Notice that $\sin(t + \pi) = \sin t \cos \pi + \cos t \sin \pi = -\sin t$. Thus,

$$\mathcal{L}\{\sin t\,\mathcal{U}(t-\pi)\} = e^{-\pi s}\mathcal{L}\{\sin(t+\pi)\}$$

$$= e^{-\pi s}\mathcal{L}\{-\sin t\}$$

$$= -e^{-\pi s} \times \frac{1}{s^2 + 1} = -\frac{e^{-\pi s}}{s^2 + 1}.$$

 Find $\mathcal{L}\{\cos t\,\mathcal{U}(t - \pi)\}$.

From the previous theorem, it follows:

Theorem 48. *Suppose that $F(s) = \mathcal{L}\{f(t)\}$ exists for $s > b \geq 0$. If a is a positive constant and $f(t)$ is continuous on $[0, \infty)$, then*

$$\mathcal{L}^{-1}\{e^{-as}F(s)\} = f(t-a)\mathcal{U}(t-a).$$

EXAMPLE 8.4.4

Find (a) $\mathcal{L}^{-1}\left\{\frac{e^{-4s}}{s^3}\right\}$ and (b) $\mathcal{L}^{-1}\left\{\frac{e^{-\pi s/2}}{s^2+16}\right\}$.

Solution

(a) If we write the expression $e^{-4s}s^{-3}$ in the form $e^{-as}F(s)$, we see that $a = 4$ and $F(s) = s^{-3}$. Hence, $f(t) = \mathcal{L}^{-1}\{s^{-3}\} = \frac{1}{2}t^2$ and

$$\mathcal{L}^{-1}\left\{\frac{e^{-4s}}{s^3}\right\} = f(t-4)\mathcal{U}(t-4)$$

$$= \frac{1}{2}(t-4)^2\mathcal{U}(t-4).$$

(b) In this case, $a = \pi/2$ and $F(s) = 1/(s^2 + 16)$. Then, $f(t) = \mathcal{L}^{-1}\{1/(s^2 + 16)\} = \frac{1}{4}\sin 4t$ and

$$\mathcal{L}^{-1}\left\{\frac{e^{-\pi s/2}}{s^2 + 16}\right\} = f\left(t - \frac{\pi}{2}\right)\mathcal{U}\left(t - \frac{\pi}{2}\right)$$

$$= \frac{1}{4}\sin\left[4\left(t - \frac{\pi}{2}\right)\right]\mathcal{U}\left(t - \frac{\pi}{2}\right)$$

$$= \frac{1}{4}\sin(4t - 2\pi)\mathcal{U}\left(t - \frac{\pi}{2}\right)$$

$$= \frac{1}{4}\sin 4t\,\mathcal{U}\left(t - \frac{\pi}{2}\right).$$

With the unit step function, we can solve initial-value problems that involve piecewise-continuous functions.

EXAMPLE 8.4.5

Solve $y'' + 9y = \begin{cases} 1, & 0 \le t < \pi \\ 0, & t \ge \pi \end{cases}$, subject to $y(0) = y'(0) = 0$.

Solution

To solve this initial-value problem we must compute $\mathcal{L}\{f(t)\}$, where $f(t) = \begin{cases} 1, & 0 \le t < \pi \\ 0, & t \ge \pi \end{cases}$.

Because this is a piecewise-continuous function, for $t \ge 0$ we write it in terms of the unit step function as

$$f(t) = 1 \times [\mathcal{U}(t-0) - \mathcal{U}(t-\pi)] + 0 \times \mathcal{U}(t-\pi)$$
$$= \mathcal{U}(t) - \mathcal{U}(t-\pi) = 1 - \mathcal{U}(t-\pi).$$

Then,

$$\mathcal{L}\{f(t)\} = \mathcal{L}\{1 - \mathcal{U}(t-\pi)\} = \frac{1}{s} - \frac{e^{\pi s}}{s}.$$

Hence, applying to the differential equation gives us

$$\mathcal{L}\{y''\} + 9\mathcal{L}\{y\} = \mathcal{L}\{f(t)\}$$
$$s^2 Y(s) - sy(0)$$
$$-y'(0) + 9Y(s) = \frac{1}{s} - \frac{e^{\pi s}}{s}$$
$$(s^2 + 9)Y(s) = \frac{1}{s} - \frac{e^{\pi s}}{s}$$
$$Y(s) = \frac{1}{s(s^2+9)} - \frac{e^{-\pi s}}{s(s^2+9)}.$$

Then,

$$y(t) = \mathcal{L}^{-1}\{Y(s)\} = \mathcal{L}^{-1}\left\{\frac{1}{s(s^2+9)}\right\}$$
$$-\mathcal{L}^{-1}\left\{\frac{e^{-\pi s}}{s(s^2+9)}\right\}.$$

Consider $\mathcal{L}^{-1}\left\{\frac{e^{-\pi s}}{s(s^2+9)}\right\}$. In the form of $\mathcal{L}^{-1}\{e^{-as}F(s)\}$, $a = \pi$ and $F(s) = 1/[s(s^2+9)]$. Now, $f(t) = \mathcal{L}^{-1}\{F(s)\}$ can be found with either a partial fraction expansion or with the formula

$$f(t) = \mathcal{L}^{-1}\left\{\frac{1}{s(s^2+9)}\right\} = \int_0^t \mathcal{L}^{-1}\left\{\frac{1}{s^2+9}\right\}$$

$$= \int_0^t \frac{1}{3}\sin 3\alpha \, d\alpha$$
$$= -\frac{1}{3}\frac{\cos 3\alpha}{3}\Big|_0^t = \frac{1}{9} - \frac{1}{9}\cos 3t.$$

Then with $\cos(3t - 3\pi) = \cos 3t \cos 3\pi + \sin 3t \sin 3\pi = -\cos 3t$, we have

$$\mathcal{L}^{-1}\left\{\frac{e^{-\pi s}}{s(s^2+9)}\right\} = \left[\frac{1}{9} - \frac{1}{9}\underbrace{\cos(3t-3\pi)}_{=\cos 3(t-\pi)}\right]\mathcal{U}(t-\pi)$$
$$= \left(\frac{1}{9} + \frac{1}{9}\cos 3t\right)\mathcal{U}(t-\pi).$$

Combining these results yields the solution

$$y(t) = \mathcal{L}^{-1}\{Y(s)\} = \mathcal{L}^{-1}\left\{\frac{1}{s(s^2+9)}\right\}$$
$$-\mathcal{L}^{-1}\left\{\frac{e^{-\pi s}}{s(s^2+9)}\right\}$$
$$= \frac{1}{9} - \frac{1}{9}\cos 3t - \left(\frac{1}{9} - \frac{1}{9}\cos t\right)\mathcal{U}(t-\pi).$$

Notice that without using the unit step function, $\mathcal{U}$, we can rewrite this function as a piecewise defined function as

$$y(t) = \begin{cases} \frac{1}{9} - \frac{1}{9}\cos 3t, & 0 \le t < \pi \\ -\frac{2}{9}\cos 3t, & t \ge \pi \end{cases},$$

which is graphed in Figure 8.5.

Periodic Functions

Another type of function that is encountered in many areas of applied mathematics is the *periodic* function.

Definition 36 (Periodic Function). *A function $f(t)$ is **periodic** if there is a positive number T such that $f(t+T) = f(t)$ for all $t \ge 0$. The minimum value of T that satisfies this equation is called the **period** of $f(t)$.*

The calculation of the Laplace transform of periodic functions is simplified through the use of the following theorem.

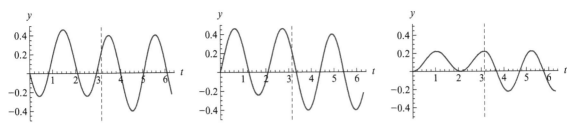

FIGURE 8.5 Graphs of three solutions of $y'' + 9y = \mathcal{U}(t) - \mathcal{U}(t - \pi)$ that satisfy $y(0) = 0$. Which one is the graph of the solution that satisfies $y'(0) = 0$?

Theorem 49 (Laplace Transform of Periodic Functions). *Suppose that $f(t)$ is a periodic function of period T and that $f(t)$ is piecewise continuous on $[0, \infty)$. Then, $\mathcal{L}\{f(t)\}$ exists for $s > 0$ and is given by the definite integral*

$$\mathcal{L}\{f(t)\} = \frac{1}{1 - e^{-sT}} \int_0^T e^{-st} f(t) \, dt.$$

Proof. We begin by writing $\mathcal{L}\{f(t)\} = \int_0^\infty e^{st} f(t) f(t) \, dt$ as the sum $\mathcal{L}\{f(t)\} = \int_0^T e^{-st} f(t) \, dt + \int_T^\infty e^{-st} f(t) \, dt$. Changing the variable in the second integral to $u = t - T$ from which it follows that $du = dt$, we obtain

$$\int_T^\infty e^{-st} f(t) \, dt = \int_0^\infty e^{-s(u+T)} \underbrace{f(u + T)}_{= f(u)} \, du$$

$$= e^{-sT} \int_0^\infty e^{-su} f(u) \, du$$

$$= e^{-sT} \mathcal{L}\{f(t)\}.$$

Then,

$$\mathcal{L}\{f(t)\} = \int_0^T e^{-st} f(t) \, dt + \int_T^\infty e^{-st} f(t) \, dt$$

$$= e^{-sT} \mathcal{L}\{f(t)\} + \int_0^T e^{-st} f(t) \, dt.$$

which is solved for $\mathcal{L}\{f(t)\}$ resulting in $\mathcal{L}\{f(t)\} = \frac{1}{1-e^{-sT}} \int_0^T e^{-st} f(t) \, dt$.

Because $\frac{1}{1-x} = \sum_{k=0}^\infty x^k$ if $|x| < 1$, $\frac{1}{1-e^{-sT}} = \sum_{k=0}^\infty e^{-skT}$ so

$$\mathcal{L}\{f(t)\} = \frac{1}{1 - e^{-sT}} \int_0^T e^{-st} f(t) \, dt$$

$$= \sum_{k=0}^\infty e^{-skT} \int_0^T e^{-st} f(t) \, dt.$$

 Is the Laplace transform of a periodic function a periodic function?

EXAMPLE 8.4.6

Find the Laplace transform of the periodic function $f(t) = t, 0 \leq t < 1, f(t + 1) = f(t), t \geq 1$.

Solution

The period of f, which is graphed in Figure 8.6, is $T = 1$. Through integration by parts,

$$\mathcal{L}\{f(t)\} = \frac{1}{1 - e^{-s}} \int_0^t e^{-st} t \, dt$$

$$= \frac{1}{1 - e^{-s}} \left\{ \left[-\frac{t e^{-st}}{s} \right]_{t=0}^{t=1} + \int_0^1 \frac{e^{-st}}{s} \, dt \right\}$$

$$= \frac{1}{1 - e^{-s}} \left\{ -\frac{e^{-s}}{s} - \left[\frac{e^{-st}}{s^2} \right]_{t=0}^{t=1} \right\}$$

$$= \frac{1}{1 - e^{-s}} \left[-\frac{e^{-s}}{s} + \frac{1 - e^{-s}}{s^2} \right]$$

$$= \frac{1 - (s+1)e^{-s}}{s^2(1 - e^{-s})}.$$

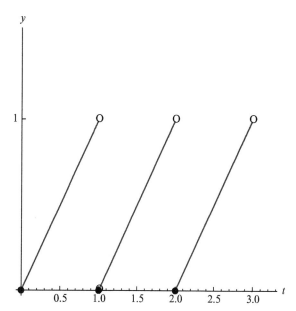

FIGURE 8.6 Although f is discontinuous, its Laplace transform is continuous for $s > 0$.

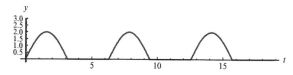

FIGURE 8.7 The half-wave rectification of $2\sin t$.

Laplace transforms can be used more easily than other methods to solve initial-value problems with periodic forcing functions.

EXAMPLE 8.4.7

Solve $y'' + y = f(t)$ subject to $y(0) = y'(0) = 0$ if $f(t) = \begin{cases} 2\sin t, & 0 \le t < \pi \\ 0, & \pi \le t < 2\pi \end{cases}$ and $f(t+2\pi) = f(t)$. ($f(t)$ is known as the *half-wave rectification* of $2\sin t$. See Figure 8.7.)

Solution

We begin by finding $\mathcal{L}\{f(t)\}$. Because the period of $f(t)$ is $T = 2\pi$, we have

$$\mathcal{L}\{f(t)\} = \frac{1}{1 - e^{-2\pi s}} \int_0^{2\pi} e^{-st} f(t)\, dt$$

$$= \frac{1}{1 - e^{-2\pi s}} \left[\int_0^{\pi} e^{-st} \times 2\sin t\, dt \right.$$

$$\left. + \int_{\pi}^{2\pi} e^{-st} \times 0\, dt \right]$$

$$= \frac{2}{1 - e^{-2\pi s}} \int_0^{\pi} e^{-st} \sin t\, dt.$$

Using integration by parts, a table of integrals, or a computer algebra system yields

$$\mathcal{L}\{f(t)\} = \frac{2}{1 - e^{-2\pi s}} \left[\frac{e^{-st}(-s\sin t - \cos t)}{s^2 + 1} \right]_0^{\pi}$$

$$= \frac{2}{1 - e^{-2\pi s}} \left[\frac{e^{-\pi s}}{s^2 + 1} + \frac{1}{s^2 + 1} \right]$$

$$= \frac{2(e^{-\pi s} + 1)}{(1 - e^{-2\pi s})(s^2 + 1)}$$

$$= \frac{2(e^{-\pi s} + 1)}{(1 - e^{-\pi s})(1 + e^{-\pi s})(s^2 + 1)}$$

$$= \frac{2}{(1 - e^{-\pi s})(s^2 + 1)}.$$

Taking the Laplace transform of both sides of the equation and solving for $Y(s)$ gives us

$$\mathcal{L}\{y''\} + \mathcal{L}\{y\} = \mathcal{L}\{f(t)\}$$

$$s^2 Y(s) - sy(0) - y'(0) + Y(s) = \frac{2}{(1 - e^{-\pi s})(s^2 + 1)}$$

$$Y(s) = \frac{2}{(1 - e^{-\pi s})(s^2 + 1)^2}.$$

Recall from your work with the geometric series that if $|x| < 1$, then

$$\frac{1}{1 - x} = \sum_{k=0}^{\infty} x^k = 1 + x + x^2 + x^3 + \cdots$$

Because we do not know the inverse Laplace transform of $\frac{2}{(1 - e^{-\pi s})(s^2 + 1)^2}$, we use a geometric series expansion of $1/(1 - e^{-\pi s})$ to obtain terms for which we can calculate the inverse Laplace transform. Replacing x with $e^{-\pi s}$ in the geometric series gives us

$$\frac{1}{1-e^{-\pi s}} = \sum_{k=0}^{\infty} e^{-\pi ks} = 1 + e^{-\pi s} + e^{-2\pi s} + e^{-3\pi s} + \cdots$$

so

$$Y(s) = \frac{2}{(s^2+1)^2} \sum_{k=0}^{\infty} e^{-\pi ks}$$

$$= (1 + e^{-\pi s} + e^{-2\pi s} + e^{-3\pi s} + \cdots) \frac{2}{(s^2+1)^2}$$

$$= 2 \sum_{k=0}^{\infty} \frac{e^{-\pi ks}}{(s^2+1)^2} = 2 \left[\frac{1}{(s^2+1)^2} + \frac{e^{-\pi s}}{(s^2+1)^2} \right.$$

$$\left. + \frac{e^{-2\pi s}}{(s^2+1)^2} + \frac{e^{-3\pi s}}{(s^2+1)^2} + \cdots \right].$$

Then,

$$y(t) = 2\mathcal{L}^{-1} \left\{ \sum_{k=0}^{\infty} \frac{e^{-\pi ks}}{(s^2+1)^2} \right\} = 2\mathcal{L}^{-1} \left\{ \frac{1}{(s^2+1)^2} \right.$$

$$+ \frac{e^{-\pi s}}{(s^2+1)^2} + \frac{e^{-2\pi s}}{(s^2+1)^2} + \cdots \Bigg\}$$

$$= 2 \sum_{k=0}^{\infty} \mathcal{L}^{-1} \left\{ \frac{e^{-\pi ks}}{(s^2+1)^2} \right\} = 2\mathcal{L}^{-1} \left\{ \frac{1}{(s^2+1)^2} \right\}$$

$$+ 2\mathcal{L}^{-1} \left\{ \frac{e^{-\pi s}}{(s^2+1)^2} \right\} +$$

$$\mathcal{L}^{-1} \left\{ \frac{e^{-2\pi s}}{(s^2+1)^2} \right\} + \cdots.$$

Notice that $\mathcal{L}^{-1}\left\{1/(s^2+1)^2\right\}$ is needed to find all the other terms. Using a computer algebra system, we have

$$\mathcal{L}^{-1} \left\{ \frac{1}{(s^2+1)^2} \right\} = \frac{1}{2}(\sin t - t \cos t)$$

and

$$\mathcal{L}^{-1} \left\{ \frac{e^{-\pi ks}}{(s^2+1)^2} \right\} = \frac{1}{2} [\sin(t-k\pi)$$

$$- (t-k\pi)\cos(t-k\pi)]\mathcal{U}(t-k\pi).$$

Then,

$$y(t) = \sum_{k=0}^{\infty} [\sin(t-k\pi) - (t-k\pi)\cos(t-k\pi)]\mathcal{U}(t-k\pi)$$

$$= (\sin t - t \cos t) + [\sin(t-\pi)$$

$$- (t-\pi)\cos(t-\pi)]\mathcal{U}(t-\pi) +$$

$$[\sin(t-2\pi) - (t-2\pi)\cos(t-2\pi)]\mathcal{U}(t-2\pi)$$

$$+ [\sin(t-3\pi) - (t-3\pi)\cos(t-3\pi)]$$

$$\mathcal{U}(t-3\pi) + \cdots.$$

We write $y = y(t)$ as the piecewise defined function

$$y(t) = \begin{cases} \sin t - t \cos t, & 0 \le t < \pi \\ -\pi \cos t, & \pi \le t < 2\pi \\ \sin t - t \cos t + \pi \cos t, & 2\pi \le t < 3\pi \\ -2\pi \cos t, & 3\pi \le t < 4\pi \\ \vdots \end{cases}$$

which is graphed in Figure 8.8. Is y a smooth function? (That is, is y' continuous?) Is y periodic?

Impulse Functions, the Delta Function

In some situations, we consider a force $f(t)$, known as an *impulse*, that acts only over a brief period of time, $t_0 < \alpha < t < t_0 + \alpha, \alpha > 0$. For example, we may strike a pendulum (or spring-mass system) or there may be a voltage surge in

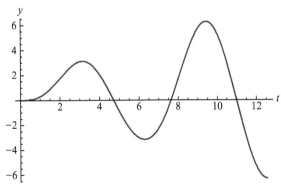

FIGURE 8.8 Even though the forcing function is discontinuous, the solution to the initial-value problem is continuous.

a circuit. In these cases, the impulse delivered by the force is given by

$$I = \int_{t_0-\alpha}^{t_0+\alpha} f(t)\, dt.$$

To better describe $f(t)$ suppose that this function has an impulse of 1 over the interval $t_0 - \alpha < t < t_0 + \alpha$, so that the impulse begins at $t = t_0 - \alpha$ and ends at $t = t_0 + \alpha$. Under these assumptions, we let

$$f(t) = \delta_\alpha(t - t_0) = \begin{cases} \dfrac{1}{2\alpha}, & t_0 - \alpha < t < t_0 + \alpha \\ 0, & \text{otherwise} \end{cases}.$$

We graph $\delta_\alpha(t - t_0)$ for several values of α in Figure 8.9.

 Graph $f(t) = \delta_{1/4}(t - 1)$. What is the maximum value of $f(t)$?

The impulse of $f(t) = \delta_\alpha(t - t_0)$ is given by

$$I = \int_{t_0-\alpha}^{t_0+\alpha} f(t)\, dt = \int_{t_0-\alpha}^{t_0+\alpha} \frac{1}{2\alpha}\, dt = \frac{1}{2\alpha}[(t_0 + \alpha)$$

$$- (t_0 - \alpha)] = \frac{1}{2\alpha} \times 2\alpha = 1.$$

Notice that the value of this integral does not depend on α as long as α is not zero. We now try to create the *idealized impulse function* by requiring

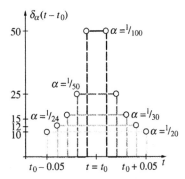

FIGURE 8.9   The area under the graph of $y = \delta_\alpha(t - t_0)$ is independent of α and t_0: the area is 1.

that $\delta_\alpha(t - t_0)$ act on smaller and smaller intervals. From the integral calculation, we have

$$\lim_{\alpha \to 0^+} I(t) = 1.$$

We also note that

$$\lim_{\alpha \to 0^+} \delta_\alpha(t - t_0) = 0, \quad t \neq t_0.$$

We use these properties to define the idealized unit impulse function. Notice that this idealized function represents an instantaneous impulse of magnitude one that acts at $t = t_0$.

Definition 37 (Unit Impulse Function). *The (idealized) unit impulse function δ satisfies*

$$\delta(t - t_0) = 0,\ t \neq t_0 \quad \text{and} \quad \int_{-\infty}^{+\infty} \delta(t - t_0)\, dt = 1.$$

We now state the following useful theorem involving the unit impulse function.

Theorem 50. *Suppose that $g(t)$ is a bounded and continuous function. Then,*

$$\int_{-\infty}^{\infty} \delta(t - t_0)\, g(t)\, dt = g(t_0).$$

The "function" $\delta(t - t_0)$, known as the *Dirac delta function*, is an example of a *generalized function*. It is not a function of the type studied in first-year calculus. The Dirac delta function is useful in the definition of impulse-forcing functions and arise in many areas of applied mathematics. Although it does not possess the properties required to apply the Laplace transform, we can determine $\mathcal{L}\{\delta(t - t_0)\}$ formally.

Theorem 51. *For $t_0 \geq 0$, $\mathcal{L}\{\delta(t - t_0)\} = e^{-st_0}$.*

Proof. Because the Laplace transform is linear, we find $\mathcal{L}\{\delta(t - t_0)\}$ through the following calculations:

$$\mathcal{L}\{\delta(t - t_0)\} = \mathcal{L}\left\{ \lim_{\alpha \to 0^+} \delta_\alpha(t - t_0) \right\}$$

$$= \lim_{\alpha \to 0^+} \mathcal{L}\{\delta_\alpha(t - t_0)\}.$$

We can represent the delta function $\delta_\alpha(t - t_0)$ in terms of the unit step function as

$$\delta_\alpha(t - t_0) = \frac{1}{2\alpha}\left[\mathcal{U}(t - (t_0 - \alpha)) - \mathcal{U}(t - (t_0 + \alpha)) \right].$$

Hence,

$$\mathcal{L}\{\delta_\alpha(t-t_0)\} = \mathcal{L}\left\{\frac{1}{2\alpha}[\mathcal{U}(t-(t_0-\alpha))\right.$$

$$\left.-\mathcal{U}(t-(t_0+\alpha))]\right\}$$

$$= \frac{1}{2\alpha}\left[\frac{e^{-s(t_0-\alpha)}}{s}-\frac{e^{-s(t_0+\alpha)}}{s}\right]$$

$$= e^{-st_0} \times \frac{e^{\alpha s}-e^{-\alpha s}}{2\alpha s}.$$

Therefore,

$$\mathcal{L}\{\delta(t-t_0)\} = \lim_{\alpha \to 0^+} \mathcal{L}\{\delta_\alpha(t-t_0)\}$$

$$= \lim_{\alpha \to 0^+} e^{-st_0} \times \frac{e^{\alpha s}-e^{-\alpha s}}{2\alpha s}.$$

Because the limit is of the indeterminant form $0/0$, we use L'Hopital's rule to find that

$$\mathcal{L}\{\delta(t-t_0)\} = \lim_{\alpha \to 0^+} e^{-st_0} \times \frac{e^{\alpha s}-e^{-\alpha s}}{2\alpha s}$$

$$= \lim_{\alpha \to 0^+} e^{-st_0} \times \frac{se^{\alpha s}+se^{-\alpha s}}{2s}$$

$$= e^{-st_0} \times 1 = e^{-st_0}.$$

EXAMPLE 8.4.8

Find (a) $\mathcal{L}\{\delta(t-1)\}$; (b) $\mathcal{L}\{\delta(t-\pi)\}$; and (c) $\mathcal{L}\{\delta(t)\}$.

Solution

(a) In this case $t_0 = 1$, so $\mathcal{L}\{\delta(t-1)\} = e^{-s}$. (b) With $t_0 = \pi$, $\mathcal{L}\{\delta(t-\pi)\} = e^{-\pi s}$. (c) Because $t_0 = 0$, $\mathcal{L}\{\delta(t)\} = e^{-s \times 0} = 1$.

EXAMPLE 8.4.9

Solve $y'' + y = 3\sin t\,\delta(t-\pi/2)$, $y(0) = 0$, $y'(0) = -1$.

Solution

We solve this initial-value problem by taking the Laplace transform of both sides of the

differential equation and applying the initial conditions. This yields

$$L\{y''\} + L\{y\} = L\{3\sin t\,\delta(t-\pi/2)\}$$

$$s^2Y(s) - sy(0) - y'(0) + Y(s) = 3e^{-\pi s/2}$$

$$(s^2+1)Y(s) + 1 = 3e^{-\pi s/2}$$

$$Y(s) = \frac{3e^{-\pi s/2}-1}{s^2+1}$$

$$= \frac{3e^{-\pi s/2}}{s^2+1} - \frac{1}{s^2+1}.$$

Because $\mathcal{L}^{-1}\{1/(s^2+1)\} = \sin t$, it follows that

$$y(t) = \mathcal{L}^{-1}\left\{\frac{3e^{-\pi s/2}}{s^2+1} - \frac{1}{s^2+1}\right\}$$

$$= -\sin t - 3\mathcal{U}(t-\pi/2)\cos t.$$

(*Note:* $\sin(t-\pi/2) = -\cos t$.) The graph of $y(t)$ is shown in Figure 8.10. Notice the sharp inward change of the velocity (y') that occurs when $t = \pi/2$.

 EXERCISES 8.4

Find the Laplace transform of the given function.

1. $-28\mathcal{U}(t-3)$
2. $-16\mathcal{U}(t-6)$
3. $3\mathcal{U}(t-8) - \mathcal{U}(t-4)$
4. $2\mathcal{U}(t-5) + 7\mathcal{U}(t-9)$
5. $-42e^{t-4}\mathcal{U}(t-4)$
6. $6e^{t-2}\mathcal{U}(t-2)$
7. $12\sinh(t-2)\mathcal{U}(t-2)$
8. $\cosh(t-3)\mathcal{U}(t-3)$
9. $-14\sin(t-2\pi/3)\mathcal{U}(t-2\pi/3)$
10. $-3\cos(t-\pi/4)\mathcal{U}(t-\pi/4)$
11. $f(t) = \begin{cases} t, & 0 \le t < 1 \\ 0, & 1 \le t < 2 \end{cases}$ and $f(t) = f(t-2)$ if $t \ge 2$

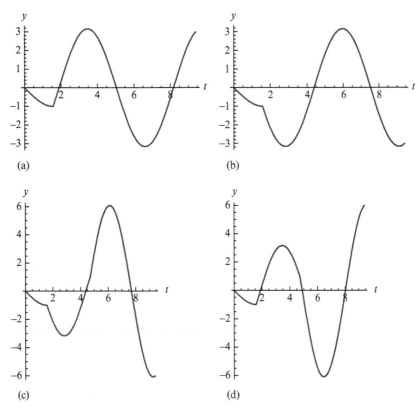

FIGURE 8.10 (a) Shows the graph of the solution in the example. In (b), (c), and (d), the forcing functions are $f(t) = -3\sin t\,\delta(t - \pi/2)$, $f(t) = -3\sin t\,(\delta(t - \pi/2) + \delta(t - 3\pi/2))$, and $f(t) = 3\sin t\,(\delta(t - \pi/2) + \delta(t - 3\pi/2))$. *Explain how to match each solution graph with its forcing function.*

12. $f(t) = \begin{cases} 1, & 0 \le t < 1 \\ 2 - t, & 1 \le t < 2 \end{cases}$ and $f(t) = f(t-2)$

    if $t \ge 2$

13. $f(t) = \begin{cases} 0, & 0 \le t < 1 \\ 1, & 1 \le t < 2 \\ 2, & 2 \le t \le 3 \end{cases}$ and $f(t) = f(t-3)$ if

    $t \ge 3$

14. $f(t) = \begin{cases} t, & 0 \le t < 1 \\ 1, & 1 \le t < 2 \\ 3 - t, & 2 \le t \le 3 \end{cases}$ and $f(t) = f(t-3)$

    if $t \ge 3$

15. $\delta(t - \pi)$

16. $\delta(t - 3)$

17. $\delta(t - 1) + 100\delta(t - 2)$

18. $2\delta(t - 3) - \delta(t - 4)$

19. $\sin t\,\mathcal{U}(t - 2\pi) + 100\delta(t - \pi/2)$

20. $\cos 2t\,\mathcal{U}(t - \pi) + 10\delta(t - \pi/4)$

21. $f(t) = \delta(t - 1)$, $0 \le t < 2$ and $f(t) = f(t-2)$ if

    $t \ge 2$

22. $f(t) = \delta(t - 2) - 2\delta(t - 1)$, $0 \le t < 3$,

    $f(t) = f(t-3)$ if $t \ge 3$

In Exercises 23–44, find the inverse Laplace transform of the given function.

23. $\dfrac{-3}{se^{\pi s}}$

24. $\dfrac{-10}{se^{4s}}$

25. $\dfrac{2e^{3s} - 3}{se^{4s}}$

26. $\dfrac{3 + e^{4s}}{se^{6s}}$

27. $\dfrac{3e^s - 4e^{3s} - 3}{se^{6s}}$

28. $\dfrac{5 - 6e^s - 3e^{2s}}{se^{4s}}$

29. $\dfrac{e^{-2s}}{s - 3}$

30. $\dfrac{e^{-s}}{s - 4}$

31. $\dfrac{se^{-3s}}{s^2 + 4}$

32. $\dfrac{e^{-5s}}{s(s^2 + 1)}$

33. $\dfrac{e^{-5s}}{s(s^2 + 4)}$

34. $\dfrac{1 - e^s + se^{2s}}{s^2(e^{2s} - 1)}$

35. $\dfrac{2(1 - 2e^{2s} - e^{4s})}{s^2 e^{4s}(1 - e^{-4s})}$

36. $\dfrac{-9}{e^{4s}(s^2 - 1)}$

37. $\dfrac{s}{e^{3s}(s^2 - 4)}$

38. $\dfrac{1}{s^3(1 + e^{-3s})}$

39. $\dfrac{1}{s^2(1 + e^{-4s})}$

40. $\dfrac{s}{(1 + e^{-2s})(s^2 + 9)}$

41. $\dfrac{1}{(1 + e^{-s})(s^2 + 4)}$

42. $\dfrac{e^{\pi s}}{e^{2\pi s} - 1}$

43. $\dfrac{e^{s/2}}{1 - e^s}$

44. $\dfrac{e^s(e^s - 2)}{e^{3s} - 1}$

In Exercises 45-63, solve the initial-value problem.

45. $y' + 3y = f(t), y(0) = 1$,
$$f(t) = \begin{cases} 1, & 0 \le t < 2 \\ 0, & t \ge 2 \end{cases}$$

46. $y' + y = f(t), y(0) = 0$,
$$f(t) = \begin{cases} \sin t, & 0 \le t < 1 \\ 0, & t \ge 1 \end{cases}$$

47. $y'' + 4y' = f(t), y(0) = y'(0) = 0$,
$$f(t) = \begin{cases} 1, & 0 \le t < 1 \\ -1, & 1 \le t < 2 \\ 0, & t \ge 2 \end{cases}$$

48. $y'' + y' = f(t), y(0) = y'(0) = 0$,
$$f(t) = \begin{cases} 1, & 0 \le t < 1 \\ 2, & 1 \le t < 2 \\ 0, & t \ge 2 \end{cases}$$

49. $x'' + x = 1 + 10\delta(t - \pi), x(0) = x'(0) = 0$

50. $x'' + 4x = 2\delta(t - \pi) - 8\delta(t - 2\pi), x(0) = 1,$
$x'(0) = 0$

51. $x'' + 4x' + 13x = 20\delta(t - \pi/2), x(0) = 0,$
$x'(0) = 10$

52. $x'' + 4x' + 13x = 5\delta(t - \pi) - 5\delta(t - \pi/2),$
$x(0) = 0, x'(0) = 10$

53. $x'' + 4x' + 13x = f(t), f(t) = \begin{cases} 1, & 0 \le t < 1 \\ 2, & t \ge 1 \end{cases}$,
$x(0) = x'(0) = 0$

54. $x'' + 4x' + 13x = f(t), f(t) = \begin{cases} t, & 0 \le t < \pi \\ 0, & t \ge \pi \end{cases}$,
$x(0) = x'(0) = 0$

55. $x' + 2x = f(t), f(t) = \begin{cases} 0, & 0 \le t < 1/2 \\ 10, & 1/2 \le t < 1 \end{cases}$ and
$f(t) = f(t - 1)$ if $t \ge 1, x(0) = 0$

56. $x' + 5x = f(t), f(t) = \begin{cases} 0, & 0 \le t < 1 \\ t - 1, & 1 \le t < 2 \end{cases}$ and
$f(t) = f(t - 2)$ if $t \ge 2, x(0) = 0$

57. $x'' = f(t), f(t) = \begin{cases} 1, & 0 \le t < 1 \\ 2 - t, & 1 \le t < 2 \end{cases}$ and
$f(t) = f(t - 2)$ if $t \ge 2, x(0) = x'(0) = 0$

58. $x'' = f(t), f(t) = \begin{cases} 0, & 0 \le t < \pi/2 \\ -\cos t, & \pi/2 \le t < 3\pi/2 \\ 0, & 3\pi/2 \le t < 2\pi \end{cases}$
and $f(t) = f(t - 2\pi)$ if $t \ge 2\pi$,
$x(0) = x'(0) = 0$

59. $x'' + x' = f(t), f(t) = \begin{cases} 1, & 0 \le t < 1 \\ -1, & 1 \le t < 2 \end{cases}$ and
$f(t) = f(t - 2)$ if $t \ge 2$, $x(0) = x'(0) = 0$

60. $x'' + 4x = f(t), f(t) = \begin{cases} t - 1, & 0 \le t < 1 \\ 0, & 1 \le t < 2 \end{cases}$ and
$f(t) = f(t - 2)$ if $t \ge 2$, $x(0) = x'(0) = 0$

61. $x'' + 4x = f(t)$,
$f(t) = \begin{cases} 0, & 0 \le t < \pi/2 \\ -\sin 2t, & \pi/2 \le t < \pi \end{cases}$ and
$f(t) = f(t - \pi)$ if $t \ge \pi$, $x(0) = x'(0) = 0$

62. $y'' + 9\pi^2 y = f(t), f(t) = \delta(t - 1) - \delta(t - 2)$,
$0 \le t < 3$, and $f(t) = f(t - 3)$ if $t \ge 3$,
$y(0) = 0, y'(0) = 1$

63. $y'' + 2y' = f(t), f(t) = \delta(t - 1), 0 \le t < 2$,
$f(t) = f(t - 2)$ if $t \ge 2$, $y(0) = y'(0) = 0$

64. Show that $\mathcal{L}\{g(t)\mathcal{U}(t - a)\} = e^{-st}\mathcal{L}\{g(t + a)\}$.
(*Hint:* Use the definition of the Laplace transform and the change of variable $u = t - a$.)

65. (*Square wave*) If $f(t) = \begin{cases} 1, & 0 \le t < a \\ -1, & a \le t < 2a \end{cases}$
$f(t + 2a) = f(t)$, show that
$F(s) = \dfrac{1 - e^{-as}}{s(1 + e^{-as})} = \dfrac{1}{s}\tanh\dfrac{as}{2}$.

66. (*Triangular wave*) Consider the function
$g(t) = \begin{cases} x, & 0 \le t < a \\ 2a - x, & a \le t < 2a \end{cases}, g(t + 2a) = g(t)$.
Show that $G(s) = \dfrac{1}{s^2}\tanh\dfrac{as}{2}$. (*Hint:* Use the square wave function and the relationship $g'(t) = f(t)$.)

67. (*Sawtooth wave*) If $f(t) = t/a, 0 \le t < a$,
$f(t + a) = f(t)$, show that
$F(s) = \dfrac{1}{as^2} - \dfrac{e^{-as}}{s(1 - e^{-as})}$.

68. (*Rectified sine wave*) Let $f(t) = |\sin(\pi t/a)|$.
Show that $F(s) = \dfrac{\pi a}{a^2 s^2 + \pi^2}\coth\dfrac{as}{2}$.

69. (*Half-rectified sine wave*) If
$f(t) = \begin{cases} \sin(\pi t/a), & 0 \le t < a \\ 0, & a \le t < 2a \end{cases}$
$f(t + 2a) = f(t)$, show that
$F(s) = \dfrac{\pi a}{(a^2 s^2 + \pi^2)(1 - e^{-as})}$.

70. (*Stair step*) Consider the function
$g(t) = \sum_{n=1}^{\infty}\mathcal{U}(t - na)$. Sketch the graph of
$g(t)$ on $[0, 4]$. Show that $G(s) = \dfrac{1}{s(1 - e^{-as})}$.

8.5 THE CONVOLUTION THEOREM

- THE CONVOLUTION THEOREM
- INTEGRAL AND INTEGRODIFFERENTIAL EQUATIONS

The Convolution Theorem

In many cases, we are required to determine the inverse Laplace transform of a product of two functions. Just as in integral calculus when the integral of the product of two functions did not produce the product of the integrals, neither does the inverse Laplace transform of the product yield the product of the inverse Laplace transforms. We had a preview of this when we found $\mathcal{L}^{-1}\left\{\dfrac{F(s)}{s}\right\} = \mathcal{L}^{-1}\left\{\dfrac{1}{s}F(s)\right\}$ earlier in this chapter. We recall that $\mathcal{L}^{-1}\{F(s)/s\} = \int_0^t f(\alpha)\, d\alpha$, where $f(t) = \mathcal{L}^{-1}\{F(s)\}$ so we realize that

$$\mathcal{L}^{-1}\left\{\frac{F(s)}{s}\right\} \neq \mathcal{L}^{-1}\left\{\frac{1}{s}\right\} \times \mathcal{L}^{-1}\{F(s)\} = f(t).$$

The *convolution theorem* gives a relationship between the inverse Laplace transform of the product of two functions, $\mathcal{L}^{-1}\{F(s)\,G(s)\}$, and the inverse Laplace transform of each function, $\mathcal{L}^{-1}\{F(s)\}$ and $\mathcal{L}^{-1}\{G(s)\}$.

Theorem 52 (Convolution Theorem). *Suppose that $f(t)$ and $g(t)$ are piecewise continuous on $[0, \infty)$ and both of exponential order b. Further, suppose that $\mathcal{L}\{f(t)\} = F(s)$ and $\mathcal{L}\{g(t)\} = G(s)$. Then,*

$$\mathcal{L}^{-1}\{f(s)G(s)\} = \mathcal{L}^{-1}\{\mathcal{L}\{(f * g)(t)\}\}$$

$$= (f * g)(t) = \int_0^t f(t - v)g(v)\,dv.$$

Note that $(f * g)(t) = \int_0^t f(t - v)g(v)\,dv$ is called the **convolution integral.**

Proof. We prove the convolution theorem by computing the product $F(s)G(s)$ with the definition of the Laplace transform. This yields

$$F(s)G(s) = \int_0^\infty e^{-sx}f(x)\,dx \times \int_0^\infty e^{-sv}g(v)\,dv,$$

which can be written as the iterated integral

$$F(s)G(s) = \int_0^\infty \int_0^\infty e^{-s(x+v)}f(x)g(v)\,dx\,dv.$$

Changing variables with $x = t - v$ (so that $t = x + v$ and $dx = dt$) then yields

$$F(s)G(s) = \int_0^\infty \int_v^\infty e^{-st}f(t - v)g(v)\,dt\,dv,$$

where the region of integration R is the unbounded triangular region shown in Figure 8.11. Recall from multivariable calculus that the order of integration can be interchanged. Then, we obtain

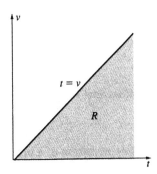

FIGURE 8.11 Observe that $R = \{(t, v) : v \le t < \infty, 0 \le v < \infty\} = \{(t, v) : 0 \le v \le t, 0 \le t < \infty\}$.

$$F(s)G(s) = \int_0^\infty \int_0^t e^{-st}f(t - v)g(v)\,dv\,dt$$

(why?), can be written as

$$F(s)G(s) = \int_0^\infty e^{-st} \int_0^t f(t - v)g(v)\,dv\,dt$$

$$= \mathcal{L}\left\{\int_0^t f(t - v)g(v)\,dv\right\} = \mathcal{L}\{(f * g)(t)\}.$$

Therefore, $\mathcal{L}^{-1}\{f(s)G(s)\} = \mathcal{L}^{-1}\{\mathcal{L}\{(f * g)(t)\}\} = (f * g)(t)$.

EXAMPLE 8.5.1

Compute $(f * g)(t)$ and $(g * f)(t)$ if $f(t) = e^{-t}$ and $g(t) = \sin t$. Verify the convolution theorem for these functions.

Solution

We use the definition and a computer algebra system (or table of integrals) to obtain

$$(f * g)(t) = \int_0^t f(t - v)g(v)\,dv = \int_0^t e^{-(t-v)}\sin v\,dv$$

$$= e^{-t}\int_0^t e^v \sin v\,dv$$

$$= e^{-t}\left[\frac{1}{2}e^v(\sin v - \cos v)\right]_0^t$$

$$= \frac{1}{2}e^{-t}\left[e^t(\sin t - \cos t) - (\sin 0 - \cos 0)\right]$$

$$= \frac{1}{2}(\sin t - \cos t + e^{-t}).$$

With a computer algebra system, we find that

$$(g * f)(t) = \int_0^t g(t - v)f(v)\,dv = \int_0^t \sin(t - v)e^v\,dv$$

$$= \frac{1}{2}(\sin t - \cos t + e^{-t}).$$

Now, according to the convolution theorem, $\mathcal{L}\{f(t)\}\mathcal{L}\{g(t)\} = \mathcal{L}\{(f * g)(t)\}$. In this example, we have

$$F(s) = \mathcal{L}\{f(t)\} = \mathcal{L}\{e^{-t}\} = \frac{1}{s + 1} \quad \text{and}$$

$$G(s) = \mathcal{L}\{g(t)\} = \mathcal{L}\{\sin t\} = \frac{1}{s^2 + 1}.$$

Hence, $\mathcal{L}^{-1}\{F(s)G(s)\} = \mathcal{L}^{-1}\left\{\frac{1}{s+1} \times \frac{1}{s^2+1}\right\}$ should equal $(f * g)(t)$. We compute $\mathcal{L}^{-1}\left\{\frac{1}{s+1} \times \frac{1}{s^2+1}\right\}$ through the partial fraction decomposition

$$\frac{1}{s+1} \times \frac{1}{s^2+1} = \frac{A}{s+1} + \frac{Bs+C}{s^2+1}$$

and find that $A = C = 1/2$ and $B = -1/2$. Therefore,

$$\mathcal{L}^{-1}\left\{\frac{1}{s+1} \times \frac{1}{s^2+1}\right\} = \frac{1}{2}\mathcal{L}^{-1}\left\{\frac{1}{s+1} + \frac{-s+1}{s^2+1}\right\}$$

$$= \frac{1}{2}\mathcal{L}^{-1}\left\{\frac{1}{s+1}\right\}$$

$$-\frac{1}{2}\mathcal{L}^{-1}\left\{\frac{s}{s^2+1}\right\}$$

$$+\frac{1}{2}\mathcal{L}^{-1}\left\{\frac{1}{s^2+1}\right\}$$

$$= \frac{1}{2}e^{-t} - \frac{1}{2}\cos t + \frac{1}{2}\sin t,$$

which is the same result as that obtained for $(f * g)(t)$.

Notice that in Example 8.5.1, $(g * f)(t) = (f * g)(t)$. This is no coincidence. With a straightforward change of variables, we can prove this relationship in general to see that the convolution integral operation on two functions is *commutative* (see the exercises at the end of the section).

EXAMPLE 8.5.2

Use the convolution theorem to find the Laplace transform of $h(t) = \int_0^t \cos(t - v) \sin v \, dv$.

Solution

Notice that $h(t) = (f * g)(t)$, where $f(t) = \cos t$ and $g(t) = \sin t$. Therefore, by the convolution theorem, $\mathcal{L}\{(f * g)(t)\} = F(s)G(s)$. Hence,

$$\mathcal{L}\{h(t)\} = \mathcal{L}\{f(t)\}\mathcal{L}\{g(t)\} = \mathcal{L}\{\cos t\}\mathcal{L}\{\sin t\}$$

$$= \frac{s}{s^2+1} \times \frac{1}{s^2+1} = \frac{s}{(s^2+1)^2}.$$

 Find the Laplace transform of $h(t) = \int_0^t e^{t-v} \sin v \, dv$.

Integral and Integrodifferential Equations

The convolution theorem is useful in solving numerous problems. In particular, this theorem can be used to solve *integral equations*, which are equations that involve the integral of the unknown function.

EXAMPLE 8.5.3

Use the convolution theorem to solve the integral equation

$$h(t) = 4t + \int_0^t h(t - v) \sin v \, dv.$$

Solution

We need to find $h(t)$, so we first note that the integral in this equation represents $(h * g)(t)$ with $g(t) = \sin t$. Therefore, applying the Laplace transform to both sides of the equation gives us

$$\mathcal{L}\{h(t)\} = \mathcal{L}\{4t\} + \mathcal{L}\{h(t)\}\mathcal{L}\{\sin t\} \quad \text{or}$$

$$H(s) = \frac{4}{s^2} + H(s)\frac{1}{s^2+1},$$

where $\mathcal{L}\{h(t)\} = H(s)$. Solving for $H(s)$, we have

$$H(s)\left(1 - \frac{1}{s^2+1}\right) = \frac{4}{s^2} \quad \text{so}$$

$$H(s) = \frac{4(s^2+1)}{s^4} = \frac{4}{s^2} + \frac{4}{s^4}.$$

Then by computing the inverse Laplace transform, we find that

$$h(t) = \mathcal{L}^{-1}\left\{\frac{4}{s^2} + \frac{4}{s^4}\right\} = 4t + \frac{2}{3}t^3.$$

Laplace transforms are also helpful in solving *integrodifferential equations*, which are equations that involve a derivative as well as an integral of the unknown function.

EXAMPLE 8.5.4

Solve $\dfrac{dy}{dt} + y + \int_0^t y(u)\,du = 1$ subject to $y(0) = 0$.

Solution

Because we must take the Laplace transform of both sides of this integrodifferential equation, we first compute

$$\mathcal{L}\left\{\int_0^t y(u)\,du\right\} = \mathcal{L}\{(1 * y)(t)\} = \mathcal{L}\{1\}\mathcal{L}\{y\} = \frac{Y(s)}{s}.$$

Differentiate this equation to show that this initial-value problem is equivalent to $y'' + y' + y = 0$, $y(0) = 0$, $y'(0) = 1$. Why must $y'(0) = 1$?

Hence,

$$\mathcal{L}\left\{\frac{dy}{dt}\right\} + \mathcal{L}\{y\} + \mathcal{L}\left\{\int_0^t y(u)\,du\right\} = \mathcal{L}\{1\}$$

$$sY(s) - y(0) + Y(s) + \frac{Y(s)}{s} = \frac{1}{s}$$

$$s^2Y(s) + sY(s) + Y(s) = 1$$

$$Y(s) = \frac{1}{s^2 + s + 1}.$$

Completing the square in the denominator shows us that

$$Y(s) = \frac{1}{s^2 + s + 1} = \frac{1}{(s + 1/2)^2 + (\sqrt{3}/2)^2} \quad \text{so}$$

$$y(t) = \frac{2}{\sqrt{3}} e^{-t/2} \sin \frac{\sqrt{3}}{2} t.$$

EXERCISES 8.5

In Exercises 1-6, compute the convolution $(f * g)(t)$ using the indicated pair of functions.

1. $f(t) = 1$, $g(t) = t^2$
2. $f(t) = e^{-3t}$, $g(t) = e^{-3t}$

3. $f(t) = t$, $g(t) = t^2$
4. $f(t) = \sin 2t$, $g(t) = e^{-t}$
5. $f(t) = t^3$, $g(t) = \sin 4t$
6. $f(t) = \sin 2t$, $g(t) = e^{-4t}$

In Exercises 7-15, find the Laplace transform of h.

7. $h(t) = \int_0^t e^{-v}\,dv$
8. $h(t) = \int_0^t \sin v\,dv$
9. $h(t) = \int_0^t (t - v)\sin v\,dv$
10. $h(t) = \int_0^t v\cos(t - v)\,dv$
11. $h(t) = \int_0^t v e^{t-v}\,dv$
12. $h(t) = \int_0^t e^v (t - v)^2\,dv$
13. $h(t) = \int_0^t \mathcal{U}(t - v - 1)\mathcal{U}(v - 2)\,dv$
14. $h(t) = \int_0^t (t - v - 1)(\mathcal{U}(v - 1) - \mathcal{U}(v - 2))\,dv$
15. $h(t) = \int_0^t \cos(t - v)\,(\mathcal{U}(t - \pi/2) - \mathcal{U}(t - 3\pi/2))\,dv$

In Exercises 16-22, find the inverse Laplace transform of each function using the convolution theorem.

16. $\dfrac{1}{s^2(s + 1)}$

17. $\dfrac{1}{s^3(s + 3)}$

18. $\dfrac{s}{(s^2 + 1)^2}$

19. $\dfrac{1}{(s^2 + 4)(s + 1)}$

20. $\dfrac{1}{(s^2 + 9)(s - 2)}$

21. $\dfrac{e^{-\pi s}}{s^3 + s}$

22. $\dfrac{e^{1-s}}{s^2 - 1}$

In Exercises 23-26, solve the given integral equation using Laplace transforms.

23. $g(t) - t = -\int_0^t (t - v)g(v)\,dv$
24. $h(t) = 2 - \int_0^t h(v)\,dv$
25. $h(t) - 4e^{-2t} = \cos t - \int_0^t \sin(t - v)\,h(v)\,dv$
26. $f(t) = 3 - \int_0^t v f(t - v)\,dv$

In Exercises 27 and 28, solve the given integrod-ifferential equation using Laplace transforms.

27. $\dfrac{dy}{dt} - 4y + 4\int_0^t y(v)\,dv = t^3 e^{2t}$, $y(0) = 0$

28. $\dfrac{dx}{dt} + 16\int_0^t x(v)\,dv = \sin 4t$, $x(0) = 0$

29. Show that $(f * g)(t) = (g * f)(t)$ by verifying that $\int_0^t f(v)g(t-v)\,dv = \int_0^t g(v)g(t-v)\,dv$. Therefore, the convolution integral is *commutative*.

30. Show that the convolution integral is *associative* by proving that $(f * (g * h))(t) = ((f * g) * h)(t)$.

31. Show that the convolution integral satisfies the *distributive property* $(f * (g + h))(t) = (f * g)(t) + (f * h)(t)$.

32. Show that for any constant k, $((kf) * g)(t) = k(f * g)(t)$.

33. Show that $(\sin t) * (\cos kt) = (\cos t) * (\frac{1}{k}\sin kt)$.

34. Show that $t^{-1/2} * t^{-1/2} = \pi$.

35. Express the integrodifferential equation in Exercise 27 as an equivalent second-order initial-value problem and solve this problem.

36. Express the integrodifferential equation in Exercise 28 as an equivalent second-order initial-value problem and solve this problem.

37. Verify each of the following:

(a) $(\sin kt) * (\cos kt) = \dfrac{1}{2}t\sin kt$

(b) $(\sin kt) * (\sin kt) = \dfrac{1}{2k}\sin kt - \dfrac{1}{2}t\cos kt$

(c) $(\cos kt) * (\cos kt) = \dfrac{1}{2k}\sin kt + \dfrac{1}{2}t\cos kt$

(d) Use these results to very that
$$\mathcal{L}^{-1}\left\{\dfrac{ks}{(s^2+k^2)^2}\right\} = \tfrac{1}{2}t\sin kt,$$
$$\mathcal{L}^{-1}\left\{\dfrac{k^2}{(s^2+k^2)^2}\right\} = \tfrac{1}{2k}\sin kt - \tfrac{1}{2}t\cos kt, \text{ and}$$
$$\mathcal{L}^{-1}\left\{\dfrac{s^2}{(s^2+k^2)^2}\right\} = \tfrac{1}{2k}\sin kt + \tfrac{1}{2}t\cos kt.$$

8.6 LAPLACE TRANSFORM METHODS FOR SOLVING SYSTEMS

In many cases, Laplace transforms can be used to solve initial-value problems that involve a system of linear differential equations. This method is applied in much the same way that it was in solving initial-value problems involving higher order differential equations, except that a system of algebraic equations is obtained after taking the Laplace transform of each equation. After solving for the Laplace transform of each of the unknown functions, the inverse Laplace transform is used to find each unknown function in the solution of the system.

EXAMPLE 8.6.1

Solve $\mathbf{X}' = \begin{pmatrix} 0 & -1 \\ 1 & 0 \end{pmatrix}\mathbf{X} + \begin{pmatrix} \sin t \\ 2\cos t \end{pmatrix}$ subject to

$\mathbf{X}(0) = \begin{pmatrix} 0 \\ 0 \end{pmatrix}$.

Solution

Let $\mathbf{X}(t) = \begin{pmatrix} x(t) \\ y(t) \end{pmatrix}$. Then, we can rewrite this problem as

$$x' = -y + \sin t$$
$$y' = x + 2\cos t, \quad x(0) = 0, \ y(0) = 0.$$

Taking the Laplace transform of both sides of each equation yields the system

$$sX(s) - x(0) = -Y(s) + \dfrac{1}{s^2+1}$$
$$sY(s) - y(0) = X(s) + \dfrac{2s}{s^2+1},$$

which is equivalent to

$$sX(s) + Y(s) = \dfrac{1}{s^2+1}$$
$$-X(s) + sY(s) = \dfrac{2s}{s^2+1}.$$

> You could also use Cramer's rule to solve a system of equations such as this one.

Multiplying the second equation by s, adding the two equations, and finding the partial fraction decomposition, we have

$$(s^2 + 1)Y(s) = \frac{2s^2}{s^2 + 1} + \frac{1}{s^2 + 1}$$

$$Y(s) = \frac{2s^2}{(s^2 + 1)^2} + \frac{1}{(s^2 + 1)^2}$$

$$Y(s) = \frac{2}{s^2 + 1} - \frac{1}{(s^2 + 1)^2}$$

Taking the inverse Laplace transform then yields

$$y(t) = \frac{1}{2}(t\cos t + 3\sin t).$$

At this point, we can solve the system for $X(s)$ and compute $x(t)$ with the inverse Laplace transform. However, we can find $x(t)$ more easily by substituting $y(t)$ into the second differential equation $y' = x + 2\cos t$. Because $y' = \frac{1}{2}(4\cos t - t\sin t)$ and $x = y' - 2\cos t$, we have

$$x(t) = y'(t) - 2\cos t = -\frac{1}{2}\sin t.$$

In Figure 8.12, we graph $x(t)$, $y(t)$, and $(x(t), y(t))$. What is the orientation of $(x(t), y(t))$?

Linear problems that involve higher order differential equations can be solved with Laplace transforms as well.

EXAMPLE 8.6.2

Solve $x'' - y'' = y + \sin t$, $y' - y - x' = 0$, $x(0) = x'(0) = y(0) = y'(0) = 0$.

Solution

We begin by taking the Laplace transform of both equations and applying the initial conditions. For the first equation, this yields

$$s^2 X(s) - s^2 Y(s) = Y(s) + \frac{1}{s^2 + 1}$$

$$s^2 X(s) - (s^2 + 1)Y(s) = \frac{1}{s^2 + 1},$$

and for the second equation,

$$sX(s) - (s - 1)Y(s) = 0.$$

To use Cramer's rule we first compute $W = \begin{vmatrix} s^2 & -(s^2 + 1) \\ s & -(s - 1) \end{vmatrix} = s(s - 1)$. Then,

$$X(s) = \frac{1}{s(s-1)} \begin{vmatrix} \frac{1}{s^2 + 1} & -(s^2 + 1) \\ 0 & -(s - 1) \end{vmatrix}$$

$$= \frac{1 - s}{s(s + 1)(s^2 + 1)} = \frac{1}{s} - \frac{1}{s + 1} - \frac{1}{s^2 + 1}$$

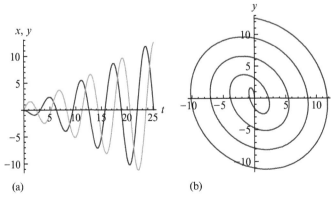

(a)   (b)

FIGURE 8.12 (a) Graph of $x(t)$ and $y(t)$ for $0 \leq t \leq 8\pi$. (b) Graph of $(x(t), y(t))$ for $0 \leq t \leq 8\pi$.

and

$$Y(s) = \frac{1}{s(s-1)} \begin{vmatrix} s^2 & \frac{1}{s^2+1} \\ s & 0 \end{vmatrix} = -\frac{1}{(s+1)(s^2+1)}$$

$$= -\frac{1}{s(s+1)} + \frac{s-1}{2(s^2+1)}.$$

Taking the inverse Laplace transform of $X(s)$ and $Y(s)$ gives us

$$x(t) = 1 - e^{-t} - \sin t \quad \text{and}$$

$$y(t) = \frac{1}{2}(-e^{-t} + \cos t - \sin t).$$

In Figure 8.13 we graph $x(t)$, $y(t)$, and $(x(t), y(t))$. What is the orientation of $(x(t), y(t))$?

When dealing with periodic and impulse forcing functions, computer algebra systems are of great use in dealing with the algebra.

EXAMPLE 8.6.3

 Solve

$$x' = -2y + \delta(t - \pi/2) - \delta(t - 3\pi/2)$$
$$y' = 8x + \sin 4t - \delta(t - \pi) + \delta(t - 2\pi)$$

subject to $x(0) = y(0) = 0$.

Solution

Taking the Laplace transform and applying the initial conditions results in the system

$$sX(s) = -2Y(s) + e^{-\pi s/2} - e^{-3\pi s/2}$$
$$sY(s) = 8X(s) + \frac{4}{s^2+16} - e^{-\pi s} + e^{-2\pi s}$$

or

$$sX(s) + 2Y(s) = e^{-\pi s/2} - e^{-3\pi s/2}$$
$$-8X(s) + sY(s) = \frac{4}{s^2+16} - e^{-\pi s} + e^{-2\pi s}.$$

To use Cramer's rule, we first compute $\begin{vmatrix} s & 2 \\ -8 & s \end{vmatrix} = s^2 + 16$. Then,

$$X(s) = \frac{1}{s^2+16} \begin{vmatrix} e^{-\pi s/2} - e^{-3\pi s/2} & 2 \\ \frac{4}{s^2+16} - e^{-\pi s} + e^{-2\pi s} & s \end{vmatrix}$$

$$= \frac{-2e^{-2\pi s} + 2e^{-\pi s} - se^{-3\pi s/2} + se^{-\pi s/2}}{s^2+1}$$

$$- \frac{8}{(s^2+1)(s^2+16)}$$

and

$$Y(s) = \frac{1}{s^2+16} \begin{vmatrix} s & e^{-\pi s/2} - e^{-3\pi s/2} \\ -8 & \frac{4}{s^2+16} - e^{-\pi s} + e^{-2\pi s} \end{vmatrix}$$

$$= \frac{-8e^{-3\pi s/2} + 8e^{-\pi s/2} + se^{-2\pi s} - se^{-\pi s}}{s^2+1}$$

$$+ \frac{4s}{(s^2+1)(s^2+16)}.$$

Taking the inverse Laplace transform yields the solution of the initial-value problem:

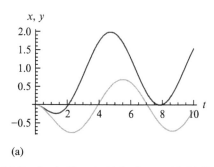

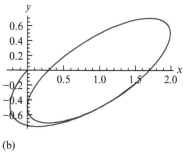

(a)   (b)

FIGURE 8.13   (a) Graph of $x(t)$ and $y(t)$ for $0 \le t \le 10$. (b) Graph of $(x(t), y(t))$ for $0 \le t \le 10$.

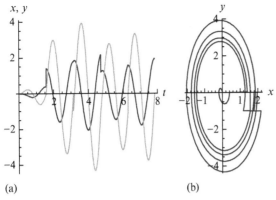

(a)   (b)

FIGURE 8.14   (a) Graph of $x(t)$ and $y(t)$ for $0 \le t \le 5\pi/2$.
(b) Graph of $(x(t), y(t))$ for $0 \le t \le 5\pi/2$.

$$x(t) = \frac{1}{15}(-8 + 4\cos t + 4\cos 3t - 30\mathcal{U}(t - 2\pi)$$
$$+15\mathcal{U}(t - 3\pi/2) - 30\mathcal{U}(t - \pi)$$
$$+15\mathcal{U}(t - \pi/2))\sin t$$

and

$$y(t) = \frac{1}{15}[-4\cos 4t + (4 + 15\mathcal{U}(t - 2\pi)$$
$$-120\mathcal{U}(t - 3\pi/2)$$
$$+15\mathcal{U}(t - \pi)12015\mathcal{U}(t - \pi/2))\cos t].$$

In Figure 8.14, we graph $x(t)$, $y(t)$, and $(x(t), y(t))$.
What is the orientation of $(x(t), y(t))$?

 EXERCISES 8.6

In Exercises 1-25, use Laplace transforms to
solve each initial-value problem. Describe other
methods that could be used to solve the system.

1. $x' - 2x + 3y = 0$, $y' + 9x + 4y = 0$, $x(0) = 0$,
   $y(0) = 4$
2. $x' + 9x - 2y = 0$, $y' + 10x - 3y = 0$,
   $x(0) = -2$, $y(0) = 0$
3. $x' - 5x - 5y = 0$, $y' + 4x + 3y = 0$, $x(0) = 2$,
   $y(0) = 0$

4. $x' + 2x - 4y = 0$, $y' + 2x - 2y = 0$, $x(0) = 0$,
   $y(0) = 5$
5. $x' = 2x - y + 3z$, $y' = 6x - 2y + 6z$,
   $z' = -2x + y - 3x$, $x(0) = 1$, $y(0) = z(0) = 0$
6. $x' = -z$, $y' = x - z$, $z' = -x$, $x(0) = y(0) = 0$,
   $z(0) = 1$
7. $x' = 2x - 3y + e^t$, $y' = 4x - 4y$,
   $x(0) = y(0) = 0$
8. $x' = x + 6y - 2\cos t$, $y' = -x - 4y$,
   $x(0) = y(0) = 0$
9. $x' = 2x - 6y$, $y' = x - 3y + e^{-t}$,
   $x(0) = y(0) = 0$
10. $x' = -2x + y - t$, $y' = -3x + 2y + 1$,
    $x(0) = y(0) = 0$
11. $x' = -2x - 4y + e^{-t}$, $y' = 5x + 2y - e^{-t}$,
    $x(0) = y(0) = 0$
12. $x' = x - 4y - 1$, $y' = 2x - 3y + t$,
    $x(0) = y(0) = 0$
13. $x' = x - 5y + 1$, $y' = x - y + t$, $x(0) = y(0) = 0$
14. $x' = -2x + 4y - e^{-t}$, $y' = -2x + 2y + e^{-t}$,
    $x(0) = y(0) = 0$
15. $x' - y = 0$, $y' + x = f(t)$, $x(0) = y(0) = 0$,
    where $f(t) = \begin{cases} \sin t, & 0 \le t < \pi \\ 0, & t \ge \pi \end{cases}$
16. $x' + 7x + 4y = f(t)$, $y' + 6x - 3y = 0$,
    $x(0) = y(0) = 0$, where $f(t) = \begin{cases} 1, & 0 \le t < 1 \\ 0, & t \ge 1 \end{cases}$
17. $x' = -2y + \delta(t - \pi/8)$, $y' = 8x - \delta(t - \pi/4)$,
    $x(0) = 1$, $y(0) = 0$
18. $x' = -2x - 8y - \delta(t - \pi/2)$, $y' = x + 2y +$
    $(\mathcal{U}(t) - \mathcal{U}(t - \pi))$, $x(0) = y(0) = 0$
19. $x' = x + y + f(t)$, $y' = -2x - 2y$,
    $x(0) = y(0) = 0$, where
    $f(t) = \begin{cases} 1, & 0 \le t < 1 \\ -1, & 1 \le t < 2 \end{cases}$ and $f(t) = f(t - 2)$ if
    $t \ge 2$
20. $x' = x - 2y$, $y' = 3x - 4y + f(t)$,
    $x(0) = y(0) = 0$, where $f(t) = t$, $0 \le t < 1$
    and $f(t) = f(t - 1)$ if $t \ge 1$
21. $x' = x - y + f(t)$, $y' = 2x - y$, $x(0) = y(0) = 0$,
    where $f(t) = \begin{cases} 1, & 0 \le t < \pi \\ 0, & \pi \le t < 2\pi \end{cases}$ and
    $f(t) = f(t - 2\pi)$ if $t \ge 2\pi$

22. $x'' - y'' = 2x + 1,\ y' + x' = 0,$
$\quad x(0) = x'(0) = y(0) = y'(0) = 0$

23. $x'' = 3x' - y' - 2x + y,\ x' + y' = 2x - y,$
$\quad x(0) = x'(0) = 0,\ y(0) = -1$

24. $x'' + y'' = y + 1,\ y' - y - x' = 0,$
$\quad x(0) = x'(0) = y(0) = y'(0) = 0$

25. $x'' + y'' = x + t,\ y' - x + x' = 0,$
$\quad x(0) = x'(0) = y(0) = y'(0) = 0$

8.7 SOME APPLICATIONS USING LAPLACE TRANSFORMS

- L-R-C Circuits Revisited
- Delay Differential Equations
- Coupled Spring-Mass Systems
- The Double Pendulum

Laplace transforms are useful in solving the differential equations associated with many of the applications that were discussed in earlier sections. However, because this method is most useful in alleviating the difficulties associated with problems involving piecewise defined functions, impulse functions, and periodic functions, we focus most of our attention on solving spring-mass systems, L-R-C circuit problems, and population problems that include functions of this type.

L-R-C Circuits Revisited

Laplace transforms can be used to solve circuit problems (see Figure 8.15(a) and (b)) that were introduced earlier in the text. Recall that the initial-value problems that model L-R-C and L-R circuits are

$$L\frac{d^2Q}{dt^2} + R\frac{dQ}{dt} + \frac{1}{C}Q = E(t)$$

$$Q(0) = Q_0, \quad I(0) = \frac{dQ}{dt}(0) = I_0 \quad \text{and}$$

$$L\frac{dI}{dt} + RI = E(t)$$

$$I(0) = I_0,$$

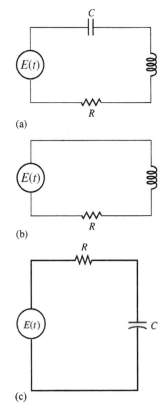

(a)

(b)

(c)

FIGURE 8.15 (a) L-R-C circuit, (b) L-R circuit, and (c) an R-C circuit with the property that $R = C = 1$.

where L, R, and C represent the inductance, resistance, and capacitance, respectively. $Q(t)$ is the charge of the capacitor and $dQ/dt = I(t)$, where $I(t)$ is the current and $E(t)$ is the voltage supply.

In circuit analysis, the Laplace transform from the t-domain (time) to the s-domain (frequency) corresponds to transforming circuit elements to impedances.

EXAMPLE 8.7.1

Consider the R-C circuit in Figure 8.15(c) with the property that $R = C = 1$.

If $E(t) = \begin{cases} 1\,V, & 0 \le t < a \\ 0, & a \le t < 2a \end{cases}$ and $E(t + 2a) = E(t)$, find the charge $Q(t)$ if $Q(0) = 0$. Determine the values of a so that the output voltage, $Q(t)/C$ (the voltage across the capacitor), matches the input voltage, $E(t)$.

Solution

This circuit is modeled with $R\frac{dQ}{dt} + \frac{1}{C}Q = E(t)$ because $L = 0$. We also are given that $Q(0) = 0$. The Laplace transform of $E(t)$ is

$$\mathcal{L}\{E(t)\} = \frac{1}{1 - e^{-2as}} \int_0^a e^{-st}\,dt$$

$$= \frac{1}{1 - e^{-2as}} \left[\frac{e^{-st}}{-s}\right]_0^a = \frac{1}{s(1 + e^{-as})}$$

$$= \frac{1}{s} \sum_{k=0}^{\infty} (-1)^k e^{-aks}$$

$$\left(\frac{1}{1 + e^{-as}} = \sum_{k=0}^{\infty} (-1)^k e^{-aks}\right)$$

Taking the Laplace transform of each side of the differential equation, applying the initial condition, and solving for $\mathcal{L}\{Q(t)\}$ gives us

$$s\mathcal{L}\{Q(t)\} - Q(0) + \mathcal{L}\{Q(t)\} = \frac{1}{s(1 + e^{-as})}$$

$$\mathcal{L}\{Q(t)\} = \frac{1}{s(s+1)(1 + e^{-as})}$$

$$= \frac{1}{s(s+1)} \sum_{k=0}^{\infty} (-1)^k e^{-aks}.$$

Because $\mathcal{L}^{-1}\{1/[s(s+1)]\} = 1 - e^{-t}$,

$$Q(t) = \sum_{k=0}^{\infty} (-1)^k (1 - e^{-(t-ak)}) \mathcal{U}(t - ak)$$

$$= (1 - e^{-t}) - (1 - e^{-(t-a)}) \mathcal{U}(t - a)$$

$$+ (1 - e^{-(t-2a)}) \mathcal{U}(t - 2a)$$

$$- (1 - e^{-(t-3a)}) \mathcal{U}(t - 3a) + \cdots.$$

Figure 8.16 shows the graph of the solution when $a = 1$, $a = 10$, $a = 20$, and $a = 50$, along with

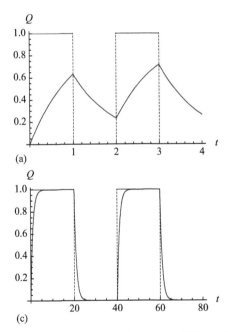

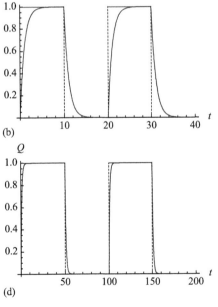

FIGURE 8.16   (a) $a = 1$, (b) $a = 10$, (c) $a = 20$, and (d) $a = 50$.

the input function $E(t)$. We notice that the output voltage, $Q(t)/C$, moves closer to $E(t)$ as a increases. Notice that if t is measured in milliseconds, then the input with $a = 1$ has frequency 500 Hz (cycles per second); while the input with $a = 50$ has frequency 10 Hz. Therefore, low frequency square waves are least altered by the circuit.

Delay Differential Equations

In some applications, there is a delay (or shift) in the argument of the dependent variable. A differential equation that involves this type of shift is called a *delay equation*. Consider the mixture problems from Section 3.3. Now, we assume that there is a delay in the effect that the salt solution flowing into the tank has on the salt concentration flowing out of the tank. To include this assumption in the mathematical model of the situation, we assume that the concentration of the salt solution flowing out equals the salt concentration in the tank at an earlier time.

EXAMPLE 8.7.2

A tank contains 100 gal of water. A salt solution with concentration 2 lb/gal flows into the tank at a rate of 4 gal/min, and the well-stirred mixture flows out of the tank at the same rate. If the salt concentration of the solution flowing out of the tank equals the average concentration 1 min earlier, determine the amount of salt in the tank at any time t.

Solution

If we let $y(t)$ represent the amount of salt in the tank at any time t, then we determine the net change in the amount of salt in the tank with the differential equation

$$\frac{dy}{dt} = \left(2\frac{\text{lb}}{\text{gal}}\right)\left(4\frac{\text{gal}}{\text{min}}\right) - \left(\frac{y(t-1)}{100}\frac{\text{lb}}{\text{gal}}\right)\left(4\frac{\text{gal}}{\text{min}}\right).$$

Therefore, because there is no salt in the tank initially (at $t = 0$ $y(t) = 0$), we solve the problem

$$\frac{dy}{dt} = 8 - \frac{1}{25}y(t-1), \quad y(t) = 0, -1 \le t \le 0.$$

 Show that $\mathcal{L}\{y(t-1)\} = e^{-s}Y(s)$.

As an exercise, you should show that $\mathcal{L}\{y(t-1)\} = e^{-s}Y(s)$. Using the Laplace transform with this property we find that

$$sY(s) - y(0) = \frac{8}{s} - \frac{1}{25}e^{-s}Y(s)$$

$$\left(s + \frac{1}{25}e^{-s}\right)Y(s) = \frac{8}{s}$$

$$Y(s) = \frac{8}{s\left(s + \dfrac{1}{25}e^{-s}\right)}$$

$$= \frac{8}{s^2\left(1 + \dfrac{1}{25s}e^{-s}\right)}.$$

With the power series expansion $\frac{1}{1+x} = \sum_{k=0}^{\infty}(-1)^k x^k = 1 - x + x^2 - x^3 + \cdots$ where we replace x with $-\frac{1}{25s}e^{-s}$, we have

$$Y(s) = \frac{8}{s^2}\sum_{k=0}^{\infty}(-1)^k\left(\frac{1}{25s}e^{-s}\right)^k = \frac{8}{s^2}\sum_{k=0}^{\infty}\frac{(-1)^k}{25^k}\frac{e^{-ks}}{s^k}$$

$$= 8\sum_{k=0}^{\infty}\frac{(-1)^k}{25^k}\frac{e^{-ks}}{s^{k+2}}.$$

Therefore, $y(t) = 8\sum_{k=0}^{\infty}\frac{(-1)^k}{25^k(k+1)!}(t-k)^{k+1}\mathcal{U}(t-k)$. In Figure 8.17(a), we graph the solution for $0 \le t \le 150$. It appears that $\lim_{t\to\infty} y(t) = 200$. We may ask how this solution differs from that of the problem that does not include a delay, $dz/dt = 8 - z/25$, $z(0) = 0$. See Figure 8.17(b). Solving this problem as a first-order linear equation, we find that $z(t) = 200 - 200e^{-t/25}$, which has the same limiting value as $y(t)$ as $t \to \infty$. When is the difference between the two functions greatest?

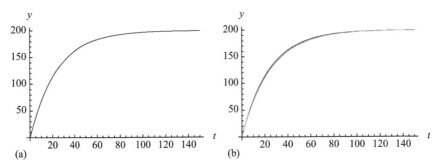

FIGURE 8.17 (a) The graph of $y(t)$. (b) The graph of $y(t)$ together with the graph of the solution to the problem that does not include a delay, $z(t)$.

Coupled Spring-Mass Systems

The displacement of a mass attached to the end of a spring was modeled with a second-order linear differential equation with constant coefficients in Chapter 5. Similarly, if a second spring and mass are attached to the end of the first mass as shown in Figure 8.18(a), the model becomes that of a system of second-order equations. To more precisely state the problem, let masses m_1 and m_2 be attached to the ends of springs S_1 and S_2 having spring constants k_1 and k_2, respectively. Spring S_2 is then attached to the base of mass m_1. Suppose that $x(t)$ and $y(t)$ represent the vertical displacement from the

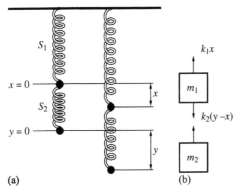

FIGURE 8.18 (a) A coupled spring-mass system. (b) Force diagram for a coupled spring-mass system.

equilibrium position of springs S_1 and S_2, respectively. Because spring S_2 undergoes both elongation and compression when the system is in motion (due to the spring S_1 and the mass m_2), according to Hooke's law, S_2 exerts the force $k_2(y - x)$ on m_2 and S_1 exerts the force $-k_1x$ on m_1. Therefore, the force acting on mass m_1 is the sum $-k_1x + k_2(y - x)$ and the force acting on m_2 is $-k_2(y - x)$. (In Figure 8.18(b), we show the forces acting on the two masses where up is the negative direction and down is positive.) Using Newton's second law ($F = ma$) with each mass, we have the system

$$m_1 \frac{d^2x}{dt^2} = -k_1x + k_2(y - x)$$

$$m_2 \frac{d^2y}{dt^2} = -k_2(y - x).$$

The initial displacement and velocity of the two masses m_1 and m_2 are given by $x(0)$, $x'(0)$, $y(0)$, and $y'(0)$, respectively. Because this system involves second-order linear equations, we can use Laplace transforms to solve problems of this type. Recall the following property of the Laplace transform: $\mathcal{L}\{f(t)\} = s^2F(s) - sf(0) - f'(0)$, where $F(s)$ is the Laplace transform of $f(t)$. This property is of great use in solving this problem because both equations involve second derivatives.

EXAMPLE 8.7.3

Consider the coupled spring-mass system with $m_1 = m_2 = 1$, $k_1 = 3$, and $k_2 = 2$. Find the position functions $x(t)$ and $y(t)$ if $x(0) = y'(0) = 0$ and $x'(0) = y(0) = 1$.

Solution

To find $x(t)$ and $y(t)$, we must solve the initial-value problem

$$\frac{d^2x}{dt^2} = -5x + 2y$$
$$\frac{d^2y}{dt^2} = 2x - 2y$$, $x(0) = y'(0) = 0$, $x'(0) = y(0) = 1$.

Taking the Laplace transform of both sides of each equation, we have

$$s^2X(s) - sx(0) - x'(0) = -5X(s) + 2Y(s)$$
$$s^2Y(s) - sy(0) - y'(0) = 2X(s) - 2Y(s),$$

which is simplified to the system

$$(s^2 + 5)X(s) - 2Y(s) = 1$$
$$-2X(s) + (s^2 + 2)Y(s) = s.$$

We use Cramer's rule to solve for $X(s)$:

$$X(s) = \frac{\begin{vmatrix} 1 & -2 \\ s & s^2+2 \end{vmatrix}}{\begin{vmatrix} s^2+5 & -2 \\ -2 & s^2+2 \end{vmatrix}} = \frac{s^2 + 2s + 2}{s^4 + 7s^2 + 6}$$

$$= \frac{s^2 + 2s + 2}{(s^2 + 6)(s^2 + 1)}.$$

Then, the partial fraction decomposition of $X(s)$ is

$$X(s) = \frac{As + B}{s^2+1} + \frac{Cs+D}{s^2+6} = \frac{2s+1}{5(s^2+1)} + \frac{2(2-s)}{5(s^2+6)}.$$

Taking the inverse Laplace transform then yields

$$x(t) = \frac{1}{15}\left(6\cos t - 6\cos\sqrt{6}t + 3\sin t + 2\sqrt{6}\sin\sqrt{6}t\right).$$

Instead of solving the system to find $Y(s)$, we use the differential equation $x'' = -5x + 2y$ to find $y(t)$. Because $y = \frac{1}{2}(x'' + 5x)$, and

$$x'(t) = \frac{1}{5}\left(\cos t + 4\cos\sqrt{6}t - 2\sin t + 2\sqrt{6}\sin\sqrt{6}t\right)$$

and

$$x''(t) = \frac{1}{5}\left(-2\cos t + 12\cos\sqrt{6}t - \sin t - 4\sqrt{6}\sin\sqrt{6}t\right),$$

it follows that

$$y = \frac{1}{15}\left(12\cos t + 3\cos\sqrt{6}t + 6\sin t - \sqrt{6}\sin\sqrt{6}t\right).$$

In Figure 8.19(a), we graph $x(t)$ and $y(t)$ simultaneously. Note that the initial point of $y(t)$ is $(0, 1)$, whereas that of $x(t)$ is $(0, 0)$. Of course, these functions can be graphed parametrically in the xy-plane as shown in Figure 8.19(b). Notice that this phase plane is different from those discussed in previous sections. One of the reasons is that the equations in the systems of differential equations are second-order instead of first-order. Finally, in Figure 8.20, we illustrate the motion of the spring-mass system by graphing the springs for several values of t.

In Example 8.7.3, what is the maximum displacement of each spring? Does one of the springs always attain its maximum displacement before the other?

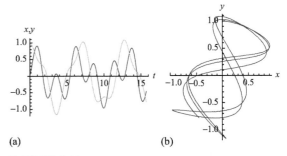

FIGURE 8.19 (a) simultaneous plot and (b) parametric plot.

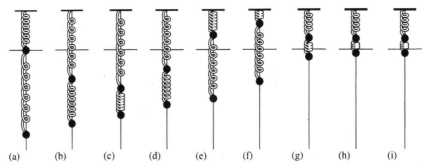

FIGURE 8.20   (a) $t = 0$, (b) $t = 1/2$, (c) $t = 1$, (d) $t = 3/2$, (e) $t = 2$, (f) $t = 5/2$, (g) $t = 3$, (h) $t = 7/2$, and (i) $t = 4$.

If external forces $F_1(t)$ and $F_2(t)$ are applied to the masses, the system of equations becomes

$$m_1 \frac{d^2x}{dt^2} = -k_1x + k_2(y - x) + F_1(t)$$

$$m_2 \frac{d^2y}{dt^2} = -k_2(y - x) + F_2(t),$$

which is again solved through the method of Laplace transforms. We investigate the effects of these external forcing functions in the exercises.

The previous situation can be modified to include a third spring with spring constant k_3 between the base of the mass m_2 and a lower support as shown in Figure 8.21. The motion of the spring-mass system is affected by the third spring. Using the techniques of the earlier case, this model becomes

$$m_1 \frac{d^2x}{dt^2} = -k_1x + k_2(y - x) + F_1(t)$$

$$m_2 \frac{d^2y}{dt^2} = -k_3y - k_2(y - x) + F_2(t).$$

The Double Pendulum

In a method similar to that of the simple pendulum in Chapter 5 and that of the coupled spring system, the motion of a double pendulum (see Figure 8.22) is modeled by the following system of equations using the approximation $\sin\theta \approx \theta$ for small displacements:

$$(m_1 + m_2)\ell_1^2\theta_1'' + m_2\ell_1\ell_2\theta_2''$$
$$+(m_1 + m_2)\ell_1 g\theta_1 = 0$$
$$m_2\ell_2^2\theta_2'' + m_2\ell_1\ell_2\theta_1'' + m_2\ell_2 g\theta_2 = 0,$$

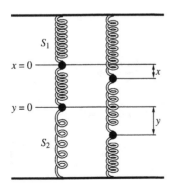

FIGURE 8.21   Applying external forces to a spring-mass system.

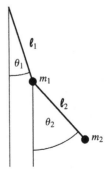

FIGURE 8.22   A double pendulum.

where θ_1 represents the displacement of the upper pendulum and θ_2 that of the lower pendulum. Also, m_1 and m_2 represent the mass attached to the upper and lower pendulums, respectively, and the length of each is given by ℓ_1 and ℓ_2.

EXAMPLE 8.7.4

Suppose that $m_1 = 3$, $m_2 = 1$, and each pendulum has length 16. If $g = 32$, determine $\theta_1(t)$ and $\theta_2(t)$ if $\theta_1(0) = 1$, $\theta_1'(0) = \theta_2(0) = 0$, and $\theta_2'(0) = -1$.

Solution

In this case, the system is

$$4 \times 16^2 \theta_1'' + 16^2 \times \theta_2'' + 4 \times 16 \times 32 \theta_1 = 0$$
$$16^2 \times \theta_2'' + 16^2 \times \theta_1'' + 16 \times 32 \times \theta_2 = 0,$$

which can be simplified to obtain

$$4\theta_1'' + \theta_2'' + 8\theta_1 = 0$$
$$\theta_2'' + \theta_1'' + 2\theta_2 = 0.$$

If we let $\mathcal{L}\{\theta_1(t)\} = X(s)$ and $\mathcal{L}\{\theta_2(t)\} = Y(s)$, taking the Laplace transform of each side of each equation gives us

$$4(s^2 X(s) - s\theta_1(0) - \theta_1'(0)) + (s^2 Y(s)$$
$$- s\theta_2(0) - \theta_2'(0)) + 8X(s) = 0$$
$$(s^2 Y(s) - s\theta_2(0) - \theta_2'(0)) + (s^2 X(s)$$
$$- s\theta_1(0) - \theta_1'(0)) + 2Y(s) = 0$$

or

$$4(s^2 + 2)X(s) + s^2 Y(s) = 4s - 1$$
$$s^2 X(s) + (s^2 + 2)Y(s) = s - 1.$$

Solving this system for $X(s)$, we obtain

$$X(s) = \frac{\begin{vmatrix} 4s - 1 & s^2 \\ s - 1 & s^2 + 2 \end{vmatrix}}{\begin{vmatrix} 4(s^2 + 2) & s^2 \\ s^2 & s^2 + 2 \end{vmatrix}} = \frac{3s^2 + 8s - 2}{3s^4 + 16s^2 + 16}$$

$$= \frac{3s^2 + 8s - 2}{(3s^2 + 4)(s^2 + 4)}.$$

After finding the partial fraction decomposition,

$$\frac{3s^2 + 8s - 2}{(3s^2 + 4)(s^2 + 4)} = \frac{As + B}{3s^2 + 4} + \frac{Cs + D}{s^2 + 4} = \frac{3}{4}\frac{2s - 1}{3s^2 + 4}$$
$$+ \frac{1}{4}\frac{2s + 1}{s^2 + 4},$$

we compute the inverse Laplace transform to find that

$$\theta_1(t) = \frac{1}{8}\left(4\cos 2t + 4\cos \frac{2}{\sqrt{3}}t + \sin 2t\right.$$
$$\left. - \sqrt{3}\sin \frac{2}{\sqrt{3}}t\right).$$

Differentiating, we have

$$\theta_1'(t) = \frac{1}{8}\left(2\cos 2t - 2\cos \frac{2}{\sqrt{3}}t - 8\sin 2t\right.$$
$$\left. - \frac{8}{\sqrt{3}}\sin \frac{2}{\sqrt{3}}t\right)$$

and

$$\theta_1''(t) = \frac{1}{6}\left(-12\cos 2t - 4\cos \frac{2}{\sqrt{3}}t - 3\sin 2t\right.$$
$$\left. + \sqrt{3}\sin \frac{2}{\sqrt{3}}t\right).$$

Using the differential equation $4\theta_1'' + \theta_2'' + 8\theta_1 = 0$ yields

$$\theta_2''(t) = -4\theta_1''(t) - 8\theta_1(t) = 4\cos 2t - \frac{4}{3}\cos \frac{2}{\sqrt{3}}t$$
$$+ \sin 2t + \frac{1}{\sqrt{3}}\sin \frac{2}{\sqrt{3}}t.$$

Integrating, we have

$$\theta_2'(t) = \frac{1}{6}\left(-3\cos 2t - 3\cos \frac{2}{\sqrt{3}}t + 12\sin 2t\right.$$
$$\left. - 4\sqrt{3}\sin \frac{2}{\sqrt{3}}t\right) + c_1.$$

Applying the initial condition $\theta_2'(0) = -1$ yields $\theta_2'(0) = -1/2 - 1/2 + c_1 = -1$, so $c_1 = 0$. Integrating again, we obtain

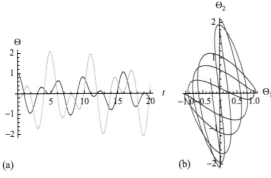

(a)   (b)

FIGURE 8.23   (a) simultaneous plot; and (b) parametric plot.

$$\theta_2(t) = -\cos 2t + \cos \frac{2}{\sqrt{3}}t - \frac{1}{4}\sin 2t$$

$$-\frac{1}{4}\sqrt{3}\sin \frac{2}{\sqrt{3}}t + c_2.$$

Then, because $\theta_2(0) = 0$, $\theta_2(0) = 1 - 1 + c_2 = 0$ so $c_2 = 0$ and it follows that

$$\theta_2(t) = -\cos 2t + \cos \frac{2}{\sqrt{3}}t - \frac{1}{4}\sin 2t - \frac{1}{4}\sqrt{3}\sin \frac{2}{\sqrt{3}}t.$$

These two functions are graphed together in Figure 8.23(a) and parametrically in Figure 8.23(b) to show the solution in the phase plane. We also show the motion of the double pendulum for several values of t in Figure 8.24.

Does the system in Example 8.7.4 come to rest? Which pendulum experiences a greater displacement from equilibrium?

EXERCISES 8.7

1. Suppose that we consider a circuit with a capacitor C, a resistor R, and a voltage supply $E(t) = \begin{cases} 100, & 0 \le t < 1 \\ 0, & t \ge 1 \end{cases}$. If $L = 0$,

find $Q(t)$ and $I(t)$ if $Q(0) = 0$, $C = 50^{-1}$ farads, and $R = 50\,\Omega$.

2. Suppose that we consider a circuit with a capacitor C, a resistor R, and a voltage supply $E(t) = \begin{cases} 100\sin t, & 0 \le t < \pi \\ 0, & t \ge \pi \end{cases}$. If $L = 0$, find $Q(t)$ and $I(t)$ if $Q(0) = 0$, $C = 10^{-2}$ farads, and $R = 100\,\Omega$.

3. Consider the circuit with no capacitor, $R = 100\,\Omega$, and $L = 100\,\text{H}$ if $E(t) = \begin{cases} 50, & 0 \le t < 1 \\ 0, & 1 \le t < 2 \end{cases}$ and $E(t+2) = E(t)$. Find the current $I(t)$ if $I(0) = 0$.

4. Consider the circuit with no capacitor, $R = 100\,\Omega$, and $L = 100\,\text{H}$ if $E(t) = \begin{cases} 100\sin t, & 0 \le t < \pi \\ 0, & \pi \le t < 2\pi \end{cases}$ and $E(t+2\pi) = E(t)$. Find the current $I(t)$ if $I(0) = 0$.

5. Consider the circuit with no capacitor, $R = 100\,\Omega$, and $L = 100\,\text{H}$ if $E(t) = \begin{cases} 100t, & 0 \le t < 1 \\ 0, & 1 \le t < 2 \end{cases}$ and $E(t+2) = E(t)$. Find the current $I(t)$ if $I(0) = 0$.

6. Consider the circuit with no capacitor, $R = 100\,\Omega$, and $L = 100\,\text{H}$ if $E(t) = \begin{cases} 100t, & 0 \le t < 1 \\ 100(2 - t), & 1 \le t < 2 \end{cases}$ and $E(t+2) = E(t)$. Find the current $I(t)$ if $I(0) = 0$.

7. Solve the L-R-C circuit equation that was derived before Example 8.7.1 for $I(t)$ if $L = 1$ Henry, $R = 6\,\Omega$, $C = 1/9$ farads, $E(t) = 100$ V, and $I(0) = 0$.

8. Solve the L-R-C equation for $I(t)$ if $L = 1$ Henry, $R = 6\,\Omega$, $C = 1/9$ farads, $E(t) = 100\sin t$ volts, and $I(0) = 0$.

9. Solve the L-R-C circuit equation for $I(t)$ using the parameter values $L = C = R = 1$ and $E(t) = \begin{cases} 1, & 0 \le t < \pi \\ 0, & t \ge \pi \end{cases}$, $I(0) = 0$.

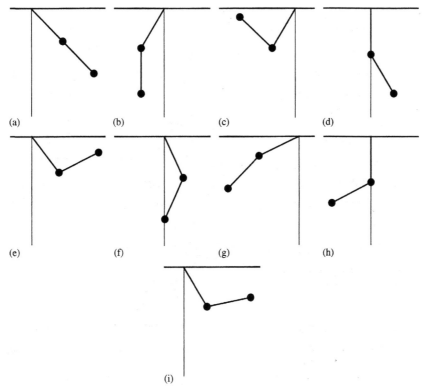

FIGURE 8.24   (a) $t = 0$, (b) $t = 5/4$, (c) $t = 5/2$, (d) $t = 15/4$, (e) $t = 5$, (f) $t = 25/4$, (g) $t = 15/2$, (h) $t = 35/4$, and (i) $t = 10$.

10. Solve the L-R-C circuit equation for $I(t)$ using the parameter values $L = C = R = 1$ and $E(t) = \begin{cases} \cos t, & 0 \le t < \pi \\ 0, & t \ge \pi \end{cases}$, $I(0) = 0$.

11. Consider a circuit with $L = 1$, $R = 4$, $E(t) = \delta(t - 1)$, and no capacitor. Determine the current $I(t)$ if (a) $I(0) = 0$ and (b) $I(0) = 1$.

12. Solve the initial-value problems in Exercise 11 using $E(t) = \delta(t - 1) + \delta(t - 2)$.

13. Consider an L-R circuit in which $L = R = 1$ and $E(t) = \begin{cases} t, & 0 \le t < 1 \\ 1, & t \ge 1 \end{cases}$. Find the current if $I(0) = 0$.

14. Consider an L-R circuit in which $L = R = 1$ and $E(t) = \begin{cases} t, & 0 \le t < 1 \\ 0, & t \ge 1 \end{cases}$. Find the current if $I(0) = 0$.

15. Consider an L-R circuit in which $L = R = 1$ and $E(t) = \delta(t - 2) + \delta(t - 6)$. Find the current if $I(0) = 0$.

16. Consider an L-R circuit in which $L = R = 1$ and $E(t) = 120\delta(t - 1)$. Find the current if $I(0) = 100$.

17. Consider the R-C circuit in Figure 8.25 in which $R = 2.5 \times 10^4\ \Omega$, $C = 10^{-5}$ farad, and $E(t) = \begin{cases} 0, & 0 \le t < 1 \\ 1, & 1 \le t < 2 \\ 0, & t \ge 2 \end{cases}$ volts. Determine the output voltage $I(t)R$ (the voltage across the resistor), by solving the appropriate integral equation. (*Note*: This equation cannot be solved as a differential equation because $E(t)$ is discontinuous at $t = 1$ and $t = 2$.)

18. (*Spring-mass systems revisited*) Just as with systems, we may use Laplace transforms to solve the initial-value problem

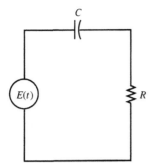

FIGURE 8.25   See Exercise 17.

$$m\frac{d^2x}{dt^2} + c\frac{dx}{dt} + kx = f(t),$$

$$x(0) = \alpha, \quad \frac{dx}{dt}(0) = \beta,$$

which models a spring-mass system involving one spring. Suppose that $m = 1$ slug, $k = 1$ lb/ft., and that there is no resistance due to damping. Further suppose that the object is released from its equilibrium position with zero initial velocity and the object is subjected to the

external force $f(t) = \begin{cases} \cos t, & 0 \le t < \pi/2 \\ 0, & t \ge \pi/2 \end{cases}$.

Show that $\mathcal{L}\{f(t)\} = \frac{s}{s^2+1} + \frac{1}{s^2+1}e^{-s\pi/2}$.

19. Given that $X(s) = \mathcal{L}\{f(t)\} = \frac{s}{s^2+1} + \frac{1}{s^2+1}e^{-s\pi/2}$, show that $x(t) = \mathcal{L}^{-1}\{X(s)\} = \frac{1}{2}t\sin t + \frac{1}{2}(-\cos t - (t-\pi/2)\sin t)\mathcal{U}(t-\pi/2)$.

In Exercises 20 and 21, use the given initial conditions to determine the displacement of the object of mass m attached to a spring with spring constant k.

20. $m = 4, k = 16, x(0) = 1, x'(0) = 0$
21. $m = 1, k = 9, x(0) = 3, x'(0) = -2$

In Exercises 22-25, determine the displacement of the object of mass m attached to a spring with spring constant k if the damping is given by $c\,dx/dt$. Use the initial conditions $x(0) = 1$, $x'(0) = 0$ in each case.

22. $m = 1, k = 6, c = 5$
23. $m = 1/2, k = 2, c = 2$
24. $m = 1, k = 13, c = 4$
25. $m = 1, k = 4, c = 5$

In Exercises 26-29, determine the displacement of the object of mass m attached to a spring with spring constant k if the damping is given by $c\,dx/dt$ and there is an external force $f(t)$. In each case, use the initial conditions $x(0) = x'(0) = 0$.

26. $m = 1, k = 6, c = 5, f(t) = \frac{1}{4}\sin t$
27. $m = 1/2, k = 2, c = 2, f(t) = te^{-t}$
28. $m = 1, k = 13, c = 4, f(t) = e^{-2t}$
29. $m = 1, k = 4, c = 5, f(t) = e^{-4t} + 2e^{-t}$

In Exercises 30-37, determine the displacement of the object of mass m attached to a spring with spring constant k if the damping is given by $c\,dx/dt$ and there is a piecewise defined external force $f(t)$. In each case, use the initial conditions $x(0) = x'(0) = 0$.

30. $m = 1, k = 6, c = 5$,

$f(t) = \begin{cases} \sin 2t, & 0 \le t < \pi/2 \\ 0, & t \ge \pi/2 \end{cases}$

31. $m = 1, k = 6, c = 5$,

$f(t) = \begin{cases} \cos \pi t, & 0 \le t < 1 \\ 0, & t \ge 1 \end{cases}$

32. $m = 1/2, k = 2, c = 2$,

$f(t) = \begin{cases} e^{-2t}, & 0 \le t < 1 \\ 0, & t \ge 1 \end{cases}$

33. $m = 1/2, k = 2, c = 2$,

$f(t) = \begin{cases} e^{-2t}, & 0 \le t < 1 \\ 1, & 1 \le t < 2 \\ 0, & t \ge 2 \end{cases}$

34. $m = 1, k = 13, c = 4$,

$f(t) = \begin{cases} e^{-2t}\cos 3t, & 0 \le t < \pi \\ 0, & t \ge \pi \end{cases}$

35. $m = 1, k = 13, c = 4,$

$$f(t) = \begin{cases} e^{-2t}\cos 3t, & 0 \le t < \pi \\ e^{-2t}\sin 3t, & \pi \le t < 2\pi \\ 0, & t \ge 2\pi \end{cases}$$

36. $m = 1, k = 4, c = 5, f(t) = \begin{cases} e^{-t}, & 0 \le t < 1 \\ 0, & t \ge 1 \end{cases}$

37. $m = 1, k = 4, c = 5,$

$$f(t) = \begin{cases} \cos \pi t, & 0 \le t < 1 \\ 0, & t \ge 1 \end{cases}$$

38. Show that the initial-value problems
$x'' + x = \delta(t), x(0) = x'(0) = 0$ is equivalent
to $x'' + x = 0, x(0) = x'(0) = 0$. (*Hint*: Solve
each equation.)

In Exercises 39-41, assume that the mass is released with zero initial velocity from its equilibrium position.

39. Suppose that an object with mass $m = 1$ is
attached to the end of a spring with spring
constant 16. If there is no damping and the
spring is subjected to the forcing function
$f(t) = \sin t$, determine the motion of the
spring if at $t = 1$, the spring is supplied with
an upward shock of 4 units.

40. An object of mass $m = 1$ is attached to a
spring with $k = 13$ and is subjected to
damping equivalent to $4\,dx/dt$. Find the
motion of the mass if the spring is supplied
with a downward shock of 1 unit at $t = 2$.

41. An object of mass $m = 1$ is attached to a
spring with $k = 13$ and is subjected to
damping equivalent to $4\,dx/dt$. Find the
motion if $f(t) = \delta(t - 1) + \delta(t - 3)$.

42. (*Population problems revisited*) Let $x(t)$
represent the population of a certain
country in which the rate of population
increase depends on the growth rate of the
population of the country as well as the rate
at which people are being added to or
subtracted from the population because of
immigration, emigration, or both. Consider
the initial-value problem $x' + kx =$
$1000(1 + a \sin t), x(0) = x_0$. (a) Show that

$$x(t) = \frac{1000}{k}(1 - e^{-kt})$$

$$+ \frac{1000a}{1 + k^2}(e^{-kt} - \cos t + k \sin t) + x_0 e^{-kt}.$$

(b) If $k = 3$, graph $x(t)$ over $0 \le t \le 10$ for
$a = 0.2, 0.4, 0.6, 0.8$ and $x_0 = 0, 1$. Describe
how the value of a affects the solution.

In Exercises 43-48, solve the initial-value problem using Laplace transforms. Interpret each as a
population problem. Is the population bounded?

43. $x' + 5x = 500(2 - \sin t), x(0) = 10,000$
44. $x' + 5x = 500(2 + \cos t), x(0) = 5000$
45. $x' + 5x = 500(2 - \cos t), x(0) = 5000$
46. $x' + 5x = 500(2 - \sin t), x(0) = 5000$
47. $x' - 2x = 500(2 + \sin t), x(0) = 5000$
48. $x' - 2x = 500(1 + \cos t), x(0) = 5000$
49. Suppose that the emigration function is

$$f(t) = \begin{cases} 5000(2 - \sin t), & 0 \le t < 5 \\ 0, & t \ge 5 \end{cases} \text{Solve}$$

$x' - x = f(t), x(0) = 10,000$. Determine
$\lim_{t \to \infty} x(t)$.

50. Suppose that the emigration function is

$$f(t) = \begin{cases} 5000(1 + \cos t), & 0 \le t < 10 \\ 0, & t \ge 10 \end{cases} \text{Solve}$$

$x' - x = f(t), x(0) = 5000$. Determine
$\lim_{t \to \infty} x(t)$.

51. If the emigration function is the periodic
function

$$f(t) = \begin{cases} 5000(1 + \cos t), & 0 \le t < 1 \\ 0, & 1 \le t < 2 \end{cases}$$

$f(t + 2) = f(t)$, solve $x' - x = f(t)$,
$x(0) = 5000$.

52. If the emigration function is the periodic
function

$$f(t) = \begin{cases} 5000(2 - \sin t), & 0 \le t < 1 \\ 0, & 1 \le t < 2 \end{cases}$$

$f(t + 2) = f(t)$, solve $x' - x = f(t)$,
$x(0) = 5000$.

53. Suppose that a bacteria population satisfies
the differential equation $dx/dt = x +$
$20\delta(t - 2)$ with $x(0) = 100$. What is the
population at $t = 5$?

54. Suppose that a patient receives glucose through an IV tube at a constant rate of c grams per minute. If at the same time the glucose is metabolized and removed from the bloodstream at a rate that is proportional to the amount of glucose present in the bloodstream, the rate at which the amount of glucose changes is modeled by $dx/dt = c - kx$, where $x(t)$ is the amount of glucose in the bloodstream at time t and k is a constant. If $x(0) = x_0$, use Laplace transforms to find $x(t)$. Does $x(t)$ approach a limit as $t \to \infty$?

55. Suppose that in Exercise 54, the patient receives glucose at a variable rate $c(1 + \sin t)$. Therefore, the rate at which the amount of glucose changes is modeled by $dx/dt = c(1 + \sin t) - kx$. If $x(0) = x_0$, use Laplace transforms to find $x(t)$. How does the solution of this problem differ from that found in Exercise 54?

56. If the person in Exercise 54 receives glucose periodically according to the function
$$f(t) = \begin{cases} c, & 0 \le t < 1 \\ 0, & 1 \le t < 2 \end{cases}, f(t+2) = f(t), \text{ solve}$$
the initial-value problem $dx/dt = f(t) - kx$, $x(0) = x_0$, and compare the solution with that of Exercise 55.

57. (*Drug dosage problem*) The drug dosage problems considered in Exercises 54-56 may be thought of as a two-tank (compartment) problem. Suppose that we administer a drug once every 4 h, where the drug moves from the gastrointestinal (GI) tract (tank A) into the circulatory system (tank B). Let $x(t)$ represent the amount of the drug in the GI tract at time t hours, $y(t)$ be the amount of the drug in the circulatory system, and $c(t)$ be the dosage of the drug. If a and b are the rates at which the drug is consumed in the GI tract and the circulatory system, respectively, and if initially neither the GI tract nor the circulatory system contain the drug, we model this situation with the system

$$\frac{dx}{dt} = c(t) - ax, \quad x(0) = 0$$
$$\frac{dy}{dt} = ax - by, \quad y(0) = 0.$$

Suppose that $c(t) = \begin{cases} c_0, & 0 \le t < 1/2 \\ 0, & 1/2 \le t < 4 \end{cases}$,
$c(t - 4) = c(t)$. (a) Show that for $0 \le t \le 24$ (over 1 day),

$$c(t) = \sum_{n=0}^{6} c_0 \left[\mathcal{U}(t - 4n) - \mathcal{U}(t - 4n - 1/2) \right]$$

and

$$\mathcal{L}\{c(t)\} = \sum_{n=0}^{6} \frac{c_0}{s} \left(e^{-4ns} - e^{-(4n+1/2)s} \right).$$

(b) Use the Laplace transform to solve $dx/dt = c(t) - ax$, $x(0) = 0$. (*Note:*
$X(s) = \sum_{n=0}^{6} \frac{c_0}{s(s+a)} \left(e^{-4ns} - e^{-(4n+1/2)s} \right)$.)
(c) Use the Laplace transform and $x(t)$ to solve $dy/dt = ax - by$, $y(0) = 0$. (*Note:*

$$Y(s) = \sum_{n=0}^{6} \frac{c_0}{s+b} \left[\left(\frac{1}{s} - \frac{1}{s+a} \right) e^{-4ns} \right.$$
$$\left. - \left(\frac{1}{s} - \frac{1}{s+a} \right) e^{-(4n+1/2)s} \right].)$$

(d) If the half-life of the drug in the GI tract is 1 h and that in the circulatory system is 4 h, then graph the amount of the drug in the GI tract and the circulatory system over $[0, 24]$. (Recall that we discussed half-life with regard to exponential decay in Chapter 3.) We note that if the half-life is 1 h, the one half of the initial dosage remains after 1 h. Therefore, we solve $x_0 e^{-a \times 1} = x_0/2$ for a, which yields $a = \ln 2$. Similarly, we solve $y_0 e^{-b \times 4} = y_0/2$ to find that $b = (\ln 2)/4$. Describe what happens to the drug concentration in each case. Assume that $c_0 = 2$. (e) Suppose that the half-life of the drug in the GI is 30 min and that in the circulatory system it is 4 h. How does this change affect the drug concentration in the circulatory system? (f) Suppose that the half-life of the drug in the circulatory system in part (d) is 5 h. How does this change affect the drug concentration in the circulatory system? (g) How does adding a buffer to the drug increase the half-life of the drug in the GI tract affect the drug concentration in the circulatory system?

58. (*The stiffness matrix*) Show that we can write the system in matrix form

$$m_1 \frac{d^2x}{dt^2} = -k_1 x + k_2(y - x),$$

$$m_2 \frac{d^2y}{dt^2} = -k_2(y - x)$$

in matrix form $\mathbf{Mx''} = \mathbf{Kx}$, where

$\mathbf{M} = \begin{pmatrix} m_1 & 0 \\ 0 & m_2 \end{pmatrix}$ is the mass matrix and

$\mathbf{K} = \begin{pmatrix} -(k_1 + k_2) & k_2 \\ k_2 & -k_2 \end{pmatrix}$ is the *stiffness matrix*.

(b) Show that we can write $\mathbf{Mx''} = \mathbf{Kx}$ as $\mathbf{x''} = \mathbf{Ax}$, where $\mathbf{A} = \mathbf{M}^{-1}\mathbf{K}$. (Why does $\mathbf{M}^{-1}$ exist?) (c) Assume that $\mathbf{x}(t) = \mathbf{v}e^{\alpha t}$ is a solution of $\mathbf{x''} = \mathbf{Ax}$. Show that $\mathbf{x}(t) = \mathbf{v}e^{\alpha t}$ satisfies the system if and only if $\alpha^2 = \lambda$, where λ is an eigenvalue of $\mathbf{A}$ and $\mathbf{v}$ is an associated eigenvector. (d) The eigenvalues of $\mathbf{A}$ are typically negative real numbers. Show that two solutions of the system are $\mathbf{x}_1 = \mathbf{v} \cos \omega t$ and $\mathbf{x}_2 = \mathbf{v} \sin \omega t$, where $\alpha^2 = \lambda = -\omega^2$ or $\alpha = \pm\omega i$.

59. Let $m_1 = 2$, $m_2 = 1/2$, $k_1 = 75$, and $k_2 = 25$ in the previous system. (a) Show that the eigenvalues of $\mathbf{A} = \mathbf{M}^{-1}\mathbf{K} = \begin{pmatrix} -50 & 25/2 \\ 50 & -50 \end{pmatrix}$

are $\lambda_1 = -75$ and $\lambda_2 = -25$. Then, $\alpha_1^2 = \lambda_1 = -\omega_1^2$ implies that $\omega_1 = \pm 5\sqrt{3}$; $\alpha_2^2 = \lambda_2 = -\omega_2^2$ implies that $\omega_2 = \pm 5i$. (b) Show that an eigenvector corresponding to $\lambda_1 = -75$ is $\mathbf{v}_1 = \begin{pmatrix} 1 \\ -2 \end{pmatrix}$; that corresponding to $\lambda_2 = -25$ is $\mathbf{v}_2 = \begin{pmatrix} 1 \\ 2 \end{pmatrix}$. (c) Show that two

linearly independent solutions corresponding to $\omega_1 = \pm 5\sqrt{3}$ are $\begin{pmatrix} \cos 5\sqrt{3}t \\ -2\cos 5\sqrt{3}t \end{pmatrix}$ and $\begin{pmatrix} \sin 5\sqrt{3}t \\ -2\sin 5\sqrt{3}t \end{pmatrix}$; those corresponding to $\omega_2 = \pm 5i$ are $\begin{pmatrix} \cos 5t \\ 2\cos 5t \end{pmatrix}$ and $\begin{pmatrix} \sin 5t \\ 2\sin 5t \end{pmatrix}$. (c) Form a general solution to the system by taking a linear combination of the four solutions. (d) If the initial conditions are $x(0) = 0$, $y(0) = 4$, $x'(0) = 0$, and $y'(0) = 0$, determine the solution to the initial-value problem.

60. (*Forced system method of undetermined coefficients*) Suppose that a system can be written as $\mathbf{x''} = \mathbf{Ax} + \mathbf{f}$, where $\mathbf{f} = \mathbf{F}_0 \cos \omega t$. Show that if a particular solution has the form $\mathbf{x}_p(t) = \mathbf{c} \cos \omega t$, then $\mathbf{c}$ satisfies the system $(\mathbf{A} + \omega^2 \mathbf{I})\mathbf{c} = -\mathbf{F}_0$. If $\omega \neq 5$, $5\sqrt{3}$, find a particular solution of the system in Exercise 59 if $\mathbf{F}_0 = \begin{pmatrix} 0 \\ 50 \end{pmatrix}$. (*Note:* The constants will depend on ω.)

61. (*Delay equation*) According to the definition, $\mathcal{L}\{y(t - t_0)\} = \int_0^\infty e^{-st}y(t - t_0)\,dt$. If $y(0) = 0$ for $-t_0 \leq t \leq 0$, use the change of variables $v = t - t_0$ to show that $\mathcal{L}\{y(t - t_0)\} = e^{-t_0 s}Y(s)$.

In Exercises 62–67, solve the coupled spring-mass system modeled by

$$m_1 \frac{d^2x}{dt^2} = -k_1 x + k_2(y - x), \quad m_2 \frac{d^2x}{dt^2} = -k_2(y - x)$$

using the indicated parameter values and initial conditions.

62. $m_1 = 2$, $m_2 = 2$, $k_1 = 6$, $k_2 = 4$, $x(0) = 0$, $x'(0) = -1$, $y(0) = 0$, $y'(0) = 0$
63. $m_1 = 2$, $m_2 = 2$, $k_1 = 6$, $k_2 = 4$, $x(0) = 0$, $x'(0) = 0$, $y(0) = 1$, $y'(0) = 0$
64. $m_1 = 2$, $m_2 = 1$, $k_1 = 4$, $k_2 = 2$, $x(0) = 0$, $x'(0) = 0$, $y(0) = 0$, $y'(0) = -1$
65. $m_1 = 2$, $m_2 = 1$, $k_1 = 4$, $k_2 = 2$, $x(0) = 1$, $x'(0) = 0$, $y(0) = 0$, $y'(0) = 0$

66. $m_1 = 2, m_2 = 2, k_1 = 3, k_2 = 2, x(0) = 0,$
$x'(0) = 1, y(0) = 0, y'(0) = 0$

67. $m_1 = 2, m_2 = 2, k_1 = 3, k_2 = 2, x(0) = 1,$
$x'(0) = 0, y(0) = 0, y'(0) = 0$

In Exercises 68-71, solve the coupled spring-mass system modeled by

$$m_1\frac{d^2x}{dt^2} = -k_1x + k_2(y - x) + F_1(t)$$

$$m_2\frac{d^2x}{dt^2} = -k_2(y - x) + F_2(t)$$

where $m_1 = 2, m_2 = 1, k_1 = 4, k_2 = 2, x(0) = 1,$
$x'(0) = y(0) = y'(0) = 0$ using the given forcing functions.

68. $F_1(t) = 1, F_2(t) = 1$
69. $F_1(t) = 1, F_2(t) = \sin t$
70. $F_1(t) = \cos t, F_2(t) = \sin t$
71. $F_1(t) = \cos 2t, F_2(t) = 0$

In Exercises 72-79, solve the coupled pendulum problem using the parameter values $m_1 = 3$ slugs, $m_2 = 1$ slug, $\ell_1 = \ell_2 = 16$ ft., and $g = 32$ ft./s^2 and the indicated initial conditions.

72. $\theta_1(0) = 1, \theta_1'(0) = 0, \theta_2(0) = 0, \theta_2'(0) = 1$
73. $\theta_1(0) = 0, \theta_1'(0) = 0, \theta_2(0) = 0, \theta_2'(0) = -1$
74. $\theta_1(0) = 1, \theta_1'(0) = 1, \theta_2(0) = 0, \theta_2'(0) = 0$
75. $\theta_1(0) = 1, \theta_1'(0) = 0, \theta_2(0) = 0, \theta_2'(0) = 0$
76. $\theta_1(0) = 0, \theta_1'(0) = 1, \theta_2(0) = -1, \theta_2'(0) = 0$
77. $\theta_1(0) = 0, \theta_1'(0) = 1, \theta_2(0) = 0, \theta_2'(0) = 0$
78. $\theta_1(0) = 0, \theta_1'(0) = 0, \theta_2(0) = -1, \theta_2'(0) = -1$
79. $\theta_1(0) = 0, \theta_1'(0) = 0, \theta_2(0) = -1, \theta_2'(0) = 0$
80. Consider the physical situation of two pendulums coupled with a spring as shown in Figure 8.26. The motion of this pendulum-spring system is approximated by solving the second-order system

$$mx'' + m\omega_0^2x = -k(x - y)$$

$$my'' + m\omega_0^2y = -k(y - x),$$

where L is the length of each pendulum, g is the gravitational constant, and $\omega_0^2 = g/L$. Use the method of Laplace transforms to solve this system if the initial conditions are $x(0) = a, x'(0) = b, y(0) = c, y'(0) = d.$

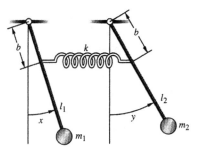

FIGURE 8.26 Two pendulums coupled with a spring.

81. Solve the initial-value problem

$$mx'' + m\omega_0^2x = -k(x - y),$$

$$my'' + m\omega_0^2y = -k(y - x), \ x(0) = -1,$$

$$y(0) = 1, \ x'(0) = y'(0) = 0.$$

Verify that the initial conditions are satisfied.

82. Solve the initial-value problem

$$mx'' + m\omega_0^2x = -k(x - y),$$

$$my'' + m\omega_0^2y = -k(y - x),$$

$$x(0) = y(0) = y'(0) = 0, \ x'(0) = 1.$$

Verify that the initial conditions are satisfied.

83. The physical situation shown in Figure 8.27 is modeled by the system of differential equations

$$m\frac{d^2x}{dt^2} + 3kx - ky = 0,$$

$$2m\frac{d^2y}{dt^2} + 3ky - kx = 0.$$

If $x(0) = y(0) = y'(0) = 0$ and $x'(0) = 1$, find x and y. Compare these results to those found if $x(0) = y(0) = x'(0) = 0$ and

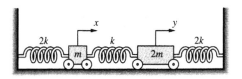

FIGURE 8.27 Two objects connected by three springs.

$y'(0) = 1$. Does this model resemble one that was introduced earlier in the section?

84. Write the system

$$m_1 \frac{d^2x}{dt^2} = -k_1 x + k_2(y - x),$$

$$m_2 \frac{d^2y}{dt^2} = -k_3 y - k_2(y - x)$$

as a system of four first-order equations with the substitutions $x = x_1$, $x_1' = x_2$, $y = y_1$, $y_1' = y_2$.

85. Show that the eigenvalues of the 4×4 coefficient matrix in the system of equations obtained in Exercise 84 are two complex conjugate pairs of the form $\pm q_1 i$ and $\pm q_2 i$, where q_1 and q_2 are real numbers. What does this tell you about the solutions to this system? Does this agree with the solution obtained with Laplace transforms?

86. In (a)-(d), determine the motion of the object of mass m attached to a spring with spring constant k if the damping is $c\, dx/dt$ and there is a piecewise defined periodic external force $f(t)$. Use the initial conditions $x(0) = x'(0) = 0$ in each problem.

(a) $m = 1, k = 6, c = 5,$

$$f(t) = \begin{cases} 1, & 0 \le t < 1 \\ 0, & 1 \le t < 2 \end{cases}, f(t + 2) = f(t)$$

(b) $m = 1, k = 6, c = 5,$

$$f(t) = \begin{cases} 1, & 0 \le t < 1 \\ -1, & 1 \le t < 2 \end{cases}, f(t + 2) = f(t)$$

(c) $m = 1, k = 13, c = 4,$

$$f(t) = \begin{cases} \sin t, & 0 \le t < \pi \\ 0, & \pi \le t < 2\pi \end{cases}$$

$f(t + 2\pi) = f(t)$

(d) $m = 1, k = 13, c = 4,$

$$f(t) = \begin{cases} 1, & 0 \le t < 1 \\ -1, & 1 \le t < 2 \end{cases}, f(t + 2) = f(t)$$

87. Consider the L-R-C circuit in which $R = 100\,\Omega$, $L = 1/10$ Henry, $C = 10^{-3}$, and $E(t) = 155 \sin 377t$. (Notice that the frequency of the voltage source is $377/(2\pi) \approx 60\,Hz = 60$ cycles/s.) (a) Find the current $I(t)$. (b) What is the maximum value of the current and where does it occur? (c) Find $\lim_{t\to\infty} I(t)$ (the steady state current). (d) What is the frequency of the steady-state current?

88. Suppose that the L-R-C circuit in Exercise 87 has the voltage source

$$E(t) = \begin{cases} 155 \cos 377t, & 0 \le t < 1 \\ 0, & t \ge 1 \end{cases}. \text{ (a) Find}$$

the current $I(t)$. (b) What is the maximum value of the current and where does it occur? (c) Find $\lim_{t\to\infty} I(t)$ (the steady state current). (d) What is the frequency of the steady-state current?

89. Suppose that the L-R-C circuit in Exercise 87 has the periodic voltage source

$$E(t) = \begin{cases} 155, & 0 \le t < 1 \\ 0, & 1 \le t < 2 \end{cases}, E(t + 2) = E(t).$$

(a) Find the current $I(t)$. (b) What is the maximum value of the current and where does it occur? (c) Find $\lim_{t\to\infty} I(t)$ (the steady state current). (d) What is the frequency of the steady-state current?

90. Solve the problem of the forced coupled spring-mass system with $m_1 = m_2 = 1$, $k_1 = 3$, and $k_2 = 2$ if the forcing functions are $F_1(t) = 1$ and $F_2(t) = \sin t$ and the initial conditions are $x(0) = y'(0) = 0$, $x'(0) = y(0) = 1$. Graph the solution parametrically as well as simultaneously. How does the motion differ from that of Example 8.7.3? What eventually happens to this system? Will the objects eventually come to rest?

91. Consider the three-spring problem shown in Figure 8.21 with $m_1 = m_2 = 1$ and $k_1 = k_2 = k_3 = 1$. Determine $x(t)$ and $y(t)$ if $x(0) = y(0) = 0$, $x'(0) = -1$, and $y'(0) = 1$. Graph the solution parametrically as well as simultaneously. When does the object attached to the top spring first pass through its equilibrium position? When does the object attached to the second spring first pass through its equilibrium position?

92. Solve Exercise 91 with the forcing functions $F_1(t) = 1$ and $F_2(t) = \cos t$. Graph the solution parametrically as well as simultaneously. When does the object attached to the top spring first pass through its equilibrium position? When does the object attached to the second spring first pass through its equilibrium position? How does the motion differ from that in Exercise 91?

93. Consider the double pendulum with $m_1 = 3$ slugs, $m_2 = 1$ slug, $\ell_1 = \ell_2 = 16$ ft. If $g = 32$ ft./s^2, determine $\theta_1(t)$ and $\theta_2(t)$ if $\theta_1(0) = 1$, $\theta_1'(0) = \theta_2(0) = 0$, $\theta_2'(0) = -1$. Graph the solution parametrically and simultaneously.

94. Suppose that in the double pendulum $m_1 = 3$ slugs, $m_2 = 1$ slug, $\ell_1 = \ell_2 = 4$ ft. If $g = 32$ ft./s^2, determine $\theta_1(t)$ and $\theta_2(t)$ if $\theta_1(0) = 1$, $\theta_1'(0) = \theta_2(0) = 0$, $\theta_2'(0) = -1$. Graph the solution parametrically and simultaneously. How does the length of each pendulum affect the motion as compared to that of Exercise 93?

95. Suppose that in the double pendulum $m_1 = m_2 = 1$ slug, $\ell_1 = \ell_2 = 4$ ft. If $g = 32$ ft./s^2, determine $\theta_1(t)$ and $\theta_2(t)$ if $\theta_1(0) = 1$, $\theta_1'(0) = \theta_2(0) = 0$, $\theta_2'(0) = -1$. Graph the solution parametrically and simultaneously. How does the mass of the first object affect the motion as compared to that of Exercise 94?

96. Consider the physical situation shown in Figure 8.28, where the uniform bar has mass m_1, a centroidal motion of inertial $\bar{I}$, and is supported by two springs each with spring constant k. A mass m_3 is attached at the center of gravity (Point G) of the bar by another spring with spring constant k. Using the coordinates q_1, q_2, and q_3 as shown in Figure 8.28, we find that the motion is modeled with

$$m_1 \frac{d^2 q_1}{dt^2} + 3q_1 - kq_3 = 0$$

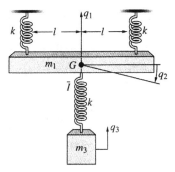

FIGURE 8.28 A mass connected to a bar that is connected to its support by two springs.

$$\bar{I} \frac{d^2 q_2}{dt^2} + 2kl^2 q_2 = 0$$

$$m_3 \frac{d^2 q_3}{dt^2} - kq_1 + kq_3 = 0.$$

(a) Determine the motion of the system if $k = 2$, $m_1 = 2$, $m_3 = 1$, $\bar{I} = 1$, $l = 1$ and the initial conditions are $q_1(0) = q_2(0) = q_2'(0) = q_3(0) = q_3'(0) = 0$, and $q_1'(0) = 1$. What is the maximum displacement of q_1, q_2, and q_3? (b) If the initial conditions are $q_1(0) = q_1'(0) = q_2(0) = q_2'(0) = q_3'(0) = 0$, and $q_3(0) = 1$, determine the motion of the system. What is the maximum displacement of q_1, q_2, and q_3? (c) How do the changes in the initial conditions affect the maximum displacement of each component of the system?

97. Consider the system of three springs shown in Figure 8.29, where springs S_1, S_2, and S_3 have spring constants k_1, k_2, and k_3, respectively, and have objects of mass m_1, m_2, and m_3, respectively, attached to them. Summing the forces acting on each mass and applying Newton's second law of motion yields the system of differential equations

$$m_1 x_1'' = -k_1 x_1 + k_2(x_2 - x_1)$$
$$m_2 x_2'' = -k_2(x_2 - x_1) + k_3(x_3 - x_2)$$
$$\text{(' denotes d/dt)}$$
$$m_3 x_3'' = -k_3(x_3 - x_2)$$

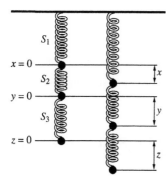

FIGURE 8.29 Two pairs of three masses connected by three springs.

Solve this system if $m_1 = 2$, $m_2 = m_3 = 1$, $k_1 = 1/2$, $k_2 = 1$, and $k_3 = 2$, using the initial conditions $x_1(0) = x_2(0) = x_3(0) = 0$, $x_1'(0) = 2$, $x_2'(0) = 1$, and $x_3'(0) = 2$. When does each mass first pass through its equilibrium position?

98. Consider the problem $y'(t) - 2y(t-1) = t$ subject to the condition $y(t) = y(0)$ for $-1 \le t \le 0$. Laplace transforms can be used to solve problems of this type with the following procedure.

(a) Take the Laplace transform of both sides of the equation to obtain

$$sY(s) - y(0) - 2\int_0^\infty e^{-st}y(t-1)\,dt = \frac{1}{s^2}.$$

(b) Evaluate $\int_0^\infty e^{-st}y(t-1)\,dt$ by letting $u = t - 1$ to obtain

$$\int_0^\infty e^{-st}y(t-1)\,dt$$
$$= e^{-s}\left(\int_{-1}^0 e^{-su}y(u)\,du + \int_0^\infty e^{-su}y(u)\,du\right).$$

(c) Use the condition $y(t) = y(0)$ for $-1 \le t \le 0$ to simplify the expression in (b). Substitute this result into the expression in (b). Substitute that result into the expression in (a) and solve for $Y(s)$ to find that

$$Y(s) = \frac{-e^s + 2sy(0) - 2e^s sy(0) - e^s s^2 y(0)}{s^3 e^s - 2s^2}.$$

(d) Use partial fractions to find that

$$Y(s) = \frac{2(1 + 2s\,y(0))}{s^3(se^s - 2)} + \frac{1 + 2sy(0) + s^2 y(0)}{s^3}.$$

(e) Find $\mathcal{L}^{-1}\left\{\dfrac{1 + 2sy(0) + s^2 y(0)}{s^3}\right\}$.

(f) Find $\mathcal{L}^{-1}\left\{\dfrac{2(1 + 2s\,y(0))}{s^3(se^s - 2)}\right\}$ by rewriting this expression as

$$\frac{2(1 + 2s\,y(0))}{s^3(se^s - 2)} = \frac{2(1 + 2s\,y(0))}{s^4 e^s}\frac{1}{1 - 2s^{-1}e^{-s}}$$

$$= \frac{2(1 + 2s\,y(0))}{s^4 e^s}\sum_{n=0}^\infty \left(2s^{-1}e^{-s}\right)^n$$

$$= \sum_{n=0}^\infty \left(2^{n+1}\frac{e^{-ns}}{s^{n+4}} + 2^{n+2}y(0)\frac{e^{-ns}}{s^{n+3}}\right).$$

(g) Show that

$$\mathcal{L}^{-1}\left\{2^{n+1}\frac{e^{-ns}}{s^{n+4}} + 2^{n+2}y(0)\frac{e^{-ns}}{s^{n+3}}\right\}$$

$$= \frac{2^{n+1}}{(n+3)!}(t-n)^{n+2}\mathcal{U}(t-n)$$

$$\times \left[t + 6y(0) + (2y(0) - 1)n\right]$$

so that

$$\mathcal{L}^{-1}\left\{\frac{2(1 + 2s\,y(0))}{s^3(se^s - 2)}\right\}$$

$$= \sum_{n=0}^\infty \left[\frac{2^{n+1}}{(n+3)!}(t-n)^{n+2}\mathcal{U}(t-n)\right.$$

$$\times \left[t + 6y(0) + (2y(0) - 1)n\right]\Big].$$

(h) Find $y(t)$.

CHAPTER 8 SUMMARY: ESSENTIAL CONCEPTS AND FORMULAS

Definition of the Laplace Transform:
$\mathcal{L}\{f(t)\} = \int_0^\infty e^{-st}f(t)\,dt.$

Linearity of the Laplace Transform:
$\mathcal{L}\{af(t) + bg(t)\} = a\mathcal{L}\{f(t)\} + b\mathcal{L}\{g(t)\}.$

Exponential Order: *A function f is of exponential order (of order b) if there are numbers b, C > 0, and T > 0 such that* $|f(t)| \le Ce^{bt}$ *for t > T.*

Shifting Property of the Laplace Transform:
$\mathcal{L}\{e^{at}f(t)\} = F(s - a).$

Inverse Laplace Transform: $f(t) = \mathcal{L}^{-1}\{f(s)\}$ if $\mathcal{L}\{f(t)\} = F(s).$

Laplace Transform of the First Derivative:
$\mathcal{L}\{f'(t)\} = s\mathcal{L}\{f(t)\} - f(0).$

Laplace Transform of Higher Order Derivative: $\mathcal{L}\{f^{(n)}(t)\} =$
$s^n\mathcal{L}\{f(t)\} - s^{n-1}f(0) - \cdots - f^{(n-1)}(0).$

Derivatives of the Laplace Transform:
$\mathcal{L}\{t^nf(t)\} = (-1)^n\dfrac{d^nF}{ds^n}(s)$ (Note that
$\mathcal{L}\{f(t)\} = F(s)$).

Laplace Transform of a Periodic Function:
$\mathcal{L}\{f(t)\} = \dfrac{1}{1 - e^{-sT}}\int_0^T e^{-st}f(t)\,dt.$

Unit Impulse Function: $\delta(t - t_0) = 0,$
$t \ne t_0, \int_{-\infty}^{\infty}\delta(t - t_0)\,dt = 1,$
$\mathcal{L}\{\delta(t - t_0)\} = e^{-st_0}.$

Convolution theorem:
$\mathcal{L}^{-1}\{F(s)G(s)\} = \mathcal{L}^{-1}\{\mathcal{L}\{(f * g)(t)\}\} =$
$(f * g)(t) = \int_0^t f(t - v)g(v)\,dv.$

Coupled Spring-Mass System: Two Springs:

$$m_1\frac{d^2x}{dt^2} = -k_1x + k_2(y - x) + F_1(t),$$

$$m_2\frac{d^2y}{dt^2} = -k_2(y - x) + F_2(t)$$

Coupled Spring-Mass System: Three Springs:

$$m_1\frac{d^2x}{dt^2} = -k_1x + k_2(y - x) + F_1(t),$$

$$m_2\frac{d^2y}{dt^2} = -k_3y - k_2(y - x) + F_2(t)$$

The Double Pendulum:

$$(m_1 + m_2)\ell_1{}^2\theta_1'' + m_2\ell_1\ell_2\theta_2''$$
$$+(m_1 + m_2)\ell_1g\theta_1 = 0$$
$$m_2\ell_2{}^2\theta_2'' + m_2\ell_1\ell_2\theta_1'' + m_2\ell_2g\theta_2 = 0$$

CHAPTER 8 REVIEW EXERCISES

In Exercises 1-4, find the Laplace transform of each function using the definition of the Laplace transform.

1. $f(t) = 1 - t$
2. $f(t) = te^{-4t}$
3. $f(t) = \begin{cases} 1, & 0 \le t < 5 \\ 0, & t \ge 5 \end{cases}$
4. $f(t) = \begin{cases} t, & 0 \le t < 1 \\ 0, & t \ge 1 \end{cases}$

In Exercises 5-26, find the Laplace transform of each function.

5. $t^2 + 5$
6. $2\sinh 4t$
7. te^{2t}
8. t^3
9. t^3e^t
10. $2e^t\sin 5t$
11. $t\cos 3t$
12. $t\sin 2t$
13. $e^{-5t}\cos 3t$
14. $\delta(t - 2\pi)$
15. $\delta(t - 3\pi/2)$
16. $7\mathcal{U}(t - 7)$
17. $6\mathcal{U}(t - 7) - 4\mathcal{U}(t - 4)$
18. $36e^{-4t}\mathcal{U}(t - 2)$
19. $-42e^{5t}\mathcal{U}(t - 1)$
20. $5\sin(t - 5)\mathcal{U}(t - 5)$
21. $t^2\mathcal{U}(t - 2)$
22. $f(t) = \begin{cases} 2, & 0 \le t < 1 \\ 1, & 1 \le t < 2 \end{cases}, f(t + 2) = f(t)$ if $t \ge 2$
23. $f(t) = 1 - t, 0 \le t < 1, f(t + 1) = f(t)$ if $t \ge 1$

24. $f(t) = \sin \pi t, 0 \le t < 1, f(t+1) = f(t)$ if $t \ge 1$

25. $f(t) = \begin{cases} \cos \frac{\pi}{2}t, & 0 \le t < 1 \\ 0, & 1 \le t < 2 \end{cases}, f(t) = f(t-2)$ if $t \ge 2$

26. $f(t) = \begin{cases} t, & 0 \le t < 2 \\ 2, & 2 \le t < 4, f(t) = f(t-6) \text{ if} \\ 6 - t, & 4 \le t < 6 \end{cases}$ $t \ge 6$

27. $f(t) = \int_0^t \frac{\sin \tau}{\tau} \, d\tau$

28. $\dfrac{e^{-at} - e^{-bt}}{t}$

29. $\dfrac{1 - e^{-t}}{t}$

In Exercises 30-41, find the inverse Laplace transform of each function.

30. $\dfrac{-10}{s^2 - 25}$

31. $-\dfrac{2s}{(s^2 + 1)^2}$

32. $\dfrac{3(40 - s^5)}{s^6}$

33. $\dfrac{2s^2 - 7s + 20}{s(s^2 - 2s + 10)}$

34. $\dfrac{-s^2 - 15s - 32}{s(s^2 + 10s + 26)}$

35. $-\dfrac{14}{se^{2s}}$

36. $\dfrac{6e^{4s} + 7}{se^{7s}}$

37. $\dfrac{3e^{6-s}}{6 - s}$

38. $\dfrac{8s}{(s^2 + 1)e^{3s}}$

39. $\dfrac{-18}{(s^2 + 1)(1 - e^{-3s})}$

40. $\dfrac{s \sin \phi + \omega \cos \phi}{s^2 + \omega^2}$

41. $\dfrac{s \cos \phi - \omega \sin \phi}{s^2 + \omega^2}$

In Exercises 42 and 43, compute the convolution $(f * g)(t)$ using the given pair of functions.

42. $f(t) = \sqrt{t}, g(t) = t^2$

43. $f(t) = \cos t, g(t) = \sin 2t$

In Exercises 44-54, solve the initial-value problem.

44. $y'' + 6y' + 10y = 0, y(0) = 0, y'(0) = 1$

45. $y'' - 4y' + 5y = 0, y(0) = 1, y'(0) = 0$

46. $y' + y = f(t), f(t) = \begin{cases} \sin \pi t, & 0 \le t < 1 \\ 0, & t \ge 1 \end{cases}$, $y(0) = 0$

47. $y' + 2y = \delta(t - 1), y(0) = 1$

48. $y'' - y = f(t), f(t) = \begin{cases} t, & 0 \le t < 1 \\ 2 - t, & 1 \le t < 2, \\ 0, & t \ge 2 \end{cases}$ $y(0) = y'(0) = 0$

49. $y'' + 12y' + 32y = f(t)$, $f(t) = \begin{cases} t, & 0 \le t < 1 \\ 2 - t, & 1 \le t < 2 \end{cases}, f(t) = f(t-2)$ if $t \ge 2, y(0) = y'(0) = 0$

50. $y'' + 6y' + 8y = f(t), f(t) = \begin{cases} 2, & 0 \le t < 1 \\ -1, & 1 \le t < 2 \end{cases}$ $f(t) = f(t-2)$ if $t \ge 2, y(0) = y'(0) = 0$

51. $x'' + 3x' + 2x = \delta(t - 1) + \delta(t - 2)$, $x(0) = x'(0) = 0$

52. $x'' + 9x = \cos 3t + \delta(t - \pi/3), x(0) = x'(0) = 0$

53. $g(t) = \sin t + \int_0^t g(t - v)e^{-v} \, dv$

54. $g(t) = 5 + t - \int_0^t (t - v) g(v) \, dv$

55. Use the Maclaurin series expansion
$\tan^{-1} x = x - \frac{1}{3}x^3 + \frac{1}{5}x^5 - \frac{1}{7}x^7 + \frac{1}{9}x^9 - \cdots$ to find the Maclaurin series expansion for $\tan^{-1}(1/s)$. Use this expansion to show that $\mathcal{L}^{-1}\{\tan^{-1}(1/s)\} = \dfrac{\sin t}{t}$. (*Hint*: Use the Maclaurin series expansion of $\sin x$: $\sin x = \sum_{k=0}^{\infty} \dfrac{(-1)^k}{(2k + 1)!} x^{2k+1} = x - \frac{1}{3!}x^3 + \frac{1}{5!}x^5 - \cdots$.)

56. Use $\mathcal{L}\{t^n f(t)\} = (-1)^n \dfrac{d^n}{ds^n} F(s)$ with $n = 1$:

$$f(t) = -\frac{1}{t} \mathcal{L}^{-1}\left\{\frac{d}{ds} F(s)\right\} \text{ to compute}$$

(a) $\mathcal{L}^{-1}\left\{\ln \dfrac{s-5}{s+2}\right\}$ and (b) $\mathcal{L}^{-1}\left\{\ln \dfrac{s^2+4}{s^2+9}\right\}$.

In Exercises 57 and 58, solve the initial-value problem $mx'' + cx' + kx = f(t)$, $x(0) = a$, $x'(0) = b$ to determine the motion of an object attached to the end of a spring using the given parameter values.

57. $m = 4, c = 0, k = 1, a = 0, b = -1$,

$$f(t) = \begin{cases} \cos 2t, & 0 \le t < \pi \\ 0, & t \ge \pi \end{cases}$$

58. $m = 1, c = 1/2, k = 145/16, a = b = 0$,
$f(t) = \delta(t - \pi)$

In Exercises 59 and 60 solve the L-R-C series circuit modeled by

$$L\frac{d^2Q}{dt^2} + R\frac{dQ}{dt} + \frac{1}{C}Q = E(t),$$

$$Q(0) = Q_0, \quad I(0) = \frac{dQ}{dt}(0) = I_0$$

using the given parameter values and functions. (Assume that the units are Henry, Ohm, farad, and volt are used, respectively, for L, R, C, and $E(t)$.)

59. $L = 1, R = 0, C = 10^{-4}$,

$$E(t) = \begin{cases} 220, & 0 \le t < 2 \\ 0, & t \ge 2 \end{cases}, Q_0 = I_0 = 0$$

60. $L = 4, R = 20, C = 1/25$,

$$E(t) = \begin{cases} 0, & 0 \le t < 1 \\ 50, & 1 \le t < 2 \end{cases}, E(t+2) = E(t) \text{ if}$$
$t \ge 2, Q_0 = I_0 = 0$

In Exercises 61 and 62, interpret the initial-value problem as a population problem. In each case, determine if the population approaches a limit as $t \to \infty$.

61. $x' - 2x = 100\,\delta(t-1), x(0) = 10,000$

62. $x' - 2x = \begin{cases} 100t, & 0 \le t < 1 \\ 100, & t \ge 1 \end{cases}, x(0) = 10,000$

63. Suppose that the vertical displacement of a horizontal beam is modeled by the boundary-value problem

$$d^4y/dx^4 = W(x), \quad y(0) = y'(0) = 0,$$
$$y(1) = y'(1) = 0,$$

where $W(x)$ is the constant load that is uniformly distributed along the beam. Use the two conditions at $x = 0$ with the method of Laplace transforms to obtain a solution that involves the arbitrary constants $A = y''(0)$ and $B = y'''(0)$. Then apply the conditions at $x = 1$ to find A and B.

64. Suppose that the load in Exercise 63 is not constant. Instead, it is given by

$$W(x) = \begin{cases} 10, & 0 \le x < 1/2 \\ 0, & 1/2 \le x < 1 \end{cases}. \text{ Solve}$$

$$d^4y/dx^4 = W(x), \quad y(0) = y'(0) = 0,$$
$$y(1) = y'(1) = 0,$$

to find the displacement of the beam.

In Exercises 65-68, solve each system.

65. $y'' = -x - 2y, x'' = -2x - 4y, x'(0) = 1$,
$x(0) = y(0) = y'(0) = 0$

66. $y'' - 3y' = -x' - 2y + x, x' + y' = 2y - x$,
$x(0) = y(0) = 0, x'(0) = 1, y'(0) = -1$

67. $x' = -2x - 4y, y' = \frac{1}{4}x + f(t)$,

$$f(t) = \begin{cases} 1, & 0 \le t < 2 \\ 0, & t \ge 2 \end{cases}, x(0) = y(0) = 0$$

68. $x' = -y + f(t), y' = x$,

$$f(t) = \begin{cases} -1, & 0 \le t < \pi \\ 1, & \pi \le t < 2\pi, x(0) = y(0) = 0 \\ 0, & t \ge 2\pi \end{cases}$$

69. Solve the spring-mass system with two springs, as shown in Figure 8.18(a), using $m_1 = m_2 = k_1 = 6, k_2 = 4$, and the initial conditions $x(0) = y(0) = y'(0) = 0$ and $x'(0) = 2$. (Assume no external forcing functions.)

70. Solve the spring-mass system with two springs as shown in Figure 8.18(a), if the forcing functions $F_1(t) = 2\cos t$ and $F_2(t) = 0$ are included with the parameters $m_1 = m_2 = 1, k_1 = 3, k_2 = 2$, and the initial conditions are $x(0) = x'(0) = y(0) = y'(0) = 0$. What physical phenomenon occurs in $y(t)$?

71. Solve the spring-mass system with three springs (as shown in Figure 8.21) if the forcing functions $F_1(t) = 2\cos t$ and $F_2(t) = 0$ and included with the parameters $m_1 = m_2 = k_1 = k_2 = k_3 = 1$ and the initial conditions are $x(0) = x'(0) = y(0) = y'(0) = 0$.

72. Solve the double pendulum with $m_1 = 3$, $m_2 = 1, \ell_1 = \ell_2 = 16$ ft., $g = 32$ ft./s^2, and the initial conditions $\theta_1(0) = \theta_1'(0) = 0$ and $\theta_2(0) = \theta_2'(0) = -1$.

73. Solve the double pendulum with $m_1 = 3$, $m_2 = 1, \ell_1 = \ell_2 = 16$ ft., $g = 32$ ft./s^2, and the initial conditions $\theta_1(0) = 0, \theta_1'(0) = -1$ and $\theta_2(0) = 0, \theta_2'(0) = 1$.

DIFFERENTIAL EQUATIONS AT WORK

A. The Tautochrone

From rest, a particle slides down a frictionless curve under the force of gravity as illustrated in Figure 8.30. What must the shape of the curve be for the time of descent to be independent of the starting position of the particle?

The shape of the curve is found through the use of the Laplace transform. Suppose that the particle starts at height y and that its speed is v when it is at a height of z. If m is the mass of the particle and g is the acceleration of gravity, then the speed can be found by equating the kinetic and potential energies of the particle with

$$\frac{1}{2}mv^2 = mg(y - z) \quad \text{or} \quad v = \sqrt{2g}\sqrt{y - z}.$$

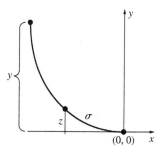

FIGURE 8.30 The tautochrone.

Christiaan Huygens (1629-1695), Dutch mathematician astronomer, and physicist, originally solved this problem in 1673 as he studied the mathematics associated with pendulum clocks. Huygens greatly influenced Leibniz. Huygens designed an internal combustion engine powered by gunpowder. Unfortunately, he was never able to build a working one.

To avoid confusion with the s that is usually used in the Laplace transform of functions, let σ denote the arc length along the curve from its lowest point to the particle. Therefore, the time required for the descent is

$$\text{Time} = \int_0^{\sigma(y)} \frac{1}{v}\,d\sigma = \int_0^y \frac{1}{v}\frac{d\sigma}{dz}\,dz$$
$$= \int_0^y \frac{1}{v}\phi(z)\,dz,$$

where $\phi(y) = d\sigma/dy$, which means that $\phi(z)$ is the value of $d\sigma/dy$ at $y = z$. Now, because the time is constant and $v = \sqrt{2g}\sqrt{y-z}$, we have

$$\int_0^y \frac{\phi(z)}{\sqrt{y-z}}\,dz = c_1,$$

where c_1 is a constant. In an attempt to use a convolution, we multiply by $e^{-sy}\,dy$ and integrate. Therefore,

$$\int_0^\infty e^{-sy}\int_0^y \frac{\phi(z)}{\sqrt{y-z}}\,dz\,dy = \int_0^\infty e^{-sy}c_1\,dy$$

$$\mathcal{L}\{\phi * y^{-1/2}\} \qquad = \mathcal{L}\{c_1\}.$$

By the convolution theorem, this simplifies to

$$\mathcal{L}\{\phi\}\,\mathcal{L}\{y^{-1/2}\} = \frac{c_1}{s}.$$

Then, because $\mathcal{L}\{t^{-1/2}\} = \sqrt{\pi/s}$, we have

$$\mathcal{L}\{\phi\}\sqrt{\frac{\pi}{s}} = \frac{c_1}{s} \quad \text{or} \quad \mathcal{L}\{\phi\} = \frac{c_1}{\sqrt{\pi}}\frac{1}{\sqrt{s}}.$$

Applying the inverse Laplace transform with $\mathcal{L}^{-1}\{s^{-1/2}\} = t^{-1/2}/\sqrt{\pi}$ then yields

$$\phi = \frac{c_1}{\sqrt{\pi}}y^{-1/2} = ky^{-1/2}.$$

Recall that $\phi(y) = d\sigma/dy$ represents arc length. Hence $\phi(y) = d\sigma/dy = \sqrt{1+(dx/dy)^2}$, so substitution into the previous equation gives

$$\sqrt{1+\left(\frac{dx}{dy}\right)^2} = ky^{-1/2} \quad \text{or} \quad 1+\left(\frac{dx}{dy}\right)^2 = \frac{k^2}{y}.$$

Solving for dx/dy then yields $dx/dy = \sqrt{(k^2/y)-1}$, which can be integrated with the substitution $y = k^2\sin^2\theta$.

1. Find $x(\theta)$ using this substitution.
2. Use the identity $\sin^2\theta = \frac{1}{2}(1-\cos 2\theta)$ to obtain a similar formula for $y(\theta)$.
3. Graph the curve $(x(\theta), y(\theta))$ for $-\pi/2 \le \theta \le 0$ for $k = 1,2,3$. How does the value of k affect the curve?

4. Find the time required for an object located at $(x(-\pi/2), y(-\pi/2))$ to move along the curve to the point $(x(0), y(0))$. Does this result depend on the value of θ?

B. Vibration Absorbers

Vibration absorbers can be used to virtually eliminate vibration in systems in which it is particularly undesirable and to reduce excessive amplitudes of vibrations in others. A typical type of vibration absorber consists of a spring-mass system constructed so that its natural frequency is easily varied. This system is then attached to the principal system that is to have its amplitude of vibration reduced, and the frequency of the absorber system is then adjusted until the desired result is achieved (see Figure 8.31(a)).

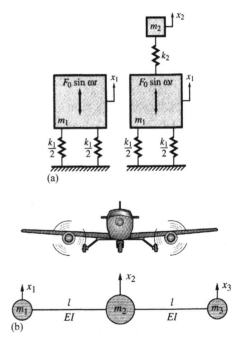

FIGURE 8.31   (a) The principal system and the vibration absorber attached to the principal system. (b) A simple model of an airplane.

If the frequency ω of the disturbing force $f_0 \sin \omega t$ is near or equal to the natural frequency $\omega_n = \sqrt{k_1/m_1}$ of the system, the amplitude of vibration of the system could become very large due to resonance. If the absorber spring-mass system, made up of components with k_2 and m_2, is attached to the principal system, the amplitude of the mass can be reduced to almost zero if the natural frequency of the absorber is adjusted until it equals that of the disturbing force $\omega = \sqrt{k_2/m_2}$. These types of absorbers are designed to have little damping and are "tuned" by varying k_2, m_2, or both. This problem is modeled by the system of differential equations

$$m_1 \frac{d^2 x_1}{dt^2} + (k_1 + k_2)x_1 - k_2 x_2 = F_0 \sin \omega t$$

$$m_2 \frac{d^2 x_2}{dt^2} - k_2 x_1 + k_2 x_2 = 0.$$

1. Solve the system for x_1 and x_2.
2. When is the amplitude of x_1 equal to zero? What does this represent?
   The mass ratio m_2/m_1 is an important parameter in the design of the absorber. To see the effect on the response of the system, transform these properties to the nondimensional form

$$w_{ss}{}^2 = \frac{k_2}{m_2} = \frac{k_1}{m_1} \quad \text{and} \quad \mu = \frac{m_2}{m_1} = \frac{k_2}{k_1}.$$

3. If A_1 is the amplitude of x_1, express $A_1/(F_0/k_1)$ as a function of ω/ω_{22}.
4. Plot the absolute value of $A_1/(F_0/k_1)$ for $\mu = 0.2$.
5. When does $A_1/(F_0/k_1)$ become infinite? Do these values correspond to resonance?
6. In designing the vibration absorber, what frequencies should the absorber *not* be "tuned" in order to avoid resonance?

C. Airplane Wing

A small airplane is modeled using three lumped masses as shown in Figure 8.31(b). We assume in this simplified model of the airplane that the wings are uniform beams of length ℓ and stiffness factor EI where E and I are constants. (E depends on the material from which the beam is made and I depends on the shape and size of the beam.) We also assume that $m_1 = m_3$ and $m_2 = 4m_1$, where m_1 and m_3 represent the mass of each wing and m_2 represents the mass of the body of the airplane.

We find the displacements x_1, x_2, and x_3 by solving the system

$$M \begin{pmatrix} x_1'' \\ x_2'' \\ x_3'' \end{pmatrix} + K \begin{pmatrix} x_1 \\ x_2 \\ x_3 \end{pmatrix} = \begin{pmatrix} 0 \\ 0 \\ 0 \end{pmatrix},$$

where $K = \dfrac{EI}{\ell^3} \begin{pmatrix} 3 & -3 & 0 \\ -3 & 6 & -3 \\ 0 & -3 & 3 \end{pmatrix}$ and $M = m_1 \begin{pmatrix} 1 & 0 & 0 \\ 0 & 4 & 0 \\ 0 & 0 & 1 \end{pmatrix}$. We can then write the system as

$$\begin{pmatrix} 1 & 0 & 0 \\ 0 & 4 & 0 \\ 0 & 0 & 1 \end{pmatrix} \begin{pmatrix} x_1'' \\ x_2'' \\ x_3'' \end{pmatrix} + \frac{EI}{m_1 \ell^3} \begin{pmatrix} 3 & -3 & 0 \\ -3 & 6 & -3 \\ 0 & -3 & 3 \end{pmatrix} \begin{pmatrix} x_1 \\ x_2 \\ x_3 \end{pmatrix}$$

$$= \begin{pmatrix} 0 \\ 0 \\ 0 \end{pmatrix}. \tag{8.9}$$

1. Solve system (8.9) subject to the initial conditions $x_1(0) = x_2(0) = x_3(0) = 0$, $x_1'(0) = 1$, $x_2'(0) = -1/2$, and $x_3'(0) = 1$.[1]
2. Determine the period of x_1, x_2, and x_3 for $c = 10^{-4}, 10^{-2}, 1/10, 1$, and 2, where $c = EI/(m_\ell^3)$.
3. Graph x_1, x_2, and x_3 over several periods for $c = 10^{-4}, 10^{-2}, 1/10, 1$, and 2.
4. Illustrate the motion of the components of the airplane under these conditions.

---

[1] James, M.L., Smith, G.M., Wolford, J.C., and Whaley, P.W., *Vibration of Mechanical and Structural Systems*, Harper and Row, New York, 1989, pp. 346-349.

D. Free Vibration of a Three-Story Building

If you have ever gone to the top of a tall building, such as the Sears Tower or Empire State Building on a windy day, you may have been acutely aware of the *sway* of the building. In fact, all buildings sway (or vibrate) naturally. Usually, we are only aware of the sway of a building when we are in a very tall building or in a building during an event such as an earthquake. In some tall buildings, like the Park Tower in Chicago, the sway of the building during high winds is reduced by having a tuned mass damper at the top of building that oscillates at the same frequency as the building but out of phase. We investigate the sway of a three-story building and then try to determine how we would investigate the sway of a building with more stories.

The Park Tower in Chicago was the first building designed with a tuned mass damper. The damper in the building weighs 300 tons. Taipei 101's 730 ton tuned mass damper was featured in *Popular Mechanics* in May, 2005.

We make two assumptions to solve this problem. First, we assume that the mass distribution of the building can be represented by the lumped masses at the different levels. Second, we assume that the girders of the structure are infinitely rigid compared with the supporting columns. With these assumptions, we can determine the motion of the building by interpreting the columns as springs in parallel.

Assume that the coordinates x_1, x_2, and x_3 as well as the velocities and acceleration are positive to the right. If we assume that $x_3 > x_2 > x_1$, the forces that the columns exert on the masses are shown in Figure 8.32.

One of the world's tallest building is Taipei 101, which has one of the world's largest tuned mass dampers. The damper weighs 730 tons!

In applying Newton's second law of motion, recall that we assumed that acceleration is in the *positive* direction. Therefore, we sum the forces in the same direction as the acceleration positively, and other negatively. With this configuration, Newton's second law on each of the three masses yields the following system of differential equations:

$$-k_1 x_1 + k_2(x_2 - x_1) = m_1 \frac{d^2 x_1}{dt^2}$$

$$-k_2(x_2 - x_1) + k_3(x_3 - x_2) = m_2 \frac{d^2 x_2}{dt^2}$$

$$-k_3(x_3 - x_2) = m_3 \frac{d^2 x_3}{dt^2},$$

which we write as

$$m_1 \frac{d^2 x_1}{dt^2} + (k_1 + k_2)x_1 - k_2 x_2 = 0$$

$$m_2 \frac{d^2 x_2}{dt^2} - k_2 x_1 + (k_2 + k_3)x_2 - k_3 x_3 = 0$$

$$m_3 \frac{d^2 x_3}{dt^2} - k_3 x_2 + k_3 x_3 = 0,$$

where m_1, m_2, and m_3 represent the mass of the building on the first, second, and third levels,

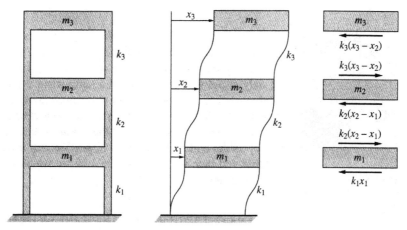

FIGURE 8.32 Diagram used to model the sway of a three-story building.

and k_1, k_2, and k_3, corresponding to the spring constants, represent the total stiffness of the columns supporting a given floor.[2]

1. Show that this system can be written in matrix form as

$$\begin{pmatrix} m_1 & 0 & 0 \\ 0 & m_2 & 0 \\ 0 & 0 & m_3 \end{pmatrix}\begin{pmatrix} x_1'' \\ x_2'' \\ x_3'' \end{pmatrix}$$
$$+ \begin{pmatrix} k_1+k_2 & -k_2 & 0 \\ -k_2 & k_2+k_3 & -k_3 \\ 0 & -k_3 & k_3 \end{pmatrix}\begin{pmatrix} x_1 \\ x_2 \\ x_3 \end{pmatrix} = \begin{pmatrix} 0 \\ 0 \\ 0 \end{pmatrix}.$$

The matrix $\begin{pmatrix} k_1+k_2 & -k_2 & 0 \\ -k_2 & k_2+k_3 & -k_3 \\ 0 & -k_3 & k_3 \end{pmatrix}$ is called the *stiffness matrix* of the system.

2. Find a general solution to the system. What can you conclude from your results? Suppose that $m_2 = 2m_1$ and $m_3 = 3m_1$. Then, we can write this system in the form

$$\begin{pmatrix} 1 & 0 & 0 \\ 0 & 2 & 0 \\ 0 & 0 & 3 \end{pmatrix}\begin{pmatrix} x_1'' \\ x_2'' \\ x_3'' \end{pmatrix} + \frac{1}{m_1}\begin{pmatrix} k_1+k_2 & -k_2 & 0 \\ -k_2 & k_2+k_3 & -k_3 \\ 0 & -k_3 & k_3 \end{pmatrix}$$
$$\begin{pmatrix} x_1 \\ x_2 \\ x_3 \end{pmatrix} = \begin{pmatrix} 0 \\ 0 \\ 0 \end{pmatrix}.$$

3. Find exact and numerical solutions to the system subject to the initial conditions $x_1(0) = 0$, $x_1'(0) = 1/4$, $x_2(0) = 0$, $x_2'(0) = -1/2$, $x_3(0) = 0$, $x_3'(0) = 1$ if $k_1 = 3$, $k_2 = 2$, and $k_3 = 1$, for $m_1 = 1, 10, 100$, and 1000.

4. Determine the period of x_1, x_2, and x_3 for $m_1 = 1, 10, 100$, and 1000.

5. Graph x_1, x_2, and x_3 for $m_1 = 1, 10, 100$, and 1000.

6. Illustrate the motion of the building under these conditions.

7. How does the system change if we consider a 5-story building? a 50-story building? a 100-story building?

[2]James, M.L., Smith, G.M., Wolford, J.C., and Whaley, P.W., *Vibration of Mechanical and Structural Systems with Microcomputer Applications*, Harper and Row, New York, 1989, pp. 282-286. Vierck, R.K., *Vibration Analysis*, Second Edition, Harper Collins, New York, 1979, pp. 266-290.

E. Control Systems

Consider the spring-mass system or circuit described by the IVP[3]

$$a\frac{d^2x}{dt^2} + b\frac{dx}{dt} + cx = f(t), \quad x(0) = x'(0) = 0.$$
$$(8.10)$$

Many times the constant coefficients a, b, and c are unknown, and it is the job of the engineer to determine the response (solution) to the input $f(t)$.

1. Show that $X(s) = F(s)/(as^2 + bs + c)$. Let $P(s) = 1/(as^2 + bs + c)$. This function is known as the *transfer function* of the system and $p(t) = \mathcal{L}^{-1}\{P(s)\}$ is called the *weight function* of the system.
2. Use the convolution theorem to show that the solution of $X(s) = P(s)F(s)$ is $x(t) = \int_0^t p(v)f(t - v)\,dv$, called *Duhamel's principle* for the system.
3. Find the response to the spring-mass system

$$\frac{d^2x}{dt^2} + 2\frac{dx}{dt} + 2x = f(t), \quad x(0) = x'(0) = 0.$$

4. Using the fact that $\mathcal{L}\{\delta(t)\} = 1$, show that $P(s) = \mathcal{L}\{\delta(t)\}/(as^2 + bs + c)$. Therefore, the solution to initial-value problem (8.10) is $p(t) = \mathcal{L}^{-1}\{P(s)\}$, called the *unit impulse response*.
5. Let $h(t)$ be the solution to $a\,d^2x/dt^2 + b\,dx/dt + cx = \mathcal{U}(t)$. Show that $H(s) = P(s)/s$, so that by the convolution

theorem, $h(t) = \int_0^t p(v)\,dv$. Therefore, $h'(t) = p(t)$. Use Duhamel's principle to show that $x(t) = \int_0^t h'(v)f(t - v)\,dv$.

6. Suppose that we differentiate the equation $L\,d^2Q/dt^2 + R\,dQ/dt + (1/C)\,Q = E(t)$ with respect to t. Use the relationship $dQ/dt = I$ to obtain $L\,d^2I/dt^2 + R\,dI/dt + (1/C)I = E'(t)$. If we do not know the values of L, R, and C, we may select the voltage source $E(t) = t$ to determine the response, as in 5. In this case, show that $I(t) = \int_0^t h'(v)E'(t - v)\,dv = \int_0^t h'(v)\,dv$, so we can use an ammeter to measure the response, $h(t)$, when $E'(t) = 1$.

Suppose that the readings on the ammeter given as ordered pairs $(t, h(t))$ are: $(0, 3.)$, $(0.299, 3.07)$, $(0.598, 2.22)$, $(0.898, 0.896)$, $(1.2, -0.337)$, $(1.5, -0.972)$, $(1.8, -0.753)$, $(2.09, 0.232)$, $(2.39, 1.58)$, $(2.69, 2.74)$, $(2.99, 3.16)$, $(3.29, 2.57)$, $(3.59, 1.)$, $(3.89, -1.14)$, $(4.19, -3.23)$, $(4.49, -4.65)$, $(4.79, -4.96)$, $(5.09, -4.06)$, $(5.39, -2.23)$, $(5.68, -0.0306)$, $(5.98, 1.89)$, $(6.28, 3.)$, $(6.58, 3.07)$, $(6.88, 2.22)$, $(7.18, 0.896)$, $(7.48, -0.337)$, $(7.78, -0.972)$, $(8.08, -0.753)$, $(8.38, 0.232)$, $(8.68, 1.58)$, $(8.98, 2.74)$, $(9.28, 3.16)$, $(9.57, 2.57)$, $(9.87, 1.)$, $(10.2, -1.14)$, $(10.5, -3.23)$, $(10.8, -4.65)$, $(11.1, -4.96)$, $(11.4, -4.06)$, $(11.7, -2.23)$, $(12, -0.0306)$, $(12.3, 1.89)$. Fit this data with a trigonometric function to determine $h(t)$ to use $h'(t)$ to find $I(t)$ if

$$E(t) = \frac{1}{10}\sin 10t.$$

---

[3] DiStephano, J.J., Stubberud, A.R., and Williams, I.J., *Feedback and Control Systems*, Schaum's Outline Series, New York, 1967.

Answers to Selected Exercises

Exercises 1.1

1. Second-order linear ordinary differential equation. The forcing function is $f(x) = x^3$ so the equation is nonhomogeneous.

3. Second-order linear partial differential equation.

5. This is a first-order ordinary differential equation. It is nonlinear because the derivative dy/dx is squared.

7. Second-order linear partial differential equation.

9. Nonlinear second-order ordinary differential equation.

11. This is a second-order partial differential equation. It is nonlinear because of the product, $uu_x = u\partial u/\partial x$, of functions involving the dependent variable, $u = u(x,t)$.

13. This is a first-order ordinary differential equation. If we write it as $(2t-y)\,dt/dy - 1 = 0$, y is independent, $t = t(y)$ is dependent, and the equation is nonlinear. If we write the equation as $dy/dt + y = 2t$, t is independent, $y = y(t)$ is dependent, and the equation is linear.

15. This is a first-order ordinary differential equation. It is nonlinear in both x (because of the $2x\,dx$ term) and y (because of the $-y\,dy$ term.

45. $y = \ln(\csc x + \cot x) + \cos x + C$

47. $y(x) = 2e^{-2x}$

49. $y(x) = \dfrac{5}{7}e^{-3x} + \dfrac{2}{7}e^{4x}$

51. $y(x) = -\dfrac{1}{4}\left(-3 + 3e^{2x} - 2x\right)$

53. $y(t) = -t^7 + t^6$

55. $y(x) = x^4 - 1/2\,x^2 + 2x + 1$

57. $y(x) = -\sin\left(x^{-1}\right) + 2$

67. $y(x) = c_1 x + c_2 x^2$

69. $y(x) = (e^x + C)\,e^{-2x}$

81. $y = -\dfrac{1}{208}e^{-x/2}(74e^x \cos 2x - 74 \cos 3x - 111e^x \sin 2x - 20 \sin x)$

83. $x = \dfrac{4}{9}e^{-6t}\left(e^{9t} + 8\right),\ y = \dfrac{8}{9}e^{-6t}\left(e^{9t} - 1\right)$

Exercises 1.2

1. No

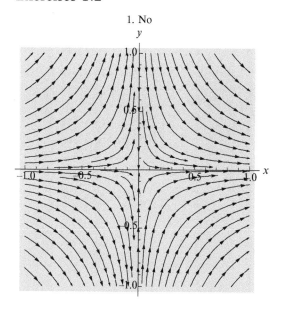

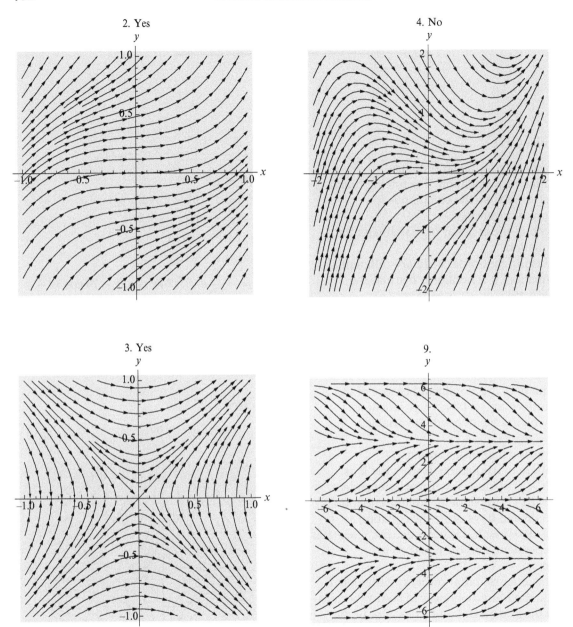

2. Yes

3. Yes

4. No

9.

10.

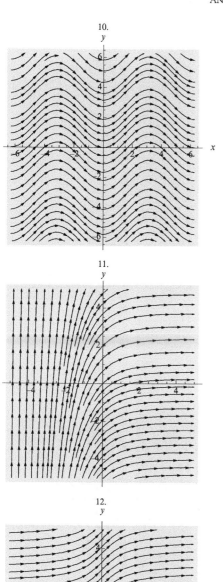

11.

12.

15. $x' = y$ so $y' = x'' = -4x$: $\{x' = y,$
$y' = -4x\}$

17. $\{x' = y, y' = -13x - 4y\}$

19. $\{x' = y, y' = -16x + \sin t\}$

21. $y = C + \dfrac{\sqrt{\pi}}{2}\,\text{erf}\,(x)$, where

$\text{erf}\,(x) = \dfrac{2}{\sqrt{\pi}}\int_0^x e^{-t^2}\,dt.$

21.

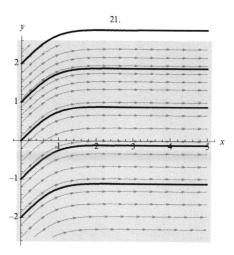

22.

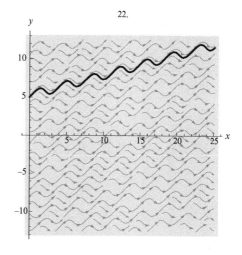

23

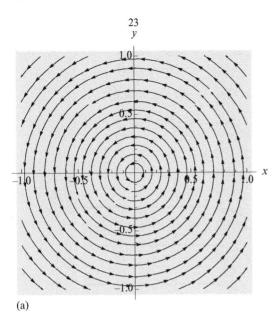

(a)

25

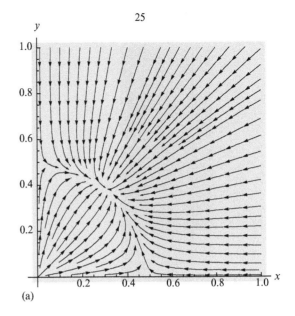

(a)

23

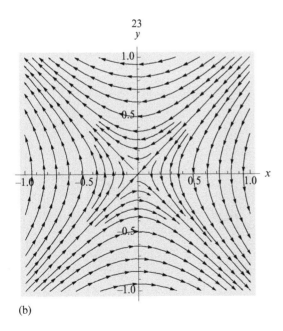

(b)

25

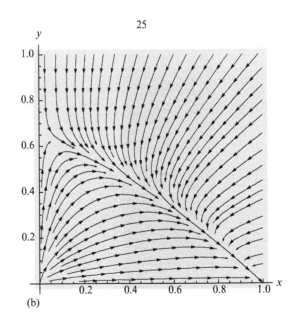

(b)

Chapter 1 Review Exercises

1. First-order ordinary linear homogeneous differential equation.
3. Second-order linear homogeneous differential equation.
5. Second-order nonlinear partial differential equation.
17. $y = (2 - x^2) \cos x + 2x \sin x$.
19. $y = \frac{1}{2}(x\sqrt{x^2 - 1} + \ln(2x + 2\sqrt{x^2 - 1}))$.
21. $y = \frac{1}{4}(x + 4 \cos 2x)$.
23. $y = \frac{1}{3}(1 - \cos^3 x)$.

Exercises 2.1

1. (a) yes; (b) no; (c) no.
3. $f(x, y) = y^{1/5}$, so $f_y(t, y) = \frac{1}{5}y^{-4/5}$ is not continuous at $(0, 0)$ so uniqueness is not guaranteed. Solutions: $y = \left(\frac{4}{5}t\right)^{5/4}$, $y = 0$.
5. $f(t, y) = 2\sqrt{|y|} = \begin{cases} 2\sqrt{y}, y \geq 0 \\ 2\sqrt{-y}, y < 0 \end{cases}$, so

   $f_y(t, y) = \begin{cases} y^{-1/2}, y\, 0 \\ -(-y)^{1/2}, y < 0 \end{cases}$ is not

   continuous at $(0, 0)$. Therefore, the hypotheses of the Existence and Uniqueness Theorem are not satisfied.
7. Yes. $y = \exp\left(\frac{2}{3}(t^{3/2} - 1)\right)$.
9. Yes. $f(t, y) = \sin y - \cos t$ and $f_y(t, y) = \cos y$ are continuous on a region containing $(\pi, 0)$.
11. $y = \sec t$ so $y' = \sec t \tan t = y \tan t$ and $y(0) = 1$. $f_y(t, y) = t$ is continuous on $-\pi/2 < t < \pi/2$ and $f(t, y) = \sec t$ is continuous on $-\pi/2 < t < \pi/2$ so the largest interval on which the solution is valid is $-\pi/2 < t < \pi/2$.

13. $f(t, y) = \sqrt{y^2 - 1}$ and $f_y(t, y) = \frac{1}{2}(y^2 - 1)^{-1/2}$; unique solution guaranteed for (a) only.
15. $(0, \infty)$.
17. $(0, \infty)$.
19. $(-\infty, 1)$.
21. $(-2, 2)$.
23. $t > 0$, $y = t^{-1} \sin t - \cos t$, $-\infty < t < \infty$.
25. $t > 1/a$ or $t < 1/a$.
27. $|t| < a$.

Exercises 2.2

1. $y = \left(\frac{3}{2}x^2 + C\right)^{1/3}$.
3. $y = (1 - 2Cx + C^2x^2)/4x^2$.
5. $3t + \frac{1}{2}t^4 + \frac{5}{2}y - \frac{9}{14}y^{-7} = C$.
7. $\sinh 3x - \frac{1}{2}\cosh 4y = C$.
9. $y = -2 + C\sqrt{2t + 1}$.
11. $y = \sin^{-1}\left(-\frac{3}{4}\cos x + C\right)$.
13. $y = Ce^{-kt}$.
15. $y = \cosh^{-1}\left(-\frac{1}{120}\sinh 6t\right.$

    $\left. -\frac{1}{16}\cosh 4t + C\right)$.
17. $y = -\frac{1}{3}\ln\left(-\frac{3}{2}e^{2t} + C\right)$.
19. $\frac{3}{4}\cos\theta - \frac{1}{12}\cos 3\theta + \frac{1}{16}\sin 4y - \sin y = C$.
21. $y \sin 2x + 2xy + 2y^2 + 5 + 4Cy = 0$.
23. $-\frac{1}{64}(1/\cos(2y) + 1)(-64C\cos(2y(x)) - 64C + \cos(8x - 2y) + \cos(8x + 2y) + 2\cos(8x) + 4\cos(4x - 2y) + 4\cos(4x + 2y) + 8\cos(4x) - 128) = 0$.
25. $-\frac{1}{4}x \sin 2x + \frac{1}{4}x^2 - \frac{1}{8}\cos 2x - 2\sin\sqrt{y} = C$.

27. $\frac{1}{64}(1/(-1+\sin(y)))(58+6\sin(y)-2\cos(4x)+\sin(y+4x)-\sin(-y+4x)+8\cos(2x)-4\sin(y+2x)+4\sin(-y+2x)-64C+64C\sin(y))=0.$

29. $\frac{2}{3}(\ln x)^{3/2}+\frac{1}{3}e^{3/y}=C.$

31. $y=\dfrac{1+Ce^{\frac{2}{3}t^2+2t}}{1-Ce^{\frac{2}{3}t^2+2t}}.$

33. $y=\dfrac{13+12C-4x(3C+1)}{4C(x-1)-3}.$

35. $-3t+\sqrt{3}\tan^{-1}\left(\dfrac{2y-1}{\sqrt{3}}\right)+$
$3\ln\left|\dfrac{(y+1)^{1/3}}{(y^2-y+1)^{1/6}}\right|=C.$

37. $y=\pm\sqrt{\dfrac{Ce^{2t}}{1-Ce^{2t}}}.$

39. $y=\pm\sqrt{\dfrac{1}{1-Ce^{2t}}}.$

41. $y(x)=1/4x^4.$

43. $x(y)=\sin(y)+2.$

45. $y(t)=2/3\sqrt{9+3t^{3/2}}.$

47. $y(t)=-1-\sqrt{-1+2e^t}.$

49. $y(x)=e^{\sqrt{2}\sqrt{x}},\ y(x)=e^{-\sqrt{2}\sqrt{x}}.$

51. $y(x)=\arctan(x)+1.$

53. $y=-3+4(3x+1)^{1/3}.$

55. $y=\ln\left(\dfrac{1}{2}e^{2x}-\dfrac{1}{2}+e\right).$

57. The solution for (a) is $y=e^{\sin t}$, for (b) it is
$y=\dfrac{1}{1-\sin t}$, and for (c) it is
$y=\dfrac{1}{4}\left(4+r\sin t+\sin^2 t\right).$

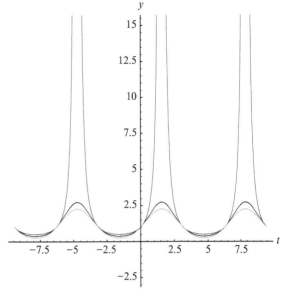

59. $y=2x(x-2)^{-1}.$

67. $y=2\dfrac{3Ce^{8t}+1}{Ce^{8t}-1}.$

71. $y=\exp\left(\dfrac{c}{t}(t-1)\right).$

73. $L(t)=L_\infty-e^{-r_Bt}.$

Exercises 2.3

1. $y=-10+Ce^t.$

3. $y=Ce^t-\cos t+\sin t.$

5. $y=Ce^t-2te^{-t}-e^{-t}.$

7. $y=\dfrac{1}{2}t+Ct^{-1}.$

9. $y=-x^{-1}e^{-x}+Cx^{-1}.$

11. $y=4t^2+1+C\sqrt{4t^2+1}.$

13. $y=-\dfrac{1}{2}\cos t\cot t+C\csc t.$

15. $y=\dfrac{1}{4}(4t^2-9)+C\sqrt{4t^2-9}.$

17. $y=\dfrac{1}{3}\sin x+C\csc^2 x.$

19. $\theta = -1 + Ce^{r^2/2}$.

21. $x(y) = -1 - y + Ce^y$.

23. $x(t) = C/(t^3 - 1)$.

25. $v(s) = se^{-s} + Ce^{-s}$.

27. $y = e^{-t}(t - 1)$.

29. $y = 1 - 2e^{-t^2}$.

31. $y = (2te^t - 2e^t - 1)/t$.

33. $y = \dfrac{t^2 - 16}{t^2 + 4}$.

35. $y = te^{2t}$.

39. (a) $y = Ce^t + 1$; (b) $y = Ce^{-t} + t$;
(c) $y = Ce^{-t} + \sin t$; (d) $y = Ce^t + e^{-t}$.

43. $y(t) = \begin{cases} t - 1 + 2e^{-t}, 0 \le t < 1 \\ 2e^{-t}, t \ge 1 \end{cases}$.

45. $y(t) = \begin{cases} e^{-2t}, 0 \le t < 1 \\ e^{2-4t}, t \ge 1 \end{cases}$.

47. $y(t) = -2/5 \cos(2t) - 1/5 \sin(2t) + Ce^t$.

49. $y(t) = (t + C)e^{-t}$.

51. $y(t) = -1/25 - 1/5t + Ce^{5t}$.

53. $y(t) = -2 \cos(t) + 4 \sin(t) + 4e^t + Ce^{1/2t}$.

55. $y(t) = 2/11e^t + Ce^{-10t}$.

57. $y(t) = (2t + C)e^t$.

59. $y(t) = \cos(t) + \sin(t) + t - 1 + Ce^{-t}$.

63. $y(t) = t - 1 + Ce^{-t}$,
$y(t) = -1/2 \cos(t) + 1/2 \sin(t) + Ce^{-t}$,
$y(t) = 1/2 \cos(t) + 1/2 \sin(t) + Ce^{-t}$,
$y(t) = 1/2e^t + Ce^{-t}$.

65. $y(t) = t - 1 - e^{-t}, y(t) = t - 1$,
$y(t) = t - 1 + e^{-t}, y(t) = t - 1 + 2e^{-t}$,
$y(t) = t - 1 + 3e^{-t}$.

Exercises 2.4

1. $M_y(t, y) = 2y - \dfrac{1}{2}t^{-1/2} = N_t(t, y)$, exact.

3. $M_y(t, y) = \cos ty - ty \sin ty = N_t(t, y)$, exact.

5. The equations is exact because
$\partial_t(sty^2) = 3y^2 = \partial_y(y^3)$.

7. $M_y(t, y) = \sin 2t \ne 2 \sin 2t = N_t(t, y)$,
not exact.

9. $M_y(t, y) = y^{-1} = N_t(t, y)$, exact.

11. $y = C + t^3$.

13. $y = 0, ty^2 = C$.

15. $t^2 + ty^3 + 4y(t) = C$.

17. $-1/3 \ln\left(\dfrac{y(3t^2 + y^2)}{t^3}\right) - \ln(t) = C$.

19. $\ln t + 2 \ln \sin y = C$.

21. $e^t \sin y + y = C$.

23. $y = 0, y = C\sqrt{\sec t^2 + \tan t^2}$.

25. $t + y \sin(yt) = C$.

27. $3 \sin(t + y) + t \sin(t + y) = C$.

29. $ye^{y/t} + t = C$.

31. $y^2t^2 - 1 = 0$.

33. $y = \dfrac{-t^3 - 1}{(t - 1)(t + 1)}$.

35. $te^y - t^2y = 0$.

37. $t(y)^2 + \cos(2t) + y - 2 = 0$.

39. $y^2 - \dfrac{\arctan(t)}{t} = 0$.

41. $y^2 - x^2 - \sin(xy) = C$.

43. $-\dfrac{1}{2}x^2 + xy^2 = C$.

49. $y = \dfrac{C}{(e^t + t^2)^{2/3}}$.

51. $y^{-1} - \dfrac{-t + C}{t^2} = 0$.

53. $yt^5 + t^4(y)^2 + 1/4t^4 = C$.

55. $\cos(y(t))t^2 + \sin(y)t = C$.

57. $t^2 + y \cos ty = C$.

59. (a) (i) $y = -t, t^2 + 2ty + y^2 = C$, (ii)
$y = \dfrac{1}{C}\left(-Ct \pm \dfrac{1}{10}\sqrt{10C^2t^2 + 10}\right)$, (iii)
$y = \dfrac{1}{C}\left(-\dfrac{19}{20}Ct \pm \dfrac{1}{20}\sqrt{-39C^2t^2 + 40}\right)$.

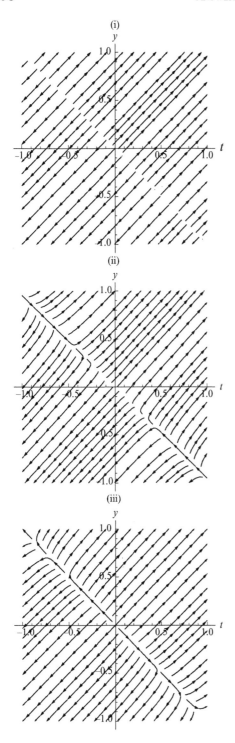

(i)

(ii)

(iii)

Exercises 2.5

1. $y = \pm\sqrt{-2 - 2t + Ce^t}$.

3. $y = \pm\dfrac{\sqrt{-(2\cos(t) + 2t\sin(t) - C)\,t}}{2\cos(t) + 2t\sin(t) - C}$.

5. $y^{3/2} + \dfrac{9}{20}\cos(t) - \dfrac{3}{20}\sin(t) - Ce^{3t} = 0$.

7. $y = \dfrac{3t}{t^3 - 3C}$.

9. $y = 1/(Ct^{-1} - 1)$.

11. Homogeneous of degree 0.

13. Not homogeneous.

15. Not homogeneous.

17. $-2\ln\left(-\dfrac{-y+t}{t}\right) + \ln\left(\dfrac{y-2t}{t}\right) - \ln(t) = C$.

19. $-1/4\ln\left(\dfrac{-t+2y}{t}\right) - 1/2\ln\left(\dfrac{y}{t}\right) - \ln(t) = C$.

21. $y^3 - \left(3\ln(t) + C\right)t^3 = 0$.

23. $y = \left(-\ln(t) + C\right)t$.

25. $-2/3\ln\left(\dfrac{3y-2t}{t}\right) + 1/2\ln\left(\dfrac{-t+2y}{t}\right) -$
$\ln(t) = C$.

27. $1/4\dfrac{y}{t} - \ln\left(\dfrac{y}{t}\right) - \ln(t) = C$.

29. $y = -t,\ -1/2\ln\left(\dfrac{t^2 - ty + y^2}{t^2}\right) +$
$1/3\sqrt{3}\arctan\left(1/3\dfrac{(t-2y)\sqrt{3}}{t}\right) - \ln(t) = C$.

31. $1/2\left(-y + t\right)y^{-1}\left(e^{-t/y}\right)^{-1} - \ln(y) = C$.

33. $y = \dfrac{1}{4}(4 - 8t + 8t^2 - 4t^3 + t^4)$.

35. $y = 1/2\sqrt{2 + 2t^2}\,t$.

37. $y = -1/2 + 1/2t^2$.

39. $y = \sqrt[3]{-3\ln(t) + 27}\,t$.

41. $\dfrac{1}{3}\dfrac{y^3}{t^3} - \ln\left(\dfrac{y}{t}\right) - \ln t - 8/3 + \ln 2 = 0$.

43. $y = \sqrt{2t}\sqrt{\ln t}$.

45. $y = \pm\sqrt{-x^2 \pm \sqrt{x^4 + C}}$.

47. The general solution is $y^{-3} = (3\cot x + C)$ $\sin^3 x$ or $y = 0$. So, the solution to the initial value problem (IVP) is $y = 0$.

53. $y = 1/4t^2 + 1/2t - 3/4, y = Ct - 1 - C^2 + C.$

55. $y = 3/2t^{2/3} - 1/2.$

Exercises 2.6

1. 47.3742, 63.2572.
3. 1.8857, 2.09847.
5. 79.8458, 123.048.
7. 1.95109, 1.95388.
9. 83.6491, 88.6035.
11. 2.37754, 2.41897.
13. 185.34, 206.981.
15. 1.95547, 1.95609.
17. 90.6405, 90.6927.
19. 216.582, 216.992.
23. 1.95629, 1.95629.
25-27. (a) $y(t) = e^{-t}, y(1) = 1/e \approx 0.367879x.$
  $h = 0.1, h = 0.005, h = 0.025.$
25. (Euler's) 0.348678, 0.358486, 0.363232.
26. (Improved) 0.368541, 0.368039, 0.367918.
27. (4th-Order RK) 0.429069, 0.414831, 0.36788.
29. $y(0.5) \approx 0.566144, 1.12971, 1.68832, 2.23992, 2.78297.$

Chapter 2 Review Exercises

1. $y = 1/5\sqrt[3]{25t^6 + 125C}.$
3. $y = \dfrac{t}{-t + C}.$
5. $y = (e^{5t} + C)^{1/5}.$
7. $y(t) = e^{\pm\sqrt{-e^{-2t}+2C}}.$
9. $y = \cos t - 3\sin t + Ce^{-3t}.$
11. $y = -t \pm \dfrac{1}{C}\sqrt{2C^2t^2 + 1}.$
13. $\dfrac{ty^2}{t + 2y} = C$ or $y = \dfrac{C \pm \sqrt{C^2 + Ct^2}}{t}.$
15. $-\dfrac{5}{8}\ln\left(\dfrac{-x + 2t}{t}\right) - \dfrac{5}{8}\ln\left(\dfrac{x + 2t}{t}\right) + \dfrac{1}{4}\ln\left(\dfrac{x}{t}\right) - \ln(t) = C.$

17. $\dfrac{1}{3}t^3y - \cos t - \sin y = C.$
19. $\dfrac{1}{2}t^2\ln y + y = C.$
21. $y = 1 + Ce^{-t^2/2}.$
23. $r = t^{-1}(\cos t + t\sin t + C).$
25. $y = t^{-1}(6e^t - 6te^t + 3t^2e^t + C)^{1/3}.$
27. $y = -2 + 2\ln(-2/t), y = Ct + 2\ln C.$
29. $y = -1/3t\sqrt{6t - 6\sqrt{t^2 - 12C}} + \dfrac{1}{54}\left(6t - 6\sqrt{t^2 - 12C}\right)^{3/2},$
$y = 1/3t\sqrt{6t - 6\sqrt{t^2 - 12C}} - \dfrac{1}{54}\left(6t - 6\sqrt{t^2 - 12C}\right)^{3/2},$
$y = -1/3t\sqrt{6t + 6\sqrt{t^2 - 12C}} + \dfrac{1}{54}\left(6t + 6\sqrt{t^2 - 12C}\right)^{3/2},$
$y = 1/3t\sqrt{6t + 6\sqrt{t^2 - 12C}} - \dfrac{1}{54}\left(6t + 6\sqrt{t^2 - 12C}\right)^{3/2}.$
31. $y + \sin(t - y) = \pi.$
33. $t\sin y - y\sin t = 0.$
35. $t\ln y + y\ln t = 0.$
37.

n	X_n	Y_n (Euler's)	Y_n(Improved Euler's)	Y_n(Runge-Kutta of order 4)
0	1	1	1	1
1	1.05	1	1.00559	1.00678
2	1.1	1.01118	1.0181	1.01921
3	1.15	1.02608	1.03383	1.03486
4	1.2	1.04368	1.052	1.05296
5	1.25	1.06345	1.07219	1.07309
6	1.3	1.08505	1.0941	1.09494
7	1.35	1.10823	1.11752	1.11831
8	1.4	1.13281	1.14228	1.14303
9	1.45	1.15866	1.16825	1.1696
10	1.5	1.18565	1.19533	1.19601
11	1.55	1.21368	1.22343	1.22407
12	1.6	1.24268	1.25247	1.25307
13	1.65	1.27256	1.28237	1.28295
14	1.7	1.30328	1.31309	1.31364
15	1.75	1.33478	1.34456	1.34509
16	1.8	1.36699	1.37675	1.37726
17	1.85	1.3999	1.40961	1.4101
18	1.9	1.43344	1.44311	1.44357
19	1.95	1.46759	1.4772	1.47764
20	2.	1.50232	1.51186	1.51229

39.

n	X_n	Y_n (Euler's)	Y_n (Improved Euler's)	Y_n (Runge-Kutta of order 4)
0	1	1	1	1
1	0.05	1	1.00125	1.00125
2	0.1	1.0025	1.00501	1.00501
3	0.15	1.0075	1.01129	1.01129
4	0.2	1.01503	1.02012	1.02013
5	0.25	1.02511	1.03156	1.03157
6	0.3	1.03779	1.04564	1.04565
7	0.35	1.0531	1.06242	1.06245
8	0.4	1.07112	1.08198	1.08201
9	0.45	1.09189	1.10437	1.10441
10	0.5	1.11548	1.12966	1.12971
11	0.55	1.14194	1.15789	1.15795
12	0.6	1.17132	1.1891	1.18918
13	0.65	1.20364	1.22329	1.2234
14	0.7	1.23889	1.26043	1.26056
15	0.75	1.27702	1.30042	1.30058
16	0.8	1.31791	1.34309	1.34329
17	0.85	1.36139	1.38818	1.38842
18	0.9	1.40717	1.43532	1.43561
19	0.95	1.45488	1.48403	1.48438
20	1.	1.50399	1.53369	1.5341

Exercises 3.1

1. $y(t) = 100e^{kt}$. With $t = 3$, $y(3) = 100e^{3k}$ $= 200$ so $e^{3k} = 2$ which means that $e^k = 2^{1/3}$ and $y(t) = 2^{t/3}$. $y(30) = 2^{10} = 1024$; while $y(t) = 2^{t/3} = 4250$ when $t = 3\ln(4250)/\ln(2) \approx 36.16$.

3. $y(t) = e^{kt}$ and $y(1) = e^k = 2/3$ so $y(t) = (3/2)^{-t}$. $y(t) = (3/2)^{-t} = 1/3$ when $t = \ln(3)/\ln(3/2) \approx 2.71$.

5. $y(1000) = 100e^{1000k} = 50$ so $e^k = 2^{-1/1000}$ and $y(t) = 100 \times 2^{-t/1000}$. $y(1) = 100 \times 2^{-1/1000} \approx 99.93$ and $y(500) \approx 70.71$.

7. $y(t) = 500e^{kt}$ and $y(6) = 500e^{6k} = 600$ so $e^k = (6/5)^{1/6}$ and $y(t) = 500 \times (6/5)^{t/6}$. $y(24) = 500 \times (6/5)^4 = 5184/5 \approx 1036.8$. $500 \times (6/5)^{t/6} = 1000$ when $t = 6\ln(2)/\ln(6/5) \approx 22.81$.

9. $y(5) = e^{5k} = 1/2$ so $e^k = 2^{-1/5}$ and $y(t) = 2^{-t/5}$. Then $y(t) = 2^{-t/5} = 1/6$ when $t = 5\ln(6)/\ln(2) \approx 12.92$ and $y(15) = 2^{-3} = 1/8$.

13. 9.72 days.

15. $y(100) \approx 0.96 y_0$.

17. $t_{\text{tool}} \approx 3561.13$ years old, $t_{\text{fossil}} \approx 4222.81$ years old, $t_{\text{fossil}} - t_{t00} \approx 661.68$ years. No.

21. $y = 100,000 \left(1 + 9e^{-t/100}\right)^{-1}$; $y(25) \approx 1.25 \times 10^5$; $\lim_{t \to \infty} y(t) = 1,000,000$.

23. $y = 200 \left(1 + 199 \left(\frac{3}{199}\right)^t\right)^{-1}$; $y(2) \approx 191$; because there is no t so that $y = 200$, all students do not theoretically learn of the rumor.

25. $y(t) = 1 + 499e^{-5t}$; $y(20) \approx 1$; quickly.

27. $\lim_{m \to \infty} S_0 \left(1 + k/m\right)^{mt} = S_0 e^{kt}$.

29. (c) $y(t) = 2\left(1 + e^{rt}\right)^{-1}$, $\lim_{t \to \infty} y(t) = 0$; (d) $y(t) = 6\left(3 - e^{rt}\right)^{-1}$, $\lim_{t \to \infty} y(t) = 0$.

31. (a) $y = 2$, semistable; $y = 0$ unstable; (b) $y = 0$, unstable; $y = 1$, unstable; (c) $y = 0$, semistable; $y = 1$, unstable; (d) $y = 3$, stable; $y = 0$, semistable, $y = -3$, unstable; (e) $y = -1$, unstable; $y = 0$, semistable; (f) $y = -1$, stable; $y = 0$, unstable; $y = 1$ stable. See Figure 8.33.

33. $y(t) = \left(100 - 99e^{-t/100}\right)^{-1}$, $y(t) = \left(20 - 19e^{-t/20}\right)^{-1}$, $y(t) = \left(10 - 9e^{-t/10}\right)^{-1}$, $y(t) = \left(2 - e^{-t/2}\right)^{-1}$. See Figure 8.34.

Exercises 3.2

1. $t \approx 55.85$ min.

3. 12.59 min.

5. $t \approx -2.45$ h, 12:30 p.m.

7. $T(5) \approx 76.3\,°\text{F}$.

9. $75\,°\text{F}$.

11. $t \approx 4.7$ min.

13. $u(t) = -5(9 + \pi^2)^{-1}(-8\pi^2 - (2\pi^2 + 27)e^{-t/4} + 3\pi \sin(\pi t/12) + 9\cos(\pi t/12) - 72)$.

15. $u(t) = -5(9 + \pi^2)^{-1}(-14\pi^2 + \pi^2 e^{-t/4} + 3\pi \sin(\pi t/12) + 9\cos(\pi t/12) - 126)$.

17. If $R_1 = R_2$, the volume remains constant, so $V(t) = V_0$. If $R_1 > R_2$, V increases. If $R_1 < R_2$, V decreases.

19. $y = 2t + 400 - 390 \times 200^3(t + 200)^{-3}$.

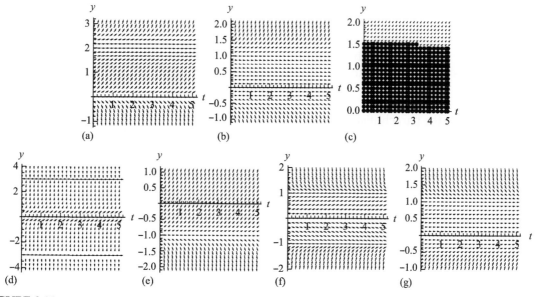

FIGURE 8.33   Using direction fields to investigate stability

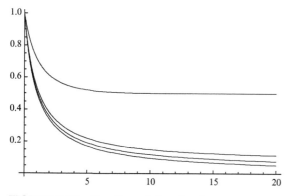

FIGURE 8.34   The effects of changing the carrying capacity on solutions of the logistic equation

21. (a) $u(t) = e^{-t/4}(-6 + 76e^{t/4})$; (b) $u(t) = (9 + \pi^2)^{-1}(639 + 71\pi^2 + (81 - \pi^2)e^{-t/4} - 90\cos(\pi t/12) - 30\pi\sin(\pi t/12))$; (c) $u(t) = 2(9 + \pi^2)^{-1}(333 + 37\pi^2 + (27 - 2\pi^2)e^{-t/4} - 45\cos(\pi t/12) - 15\pi\sin(\pi t/12))$.

23. $u_d \approx 69.8273$.

Exercises 3.3

1. $v(t) = 32 - 32e^{-t}$; $v(2) \approx 27.67$ ft./s.

3. $v(t) = \dfrac{1}{2}(1 + 15e^{-64t})$; $v(1) \approx 0.5$ ft./s.

5. $v(t) = 4 - 68e^{-8t}$; $v(0) = 0$ when $t = \dfrac{1}{8}\ln 17 \approx 0.354$s.

7. $dv/dt = 32 - v$, $v(0) = 0$ has solution $v(t) = 32 - 32e^{-t}$; $ds/dt = v(t)$, $s(0) = 0$ has solution $s(t) = 32t + 32e^{-t} - 32$; $s(4) \approx 96.59$ ft. < 300 so about 203.41 ft. above the ground.

9. $v(t) = \dfrac{49}{5}(1 - e^{-t})$; $\lim_{t\to\infty} v(t) = 49/5$.

11. $v(t) = -9800 + 9900e^{-t/1000}$; $v(t) = 0$ when $t \approx 10.152$ s; $s(t) = -9800t + 9,900,000(1 - e^{-t/1000})$; $s(10.152) \approx 506.76$ m.

13. $dv/dt = -g$, $v(0) = v_0$ has solution $v(t) = -gt + v_0$; $ds/dt = v(t)$, $s(0) = s_0$ has solution $s(t) = -\dfrac{1}{2}gt^2 + v_0t + s_0$.

15. Because the object reaches its maximum height when $v = -gt + v_0 = 0$, $t = v_0/g$ and the air resistance is ignored, the object hits the ground when $t = 2v_0/g$. The velocity at this time is $v(2v_0/g) = -g(2v_0/g) + v_0 = -v_0$.

17. $c = 5$.

19. The velocity of the parachutist after the parachute is opened is given by
$$v(t) = 8\frac{17e^{8t} + 13}{17e^{8t} - 13};$$ the limiting velocity is $\lim_{t\to\infty} v(t) = 8$.

21. (b) $\dfrac{dv}{dt} = \dfrac{dv}{dr}\dfrac{dr}{dt} = v\dfrac{dv}{dr}$;
(c) $\lim_{t\to\infty} v^2 = v_0^2 - 2gR$.

23. $g \approx 32$ ft./s$^2 \approx 0.006$ mi/s^2; $v_0 = \sqrt{2gR} = \sqrt{2 \times 0.165 \times 0.006 \times 1080} \approx 1.46$ m/s.

25. $Q(t) = E_0C + e^{-t/(RC)}(-E_0C + Q_0)$;
$I(t) = -\dfrac{1}{RC}e^{-t/(RC)}(-E_0C + Q_0)$.

27. $v(t) = 12(4 - \sqrt{3}) - 12(4 - \sqrt{3})e^{-t/3}$;
$x(t) = 12(4 - \sqrt{3})t + 36(4 - \sqrt{3})e^{-t/3} - 36(4 - \sqrt{3})$.

29. $v(t) = 32$, $\lim_{t\to\infty}(32 + Ce^{-t}) = 32$.

31. $v(t) = -gm/c$, $\lim_{t\to\infty}(-gm/c + Ce^{-ct/m}) = -gm/c$.

33. $c = 1/2$: $v(t) = 64e^{-t/2}(-1 + e^{t/2})$; $c = 1$:
$v(t) = 32e^{-t}(-1 + e^t)$; $c = 2$:
$v(t) = 16e^{-2t}(-1 + e^{2t})$.

37. approximately 300.772 s.

Chapter 3 Review Exercises

1. $y = 0$ is unstable; $y = 1/2$ is stable.

3. $y = 0$ is unstable; $y = 4$ is stable.

5. $y = P_0 3^{t/4}$; $t = (4\ln 5)/\ln 3 \approx 5.86$ days.

7. $y = y_0 2^{-t/1700}$, $y(50) \approx 0.9798 y_0$ (97.98% of y_0).

9. $y = 1000(1 + 3^{1-t})^{-1}$; $y = 750$ implies $t = (\ln 9)/(\ln 3) \approx 2$ days.

11. $T(t) = 90 - 50 \times 2^{-t/20}$;
$T(30) = 90 - 50 \times 2^{-3/2} \approx 72.3$ °F.

13. $T(t) = -225 \times (7/9)^{t/45} + 325$ so
$T(-60) \approx 10.43$ °F.

15. $v(t) = 4 - 4e^{-8t}$; $v(3) \approx 4$ ft./s;
$s(t) = 4t + \dfrac{1}{2}e^{-8t} - \dfrac{1}{2}$; $s(3) \approx 11.5$ ft.

17. $dv/dt = -9.8 - v$, $v(0) = 40$ has solution
$v(t) = \dfrac{1}{5}(-99 + 299e^{-t})$; $v(t) = 0$ when
$t = \ln(299/99) \approx 1.11$s
$ds/dt = v$, $s(0) = 0$ has solution
$s(t) = \dfrac{1}{5}(-99t - 299e^{-t} + 299)$;
$s(\ln(299/99)) \approx 18.11$ ft.

19. $dv/dt = 32 - \dfrac{1}{2}v^2$, $v(0) = 30$ has solution
$v(t) = 8\dfrac{19e^{8t} + 11}{19e^{8t} - 11}$; $\lim_{t\to\infty} v(t) = 8$ ft./s.

21. With $m = 230$ kg,
$dv/dt = 82/115 - 637v/230,000$, $v(0) = 0$,
which has solution
$v(t) = \dfrac{164,000}{637}(1 - e^{-637t/230,000})$; $v(t) = 12$
when $t = \dfrac{230,000}{637}\ln\dfrac{41,000}{39,089} \approx 17.23$s
$dy/dt = v$, $y(0) = 0$ has solution $y =$
$\dfrac{164,000}{637}t + \dfrac{37,720,000,000}{405,769}(e^{-637t/230,000} - 1)$
$H = y(17.23) \approx 104.17,798$.

23. (a) $4r^2 = 32\cos^2\theta - 16$; (b) $r = 2\sec\theta + 2$;
(c) $r = -6\cos^2(\theta/2) + 6$.

29. $y = \dfrac{1}{2}x + k$.

31. $xy = k$.

33. Yes. The orthogonal trajectories of
$y^2 - 2cx = c^2$ satisfy $\dfrac{dy}{dx} = \dfrac{y}{x \pm \sqrt{x^2 + y^2}}$.
One solution of this equation can be written as $y^2 + 2Cx = C^2$. Now replace C by $-c$.

35. Equipotential lines: $x^2 + y^2 = c$; orthogonal trajectories: $y = kx$.

37. (c) $\dfrac{1}{2}x^2 + xy - \dfrac{1}{2}y^2 = k_1$ and
$-\dfrac{1}{2}x^2 + xy + \dfrac{1}{2}y^2 = k_2$.

39. 1. $V(x) = \int_0^x \pi y^2 dt,$

$\qquad W(x) = \rho V(x) = \rho \int_0^x \pi y^2 dt$

2. $\sigma(x) = \dfrac{F(x)}{A(x)}$ where $F(x) = W(x) + L$ and

$\qquad A(x) = \pi y^2.$

3. $y(x) = K \exp\left(\dfrac{\rho x}{2\sigma}\right).$ If $y(0) = 1$, then

$\qquad K = 1.$

41. $\dfrac{dx}{dt} = kx^2,\ x(0) = 100,\ x(1) = 60.$

$\qquad x(t) = \dfrac{-1}{kt + C}$ so that $C = -1/100$ and

$\qquad k = -1/150.$ When $t = 3$,

$\qquad x(3) = \left(\dfrac{3}{150} + \dfrac{1}{100}\right)^{-1} = \dfrac{100}{3} \approx 33.33\text{g}.$

Exercises 4.1

1. $W(s) = \begin{vmatrix} t & 4t - 1 \\ 1 & 4 \end{vmatrix} = 1$; linearly independent.

3. $W(s) = \begin{vmatrix} e^{-6t} & e^{-4t} \\ -6e^{-6t} & -4e^{-4t} \end{vmatrix} = 2e^{-10t}$; linearly independent.

5. Linearly independent;

$\qquad W(S) = \begin{vmatrix} t^{-1} & t^{-2} \\ -t^{-2} & -2t^{-3} \end{vmatrix} = -2t^{-4} + t^{-4} = -t^{-4}.$

7. Linearly independent; $W(S) = 0.$

9. $W(\{e^t, e^{-t}\}) = -2.$

11. $W(\{t^{-1/2}, t^3\}) = \dfrac{7}{2} t^{3/2}.$

13. $y(t) = -2e^{2t} + e^{-t}.$

15. $y = -t^{-1/3} + 2t^3.$

17. $y = \sin t + \cos t + t \sin t.$

21. $W(S) = \begin{vmatrix} \cos 4t & \sin 4t \\ -4\sin 4t & 4\cos 4t \end{vmatrix} = 4\cos^2 4t + 4\sin^2 4t = 4.$

23. $W(S) = \begin{vmatrix} t^{-1} & t \\ -t^{-2} & 1 \end{vmatrix} = 2t^{-1}.$

25. $S = \{e^{2t}, e^{3t}\},\ y = c_1 e^{2t} + c_2 e^{3t}.$

27. $S = \{e^{2t}, te^{2t}\},\ y = e^{2t}(c_1 + c_2 t),\ y = te^{-2t}.$

29. $S = \{\cos 3t, \sin 3t\},\ y = c_1 \cos 3t + c_2 \sin 3t,$

$\qquad y = \cos 3t - \dfrac{4}{3} \sin 3t.$

31. $S = \{t^{-4}, t\},\ y = c_1 t^{-4} + c_2 t.$

33. For $t > 0$, $y_2(t) = \dfrac{\sin t}{\sqrt{t}}$ so a general solution is $y = t^{-1/2}(c_1 \cos t + c_2 \sin t).$

35. Substituting $y = e^{-t/2}$ and $y = e^{t/3}$ into the equation $y'' + (b/a)y' + (c/a)y = 0$ results in the system of equations $2b/a - 4c/a = 1$ and $3b/a - 9c/a = 1$ so $b/a = 5/6$ and $c/a = 1/6.$ Hence, the equation is $6y'' + 5y' + y = 0.$

37. Substitution of either function into the differential equation and equating coefficients gives us $a - b = -3$ and $2a = 4$ so $a = 2$ and $b = 5.$

39. $W\left(\left\{f(t), \int \dfrac{1}{(f(t))^2} e^{-\int p(t)dt} dt\right\}\right) = e^{-\int p(t)dt}.$

41. $y = c_1 + c_2 \tan(c_3 + c_4 \ln t)$ is a solution of the equation if (a) $y = c_1$ (note that if $c_2 = 0$ or $c_4 = 0$, y is a constant function); or

$\qquad$ (b) $y = -\dfrac{1}{2} + c_2 \tan(c_3 + c_2 \ln t).$ The Principle of Superposition does not hold.

43. $y = t^{-1} \cos 4t.$

45. $y = -\pi t^{-1} \cos 4t.$

47. $y = t^{-3}(a \cos t + b \sin t);\ y(\pi = -a\pi^{-3} = 0$ implies $a = 0;\ y(2\pi) = \dfrac{18}{a}\pi^{-3} = 0$ implies that $a = 0.$ $y = t^{-3} \sin t,\ C$ arbitrary.

Exercises 4.2

1. $y'' = 0$ has characteristic equation $r^2 = 0$ so $r = 0$ has multiplicity two. Two linearly independent solutions to the equation are $y_1 = 1$ and $y_2 = t$; a fundamental set of solutions is $S = \{1, t\}$; and a general solution is $y = c_1 + c_2 t.$

3. $y'' + y' = 0$ has characteristic equation $r^2 + r = 0$, which has solutions $r_1 = 0$ and $r_2 = -1.$ Two linearly independent solutions to the equation are $y_1 = 1$ and $y_2 = e^{-t}$; a fundamental set of solutions is $S = \{1, e^{-t}\}$; and a general solution is $y = c_1 + c_2 e^{-t}.$

5. $y = c_1 e^{-6t} + c_2 e^{-2t}$.

6. $6r^2 + 5r + 1 = (3r + 1)(2r + 1) = 0$ so
$r_1 = -1/3$ and $r_2 = -1/2$. Thus, two linearly
independent solutions to the equation are
$y_1 = e^{-t/3}$ and $y_2 = e^{-t/2}$; a fundamental set
of solutions is $S = \{e^{-t/3}, e^{-t/2}\}$; and a
general solution is $y = c_1 e^{-t/3} + c_2 e^{-t/2}$.

7. $y = c_1 e^{-t/4} + c_2 e^{-t/2}$.

9. $y = c_1 \cos 4t + c_2 \sin 4t$.

11. $y = c_1 \cos(\sqrt{7}t) + c_2 \sin(\sqrt{7}t)$.

13. $y = c_1 e^{-t} + c_2 e^{3t/7}$.

15. $y = c_1 e^{3t} + c_2 t e^{3t}$.

17. General: $y = c_1 + c_2 e^{t/3}$; IVP:
$y = -21(1 - e^{t/3})$.

19. General: $y = c_1 e^{3t} + c_2 e^{4t}$; IVP:
$y = 14 e^{3t} - 11 e^{4t}$.

21. General: $y = c_1 e^{5t} + c_2 e^{2t}$; IVP: $y = e^{5t}$.

23. General: $y = c_1 \cos 10t + c_2 \sin 10t$; IVP:
$y = \cos 10t + \sin 10t$.

25. General: $y = e^{-2t}(c_1 + c_2 t)$; IVP:
$y = e^{-2t}(1 + 5t)$.

27. General: $y = e^{-2t}(c_1 \cos 4t + c_2 \sin 4t)$; IVP:
$y = e^{-2t}(2 \cos 4t + \sin 4t)$.

29. $y = e^{-t/2} \cos\left(\dfrac{\sqrt{3}}{2} t\right) + \dfrac{1}{\sqrt{3}} e^{-t/2} \sin\left(\dfrac{\sqrt{3}}{2} t\right)$.

31. $y = -\dfrac{e^{\left(\frac{1}{2} - \frac{\sqrt{5}}{2}\right)t} - e^{\left(\frac{1}{2} + \frac{\sqrt{5}}{2}\right)t}}{\sqrt{5}}$.

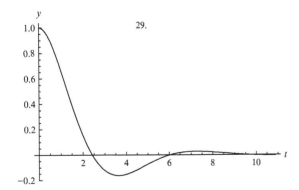

29.

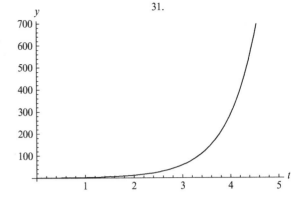

31.

33. $y = 9 e^{-t/3} - 8 e^{-t/2}$.

35. $y = \dfrac{3}{4} e^{-2t} \sin(4t) + e^{-2t} \cos(4t)$.

37. (a) $y = c_1 t^{2/3} + c_2 t$; (b) $y = t(c_1 + c_2 \ln t)$.

41. $y = Ce^{-t} \sin 2t, y = 0, y = e^{-t} \cos 2t$.

43. $y = \left(\dfrac{3}{2} a + \dfrac{1}{2} b\right) e^{-t} + \left(-\dfrac{1}{2} a - \dfrac{1}{2} b\right) e^{-3t}$ so

$y' = -\left(\dfrac{3}{2} a + \dfrac{1}{2} b\right) e^{-t} - 3\left(-\dfrac{1}{2} a - \dfrac{1}{2} b\right) e^{-3t}$;

$y' = 0$ if $t = -\ln\left(\pm \dfrac{1}{3} \dfrac{\sqrt{3}\sqrt{(b+a)(3a+b)}}{b+a}\right)$;

For none, $(b + a)(3a + b) \le 0$ while for one,
$(b + a)(3a + b) > 0$.

45. (a) No (b) To be a general solution, a
fundamental set for the equation is
$S = \{t \cos t, t \sin t\}$. Now substitute each of
these functions into the differential equation
and set the result equal to zero. Solve the
resulting system for $p(t)$ and $q(t)$ to obtain
$p(t) = -2/t$ and $q(t) = (t^2 + 2)/t^2$.

Exercises 4.3

1. $F = \{t, 1\}$.

3. $F_1 = \{e^{2t}\}, F_2 = \{1\}$.

5. $F_1 = \{e^{-t}\}, F_2 = \{t^4, t^3, t^2, t, 1\}$.

7. $F_1 = \{\cos 2t, \sin 2t\}, F_2 = \{e^{-4t}\}$.

9. $F_1 = \{\cos 3t, \sin 3t\}, F_2 = \{\cos 2t, \sin 2t\}$.

11. $F_1 = \{e^{-t} \cos 2t, e^{-t} \sin 2t\}, F_2 = \{1\}$.

13. $y_p = Ae^{2t}$ $(y = c_1 \cos t + c_2 \sin t + \frac{8}{5}e^{2t})$.

15. $y_p = Ate^{3t}$ $(y = c_1 e^{3t} + c_2 e^t + te^{3t})$.

17. $y_p = At^2 + Bt + C$
$(y = e^t(c_1 + c_2 t) + t^2 + 4t + 6)$.

19. $y_p = A \cos 2t + B \sin 2t$

$(y = c_1 \cos t + c_2 \sin t - \frac{1}{3} \cos 2t)$.

21. $y_p = At \cos 2t + Bt \sin 2t + Ct + D$

$(y = c_1 \cos 2t + c_2 \sin 2t + \frac{1}{4} t \sin 2t + \frac{1}{4} t)$.

23. $y_p = At^6 + Bt^5 + Ct^4 + Dt^3 + Et^2$

$(y = c_1 + c_2 t + \frac{1}{10} t^6 - \frac{1}{3} t^3)$.

25. $y = c_1 e^t + c_2 e^{-2t} + 1/2$.

27. $y = c_1 e^{4t} + c_2 e^{-2t} + 1 - 4t$.

29. $y = c_1 e^{-t} \sin (5t) + c_2 e^{-t} \cos (5t) + 1 - 13t$.

31. $y = 280 - 60t + 5t^2 + c_1 e^{-1/2t} + c_2 e^{-1/4t}$.

33. $y = -4 \sin (3t) + 8/3 \cos (3t) + c_1 e^{2t} + c_2$.

35. $y = c_1 e^{-3t} + c_2 e^{3t} - \frac{216}{169} \cos (2t) - \frac{54}{13} \sin (2t) t$.

37. $y = c_1 e^{-2t} + c_2 e^{-2t} t + (-4t + 3) \cos (2t) - 4t \sin (2t)(t-1)$.

41. $y = c_1 + c_2 e^{-4t} + \frac{1}{4} e^{4t} + te^{-4t}$.

43. $y = y_h + y_p = c_1 + c_2 e^{-4t} - 3t^2 + \frac{1}{2} t^2 e^{-4t}$.

45. $y = c_1 t + c_2 + \frac{1}{12} t^4 + e^t - \sin t$.

47. $y = 2e^{-t} + 2e^t - 4$.

49. $y = 2e^t - 2e^{-3t} + 2/3$.

51. $y = e^{-4t} + 4e^{-4t} t + 1/4$.

53. $y = 7/4 e^{-3t} \sin (4t) - 1/25$.

55. $y = -2t^2 e^{2t} + 6e^{2t} t - 7e^{2t} + 3/2 t^2 + 5e^t + 3t - 3/2$.

57. $y = -3t^2 + \frac{1}{2} t^2 e^{-4t}$.

59. $y = \begin{cases} \frac{2}{9} t - \frac{2}{27} \sin 3t, 0 \le t < \pi \\ -\frac{2}{9} \cos 3t - \frac{4}{27} \sin 3t, t \ge \pi \end{cases}$ Note: A
CAS was used to construct the solution.

61.

$y = \begin{cases} \sin 2t, 0 \le t < \pi \\ \frac{1}{2} (5 - 5 \cos 2t + 2 \sin 2t), \pi \le t < 2\pi. \\ \sin 2t, t \ge 2\pi \end{cases}$

Note: A CAS was used to construct the solution.

67. $\omega = 2$.

Exercises 4.4

1. $y = c_1 \sin (2t) + c_2 \cos (2t) + 1/4$.

3. $y = c_1 e^{5t} + c_2 e^{2t} - \frac{1}{2} e^{3t}$.

5. $y = c_1 e^{-2t} \sin (4t) + c_2 e^{-2t} \cos (4t) + \frac{1}{8} te^{-2t}$.

7. $y = c_1 \sin (4t) + c_2 \cos (4t) + 1/16 \ln (\sin (4t)) \sin (4t) - 1/4t \cos (4t)$.

9. $y(t) = 1/49 e^{-t}(c_1 \sin (7t) + c_2 \cos (7t) + \ln (\sin (7t)) \sin (7t) - 7t \cos (7t))$.

11. $y = c_1 e^t \sin (5t) + c_2 e^t \cos (5t) - 1/5(-1/5 \cos (5t) \ln (\cos (5t)) - 1/5 \sin (5t) \ln (\sin (5t)) + t(- \sin (5t) + \cos (5t)))e^t$.

13. $y = c_1 e^{3t} \sin (5t) + c_2 e^{3t} \cos (5t) - 1/25 e^{3t} \cos (5t) \ln \left(\frac{1 + \sin (5t)}{\cos (5t)} \right)$.

15. $y = c_1 e^{6t} \sin (t) + c_2 e^{6t} \cos (t) + e^{6t} (t \sin (t) + \ln (\cos (t)) \cos (t))$.

17. $y = c_1 e^{3t} + c_2 e^{-3t} + \frac{1}{18} e^{-3t}(-e^{3t} + e^{6t} \ln (e^{-3t} + 1) - \ln (e^{3t} + 1))$.

19. $y = c_1 e^t + c_2 e^{-t} + 1/2 e^{-t}((t + 1/2 \cosh (2t) - 1/2 \sinh (2t)) e^{2t} - 1/2 \cosh (2t) - 1/2 + t - 1/2 \sinh (2t))$.

21. $y = c_1 e^{2t} + c_2 e^{2t} - e^{2t} (\ln (t) + 1)$.

23. $y = c_1 e^{-3t} + c_2 e^{-3t} t + t (-1 + \ln (t)) e^{-3t}$.

25. $y = \left(2 \frac{(\tan (1/2e^t))^2}{\left(1 + (\tan (1/2e^t))^2 \right) e^t} - c_1 e^{-t} + c_2 \right) e^{-t}$.

27. $y = c_1 e^t + c_2 e^t t +$
$\frac{1}{6} e^t \left(2\sqrt{1-t^2} + \sqrt{1-t^2}\, t^2 + 3t \arcsin(t) \right).$

29. $y = c_1 e^{2t} + c_2 e^{2t} t + \frac{1}{2} e^{2t} (\arctan(t)\, t^2 + t -$
$\arctan(t) - t \ln(t^2 + 1)).$

31. $y = c_1 \sin(1/2t) + c_2 \cos(1/2t) +$
$\cos(1/2t) \ln(\cos(1/2t)) +$
$\sin(1/2t) \ln(\sin(1/2t)) -$
$1/2t\, (-\sin(1/2t) + \cos(1/2t)).$

33. $y = c_1 \cos 3t + c_2 \sin 3t + \frac{1}{9} \cos 3t \ln \cos$
$3t + \frac{1}{3} t \sin 3t.$

35. $y = c_1 \cos 2t + c_2 \sin 2t -$
$\frac{1}{4} \cos 2t \left(\ln(\sec 2t + \tan 2t) - \sin 2t \right)$
$+ \frac{1}{2} \sin 2t \cos^2 2t.$

37. $y = c_2 \cos 2t + c_2 \sin 2t + \frac{1}{4} \cos 2t (\sin 2t - 2t)$
$+ \frac{1}{2} \sin 2t \left(\ln \cos t - \frac{1}{2} \cos 2t \right).$

39. $y = c_1 \cos 2t + c_2 \sin 2t + \frac{1}{2} t \cos 2t$
$+ \frac{1}{4} \sin 2t \ln \sec 2t.$

41. $y = \frac{1}{4} \cos 2t + \frac{1}{2} \sin 2t - \frac{1}{4} +$
$\frac{1}{4} \sin 2t \ln(\sec 2t + \tan 2t).$

43. $y = 2\cos t + \sin t - 2 + \sin t - \ln(\sec t + \tan t).$

45. $y = \left(-\frac{1}{6}\sqrt{2} + \frac{1}{36}\pi \right) \cos 3t + \frac{1}{36}\sqrt{2}(6 +$
$\sqrt{2} \ln 2) \sin 3t - \frac{1}{3} t \cos 3t + \frac{1}{9} \sin 3t \ln \sin 3t.$

51. $y = \frac{1}{k} \left[\left(\int_0^t f(z) \cos kz\, dz \right) \sin kt \right.$
$\left. - \left(\int_0^t f(z) \sin kz\, dz \right) \cos kt \right].$

53. $y = \frac{1}{2k} \left[\left(\int_0^t e^{-kz} f(z) dz \right) e^{kt} \right.$
$\left. - \left(\int_0^t e^{kz} f(z) dz \right) e^{-kt} \right].$

55. $y = c_1 t^{-1} + c_2 t^{-1} \ln t + \ln t - 2.$

57. $y = c_1 t^6 + c_2 t^{-1} - \frac{1}{3} \ln t + \frac{5}{18}.$

59. $y = \exp\left(\frac{1}{2} t^2 e^{2t} - \frac{1}{2} t e^{2t} + \frac{1}{4} e^{2t} + c_1 t + c_2 \right).$

61. (a) $y = t^2 (c_1 \cos t + c_2 \sin t) + t$ (b)
$y = t^2 (c_1 \cos t + c_2 \sin t) + t$ (infinitely
many solutions).

63. (a) $y = t^{-1/2}(c_1 \cos 2t + c_2 \sin 2t) + t^{-1/2}$
(b) No solution (c) $y = t^{-1/2}(1 - \cos 2t).$

65. A general solution is $y = c_1 t + c_2 \sin t + \cos t$
and the solution to the IVP is
$$y = \frac{4\sqrt{2}}{\pi - 4} t + \frac{\pi + 4}{\pi - 4} \sin t + \cos t.$$

Exercises 4.5

1. $W(S) = \begin{vmatrix} 3t^2 & t & 2t - 2t^2 \\ 6t & 1 & 2 - 4t \\ 6 & 0 & -4 \end{vmatrix} = 0$, linearly
dependent.

3. $W(S) = \begin{vmatrix} e^{-t} & e^{3t} & te^{3t} \\ -e^{-t} & 3e^{3t} & e^{3t} + 3te^{3t} \\ e^{-t} & 9e^{3t} & 6e^{3t} + 9te^{3t} \end{vmatrix} = 16e^{5t},$
linearly independent.

5. $W(S) = \begin{vmatrix} 1 & t & t^2 & t^3 & t^4 \\ 0 & 1 & 2t & 3t^2 & 4t^3 \\ 0 & 0 & 2 & 6t & 12t^2 \\ 0 & 0 & 0 & 6 & 24t \\ 0 & 0 & 0 & 0 & 24 \end{vmatrix} = 288,$ linearly
independent.

7. Fourth order, $y = (c_1 + c_2 t + c_3 t^2)e^{-2t} + c_4 e^{2t}.$

9. Fourth order, $y = c_1 + c_2 t + c_3 \cos 3t$
$+ c_4 \sin 3t.$

11. Sixth order, $y = e^{-3t}[(c_1 + c_2 t) \cos 4t$
$+ (c_3 + c_4 t) \sin 4t] + c_5 e^{-5t} + c_6 e^{-t/3}.$

13. All are solutions of $y''' = 0$ but the Wronskian of the set is 0 so they cannot be a fundamental set.

15. Yes: $y^{(4)} - \frac{1}{4}y''' - 2y'' - \frac{7}{2}y' + y = 0.$

17. $k = 0$ with multiplicity 3 is the solution of the characteristic equation so $S = \{1, t, t^2\}$ is a fundamental set of solutions and, thus, $y = c_0 + c_1 t + c_2 t^2$ is a general solution.

19. $y = c_1 + c_2 e^{5t} + c_3 e^{5t}t.$

21. The characteristic equation is $k^4 + 16k^2 = 0$ and the left-hand side of the equation factors resulting in $k^2(k^2 + 16) = 0$ so $k_{1,2} = 0$ is a 0 of multiplicity 2 and $k^2 + 16 = 0$ has roots $k_{3,4} = \pm 4i$. Thus, a fundamental set of solutions is $S = \{1, t, \cos 4t, \sin 3t\}$ and a general solution of the equation is $y = c_1 + c_2 t + c_3 \cos 4t + c_4 \sin 4t.$

23. $y = c_1 e^t + c_2 e^{t/3} + c_3 t/2.$

25. $y = c_1 e^{2t} + c_2 e^{(-\frac{1}{2}-\sqrt{3})t} + c_3 e^{(-\frac{1}{2}+\sqrt{3})t}.$

27. The characteristic equation is $k^4 + k^3 = k^3(k + 1) = 0$ so $k_{1,2,3} = 0$ is a zero of multiplicity three and $k_4 = -1$ is a zero of multiplicity one. A fundamental set is $S = \{1, t, t^2, e^{-t}\}$; a general solution is $y = c_1 + c_2 t + c_3 t^2 + c_4 e^{-t}.$

29. $y = c_1 e^{-2t} + c_2 e^{2t} + c_3 \sin(2t) + c_4 \cos(2t).$

31. $y = c_1 e^{2t} + c_2 e^{-t} + c_3 e^{-4t} + c_4 e^{-4t} t.$

33. $y = c_1 + c_2 t + c_3 t^2 + c_4 t^3 + c_5 e^{-4t}.$

35. $y = c_1 + c_2 t + c_3 e^{-t} + c_4 t e^{-t} + c_5 t^2 e^{-t}.$

37. $y = (c_1 + c_2 t) \cos 2t + (c_3 + c_4 t) \sin 2t.$

39. $y = (c_1 + c_2 t + c_3 t^2) \cos 2t + (c_4 + c_5 t + c_6 t^2) \sin 2t.$

41. $y = e^t - \sqrt{3} e^{-1/2t} \sin\left(1/2\sqrt{3}t\right) - e^{-1/2t} \cos\left(1/2\sqrt{3}t\right).$

43. $y = t(e^{2t} - e^{-2t}).$

45. $y = e^{-t} - 2e^{-2t}.$

47. $y = e^{-t/2} (1 + 3 \cos 3t - 4 \sin 3t)$ *Note: A CAS was used to assist in constructing the solution.*

49. $y = 8t^3 + 4t + 9.$

53. (a) 3, (b) $t - 1$, (c) 2, (d) 0.

59. $W(S) = \begin{vmatrix} e^{-3t} & e^{-t} \\ -3e^{-3t} & -e^{-t} \\ 9e^{-3t} & e^{-t} \\ -27e^{-3t} & -e^{-t} \end{vmatrix}$

$\begin{vmatrix} e^{-4t} \cos(3t) \\ -4e^{-4t} \cos(3t) - 3e^{-4t} \sin(3t) \\ 7e^{-4t} \cos(3t) + 24e^{-4t} \sin(3t) \\ 44e^{-4t} \cos(3t) - 117e^{-4t} \sin(3t) \end{vmatrix}$

$\begin{vmatrix} e^{-4t} \sin(3t) \\ -4e^{-4t} \sin(3t) + 3e^{-4t} \cos(3t) \\ 7e^{-4t} \sin(3t) - 24e^{-4t} \cos(3t) \\ 44e^{-4t} \sin(3t) + 117e^{-4t} \cos(3t) \end{vmatrix} = 1080e^{-12t}.$

61. (b) $W(S) = \begin{vmatrix} f(t) & tf(t) \\ \frac{d}{dt}f(t) & f(t) + t\frac{d}{dt}f(t) \\ \frac{d^2}{dt^2}f(t) & 2\frac{d}{dt}f(t) + t\frac{d^2}{dt^2}f(t) \end{vmatrix}$

$\begin{vmatrix} t^2 f(t) \\ 2tf(t) + t^2\frac{d}{dt}f(t) \\ 2f(t) + 4t\frac{d}{dt}f(t) + t^2\frac{d^2}{dt^2}f(t) \end{vmatrix} = 2(f(t))^3.$

63. (a) $y = -0.09090909091e^{-3.0t} + 0.5142594770 \sin(1.414213562t) + 0.09090909091 \cos(1.414213562t)$
(b) $y = 2.020725943e^{2.0t} \sin(1.732050808t) + e^{2.0t} \cos(1.732050808t) - 0.5773502693e^{2.0t} \sin(1.732050808t)t - 3.500000000e^{2.0t} \cos(1.732050808t)t$
(c) $y = 0.0002889303734e^{-36.99601821t} - 1.061192194e^{0.4334929696t} \sin(0.5236492683t) - 1.000288930e^{0.4334929696t} \cos(0.5236492683t)].$

65. $y = 0.05460976246y_0e^{0.4712630674t} +$
$0.8550147286y_0e^{-0.2076539530t}$
$\sin(1.126208087t) -$
$0.05460976246y_0e^{-0.2076539530t}$
$\cos(1.126208087t).$

Exercises 4.6

1. $y = c_1 + c_2t + c_3e^{-t} + \dfrac{1}{2}e^t.$

3. $y = c_1 + c_2t + c_3t^2 + c_4t^3 + c_5e^t - \dfrac{1}{24}t^4.$

5. $y = c_1 + c_2t + c_3e^{-3t} + c_4e^{3t} + \dfrac{1}{6}te^{3t}.$

7. $y = e^{-3t}t + 3/4te^{-t} - \dfrac{7}{8}e^{-t} + 3/2e^{-3t} -$
$1/4e^{-t}t^2 + c_1e^{-3t} + c_2e^{-2t} + c_3e^{-t}.$

9. $y = e^t + c_1e^{4t} + c_2e^{-5t}\cos(t) + c_3e^{-5t}\sin(t).$

11. $y = c_1 + c_2\cos 2t + c_3\sin 2t + \dfrac{1}{8}\ln\sec 2t +$
$\dfrac{1}{8}\cos 2t(4t - \sin 4t) - \dfrac{1}{8}\sin 2t\ln\sec 2t.$

13. $y = y_h + y_p = c_1 + c_2t + c_3\cos 2t + c_4\sin 2t +$
$\dfrac{1}{16}\left(\ln\sec 2t - \sin 2t\ln(\sec 2t + \tan 2t)\right).$

15. $y = \dfrac{1}{27}[(c_1 + \ln(\cos(3t/2) - \sin(3t/2)) -$
$\ln(\cos(3t/2)+\sin(3t/2)))\cos 3t + c_2\sin 3t + c_3].$

17. $y = \dfrac{1}{8}(c_1 + c_2\cos^2 t - 2t\cos 2t$
$- \ln(\cos t - \sin t) + \ln(\cos t + \sin t) +$
$c_3\sin 2t + \ln(\cos 2t)\sin 2t).$

19. $y = -3/4t^2e^t + 1/2\ln(t)t^2e^t + c_1e^t +$
$c_2te^t + c_3t^2e^t.$

21. $y = 1/6e^{-3t} + c_1e^{-4t} + c_2e^{-2t} + c_3e^{3t}.$

23. $y = 1/10\sin(t) - 3/10\cos(t) + c_1e^{-2t}$
$+ c_2e^{-t} + c_3.$

25. $y = -\dfrac{1}{2}t^2 + c_1 + c_2\cos t - \ln(-\cos(t/2)$
$- \sin(t/2)) - \ln(-\cos(t/2) + \sin(t/2))$
$- (c_3 - \ln(\cos(t/2) - \sin(t/2)) +$
$\ln(\cos(t/2) + \sin(t/2)))\sin t.$

27. $y = 1 + t - \cos t - \ln(\cos t)$
$- 2\tanh^{-1}(\tan(t/2))\sin t.$

29. $y = 2 - 2\cos t - \dfrac{1}{2}t\sin t.$

31. $y = c_1 + c_2t^2 + c_3t\ln t + t$

32. $y = c_1 + c_2e^t + c_3\ln t + t.$

33. A general solution is $y = c_1t + c_2t\ln t$
$+ c_3\sqrt{t} - t^2$; the solution of the IVP is
$y = t - t^2 + 2t\ln t.$

35. A general solution is $y = c_1 + c_2t$
$+ c_3t\ln t + c_4t^2 + 2t^{-1/2}$; the solution of the
IVP is $y = \dfrac{1}{8}(-41 + 16t^{-1/2} + 12t$
$+ 13t^2 - 30t\ln t).$

Exercises 4.7

1. $y = c_1x^{1/2} + c_2x^{5/2}.$

3. $y = c_1x + c_2x^4.$

5. $y = \sqrt{x}(c_1\cos(2\ln x) + c_2\sin(2\ln x)).$

7. $y = x(c_1\cos(3\ln x) + c_2\sin(3\ln x)).$

9. $y = x^{-1/2}(c_1 + c_2\ln x).$

11. $y = x^3(c_1 + c_2\ln x).$

13. $y = c_1x^{-10} + c_2x^{-7} + c_3x^{-2}.$

15. $y = c_1x^{-2} + c_2x + c_3x^3.$

17. $y = c_1x + x(c_2\cos(\ln x) + c_3\sin(\ln x)).$

19. $y = x^{-1}(c_1 + c_2\ln x + c_3(\ln x)^2).$

21. $y = \dfrac{c_1}{x^2} + \dfrac{c_2\ln(x)}{x^2} + 1/9x^{-5}.$

22. $y = c_1x^3 + c_2x^3\ln(x) + 1/2\left(\ln(x)\right)^2x^3.$

23. $y = c_1\sin\left(\ln(x)\right) + c_2\cos\left(\ln(x)\right) + 1/5x^{-2}.$

25. $y = \dfrac{c_1}{x^3} + c_2x^2 - 1/2x.$

27. $y = c_1\sin\left(2\ln(x)\right) + c_2\cos\left(2\ln(x)\right) + 2.$

29. $y = 1/25x^{-3} + c_1x^2 + \dfrac{c_2}{x^4} + c_3x^2\ln(x).$

31. $y = 9/5\sqrt[3]{x} + 1/5x^2.$

33. $y = \cos\left(2\ln(x)\right).$

35. $y = -3/4x^2 + 1/44x^{-10} + \dfrac{8}{11}x.$

37. $y = 3/2x + 1/2x^2 \sin\left(\ln(x)\right)$
$- 3/2x^2 \cos\left(\ln(x)\right).$

39. $y = \dfrac{22}{15}\sqrt{x} + 1/5x^{-2} - 5/3x^{-1}.$

41. $y = \dfrac{24}{25}\sqrt{x} - 8/5\sqrt{x}\ln(x) + 1/25x^3.$

43. $W(S) = \begin{vmatrix} x^{r_1} & x^{r_2} \\ x^{r_1-1}r_1 & x^{r_2-1}r_2 \end{vmatrix} =$
$-x^{r_1+r_2-1}(-r_2 + r_1).$

49. $y = c_1 + c_2\cos(6\ln x) + c_3\sin(6\ln x).$

51. $y = x(c_1 + c_2\ln x + c_3(\ln x)^2).$

53. (c) $y = c_1\sin(2\arctan(x))$
$+ c_2\cos(2\arctan(x))$
(d) $y = c_1\sin(2\arctan(x)) + c_2$
$\cos(2\arctan(x)) + 1/4\arctan(x)$
(e) $y = 1/2\sin(2\arctan(x))$ (f) $y = 3/8$
$\sin(2\arctan(x)) + 1/4\arctan(x).$

67. $y = \pm\sqrt{2c_1\ln(x) + 2c_2}.$

Exercises 4.8

1. $x = 0, R \geq 1.$

3. $x = \pm 2, R \geq 1.$

5. $y = a_0 + a_1x + (11/2a_1 - 15a_0)x^2 +$
$\left(\dfrac{91}{6}a_1 - 55a_0\right)x^3 + \left(\dfrac{671}{24}a_1 - \dfrac{455}{4}a_0\right)x^4 +$
$\left(\dfrac{4651}{120}a_1 - \dfrac{671}{4}a_0\right)x^5 + \cdots.$

7. $y = a_0 + a_1x + \left(1/2a_1 + a_0 + \dfrac{1}{2}\right)x^2 +$
$\left(1/2a_1 + 1/3a_0 + \dfrac{1}{3}\right)x^3 +$
$\left(\dfrac{5}{24}a_1 + 1/4a_0 + \dfrac{5}{24}\right)x^4 +$
$\left(\dfrac{11}{120}a_1 + 1/12a_0 + \dfrac{1}{12}\right)x^5 + \cdots.$

9. $y = a_0 + a_1x - 1/4a_1x^3 + 3/16a_1x^4$
$- \dfrac{9}{80}a_1x^5 + \cdots.$

11. $y = a_0 + a_1x + (1/2a_1 - a_0)x^2 +$
$(-1/3a_1 - 1/3a_0)x^3 + \left(-\dfrac{5}{24}a_1 + 1/6a_0\right)x^4 +$
$\left(-\dfrac{1}{120}a_1 + 1/15a_0\right)x^5 + \cdots.$

13. $y = a_0 + a_1x + 3/4a_0x^2 + 1/12a_1x^3 -$
$\dfrac{5}{32}a_0x^4 - 1/32a_1x^5 + \cdots.$

15. $y = 1 + \dfrac{1}{3}x^4 + \dfrac{1}{42}x^8 + \dfrac{1}{1386}x^{12} + \cdots.$

17. $y = a_1 + a_2(x-1) + 1/8a_2(x-1)^2 +$
$\dfrac{5}{96}a_2(x-1)^3 + \dfrac{15}{512}a_2(x-1)^4 +$
$\dfrac{39}{2048}a_2(x-1)^5 + O((x-1)^6).$

19. $y = 1 + \dfrac{1}{6}x^3 - \dfrac{1}{30}x^5 + \dfrac{1}{240}x^7 - \dfrac{37}{90,720}x^9 + \cdots.$

21. $y = x + \dfrac{1}{2}x^2 - \dfrac{1}{8}x^4 - \dfrac{1}{8}x^5 - \dfrac{7}{120}x^6 +$
$\dfrac{11}{1680}x^7 + \dfrac{89}{2688}x^8 + \dfrac{79}{2880}x^9 + \cdots.$

23. (a) $y = a_1 + a_2x - 1/2a_1x^2 + 1/6a_2x^3$
$- 1/8a_1x^4 + 1/24a_2x^5 - \dfrac{7}{240}a_1x^6 + \dfrac{1}{112}a_2x^7$
$- \dfrac{11}{1920}a_1x^8 + \dfrac{13}{8064}a_2x^9 + \cdots$ (b) $y = a_1$
$+ a_2x - 3/2a_1x^2 - 1/6a_2x^3 - 1/8a_1x^4$
$- 1/40a_2x^5 - 1/48a_1x^6 - \dfrac{1}{240}a_2x^7 - \dfrac{3}{896}a_1x^8$
$- \dfrac{11}{17,280}a_2x^9 + \cdots; y = a_1 + a_2x - 1/2ka_1x^2$
$+ (1/3a_2 - 1/6ka_2)x^3 + (-1/6ka_1$
$+ 1/24k^2a_1)x^4 + \left(-1/15ka_2 + \dfrac{1}{120}k^2a_2\right.$
$\left. + 1/10a_2\right)x^5 + \left(\dfrac{1}{60}k^2a_1 - \dfrac{1}{720}k^3a_1\right.$
$\left. - \dfrac{2}{45}ka_1\right)x^6 + \left(-\dfrac{23}{1260}ka_2 + \dfrac{1}{280}k^2a_2\right.$
$\left. + 1/42a_2 - \dfrac{1}{5040}k^3a_2\right)x^7 + \left(-\dfrac{1}{1680}k^3a_1\right.$
$\left. + \dfrac{1}{40,320}k^4a_1 + \dfrac{11}{2520}k^2a_1 - \dfrac{1}{105}ka_1\right)x^8 +$

$$\left(\frac{43}{45,360}k^2a_2 - \frac{1}{11,340}k^3a_2 - \frac{11}{2835}ka_2 \right.$$
$$\left. + \frac{1}{362,880}k^4a_2 + \frac{1}{216}a_2\right)x^9 + \cdots.$$

25. $2, \frac{2}{3}, \frac{2}{5}, \frac{2}{7}, \frac{2}{9}, \frac{2}{11}.$

27. $y = 1 - \frac{1}{2}x^2 + \frac{1}{3}x^3 - \frac{1}{24}x^4 - \frac{1}{24}x^5 + \frac{1}{40}x^6 -$
$$\frac{17}{5040}x^7 - \frac{53}{20,160}x^8 + \frac{11}{7560}x^9 -$$
$$\frac{41}{302,400}x^{10} - \frac{3097}{19,958,400}x^{11} +$$
$$\frac{8219}{119,750,400}x^{12} - \frac{3431}{2,075,673,600}x^{13} + \cdots.$$

Exercises 4.9

1. $x = 0$, regular.
3. $x = 4$, regular.
5. $y = c_1 x^{7/2}\left(1 + \frac{1}{3}x + \frac{1}{22}x^2 + \frac{1}{286}x^3\right.$
$$+ \frac{1}{5720}x^4 + \frac{3}{486,200}x^5 + \cdots\right)$$
$$+ c_2\left(1 - \frac{3}{5}x + \frac{3}{10}x^2 - \frac{3}{10}x^3\right.$$
$$\left. - \frac{9}{40}x^4 - \frac{9}{200}x^5 + \cdots\right).$$

7. $y = c_1\left(1 + \frac{1}{4}x - \frac{3}{104}x^2 - \frac{29}{6864}x^3 +$
$$\frac{13}{65,472}x^4 + \frac{251}{11,348,480}x^5 + \cdots\right)x^{-5/9} +$$
$$c_2\left(1 + \frac{1}{14}x - \frac{13}{644}x^2 - \frac{59}{61,824}x^3 +$$
$$\frac{29}{247,296}x^4 + \frac{53}{12,364,800}x^5 + \cdots\right).$$

9. $y = c_1 x\left(1 - \frac{1}{3}x + \frac{1}{24}x^2 - \frac{1}{360}x^3 + \frac{1}{8640}x^4 -$
$$\frac{1}{302,400}x^5 + \cdots\right) + c_2\left(\ln(x)\left(x^2 - \frac{1}{3}x^3 +$$
$$\frac{1}{24}x^4 - \frac{1}{360}x^5 + \cdots\right)x^{-1} + (-2 - 2x + \frac{4}{9}x^3 -$$
$$\frac{25}{288}x^4 + \frac{157}{21,600}x^5 + \cdots)x^{-1}\right).$$

11. $y = c_1\left(1 + \frac{3}{2}x^2 - \frac{9}{40}x^4 + \cdots\right)x^{-2} +$
$$c_2\sqrt[3]{x}\left(1 - \frac{3}{26}x^2 + \frac{9}{1976}x^4 + \cdots\right).$$

13. $y = c_1 x^{7/4}\left(1 + \frac{7}{8}x + \frac{77}{160}x^2 + \frac{77}{384}x^3 + \frac{209}{3072}x^4 +$
$$\frac{4807}{245,760}x^5 + \cdots\right) + c_2\left(\ln(x)\left(\frac{15}{8}x^3 +$$
$$\frac{105}{64}x^4 + \frac{231}{256}x^5 + \cdots\right)x^{-5/4} + \left(12 + 15x +$$
$$\frac{15}{4}x^2 - \frac{13}{2}x^3 - \frac{1741}{256}x^4 - \frac{4141}{1024}x^5 + \cdots x^{-5/4}\right).$$

15. $y = c_1 x^{-7} + c_2 x.$
17. $y = c_1 x^2 + c_2 x^2 \ln x.$

21. $y = \left(c_1\left(-2\frac{-1 + x}{\sqrt{x(-1 + x)}} + \ln(-1/2 + x +$
$$\sqrt{-x + x^2})\right) + c_2\right)\sqrt{x}(-1 + x)^{-3/2}.$$

23. $y = c_1(1 - 2x) + c_2((-1 + 2x)\ln(x) - 2 +$
$(1 - 2x)\ln(-1 + x)).$

27. (a) $= x^{-1/2}(c_1\cos x + c_2\sin x)$
(b) $y = c_1 J_5(4x) + Y_5(4x).$

33. $y = 1 + (1 - x) + \frac{3}{8}(-1 + x)^2 - \frac{1}{3}(-1 + x)^3 +$
$$\frac{101}{384}(-1 + x)^4 - \frac{29}{128}(-1 + x)^5 + \cdots.$$

Chapter 4 Review Exercises

1. Both are solutions of $y''' - 25y' = 0$ and
$$W(\{e^{5t}, 1\}) = \begin{vmatrix} e^{5t} & 1 \\ 5e^{5t} & 0 \end{vmatrix} = -5e^{5t} \neq 0;\text{ linearly}$$
independent.

3. $W(S) = \begin{vmatrix} t & t\ln(t) \\ 1 & \ln(t) + 1 \end{vmatrix} = t;$ linearly
independent.

5. All three are solutions of $y'''' + y'' = 0$ and
$$W(S) = \begin{vmatrix} t & \cos(t) & \sin(t) \\ 1 & -\sin(t) & \cos(t) \\ 0 & -\cos(t) & -\sin(t) \end{vmatrix} = t;$$ linearly
independent.

11. $y = t^{-1}(c_1\cos t + c_2\sin t).$
13. $y = c_1 e^{1/2t} + c_2 e^{-4/3t}.$

15. $y = c_1 e^{-2t} + c_2 e^{-t}$.

17. $y = c_1 e^{1/2t} + c_2 e^{2t}$.

19. $y = c_1 e^{-1/4t} + c_2 e^{1/5t}$.

21. $y(t) = c_1 + c_2 e^{-1/2t} + c_3 e^{-t}$.

23. $y = c_1 + c_2 e^{-2/3t} \sin(t) + c_3 e^{-2/3t} \cos(t)$.

25. $y = -1/5t^2 + 1/3t^3 - 1/5c_1 e^{-5t} + \dfrac{2}{25}t + c_2$.

27. $y = c_1 e^{-t} \sin(2t) + c_2 e^{-t} \cos(2t) + \dfrac{3}{17}\sin(2t) - \dfrac{12}{17}\cos(2t)$.

29. $y = 1/4 \ln\left(1 + e^{2t}\right) + 1/4e^{2t} \ln\left(1 + e^{2t}\right) - 1/4 - 1/2t - 1/2te^{2t} + c_1 e^{2t} + c_2$.

31. $y = e^{3t}(c_1 \cos 2t + c_2 \sin 2t) + \dfrac{3}{29}e^{-2t}$.

33. $y = c_1 e^{-3t} + c_2 e^{-4t} - t\left(6 + 3t + t^2\right)e^{-4t}$.

35. $y = \left(1/36te^{10t} - \dfrac{1}{108}e^{10t} + 1/12t^2 e^{4t} + 1/36te^{4t} + \dfrac{1}{216}e^{4t}\right)e^{-6t} + c_1 e^{-2t} + c_2 e^{4t} + c_3 e^{-2t}t$.

37. $y = \dfrac{4}{625} - \dfrac{8}{125}t + 1/25t^2 + c_1 e^{-t} \cos(2t) + c_2 e^{-t} \sin(2t) + c_3 e^{-t} \cos(2t)\, t + c_4 e^{-t} \sin(2t)\, t$.

39. $y = 2/3e^{-2t} - 2/3e^{-8t}$.

41. $y = \cos(5t)$.

43. $y = \dfrac{19}{25}e^t - \dfrac{19}{25}e^{-4t} + 1/5te^t$.

45. $y = 1/2 \sin(t)\, t$.

47. $y = c_1 \sin(t) + c_2 \cos(t) + \ln(\sin(t)) \sin(t) - t \cos(t)$.

49. $y = \dfrac{1}{20}e^{4t}\left(t^5 + 20t\right)$.

51. $y = \dfrac{1}{4}e^{t-1}\left(8 - e - 4t + 4et - 3et^2 + 2et^2 \ln t\right)$.

55. (a) $y = -e^{-t-1} + \dfrac{1}{2}e^{t-1}$.

57. $y = c_1 x^2 + c_2 x^3$.

59. $y = -x^{-1} + 2x^{-1/2}$.

61. $y = x^4(c_1 \cos(3 \ln x) + c_2 \sin(3 \ln x))$.

62. $y = c_1 x^3 + c_2 x^5 + x$.

63. $y = a_1 + a_2 x + (2a_2 - 2a_1)x^2 + (2a_2 - 8/3a_1)x^3 + (4/3a_2 - 2a_1)x^4 + \left(2/3a_2 - \dfrac{16}{15}a_1\right)x^5 + \cdots$.

65. $y = a_1 + a_2 x - 3/2a_1 x^2 - 1/6a_2 x^3 - 5/8a_1 x^4 - 1/8a_2 x^5 + \cdots$.

66. $y = c_1\left(1 - \dfrac{1}{5}x + \dfrac{1}{20}x^2 + \dfrac{1}{60}x^3 + \dfrac{1}{960}x^4 + \dfrac{1}{33,600}x^5 + \cdots\right)x^{-8/3} + c_2\left(1 + \dfrac{1}{11}x + \dfrac{1}{308}x^2 + \dfrac{1}{15,708}x^3 + \dfrac{1}{1,256,640}x^4 + \dfrac{1}{144,513,600}x^5 + \cdots\right)$.

67. $y = c_1 x^{-2} + c_2 x^{1/2}$.

69. $y = c_1 (-1 + x)^4 (x - 1/6) + c_2\Big(-6(-1 + x)^4 (x - 1/6) \ln(-1 + x) + 6(-1 + x)^4 (x - 1/6)\ln(x) + \dfrac{101}{6}x - \dfrac{59}{2}x^2 - \dfrac{197}{60} + 22x^3 - 6x^4\Big)$.

73. $y = \sqrt{2t - 4 \ln t - 1}, t \ge 1$.

Exercises 5.1

1. $m = 4$ slugs, $k = 9$ lb/ft.; released 1 ft. above equilibrium with zero initial velocity.

3. $m = 1/4$ slugs, $k = 16$ lb/ft.; released 3/4 ft. (8 in.) below equilibrium with an upward initial velocity of 2 ft./s.

5. $x(t) = 5\cos(t - \phi)$, $\phi = -\cos^{-1}(3/5) \approx 0.93$rad; period $= 2\pi$, amplitude $= 5$.

7. $x(t) = \dfrac{\sqrt{65}}{4}\cos(t - \phi)$; $\phi = \cos^{-1}(-8/\sqrt{65}) \approx 3.02$ rad; period $= \pi/2$; amplitude $= \sqrt{65}/4$.

9. $x(t) = \dfrac{\sqrt{2}}{3}\cos(3t - \phi)$, $\phi = -\cos^{-1}(1/\sqrt{2}) = -\pi/4$ rad; period $= 2\pi/3$; amplitude $= \sqrt{2}/3$.

11. $x(t) = \cos 8t$, maximum displacement $= 1$ ft. when $-8\sin 8t = 0$ or $8t = n\pi$ so that $t = n\pi/8$, $n = 0, 1, 2, \ldots$.

13. $x(t) = -\dfrac{1}{4}\cos 8t$; $t = \pi/16$ sec; $x(5) \approx 0.167$ ft.; $x(t) = \dfrac{1}{8}\sin 8t$; $t = \pi/8$ s.

15. As k increases, the frequency at which the spring-mass system passes through equilibrium increases.

17. $b = \sqrt{1023}/16$.

19. $m = 2$ slugs.

21. $\omega\sqrt{\alpha^2 + \beta^2/\omega^2}$.

23. $y(t) = 1.61\cos 3.5t - 0.856379\sin 3.5t$;
maximum displacement
$= \sqrt{(1.61)^2 + (0.856379)^2} = 1.8236$ ft.

27. $\alpha = 1$: $x = \cos 2t$; $\alpha = 4$: $x = 4\cos 2t$; $\alpha = -2$:
$x = -2\cos 2t$; $\beta = 1$: $x = \dfrac{1}{2}\sin 2t$; $\beta = 4$:
$x = 2\sin 2t$; $\beta = -2$: $x = -\sin 2t$.

29. $\dfrac{1}{6}\sqrt{137/2} \approx 1.37941$, $\dfrac{1}{12}\sqrt{199} \approx 1.17556$.

31. $\dfrac{1}{4}\sqrt{199/15} \approx 0.910586$.

Exercises 5.2

1. $m = 1, c = 4, k = 3$; released from
equilibrium with an upward initial
velocity of 4 ft./s.

3. $m = 1/4, c = 2, k = 1$; released 6 in. above
equilibrium with a downward initial
velocity of 1 ft./s.

5. $x(t) = \dfrac{\sqrt{10}}{3}e^{-2t}\cos(3t - \phi)$,
$\phi = \cos^{-1}(3/\sqrt{10}) \approx 0.32$ rad, Q.P.: $2\pi/3$,
$t \approx 0.63$.

7. $x(t) = \dfrac{\sqrt{29}}{5}e^{-t}\cos(5t - \phi)$,
$\phi = \cos^{-1}(5/\sqrt{29}) \approx 0.38$ rad, Q.P.: $2\pi/5$,
$t \approx 0.39$.

9. $x(t) = -\dfrac{1}{2}e^{-5t} + \dfrac{1}{2}e^{-3t}$; overdamped; does
not pass through equilibrium;
$x\left(\dfrac{1}{2}\ln(5/3)\right) \approx 0.093$.

11. $x(t) = -3e^{-t} + 2e^{-t/2}$; overdamped; $x(t) = 0$
implied $t = 2\ln(3/2) \approx 0.811$; maximum
displacement $= 1$ at $t = 0$.

13. $x(t) = 4e^{-4t} + 14te^{-4t}$; critically damped;
does not pass through equilibrium;
maximum displacement $= 4$ at $t = 0$.

15. $x(t) = -5e^{-5t} - 24te^{-5t}$; critically damped;
does not pass through equilibrium;
maximum displacement $= 5$ at $t = 0$.

17. $c = 2$.

19. $x(t) = -\dfrac{3}{2}e^{-4t} + e^{-6t}$; does not pass through
equilibrium; maximum displacement $= 1/2$
at $t = 0$.

21. $x(t) = e^{-2t}\left(-3\cos(4\sqrt{11/5}t) - \dfrac{3}{2}\sqrt{5/11}\right.$
$\left.\sin(4\sqrt{11/5}t)\right)$ or $x(t) = \dfrac{21}{2\sqrt{11}}$
$\cos(4\sqrt{11/5}t - \phi)$, where
$\phi = \cos^{-1}(-2\sqrt{11}/7) \approx 2.81646$; $x(t) = 0$
when $t = \dfrac{1}{4}\sqrt{5/11}\left(\dfrac{1}{2}(2n+1)\pi + \phi\right)$, n an
integer or $t \approx 0.739471, 1.26899, 1.7985,$
$2.32802, 2.85753, \ldots$.

23. $c = 32, 0 < c < 32$.

25. $c = 10/13$.

29. $x(t) = c_1 e^{-\rho t} + c_2 t e^{-\rho t}$, $\rho = c/(2m)$; $x(0) = \alpha$,
$x'(0) = \beta$ implies that $x(t) = \alpha e^{-\rho t}(1 + \rho t)$;
$x(t) = 0$ implies $t = -1/\rho < 0$.

35. $c = 2\sqrt{6}$: $x = te^{-t\sqrt{6}}$; $c = 4\sqrt{6}$:
$x = \dfrac{1}{6\sqrt{2}}e^{-3\sqrt{2}(1+\sqrt{3})t}(e^{6t\sqrt{2}} - 1)$; $c = \sqrt{6}$:
$x = \dfrac{1}{3}\sqrt{2}e^{-t\sqrt{3/2}}\sin\left(\dfrac{3}{\sqrt{2}}t\right)$.

37. The object does not come into contact with
the floor in any of the cases.

39. $c = 2$: $x = e^{-t}(2t - 1)$; $c = \sqrt{8}$:
$x = (1/\sqrt{2} - 2)e^{-(1+\sqrt{2})t} + (1 - 1/\sqrt{2})e^{(1-\sqrt{2})t}$

Exercises 5.3

1. $x(t) = -\dfrac{2}{7}\cos 4t + \dfrac{1}{2}\sin 4t + \dfrac{2}{7}\cos 3t$, $\omega = 4$.

3. $x(t) = \dfrac{35}{141}\cos(4\sqrt{3}t) + \dfrac{4}{47}\cos t$.

5. $x(t) = \begin{cases} \dfrac{4}{\omega^2 - 9}(\cos 3t - \cos \omega t), \omega \neq 3 \\ \dfrac{2}{3}t\sin 3t, \omega = 3 \end{cases}$;
resonance occurs if $\omega = 3$.

7. $x(t) = -e^{-4t}\left(\dfrac{1}{20}\cos 3t + \dfrac{7}{120}\sin 3t\right) +$
$\dfrac{1}{40}(2\cos t - \sin t)$.

9. $x(t) = \dfrac{1}{3}(\cos t - \cos 2t); \pm\dfrac{2}{3}\sin\left(\dfrac{1}{2}t\right);$ decreases.

11. $x(t) = e^{-t/2}\left(2\cos\left(\dfrac{5}{2}t\right) - 4\sin\left(\dfrac{5}{2}t\right)\right) -$ $2\cos t + 11\sin t$; transient:

$e^{-t/2}\left(2\cos\left(\dfrac{5}{2}t\right) - 4\sin\left(\dfrac{5}{2}t\right)\right);$

steady-state: $-2\cos t + 11\sin t.$

13. (a) $x(t) = \alpha\cos(\sqrt{k/m}t)$

$-\dfrac{F\omega}{k-m\omega^2}\sqrt{m/k}\sin(\sqrt{k/m}t) +$

$\dfrac{F}{k-m\omega^2}\sqrt{m/k}\sin\omega t;$

(b) $x(t) = \left(\beta - \dfrac{F\omega}{k-m\omega^2}\right)$

$\sqrt{m/k}\sin(\sqrt{k/m}t) + \dfrac{F}{k-m\omega^2}\sqrt{m/k}\sin\omega t.$

15. $x(t) = \begin{cases} 1 - \cos t, 0 \le t \le \pi \\ -2\cos t, t > \pi \end{cases}.$

17. $x(t) =$

$\begin{cases} t - \sin t, 0 \le t \le 1 \\ -2\sin 1\cos t + (2\cos 1 - 1)\sin t + 2 - t, \\ \quad 1 < t \le 2 \\ (\sin 2 - 2\sin 1)\cos t + 4\cos 1\sin^2\dfrac{1}{2}\sin t, \\ \quad t > 2 \end{cases}.$

21. $b = 0: x = \dfrac{1}{2}t\sin t; b = 1:$

$x = \dfrac{1}{2}(2\sin t + t\sin t).$

23. (a) $x(t) = \dfrac{100}{39}(\cos(19t/10) - \cos 2t);$ (b)

$x(t) = \dfrac{100}{41}(\cos 2t - \cos(21t/10)).$

25. $x(t) =$

$\begin{cases} t - \sin t + a\cos t + b\sin t, 0 \le t \le 1 \\ -t - \sin t + 2\sin(t-1) + 2 + a\cos t \\ \quad +b\sin t, 1 < t \le 2 \\ -\sin t + 2\sin(t-1) - \sin(t-2) + a\cos t \\ \quad +b\sin t, t > 1 \end{cases}.$

27. $x(t) = -\dfrac{5000}{89,733,775,561}(\cos(299,581t/50) -$ $\cos(20\sqrt{15}t)).$

Exercises 5.4

1. $Q(t) = \dfrac{55}{8}(1 - \cos 4t).$

3. $Q(t) = \dfrac{1}{64}(16t - \sin 16t).$

5. $Q(t) = \dfrac{8\sqrt{7}}{175}e^{-125t/2}\sin\left(\dfrac{25\sqrt{7}}{2}t\right);$ max.: 0.0225 at $t \approx 0.0147.$

7. $Q(t) = e^{-125t/2}\left(-\dfrac{12,185}{12,502,813}\cos\left(\dfrac{25\sqrt{7}}{2}t\right) + \dfrac{26,554,371}{312,570,325\sqrt{7}}\sin\left(\dfrac{25\sqrt{7}}{2}t\right)\right) + \dfrac{12,185}{12,502,813}\cos t + \dfrac{62,526,875}{12,502,813}\sin t;$ steady-state charge: $\lim_{t\to\infty}Q(t) = \dfrac{12,185}{12,502,813}\cos t + \dfrac{62,526,875}{12,502,813}\sin t;$ steady-state current: $-\dfrac{12,185}{12,502,813}\sin t + \dfrac{62,526,875}{12,502,813}\cos t.$

9. (a) $s(x) = \dfrac{1}{300}x^4 + \dfrac{1}{3}x^2 - \dfrac{1}{15}x^3;$

(b) $s(x) = \dfrac{1}{30}x^4 + \dfrac{10}{3}x^2 - \dfrac{2}{3}x^3;$

(c) $s(x) = \dfrac{1}{3}x^4 + \dfrac{100}{3}x^2 - \dfrac{20}{3}x^3.$

11. (a) $s(x) = \dfrac{1}{300}x^4 + \dfrac{10}{3}x - \dfrac{1}{15}x^3;$

(b) $s(x) = \dfrac{1}{30}x^4 + \dfrac{100}{3}x - \dfrac{2}{3}x^3;$

(c) $s(x) = \dfrac{1}{3}x^4 + \dfrac{1000}{3}x - \dfrac{20}{3}x^3;$ simple support leads to larger maximum displacement.

13. (a) $s(x) = \dfrac{1}{360}x^6 + \dfrac{10,000}{9}x - \dfrac{125}{9}x^3;$

(d) $s(x) = \dfrac{1}{360}x^6 + \dfrac{1250}{3}x - \dfrac{125}{18}x^3.$

15. $Q(t) = Q_0\cos(t/\sqrt{LC});$ $I(t) = -Q_0/\sqrt{LC}\sin(t/\sqrt{LC});$ maximum $Q = Q_0;$ maximum $I = Q_0/\sqrt{LC}.$

17. $E_0((L\omega - 1/(C\omega)^2 + R^2))^{-1}(R\sin\omega t$ $-(L\omega - 1/(C\omega))\cos\omega t).$

21. $s(x) = \dfrac{1}{360}x^6 - \dfrac{175}{9}x^3 + \dfrac{500}{3}x^2;$

(b) $s(x) = \dfrac{1}{360}x^6 - \dfrac{500}{9}x^3 + 1250x^2;$

(c) $s(x) = \dfrac{1}{360}x^6 - \dfrac{500}{9}x^3 + 750x^2;$

(d) $s(x) = \dfrac{1}{360}x^6 - \dfrac{175}{9}x^3 + \dfrac{500}{3}x^2.$

23. (a) $s(x) = 480,000\pi^{-4}\sin(\pi x/10)$; (b) no solution; (c) $s(x) = -80\pi^{-4}(-600\pi x - 300\pi^3 x - +\pi^3 x^3 - 6000\sin(\pi x/10))$ (d) $s(x) = 240\pi^{-4}(-100\pi x + \pi x^3 + 2000\sin(\pi x/10)).$

25. (a) $s(x) = \dfrac{1}{36,000}x^6 - \dfrac{5}{36}x^3 + \dfrac{100}{9}x$; (b) no solution; (c) $s(x) = \dfrac{1}{36,000}x^6 - \dfrac{5}{9}x^3 + 150x;$

(d) $s(x) = \dfrac{1}{36,000}x^6 - \dfrac{5}{72}x^3 + \dfrac{25}{6}x.$

Exercises 5.5

1. (a) $\theta(t) = \dfrac{1}{20}\cos 4t;$

(b) $\theta(t) = \dfrac{1}{20}\cos 4t + \dfrac{1}{4}\sin 4t;$

(c) $\theta(t) = \dfrac{1}{20}\cos 4t - \dfrac{1}{4}\sin 4t$; maximum displacement: (a) $1/20$; (b) and (c): $\sqrt{26}/20 \approx 0.255.$

3. (a) $\theta(t) = \dfrac{1}{60}e^{-t\sqrt{7}}(\sqrt{7}\sin 3t + 3\cos 3t);$

(b) $\theta(t) = \dfrac{1}{60}e^{-t\sqrt{7}}\Big((\sqrt{7}+20)\sin 3t + 3\cos 3t\Big)$; (c) $\theta(t) = \dfrac{1}{60}e^{-t\sqrt{7}}\Big((\sqrt{7}-20)\sin 3t + 3\cos 3t\Big).$

9. $T = 2\pi\sqrt{2/9.8} \approx 2.83\,\text{s}.$

11. $T = 2\pi\sqrt{8/32} \approx 3.14\,\text{s}.$

13. $2\pi\sqrt{L/9.8} = 1$ if $L = 0.248\,\text{m}.$

17. $b^2 - 4Lg > 0$, overdamped; $b^2 - 4Lg = 0,$ critically damped; $b^2 - 4Lg < 0,$ underdamped.

19. (c) yes; (d) no.

21. (a) $\theta(t) = 2\sin t$; (b) $\theta(t) = 2\cos t$; (c) $\theta(t) = -2\cos t$; (d) $\theta(t) = -\sin t$; (e) $\theta(t) = -2\sin t$; (f) $\theta(t) = \cos t - \sin t$; (g) $\theta(t) = -\cos t + \sin t.$

Chapter 5 Review Exercises

1. $x(t) = \dfrac{1}{3}\cos 8t$; maximum displacement $= 1/3; t = \pi/16, \pi/8.$

3. $x(t) = -\dfrac{1}{3}e^{-2t}\sin 3t$, $\lim_{t\to\infty} x(t) = 0$; quasiperiod $= 2\pi/3$; maximum displacement $= \left| x(\dfrac{1}{3}\tan^{-1}(3/2)) \right| \approx 0.144$; $t = \pi/3.$

5. $x(t) = \dfrac{1}{4}(1 - \cos 2t)$; maximum displacement is $1/2$ and occurs when $t = \pi/2.$

7. $x(t) = \dfrac{1}{3}\cos t - \dfrac{1}{3}\cos 2t$; beats; $\pm\dfrac{2}{3}\sin(t/2).$

9. $Q(t) = \dfrac{25}{327}[3 - 3e^{-10t}(\cos 3t - 10\sin 3t)];$ $I(t) = \dfrac{25}{3}e^{-10t}\sin 3t$; $\lim_{t\to\infty} Q(t) = 25/109;$ $\lim_{t\to\infty} I(t) = 0.$

11. $Q(t) = \dfrac{11}{500}(1 - \cos 100t)$; $I(t) = \dfrac{11}{5}\sin 100t$; $\lim_{t\to\infty} Q(t)$ and $\lim_{t\to\infty} I(t)$ do not exist.

13. $s(x) = \dfrac{250}{3}x^2 - \dfrac{125}{9}x^3 + \dfrac{1}{12}x^5 - \dfrac{1}{360}x^6.$

15. $s(x) = \dfrac{1250}{3}x^2 - \dfrac{250}{9}x^3 + \dfrac{1}{12}x^5 - \dfrac{1}{360}x^6.$

17. $\theta(t) = \cos 8t$; maximum displacement $= 1$; $t = \pi/16.$

19. $\theta(t) = e^{-8t} + 8te^{-8t}$; motion is not periodic.

21. $\theta(t) = 0.2168\cos 3.7t + 0.0471622\sin 3.7t.$

23. $y(t) = \begin{cases} 3 - t^2 - 3\cos t, 0 \le t \le 1 \\ (-3 + 2\cos 1 + 2\sin 1)\cos t - \\ \quad (2\cos 1 - 2\sin 1 + \sin 2)\sin t, t > 1 \end{cases}$

25. $x(t) = 25\cos\left(\dfrac{7\sqrt{5}}{25}t\right).$

27. $x(t) = A\sin \omega_n t + B\cos \omega_n t + F/k,$ $x(t) = (x_0 - F/k)\cos \omega_n t + F/k.$

Exercises 6.1

1. $\{x(t) = 6t + c_2, y(t) = \sin(t) + c_1\}$.
3. $\{x(t) = c_2, y(t) = c_1 e^{-2t}\}$.
5. $\{x_1(t) = -e^{-3t}, x_2(t) = e^t\}$.
7. $\left\{x(t) = -\dfrac{3}{2}c_1 e^{-7t} + c_2 e^{3t}, y(t) = \right.$

$\left. c_1 e^{-7t} + c_2 e^{3t}\right\}$.

9. $\{x(t) = c_1 \sin(t) + c_2 \cos(t), y(t) = -1/2 c_1 \cos(t) + 1/2 c_2 \sin(t) - 1/2 c_1 \sin(t) - 1/2 c_2 \cos(t)\}$.
11. $\{x(t) = -c_1 \cos(t) + c_2 \sin(t) + 1, y(t) = c_1 \sin(t) + c_2 \cos(t)\}$.
13. $\{x' = y, y' = -4x + 3y\}$.
15. $\{x' = y, y' = -16x + t \sin t$.
17. $\{y' = x, x' = z, z' = -6x - 3y - 3z + t\}$.
19. Write the equation as

$$x_1'' + \mu \left(\frac{1}{3}x_1^2\right) x_1' + x_1 = 0. \text{ Now}$$

let $x_1' = x_2$. Then, $x_2' = x_1''$

$$= -x_1 - \mu \left(\frac{1}{3}x_2^2\right) x_2.$$

21. $\left\{x = c_2 e^{-4t} + c_1 e^{3t} + \dfrac{1}{3}, y = \right.$

$\left. c_2 e^{-4t} - \dfrac{5}{2}c_1 e^{3t} + \dfrac{5}{3}\right\}$.

23. $\{x = 2/9 + 2e^{-3t}c_2 t - 1/4e^t + e^{-3t}c_1,$
$y = 1/3 + e^{-3t}c_2\}$.
25. $x = c_1 e^t + c_2 e^t t + c_3 \sin(t) + c_4 \cos(t),$
$y = -c_1 e^t - c_2 e^t t - c_3 \sin(t) - c_4 \cos(t) + c_3 \cos(t) - c_4 \sin(t)$.
27. $x = -3/10 \cos(t) - 1/10 \sin(t) - 2c_1 e^{-t} - 2c_2 e^{2t} - 2c_3 e^{-t}t + 3c_3 e^{-t}, y = 1/2 \sin(t) + 3c_1 e^{-t} + 3c_3 e^{-t}t - 4c_3 e^{-t}, z = 3/10 \sin(t) + 2/5 \cos(t) + c_1 e^{-t} + c_2 e^{2t} + c_3 e^{-t}t$.
29. $\{x = -20 + e^t(20 - 20t), y = -30 + 1/2e^t(60 - 40t)\}$.
33. $\{x = x, y = -x + 1/12t^4 + tc_1 + c_2\}$.
35. $\{x = -1/2t^2 + 1/8t^2 k + c_2, y = -1/6t^3 + 1/24kt^3 + c_2 t + 1/8t^2 k + c_1\}$.
37. $[\{x(t) = (t + c_1) e^{-t}, y(t) = 0\}$.

Exercises 6.2

1. $\begin{pmatrix} -3 & 5 \\ 3 & 2 \end{pmatrix}$.

3. $\begin{pmatrix} -5 & 21 \\ 1 & -3 \end{pmatrix}$.

5. $\mathbf{AB} = \begin{pmatrix} -1 & -11 \\ -2 & -28 \end{pmatrix}; \mathbf{BA} = \begin{pmatrix} -2 & -3 \\ -16 & -27 \end{pmatrix}$.

7. $\mathbf{AB} = \begin{pmatrix} 22 & -5 & -13 \\ 22 & -24 & -17 \\ -31 & -4 & 37 \end{pmatrix}$ and

$\mathbf{BA} = \begin{pmatrix} 1 & 9 & 16 & -35 \\ -25 & 20 & 20 & 5 \\ 0 & -9 & -8 & -28 \\ -27 & -11 & -22 & 22 \end{pmatrix}$.

9. 17.
11. -3.
13. 1.

15. $\begin{pmatrix} 1 & \frac{1}{2} \\ -1 & 0 \end{pmatrix}$.

17. $\begin{pmatrix} -1 & -3 & 5 & -2 \\ 0 & 0 & 1 & 0 \\ 1 & 1 & -3 & 1 \\ 1 & 2 & -4 & 2 \end{pmatrix}$.

19. $\lambda_1 = -5, \mathbf{v}_1 = \begin{pmatrix} 1 \\ 1 \end{pmatrix}; \lambda_2 = -4, \mathbf{v}_1 = \begin{pmatrix} 1 \\ 2 \end{pmatrix}$.

21. $\lambda_{1,2} = -2 \pm 2i, \mathbf{v}_{1,2} = \begin{pmatrix} 1 \pm 2i \\ 5 \end{pmatrix}$.

23. $\lambda_1 = -2$ has multiplicity one and $\lambda_{2,3} = 1$ has multiplicity two. $\mathbf{v}_1 = \begin{pmatrix} 1 \\ -1 \\ 1 \end{pmatrix};$

$\mathbf{v}_2 = \begin{pmatrix} -1 \\ 2 \\ 0 \end{pmatrix}$ and $\mathbf{v}_3 = \begin{pmatrix} -1 \\ 0 \\ 1 \end{pmatrix}$.

25. $\lambda_1 = -1$ and $\lambda_{2,3} = -3 \pm i, \mathbf{v}_1 = \begin{pmatrix} 0 \\ 0 \\ 1 \end{pmatrix},$

$\mathbf{v}_2 = \begin{pmatrix} -i \\ 1+i \\ 1 \end{pmatrix} = \begin{pmatrix} 0 \\ 1 \\ 1 \end{pmatrix} + \begin{pmatrix} -1 \\ 1 \\ 0 \end{pmatrix} i.$ Since

eigenvectors of complex conjugate

eigenvalues are also complex conjugates,

$$\mathbf{v}_3 = \begin{pmatrix} 0 \\ 1 \\ 1 \end{pmatrix} - \begin{pmatrix} -1 \\ 1 \\ 0 \end{pmatrix} i.$$

27. $\lambda_1 = -2$ and $\lambda_{2,3,4} = -1$; $\mathbf{v}_1 = \begin{pmatrix} 1 \\ -2 \\ 1 \\ 1 \end{pmatrix}$,

$$\mathbf{v}_2 = \begin{pmatrix} 1 \\ 0 \\ 0 \\ 1 \end{pmatrix}, \text{ and } \mathbf{v}_{3,4} = \begin{pmatrix} -1 \\ 1 \\ 0 \\ 0 \end{pmatrix}.$$

29. $\begin{pmatrix} -2e^{-2t} \\ -5e^{-5t} \end{pmatrix}; \begin{pmatrix} -\frac{1}{2}e^{-2t} \\ -\frac{1}{5}e^{-5t} \end{pmatrix}.$

31. $\begin{pmatrix} -\sin t \cos t - t\sin t \\ \cos t \quad t\cos t + \sin t \end{pmatrix};$

$\begin{pmatrix} \sin t \quad \cos t + t\sin t \\ -\cos t \ \sin t - t\cos t \end{pmatrix}.$

33. $\begin{pmatrix} 4e^{4t} \\ -3\sin 3t \\ 3\cos 3t \end{pmatrix}; \begin{pmatrix} \frac{1}{4}e^{4t} \\ \frac{1}{3}\sin 3t \\ -\frac{1}{3}\cos 3t \end{pmatrix}.$

Exercises 6.3

1. $\begin{pmatrix} x \\ y \end{pmatrix}' = \begin{pmatrix} 1 & 4 \\ 2 & -1 \end{pmatrix} \begin{pmatrix} x \\ y \end{pmatrix}.$

3. $\begin{pmatrix} x \\ y \end{pmatrix}' = \begin{pmatrix} 1 & 0 \\ 1 & 1 \end{pmatrix} \begin{pmatrix} x \\ y \end{pmatrix} + \begin{pmatrix} e^t \\ 0 \end{pmatrix}.$

5. $\begin{pmatrix} t(t^3 - 1)x' \\ t(t^3 - 1)y' \end{pmatrix} = \begin{pmatrix} 1 & -t \\ -2t^2 & t^3 + 1 \end{pmatrix} \begin{pmatrix} x \\ y \end{pmatrix}.$

7. $\begin{pmatrix} x \\ y \\ z \end{pmatrix}' = \begin{pmatrix} 1 & 1 & 1 \\ 2 & 0 & 1 \\ 0 & -2 & -1 \end{pmatrix} \begin{pmatrix} x \\ y \\ z \end{pmatrix} + \begin{pmatrix} \cos t \\ 0 \\ \sin t \end{pmatrix}.$

9. -5, linearly independent.
11. $2 - 4\cos^2 2t$, linearly independent.
13. $2e^{2t}$, linearly independent.

15-22. $\mathbf{X}(t)\Phi(t)\mathbf{C}$ where $\mathbf{C} = \begin{pmatrix} c_1 \\ c_2 \end{pmatrix}$ for 2×2 or

$\begin{pmatrix} c_1 \\ c_2 \\ c_3 \end{pmatrix}$ for 3×3.

23. $\{x(t) = e^{-t} + 2e^{4t}, y(t) = -e^{-t} + 3e^{4t}\}.$

25. $\{x(t) = \sin(t), y(t) = -\sin(t) + \cos(t)\}$

27. $x = 4t^{-1} - 1, y = 4 - t^2.$

29. $\{x(t) = t, y(t) = \sin(3t) + \cos(3t), z(t) = -\cos(3t) + \sin(3t)\}.$

33. $\Phi(t) = \begin{pmatrix} e^t & e^{-t} \\ 2e^t & e^{-t} \end{pmatrix}.$

Exercises 6.4

1. $\mathbf{X} = \begin{pmatrix} e^{4t} & e^{2t} \\ e^{4t} & 3e^{2t} \end{pmatrix} \mathbf{C}.$

3. $\mathbf{X} = \begin{pmatrix} 6e^{-4t} & 0 \\ e^{-4t} & e^{2t} \end{pmatrix} \mathbf{C}.$

5. $\mathbf{X} = \begin{pmatrix} e^{2t} & e^{2t}t \\ e^{2t} & e^{2t}t + e^{2t} \end{pmatrix} \mathbf{C}.$

7. $\mathbf{X} = \begin{pmatrix} e^t \sin(2t) & e^t \cos(2t) \\ -1/2e^t \sin(2t) + 1/2e^t \cos(2t) & -1/2e^t \cos(2t) - 1/2e^t \sin(2t) \end{pmatrix} \mathbf{C}.$

9. $\mathbf{X} = \begin{pmatrix} e^{3t} & 0 & 0 \\ 0 & e^{2t} & e^{2t}t \\ 0 & 0 & e^{2t} \end{pmatrix} \mathbf{C}.$

11. $\mathbf{X} = \begin{pmatrix} \sin(t) & \cos(t) & 0 \\ -\cos(t) & \sin(t) & 0 \\ 0 & 0 & e^{2t} \end{pmatrix} \mathbf{C}.$

13. $\mathbf{X} = \begin{pmatrix} e^{-4t} & e^{15t} \\ 1/2e^{-4t} & -7/5e^{15t} \end{pmatrix} \mathbf{C}.$

15. $\mathbf{X} = \begin{pmatrix} e^t & e^{5t} \\ 5e^t & e^{5t} \end{pmatrix} \mathbf{C}.$

17. $\mathbf{X} = \begin{pmatrix} 0 & 3e^{7t} \\ e^{-8t} & e^{7t} \end{pmatrix} \mathbf{C}.$

19. $\mathbf{X} = \begin{pmatrix} e^{-4t} & e^{-11t} \\ 1/3e^{-4t} & -2e^{-11t} \end{pmatrix} \mathbf{C}.$

21. $\mathbf{X} = \begin{pmatrix} e^{-4t} & e^{-12t} \\ -1/2e^{-4t} & 3/2e^{-12t} \end{pmatrix} \mathbf{C}.$

23. $X = \begin{pmatrix} e^{-8t} & e^{-8t}t \\ -e^{-8t} & -e^{-8t}t + 1/2e^{-8t} \end{pmatrix} C.$

25. $X = \begin{pmatrix} \sin(4t) & \cos(4t) \\ 1/2\cos(4t) & -1/2\sin(4t) \end{pmatrix} C.$

27. $X = \begin{pmatrix} e^{4t} & e^{-t} & 0 \\ 0 & 0 & e^{-2t} \\ 0 & -5e^{-t} & 0 \end{pmatrix} C.$

29. $X = \begin{pmatrix} e^{-t} & e^{3t} & e^{-t}t \\ 0 & -4/7e^{3t} & -e^{-t} \\ -2e^{-t} & -6/7e^{3t} & -2e^{-t}t \end{pmatrix} C.$

31. $x = \begin{pmatrix} e^t & e^t\cos(2t) & e^t\sin(2t) \\ 0 & \frac{6}{13}e^t\cos(2t) - \frac{4}{13}e^t\sin(2t) & \frac{6}{13}e^t\sin(2t) + \frac{4}{13}e^t\cos(2t) \\ 0 & -\frac{4}{13}e^t\cos(2t) - \frac{6}{13}e^t\sin(2t) & \frac{6}{13}e^t\cos(2t) - \frac{4}{13}e^t\sin(2t) \end{pmatrix} C.$

33. $X = \begin{pmatrix} -2e^t & -2e^t & 1 & 4 \\ 0 & -3e^t & 1 & 1 \\ 0 & 3e^t & 0 & 2 \\ 3e^t & 0 & 1 & 0 \end{pmatrix} \begin{pmatrix} c_1 \\ c_2 \\ c_3 \\ c_4 \end{pmatrix}.$

35. $x = 4e^t(-1 + e^t), y = 4e^{2t}.$

37. $x = 8e^{4t}, y = 16e^{4t}t.$

39. $x = -8\sin 4t, y = 8\cos 4t.$

41. $x = -e^{2t}(-1 + 2e^t), y = 2e^t(1 - e^t + e^{2t}),$
$z = 2e^{2t}(-1 + e^t).$

43. $x = 9e^{-t}(-1 + e^t), y = e^{-t}(9 - 15e^t + 7e^{2t}),$
$z = -\dfrac{1}{2}e^{-t}(9 - 18e^t + 7e^{2t}),$
$w = -7(-1 + e^t).$

Exercises 6.5

1. $X = \begin{pmatrix} 2e^{-t} & 1 \\ e^{-t} & 1 \end{pmatrix}\begin{pmatrix} c_1 \\ c_2 \end{pmatrix} + \begin{pmatrix} -e^t \\ e^t \end{pmatrix}.$

3. $x = -c_1e^{-t} - 2c_2e^t - t + 1,$
$y = c_1e^{-t} + 3c_2e^t - 2t.$

5. $X = \begin{pmatrix} e^{-2t} & 3e^{-t} \\ e^{-2t} & 2e^{-t} \end{pmatrix}\begin{pmatrix} c_1 \\ c_2 \end{pmatrix} + \begin{pmatrix} -\cos t \\ \sin t - \cos t \end{pmatrix}.$

7. $X = \begin{pmatrix} -e^{-2t} & -2e^{-t} \\ 2e^{-2t} & 3e^{-t} \end{pmatrix}\begin{pmatrix} c_1 \\ c_2 \end{pmatrix} + \begin{pmatrix} -te^{-2t} \\ 2te^{-2t} - e^{-2t} \end{pmatrix}.$

9. $X = \begin{pmatrix} 3e^{-2t} & e^{-t} \\ 2e^{-2t} & e^{-t} \end{pmatrix}\begin{pmatrix} c_1 \\ c_2 \end{pmatrix} + \begin{pmatrix} 0 \\ -2e^{-t} \end{pmatrix}.$

11. $X =$
$\begin{pmatrix} 2te^{-t} + e^{-t} & -2te^{-t} \\ 2te^{-2t} & -2te^{-t} + e^{-t} \end{pmatrix}\begin{pmatrix} c_1 \\ c_2 \end{pmatrix} + \begin{pmatrix} e^{-t} \\ 0 \end{pmatrix}.$

13. (a)

$x = 2c_1e^{-2t} + 3c_2e^{-t},$
$y = -c_1e^{-2t} + c_2e^{-t}.$

(b)

$x = 2c_1e^{-2t} + 3c_2e^{-t} + 2e^t,$
$y = -c_1e^{-2t} + c_2e^{-t}.$

(c)

$x = 2c_1e^{-2t} + 3c_2e^{-t} + e^{-t},$
$y = -c_1e^{-2t} + c_2e^{-t}.$

15. $X = \begin{pmatrix} -e^{-t} & -te^{-t} - \frac{1}{3}e^{-t} \\ e^{-t} & te^{-t} \end{pmatrix}\begin{pmatrix} c_1 \\ c_2 \end{pmatrix},$

$X_p = \begin{pmatrix} te^{-t} \\ -te^{-t} \end{pmatrix}$ and

$X = \begin{pmatrix} -e^{-t} & -te^{-t} - \frac{1}{3}e^{-t} \\ e^{-t} & te^{-t} \end{pmatrix}\begin{pmatrix} c_1 \\ c_2 \end{pmatrix} + \begin{pmatrix} te^{-t} \\ -te^{-t} \end{pmatrix},$

$X_p = \begin{pmatrix} \frac{3}{2}t^2e^{-t} + 2te^{-t} \\ -\frac{3}{2}t^2e^{-t} - te^{-t} \end{pmatrix}$ and

$X = \begin{pmatrix} -e^{-t} & -te^{-t} - \frac{1}{3}e^{-t} \\ e^{-t} & te^{-t} \end{pmatrix}\begin{pmatrix} c_1 \\ c_2 \end{pmatrix} +$

$\begin{pmatrix} \frac{3}{2}t^2e^{-t} + 2te^{-t} \\ -\frac{3}{2}t^2e^{-t} - te^{-t} \end{pmatrix}.$

17. $x(t) = \cos(t)c_2 - \sin(t)c_1 + t\cos(t) - \ln(\cos(t))\sin(t), y(t) = c_2\sin(t) + c_1\cos(t) + t\sin(t) + \ln(\cos(t))\cos(t).$

19. $x(t) = c_2 \sin(t) + c_1 \cos(t) -$
$\qquad 2\sin(t)\ln\left(-\dfrac{-1+\cos(t)}{\sin(t)}\right),$
$\qquad y(t) = c_2\cos(t) -$
$\qquad 2\cos(t)\ln\left(-\dfrac{-1+\cos(t)}{\sin(t)}\right) - c_1\sin(t) - 2$

21. $x(t) = -\dfrac{83}{16} + 11/4t - t^2 + c_1 e^{-4t} + c_2 e^{-t} + c_3 e^{2t},$
$\qquad y(t) = -\dfrac{125}{32} + 3/2 c_1 e^{-4t} + 3/4 c_2 e^{-t} + \dfrac{9}{8} c_3 e^{2t}$
$\qquad + \dfrac{15}{8} t - t^2, z(t) = \dfrac{11}{16} - 3/4t - c_1 e^{-4t}$
$\qquad - 1/4 c_2 e^{-t} + 1/2 c_3 e^{2t}.$

23.

$$\mathbf{X} = \begin{pmatrix} -e^{-t} & -e^{-t}\cos 2t \\ 0 & e^{-t}(\cos 2t + 2\sin 2t) \\ e^{-t} & 2e^{-t}\cos 2t \end{pmatrix}$$

$$\begin{pmatrix} -e^{-t}\sin 2t \\ e^{-t}(-2\cos 2t + \sin 2t) \\ 2e^{-t}\sin 2t \end{pmatrix}\begin{pmatrix} c_1 \\ c_2 \\ c_3 \end{pmatrix},$$

$\mathbf{X} = \mathbf{X}_h + \mathbf{X}_p$

$$= \begin{pmatrix} -e^{-t} & -e^{-t}\cos 2t \\ 0 & e^{-t}(\cos 2t + 2\sin 2t) \\ e^{-t} & 2e^{-t}\cos 2t \end{pmatrix}$$

$$\begin{pmatrix} -e^{-t}\sin 2t \\ e^{-t}(-2\cos 2t + \sin 2t) \\ 2e^{-t}\sin 2t \end{pmatrix}\begin{pmatrix} c_1 \\ c_2 \\ c_3 \end{pmatrix} + \begin{pmatrix} e^t \\ \cos 2t \\ \sin 2t. \end{pmatrix}$$

25. $x(t) = -1/2\cos(2t)c_2 + 1/2\sin(2t)c_1$
$\qquad + 1/8\cos(2t) + 1/4\sin(2t)t + c_3,$
$\qquad y(t) = 1/2 c_2\cos(2t) - 1/2 c_2\sin(2t)$
$\qquad - 1/8\cos(2t) - 1/4\sin(2t)t - 2c_3,$
$\qquad z(t) = \sin(2t)c_2 + \cos(2t)c_1 + 1/2t$
$\qquad \cos(2t) + 3c_3.$

27. $x(t) = -1 + t + e^{-t}, y(t) = 0.$

29. $x(t) = -2/3e^{-t} + 1/6e^{5t} - 1/2e^t,$
$\qquad y(t) = 2/3e^{-t} + 1/12e^{5t} + 1/4e^t.$

31. $x(t) = \dfrac{20}{13} + e^{2t}\left(1/39\sin(3t) - \dfrac{7}{13}\cos(3t)\right),$
$\qquad y(t) = -\dfrac{30}{13} - 1/2e^{2t}\left(-\dfrac{62}{39}\sin(3t)\right.$
$\qquad \left. - \dfrac{8}{13}\cos(3t)\right).$

33. $x(t) = -1/4\sin(2t) + 1/2\cos(2t)$
$\qquad -1/2\sin(2t)\ln\left(\dfrac{1+\sin(2t)}{\cos(2t)}\right)$
$\qquad -1/4\cos(2t)\ln\left(\dfrac{1+\sin(2t)}{\cos(2t)}\right)$
$\qquad +5/4\cos(2t)\ln(\cos(2t)) + 5/2$
$\qquad \sin(2t)t + 1/2, y(t) = -1/4\sin(2t)$
$\qquad -1/4\cos(2t)\ln\left(\dfrac{1+\sin(2t)}{\cos(2t)}\right)$
$\qquad -1/2\ln(\cos(2t))\sin(2t)$
$\qquad +1/4\cos(2t)\ln(\cos(2t)) + \cos(2t)t$
$\qquad +1/2\sin(2t)t.$

35.
$$x = \frac{1}{4}(5\ln(\cos(t) - \sin(t))\cos(2t)$$
$$- 5\ln(\cos(t) + \sin(t))\cos(2t)$$
$$+ 4\cos(2t) - 3\sin(2t))$$

and
$$y = \frac{1}{2}\bigl(-\cos^2(2t) + 2\ln(\cos(t) - \sin(t))$$
$$\cos(2t) - 2\ln(\cos(t) + \sin(t))\cos(2t)$$
$$+ \cos(2t) - \sin^2(2t) - \ln(\cos(t)$$
$$- \sin(t))\sin(2t) + \ln(\cos(t)$$
$$+ \sin(t))\sin(2t) - 2\sin(2t)\bigr).$$

37. $\mathbf{X} = \begin{pmatrix} 1 & \sqrt{t} \\ t & t \end{pmatrix}\begin{pmatrix} c_1 \\ c_2 \end{pmatrix} + \begin{pmatrix} t \\ 1 \end{pmatrix}.$

39. $\mathbf{X} = \begin{pmatrix} 1 & f(t) \\ f(t) & 1 \end{pmatrix}\begin{pmatrix} c_1 \\ c_2 \end{pmatrix} + \begin{pmatrix} f(t) \\ f(t) \end{pmatrix}.$

41. (a) $x(t) = -\dfrac{199}{343}e^{-t} + \dfrac{453}{1372}e^{-8t} + 1/4$
$\qquad -3/14e^{-t}t^2 + \dfrac{3}{49}te^{-t}, y(t) =$
$\qquad \dfrac{391}{343}e^{-t} + \dfrac{151}{1372}e^{-8t} - 1/4 + 1/49te^{-t} + 3/7e^{-t}t^2$

$\qquad$ (b) $x(t) = \dfrac{260}{81}e^t - \dfrac{179}{81}e^{-2t} + \dfrac{10}{9}te^t$
$\qquad -5/3t^2e^t - \dfrac{20}{27}e^{-2t}t - \dfrac{10}{9}t^2e^{-2t} - \dfrac{7}{9}e^{-2t}t^3,$
$\qquad y(t) = -\dfrac{173}{81}e^t + \dfrac{173}{81}e^{-2t} - \dfrac{10}{9}te^t$
$\qquad +7/6t^2e^t + \dfrac{7}{9}t^2e^{-2t} + \dfrac{14}{27}e^{-2t}t + \dfrac{7}{9}e^{-2t}t^3;$

(c) $x(t) = \dfrac{279}{289}\cos(4t) - \dfrac{3479}{4624}\sin(4t) +$

$\dfrac{10}{289}e^{-t} + 1/2\sin(4t)\,t - 1/4\cos(4t)\,t + \dfrac{5}{17}te^{-t}$,

$y(t) = \dfrac{1321}{2312}\sin(4t) + \dfrac{300}{289}\cos(4t) -$

$\dfrac{11}{289}e^{-t} + \dfrac{3}{17}te^{-t} - 1/2\cos(4t)\,t$.

2.

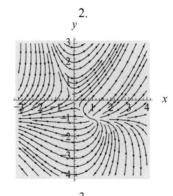

Exercises 6.6

1. $\lambda_{1,2} = 1, -11$; Saddle point, unstable.
3. $\lambda_{1,2} = 2$; Deficient node, unstable.
5. $\lambda_{1,2} = -8, -12$; Improper node, asymptotically stable.
7. $\lambda_{1,2} = -1$; Star node, asymptotically stable.
9. $\lambda_{1,2} = 2 \pm 5i\sqrt{2}$; Spiral point, unstable.
11. $\lambda_{1,2} = \pm i\sqrt{41}$; Center, stable.
17. $x'' - x' - 2x = 0$, $\mathbf{X}' = \begin{pmatrix} 0 & 1 \\ 2 & 1 \end{pmatrix}\mathbf{X}$, Saddle point, unstable.
19. $x'' - 6x' + 9x = 0$, $\mathbf{X}' = \begin{pmatrix} 0 & 1 \\ -9 & 6 \end{pmatrix}\mathbf{X}$, Deficient node, unstable.

3.

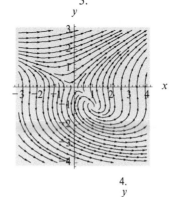

4.

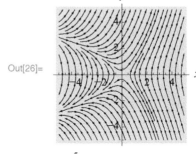

Exercises 6.7

1. $(0,0)$, saddle point, unstable; $(1,0)$ inconclusive-center or spiral point.
3. $(0,0)$, saddle point, unstable; $(1,-1)$, spiral point, unstable.
5. $(0,0)$, inconclusive-center or spiral point.

1.

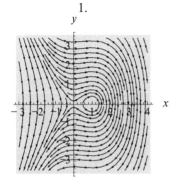

5.

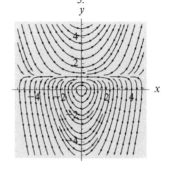

6.

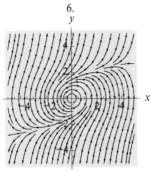

7. $(0,0)$, node or spiral point, unstable.

9. $(0,0)$, improper node, asymptotically stable; $(0,4)$, saddle point, unstable; $(2,0)$, saddle point, unstable; $(3,1)$, spiral point, unstable.

7.

8.

9.

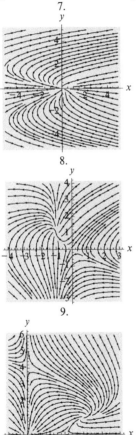

11. $(-4,4)$, saddle point, unstable; $(1,-1)$, saddle point unstable.

13. $(2,-1)$: $\lambda_{1,2} = -6, 1$, saddle; $(8,-1)$: $\lambda_{1,2} = 6, 1$, unstable node.

15. $(0,0)$, $\lambda_{1,2} = \pm\sqrt{2}$, saddle; $(1,1)$, $\lambda_{1,2} = -1 \pm i$, stable spiral.

10.

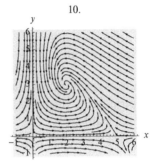

11.

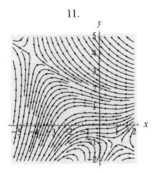

12.

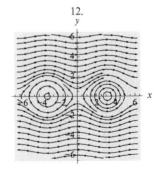

13.

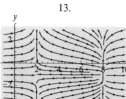

16.

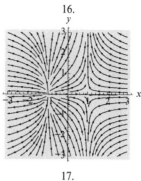

14.

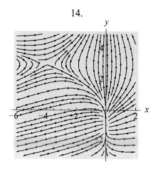

17.

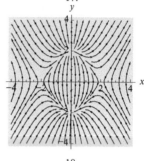

15.

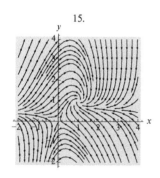

18.

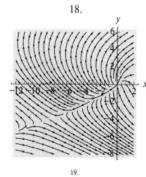

19.

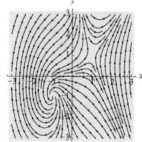

17. $(0, 2)$, $\lambda_{1,2} = -6, 1$, unstable node; $(0, -2)$, $\lambda_{1,2} = -2, -4$, stable node; $(2, 0)$, $\lambda_{1,2} = \pm 2\sqrt{2}$, saddle; $(-2, 0)$, $\lambda_{1,2} = \pm 2\sqrt{2}$, saddle.

19. $(0, 0)$, $\lambda_{1,2} = \pm 2$, saddle; $(-1, -1)$, $\lambda_{1,2} = 1 \pm i\sqrt{3}$, unstable spiral.

20.

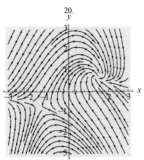

21. $(0,0)$, $\lambda_{1,2} = a_1, -a_2$, saddle; $(a_1/b_1, 0)$,
$\lambda_{1,2} = -a_1, a_1c_2/b_1 - a_2$, stable node;
$(-a_2/c_2, (a_1c_2 + a_2b_1)/(c_1c_2))$,

$$\lambda_{1,2} = \frac{a_2b_1}{c_2} \pm \sqrt{\left(\frac{a_2b_1}{c_2}\right)^2 - \frac{4a_2(a_1c_2 + a_2b_1)}{b_1}},$$

unstable node if
$\left(\dfrac{a_2b_1}{c_2}\right)^2 - \dfrac{4a_2(a_1c_2 + a_2b_1)}{b_1} \geq 0$; unstable

spiral if $\left(\dfrac{a_2b_1}{c_2}\right)^2 - \dfrac{4a_2(a_1c_2 + a_2b_1)}{b_1} < 0$.

25. $H(x,y) = \int (2y + x^{-1}y^{-2})\, dy + g(x)$;
$g(x) = -\int -2xx$ so
$H(x,y) = x^2 - (xy)^{-1} + y^2$.

25.

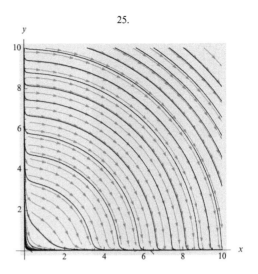

27. $H(x,y) = \int (x \cos xy - \sin 2y)dy + g(x)$;
$g(x) = -\int \sin 2x\, dx = \dfrac{1}{2}\cos 2x$ so
$H(x,y) = \dfrac{1}{2}\cos 2y + \dfrac{1}{2}\cos 2x + \sin xy$.

24.

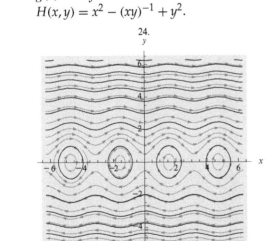

26.

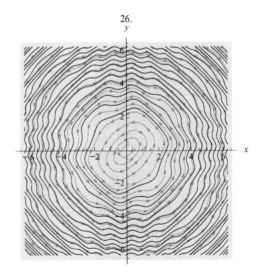

27.

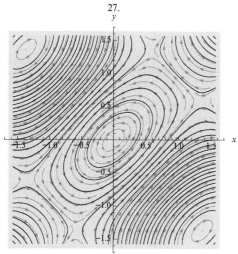

29. $x(t) = c_1 e^t$, $y(t) = -\dfrac{1}{3} c_1{}^2 e^{2t} + c_2 e^t$.

29.

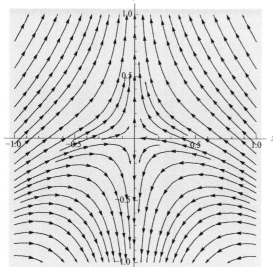

31. 1. Take the appropriate partial derivatives.

    2. Substitute $x = y$ and $z = \dfrac{y^2}{b}$ into the middle equation.

    3. The characteristic equation is
$(-b - \lambda)\left(\lambda^2 + (a - k_2 + 1)\lambda\right.$
$\left. - a(k_2 - 1 + c + k_1)\right) = 0$, which does not depend on k_3.

4. Since the eigenvalues do not depend on k_3, we can simplify the problem with $k_3 = 0$. Substitute the values of k_1, k_2 and $k_3 = 0$ into the controlled system.

5. The equilibrium points are
$\left(3\sqrt{7}, 3\sqrt{7}, 21\right)$ and $\left(-3\sqrt{7}, -3\sqrt{7}, 21\right)$
with corresponding eigenvalues
$\lambda_1 \approx -18.43$, $\lambda_{2,3} \approx 4.2 \pm 14.88i$. The positive real components of $\lambda_{2,3}$ cause the equilibria to be unstable.

33.

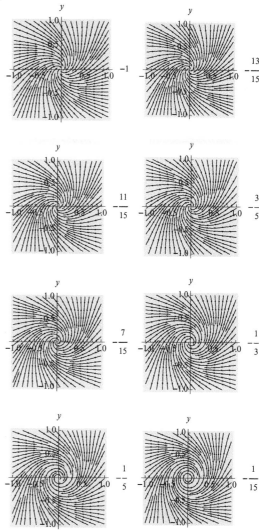

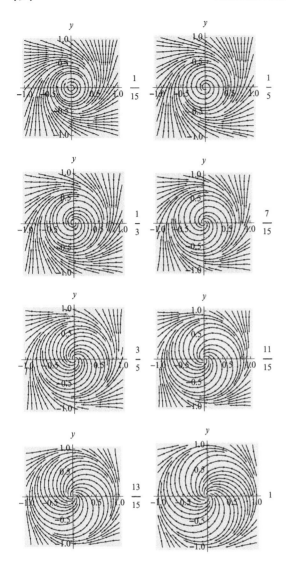

13. $x' = y, y' = -2x - 3y, x(0) = 0, y(0) = -3$;
Euler's method yields $x(1) \approx -0.723913$,
$y(1) \approx 0.40179$; Exact solution is
$x(t) = 3e^{-2t} - 3e^{-t}$ so
$x(1) = 3e^{-2}(1 - e) \approx -0.697632$.

15. $x' = y, y' = -9x, x(0) = 0, y(0) = 3$; Euler's
method yields $x(1) \approx 0.346313$,
$y(1) \approx -4.49743$; Exact solution is
$x(t) = \sin 3t$ so $x(1) = \sin 3 \approx 0.14112$.

17. $x' = y, y' = -t^{-2}(16x + ty), x(1) = 0$,
$y(1) = 4$; Euler's method yields
$x(2) \approx 0.354942, y(2) \approx -2.90834$ / Exact
solution is $x(t) = \sin(4 \ln t)$ so
$x(2) = \sin(4 \ln 2) \approx 0.360687$.

19. (See 13) Runge-Kutta yields
$x(1) \approx -0.697621, y(1) \approx 0.291602$.

21. (See 15) Runge-Kutta yields
$x(1) \approx 0.141307, y(1) \approx -2.96975$.

23. (See 17) Runge-Kutta yields
$x(1) \approx 0.360845, y(1) \approx -1.86541$.

25. $x(t) = \dfrac{1}{3}e^{-2t}(e^{3t} - 1), y(t) = \dfrac{1}{3}e^{-2t}(e^{3t} + 2)$.

27. $x(t) = 2e^t(t + 1), y(t) = -2te^t$.

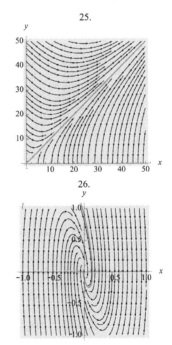

25.

26.

Exercises 6.8

1. $x(1) \approx 8.64479, y(1) \approx 8.29263$.
3. $x(1) \approx -7.2362, y(1) \approx -1.5998$.
5. $x(1) \approx -0.113115, y(1) \approx -1.06576$.
7. $x(1) \approx -326.204, y(1) \approx 608$.
9. $x(1) \approx 0.164251, y(1) \approx -0.504587$.
11. $x(1) \approx 1.52319, y(1) \approx 0.419975$.

27.
y

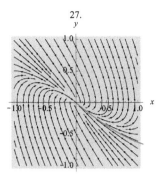

Chapter 6 Review Exercises

1. $\lambda_1 = 11$, $\lambda_2 = -4$, $\mathbf{v}_1 = \begin{pmatrix} 1 \\ 2 \end{pmatrix}$, $\mathbf{v}_2 = \begin{pmatrix} -2 \\ 1 \end{pmatrix}$.

3. $\lambda_{1,2} = -4$, $\mathbf{v} = \begin{pmatrix} 1 \\ 1 \end{pmatrix}$.

5. $\lambda_{1,2} = \pm 3i$, $\lambda_3 = -1$, $\mathbf{v}_{1,2} = \begin{pmatrix} -1 \mp \frac{1}{2}i \\ -1/2 \\ 1 \end{pmatrix}$,
$\mathbf{v}_3 = \begin{pmatrix} -2 \\ -1 \\ 1 \end{pmatrix}$.

7. $\lambda_{1,2,3} = 0$, $\mathbf{v} = \begin{pmatrix} 0 \\ 1 \\ 1 \end{pmatrix}$.

9. $x = \frac{1}{5}e^t \left((4 + e^{5t}) c_1 - 4 (-1 + e^{5t}) c_2 \right)$,
$y = \frac{1}{5}e^t \left(-e^{5t}c_1 + c_1 + 4e^{5t}c_2 + c_2 \right)$.

11. $x = -e^{-2t} + 2e^t$, $y = 3e^{-2t} - 3e^t$, $x(0) = 1$,
$y(0) = 0$.

13. $x = 2 \sin 4t$, $y = \cos 4t + \sin 4t$.

15. $x = \frac{17}{4}e^{-2t} \sin 4t$,
$y = \frac{1}{4}e^{-2t}(4 \cos 4t - \sin 4t)$.

17. $x = e^{-t} (2tc_1 + c_1 + 2tc_2)$,
$y = e^{-t} (c_2 - 2t (c_1 + c_2))$.

19. $x = e^{-t} ((5 - 4e^t) c_1$
$+ 2 (-1 + e^t) (3c_2 - 2c_3))$,
$y = (-2 + 2e^{-2t}) c_1 + (3 - 2e^{-2t}) c_2 +$
$2 (-1 + e^{-2t}) c_3$, $z = e^{-2t} \left(-3c_2 (-1 + e^t)^2 \right.$
$+ (3 - 5e^t + 2e^{2t}) c_1 + (3 - 4e^t + 2e^{2t}) c_3 \left. \right)$.

21. $x = \cos 3t + \dfrac{11}{3} \sin 3t$,
$y = \dfrac{1}{3}(-5 + 5 \cos 3t + \sin 3t)$,
$z = -\cos 3t + 5 \sin 3t$.

23. $x(t) = -e^{7t} + c_2 e^{-t}$,
$y(t) = \dfrac{7}{9}e^{7t} + c_2 e^{-t} + c_1 e^{-2t}$.

25. $x(t) = c_2 e^{-t} + c_1 e^{5t} + 1$,
$y(t) = -c_2 e^{-t} + 1/2 c_1 e^{5t} - 1$.

27. $x(t) = c_2 \sin(t) + c_1 \cos(t) - \cos(t)$
$-t \sin(t)$, $y(t) = -c_2 \cos(t)$
$+c_1 \sin(t) + t \cos(t)$
$- \sin(t)$.

29. $x(t) = 15/2t^2 \ln(t) - \dfrac{45}{4}t^2 - t \ln$
$(t) + t + c_1 t + c_2$, $y(t) = t \ln(t)$
$- t + 1/9c_1 + 5t^2 \ln(t) - 15/2t^2$
$+ 2/3c_1 t + 2/3c_2$.

31. $x = \dfrac{1}{210} \left(140t^{3/2} + 16t^{7/2} \right)$,
$y = \dfrac{1}{15} \left(-10t^{3/2} - 12t^{5/2} \right)$,
$z = \dfrac{1}{210} \left(140t^{3/2} + 56t^{5/2} + 16t^{3/2} \right)$.

32.
y

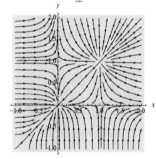

33.
y

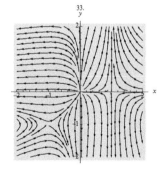

Exercises 7.1

1. (a) $Q(t) = \dfrac{3\sqrt{5}}{5}e^{-5t/3}\sin\left(\dfrac{\sqrt{5}}{3}t\right)$,

$I(t) = e^{-5t/3}\left(\cos\left(\dfrac{\sqrt{5}}{3}t\right) - \sqrt{5}\sin\left(\dfrac{\sqrt{5}}{3}t\right)\right)$;

(b) $Q(t) = \dfrac{1}{3}e^{-t} + \dfrac{1}{15}e^{-5t/3}$

$\left(-5\cos\left(\dfrac{\sqrt{5}}{3}t\right) + 7\sqrt{5}\sin\left(\dfrac{\sqrt{5}}{3}t\right)\right)$,

$I(t) = -\dfrac{1}{3}e^{-t} -$

$\dfrac{2}{3}e^{-5t/3}\left(2\cos\left(\dfrac{\sqrt{5}}{3}t\right) - \sqrt{5}\sin\left(\dfrac{\sqrt{5}}{3}t\right)\right)$.

3. (a) $Q(t) = 10^{-6}e^{-t}\cos(\sqrt{2}t)$

(b) $Q(t) = \dfrac{2}{9}e^{-t/2} + \dfrac{4\sqrt{2}}{9}e^{-t}\sin(\sqrt{2}t)$

$- \dfrac{1,999,991}{9,000,000}e^{-t}\cos(\sqrt{2}t)$,

$I_2(t) = \dfrac{8}{9}e^{-t/2} - \dfrac{1,999,991}{4,500,000\sqrt{2}}e^{-t}\sin(\sqrt{2}t)$

$- \dfrac{8}{9}e^{-t}\cos(\sqrt{2}t)$.

5. $Q(t) = 10^{-6}e^{-2t}(1+t)$, $I_2(t) = 10^{-6}te^{-2t}$ so
$I(t) = -10^{-6}e^{-2t}(1+2t)$ and then
$I_1(t) = -10^{-6}e^{-2t}(1+t)$.

7. (a) $Q(t) = 10^{-6}e^{-t} - \dfrac{1}{2} \times 10^{-6}t^2e^{-t}$,

$I_2(t) = 10^{-6}te^{-t}$, $I_3(t) = \dfrac{1}{2} \times 10^{-6}t^2e^{-t}$;

(b) $Q(t) = 10^{-6}e^{-t} + te^{-t} - \dfrac{1}{2} \times 10^{-6}t^2e^{-t}$,

$I_2(t) = 10^{-6}te^{-t}$, $I_3(t) = -te^{-t}$
$+ \dfrac{1}{2} \times 10^{-6}t^2e^{-t}$.

9. (a) $Q(t) = 90 - te^{-t} - 90e^{-t}$, $I_2(t) = e^{-t}$,
$I_3(t) = -90 + te^{-t} + 90e^{-t}$;
(b) $Q(t) = -te^{-t} + 45e^{-t} + 45\sin t - 45\cos t$,
$I_2(t) = e^{-t}$, $I_3(t) = te^{-t} - 45e^{-t}$
$- 45\sin t + 45\cos t$.

11. (a) $Q(t) = 90 - \dfrac{4}{3}e^{-t/4} - \dfrac{266}{3}e^{-t}$,

$I_2(t) = e^{-t/4}$, $I_3(t) = -90 + \dfrac{4}{3}e^{-t/4} + \dfrac{266}{3}e^{-t}$;

(b) $Q(t) = -45\cos t + 45\sin t$
$- \dfrac{4}{3}e^{-t/4} + \dfrac{139}{3}e^{-t}$, $I_2(t) = e^{-t/4}$,

$I_3(t) = 45\cos t - 45\sin t + \dfrac{4}{3}e^{-t/4} - \dfrac{139}{3}e^{-t}$.

13. $x' = y$, $y' = -9x$, $\lambda = \pm 3i$; undamped.
15. $x' = y$, $y' = -9x - 10y$, $\lambda_1 = -1$, $\lambda_2 = -9$; overdamped.
17. $x' = y$, $y' = -50x - 10y$, $\lambda_{1,2} = -5 \pm 5i$; underdamped.
19. $x' = y$, $y' = -25x - 10y$, $\lambda_{1,2} = -5$; critically damped.
21. (13) $x(t) = \cos 3t$, $y(t) = -3\sin 3t$; (15)

$x(t) = \dfrac{9}{8}e^{-t} - \dfrac{1}{8}e^{-9t}$, $y(t) = -\dfrac{9}{8}e^{-t} + \dfrac{9}{8}e^{-9t}$.

23. $\lambda = (2m)^{-1}\left(-c \pm \sqrt{c^2 - 4km}\right)$, overdamped if $c^2 > 4km$; critically damped if $c^2 = 4km$; underdamped if $c^2 < 4km$.
25. $\lim_{t\to\infty}Q(t)$ does not exist; $\lim_{t\to\infty}I(t)$ does not exist.
27. $\lim_{t\to\infty}Q(t)$ does not exist; $\lim_{t\to\infty}I_2(t)$ does not exist.
29. $\lim_{t\to\infty}Q(t) \approx 91$; $\lim_{t\to\infty}I_2(t) = 0$; $\lim_{t\to\infty}I_3(t) \approx -43.5$.

31. $x(t) = \dfrac{1}{2}e^{-3t/2}((3e^t - 1)x_0 + 2(e^t - 1)y_0)$,

$y(t) = -\dfrac{1}{4}e^{-3t/2}(3(e^t - 1)x_0 + 2(e^t - 3)y_0)$;
stable node.

Exercises 7.2

1. $x(t) = \dfrac{3}{2}(1 - 3e^{-t})$, $y(t) = \dfrac{3}{2}(1 + 3e^{-t})$,
$\lim_{t\to\infty}(x(t), y(t)) = (3/2, 3/2)$.

3. $x(t) = \dfrac{8}{3}(1 - e^{-3t/2})$, $y(t) = \dfrac{4}{3}(1 + 2e^{-3t/2})$,
$\lim_{t\to\infty}(x(t), y(t)) = (8/3, 4/3)$.

5. $x(t) = 1 + 3e^{-15t/4}$, $y(t) = 4 - 3e^{-15t/4}$,
$\lim_{t\to\infty}(x(t), y(t)) = (1, 4)$.

7. $x(t) = 10(1 - e^{-t/5})$,
$y(t) = 10(1 - e^{-t/5}) - 2te^{-t/5}$,
$\lim_{t\to\infty}(x(t), y(t)) = (10, 10)$, $x(t)$.

9. $x(t) = 4e^{-t/5}$, $y(t) = 4e^{-t/5} + \frac{4}{5}te^{-t/5}$,

$\lim_{t\to\infty}(x(t), y(t)) = (0, 0)$, $x(t)$.

11. (a) $x(t) = 300(1 - e^{-t/20})$,
$y(t) = 150(1 + e^{-t/10} - 2e^{-t/20})$, (b)
$\lim_{t\to\infty}(x(t), y(t)) = (300, 150)$.

13. $x(t) = \frac{325}{6} - \frac{125}{12}e^{-3t/25} - \frac{174}{5}e^{/-t/5}$,

$y(t) = \frac{200}{3} + \frac{125}{6}e^{-3t/25} - \frac{175}{2}e^{-t/25}$;

$\lim_{t\to\infty}(x(t), y(t)) = (325/6, 200/3)$;

$x(t) = y(t)$ at $t \approx 29.1$ min.

15. System: $dx/dt = 2y - 2x + 1$,
$dy/dt = x - 3y + 1$;

$x(t) = \frac{5}{4} - \frac{23}{12}e^{-4t} + \frac{8}{3}e^{-t}$,

$y(t) = \frac{3}{4} + \frac{23}{12}e^{-4t} + \frac{4}{3}e^{-t}$;

$\lim_{t\to\infty}(x(t), y(t)) = (4/3, 3/4)$; maximum is
$x \approx 4$ and $y \approx 1$.

17. System: $dx/dt = -\frac{1}{10}x + 10$,

$dy/dt = \frac{1}{10}x - \frac{1}{10}y$, $dz/dt = \frac{1}{10}y - \frac{1}{10}z$;

$x(t) = 100(1 - e^{-t/10})$,
$y(t) = 10(10 - 10e^{-t/10} - te^{-t/10})$,

$z(t) = 100 - 100e^{-t/10} - 10te^{-t/10} - \frac{1}{2}t^2e^{-t/10}$,

$\lim_{t\to\infty}(x(t), y(t), z(t)) = (100, 100, 100)$.

19. $x(t) = \frac{25}{3}e^{4t} + \frac{5}{3}e^{-2t}$, $y(t) = 25e^{4t} - 5e^{-2t}$.

21. $x(t) = 15e^{2t} - 10e^{-t}$, $y(t) = 10e^{-t}$.

23. $x(t) = -\frac{67}{7}e^{-10t} + \frac{171}{10}e^{7t} + \frac{663}{70}e^{-3t}$,

$y(t) = 5e^{-10t} + \frac{171}{10}e^{7t} - \frac{221}{10}e^{-3t}$,

$z(t) = \frac{52}{7}e^{-10t} + \frac{171}{10}e^{7t} + \frac{663}{70}e^{-3t}$;

$x(1) \approx z(1) \approx 18,752.89$.

25. $x(t) = \frac{43}{7} + \frac{46}{21}e^{7t} - \frac{1}{3}e^{-8t}$,

$y(z) = \frac{4}{3}e^{7t} + \frac{2}{3}e^{-8t}$,

$z(t) = -\frac{43}{7} + \frac{136}{21}e^{7t} - \frac{1}{3}e^{-8t}$; $z(1) \approx 7095.86$.

29. $x(t) = x_0e^{-at}$, $y(t) = y_0e^{-at} + ax_0te^{-at}$,
$z(t) = x_0 + y_0 + z_0 - (x_0 + y_0)e^{-at} - ax_0te^{-at}$;

$\lim_{t\to\infty} x(t) = \lim_{t\to\infty} y(t) = 0$,
$\lim_{t\to\infty} z(t) = x_0 + y_0 + z_0$.

33. $x(t) = 7e^{-6t}$, $y(t) = \frac{47}{5}e^{-t} - \frac{42}{5}e^{-6t}$,

$z(t) = 16 - \frac{47}{5}e^{-t} + \frac{7}{5}e^{-6t}$; $\lim_{t\to\infty} z(t) = 16$.

35. $x(t) = e^{-4t}$, $y(t) = 3e^{-2t} - 2e^{-4t}$,
$z(t) = 3 - 3e^{-2t} + e^{-4t}$; $\lim_{t\to\infty} z(t) = 3$.

37. $x(t) = 2$, $y(t) = 2 - e^{-t}$, $z(t) = t + 1 + e^{-t}$;
$\lim_{t\to\infty} x * t) = \lim_{t\to\infty} y(t) = 2$,
$\lim_{t\to\infty} z(t) = \infty$.

39. $x(t) = 10 - 2e^{-t}$, $y(t) = 10 - 8e^{-t} - 2te^{-t}$,
$z(t) = 10t - 8 + 10e^{-t} + 2te^{-t}$;
$\lim_{t\to\infty} x * t) = \lim_{t\to\infty} y(t) = 10$,
$\lim_{t\to\infty} z(t) = \infty$.

41. $x(t) = (v_1 + v_2)^{-1}$
$[bV_1\left(1 - \exp(-Pt(v_1 + v_2)/(V_2V_2))\right)$
$+a\left(V_1 + V_2\exp(-Pt(v_1 + v_2)/(V_2V_2))\right)]$,
$y(t) = (v_1 + v_2)^{-1}$
$[aV_2\left(1 - \exp(-Pt(v_1 + v_2)/(V_2V_2))\right)$
$+b\left(V_2 + V_1\exp(-Pt(v_1 + v_2)/(V_2V_2))\right)]$
(a) $\lim_{t\to\infty} x(t) = (a + b)V_1/(V_1 + V_2)$,
$\lim_{t\to\infty} y(t) = (a + b)V_2/(V_1 + V_2)$
(b) $V_1 > V_2$, $V_1 = V_2$ (c) $bV_1 - aV_2 > 0$,
$aV_2 - bV_1 > 0$, no.

43. The problem can be solved as a system or
the system can be written as a second-order
ODE, where $x'' = (a_1 - a_2)x' + b_1y'$
(derivative of first equation) or
$y' = 1/b_1(x'' - (a_1 - a_2)x')$. Substitution of y'
and $y = 1/b_1(x' - (a_1 - a_2)x)$ into the second
equation and simplification yields
$x'' - [(a_1 - a_2) + (b_1 - b_2)]x' + [(b_1 - b_2)(a_1 - a_2) - a_2b_1]x = 0$.

We consider the characteristics of this equation
(same as the eigenvalues of the system).
Periodic if $a_1 + b_1 = a_2 + b_2$ and $a_1b_1 - 2a_2b_1 - a_1b_2 + a_2b_2 > 0$. For example, if $a_1 = a_2 = -1$,
$b_1 = b_2 = 1$, $x_0 = 1$, and $y_0 = 0$, then $x(t) = \cos t$,
$y(t) = -\sin t$.
Exponential decay if $b_2 - b_1 > a_1 - a_2$ and
$(b_1 - b_2)(a_1 - a_2) - a_2b_1 \geq 0$. For example, if $a_1 = -1$, $a_2 = 0$, $b_1 = b_2 = 1$, $x_0 = 1$, $y_0 = 1$, then
$x(t) = e^{-t}$, $y(t) = 0$.

Exponential growth if $a_1 - a_2 > b_2 - b_1$ (will have at least one positive eigenvalue). For example, if $a_1 = 2$, $a_2 = 0$, $b_1 = b_1 = 1$, $x_0 = 1$, $y_0 = 0$, then $x(t) = e^{2t}$, $y(t) = 0$.

Exercises 7.3

1. $-a \ln|y| + by - c \ln|x| + dx = C$.
3. (a) $(2k, k)$ stable spiral; (b) $(2k, k)$ center.
5. $dx/dt = y$, $dy/dt = -\sin x$, $(k\pi, 0)$, $k = 0$, $\pm 1, \pm 2, \ldots$.
7. $dx/dt = y$, $dy/dt = -k^2 x$,
   $dy/dx = (dy/dt)/(dx/dt) = -k^2 x/y$;
   $\frac{1}{2}y^2 + \frac{1}{2}k^2 x^2 = C$; $(0,0)$ center.
9. $dx/dt = y$,
   $dy/dt = -gM_1 x^{-1} + gM_2(R-x)^{-2}$;
   $(x_0, y_0) =$
   $(R(M_1 + \sqrt{M_1 M_2})/(M_1 - M_2), 0) =$
   $(R\sqrt{M_1}/(\sqrt{M_1} - \sqrt{M_2}), 0)$; saddle.
11. If $dx/dt = y - F(x) = y - \int_0^x f(u)du$ and
    $dy/dt = -g(x)$, then $d^2x/dt^2 =$
    $dy/dt - f(x)dx/dt = -g(x) - f(x)dx/dt$,
    which is equivalent to
    $d^2x/dt^2 + f(x)dx/dt + g(x) = 0$.
13. In each case, no limit cycle in the xy-plane:
    (a) $f_x(x, y) + g_y(x, y)$
    $= 3x^2 + 1 + x^2 = 4x^2 + 1 > 0$;
    (b) $f_x(x, y) + g_y(x, y) = -1 - 1 = -2 < 0$;
    (c) $f_x(x, y) + g_y(x, y) = y^2 + 8 > 0$.
15. $du/d\theta = y$, $dy/d\theta = \alpha - u = ku^2$; $y = 0$ and
    $\alpha - u - ku^2 = 0$ so $u = \frac{1}{2k}(1 \pm \sqrt{1 - 4k\alpha})$. We
    choose $u = \frac{1}{2k}(1 - \sqrt{1 - 4k\alpha})$ because it is
    closer to $u = 0$ than $u = \frac{1}{2k}(1 + \sqrt{1 - 4k\alpha})$,
    and we are considering a small change in
    the orbit. The eigenvalues of
    $J\left(\frac{1}{2k}(1 - \sqrt{1 - 4k\alpha}), 0\right) = \begin{pmatrix} 0 & 1 \\ -\sqrt{1 - 4k\alpha} & 0 \end{pmatrix}$
    satisfy $\lambda^2 + \sqrt{1 - 4k\alpha} = 0$; center.
17. Differentiating $\frac{1}{2}m(x)(dx/dt)^2 + V(x) = E$
    with respect to t yields $\frac{1}{2}m'(x)(dx/dt)^3 +$

$m(x)dx/dtd^2x/dt^2 + V'(x)dx/dt = dx/dt$
$\left[\frac{1}{2}m'(x)(dx/dt)^2 + m(x)d^2x/dt^2 + V'(x)\right]$
$= 0$, so

$$\frac{1}{2}m'(x)\left(\frac{dx}{dt}\right)^2 + m(x)\frac{d^2x}{dt^2} + V'(x) = 0,$$

$$\frac{1}{\sqrt{m(x)}}\left[\frac{1}{2}m'(x)\left(\frac{dx}{dt}\right)^2 + m(x)\frac{d^2x}{dt^2} + V'(x)\right]$$
$= 0$,

$$\frac{1}{2}\frac{m'(x)}{\sqrt{m(x)}}\left(\frac{dx}{dt}\right)^2 + \sqrt{m(x)}\frac{d^2x}{dt^2} + \frac{V'(x)}{\sqrt{m(x)}} = 0,$$

$$\frac{d}{dt}\left[\sqrt{m(x)}\frac{dx}{dt}\right] + \frac{V'(x)}{\sqrt{m(x)}} = 0.$$

Notice that $du/dx = \sqrt{m(x)}$, so the equation is

$$\frac{d}{dt}\left[\frac{du}{dx}\frac{dx}{dt}\right] + \frac{V'(x)}{\sqrt{m(x)}} = 0.$$

Notice also that $du/dt = (du/dx)(dx/dt)$, so we have

$$\frac{d}{dt}\left(\frac{du}{dt}\right) + \frac{V'(x)}{\sqrt{m(x)}} = 0$$

or

$$\frac{d^2x}{dt^2} + + \frac{V'(x)}{\sqrt{m(x)}} = 0.$$

19. paths; circles; equilibrium point; center; agree.
21. (a) center if $x = 1$; saddle if $x = -1$;
    (b) saddle if $x = 0$; center if $x = 1$, $x = -1$.
23. $d\theta/dt = y$, $dy/dt = -FR\text{sgn}(y)/I$;
    equilibrium points: $(\theta, 0)$; if $d\theta/dt > 0$, then
    we integrate $I\frac{d\theta}{dt}\frac{d}{d\theta}\left(\frac{d\theta}{dt}\right) = -FR$ with
    respect to θ to obtain the parabolas
    $\frac{1}{2}I\left(\frac{d\theta}{dt}\right)^2 = -FR\theta + C$, $d\theta/dt > 0$. Similar
    calculations follow for $d\theta/dt < 0$.
25. (a) With $' = d/dt$,
    $(x^2 + y^2 + x^2)' = 2xx' + 2yy' + 2zz'$ and
    substitute for x', y', and z', where $x' = y$,

$y' = -x$ on $x^2 + y^2 + z^2 = 1$. (b) $z(t) = C_1$, so $z(t) = z_0$. (c) Use the change of variables in Exercise 10.

27. Use a numerical solver with initial guess $x_0 = 1$ and graph the solution. Observe the graph to see that $x_0 = 2$.

Chapter 7 Review Exercises

1. (a) $Q(t) = \dfrac{1}{500,000}\left(\dfrac{3}{2}e^{-2t}3 - e^{-t}\right)$,

$I(t) = \dfrac{1}{500,000}(-e^{-2t}3 + e^{-t})$; (b) $Q(t) =$

$-\dfrac{4,500,001}{500,000}e^{-t} - 3te^{-t} + \dfrac{9,000,003}{1,000,000}e^{-2t/3}$,

$I(t) =$

$\dfrac{3,000,001}{500,000}e^{-t} + 3te^{-t} - \dfrac{3,000,001}{1,000,000}e^{-2t/3}$.

3. (a) $Q(t) = \dfrac{1}{1,000,000}e^{-t}\cos t$,

$I_2(t) = \dfrac{1}{1,000,000}e^{-t}\sin t$; (b)

$Q(t) = 60 + 60e^{-t} - \dfrac{59,999,999}{1,000,000}e^{-t}\cos t$,

$I_2(t) = 60 - \dfrac{59,999,999}{1,000,000}e^{-t}\sin t - 60e^{-t}\cos t$.

5. (a) $x(t) = e^{-t} + te^{-t}$, $y(t) = -te^{-t}$ (critically damped); (b) $x(t) = \cos 2t$, $y(t) = -2\sin 2t$ (undamped).

7. $x(t) = \dfrac{5}{2}(3 - e^{-t})$, $y(t) = \dfrac{5}{2}(3 + e^{-t})$.

9. $x(t) = 15e^{3t} + 5e^t$, $y(t) = 15e^{3t} - 5e^t$.

11. $x(t) = -\dfrac{5}{3}e^{-5t} - \dfrac{128}{51}e^{-14t} + \dfrac{139}{17}e^{3t}$,

$y(t) = \dfrac{7}{6}e^{-5t} - \dfrac{64}{51}e^{-14t} + \dfrac{139}{34}e^{3t}$,

$z(t) = -\dfrac{1}{6}e^{-5t} + \dfrac{208}{51}e^{-14t} + \dfrac{139}{34}e^{3t}$.

13. $(0,0)$; $\lambda_1 - a > 0$, $\lambda_2 = -c < 0$, saddle point, unstable; $(a/k, 0)$, $\lambda_1 = -a < 0$, $\lambda_2 = -c + ad/k = d(a/k - c/d) > 0$, saddle point, unstable; $(c/d, (ad - ck/(bd))$,

$\lambda_{1,2} = \dfrac{1}{2d}(-ck \pm \sqrt{-4acd^2 + 4c^2dk + c^2k^2})$,

where $4dc(cd - ad) < 0$, so $-ck - \sqrt{-4acd^2 + 4c^2dk + c^2k^2} < 0$ if

$-4acd^2 + 4c^2dk + c^2k^2 > 0$ (improper node) and $\lambda_{1,2}$ has negative real part if $-4acd^2 + 4c^2dk + c^2k^2 < 0$ (spiral point); asymptotically stable.

Exercises 8.1

1. $21s^{-2}$.

3. $2/(s - 1)$.

5. $4/(s^2 + 4)$.

7. $e^{-s}s^{-1}$.

9. $\dfrac{s + e^{-\pi s/2}}{s^2 + 1}$.

11. $(3s - 1 + e^{-3s})s^{-2}$.

13. $(1 - 2e^{-10s})s^{-1}$.

15. $k/(s^2 + k^2)$.

21. $(s - 1)/[(s - 1)^2 + 9]$.

23. $5/[(s + 1)^2 + 25]$.

25. $16(2s + 1)^{-3}$.

27. $-18(s - 3)^{-1}$.

29. $\dfrac{s^2 - 25}{(s^2 + 25)^2}$.

31. $\dfrac{1}{s} + \dfrac{5}{s^2 + 25}$.

33. $\dfrac{s^2 + 2s - 3}{(s^2 + 2s + 5)^2}$.

35. $(s + 1)^{-8}$.

37. $s^{-6} - s^{-4} + s^{-2}$.

39. $\dfrac{24(5s^4 - 10s^2 + 1)}{(s^2 + 1)^5}$.

41. $\dfrac{6s}{(s^2 - 9)^2}$.

43. $\dfrac{14s}{(s^2 + 49)^2}$.

45. $\dfrac{2(s^3 + 27s)}{(s^2 - 9)^3}$.

47. $\dfrac{2}{s^2 + 2s - 3}$.

49. $\dfrac{s + 2}{(s + 2)^2 + 16}$.

51. $\dfrac{4}{(s + 5)^2 + 16}$.

55. (a) $\dfrac{s^2 + 2k^2}{s^3 + 4k^2s}$ (b) $\dfrac{2k^2}{s^3 + 4k^2s}$.

61. No to all.

Exercises 8.2

1. $\dfrac{1}{5040}t^7$.

3. e^{-5t}.

5. $\dfrac{1}{2}t^2 e^{-6t}$.

7. $\dfrac{1}{3}\sin 3t$.

9. $8e^{-8t} - 7e^{-7t}$.

11. $\dfrac{2}{9}e^{-2t} + \dfrac{7}{9}e^{7t}$.

13. $\dfrac{1}{2} + \dfrac{1}{2}e^{4t}$.

15. $-\dfrac{1}{8}e^{-2t} + \dfrac{1}{8}e^{6t}$.

17. $2 - \sin t$.

19. $2 + e^{2t}$.

21. $t + \cos 2t$.

23. $1 + t + t^2$.

25. $t - \dfrac{1}{2}t^2 + \dfrac{1}{3}t^3 - \dfrac{1}{4}t^4$.

27. (a) $\mathcal{L}\{a \sin bt - b \sin at\} = \dfrac{a^3 b - a b^3}{(s^2 + a^2)(s^2 + b^2)}$

and $\mathcal{L}\{\cos bt - \cos at\} = \dfrac{a^2 s - b^2 s}{(s^2 + a^2)(s^2 + b^2)}$;

(b)
$$\mathcal{L}^{-1}\left\{\frac{1}{(s^2 + a^2)(s^2 + b^2)}\right\} = \frac{a \sin bt - b \sin at}{ab(a^2 - b^2)}$$
and
$$\mathcal{L}^{-1}\left\{\frac{s}{(s^2 + a^2)(s^2 + b^2)}\right\} = \frac{\cos bt - \cos at}{a^2 - b^2}.$$

29. (b) $\mathcal{L}^{-1}\left\{\dfrac{k}{s(s^2 + k^2)}\right\} = \int_0^t \sin k \ \ \tau =$

$\dfrac{1}{k}(1 - \cos kt)$ and $\mathcal{L}^{-1}\left\{\dfrac{k}{s^2(s^2 + k^2)}\right\} =$

$\int_0^t \int_0^\tau \sin k\lambda \, d\lambda \, d\tau = \dfrac{1}{k^2}(kt - \sin kt)$.

Exercises 8.3

1. $y = \dfrac{1}{5}e^{-8t}(3 - 8e^{5t})$

3. $y = -\dfrac{1}{7}e^{-5t}(3 + 4e^{7t})$.

5. $y = te^t$.

7. $y = \dfrac{1}{2}(-2 + e^t + \cos t - \sin t)$.

9. $y = \dfrac{1}{2}(2\cos t + t^2 - 2)$.

11. $y = \dfrac{1}{2}t^2 e^{-t}$.

13. $y = \dfrac{1}{2}e^{-t}(1 - e^{2t} + t + te^{2t})$.

15. $y = \dfrac{1}{2} - \dfrac{3}{2}e^{-2t/3} + 4e^{-t/2} - 3e^{-t/3}$.

17. $y = \dfrac{3}{2}t^2$.

19. $y = t^2 + 1$.

21. $y = t^2$.

23. $Y'(s) + s^2 Y(s) = s$,
$-(s^2 + 2s)Y'(s) + (s^2 + n(n+1))Y(s) = 1$, one.

27. $y(t) = -1/24 e^{-t} \sin\left(\sqrt{3}t\right)\sqrt{3} +$

$\dfrac{35}{24}e^{-t}\cos\left(\sqrt{3}t\right) + 1/24\left(-8 + (6t - 3)e^t\right)e^{-t}$.

29. $y(t) =$
$e^{-t} - 1/6 te^{-t} + 1/6\left(-3t^2 + t + 2t^3\right)e^{-t}$.

Exercises 8.4

1. $-28e^{-s}s^{-1}$.

3. $(3e^{-8s} - e^{-4s})s^{-1}$.

5. $-\dfrac{42e^{-4s}}{s - 1}$.

7. $\dfrac{12e^{-2s}}{s^2 - 1}$.

9. $-\dfrac{14e^{-\pi s/3}}{s^2 + 1}$.

11. $\dfrac{e^s(e^s - 1 - s)}{s^2(e^{2s} - 1)}$.

13. $\dfrac{2e^s + 3}{se^{2s} + se^s + s}$.

15. $e^{-\pi s}$.

17. $100e^{-2s} + e^{-s}$.

19. $e^{-2\pi s}\left(100e^{3\pi s/2} + \dfrac{1}{s^2 + 1}\right)$.

21. $\mathcal{L}\{f(t)\} = \dfrac{1}{1 - e^{-2s}}\int_0^2 e^{-st}\delta(t - 1)$

$dt = \dfrac{e^{-s}}{1 - e^{-2s}}$.

23. $-3\mathcal{U}(t - \pi)$.

25. $-3\mathcal{U}(t - 4) + 2\mathcal{U}(t - 1)$.

27. $-3\mathcal{U}(t - 6) + 3\mathcal{U}(t - 5) - 4\mathcal{U}(t - 3)$.

29. $e^{3t-6}\mathcal{U}(t - 2)$.

31. $\cos(2t - 6)\mathcal{U}(t - 3)$.

33. $\dfrac{1}{2}\sin^2(t - 5)\mathcal{U}(t - 5)$.

35. $\dfrac{2(1 - 2e^{2s} - e^{4s})}{s^2 e^{4s}(1 - e^{-4s})}$

$= \dfrac{2(1 - 2e^{2s} - e^{-4s})}{s^2 e^{4s}} \dfrac{1}{1 - e^{-4s}}$

$= \dfrac{2(1 - 2e^{2s} - e^{-4s})}{s^2 e^{4s}} \sum_{k=0}^{\infty} e^{-4ks}$ so

$\mathcal{L}^{-1}\left\{\dfrac{2(1 - 2e^{2s} - e^{4s})}{s^2 e^{4s}(1 - e^{-4s})}\right\}$

$= \sum_{k=0}^{\infty} \mathcal{L}^{-1}\left\{\dfrac{2(1 - 2e^{2s} - e^{-4s})}{s^2 e^{4s}}e^{-4ks}\right\}$

$= 2\sum_{k=0}^{\infty}[(t - 4k - 4)\mathcal{U}(t - 4k - 4)$

$\quad - (2t - 8k - 4)\mathcal{U}(t - 4k - 2)$

$\quad - \quad (t - 4k)\mathcal{U}(t - 4k)]$.

37. $\cosh(6 - 2t)\mathcal{U}(t - 3)$.

39. $\dfrac{1}{s^2(1 + e^{-4s})} = \dfrac{1}{s^2}\sum_{k=0}^{\infty} e^{-4ks}$. Then,

$\sum_{k=0}^{\infty} \mathcal{L}^{-1}\left\{\dfrac{1}{s^2}e^{-4ks}\right\} = \sum_{k=0}^{\infty}(t - 4k)\mathcal{U}(t - 4k)$

41. $\mathcal{L}^{-1}\left\{\sum_{k=0}^{\infty}\dfrac{1}{s^2 + 4}e^{-ks}\right\} =$

$\dfrac{1}{2}\sum_{k=0}^{\infty}\sin(2t - 2k)\mathcal{U}(t - k)$.

43. $f(t) = -\delta(t - 1/2), 0 \le t < 1, f(t) = f(t - 1)$, $t \ge 1$.

45. $y = \dfrac{1}{3}e^{-3t}\left[2 + e^3 + (e^{3t} - e^3)\mathcal{U}(1 - t)\right]$.

47. $y =$
$\begin{cases} \dfrac{1}{16}(4t - 1 + e^{-4t}), 0 \le t \le 1 \\ \dfrac{1}{16}e^{-4t}(1 - 2e^4 + e^{4t}(9 - 4t)), 1 < t \le 2 \\ \dfrac{1}{16}e^{-4t}(e^4 - 1)^2, t > 2 \end{cases}$

49. $x = 1 - \cos t - 10\sin t\mathcal{U}(t - \pi)$.

51. $x = \dfrac{10}{3}e^{-2t}[\sin 3t + 2e^\pi \cos 3t\mathcal{U}(2t - \pi)]$.

53. $x = \dfrac{1}{39}e^{-2t}$
$\begin{cases} 3e^{2t} - 3\cos 3t - 2\sin 3t, 0 \le t < 1 \\ 6e^{2t} - 3\cos 3t + e^2(-3\cos(3t - 3) \\ \quad -2\sin(3t - 3)) - 2\sin 3t, t \ge 1 \end{cases}$

55.

$x(t) = \sum_{k=0}^{\infty}\mathcal{L}^{-1}\left\{\dfrac{10e^{-s}(e^{s/2} - 1)}{s(s + 2)}e^{-sk}\right\}$

$= 5e^{-2t}\sum_{k=0}^{\infty}\left[(e^{2+2k} - e^{2t})\mathcal{U}(t - k - 1)\right.$

$\left. +(e^{2t} - e^{2k+1})\mathcal{U}(t - 1/2 - k)\right]$.

57.

$x(t) = \sum_{k=0}^{\infty}\mathcal{L}^{-1}\left\{\dfrac{e^{-2s} - e^{-s} + s}{s^4}e^{-2ks}\right\}$

$= \dfrac{1}{6}\sum_{k=0}^{\infty}\left[(t - 2 - 2k)^3\mathcal{U}(t - 2 - 2k)\right.$

$\quad + (1 - t + 2k)^3\mathcal{U}(t - 1 - 2k)$

$\left. +3(t - 2k)^2\mathcal{U}(t - 2k)\right]$.

59.

$x(t) = \sum_{k=0}^{\infty}\mathcal{L}^{-1}\left\{\dfrac{2e^{-\pi s/2}}{(s^2 + 4)^2}e^{-\pi ks/2}\right\}$

$= -\dfrac{1}{8}\sum_{k=0}^{\infty}\left[(\pi - 2t + k\pi)\cos(2t - k\pi)\right.$

$\left. +\sin(2t - k\pi)\right]\mathcal{U}(t - \pi/2 - \pi k/2)$.

63. $y(t) = \dfrac{1}{2}\sum_{k=0}^{\infty}(1 - e^{2+4k-2t})\mathcal{U}(t - 1 - 2k)$.

Exercises 8.5

1. $\dfrac{1}{3}t^3$.

2. te^{-3t}.

3. $\dfrac{1}{12}t^4$.

4. $\dfrac{1}{5}(2e^{-t} - 2\cos 2t + \sin 2t)$.

5. $\dfrac{1}{128}(32t^3 - 12t + 3\sin 4t)$.

7. $\dfrac{1}{s} - \dfrac{1}{1+s}$.

9. $\dfrac{1}{s^2} - \dfrac{1}{s^2+1}$.

11. $\dfrac{1}{s-1} - \dfrac{1}{s^2} - \dfrac{1}{s}$.

13. $\dfrac{e^{-3s}}{s^2}$.

15. $\dfrac{e^{-3\pi s/2}(e^{\pi s} - 1)}{s^2+1}$.

17. $\dfrac{1}{54}(2 - 2e^{-3t} - 6t + 9t^2)$.

19. $\dfrac{1}{10}(2e^{-t} - 2\cos 2t + \sin 2t)$.

21. $(1 + \cos t)\mathcal{U}(t - \pi)$.

23. $g(t) = \sin t$.

25. $h(t) = \dfrac{5}{3}\cos(\sqrt{2}t) - \dfrac{2\sqrt{2}}{3}\sin(\sqrt{2}t) + \dfrac{10}{3}e^{-2t}$.

27. $y(t) = (\dfrac{1}{10}t^5 + \dfrac{1}{4}t^4)e^{2t}$.

Exercises 8.6

1. $x = e^{-7t} - e^{5t}, y = 3e^{-7t} + e^{5t}$.

3. $x = 2e^t(\cos 2t + 2\sin 2t), y = -4e^t \sin 2t$.

5. $x = e^{-2t}(-3 + 4e^t), y = 3e^{-2t}(-1 + e^{2t})$, $z = e^{-2t}(3 - 4e^t + e^{2t})$.

7. $x = e^{-2t} - 2e^{-t} + e^t$, $y = \dfrac{2}{3}e^{-2t}(e^t - 1)^2(2 + e^t)$.

9. $x = e^{-t}(6 - 6e^t + 6t), y' = e^{-t}(2 - 2e^t + 3t)$.

11. $x = \dfrac{1}{34}(2e^{-t} - 2\cos 4t + 9\sin 4t$, $y = \dfrac{1}{68}(16e^t - 16\cos 4t - 13\sin 4t)$.

13. $x = \dfrac{1}{8}(2 - 10t - 2\cos 2t + 9\sin 2t)$, $y = \dfrac{1}{8}(4 - 2t - 4\cos 2t + \sin 2t)$.

15. $x = \dfrac{1}{2}\begin{cases} -t\cos t + \sin t, 0 \le t < \pi \\ -\pi \cos t, t \ge \pi \end{cases}$,

$y = \dfrac{1}{2}\begin{cases} t\sin t, 0 \le t < \pi \\ \pi \sin t, t \ge \pi \end{cases}$.

17. $x = \cos 4t - \dfrac{1}{2}(\mathcal{U}(4t - \pi) - 2\mathcal{U}(8t - \pi))\sin 4t$, $y = (\mathcal{U}(4t - \pi) - 2\mathcal{U}(8t - \pi))\cos 4t + 2\sin 4t$.

19.

$$x(t) = \sum_{k=0}^{\infty}(-1)^k e^{-t}$$
$$\left[\left(-e^{k+1} + e^t(3 - 2t + 2k)\right)\mathcal{U}(t - 1 - k)\right.$$
$$\left. + \left(e^k + e^t(2t - 1 - 2k)\right)\mathcal{U}(t - k)\right]$$

and

$$y(t) = 2\sum_{k=0}^{\infty}(-1)^{k+1}\left[-\left(t - 2 - k + e^{1-t+k}\right)\right.$$
$$\mathcal{U}(t - 1 - k) + \left(t - 1 - k + e^{k-t}\right)$$
$$\left.\mathcal{U}(t - k)\right].$$

21.

$$x(t) = (\sin(t - 2k\pi) - \cos(t - 2k\pi) + 1)$$
$$\mathcal{U}(t - 2k\pi) - (\cos(t - 2k\pi)$$
$$- \sin(t - 2k\pi + 1)\mathcal{U}(t - \pi - 2k\pi)$$

and

$$y = -2(\cos(t - 2k\pi) + 1)\mathcal{U}(t - \pi - 2k\pi)$$
$$+ 4\sin^2(t/2 - k\pi)\mathcal{U}(t - 2k\pi).$$

23. $x(t) = -1 - e^{2t} + 2e^t, y(t) = -2 + e^t$.

25. $x = -1 + e^t - t, y = -\dfrac{1}{2}t^2$.

Exercises 8.7

1. $Q(t) = 2 - e^{-t} - 2(1 - e^{1-t})\mathcal{U}(t - 1)$, $I(t) = 2e^{-t} - 2(1 - e^{1-t})\delta(t - 1) - 2e^{1-t}\mathcal{U}(t - 1)$.

3. $I(t) = \dfrac{1}{2}\sum_{n=0}^{\infty}(-1)^n(1 - e^{-(t-n)})\mathcal{U}(t - n)$.

5. $I(t) = (t - 1 + e^{-t}) - (t - 1)\mathcal{U}(t - 1)$
$+ (t - 3 + e^{-(t-2)})\mathcal{U}(t - 2) - (t - 3)\mathcal{U}(t - 3)$
$+ (t - 5 + e^{-(t-4)})\mathcal{U}(t - 4) - (t - 5)\mathcal{U}(t - 5) + \cdots$.

7. $I(t) = 100te^{-3t}$.

9. $I(t) = \dfrac{2}{3}\sqrt{3}e^{-t/2}\sin\left(\dfrac{\sqrt{3}}{2}t\right)$

$- \dfrac{2}{3}\sqrt{3}e^{-t/2+\pi/2}\sin\left(\dfrac{\sqrt{3}}{2}(t - \pi)\right)\mathcal{U}(t - \pi)$.

11. (a) $I(t) = e^{-4t+4}\mathcal{U}(t - 1)$;
(b) $I(t) = e^{-4t} + e^{-4t+4}\mathcal{U}(t - 1)$.

13. $I(t) = t - 1 + e^{-t} - (t - 2 + e^{-t+1})\mathcal{U}(t - 1)$.

15. $I(t) = e^{2-t}\mathcal{U}(t - 2) + e^{6-t}\mathcal{U}(t - 6)$.

19. $x(t) = \dfrac{1}{2}((-t\cos t + \sin t)\mathcal{U}(t)$
$- ((\pi - t)\cos t + \sin t)\mathcal{U}(t - \pi))$.

21. $x(t) = 3\cos 3t - \dfrac{2}{3}\sin 3t$.

23. $x(t) = e^{-2t}(2t + 1)$.

25. $x(t) = \dfrac{1}{3}e^{-4t}(4e^{3t} - 1)$.

27. $x(t) = e^{-2t}(4 - 4e^t + 2t + 2te^t)$.

29. $x(t) = \dfrac{1}{9}e^{-4t}(1 - e^{3t} - 3t + 6te^{3t})$.

31. $x(t) = \dfrac{1}{36 + 13\pi^2 + \pi^4}[((\pi^2 - 6)\cos \pi t$
$- e^{-3t}(-12e^3 + 18e^{t+2} - 3e^3\pi^2 + 2\pi^2 e^{t+2}$
$+ 5\pi e^{3t}\sin \pi t)\mathcal{U}(t-1)) - e^{-3t}(-12 + 18e^t - 3\pi^2$
$+ 2\pi^2 e^t + e^{3t}(\pi^2 - 6)\cos \pi t -$
$5\pi e^{3t}\sin \pi t)\mathcal{U}(t)]$.

33. $x(t) = -\dfrac{1}{2}e^{-2t}((3e^4 + e^{2t} - 2te^4)\mathcal{U}(t - 2)$
$- (-2 + e^2 + e^{2t} + 4t - 2te^2 - 2t^2)\mathcal{U}(t - 1)$
$- 2t^2\mathcal{U}(t))$.

35. $x(t) = \dfrac{1}{18}e^{-2t}(3t\sin 3t\mathcal{U}(t) - (6\pi\cos$
$3t - 3t\cos 3t + \sin 3t)\mathcal{U}(t - 2\pi)$
$+ (3\pi\cos 3t - 3t\cos 3t + \sin 3t + 3\pi\sin$
$3t - 3t\sin 3t)\mathcal{U}(t - \pi))$.

37. $x(t) = \dfrac{1}{3(\pi^4 + 17\pi^2 + 16)}[(3(\pi^2 - 4)\cos$
$\pi t - e^{-4t}(15\pi e^{4t}\sin \pi t + 16e^{3t+1} - 4e^4 +$
$\pi^2 e^{3t+1} - 4\pi^2 e^4))\mathcal{U}(t - 1) - e^{-4t}(3e^{4t}$
$(\pi^2 - 4)\cos \pi t + 16e^{3t} + \pi^2 e^{3t} - 4\pi^2$
$- 15\pi e^{4t}\sin \pi t - 4)\mathcal{U}(t)]$.

39. $x(t) = \dfrac{1}{15}(\sin t - \dfrac{1}{4}\sin 4t$
$- 15\sin(4 - 4t)\mathcal{U}(t - 1))$.

41. $x(t) = \dfrac{1}{3}e^{-2t}(e^6\sin(3t - 9)\mathcal{U}(t - 3) +$
$e^2\sin(3t - 3)\mathcal{U}(t - 1))$.

43. $x(t) = 200 + \dfrac{127,150}{13}e^{-5t}$
$+ \dfrac{250}{13}\cos t - \dfrac{1250}{13}\sin t$, bounded.

45. $x(t) = 200 + \dfrac{63,650}{13}e^{-5t}$
$- \dfrac{1250}{3}\cos t - \dfrac{250}{13}\sin t$, bounded.

47. $x(t) = -250 + 5350e^{2t} - 200\sin t - 100\cos t$,
unbounded.

49. $x(t) =$
$-10,000 + 17,500e^t + 2500\cos t + 2500\sin$
$t + (10,000 - 10,000e^{t-5} + 2500e^{t-5}\cos$
$5 - 2500\cos 5\cos(t - 5) + 2500e^{t-5}\sin$
$5 - 2500\sin 5\cos(t - 5) - 2500\cos 5\sin(t - 5)$
$+ 2500\sin 5\sin(t - 5))\mathcal{U}(t - 5)$.

51. $x(t) = -5000\left\{1 - \dfrac{5}{2}e^t + \dfrac{1}{2}(\cos t - \sin t) + \dfrac{1}{2}\right.$
$(-2 + 2e^{t-1} + e^{t-1}\cos 1 - \cos t - e^{t-1}\sin 1$
$+ \sin t)\mathcal{U}(t - 1) +$
$\left(1 - \dfrac{3}{2}e^{t-2} + \dfrac{1}{2}(\cos(t - 2) - \sin(t - 2))\right)$
$\left.\mathcal{U}(t - 2) + \cdots\right\}$.

53. $x(t) = e^{t-2}(100e^2 + 200\mathcal{U}(t - 2))$ so
$x(5) = e^3(200 + 100e^2) \approx 18,858$.

55. $x(t) = \dfrac{1}{k + k^3}e^{-kt}(-c + ce^{kt} + ck - ck^2 + ck^2 e^{kt}$
$+ kx_0 + k^3 x_0 - cke^{kt}\cos t + ck^2 e^{kt}\sin t)$.

57. (b) $x(t) = \sum_{n=0}^{6}\dfrac{c_0}{a}$
$[(1 - e^{-a(t-4n)})\mathcal{U}(t - 4n)$
$- (1 - e^{-a(t-4n-1/2)})\mathcal{U}(t - 4n - 1/2)]$;
(c) $y(t) = \sum_{n=0}^{6}c_0$
$\left[\dfrac{1}{b}(1 - e^{-b(t-4n)})\mathcal{U}(t - 4n)\right.$
$\left.- \dfrac{1}{b - a}(e^{-a(t-4n)} + e^{-b(t-4n)})\mathcal{U}(t - 4n)\right]$
$- \sum_{n=0}^{6}c_0\left[\dfrac{1}{b}(1 - e^{-b(t-4n-1/2)})\right.$

$\mathcal{U}(t - 4n - 1/2)$

$-\dfrac{1}{b-a}(e^{-a(t-4n-1/2)} + e^{-b(t-4n-1/2)})$

$\mathcal{U}(t - 4n - 1/2)]$.

59. (d) $x(t) = \cos 5t - \cos 5\sqrt{3}t$,

$\quad y(t) = 2(\cos 5t + \cos 5\sqrt{3}t)$.

63. $x(t) = \dfrac{2}{5}(\cos t - \cos \sqrt{6}t)$,

$\quad y(t) = \dfrac{1}{5}(4\cos t + \cos \sqrt{6}t)$.

65. $x(t) = \dfrac{1}{3}(\cos t + 2\cos 2t)$,

$\quad y(t) = \dfrac{2}{3}(\cos t - \cos 2t)$.

67. $x(t) = \dfrac{1}{5}\left(4\cos \sqrt{3}t + \cos\left(\dfrac{1}{\sqrt{2}}t\right)\right)$,

$\quad y(t) = \dfrac{2}{5}\left(-\cos \sqrt{3}t + \cos\left(\dfrac{1}{\sqrt{2}}t\right)\right)$.

69. $x(t) = \dfrac{1}{4} + \dfrac{1}{18}\sin 2t + \dfrac{7}{12}\cos 2t + \dfrac{1}{18}\sin t$

$\quad + \dfrac{1}{6}\cos t - \dfrac{1}{6}t\cos t, \, y(t) = \dfrac{1}{4} - \dfrac{1}{18}\sin 2t$

$\quad - \dfrac{7}{12}\cos 2t + \dfrac{4}{9}\sin t + \dfrac{1}{3}\cos t - \dfrac{1}{3}t\cos t$.

71. $x(t) = \dfrac{7}{18}\cos t + \dfrac{11}{18}\cos 2t + \dfrac{1}{12}t\sin 2t$,

$\quad y(t) = \dfrac{7}{9}\cos t - \dfrac{7}{9}\cos 2t - \dfrac{1}{12}t\sin 2t$.

73. $\theta_1(t) = -\dfrac{1}{8}\sqrt{3}\sin\left(\dfrac{2}{\sqrt{3}}t\right) + \dfrac{1}{8}\sin 2t$,

$\quad \theta_2(t) = -\dfrac{1}{4}\sqrt{3}\sin\left(\dfrac{2}{\sqrt{3}}t\right) - \dfrac{1}{4}\sin 2t$.

75. $\theta_1(t) = \dfrac{1}{2}\cos\left(\dfrac{2}{\sqrt{3}}t\right) + \dfrac{1}{2}\cos 2t$,

$\quad \theta_2(t) = \cos\left(\dfrac{2}{\sqrt{3}}t\right) - \cos 2t$.

77. $\theta_1(t) = \dfrac{1}{4}\sqrt{3}\sin\left(\dfrac{2}{\sqrt{3}}t\right) + \dfrac{1}{4}\sin 2t$,

$\quad \theta_2(t) = \dfrac{1}{2}\sqrt{3}\sin\left(\dfrac{2}{\sqrt{3}}t\right) - \dfrac{1}{2}\sin 2t$.

79. $\theta_1(t) = -\dfrac{1}{4}\cos\left(\dfrac{2}{\sqrt{3}}t\right) + \dfrac{1}{4}\cos 2t$,

$\quad \theta_2(t) = -\dfrac{1}{2}\cos\left(\dfrac{2}{\sqrt{3}}t\right) - \dfrac{1}{2}\cos 2t$.

81. $x(t) = -\cos\left(\dfrac{1}{\sqrt{m}}t\sqrt{2k + k\omega^2}\right)$,

$\quad y(t) = \cos\left(\dfrac{1}{\sqrt{m}}t\sqrt{2k + k\omega^2}\right)$.

83. $x(t)$ has frequency ω_1, where

$\quad \omega_1{}^2 = 9k/(2m) - k\sqrt{17}/(2m); \, y(t)$ has

$\quad$ frequency ω_2, where

$\quad \omega_2{}^2 = 9k/(2m) + k\sqrt{17}/(2m)$.

87. (a) $Q(t) = \dfrac{292,175,000}{159,587,072,641}e^{-100(5+2\sqrt{6})t}$

$\quad - \dfrac{7,387,691,623}{1,276,696,581,128\sqrt{6}}e^{-100(5+2\sqrt{6})t}$

$\quad + \dfrac{292,175,000}{159,587,072,641}e^{400\sqrt{6}t - 100(5+2\sqrt{6})t}$

$\quad + \dfrac{7,387,691,623}{1,276,696,581,128\sqrt{6}}e^{400\sqrt{6}t - 100(5+2\sqrt{6})t}$

$\quad - \dfrac{584,350,000}{159,587,072,641}\cos 377t$

$\quad - \dfrac{204,799,950}{159,587,072,641}\sin 377t;$

$\quad$ (c) $I(t) = -\dfrac{584,350,000}{159,587,072,641}\cos 377t$

$\quad - \dfrac{204,799,950}{159,587,072,641}\sin 377t$.

91. $x(t) = -\dfrac{1}{3}\sqrt{3}\sin\sqrt{3}t, \, y(t) = \dfrac{1}{3}\sqrt{3}\sin\sqrt{3}t$.

93. $\theta_1(t) = \dfrac{1}{8}\left[4\cos 2t + 4\cos\left(\dfrac{2}{\sqrt{3}}t\right)\right.$

$\quad \left. + \sin 2t - \sqrt{3}\sin\left(\dfrac{2}{\sqrt{3}}t\right)\right], \, \theta_2(t) = -\cos 2t +$

$\quad \cos\left(\dfrac{2}{\sqrt{3}}t\right) - \dfrac{1}{4}\sin 2t - \dfrac{\sqrt{3}}{4}\sin\left(\dfrac{2}{\sqrt{3}}t\right)$.

97. $x_1(t) = -2\sin t, \, x_2(t) = \sin t, \, x_3(t) = 2\sin t$.

Chapter 8 Review Exercises

1. $\dfrac{s-1}{s^2}$.

3. $\dfrac{1 - e^{-5s}}{s}$.

5. $2s^{-3} + 5s^{-1}$.

7. $(s-2)^{-2}$.

9. $6(s-1)^{-4}$.

11. $\dfrac{s^2-9}{(s^2+9)^2}$.

13. $\dfrac{s+5}{s^2+10s+34}$.

15. $e^{-3/2s\pi}$.

17. $-2\dfrac{-3e^{-7s}+2e^{-4s}}{s}$.

19. $-42\dfrac{e^{5-s}}{s-5}$.

21. $2\dfrac{e^{-2s}\left(2s^2+2s+1\right)}{s^3}$.

23. $\mathcal{L}\{f(t)\} = \dfrac{1}{1-e^{-s}}\int_0^1 (1-t)e^{-st}dt$
$= \dfrac{(s-1)e^s+1}{s^2(e^s-1)}$.

25. $\mathcal{L}\{f(t)\} = \dfrac{1}{1-e^{-2s}}\int_0^1 \cos\dfrac{\pi}{2}te^{-st}dt$
$= \dfrac{2e^2(2se^s+\pi)}{(4s^2+\pi^2)(e^{2s}-1)}$.

27. $\mathcal{L}\left\{\int_0^t \dfrac{\sin\tau}{\tau}d\tau\right\} = \dfrac{1}{s}\tan^{-1}\dfrac{1}{s}$.

29. $\mathcal{L}\left\{\dfrac{1-e^{-t}}{t}\right\} = \ln\left(1+\dfrac{1}{s}\right)$.

31. $\mathcal{L}^{-1}\left\{-\dfrac{2s}{(s^2+1)^2}\right\} = -2\sin t$.

33. $\mathcal{L}^{-1}\left\{\dfrac{2s^2-7s+20}{s(s^2-2s+10)}\right\} = 2-e^t\sin 3t$.

35. $\mathcal{L}^{-1}\left\{-\dfrac{14}{se^{2s}}\right\} = -14\mathcal{U}(t-2)$.

37. $\mathcal{L}^{-1}\left\{\dfrac{3e^{6-s}}{6-s}\right\} = -3e^{6t}\mathcal{U}(t-1)$.

39. $\mathcal{L}^{-1}\left\{\dfrac{-18}{(s^2+1)(1-e^{-3s})}\right\} =$
$\mathcal{L}^{-1}\left\{\dfrac{-18}{s^2+1}\sum_{k=0}^{\infty}e^{-3ks}\right\} =$
$\sum_{k=0}^{\infty}18\sin(3k-t)\mathcal{U}(t-3k)$.

41. $\mathcal{L}^{-1}\left\{\dfrac{s\cos\phi-\omega\sin\phi}{s^2+\omega^2}\right\} = \cos(\omega t+\phi)$.

43. $\dfrac{2}{3}(\cos t-\cos 2t)$.

45. $y = e^{2t}(\cos t-2\sin t)$.

47. $y = e^{-2t}(1+e^2\mathcal{U}(t-1))$.

49.

$$y(t) = \mathcal{L}^{-1}\left\{\dfrac{1}{s^2(s+4)(s+8)}\sum_{k=0}^{\infty}(-1)^k e^{-sk}\right\}$$

$$-\mathcal{L}^{-1}\left\{\dfrac{1}{s^2(s+4)(s+8)}\sum_{k=0}^{\infty}(-1)^k e^{-sk-s}\right\}$$

$$=\dfrac{1}{256}e^{-8t}\sum_{k=0}^{\infty}(-1)^k$$

$$\left[\left(e^{8+8k}-4e^{4(t+1+k)}+e^{8t}(11+8k-8t)\right)\right.$$

$$\mathcal{U}(t-k-1)-$$

$$\left(e^{8k}-4e^{4(t+k)}+e^{8t}(3+8k-8t)\right)$$

$$\left.\mathcal{U}(t-k)\right].$$

51. $x(t) = e^{1-2t}\left[e(-e^2+e^t)\mathcal{U}(t-2)+(-e+e^t)\right.$
$\left.\mathcal{U}(t-1)\right]$.

52. $x(t) = \dfrac{1}{6}(t-2\mathcal{U}(3t-\pi))\sin 3t$.

53. $g(t) = 1+\sin t-\cos t$.

57. $x(t) =$
$$\begin{cases} \dfrac{1}{15}(\cos(t/2)-\cos 2t-30\sin(t/2)), \\ 0\le t<\pi \\ \dfrac{1}{15}(\cos(t/2)-31\sin(t/2)), t\ge\pi \end{cases}$$

58. $x(t) = -\dfrac{1}{3}e^{(\pi-t)/4}\mathcal{U}(t-\pi)\sin 3t$.

59. $Q(t) =$
$$\begin{cases} \dfrac{11}{250}\sin^2 50t, 0\le t<2 \\ \dfrac{11}{500}((-1+\cos 200)\cos 100t+\sin 200 \\ \sin 200t), t\ge 2 \end{cases}$$

61. $x(t) = 100e^{2t-2}(100e^2+\mathcal{U}(t-1))$.

65. $x = \dfrac{1}{2}(t+\cos t\sin t), y = \dfrac{1}{8}(-2t+\sin 2t)$.

67. $x = 4e^{-t}((e^t - e^2(-1+t))\mathcal{U}(t-2) + (1 - e^t + t)\mathcal{U}(t)), y = e^{-t}((-2e^t + te^2)\mathcal{U}(t-2) + (-2 + 2e^t - t)\mathcal{U}(t)).$

69. $x(t) = \dfrac{2}{5}\left(2\sqrt{2}\sin(\sqrt{2}t) + \sqrt{3}\sin\left(\dfrac{1}{\sqrt{3}}t\right)\right),$

$y(t) = \dfrac{1}{5}\left(-2\sqrt{2}\sin(\sqrt{2}t) + 4\sqrt{3}\sin\left(\dfrac{1}{\sqrt{3}}t\right)\right).$

71. $x(t) = \dfrac{1}{2}(\cos t - \cos\sqrt{3}t + t\sin t),$

$y(t) = \dfrac{1}{2}(-\cos t + \cos\sqrt{3}t + t\sin t).$

73. $x(t) = \dfrac{1}{8}\left(-3\sin 2t - \sqrt{3}\sin\left(\dfrac{2}{\sqrt{3}}t\right)\right),$

$y(t) = \dfrac{1}{4}\left(3\sin 2t - \sqrt{3}\sin\left(\dfrac{2}{\sqrt{3}}t\right)\right).$

Bibliography

[1] Abell, M. and Braselton, J., *Mathematica by Example*, Fourth Edition, Academic Press, Boston, MA, 2008.

[2] Apostol, T., *Mathematical Analysis*, Second Edition, Addison-Wesley, Reading, MA, 1974.

[3] Barnsley, M., *Fractals Everywhere*, Second Edition, Morgan Kaufmann, 2000.

[4] Boyce, W.E. and DiPrima, R.C., *Elementary Differential Equations and Boundary-Value Problems*, Tenth Edition, John Wiley & Sons, 2012.

[5] *CIA—The World Factbook* at twww.cia.gov/cia/publications/factbook/index.html, 2006 (active on November 21, 2006).

[6] Coddington, E. and Levinson, N., *Theory of Ordinary Differential Equations*, Robert E. Krieger Publishing Company/McGraw Hill, 1955/1984.

[7] Corduneanu, C., *Principles of Differential and Integral Equations*, Chelsea Publishing, London, 1977.

[8] Devaney, R.L. and Keen, L. (eds.), *Chaos and Fractals: The Mathematics Behind the Computer Graphics*, Proceedings of Symposia in Applied Mathematics, Volume 39, American Mathematical Society, 1989.

[9] Edwards, C.H. and Penney, D.E., *Calculus with Analytic Geometry*, Sixth Edition, Prentice Hall, 2002.

[10] Gaylord, R.J., Kamin, S.N., and Wellin, P.R., *Introduction to Programming with Mathematica*, Second Edition, TELOS/Springer-Verlag, New York, 1996.

[11] Giordano, F.R., Weir, M.D., and Fox, W.P., *A First Course in Mathematical Modeling*, Third Edition, Thomson/Brooks/Cole, Pacific Grove, CA, 2003.

[12] Graff, K.F., *Wave Motion in Elastic Solids*, Oxford University Press, Dover, 1975/1991.

[13] Gray, A., *Modern Differential Geometry of Curves and Surfaces*, Second Edition, CRC Press, Boca Raton, FL, 1997.

[14] Gray, J.W., *Mastering Mathematica: Programming Methods and Applications*, Second Edition, Academic Press, Boston, MA, 1997.

[15] Herriott, S.R., *College Algebra through Functions and Models*, Preliminary Edition, Brooks/Cole, 2002.

[16] Jordan, D.W. and Smith, P., *Nonlinear Ordinary Differential Equations: An Introduction to Dynamical Systems*, Fourth Edition, Oxford Applied and Engineering Mathematics, Oxford University Press, 2007.

[17] Kreyszig, E., *Advanced Engineering Mathematics*, Tenth Edition, John Wiley & Sons, 2011.

[18] Larson, R.E., Hostetler, R.P., and Edwards, B.H., *Calculus with Analytic Geometry*, Sixth Edition, Houghton Mifflin, 1998.

[19] O'Connor, J.J. and Robertson, E.F, *The MacTutor History of Mathematics Archive* at turnbull.mcs.st-and.ac.uk/history/, 2006 (active on November 6, 2006).

[20] Maeder, R.E., *The Mathematica Programmer II*, Academic Press, Boston, MA, 1996.

[21] Maeder, R.E., *Programming in Mathematica*, Third Edition, Addison-Wesley, Reading, 1996.

[22] *Malthus Family Homepage* at homepages.caverock.net.nz/~kh/ (active on October 30, 2006).

[23] Malthus, T.R., *An Essay on the Principle of Population* at Electronic Scholarly Publishing, www.esp.org/books/malthus/population/malthus.pdf, 1798 (active on October 30, 2006).

[24] O'Neil, D., *Early Theories of Evolution: 17th-19th Century Discoveries that Led to the Acceptance of Biological Evolution* at anthro.palomar.edu/evolve/, 1997-2006 (active on October 30, 2006).

[25] Rabenstein, A.L., *Introduction to Ordinary Differential Equations*, Academic Press, Boston, MA, 1966.

[26] Robinson, C., *Dynamical Systems: Stability, Symbolic Dynamics, and Chaos*, Second Edition, CRC Press, Boca Raton, FL, 1999.

[27] Smith, H.L. and Waltman, P., *The Theory of the Chemostat: Dynamics of Microbial Competition*, Cambridge University Press, Cambridge, 1995.

[28] Simmons, G.F. and Krantz, S.G., *Differential Equations: Theory, Technique, Practice*, McGraw-Hill, 2007.

[29] Stewart, J., *Calculus: Concepts and Contexts*, Second Edition, Brooks/Cole, 2001.

[30] Weisstein, E.W., *CRC Concise Encyclopedia of Mathematics*, CRC Press, Boca Raton, FL, 1999.

[31] Waltman, P., *A Second Course in Elementary Differential Equations*, Dover Publications, 1986.

[32] *Wikipedia: The Free Encyclopedia* at en.wikipedia.org/wiki/Main_Page (active on November 21, 2013).

[33] Wolfram, S., *A New Kind of Science*, Wolfram Media, Champaign, IL, 2002.

[34] Wolfram, S., *The Mathematica Book*, Fourth Edition, Wolfram Media, 2004.

[35] Zwillinger, D., *Handbook of Differential Equations*, Third Edition, Academic Press, Boston, MA, 1997.

Appendices

$f(t)$	$F(s) = \mathcal{L}\{f(t)\}$
1	$\dfrac{1}{s}$
t^n, n a positive integer	$\dfrac{n!}{s^{n+1}}$
e^{at}	$\dfrac{1}{s-a}$
$\sin kt$	$\dfrac{k}{s^2+k^2}$
$\cos kt$	$\dfrac{s}{s^2+k^2}$
$\sinh kt$	$\dfrac{k}{s^2-k^2}$
$\cosh kt$	$\dfrac{s}{s^2-k^2}$
$e^{at}\sin kt$	$\dfrac{k}{(s-a)^2+k^2}$
$e^{at}\cos kt$	$\dfrac{s-a}{(s-a)^2+k^2}$
$e^{at}\sinh kt$	$\dfrac{k}{(s-a)^2-k^2}$
$e^{at}\cosh kt$	$\dfrac{s-a}{(s-a)^2-k^2}$
te^{at}	$\dfrac{1}{(s-a)^2}$
$t\sin kt$	$\dfrac{2ks}{(s^2+k^2)^2}$
$t\cos kt$	$\dfrac{s^2-k^2}{(s^2+k^2)^2}$
$\sin kt - kt\cos kt$	$\dfrac{2k^3}{(s^2+k^2)^2}$
$\sin kt + kt\cos kt$	$\dfrac{2ks^2}{(s^2+k^2)^2}$
$1 - \cos kt$	$\dfrac{k^2}{s(s^2+k^2)}$
$kt - \sin kt$	$\dfrac{k^3}{s^2(s^2+k^2)}$
$(1+k^2t^2)\sin kt - kt\cos kt$	$\dfrac{8k^3s^2}{(s^2+k^2)^3}$
$\sin kt \sinh kt$	$\dfrac{2k^2s}{s^4+4k^4}$
$\cos kt \cosh kt$	$\dfrac{s^3}{s^4+4k^4}$

$f(t)$	$F(s) = \mathcal{L}\{f(t)\}$
$\sin kt \cosh kt$	$\dfrac{k(s^2+2k^2)}{s^4+4k^4}$
$\cos kt \sinh kt$	$\dfrac{k(s^2-2k^2)}{s^4+4k^4}$
$e^{at} - e^{bt}$	$\dfrac{a-b}{(s-a)(s-b)}$
$ae^{at} - be^{bt}$	$\dfrac{(a-b)s}{(s-a)(s-b)}$
$t^n e^{at}$, n a positive integer	$\dfrac{n!}{(s-a)^{n+1}}$
$t^n f(t)$, n a positive integer	$(-1)^n F^{(n)}(s)$
$f'(t)$	$sF(s) - f(0)$
$f''(t)$	$s^2 F(s) - sf(0) - f'(0)$
$f^{(n)}(t)$	$s^n F(s) - s^{n-1}f(0) - s^{n-2}f'(0) - \cdots - sf^{(n-2)}(0) - f^{(n-1)}(0)$
$e^{at}f(t)$	$F(s-a)$
$(f * g)(t) = \displaystyle\int_0^t f(t-v)g(v)\,dv$	$F(s)G(s)$
$\mathcal{U}(t-a)$, $a \geq 0$	$\dfrac{e^{-as}}{s}$
$f(t-a)\mathcal{U}(t-a)$, $a \geq 0$	$e^{-as}F(s)$
$f(t)\mathcal{U}(t-a)$, $a \geq 0$	$e^{-as}\mathcal{L}\{f(t+a)\}$
$\delta(t-t_0)$, $t_0 \geq 0$	$e^{-t_0 s}$
$f(t-T) = f(t)$	$\dfrac{1}{1-e^{-sT}}\displaystyle\int_0^T e^{-st}f(t)\,dt$
$t^{-1/2}$	$\sqrt{\dfrac{\pi}{s}}$
$t^{1/2}$	$\dfrac{\sqrt{\pi}}{2s^{3/2}}$
$t^{n-1/2}$, $n = 1, 2, \ldots$	$\dfrac{1 \cdot 3 \cdot 5 \cdots (2n-1)\sqrt{\pi}}{2^n s^{n+1/2}}$
t^α, $\alpha > -1$	$\dfrac{\Gamma(\alpha+1)}{s^{\alpha+1}}$
$\dfrac{1}{t}f(t)$	$\displaystyle\int_s^\infty F(u)\,du$
$\displaystyle\int_0^t f(\alpha)\,d\alpha$	$\dfrac{F(s)}{s}$
$\dfrac{1}{\sqrt{\pi t}}e^{at}(1+2at)$	$\dfrac{s}{(s-a)^{3/2}}$
$\dfrac{1}{2\sqrt{\pi t^3}}(e^{bt} - e^{at})$	$\sqrt{s-a} - \sqrt{s-b}$

509

$\dfrac{1}{\sqrt{\pi t}} - ae^{a^2 t}\operatorname{erfc}(a\sqrt{t})$	$\dfrac{1}{\sqrt{s}+a}$	$\dfrac{1}{\sqrt{\pi k}}\sin(2\sqrt{kt})$	$\dfrac{1}{s^{3/2}}e^{-k/s}$
$\dfrac{1}{\sqrt{\pi t}} + ae^{a^2 t}\operatorname{erf}(a\sqrt{t})$	$\dfrac{\sqrt{s}}{s-a^2}$	$\dfrac{1}{\sqrt{\pi t}}\cosh(2\sqrt{kt})$	$\dfrac{1}{\sqrt{s}}e^{k/s}$
$\dfrac{1}{a}e^{a^2 t}\operatorname{erf}(a\sqrt{t})$	$\dfrac{1}{\sqrt{s}(s-a^2)}$	$\dfrac{1}{\sqrt{\pi k}}\sinh(2\sqrt{kt})$	$\dfrac{1}{s^{3/2}}e^{k/s}$
$e^{a^2 t}\operatorname{erfc}(a\sqrt{t})$	$\dfrac{1}{\sqrt{s}(\sqrt{s}+a)}$	$\dfrac{1}{t}(e^{bt}-e^{at})$	$\ln\dfrac{s-a}{s-b}$
$J_0(kt)$	$\dfrac{1}{\sqrt{s^2+k^2}}$	$\dfrac{2}{t}(1-\cos kt)$	$\ln\dfrac{s^2+k^2}{s^2}$
		$\dfrac{2}{t}(1-\cosh kt)$	$\ln\dfrac{s^2-k^2}{s^2}$
$\dfrac{1}{\sqrt{\pi t}}\cos(2\sqrt{kt})$	$\dfrac{1}{\sqrt{s}}e^{-k/s}$	$\dfrac{1}{t}\sin kt$	$\tan^{-1}\dfrac{k}{s}$

Special Formulas

Trigonometric Identities

$\cos^2\alpha + \sin^2\alpha = 1$

$1 + \tan^2\alpha = \sec^2\alpha$

$1 + \cot^2\alpha = \csc^2\alpha$

$\cos(\alpha + \beta) = \cos\alpha\cos\beta - \sin\alpha\sin\beta$

$\cos(\alpha - \beta) = \cos\alpha\cos\beta + \sin\alpha\sin\beta$

$\sin(\alpha + \beta) = \sin\alpha\cos\beta + \cos\alpha\sin\beta$

$\sin(\alpha - \beta) = \sin\alpha\cos\beta - \cos\alpha\sin\beta$

$\sin 2\alpha = 2\sin\alpha\cos\alpha$

$\cos 2\alpha = \cos^2\alpha - \sin^2\alpha$

$\cos^2\alpha = \dfrac{1 + \cos 2\alpha}{2} \qquad \sin^2\alpha = \dfrac{1 - \cos 2\alpha}{2}$

$\tan\alpha = \dfrac{\sin\alpha}{\cos\alpha} \qquad \cot\alpha = \dfrac{\cos\alpha}{\sin\alpha}$

$\sec\alpha = \dfrac{1}{\cos\alpha} \qquad \csc\alpha = \dfrac{1}{\sin\alpha}$

$\cos(-\alpha) = \cos\alpha \qquad \sin(-\alpha) = -\sin\alpha$

Logarithmic and Exponential Function Properties

$\ln a^x = x\ln a,\ a > 0$

$\ln ab = \ln a + \ln b;\ a, b > 0$

$\ln\dfrac{a}{b} = \ln a - \ln b;\ a, b > 0$

$\ln e^x = x$

$e^{\ln x} = x,\ x > 0$

$e^{x+y} = e^x e^y$

$e^{xy} = (e^x)^y$

Hyperbolic Trigonometric Functions

$\cosh x = \dfrac{1}{2}(e^x + e^{-x})$

$\sinh x = \dfrac{1}{2}(e^x - e^{-x})$

$\tanh x = \dfrac{\sinh x}{\cosh x} = \dfrac{e^x - e^{-x}}{e^x + e^{-x}}$

$\operatorname{sech} x = \dfrac{1}{\cosh x} = \dfrac{2}{e^x + e^{-x}}$

$\operatorname{csch} x = \dfrac{1}{\sinh x} = \dfrac{2}{e^x - e^{-x}}$

$\coth x = \dfrac{1}{\tanh x} = \dfrac{e^x + e^{-x}}{e^x - e^{-x}}$

$\cosh^2 x - \sinh^2 x = 1$

Maclaurin Series

$e^x = 1 + x + \dfrac{x^2}{2!} + \dfrac{x^3}{3!} + \cdots = \sum_{n=0}^{\infty}\dfrac{x^n}{n!},$
$-\infty < x < \infty$

$\sin x = x - \dfrac{x^3}{3!} + \dfrac{x^5}{5!} - \cdots = \sum_{n=0}^{\infty}(-1)^n\dfrac{x^{2n+1}}{(2n+1)!},$
$-\infty < x < \infty$

$\cos x = 1 - \dfrac{x^2}{2!} + \dfrac{x^4}{4!} - \cdots = \sum_{n=0}^{\infty}(-1)^n\dfrac{x^{2n}}{(2n)!},$
$-\infty < x < \infty$

$\dfrac{1}{1-x} = 1 + x + x^2 + \cdots = \sum_{n=0}^{\infty}x^n,\ -1 < x < 1$

Table of Integrals

$$\int u^n \, du = \frac{1}{n+1} u^{n+1} + C, \, n \neq -1$$

$$\int \frac{1}{u} \, du = \ln|u| + C$$

$$\int e^u \, du = e^u + C$$

$$\int u e^{au} \, du = \frac{au - 1}{a^2} e^{au} + C$$

$$\int u^n e^{au} \, du = \frac{1}{a} u^n e^{au} - \frac{n}{a} \int u^{n-1} e^{au} \, du$$

$$\int a^u \, du = \frac{a^u}{\ln a} + C$$

$$\int \sin u \, du = -\cos u + C$$

$$\int \cos u \, du = \sin u + C$$

$$\int \tan u \, du = -\ln|\cos u| + C$$

$$\int \cot u \, du = \ln|\sin u| + C$$

$$\int \sec u \, du = \ln|\sec u + \tan u| + C$$

$$\int \csc u \, du = \ln|\csc u - \cot u| + C$$

$$\int \sec u \tan u \, du = \sec u + C$$

$$\int \sec^2 u \, du = \tan u + C$$

$$\int \sin^2 u \, du = \frac{u}{2} - \frac{1}{4} \sin 2u + C$$

$$\int \cos^2 u \, du = \frac{u}{2} + \frac{1}{4} \sin 2u + C$$

$$\int \sin^3 u \, du = -\frac{1}{3}(2 + \sin^2 u)\cos u + C$$

$$\int \cos^3 u \, du = \frac{1}{3}(2 + \cos^2 u)\sin u + C$$

$$\int \sin^n u \, du = -\frac{1}{n} \sin^{n-1} u \cos u + \frac{n-1}{n} \int \sin^{n-2} u \, du$$

$$\int \cos^n u \, du = \frac{1}{n} \cos^{n-1} u \sin u + \frac{n-1}{n} \int \cos^{n-2} u \, du$$

$$\int \tan^2 u \, du = \tan u - u + C$$

$$\int \tan^3 u \, du = \frac{1}{2} \tan^2 u + \ln|\cos u| + C$$

$$\int \sec^3 u \, du = \frac{1}{2} \sec u \tan u + \frac{1}{2} \ln|\sec u + \tan u| + C$$

$$\int \tan^n u \, du = \frac{1}{n-1} \tan^{n-1} u - \int \tan^{n-2} u \, du$$

$$\int \sec^n u \, du = \frac{1}{n-1} \tan u \sec^{n-2} u + \frac{n-2}{n-1} \int \sec^{n-2} u \, du + C$$

$$\int u \sin u \, du = \sin u - u \cos u + C$$

$$\int u \cos u \, du = \cos u + u \sin u + C$$

$$\int u^n \sin u \, du = -u^n \cos u + n \int u^{n-1} \cos u \, du$$

$$\int u^n \cos u \, du = u^n \sin u - n \int u^{n-1} \sin u \, du$$

$$\int \sin au \sin bu \, du = \frac{\sin(a-b)u}{2(a-b)} - \frac{\sin(a+b)u}{2(a+b)} + C$$

$$\int \cos au \cos bu \, du = \frac{\sin(a-b)u}{2(a-b)} + $$

$$\frac{\sin(a+b)u}{2(a+b)} + C$$

$$\int \sin au \cos bu \, du = -\frac{\cos(a-b)u}{2(a-b)} - $$

$$\frac{\cos(a+b)u}{2(a+b)} + C$$

$$\int \sin^n u \cos^m u \, du = -\frac{\sin^{n-1} u \cos^{m+1} u}{n+m} + $$

$$\frac{n-1}{n+m}\int \sin^{n-2} u \cos^m u \, du = \frac{\sin^{n+1} u \cos^{m-1} u}{n+m} + $$

$$\frac{m-1}{n+m} \int \sin^n u \cos^{m-2} u \, du$$

$$\int \frac{1}{\sqrt{a^2 - u^2}} \, du = \sin^{-1} \frac{u}{a} + C$$

$$\int \frac{1}{a^2 + u^2} \, du = \frac{1}{a} \tan^{-1} \frac{u}{a} + C$$

$$\int \frac{1}{u\sqrt{u^2 - a^2}} \, du = \frac{1}{a} \sec^{-1} \frac{u}{a} + C$$

$$\int \frac{1}{u^2 - a^2} \, du = \frac{1}{2a} \ln\left|\frac{u-a}{u+a}\right| + C$$

$$\int \frac{1}{a^2 - u^2} \, du = \frac{1}{2a} \ln\left|\frac{u+a}{u-a}\right| + C$$

$$\int \sqrt{a^2 + u^2} \, du = \frac{u}{2}\sqrt{a^2 + u^2} + $$

$$\frac{a^2}{2} \ln|u + \sqrt{a^2 + u^2}| + C$$

$$\int \frac{1}{\sqrt{a^2 + u^2}} \, du = \ln|u + \sqrt{a^2 + u^2}| + C$$

$$\int \sqrt{u^2 - a^2} \, du = \frac{u}{2}\sqrt{u^2 - a^2} - $$

$$\frac{a^2}{2} \ln|u + \sqrt{u^2 - a^2}| + C$$

$$\int \frac{1}{\sqrt{u^2 - a^2}} \, du = \ln|u + \sqrt{u^2 - a^2}| \div C$$

$$\int \frac{1}{(u^2 - a^2)^{3/2}} \, du = -\frac{u}{a^2\sqrt{u^2 - a^2}} + C$$

$$\int \sqrt{a^2 - u^2} \, du = \frac{u}{2}\sqrt{a^2 - u^2} + \frac{a^2}{2} \sin^{-1} \frac{u}{a} + $$

$$\int u^2\sqrt{a^2 - u^2} \, du = \frac{u}{8}(2u^2 - a^2)\sqrt{a^2 - u^2} + $$

$$\frac{a^4}{8} \sin^{-1} \frac{u}{a} + C$$

$$\int \sin^{-1} u \, du = u \sin^{-1} u + \sqrt{1 - u^2} + C$$

$$\int \cos^{-1} u \, du = u \cos^{-1} u - \sqrt{1 - u^2} + C$$

$$\int \tan^{-1} u \, du = u \tan^{-1} u - \frac{1}{2} \ln(1 + u^2) + C$$

$$\int e^{au} \sin bu \, du = $$

$$\frac{e^{au}}{a^2 + b^2}(a \sin bu - b \cos bu) + C$$

$$\int e^{au} \cos bu \, du = $$

$$\frac{e^{au}}{a^2 + b^2}(a \cos bu + b \sin bu) + C$$

$$\int \ln u \, du = u \ln u - u + C$$

$$\int u^n \ln u \, du = \frac{u^{n+1}}{(n+1)^2}[(n+1) \ln u - 1] + C$$

$$\int u \, dv = uv - \int v \, du$$

Index

Edwards Brothers Malloy
Ann Arbor MI. USA
April 2, 2015